POPULAR ENCYCLOPEDIA OF PLANTS

POPULAR ENCYCLOPEDIA OF
PLANTS

Chief Editor: Vernon H. Heywood

Associate Editor: Stuart R. Chant

CAMBRIDGE UNIVERSITY PRESS

CAMBRIDGE
LONDON   NEW YORK   NEW ROCHELLE
MELBOURNE   SYDNEY

Project Editor: Graham Bateman
Subeditor: Peter Forbes
Production: Clive Sparling
Picture Research: Christine Forth
Design: John Fitzmaurice

Published by the Press Syndicate of the
University of Cambridge,
The Pitt Building, Trumpington Street,
Cambridge CB2 1RP
35 East 57th Street, New York,
NY 10022, USA
296 Beaconsfield Parade, Middle Park,
Melbourne 3200, Australia

Planned and produced by and
© 1982 ⟶ Equinox (Oxford)
Limited, Mayfield House,
256 Banbury Road, Oxford,
England OX2 7DH

First published 1982

British Library cataloguing in publication
data
Popular encyclopedia of plants.
   1. Plants – Dictionaries
   I. Heywood, V. H.   II. Chant, S. R.
   581'.03'21   QK7

Library of Congress Cataloging in
Publication Data
Main entry under title:
Popular Encyclopedia of Plants.
   Bibliography: p.
   Includes index.
   1. Botany, Economic — dictionaries.
   2. Botany — Dictionaries.
I. Heywood, V. H.   II. Chant, S. R.
SB107.P67   630'.3'21   81-21713

ISBN 0521 24611 3   AACR2

Origination by Art Color Offset, Rome, Italy.
Filmset by Filmtype Services Limited,
Scarborough, England
Printed and bound by Graficoop, Bologna,
Italy

Contents

6–7   **Introduction**

8–351   **A–Z Dictionary**
(inclusive of special feature articles)

352–354   **Glossary**

355   **Index of Scientific Names**

356–367   **Index of Common Names**

368   **Bibliography**

SPECIAL FEATURE ARTICLES:

18   Alcoholic Beverages

20–23   Algae

26   Alkaloids

32–35   Angiosperms

50–51   Bacteria and Blue-green Algae

56   Beverage Plants

62–63   Brassicas

81   Cereals

92–93   Club Mosses and their Allies

98–99   Conifers and their Allies

122   Dyes and Tannins from Plants

132   Essential Oils

136–137   Ferns

140–141   Fibers from Plants

142–143   Flavorings from Plants

145   Fodder Crops

147–149   Fruits

152–155   Fungi

168–169   Herbs

174–175   Horsetails or Scouring Rushes

194–195   Legumes and Pulses

198–199   Lichens

216–217   Marrows, Squashes, Pumpkins and Gourds

220–221   Medicinal and Narcotic Plants

228–229   Mosses and Liverworts

238–239   Nuts

244–245   Oil Crops

290–291   Rootcrops

318   Sugar and Starch Crops

328–329   Timber

338–340   Vegetables

Introduction

Man, like all other animals, is dependent on green plants for his very existence. Plants are the primary producers, converting the sun's energy into food through the near-miracle of photosynthesis. This ultimate dependence of Man on plant life is a fundamental truth that is often overlooked in today's industrial society.

From earliest times, Man had to learn which plants would feed him, which were pleasant-flavored, which were poisonous, which would alleviate pain or other symptoms of illness, which could be used as efficient fuel, which would provide pigments for coloring and dyeing, and which were the "magic" plants that could transport him from reality. It was a long trial-and-error process, as a result of which hundreds of plants were selected in various parts of the world for their specific value to Man. Medicinal plants greatly outnumbered those grown or used as food, and the medicine man and witch-doctor were skilled botanists who had an important role and status in society. They can be considered as the first professional botanists.

As civilizations developed, more and more plants became known, and information about their properties and uses spread from society to society. This process continues even today as the secrets of the tropical forests are increasingly being discovered, although, tragically, many of the species in these tropical areas are in danger of extinction due to the widespread and accelerating destruction of the forests for timber extraction, agriculture, grazing or other uses. Most of the plants used by Man belong to the flowering plants (or angiosperms), the dominant group of plants on earth today. The gymnosperms, especially the conifers, are also economically very important in that they provide much of the timber (softwood) used by Man, as well as other products. The other plant groups supply us with some species of economic or ornamental interest and the fungi are particularly important, not only for edible purposes, fermentation and baking, but in the production of antibiotics and for their role in disease, decay and other ecologically important processes.

The number of plant species used by Man runs into the thousands. Not only do they include the major and minor food crops, the timber species, the herbs and spices, the medicinal and drug plants and halluci-nogens, the beverage plants, but also those that provide us with fibers, cork, tannins, dyes, resins, starches, oils, waxes, and countless other products used in our daily lives. And then there are the thousands of species of trees, shrubs and herbs grown in parks and gardens and in streets for ornament, decoration, display and pleasure, and the very important group of plants grown as animal fodder.

Convinced of the need to make knowledge of at least some of this enormous diversity of plants and their myriad uses accessible to as wide a public as possible, we have designed this book as an illustrated guide to the main species of plants used by Man. As we have noted, the number of different kinds of plant used by Man is vast and this has made the task of selection difficult. The main body of the text consists of a series of over 2 200 articles, arranged alphabetically under the scientific generic name or the popular name of the most important groups of plants that contain species employed by Man because of their economic importance, because of their ornamental value, or because of their special scientific interest. In addition, many plant products such as fruits, vegetables, gums, resins, herbs, spices, sweeteners and so on are included in this section. These alphabetical entries have been chosen so as to cover all the principal crop species of the world: cereals, legumes, oil plants, fruits, vegetables, beverages, stimulants, fibers, as well as timber trees and other plants of economic importance. In addition, we have included a selection of the most important genera of plants containing species cultivated in parks and gardens, in greenhouses, as amenity plants or as street trees.

In the case of the generic or group name entries we have, within the limitations of space, indicated the main area of natural distribution, habit, and the principal species of economic, agricultural, horticultural or scientific importance, together with their common names and a note of which part(s) of the plant produces items of economic importance. In the longer articles, referring to major crops, a brief discussion is usually given of the origin and history of the crop-plant species, a short description, its main areas of cultivation and special features of its culture, how it is processed or used, and an indication of world production. A number of entries refer to plant products such as rubber or resins; in such cases, information is given about the main species (often belonging to different families) which produce them in commercial quantities. Every effort has been made to give information that is as up-to-date as possible.

In the case of horticulturally important genera or groups, we have indicated the most widely cultivated species and cultivars, with a bias towards Europe and North America, where a wider range of species is grown because of the more highly developed art of gardening in these countries. On the other hand, many hundreds of tropical and subtropical cultivated ornamentals are included in the main section of the book, some being grown under glass as protection in temperate climates as well as out-of-doors in tropical and subtropical countries.

Popular names pose many problems, as the same plant can be known under different common names in different countries. Only the commonly used English names are given (including those used in North America) unless the foreign name has wide currency. These are typographically indicated in the text by the use of SMALL CAPITAL letters. Many garden plants are cultivated under the wrong names and we have taken great pains to try to give the correct scientific names, together with widely used synonyms by which they are also known. Two indexes have been compiled: one to the popular names of plants and their products and a second to the scientific names of genera that do not appear as key words in their own right.

Twenty-one special feature articles on economic plant groups or products are included (see Contents for details). Each of these features includes an introductory essay outlining the origins, uses, and nature of the plants or products concerned, plus a concise table of the main crops or products. An extensive cross-referencing system is used throughout the book, an asterisk (*) indicating a key word where further information can be obtained. This system is particularly useful with these economic plant special features which act as gateways to hundreds of more detailed entries.

Although, as we have already noted, the vast majority of plants used by Man belong to the flowering plants and to a lesser extent to the gymnosperms, a series of special feature articles is included giving a brief introduction to all the main groups that make up the Plant

World. These articles include information on the form and structure of the groups concerned, their reproduction, distribution and ecology, and their main uses. The main groups of the plant world, reflecting our latest knowledge and interpretations, belong to four separate kingdoms, one of which, the Monera, is made up of organisms known as *prokaryotes*, the cells of members of which *lack* a nuclear envelope, mitochondria, Golgi apparatus and plastids, and a complex structure of microtubules in their flagella. The Monera comprises the two divisions Bacteria (the bacteria) and the Cyanophyta (the blue-green algae). The three other kingdoms of the Plant World (and indeed all other groups of organisms) are known as *eukaryotes*; the cells of these *possess* a definite nucleus bounded by a double membrane, complex chromosomes, organelles such as mitochondria and flagella, and have cilia with a complex structure of microtubules. These three kingdoms of eukaryotes are the Protista, which contain the various groups of algae other than the blue-green; the Fungi, which are subdivided into several divisions and subdivisions, including the lichens; and the Plantae (the plant kingdom proper), which is in turn subdivided into mosses, liverworts and hornworts (division Bryophyta), and the club mosses (Lycophytina), whisk ferns (Psilophytina), horsetails (Sphenophytina), ferns (Filicophytina), cycads, conifers and allies (gymnosperms) and the angiosperms (Spermatophytina), all of which belong to the division Tracheophyta (the vascular plants) (see table).

The system of classification followed for the flowering plants is essentially that of G.L. Stebbins, *Flowering Plants—Evolution above the Species Level* (1974), as adopted in V. H. Heywood, *Flowering Plants of the World* (1978).

Illustrations play a major role in this volume. Over 700 color photographs have been used and certain of the economic plant special features are illustrated by composite artwork panels. Thus the vast majority of economically important plants are illustrated, together with several hundred ornamental species. In all, over 800 species are illustrated.

This volume, therefore, provides a wealth of information on the plants around us and the ways in which they influence our lives. Its information has been gleaned from the most up-to-date scientific sources and for this reason alone it will provide a valuable reference work for the professional botanist as well as for the general reader. The editors are indebted to the many scientists whose texts prepared for a larger encyclopedia (*Plants and Man*) formed, in part, the basis for the material presented here.

V. H. Heywood

Main Groups of The Plant World

KINGDOM MONERA

   Division Bacteria (**Bacteria**, p. 50)

   Division Cyanophyta (**Blue-green Algae**, p. 50)

KINGDOM PROTISTA (**Algae**, p. 20)

KINGDOM FUNGI (**Fungi**, p. 152, and **Lichens**, p. 198)

KINGDOM PLANTAE

   Division Bryophyta (**Mosses, Liverworts, Hornworts**, p. 228)

   Division Tracheophyta (Vascular Plants)

   Subdivision Lycophytina (**Club Mosses**, p. 92)

   Subdivision Psilophytina (Whisk Ferns, p. 92)

   Subdivision Sphenophytina (**Horsetails**, p. 174)

   Subdivision Filicophytina (**Ferns**, p. 136)

   Subdivision Spermatophytina (Seed Plants)

     Class Cycadinae (Cycads, p. 98)

     Class Coniferinae (**Conifers**, p. 98)

     Class Gnetinae (p. 98)

     Class Angiospermae (Flowering Plants, p. 32)

     Subclass Dicotyledoneae (Dicotyledons)

     Subclass Monocotyledoneae (Monocotyledons)

(Page numbers indicate the location of special feature articles including these groups.)

A

Abaca fiber or **Manila hemp** a fiber obtained from *Musa textilis*, a large banana-like perennial herb native to the Philippines and Borneo but cultivated extensively in Central America in similar soil and climatic conditions to those of the *banana, to which it is closely related. A tough light hemp fiber, used for ropes and cordage, is produced from narrow strips of bleached and dried leaf bases.
MUSACEAE.

Abelia a genus of semievergreen or deciduous, small or medium-sized shrubs native to Mexico, East and Central Asia and the Himalayas. A number of species are widely cultivated as ornamentals for their attractive flowers and foliage. One of the most popular is the Mexican *A. floribunda*, an evergreen shrub with ovate leaves and narrowly tubular carmine-rose flowers. *A. schumannii*, from China, is a deciduous species which bears tubular pink flowers.
CAPRIFOLIACEAE, about 30 species.

Abies [FIRS] a genus of coniferous evergreen trees widely distributed in the mountainous regions of the Northern Hemisphere, including central and southern Europe, Asia (from the Himalayas northward), Japan and extensive areas of North America. In the English language, the word "FIR" is now restricted to species of this genus, except that *DOUGLAS FIR is the traditional name for species of the genus *Pseudotsuga, and SCOTCH FIR is the name sometimes given to *Pinus sylvestris [SCOTS PINE].

Some of the many species that provide an important source of timber are the North American *A. balsamea* [BALSAM FIR, BALM OF GILEAD], *A. grandis* [GIANT FIR, GRAND FIR], *A. magnifica* [RED FIR] and *A. procera* [NOBLE FIR], and the European *A. alba* [COMMON SILVER FIR, WHITE WOOD FIR]. The timber of *A. concolor* [WHITE FIR, COLORADO FIR] is used as a source of wood pulp for the paper industry.

Abies wood is sold as (white) deal and varies from white to yellowish- or reddish-brown in color. It is easily worked, yielding a good surface which readily accepts paint and polish. Its main use is for indoor work but, treated with a preservative, it has been used outdoors, eg for telegraph poles. As the wood has no noticeable smell it has also been used for crating grocery and dairy products that might otherwise become tainted.

Resins are obtained from the bark of certain American firs. After steam distillation to remove turpentine, the residual solid (rosin) is used in the manufacture of products such as soaps, plastics and varnishes. Canada balsam, a permanent mounting medium for many microscopical preparations, is obtained from the BALSAM FIR and other North American species.

Some 30 species, including *A. koreana* [KOREAN FIR], *A. nordmanniana* [CAUCASIAN FIR] and *A. spectabilis* [HIMALAYAN FIR], as well as some of the North American firs, are planted outside their native regions, but more for their fine and lofty appearance than as a commercial undertaking.
PINACEAE, 40–50 species.

Abroma a small genus of shrubs native to tropical Asia and Australia. *A. fastuosa* is grown in the tropics as an ornamental for its large cordate leaves and dark purple flowers.

The banana-like herb Abaca (Musa textilis) is cultivated for the strong light fibers which are extracted from the sheathing petioles. (× 1/60)

Red Firs (Abies magnifica) growing in the Yosemite National Park, California, USA. This species is an important source of timber.

A. augusta [DEVIL'S COTTON] is economically useful because fibers obtained from the stems are used for making ropes and cordage.
STERCULIACEAE, 2 species.

Abrus a small genus of tropical and subtropical shrubs or shrublets, often climbers. *A. precatorius* [ROSARY PEA, CRAB'S EYES, INDIAN LICORICE] is widely naturalized and produces hard shiny scarlet seeds with black tips used to make ornamental jewelry such as necklaces and rosaries. They are also used as weights (rati) in India. The seeds constitute a hazard since they contain a highly toxic substance, abrin, which can cause serious poisoning when they are broken or bitten. The roots of *A. precatorius* and of other species

are used as a *licorice substitute (wild or Indian licorice).
LEGUMINOSAE, 6 species.

Abutilon a genus of herbs, soft-woody shrubs or trees from the warm temperate and tropical regions. Many species of *Abutilon* are extremely attractive garden plants, several of them with leaves that are prettily blotched with yellow as a result of a virus infection.

A. esculentum from Brazil is a shrub with clusters of purple flowers which are edible when cooked. *A. insigne* from Colombia and Venezuela is an attractive small shrub with large white, red-veined flowers borne in groups of one to three. Another widely grown ornamental species is the Brazilian *A. megapotamicum*, an evergreen shrub with pendulous red and yellow flowers. A number of *Abutilon* species, such as *A. theophrasti* (=*A. avicennae*) [VELVET LEAF], yield a tough fiber (see China jute).
MALVACEAE, 100–120 species.

Acacia [WATTLES] a very large genus of evergreen, usually xerophytic, tropical and subtropical trees or shrubs, including many plants of economic or horticultural importance. Many of the species bear attractive, silvery-gray leaves, giving a feathery appearance, and yellow flowers in globe-shaped heads as in *A. baileyana* [GOLDEN MIMOSA, BAILEY'S MIMOSA, COOTAMUNDRA WATTLE] or in cylindrical spike-like heads as in *A. retinodes*.

Acacias are an important component of the widespread scrub found in Australia, where they are called wattles, following their use by early settlers for building huts of wattlework plastered with mud.

Other species are frequently the only trees found in desert and dry areas of Africa and India. They are often called thorn-trees due to the large thorns at the leaf-base in many species, which serve as a protection against grazing animals. Examples include the South African *A. giraffae* [CAMEL THORN] and *A. karroo* [KARROO THORN], and the African and Indian *A. nilotica* [GUM ARABIC TREE, BABUL or EGYPTIAN THORN].

In some species, such as *A. cornigera* [BULL-HORN ACACIA] from Central America, a species of ant colonizes the thorns and obtains nourishment from the oils and proteins contained in the leaflets. Mutual advantage is gained from this association, as the worker ants which swarm over the tree sting and bite any potential predators, thus keeping them at bay.

Gum arabic is obtained from *A. senegal* [GUM ARABIC TREE] from tropical Africa and northern India, and from some other species, the gum exuding largely from the branches. Black catechu or *cutch, a dark extract rich in tannin, is obtained from *A. catechu* [CATECHU, CUTCH, BLACK CUTCH, KHAIR]. Other tannin-producing species include *A. mearnsii* [AUSTRALIAN BLACK WATTLE], *A. pycnantha* [GOLDEN WATTLE] and *A. dealbata* [SILVER WATTLE], which are planted in southern Europe for this purpose. The latter species is

also widely cultivated for timber, ornament and soil stabilization in areas such as the Côte d'Azur, and is the MIMOSA of florists used as a winter cut-flower. Several species, such as *A. melanoxylon* [AUSTRALIAN BLACKWOOD], yield a valuable timber used for making furniture, boats, boomerangs and spears.

Among the most popular ornamental species of *Acacia*, in addition to the SILVER WATTLE, are *A. armata* [KANGAROO THORN] and *A. longifolia* [WHITE SALLOW] including var *floribunda* [SYDNEY GOLDEN WATTLE]. Some species, such as the KARROO THORN and *A. farnesiana* [POPINAC, OPOPANAX, *CASSIE, HUISACHE, SWEET ACACIA] from tropical and

Flower of Abutilon megapotamicum *showing its scarlet calyx, yellow petals and red stamens.* (×2)

subtropical America are used for stabilizing sand dunes. The latter species also provides the cassie flowers used in perfumery.
LEGUMINOSAE, about 800 species.

Acaena [NEW ZEALAND BUR] a genus of perennial, often shrubby herbs related to *Poterium*. Most species occur in southern South America and Australasia but one or two are found in Hawaii, California and the remote islands surrounding Antarctica, as a result of dispersal by migrating birds which inadvertently carry the spiny fruits on their plumage.

Some species of *Acaena* are naturalized in Europe as a consequence of the accidental transport of seeds in wool from New Zealand. *A. magellanica*, *A. microphylla* and *A. montana*

are frequently used as ground cover in gardens.
ROSACEAE, about 100 species.

Acajou a gum used in bookbinding, produced from the stem of the tropical tree *Anacardium occidentale* [CASHEW NUT TREE]. The West Indian forest tree *Guarea guara*, which produces a useful deep bluish-red timber, is also sometimes called acajou.
ANACARDIACEAE.

Acalypha a large genus of herbs and shrubs native to the tropics and subtropics. A number are grown as ornamentals either for their attractive foliage, eg *A. wilkesiana* 'Marginata' [COPPER-LEAF], or for their showy inflorescences, eg *A. hispida* [CHENILLE, RED-HOT CATTAIL].
EUPHORBIACEAE, about 450 species.

Acantholimon [PRICKLY HEATHS, PRICKLY THRIFTS] a large genus of evergreen perennials mainly native to dry and stony places from the eastern Mediterranean to Central Asia. Some of the mat-forming species such as *A. acerosum*, *A. glumaceum* and *A. venustum*, which bear spikes of pink, rose or white starry flowers, are cultivated in rock gardens or in alpine houses.
PLUMBAGINACEAE, about 150 species.

Acanthosyris a small genus of trees and shrubs from temperate South America. *A. falcata* is a small tree whose wood is

Acacia trees are a major component of the landscape in dry, savanna and desert zones of Africa, Arabia and India, tropical America and Australia.

sometimes used for furniture-making. It also has red edible fruits.
SANTALACEAE, 3 species.

Acanthus [BEAR'S BREECHES] a genus of perennial herbs or small shrubs, some of which originate from southern Europe but mainly from tropical and subtropical Asia and Africa. Three species native to the Mediterranean region are popular hardy perennials in gardens. They are *A. balcanicus* (= *A. longifolius*), *A. mollis* (= *A. hispanicus*) and *A. spinosus*, all with handsome foliage and mauve-pink flowers. The leaf-form seems to have been used by the Greeks as the basis for decoration of their Corinthian columns, and later the same motif became very popular in furniture embellishment. *A. ebracteatus*, from tropical Asia, is used as a constituent of a local cough medicine.
ACANTHACEAE, about 50 species.

Acer [MAPLES, SYCAMORE] a large genus of deciduous or evergreen trees and shrubs ranging over almost all the north temperate region and extending into the tropics. The great majority are Asian and many are extremely rare.
Maples vary in height from the often shrubby species like the Asian *A. ginnala* [AMUR MAPLE] and *A. carpinifolium* [HORN-

BEAM MAPLE] through medium-sized trees such as the North American *A. negundo* [ASH LEAVED MAPLE, BOX-ELDER] and the European *A. opalus* [ITALIAN MAPLE] and *A. campestre* [FIELD MAPLE], to tall trees such as *A. pseudoplatanus* [SYCAMORE, GREAT MAPLE] and *A. platanoides* [NORWAY MAPLE] which are of the order of 30m tall.

Leaves vary in size from the evergreen type on *A. sempervirens* [CRETAN MAPLE], which are entire and only 3cm long, to those on *A. macrophyllum* [OREGON MAPLE], which are deeply lobed and measure 25–30cm across.

Many maples, such as *A. rubrum* [RED MAPLE], the NORWAY MAPLE and *A. palmatum* [JAPANESE MAPLE], are often grown as ornamentals in city streets and parks. In March, before the trees are in leaf, the bright red flowers of the RED MAPLE clustering along the shoots are highly decorative but less spectacular than the larger bunches of bright greenish-yellow flowers on big crowns of the NORWAY MAPLE. The best species for summer foliage include *A. buergeranum* [TRIDENT MAPLE] with small leaves and *A. pectinatum* for decorative lobing.

Most maples have particularly attractive autumn foliage as exemplified by the New England *A. saccharum* [SUGAR MAPLE] with its orange-scarlet leaves and by the RED MAPLE whose leaves run the gamut from lemon-yellow to deep purple. In Europe the best colors are on *A. palmatum* 'Osakazuki', *A. capillipes* [SNAKE BARK], *A. hersii* [HERS'S MAPLE] and *A. japonicum* [DOWNY JAPANESE MAPLE] 'Vitifolium'.

Several species, including the SYCAMORE, yield valuable hard close-grained timber used for flooring and furniture. Maple sugar is obtained from the SUGAR MAPLE and some other species by boring holes into the trunk in early spring to collect the sap.
ACERACEAE, about 200 species.

Aceras a genus represented by only one species, *A. anthropophorum* [MAN ORCHID,

*Inflorescence of the Man Orchid (*Aceras anthropophorum*), so called because each flower resembles the figure of a man. (×1)*

GREEN MAN ORCHIS], occurring in Europe, North Africa and Cyprus. Its common name is a reference to the shape of the flower, which is considered to simulate the shape of a man's body, although perhaps the French common name "THE HANGED MAN" is even more appropriate.
ORCHIDACEAE, 1 species.

Acetabularia [MERMAID'S WINE GLASS] a genus of marine green algae found in warm seas, where they grow mainly just below the low-water mark. Relatives of *Acetabularia* have a long fossil history dating as far back as the Cambrian period.
CHLOROPHYCEAE, about 20 species.

Achillea a large genus of hardy herbaceous perennials widespread in north temperate regions. They include the well-known weed species, *A. millefolium* [YARROW, MILFOIL]. One group of species (often referred to as the "Ptarmica" group) have relatively large flowering heads with distinct white ligulate ray florets, and are frequently alpine in distribution. The remaining species (sometimes referred to as the "Millefolium" group) possess small, crowded flowering heads with white, pinkish or yellow florets.

Some members of the genus, such as *A. millefolium*, have been used medicinally to prevent bleeding; hence the old common names "SOLDIER'S WOUND-WORT" and "HERBA MILITARIS". More recently certain species have been used as a substitute for tea in Europe, and as an additive in the manufacture of beer in Sweden, hence another name, "FIELD HOP". Other species have been used in liqueurs.

At least 20 species are commonly cultivated either as alpine-garden or rockery plants, often with white flowers and delicate silvery foliage, or as robust border plants with bright yellow flowers. Among the for-

The yellow flower heads of the Fernleaf Yarrow (Achillea filipendulina), a hardy perennial that flowers in late summer. ($\times \frac{1}{4}$)

The graceful, scented flowers of Acidanthera bicolor make this an attractive species, both for outdoor use and as an indoor pot plant. ($\times 1$)

mer, in the "Ptarmica" group are *A. moschata* [MUSK YARROW] and *A. ptarmica* [SNEEZE-WORT]. Members of the "Millefolium" group most commonly planted in borders include the yellow-flowered *A. filipendulina* [FERN-LEAF YARROW] and *A. tomentosa* [WOOLLY YARROW].
COMPOSITAE, about 85 species.

Achimenes a tropical Central American genus of perennial herbs, some of which are cultivated. The leaves and aerial stems of most species are covered with long, soft or stiff hairs. The flowers are usually tubular or trumpet-shaped.

Among the species cultivated as pot plants are *A. grandiflora*, which bears solitary, purplish-red flowers, and *A. longiflora*, whose large solitary flowers are white in cultivar 'Alba' and violet-blue in cultivar 'Major'.
GESNERIACEAE, about 30 species.

Achras see *Manilkara*.

Acidanthera a genus of perennial herbs from tropical and southern Africa. They produce long, narrow, iris-like leaves and loose spikes of gladiolus-like, cylindrical, scented flowers. *A. bicolor*, the only widely cultivated species, may be grown outdoors or, alternatively, indoors as a pot plant. The most popular variety is *murieliae* from Ethiopia, with white, crimson-blotched flowers.
IRIDACEAE, about 40 species.

Aciphylla a small genus of herbaceous perennials mainly from New Zealand with two species in Australia, mostly with stiff, spiny, erect leaves and dense clusters of yellowish or white flowers. A number are cultivated as rock-garden plants, such as *A. lyallii*, or as tall border perennials, such as *A. squarrosa* [SPEARGRASS, BAYONET PLANT] and *A. colensoi* [SPANIARD, WILD SPANIARD]. The latter

species is bushy in habit with massive inflorescences.
UMBELLIFERAE, about 35 species.

Acokanthera an African genus of evergreen shrubs or small trees, one of which, *A. oblongifolia* (= *Toxicophloea spectabilis*) [WINTERSWEET], is very commonly cultivated in parks and gardens throughout the subtropical areas of the world, and elsewhere as a greenhouse shrub. It bears very fragrant tubular flowers in clusters at the ends of the branches.

In their native African countries a poisonous thick tar-like substance is obtained by boiling up the wood of various species including *A. oppositifolia* (= *A. venenata*, *T. cestroides*) [BUSHMAN'S-POISON]. This is then spread over the tips and shafts of arrows. The active constituent of the poison is a very virulent cardiac glycoside, ouabain.
APOCYNACEAE, about 15 species.

Aconitum [ACONITE, MONKSHOOD, WOLF-BANE] a large genus of herbaceous perennials native to Europe, Asia and North America. The plants may be twining as in *A. ferox* [INDIAN ACONITE] but more often are erect, with palmately lobed leaves and blue, purple, white or yellow hood-shaped flowers borne in terminal inflorescences. Popular cultivated species include the violet-blue flowered *A. carmichaelii* (= *A. fischeri*) and the deep blue flowered *A. napellus* [MONKSHOOD, TURK'S CAP].

All parts of the plant contain the narcotic alkaloids, aconitine or pseudoaconitine, and are very poisonous to Man and other animals.
RANUNCULACEAE, about 200–300 species.

Acorus two species of temperate and subtropical perennial herbs, the best-known being *A. calamus* [SWEET FLAG, FLAG ROOT]. Originally from Asia, it has been widely introduced into Europe and America. It grows in mud along the banks of pools and sluggish streams, bearing long narrow leaves, which like the terminal inflor-

The attractive inflorescence of the Monkshood (Aconitum napellus). All parts of this and other aconites are very poisonous. (×¼)

escences, are aromatic, sweet-smelling.

The underground rhizome of *A. calamus* yields oil of calamus which is used in perfumery and medicine. It can also be candied as a confection, or powdered when dried for perfume sachets.
ARACEAE, 2 species.

Acrostichum a small genus of ferns found throughout the tropics and subtropics, most of which are tolerant of salt or brackish conditions. *A. aureum* [MARSH FERN, LEATHER FERN] is a cosmopolitan mangrove species often reaching 3m in height and over 1m across the tussock-forming base.
ADIANTACEAE, 3–5 species.

Actaea a small north temperate genus of herbaceous perennials, characterized by small, inconspicuous white flowers and large divided elder-like leaves. Several species are cultivated for their attractive berries, which are poisonous; they are black in *A. spicata* [BLACK SNAKEROOT, BUGBANE, BANEBERRY, BLACK COHOSH, HERB CHRISTOPHER], red in *A. rubra* and *A. erythrocarpa* and white in *A. pachypoda*.
RANUNCULACEAE, 10 species.

Actinidia a genus of hardy deciduous, climbing shrubs from East Asia to the Himalayas, possessing large, often heart-shaped leaves. Two of the better-known species, which grow best on walls, old trees and trellises, are *A. arguta* [BOWER ACTINIDIA, TARA-VINE] and *A. chinensis* [YANGTAO, CHINESE GOOSEBERRY]. The former grows vigorously, bearing white flowers and eventually greenish-yellow berries. The CHINESE GOOSEBERRY has rampant reddish hairy shoots, bearing white flowers in late sum-

mer, followed by edible greenish-brown fruits with a gooseberry flavor. It is cultivated commercially in New Zealand as KIWI FRUIT.
ACTINIDIACEAE, about 40 species.

Adansonia a genus of tropical trees native to Africa, Madagascar and Australia. One of the most important species, the African *A. digitata* [BAOBAB, MONKEY-BREAD TREE, SOUR GOURD, CREAM OF TARTAR TREE], has an extraordinarily thick, swollen trunk, which in some specimens has been reported to be 12m in diameter. Despite their great girth they are relatively short, normally about 12m in height, bearing stumpy branches which resemble roots. The white, fragrant flowers are some 15cm in diameter. The gourd-like woody fruit shells are used as pots and vessels. A glue is produced from the pollen and the young leaves are eaten by Man and other animals. The mealy pulp of the fruit of *A. madagascariensis* [also called the MONKEY-BREAD TREE] is also edible, as are the seeds of the Australian *A. gregorii* [BOTTLE TREE, GOUTY STEM].
BOMBACACEAE, 10 species.

Adenocarpus a small genus of deciduous or evergreen shrubs or small trees, native to the Canary Islands and the Mediterranean region. A number of the species are cultivated for ornament, particularly for their attractive racemes of pea-like flowers. They include the small deciduous shrub *A. complicatus* (= *A. intermedius*) with yellow, red-streaked flowers and the evergreen shrub *A. viscosus* (= *A. frankenioides*) with yellow flowers. Both species may reach a height of 1m in contrast to the deciduous *A. decorticans* which may grow up to 8m.

Another species cultivated is *A. foliosus*, an evergreen shrub similar to *A. viscosus*.
LEGUMINOSAE, about 20 species.

The flower-spike (spadix) and linear leaves of the Sweet Flag (Acorus calamus). This species is native to Asia but naturalized in Europe and America. (×2)

Foliage of the Common Maidenhair Fern (Adiantum capillus-veneris), showing the delicate fan-shaped leaf segments, bearing heart-shaped indusia. (×3)

Adiantum [MAIDENHAIR FERNS] a genus of delicate ferns with shiny black or purple stalks and thin, usually dissected fronds. They are shade-loving and mainly tropical or warm temperate plants although one species, *A. capillus-veneris* [COMMON MAIDENHAIR, VENUS HAIR, TRUE or BLACK MAIDENHAIR] occurs in Europe and another species grows as far north as Alaska.

Many are cultivated, especially in greenhouses or as houseplants in hanging baskets, including *A. capillus-veneris* with the fronds bearing delicate fan-shaped segments, *A. raddianum*, a Brazilian species with many cultivars, *A. macrophyllum* and *A. caudatum*.

Several species are used medicinally: the rhizomes of *A. aethiopicum* of tropical Africa as an abortifacient and *A. pedatum* [AMERICAN MAIDENHAIR] of North America and East Asia as a stimulant and expectorant.
ADIANTACEAE, about 200 species.

Adonis a small genus of annual or perennial herbs, native to Europe and temperate Asia, with leaves much divided into lobes, and solitary showy red or yellow flowers. Several species are cultivated in gardens, including the early yellow-flowered perennial *A. vernalis* [SPRING ADONIS], the annual, scarlet- or crimson-flowered *A. aestivalis* [SUMMER ADONIS] and *A. annua* [PHEASANT'S EYE, AUTUMN ADONIS]. Several species contain cardiac glycosides in the leaves or roots, with effects similar to those of digitalis, for which they are sometimes used as substitutes.
RANUNCULACEAE, about 20 species.

Aechmea a tropical American genus of herbs, mostly epiphytes, many of which have been introduced into cultivation as warm greenhouse or indoor plants.

Popular cultivated species include *A. fasciata* (= *Billbergia rhodocyanea*), which produces a beautiful dense conical head with rosy-pink bracts and bluish flowers, and *A. fulgens*, which bears a dense inflorescence of scarlet bracts with purplish-blue flowers,

followed later by dark red berries that remain on the plant for some time. *A. magdalenae* is cultivated in Central America for its tough fiber.
BROMELIACEAE, about 140 species.

Aegilops a genus of annual or biennial grasses native to temperate western Asia and Europe. Although such species as *A. neglecta* and *A. truncialis* are major components of Mediterranean grasslands, they have little commercial value. Some species (*A. squarrosa* and *A. umbellulata*) have been involved in the origin of the wheats.
GRAMINEAE, about 25 species.

Aegle a genus of small trees native to Indomalaysia. The best-known is *A. marmelos* [BAEL, BEL FRUIT, BENGAL QUINCE, BILVA], wild in the Indomalaysian monsoon forests but cultivated throughout Southeast Asia. This tree is armed with straight spines and has unpleasant smelling leaves and clusters of fragrant greenish-white flowers. The fruits are eaten fresh, or dried and made into a drink. Ripe fruit is said to be a tonic and laxative; boiled unripe fruit is said to be effective in the treatment of dysentery and diarrhea.
RUTACEAE, 3 species.

Aegopodium a genus of rhizomatous perennial herbs, native to northern and middle Europe, western Asia and Siberia. *A. podagraria* [GOUTWEED, BISHOP'S-WEED, GROUND ELDER] spreads rapidly by means of deep rhizomes which are difficult to eradicate and hence the plant can be an obnoxious weed in gardens. A less vigorous variegated form is occasionally planted as an ornamental. GROUND ELDER was once much cultivated as a potherb, and an infusion has been used to treat gout.
UMBELLIFERAE, 7 species.

Aeonium a genus of succulent herbs and subshrubs extending as far east as Ethiopia

The yellow-flowered perennial Spring Adonis (Adonis vernalis) is one of the most commonly cultivated species of the genus Adonis. ($\times \frac{1}{5}$)

and Arabia but with the main center of distribution in Macaronesia, especially the Canary Islands. There is an enormous range in habit from small-leaved, cushion plants to tall shrubs and cabbage-like rosette plants. Flower color varies from white through yellow and pink to bright red. *A. cuneatum,*

A. canariense and *A. arboreum* (particularly in its purple-leaved form) are often cultivated.
CRASSULACEAE, about 40 species.

Aeschynanthus [BASKET PLANT] a mainly Chinese and Indomalaysian genus of herbaceous or woody perennial evergreen trailing and climbing plants or epiphytes, some of which are cultivated as greenhouse ornamentals for their attractive tubular flowers. They include *A. radicans* (= *A. lobbianus*) [LIPSTICK PLANT], *A. pulcher* and *A. marmoratus* (= *A. zebrinus*). A particularly ornamental species is *A. parasiticus* (= *A. grandiflorus*) which bears clusters of long crimson and orange flowers at the ends of the branches.
GESNERIACEAE, about 80 species.

Aeschynomene a genus of tropical and subtropical shrubs and herbs, mainly native to India, Sri Lanka, Malaysia and tropical Africa. The stems are usually spongy, upright or floating and bear inflorescences of pea-type flowers.

The best-known species is *A. aspera* [SHOLA, PITH PLANT], a water plant with sensitive leaves whose stem pith is cut into slices and forms the basis of the industry (mainly centered in Calcutta) for the manufacture of sun helmets. The perennial herb

The Urn Plant (Aechmea fasciata), from Brazil, is one of the most popular bromeliads that are grown as house plants. ($\times \frac{1}{8}$)

A. uniflora yields a soft wood which can be used for a similar purpose as well as for the manufacture of floats for fishing nets. LEGUMINOSAE, about 150 species.

Aesculus [HORSE CHESTNUTS, BUCKEYES] a genus widely distributed in the north temperate zone in India, Europe and North America. The best-known is *A. hippocastanum* [COMMON HORSE CHESTNUT], introduced into Britain in the 16th century from its native mountain habitat in the Balkan Peninsula. They are deciduous trees or shrubs with a large spreading habit, palmate leaves and striking blossoms. In the spring the sticky, shiny buds are a favorite in floral decoration, while in the autumn the usually spherical, often spiny fruits split open to reveal the beautiful red-brown, polished seeds (chestnuts, conkers).

Most species, and many varieties of them, have become well-known ornamentals in parks and gardens in temperate regions, especially the HORSE CHESTNUT, which is the largest of these, growing to over 40m tall, and is extensively planted for ornament and as a shade tree in most of Europe. The flowers are white with a pattern of crimson and yellow spots towards the center. Two other large trees in this genus are *A. octandra* (= *A. flava*) [SWEET BUCKEYE, YELLOW BUCKEYE], from southeast USA, with yellow flowers and roundish fruits and *A. chinensis* [CHINESE HORSE CHESTNUT] from northern

The shrub-like habit and rosettes of fleshy red leaves of Aeonium arboreum ($\times \frac{1}{10}$)

The upright inflorescences of the Common Horse Chestnut (Aesculus hippocastanum) *produce a conspicuous display in May and June.* ($\times \frac{1}{6}$)

China with white flowers and rough fruits. By contrast, the white-flowered *A. parviflora* [BOTTLE-BRUSH or DWARF BUCKEYE] and the red-flowered *A. pavia* [RED BUCKEYE] from southern USA rarely exceed 3m in height. Medium-sized ornamental species which attain a height of 8–12m include *A. glabra* [OHIO BUCKEYE] and *A. californica* [CALIFORNIA BUCKEYE].

There are reports of the use of the bitter fruits for curing horses of coughs and other ailments. Bark and fruit of some species are used locally to stupefy fish. The timber is soft and not durable, but has been used for flooring, cabinet-making, boxes and charcoal.
HIPPOCASTANACEAE, 13 species.

Aethionema a genus of annual and perennial herbs widely distributed from East Asia to the Mediterranean region. Several species make attractive garden plants, perhaps the most popular being *A. coridifolium* [LEBANON CRESS], a dwarf, shrubby plant with showy terminal racemes of pale pink or pink-lilac flowers, and *A. grandiflorum* [PERSIAN STONECRESS] which grows to a height of 30cm and bears brilliant pink flowers.
CRUCIFERAE, about 50 species.

Aextoxicon a genus whose only species, *A. punctatum* [PALO MUERTO], is a large tree native to Chile. The durable wood is used for joinery.
EUPHORBIACEAE, 1 species.

Afrormosia a genus containing two extremely large tree species from the tropical rain forests of Africa. They provide a hard, close-grained yellow to brown wood, similar to teak. That of *A. elata* from Zaire, Nigeria and the Ivory Coast is exported to Europe and America for use in the veneer-making and furniture trades, while that of *A. angolensis*, from Angola and Zimbabwe, is used in furniture, boats and wagons. The genus is now generally included in *Pericopsis*.
LEGUMINOSAE, 2 species.

Afzelia a small genus of tropical African and Asian trees, some of which yield hardwood used in furniture-making and locally in construction. *A. africana* [AFRICAN MAHOGANY], from tropical Africa, has attractive durable wood, but has also been cultivated for ornament, because of its fragrant white and red-veined flowers. *A. bipindensis* and *A. pachyloba*, also from West Africa, produce commercial timber. Another ornamental species is the red-flowered *A. cuanzensis* [POD MAHOGANY] from central and southern Africa.
LEGUMINOSAE, 8 species.

Agapanthus [AFRICAN LILIES] a small genus of South African perennial herbs cultivated for their handsome foliage and flowers. The most commonly cultivated species is

A. orientalis, which has a range of cultivars with variegated foliage and double or multiple flowers of different shades of blue. *A. africanus* is often confused with *A. orientalis* but is less often cultivated. A decoction of its roots is used by several African tribes to ease childbirth.
LILIACEAE, about 9 species.

Agaricus a genus of fungi, one species of which, at least in Great Britain, is regarded as the only "true" MUSHROOM, the cultivated *A. bisporus*. Although a number of species, such as *A. arvensis* [HORSE MUSHROOM], are edible, only *A. bisporus* is economically important. It is the basis of the mushroom industry in North America and Western Europe.

No species is known to be deadly, but *A. silvicola* and *A. xanthodermus* and closely related species can cause alarming symptoms (coma) in a few susceptible people, but complete recovery is normal.
AGARICACEAE, 60 species.

Agathis see KAURI or KAURI PINES.
ARAUCARIACEAE, about 20 species.

Agave a genus of decorative and economically important fleshy rosette plants found in warm arid and semiarid regions from South America to southern North America. The very short stems bear rosettes of extremely large, tough, fleshy, wax-coated leaves, 1m or more in length in some species, and ending in a stout vicious spine. The flowers are white, yellowish or greenish and are borne in

The violet-blue inflorescence of the African Lily (Agapanthus africanus), which is known by the names Blue Agapanthus and Lily-of-the-Nile. (×¼)

large numbers on a giant, terminal, often spike-like inflorescence.

Some species flower every year but in others, such as *A. americana* [CENTURY PLANT, AMERICAN ALOE, MAGUEY], only a few leaves are produced each year and the plant builds up massive food reserves, flowering only after long intervals of anything up to 50 or 60 years. After this the plant usually dies.

The Mexican beverage commonly known as "pulque" is made from several species of *Agave* by fermenting the sap obtained from the stem. The distilled sap produces tequila. Fibers used for matting, carpets, rope and sacking are obtained from several species (see Istle fiber and Sisal hemp).

A. americana is often cultivated in gardens for ornament and is widely naturalized in the Mediterranean region.
AGAVACEAE, about 300 species.

Ageratum a genus of annual and biennial herbs native to tropical America. Many of the species are weeds but a few are cultivated as houseplants or bedding plants in gardens. These include *A. conyzoides* with a bushy habit, bearing blue or white flower clusters and the smaller, dwarf *A. houstonianum* with a wide range of pink, blue and white-flowered varieties.
COMPOSITAE, about 60 species.

Aglaia a large and important genus of shrubs and trees with the main areas of distribution in southern China, Indomalaysia, Indonesia, Australia and the Pacific Islands. Many species, including *A. ganggo* and *A. eusideroxylon*, yield valuable, strong and durable timber suitable for heavy construction work. The dark red timber of *A. argentea* is also used for furniture and

The emerging fruiting bodies of the Field Mushroom (Agaricus campestris). This is one of the best edible species. (×1)

cabinetwork. The flowers of the shrub *A. odorata*, cultivated in China, are used for scenting tea. The very fragrant flowers of the Indonesian shrub *A. odoratissima* yield an essential oil used in perfumes. *A. acida* and *A. oligantha* and some other species bear edible fruits.
MELIACEAE 250–300 species.

Agrimonia [AGRIMONY] a genus of perennial herbs found in north temperate regions. The best-known species is *A. eupatoria* [COMMON AGRIMONY], which yields a yellow dye. This species and the similar but hairy *A. repens* (= *A. odorata*) [SCENTED AGRIMONY] are occasionally planted in woodlands. The medicinal properties of *Agrimonia* were known to the Ancient Greeks, who used the plants for healing the eyes. Agrimony is still used as a source of a mild astringent.
ROSACEAE, about 15 species.

Agropyron [COUCH-GRASS, TWITCH-GRASSES, WHEATGRASSES, QUACK-GRASSES] a genus of perennial hardy grasses which occur extensively in temperate regions, occupying a wide range of habitats including near-marine situations. In this type of habitat the species most tolerant of salt is *A. junceiforme* [SAND COUCH], which can withstand short periods of immersion in sea water. It is a useful pioneer grass in that it is able to colonize sandy shores otherwise unpopulated by plants, where its extensive underground stems (rhizomes) help stabilize the sand into low dunes.

Other species of *Agropyron* are valuable forage grasses, for example *A. intermedium* [INTERMEDIATE WHEATGRASS], which grows from central Europe to Iran. The hardy and drought-resistant *A. smithii* [WESTERN WHEATGRASS] is planted on bare ground in North America to prevent soil erosion. *A. cristatum* [CRESTED WHEATGRASS] and *A. sibiricum* (= *A. desertorum*) have been very widely used for planting range and pasture land in North America. The best-known species, however, is almost certainly *A. repens* [COUCH-GRASS, QUACK-GRASS], one of the worst perennial weeds of temperate regions. It is

The inflorescence of the Common Agrimony (Agrimonia eupatoria), a plant long used as a home remedy for liver complaints. (×1¼)

mainly a nuisance on arable land where it rapidly regenerates from fragments of the creeping underground stems. The rhizomes of *A. repens* have been used under famine conditions as a source of food.
GRAMINEAE, approximately 150 species.

Agrostemma a small genus of erect annual herbs native to Europe and Asia Minor. *A. githago* (=*Lychnis githago, Githago segetum*) [CORN COCKLE] was introduced as a weed of cultivation almost throughout Europe. Although its seeds are considered poisonous, the young leaves have been used as an emergency food.
CARYOPHYLLACEAE, 3 species.

Agrostis [BENT-GRASSES, RED TOPS] a cosmopolitan genus of perennial and annual grasses, chiefly north temperate, widely distributed in a variety of habitats ranging from alpine meadows to coastal sand dunes. Many perennial species are dominant members of various grasslands and have a particular value in agriculture and horticulture because they are so hardy. *A. tenuis* [COMMON BENT, COLONIAL BENT, RHODE ISLAND BENT], for example, is widespread on the heath, moorland and wasteland grasslands of Europe and temperate Asia and has been introduced as a pasture grass in Australia, New Zealand and the Americas. In addition, its hard-wearing short turf makes it a very useful sports-field grass on bowling greens, lawns and golf courses. Another widespread species is *A. canina* which grows throughout Europe, temperate Asia and North America. There are two recognized strains which are utilized commercially – var *canina* [VELVET BENT] and var *montana* [BROWN BENT]. The former is a fine-leaved tufted grass of wet places and gives a short soft turf which is ideal for lawns. BROWN BENT on the other hand is a much

more drought-resistant grass of rough pastures and is thus used in lawns requiring a compact turf. Another important lawn grass is *A. stolonifera* [CREEPING BENT, WHITE BENT, FIORIN] a very variable, widespread species forming densely matted turf. Its dwarf coastal variety, commonly known as sea-marsh turf, produces very hard-wearing, close-cropping lawns, as required on bowling greens.

In the USA a number of the taller species are used as soil binders and hay grasses on poor soils. Notable examples of this type include *A. gigantea* [RED TOP, BLACK BENT] and *A. exarata* [WESTERN RED TOP], which are both found at varying altitudes as range grasses. GRAMINEAE, about 200 species.

Ailanthus a small genus of tall deciduous trees native to East and Southeast Asia and northern Australia. *A. altissima* [TREE OF HEAVEN] is often grown as a street tree in Europe and North America. In China the silkworm (*Attacus cynthia*) feeds on

The widespread Bent-grasses (Agrostis spp), seen here covered in cobwebs and dew, are the dominant component of many temperate grasslands. (×⅓)

A. altissima and other species and produces a silk that is cheaper and more durable than mulberry silk. *A. malabarica* is cultivated in North Vietnam for the leaves, which yield a black dye for silk and satin.

The bark of several species is tapped for resin which is used for incense and in local medicine to treat dysentery and other abdominal complaints. The hard yellowish wood is light, and in spite of its coarse grain takes a fine polish and is used for furniture, fishing boats and wooden shoes.
SIMAROUBACEAE, 8–10 species.

Ajuga a genus of annual and perennial herbs from temperate regions of the Old World. Three species, under the common name of BUGLE, are cultivated in rock gardens: the bluish-flowered *A. genevensis*,

A. pyramidalis and *A. reptans*. The dried leaves and stems of *A. chamaepitys* [GROUND-PINE], when made into an infusion, are said to stop bleeding as well as lowering the pulse-rate. LABIATAE, about 40 species.

Akebia an East Asian genus of twining climbers, the most common of which are *A. quinata* and *A. trifoliata* (= *A. lobata*), both natives of China and Japan. They are deciduous and woody with palmately divided leaves, and chocolate-purple unisexual flowers borne on the same plant. The dried leaves of *A. trifoliata* are used as tea, the purplish, sausage-shaped fruits eaten, and the stems used in basketry in Japan. LARDIZABALACEAE, 2–5 species.

Akee the fruit of an evergreen tree, *Blighia sapida* [SAPIDA], now naturalized and commonly cultivated in Jamaica, where it was taken towards the end of the 18th century, from its native West Africa. The plant bears small, greenish, fragrant flowers which ultimately give rise to fruits about 6cm long and red, or yellow tinged with red. The fleshy aril surrounding the black shiny seeds is edible and is eaten in West Africa and in the West Indies, but since a poison, hypoglycin A, occurs in unripe arils, and in part of the seed stalk, great care is necessary in the preparation of the aril for food.
SAPINDACEAE.

Albizia a large genus of deciduous trees or shrubs native to warm regions of the Old World and with one or two species in Central and South America. The plants grow up to 15m or more, bearing showy pink, yellowish or white flowers in globose heads or cylindrical spikes.

A number of species provide valuable timber, including *A. lebbeck* [EAST INDIA WALNUT, LEBBECK TREE, SIRIS TREE], *A. procera* [TALL ALBIZIA] and *A. welwitschii*, with hard, dense, close-grained, easily workable wood, rose-red or dark in color. Several species are a source of gum or resin. *A. saponaria* is used in the Philippines for washing – the bark is soapy and the fresh wood readily lathers in water. Other species such as *A. odoratissima* [FRAGRANT ALBIZIA] and *A. lebbeck* are used for shading tea or coffee shrubs in plantations.

A. julibrissin [PINK SIRIS TREE, SILK TREE, NEMU TREE] is one of several trees used for ornament.
LEGUMINOSAE, 100–150 species.

Alchemilla [LADY'S MANTLES]   a large genus of mainly perennial herbs with palmately lobed or divided leaves, often cultivated as garden plants. They are mainly native to Europe, Asia and America, particularly on mountains. Many of the European species reproduce vegetatively and most are apomictic. The whole plant is often covered with hairs, which in some species give the leaves a silvery appearance.

The very dwarf species occur in mountain snow patches and include *A. alpina* [ALPINE LADY'S MANTLE], while the largest herbaceous species generally come from northern and montane hay meadows and roadsides. Several species are widely cultivated in gardens, particularly *A. conjuncta*, which has deeply lobed leaves, and *A. mollis*, a large-leaved plant, densely covered with hairs.

A. xanthoclora (= *A. vulgaris* auct.) [LADY'S MANTLE] was once widely used medicinally, particularly for stopping bleeding.
ROSACEAE, about 350 species.

Alcoholic Beverages   see p. 18.

Alder   the common name for the genus *Alnus*, comprising some 35 species of deciduous trees and shrubs which bear flowers in catkins. They are predominantly natives of north temperate regions, but one or two species extend down the Andes of South America to Chile, Peru and Argentina. The species are mainly characteristic of damp

habitats. This tolerance of wet soils gives rise to a particular type of alder scrub or woodland known locally as "carrs", associated with the Icelandic word "kjarr", meaning fenwood.

Most of the alders in cultivation are hardy, and include *A. cordata* [ITALIAN ALDER], *A. glutinosa* [COMMON or BLACK ALDER], *A. incana* [GRAY or WHITE ALDER] and *A. oregona* [RED ALDER]. Their ability to flourish in wet soil conditions makes them popular for planting by rivers, streams and ponds. The wood is also used for furniture and clog making and some toy manufacturing. The bark of some species has long been a source of tannin and a dye (see Alder bark).

The term "alder" is also used popularly for several unrelated species from different genera such as **Frangula alnus* [ALDER BUCKTHORN].
BETULACEAE, about 35 species.

Alder bark   bark obtained from various *Alnus* species, such as *A. glutinosa* [BLACK ALDER] and *A. incana* [GRAY or WHITE ALDER], which has long been used for its tannin extract used in converting raw hides into reddish-colored leather. The tannin has also been used for dyeing linen, using a suitable mordant. Medicinally, preparations of alder bark are astringent, styptic and purgative.
BETULACEAE.

Aleurites   a genus of shrubs or small trees native to tropical Asia. The seeds of several species, in particular *A. fordii* [TUNG-OIL TREE, CHINA WOOD-OIL TREE] and *A. montana* [MU TREE], yield tung oil. *A. moluccana* [CANDLENUT, CANDLEBERRY TREE, VARNISH TREE] is the source of an inferior candle oil.
EUPHORBIACEAE, 6 species.

Algae   see p. 20.

Alchemilla glabra, a common species of Lady's Mantle, which grows at higher altitudes than A. xanthoclora. (× 1/10)

Bugle (Ajuga reptans) is a popular garden plant that spreads quickly in damp shady places by means of the creeping shoots. (× 1/4)

Algarroba or **algarrobo**   a Spanish word which once described only the pods of the eastern Mediterranean *Ceratonia siliqua* [*CAROB TREE]. The term is now applied to a number of plants and their pods containing a sweet mucilage which provides an important foodstuff for Man and cattle, the most important of these being species of **Prosopis*.
LEGUMINOSAE.

Alisma   a small genus of perennial aquatic or marsh herbs native to north temperate regions and to Australia, including *A. gramineum* [RIBBON-LEAVED WATER PLANTAIN] and *A. plantago-aquatica* [COMMON or GREAT WATER PLANTAIN]. The leaves of the latter have diuretic qualities and have been used medicinally for urinary complaints. These species and *A. lanceolatum* and *A. rariflorum* are grown at pond margins for their panicles of rose-lilac or white flowers.
ALISMATACEAE, 10 species.

Alkaloids   see p. 26.

Alkanet   the name of a number of similar perennial herbs, chiefly from the Mediterranean region, known since ancient times as the source of red dyes and pigments (alkanet, alkannin). Among them are **Anchusa officinalis* [BUGLOSS], *Pentaglottis sempervirens* and *Alkanna tinctoria* [DYER'S ALKANET].
BORAGINACEAE.

Allamanda   a small genus of mostly evergreen, climbing shrubs from the West Indies, South and Central America, which bear large, showy, usually yellow, funnel-shaped

Alcoholic Beverages

The process of fermentation whereby sugars or starches are broken down by yeasts or other microorganisms into alcohol and carbon dioxide in an aqueous environment antedates Man's evolution. Man has taken advantage of this natural production of inebriants for several thousand years: there is evidence of wine-making being practiced in the Mesolithic period (8000–6000 BC). Beer-making has probably as old a history and the fermentation of honey may be even older. The vine was the most frequent source of alcohol in the valley of the Nile and in Mesopotamia (before 2000 BC) and spread from early Hebrew sects to the Arabs and the Copts. Date-palm wine was another early popular source of alcohol.

Although the vine has historically been one of the main sources of alcoholic beverages and still retains a major role, many other plants have been used in the search for alcoholic inebriants. Any part of a plant – roots, stems, leaves, flowers and fruits – which contains sugar or starch can be used. The range of plants used is enormous and includes many local and national products.

Both wines and beer have played a major role in various civilizations, having mystical and religious significance as well as providing food and refreshment. The spread of wine and viticulture has been the more profound tradition largely due to the symbolic association between the red-colored liquid and blood and to the effect it has on Man's conscious state. Mythology is replete with allusions to the vine. For the Egyptians, Osiris, the God of the dead and underworld, personified the vine and libations were poured to him.

The soil and climate of Egypt and Mesopotamia were not ideally suited for growing vines of any quality and gradually viticulture extended to the northern Mediterranean in both the Greek and the Roman civilizations. Greece was ideally suited for growing good quality vines and viticulture became widespread. The God of the vine in Greek mythology was Dionysus who is often depicted on bas-reliefs, frescoes and amphoras bearing grapes or a wine cup in one hand and a staff (thyrsus) surmounted by a pine cone in the other.

In the Roman empire the spread of viticulture soon overtook that of the Greeks, especially after the conquest of Gaul; the God of wine was Bacchus to whom many of the traditions of Dionysus were ascribed, becoming eventually the God of pleasures and the God of civilization.

The process of brewing from grain crops to produce beer probably antedates the production of wine from grapes in the Sumerian culture of Mesopotamia. Fermentation of barley was widespread and beer was used as a food, beverage, libation and as a basis for many medicines. The use of *hops in beer-making probably derives from the Jews during their captivity in Babylonia, as a prophylactic against leprosy.

Distillation to produce strong spirits, although known for many centuries, is a relatively modern phenomenon as a large-scale industry. Whisky, gin, vodka, rum and brandy are the main commercially produced spirits as well as countless liqueurs made in smaller quantities.

Alcoholic beverages are often considered mistakenly as stimulants. On the contrary, alcohol is a protoplasmic poison and has a depressant effect on the central nervous system. It is also falsely reported to be an aphrodisiac but this is due to its mildly euphoric effect if taken in small quantities.

ALCOHOLIC BEVERAGES

Wines	Main production area	Genera utilized	Parts of plant utilized
Grape	Europe, N Africa, USA, Australia, S America, S Africa.	Vitis	Fruit
Flower	Europe (UK)	Cytisus, Primula, Tussilago, Sambucus	Flower
Fruit	Europe and USA, C Africa	Prunus, Ribes, Rubus, Fragaria, Sambucus, Sorbus, Rosa, Malus, Pyrus	Fruit
Mead	N Europe	various	Honey (nectar)
Beers			
Beer, Ale, Lager, Stout	Europe, N America, Australia, New Zealand	Hordeum, Humulus	Barley grain and hop cones
Cider, Cyder	England, France, Spain, USA	Malus	Apples
Ginger beer	England	Zingiber	Rhizome
Saké, Saki	Japan	Oryza	Rice grains
Pulque	Mexico	Agave	Sap from stems
Kvass (Quass)	USSR	Hordeum, Secale, Mentha	Barley, rye grain, leaves of peppermint
Pombe (boura)	Africa	Eleusine	Millet seed
Chicha	S America	Zea, Chenopodium	Corn grains or quinoa seeds
Sorgo	Africa, Asia	Sorghum	Grain
Kava	Pacific Islands	Piper methysticum	Roots
Palm wine (toddy)	C and S America, Africa, Orient	Borassus, Caryota, Phoenix, Nipa, Raphia, Acrocomia, Jubaea, Mauritia	Sap from decapitated stem apex or maltreated inflorescence
Spirits			
Brandy (cognac, armagnac, marc grappa, bagaceira, Hefebranntwein, pisco, etc)	Europe (France, Spain, Italy, Greece, Germany), S Africa, S America, Australia, USA	Vitis	Distilled grape-wine
Fruit brandy (slivovitz, kirsch, calvados, applejack)	France, C and E Europe, Jugoslavia, USA	Malus, Prunus	Distilled fruit-wine, sometimes with some kernels
Whisky, whiskey	Scotland, Ireland, USA, Canada	Hordeum, Zea, Secale	Barley, grain, maize, corn, rye malted, fermented and distilled
Rum	West Indies, Caribbean mainland, USA	Saccharum	Sugar-cane juice fermented and distilled
Gin	Britain, Holland, USA	Zea, Secale, Juniperus	Maize (corn) and rye, fermented and distilled flavored with Juniper and botanicals
Vodka	Russia, Poland, Finland	Secale, Solanum tuberosum or Triticum	Distilled potato or grain starch
Akvavit, Aquavit	Norway, Denmark, Sweden	Solanum tuberosum Carum carvi	Distilled potato starch flavored with caraway seed
Raki, Ouzo	Balkan peninsula, Turkey	Vitis vinifera, Triticum, Solanum or Saccharum	Distillate of wine, grain, potatoes, or molasses, sometimes flavored with aniseed
Arrack	Throughout the Orient	Oryza or Saccharum or Borassus or a mixture of Oryza and Borassus	Distillate of fermented rice and molasses or fermented juice of palms, or a mixture of palm toddy or rice
Mezcal	Mexico	Agave	Distillate of fermented agave sap and fibrous pulp
Tequila	Mexico	Agave	Distillate of fermented agave sap and fibrous pulp
Various			
Absinthe	France	Vitis vinifera, Artemisia absinthium	Spirit flavored with oil of wormwood
Vermouth	Italy, France, Germany, USA	Vitis vinifera, Artemisia absinthium	Wine, fortified, sweetened or not, and flavored with herbs
Liqueurs	Europe, S Africa	Vitis vinifera, Saccharum	Spirits, sweetened and flavored with herbs

(after Emboden, 1974)

Mature bulbs of the cultivated onion (Allium cepa). The bulk of the bulb comprises leaf-bases that are swollen with food reserves. (× ⅓)

flowers. Cultivated ornamental species include *A. cathartica*, a tall climber from Brazil, and *A. neriifolia*, an erect shrub, grown either as a bush or as a hedge.
APOCYNACEAE, 12–15 species.

All-heal the name given to a number of plants which were regarded as panaceas in the early days of herbal medicine. They include *Viscum album* [MISTLETOE], *Valeriana officinalis* [VALERIAN], whose roots and rhizomes are soporifics and tranquillizers, and also *Prunella vulgaris* [SELF-HEAL] a decoction of whose leaves has been used to treat sore throats and internal bleeding.

Allium a genus of bulbous plants which includes a number of important crop plants. Wild alliums are widely distributed in the Northern Hemisphere but the greatest diversity of species is found in Central Asia and in western North America. They are not found in the humid tropics or in dense forests.

Important crop species of Middle Eastern or Mediterranean origin are *A. cepa* [*ONION], *A. cepa* var *aggregatum* (= *A. ascalonicum*) [*SHALLOTS], *A. porrum* [*LEEK], *A. sativum* [*GARLIC] and *A. schoenoprasum* [*CHIVES]. Crops domesticated in the Far East are *A. fistulosum* [WELSH ONION, SPRING ONION], *A. tuberosum* [CHINESE CHIVES] and *A. chinense*

[RAKKYO]. Other species are recorded as being collected for food in many parts of the world. Some species are important weeds, notably *A. vineale*, a grassland weed that can cause an onion-like taint in milk.

Many alliums such as *A. aflatunense*, *A. moly* and *A. neapolitanum* are cultivated for their attractive flowers.
LILIACEAE, 450–500 species.

Allomyces a small genus of aquatic and soil fungi usually occurring in tropical and subtropical regions.
BLASTOCLADIACEAE, 6 species.

Allspice the common name given to the tropical evergreen tree *Pimenta dioica* (= *P. officinalis*) (Myrtaceae) which produces berries used as a spice. Native to the West Indies and Central America, it is so common in Jamaica that it does not have to be cultivated. The dried berries were mistaken by Spanish explorers for peppercorns ("pimienta"), hence the generic name *Pimenta* and the vernacular names PIMENTO, PIMENTA and JAMAICAN PEPPER. It is called "allspice" because its flavor resembles a combination of cloves, cinnamon and nutmeg. The fruits are picked when mature and after being dried in the sun become wrinkled and more aromatic. Allspice is used in baking, in preserves such as mincemeat, in mixed pickles and in curing meats.

The name is applied to some other plants: *Calycanthus floridus* [*CAROLINA ALLSPICE], *Chimonanthus praecox* [JAPANESE ALLSPICE] and *Lindera benzoin* [WILD SPICE, SPICEBUSH].

Almond the common name given to the important tree *Prunus amygdalus* (= *P. dulcis*) (Rosaceae), and to the edible nut that it produces. The almond is a native of western Asia although it is now grown commercially in southern Europe, parts of North Africa, South Africa and California.

P. amygdalus is closely related to *P. persica* [*PEACH] and the resemblance is particularly apparent in its pink flowers and young fruit. The blossom is an attractive sight in early spring and for this reason the tree is widely cultivated as an ornamental.
(continued on p. 24)

An ornamental, white-flowered cultivar of the Almond Tree (Prunus amygdalus).

Algae

ALGAE ARE THE MOST PRIMITIVE OF THE chlorophyll-containing plants and, with the exception of blue-green algae (see p.50), comprise a single kingdom of living organisms called the Protista. Whereas other groups are essentially land plants, algae are almost exclusively aquatic plants. They are found in virtually every watery environment from hot springs to ice-caps and from salt ponds to temporary rainwater pools. In the sea they constitute the sole primary producers of organic food materials and in freshwater lakes and ponds they are generally more important producers than the larger plants. Algae show a great range of size, from single-celled organisms invisible to the naked eye to giant multicellular seaweeds which may be over 50m in length.

It is difficult to define an alga, for the large assemblage of organisms covered by the term is united more by what they lack than by what they have in common. For example, the

algal plant body is not divisible into stem, leaf and root, as are many higher plants; instead the whole plant is referred to as a thallus. Algae do not achieve the level of reproductive specialization found in even the most lowly land plants, such as mosses or ferns.

Ecology. Algae are found in virtually all bodies of water, all watercourses and in damp sites on the land, such as marshes, bogs, moist soil and surfaces of trees and rocks where the humidity is normally high. In these environments they may be found free-floating and unattached or they may be attached in a number of ways. Free-floating algae in lakes and the sea are termed phytoplankton. Algae that are attached to the bottom of the sea and lakes are called "benthic" algae and this term is usually extended to include those in the littoral zone (attached to the shore). Rather surprisingly, most soils contain microscopic algae — diatoms and green algae. At the margins of lakes and tidal rivers there are large areas of mud, which are often extensively colonized by micro-algae. The main organisms involved here are *Euglena* and various diatoms. These are all motile and when the mud is covered by water they are able to retreat amongst the particles of sand and silt and can move out on to the surface again when they have been buried.

Most bodies of fresh water have marginal vegetation and they probably also have large submerged plants. All of these provide substrates for the epiphytic algae (mainly diatoms and filamentous green algae). A similar situation is found in marine environments where most larger algae have smaller algae attached to them.

On rocky shores, by far the most important algae are species belonging to the red, green and brown algal classes. These invariably grow in well-defined zones. The zonation is a product of a combination of factors, the most important of which are the ability of the algae to withstand desiccation during periods of exposure to the air, and competition between species. On European shores, for example, the upper zone is usually occupied by the small tough brown seaweed *Pelvetia*, which is only covered by water for a few days each month during exceptional high tides or spring tides. The middle of the shore is occupied by fucoids (brown algae of the order Fucales), which usually form zones; in descending order these are *Fucus spiralis*,

F. vesiculosus, *Ascophyllum nodosum* and *F. serratus*. These are all large plants that tend to cover the rock surfaces not already covered by animals and they leave little space for red or green algae. Another factor that can remarkably affect the zonation is the amount of wave action, for some algae are unable to tolerate exposure to this. For example, in very sheltered situations, *Ascophyllum nodosum* grows abundantly, but with increased exposure to rough water it disappears and the area it might have occupied will be taken up by other brown algae, such as the strap-shaped *Himanthalia*, or the red alga *Gigartina*. At the edge of the sea, beneath the low-water mark, the sublittoral zone commences, and this extends into the sea as long as light is available for photosynthesis. Here there is a much more stable environment than in the littoral zone and this is the habitat of the largest brown algae (known as kelps), such as *Laminaria* [OAR-WEED], *Macrocystis* [GIANT KELP] and *Nereocystis* [BULL KELP]. In addition there is usually a rich undergrowth of red algae.

Economic Importance. Algae have a greater economic importance than is generally appreciated. A variety of gelatinous materials, for example agar, alginates and carrageenin, are extracted from different algae and used in the manufacture of a wide range of foodstuffs and cosmetics. On some shores, particularly

Left *The red coloration on these boulders is caused by the filamentous alga* Trentepohlia *which belongs to the green alga class (Chlorophyceae) not the red.*
Below *The prolific Japanese seaweed* Sargassum muticum *which is threatening many indigenous species on the western coast of North America and parts of Europe.*
Opposite Page Pelvetia canaliculata *(Channelled Wrack).* (× 2)

of islands, sheep and cattle have traditionally been encouraged to eat seaweed at certain times of the year. On the Orkney island of Ronaldsay, for example, there are certain sheep that feed entirely on seaweed. In various countries, brown seaweeds have been dried and crushed and used as a supplement to animal feeding-stuffs.

Another traditional use of seaweeds has been as manure for spreading on the land. Algae were traditionally collected off the beaches following high storm tides, which tear off larger brown weeds and pile them on to the shore. The modern fertilizer industry uses similar algae but these are hydrolyzed to yield a liquid that can be more readily transported inland and more easily applied than wet algae in bulk form. The solution contains a wide range of mineral nutrients as well as organic matter. On the Atlantic coast of France subtidal calcareous red algae are collected for a fertilizer known as *maerl*. This is often powdered before application and, because of its high calcium carbonate content, is especially valuable on acid or clay soils.

Seaweeds were formerly much used for the extraction of chemicals. Brown algae would be collected from the shore and then spread over adjacent land for a few days to dry out. It was then burned in simple stone-lined pits or furnaces. The resulting ash, originally called "kelp" in Europe, contained a high concentration of sodium and potassium salts and was often used without further treatment in the manufacture of glass and glazed pottery. In the early 19th century it was discovered that the ash also contained a small percentage (1–4%) of iodine and this was extracted by further processing. In Japan alone, in the early years of this century, over 200 tonnes of iodine were produced by this means but this required the collection of up to three million tonnes of seaweed, and at that time without any mechanical assistance.

In many maritime communities around the world various seaweeds are traditional sources of food; one of the most commonly used is the red *Porphyra*. In South Wales this is made into a dish called laver bread which is fried with bacon. In Japan the same alga is cultivated, harvested and dried to make flat cakes called *nori*, which are used in various dishes. *Porphyra* has also been used in a similar way by the Maoris of New Zealand. In Ireland and North America the subtidal red algae *Rhodymenia* or "dulse" is either eaten fresh, or diced and then cooked with other food.

Brown algae have been eaten mainly in Japan where various *Laminaria* species are used for the production of *kombu*. This is mostly used in soups and stews or eaten with boiled rice. In North America, stalks of kelps such as *Nereocystis* have been candied and sold as sweets. Green algae have been much less used for food than red or brown algae but in various parts of the world *Ulva* [*SEA LETTUCE] has been collected for food, often being used as a salad. In Japan *Monostroma* is cultivated fairly extensively for the production of *aonori*, which is used as a garnish for meats and fish.

Unfortunately algae can also be a nuisance, for some, such as the red tides of certain dinoflagellates, produce toxins and others may grow abundantly where clear water is desired. Diatoms and green algae also occur in reservoirs and these block filters and actually grow in filter beds. In rivers and streams the process of artificial enrichment or eutrophication that is taking place as a result of increased sewage effluents and fertilizer run-off from the land, encourages the growth of algae. Dense tangles of the blanket weed *Cladophora* may virtually block water courses, making fishing very difficult and causing a further nuisance because of the smell and depletion of oxygen when it decays.

Classification. Many schemes have been proposed for classifying the algae within the kingdom Protista and there is no general agreement on which is the most suitable. All schemes are, however, based on the premise that the algae represent a number of fairly distinct lines of evolutionary development, there being eight divisions containing 12 classes (see accompanying table). Within the divisions the similarity of basic cell structure and biochemistry is more important than the range of shapes or ways in which the fairly uniform cells are put together to form the thallus.

The earliest classification schemes were based on the color of the plants and this approach has continued to the present day when biochemical techniques make possible precise separation and identification of many pigments found in the algae. All algae possess the green pigment chlorophyll *a* which is the basic photosynthetic pigment of all higher plants. Members of the green-colored classes Chlorophyceae, Charophyceae, Prasinophyceae and Euglenophyceae contain in addition chlorophyll *b*, which is also found in higher plants. The divisions Chrysophyta, Dinophyta and Cryptophyta contain instead chlorophyll *c*. The Rhodophyta (red algae) and Cryptophyta contain the pigments phycocyanin and phycoerythrin, which are mainly responsible for the distinctive colors of these groups. A large number of yellow-brown carotenoid pigments are found in the algae and the picture is still very confused. Most groups contain ß-carotene but fucoxanthin, for example, is found mainly in the Chrysophyta and Phaeophyta (brown algae). **Morphology.** Within the algae there is a number of distinct morphological lines, some of which are found in several of the classes. For example, in the class Chlorophyceae there is a flagellate line. There are simple single-celled flagellates such as *Chlamydomonas*; there are flat colonies with 12 or 16 similar cells united in a mucilaginous plate, as in *Gonium*. Sixteen or so cells may be united into a spherical colony which is either tightly packed, as in *Pandorina*, or with individual cells separated, as in *Eudorina*. The most elaborate member of the sequence is *Volvox* where the number of component cells is very large and "division of labor" occurs. Also in the green algae there is the "coccoid line." The simplest members are non-motile unicells, such as *Chlorella* and *Chlorococcum*. There are various types of colony, such as the aggregation of four or eight cells in *Scenedesmus* and the elaborate plate-like colonies of *Pediastrum*. The most elaborate member of this line is the "water net" *Hydrodictyon*, which consists of a large number of multinucleate cells.

From the point of view of their possible evolutionary significance, the filamentous lines are more important. A number of algae, such as *Ulothrix*, consist of simple unbranched filaments in which any cell is capable of division. Many algae have branched filaments; these either show intercalary growth, where virtually any cell can divide, or apical growth, where cell division is concentrated in the tips of the branches. In the red algae, remarkable thalli are formed by the close aggregation of much-branched filaments, which may consist basically of one, branched axis or of many closely associated together.

Algae formed of masses of rounded cells (parenchyma) are found in several groups. For example, *Porphyra* (Rhodophyceae) and *Ulva* (Chlorophyceae) are flattened multicellular plants that are usually attached to rocks. The parenchymatous construction

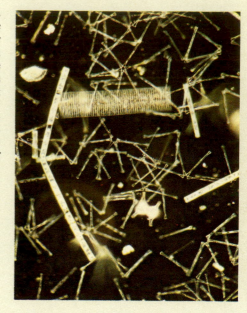

Light micrographs of microscopic, aquatic algae. Above Phytoplankton, *the free-living plants found in water. Seen here are: the star-shaped colonies of the diatom* Asterionella *with individual cells (some detached) looking like narrow threads; a chain of pennate diatoms (the horizontal, large rectangle of cells just above center); a filamentous green alga (the vertical filament left of center); and a desmid (the four-horned cell, below and right of center). (×300) Below top A sickle-shaped cell of* Closterium, *a green alga (division Chlorophyta) belonging to the group commonly known as desmids. (×400) Below Bottom The green alga* Eudorina *which is made up of a colony of 16 individual cells and inhabits fresh water. (×350)*

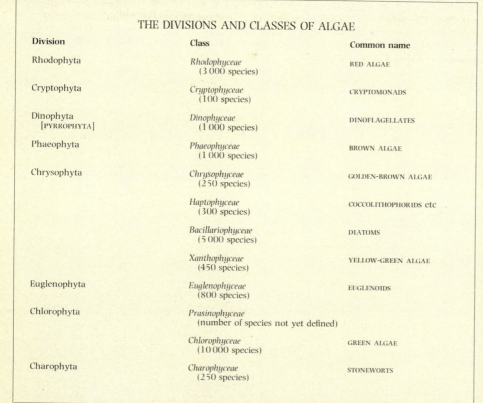

THE DIVISIONS AND CLASSES OF ALGAE

Division	Class	Common name
Rhodophyta	*Rhodophyceae* (3 000 species)	RED ALGAE
Cryptophyta	*Cryptophyceae* (100 species)	CRYPTOMONADS
Dinophyta [PYRROPHYTA]	*Dinophyceae* (1 000 species)	DINOFLAGELLATES
Phaeophyta	*Phaeophyceae* (1 000 species)	BROWN ALGAE
Chrysophyta	*Chrysophyceae* (250 species)	GOLDEN-BROWN ALGAE
	Haptophyceae (300 species)	COCCOLITHOPHORIDS etc
	Bacillariophyceae (5 000 species)	DIATOMS
	Xanthophyceae (450 species)	YELLOW-GREEN ALGAE
Euglenophyta	*Euglenophyceae* (800 species)	EUGLENOIDS
Chlorophyta	*Prasinophyceae* (number of species not yet defined)	
	Chlorophyceae (10 000 species)	GREEN ALGAE
Charophyta	*Charophyceae* (250 species)	STONEWORTS

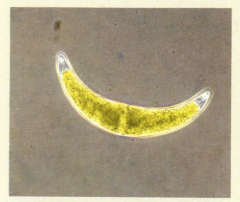

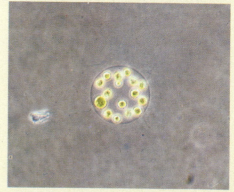

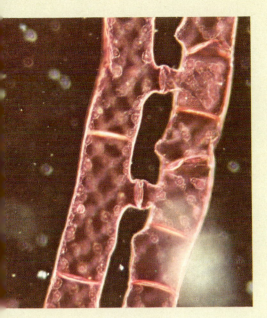

Light micrograph of the green alga Spirogyra *stained to show two filaments with conjugation tubes between. The contents of one cell will migrate to the other and fuse with it to form a zygote.*

reaches its peak in the Phaeophyceae (brown algae). Plants such as the large brown seaweeds and the giant kelps are constructed of several layers of these cells.

Another line of development is found in the siphonous or acellular algae in which there is a complete lack, or rare production of, crosswalls, the "cells" normally being multinucleate (ie containing many nuclei). Simple forms consist of branched filaments, as in *Vaucheria, or a single bladder as in *Valonia or Botrydium. More elaborate constructions are found in the green seaweed *Codium.

Reproduction and Life Histories. A number of algae are able to reproduce vegetatively from fragments of the thallus that have become detached from the parent plant, and in most algae there is some form of asexual reproduction. This is at its most simple in the flagellates of many groups and the diatoms, where binary fission of the parent cell gives two daughter cells, which then normally separate. In non-motile algae, such as filamentous and parenchymatous types, there is usually some form of motile spore (zoospore) formation. This may take place either in normal cells, as in many green algae, or in specialized bodies known as zoosporangia, as in some brown algae. The zoospores, which make possible the dispersal of the organism, have a brief period of motility before settling down on a substrate and growing into a new individual.

The most simple types of life history involve only one main vegetative phase and one sexual phase. Examples of this are common in the unicellular green algae where the ordinary cells of *Chlamydomonas or *Spirogyra, for example, may become gametes themselves and fuse with other gametes to give a thick-walled resistant zygote. Upon germination, new vegetative plants are formed. Since the vegetative cells contain just one set of chromosomes (haploid) this type of life cycle is termed haplontic, since only haploid vegetative plants are produced. In multicellular algae, such as *Volvox and *Oedogonium, the gametes produced are dissimilar (anisogametes); one gamete is motile but the other has lost its power of movement and is usually very much larger. The motile gamete swims to the other to bring about fertilization. This process is known as oogamy and is the type of reproduction found in all the more advanced algae. Very elaborate reproductive organs are found in the division Charophyta although this has a simple haplontic life history as in the other green algae described above.

In many algae there is an alternation of generations between a sexual (gametophyte) phase, which is normally haploid, and a vegetative (sporophyte) phase, which is normally diploid (ie its cells contain two sets of chromosomes – one from each parent). In what might be called the basic system, the gametophyte phase produces two sorts of gamete. The female organs (oogonia) are usually rounded or swollen structures in which the contents divide up to give one to eight eggs. The male organs (antheridia) are smaller structures in which there is considerable division to give a large number of very small flagellate antherozoids. Both gametes are released and after fertilization the zygote grows directly into a sporophyte plant, which may look exactly like the gametophyte (as in the brown alga *Dictyota) or may look quite different (as in *Laminaria). The sporophyte, when mature, forms sporangia within which the cells divide by a process of reduction division (meiosis) to form zoospores each of which now contains just the single haploid complement of chromosomes. These germinate to form new gametophytes.

In the red algae, sexual reproduction usually involves fixed carpogonia (equivalent to the oogonia) and non-motile spermatia, which are passively conveyed to the carpogonium. From this develops a parasitic carposporophyte phase of the life-cycle, which then forms carpospores that either give rise to gametophyte plants or, more usually, to sporophyte plants, which produce tetrasporangia. In these, haploid spores are formed, which then develop into new gametophyte plants.

In some algae such a complex life history appears to have been reduced to a more simple one. Thus the successful brown alga *Fucus* and its numerous relatives are diploid gametophyte plants. Reduction division takes place in the antheridia and oogonia to produce haploid gametes. After fusion the zygote immediately gives rise to a new gametophyte plant without the intervention of either a resting phase or a sporophyte.

Fossil history. The fossil history of the algae extends well into the Precambrian era where many remains of coccoid and filamentous blue-green algae have been found. In the Cambrian period (about 500–600 million years ago) green algae were well established. Also during this period, the first members of the Rhodophyta appeared. In the Silurian period the first charophytes appeared and a number of possible brown algae have been found. There is a fairly long gap until the Triassic period when both the dinoflagellates and the coccolithophorids (class Haptophyceae) make an undisputed appearance. The diatoms (class Bacillariophyceae) are very abundant from Jurassic times although some authorities allege that they first occurred much earlier. The remaining groups of algae were all probably in existence by the Cretaceous period (100 million years ago).

A drainage ditch covered with a floating mat of the green alga Enteromorpha.

Aloe saponaria *seen in cultivation in South Africa. The rosettes of fleshy leaves and spikes of showy red or yellow flowers are characteristic of aloes.* ($\times \frac{1}{5}$)

Almonds may be classified either according to the thickness of the shell, into paper-shelled, soft-shelled and thick-shelled types or, alternatively, by the presence or absence of a glucoside called amygdalin, into bitter (var *amara*) and sweet (var *dulcis*) varieties. Of the two varieties, only sweet almonds are edible; they may be roasted and salted or made into almond paste or ground for use in confectionery. Bitter almonds are inedible because of the volatile "oil of bitter almond" which contains hydrocyanic acid. Both varieties contain about 50% "oil of sweet almond", used as a flavoring and in cosmetics. Paradoxically, bitter almond is the main source of "oil of sweet almond". The residue after oil extraction is rich in protein and is used as an animal feedstuff.

Canarium commune [JAVA ALMOND], a totally unrelated species (Burseraceae), is cultivated in Malaysia. Its seeds are eaten raw or roasted and the extracted oil is used for cooking and for oil lamps. *Terminalia catappa* (Combretaceae) [COUNTRY ALMOND, INDIAN ALMOND, TROPICAL ALMOND] is culti-vated in the Old and New World for its oily seeds and for the tanning properties of the fruits, bark and roots. Its leaves are fed to silkworms.

Alnus see ALDER.
BETULACEAE, about 35 species.

Alocasia a tropical genus, all species of which are large, succulent, perennial herbs, which regenerate from corms weighing as much as 20kg. The starchy corms of some species including *A. macrorrhiza* [GIANT TARO] and *A. indica* are edible after peeling and boiling. They are used as food in various parts of Asia, including India and the Pacific Islands. The sap of some species is claimed to have medicinal properties, while that of *A. denudata*, which is native to Malaysia, contains a poison used for tipping arrows.
ARACEAE, about 70 species.

Aloe a genus of shrubby or xerophytic plants native to tropical and southern Africa, Madagascar and Arabia but introduced into many other parts of the world. A number of ornamental species are widely cultivated in Europe and the New World for their foliage and for decorative effect in gardens, parks and public buildings. Some aloes are stem-less, others grow up to 20m tall. The many leaves are spiny toothed and the flowers are showy reddish, eg *A. aristata* and *A. variegata* [PARTRIDGE BREAST ALOE], or orange, eg *A. ferox* [CAPE ALOE].

The leaves of *A. vera* (= *A. barbadensis*) [CURACAO ALOE, BARBADOS ALOE], are the source of bitter aloes, and this species is much cultivated in the tropics of the New World. After collection, the leaf sap is evaporated and cooled to form a compacted yellowish, reddish or black powder. The active principle of the powder is a glycoside used in medicine mainly as a purgative and vermifuge, and in small quantities in bitters.
LILIACEAE, about 300 species.

Alonsoa [MASKFLOWER], a small genus of herbs or shrubs native to tropical America. Cultivated ornamental species include *A. acutifolia* (= *A. myrtifolia*) with a bushy habit and deep red flowers and *A. warsce-wiczii* (= *A. grandiflora, A. compacta*) a red-stemmed compact plant with dark green ovate leaves and loose racemes of scarlet, spurred flowers.
SCROPHULARIACEAE, 6–10 species.

Alopecurus [FOXTAILS] a genus of temperate Eurasian and South American annual and perennial grasses. All species of *Alopecurus* are considered to be good forage grasses and of the perennials perhaps the best-known is *A. pratensis* [MEADOW FOX-TAIL]. It has become naturalized throughout the temperate grasslands by cultivation because it is one of the most important forage grasses in the early part of the season. Other prominent perennials include *A. geniculatus* [MARSH FOXTAIL], occurring in wet mead-ows, and *A. alpinus* [ALPINE FOXTAIL] of mountain grasslands. The well-known annuals include *A. myosuroides* [SLENDER FOXTAIL], common on dry and disturbed ground and a weed of cultivation throughout Europe and Asia, and *A. utriculatus*, a common species of the Mediterranean region. GRAMINEAE, about 50 species.

Alstroemeria a genus of perennial herbs from South America. They have leafy stems up to 1m high, bearing clusters or single lily-like flowers in brilliant colors – orange, scarlet, yellow or pink – from June to August. They make excellent cut-flowers. Well-known cultivated species include *A. aurantiaca* [PERUVIAN LILY] with orange or yellow flowers and *A. ligtu*, usually grown as the superb "hybrids" in a color range of pink, red, salmon and apricot. AMARYLLIDACEAE, 50 species.

Alternanthera a genus of mostly herbaceous species mainly native to tropical Central and South America. A number of these are propagated from cuttings as ornamental bedding plants. Most produce showy axillary clusters of small white flowers and display attractive colored blotches on the leaves.

The most commonly grown species is *A. ficoidea*, especially cultivars 'Bettzickiana', with narrow, yellow-red blotched leaves, and 'Versicolor' with wider, brown-red blotched leaves.

The tender stems and leaves of *A. triandra* are used as a vegetable and as a source of medicine in parts of Southeast Asia. AMARANTHACEAE, about 200 species.

Althaea a small genus of annual, biennial or perennial herbs found from Western Europe to northeastern Siberia and China. They have large roundish leaves and a terminal spike up to 1m long, bearing pink or yellow flowers.

The English sweetmeat, marshmallow, was originally made from the roots of *A. officinalis* [MARSH MALLOW]. It is cultivated

in France, Holland and Germany for the production of pasta althaea, a cough sweet. MALVACEAE, about 12 species.

Alyssum a large genus of annual, biennial or perennial herbs native to the Mediterranean region and east to Siberia. Most species occur wild in dry rocky or stony habitats but several are also widespread as weeds, while others are cultivated as garden plants. They are often grayish or silvery in appearance owing to the presence of hairs on the leaves or the stems.

The main cultivated species include *A. alpestre*, a procumbent perennial with pale yellow flowers, *A. argenteum*, an erect perennial with leaves silvery beneath and golden-yellow flowers, and the yellow-flowered perennials *A. corymbosum* (= *Aurinia corymbosa*) and *A. montanum*. Perhaps best-known of all is *A. saxatile* (= *Aurinia saxatilis*) [MAD-

The Peruvian Lily (Alstroemeria aurantiaca), *a native of Chile, has tall leafy stems tipped with brightly colored lily-like flowers.* ($\times \frac{1}{20}$)

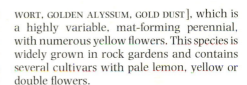

Gold Dust (Alyssum saxatile), *from the mountains of central Europe, is a splendid wall or rock-garden evergreen perennial.* ($\times \frac{1}{4}$)

WORT, GOLDEN ALYSSUM, GOLD DUST], which is a highly variable, mat-forming perennial, with numerous yellow flowers. This species is widely grown in rock gardens and contains several cultivars with pale lemon, yellow or double flowers.

The plant most commonly grown as alyssum (or alysson) is now placed in the related genus *Lobularia* as *L. maritima* [SWEET ALYSSUM]. CRUCIFERAE, about 150 species.

Amanita a genus of gill fungi which includes many common mushrooms and toadstools. It contains some of the most sought-after edible species, including *A. caesarea*, *A. ovoidea* and *A. rubescens* as well as some of the most deadly of all, for example *A. phalloides* [DEATH CAP], *A. virosa* [DESTROYING ANGEL], *A. verna* [FOOL'S MUSHROOM] and *A. pantherina* [PANTHER CAP]. *A. phalloides* is not uncommon in Europe and is responsible for nearly all fatalities by toadstool poisoning: symptoms do not occur until four to six hours, or even longer, after ingestion, and only about 50% of those who eat it recover. In North America it is rare, but is replaced by an equally poisonous species, *A. brunnescens*. The well-known *A. muscaria* [FLY AGARIC], although a poisonous species, is not usually fatal. Consumption of *A. muscaria* can cause symptoms of intoxication and high degrees of excitement, which may have serious social consequences. AGARICACEAE, about 50 species.

Amaranthus a genus of tropical and temperate, usually annual, herbs found in many parts of the world, often occurring as weeds in Europe. Some are cultivated as garden ornamentals for their showy leaves and brightly colored spikes of flowers, others as vegetables or for their seeds.

The best-known is *A. caudatus* [LOVE LIES

Alkaloids

Some of the alkaloids are of enormous economic importance, eg the entire tobacco industry is ultimately built on nicotine production and its physiological effects on the smoker, more than 70 000 tonnes of the alkaloid being produced annually. Another stimulant of worldwide significance is caffeine, which occurs in coffee and tea and also in cocoa and cola together with other alkaloids such as theobromine and cocaine.

Alkaloids are basic plant constituents containing amine nitrogen, and include a wide range of chemical structures, from simple monocyclic compounds such as coniine of *HEMLOCK to hexa- and hepta-cyclic forms such as solanine of the *POTATO and strychnine of *Strychnos nux-vomica. They produce a pronounced, usually quite specific action on different areas of the nervous system in Man and animals and are of great importance because of their poisonous properties and medicinal uses, properties that have often been known since antiquity.

Alkaloids are of wide occurrence in the plant kingdom but about 95% of the 5 500 types so far described occur in the flowering plants, the remainder being found in the fungi, algae, cycads, horsetails, club mosses, and gymnosperms. In the monocotyledons the richest alkaloid-containing families are the Amaryllidaceae and the Liliaceae, while in the dicotyledons their distribution is complex: at least one family – the Papaveraceae – is 100% alkaloid-containing, while other families such as the Rosaceae and Labiatae are apparently alkaloid-free. Only one species in the very large family Umbelliferae contains alkaloids – *Conium maculatum [HEMLOCK]. The family Apocynaceae accounts for about 18% of all known alkaloids, including reserpine obtained from species of *Rauvolfia.

It was thought until recently that alkaloids were found exclusively in plants, but the discovery of alkaloids in several marine organisms and in a number of arthropods (such as millipedes, ladybirds and water beetles) clearly indicates that alkaloid synthesis is not confined to plants.

The production of alkaloids is frequently localized in certain tissues and they are then translocated to other parts of the plant. Thus, nicotine is formed in the root and if a tobacco plant is grafted on to a tomato root, the leaves will be free of nicotine. The yellow color of *Berberis* roots is due to the presence of the alkaloid berberidine, and an important class of yellow alkaloids are the betaxanthins of the Centrospermae. In other cases alkaloids are found in the milky sap of the laticiferous vessels, eg *Papaver [*POPPIES] and *Chelidonium*.

The function of alkaloids in the plants themselves is still very much in doubt. Some authors regard the alkaloids as waste products of the plant's secondary metabolism. Although many alkaloids are toxic to fungi, only in a few instances has disease resistance been linked to a particular alkaloid produced by the host plant.

SOME COMMON ALKALOIDS, THEIR SOURCES AND CHARACTERISTICS

Alkaloid	Main sources	Pharmaceutical and other characteristics
Aconitine	*Aconitum napellus [MONKSHOOD]	Highly toxic to the central nervous system. Used as a heart and nerve sedative and for pain relief.
Anabasine	Anabasis aphylla	Also occurring in tobacco; has been of minor importance as an insecticide.
Arecoline	Areca catechu [*BETEL NUT PALM]	Used to cure tape worm infections, to reduce fever and as an astringent.
Atropine	*Atropa belladonna [BELLADONNA, DEADLY NIGHTSHADE]	Racemate of hyoscyamine. Enlarges pupil of eye, paralyzes nerve endings, antidote to toadstool poisons, nicotine etc.
Berberine	Berberis vulgaris [*BARBERRY] Hydrastis canadensis [GOLDEN SEAL]	Yellow alkaloid, abundant in plants, especially in roots. Used as a tonic.
Brucine	Strychnos nux-vomica [STRYCHNINE TREE]	Extremely toxic.
*Caffeine (theine)	Coffea arabica, C. canephora [*COFFEE] Camellia sinensis [*TEA] Ilex paraguariensis [PARAGUAY TEA, *YERBA DE MATÉ] Theobroma cacao [CACAO] Cola acuminata [*KOLA] Paullinia cupana [GUARANA] *Annona cherimolia [CHERIMOYA]	Heart stimulant and diuretic. Widely occurs in drinks produced from these plants.
Cinchonine	*Cinchona spp	Produced in bark of trees. Used to be important for treating malaria.
*Cocaine	*Erythroxylum coca [COCA TREE]	Paralyzes peripheral nerves, dilates pupil of eye, increases body temperature and is a powerful anaesthetic.
Codeine	Papaver somniferum [*OPIUM POPPY]	A narcotic closely related to morphine.
Colchicine	*Colchicum autumnale [AUTUMN CROCUS]	Highly toxic; affects spindle formation in nuclear division. Used to induce polyploidy.
Coniine	Conium maculatum [*HEMLOCK]	Toxin causing paralysis of motor nerve ends and muscles.
Curarine	Strychnos toxifera [*CURARE POISON NUT]	Extremely poisonous; used medicinally as a muscle and nerve tonic, and locally in South America to tip arrows.
Emetine	Psychotria ipecacuanha	Powerful emetic (ipecac) produced by the root.
Hygrine	*Erythroxylum coca [COCA TREE]	Used as a masticatory stimulant.
Hyoscyamine Hyoscine	*Hyoscyamus niger [HENBANE]	Closely related to atropine. Used for pain relief, as a narcotic and hypnotic.
Laudanosine	*Papaver somniferum	Constituent of opium, related to papaverine.
Lupinine	*Lupinus luteus [YELLOW LUPIN]	Toxic, bitter substance causing lupinosis.
Morphine	*Papaver somniferum [OPIUM POPPY]	Important narcotic and drug, which acts on the central nervous system.
Narcotine	Papaver somniferum [OPIUM POPPY]	A powerful narcotic, analgesic and drug, which is the chief constituent of opium.
Nicotine	Nicotiana tabacum, N. rustica [*TOBACCO]	Contained in all kinds of tobacco. Also used in the manufacture of insecticides and nicotinic acid (stimulates peripheral nerves).
Opium	Papaver somniferum [OPIUM POPPY]	Mixture of some 25 alkaloids including morphine and codeine, from juice of unripe capsules. Dangerous habit-forming drug.
Papaverine	Papaver somniferum [OPIUM POPPY]	Constituent of opium.
Pilocarpine	*Pilocarpus pennatifolius	Increases glandular secretions and peristaltic action of intestine. Contracts pupil of eye.
Piperidine Piperine	Piper nigrum, P. longum [*PEPPER]	Constituents of black, white and long pepper, responsible for peppery taste.
*Quinine	Cinchona spp, especially C. ledgeriana	Occurs in bark of trees. Very important in the treatment of malaria.
Ricinine	Ricinus communis [*CASTOR OIL PLANT]	Weakly toxic.
Scopolamine	*Atropa belladonna Scopolia atropoides	Narcotic and produces dilation of pupils.
Strychnine	*Strychnos nux-vomica [STRYCHNINE TREE]	Virulent poison, used as tonic in very small doses.
Thebaine	Papaver somniferum [*OPIUM POPPY]	Minor constituent of opium.
Theobromine	*Theobroma cacao	Heart stimulant and diuretic.
Trigonelline	*Trigonella foenum-graecum [FENUGREEK] Strophanthus spp	Used medicinally to relieve stomach upsets. Frequently found in human urine.
Yohimbine	Corynanthe yohimbe Pausinystalia yohimbe	Enlarges blood vessels. Used in veterinary medicine as aphrodisiac.

The Blusher (Amanita rubescens) is edible when cooked but poisonous raw, so is best avoided in case it is undercooked or misidentified. ($\times \frac{3}{4}$)

BLEEDING, TASSEL FLOWER], a species native to South America, with long pendent red, or rarely green, trailing tassels of flowers. The seed (called Inca wheat or quihuicha), like that of *A. hypochondriacus*, *A. cruentus* and other species in the tropics of Asia and Central America, is used to make bread (see Grain amaranth). The leaves or young plants of several species such as *A. gangeticus* and *A. mangostanus* are used like spinach, as potherbs or as salads in various parts of the tropics.
AMARANTHACEAE, about 60 species.

Amaryllis a genus represented by one species of great horticultural value, *A. belladonna* [BELLADONNA LILY], native to the Cape Province of South Africa. The large, scaled bulb gives rise to a flower stem bearing three or four usually pale pink trumpet-shaped short-tubed flowers. The name *Amaryllis* is widely applied in horticulture to species of the genus *Hippeastrum*.
AMARYLLIDACEAE, 1 species.

Ambrosia [RAGWEEDS] a genus of mainly southwest North American annual or perennial herbs many of which are aromatic, eg *A. artemisiifolia* whose fruits yield a drying oil. Some species (eg *A. hispida*, *A. maritima* and *A. peruviana*) produce extracts used locally for treating various internal and external ailments. *A. artemisiifolia* and other species are major causes of hay fever.
COMPOSITAE, 35–40 species.

Amburana a small Brazilian and Argentinian genus of trees which bear axillary racemes of yellow and white fragrant flowers. Coumarin extracted from the seeds of *A. cearensis* (= *Torresea cearensis*) is used in the perfume industry, in soap manufacture and in aromatic tobacco. A medicinal oil is derived from the resin. *Amburana* timber is used for general construction work.
LEGUMINOSAE, 3 species.

Amelanchier [SHADBUSH, SHADBLOW, SERVICEBERRY, JUNEBERRY a genus of deciduous trees or shrubs often cultivated for their abundant white flowers in spring and brilliant red autumn foliage. Most species are found in North America, extending as far south as Mexico, with the remainder distributed throughout central and southern Europe and parts of Asia.

In addition to their ornamental value some species such as *A. canadensis*, *A. spicata* and *A. stolonifera* bear small reddish-black or purple fruits which are sweet and can be used to make jellies.
ROSACEAE, about 25 species.

Amherstia is represented by only a single species, *A. nobilis*, which is a rare native of southern Burma but commonly cultivated in many tropical gardens as it is a beautiful tree bearing long dangling inflorescences of yellow and red flowers.
LEGUMINOSAE, 1 species.

Ammi a small genus of annual or biennial herbs native to southern Europe, tropical Africa, Asia Minor and Macaronesia. The ovoid, ridged, dry fruits of *A. visnaga* are used in folk medicine particularly for the treatment of urinary ailments. Those of *A. majus* (= *Apium ammi*) [BISHOP'S WEED] have been used to treat intestinal and respiratory ailments as well as angina pectoris. A modern drug derived from the fruits of *A. majus*, methoxsalen, is used to treat skin complaints. Both species are grown for ornament and *A. majus* is widely cultivated for use in the cut-flower trade.
UMBELLIFERAE, about 10 species.

Ammophila an important grass genus, one species of which, *A. breviligulata* [AMERICAN BEACHGRASS], is native to America while the closely related *A. arenaria* [EUROPEAN BEACHGRASS, MARRAM GRASS] occurs naturally on maritime sands throughout Europe and in the Mediterranean region. It has been introduced onto the Atlantic seaboard of America and also Australia.

A. arenaria has extensive, spreading rhizomes which rapidly grow through loose sand. The aerial leaves and stems, which grow to 1.2m, break the force of wind and cause wind-blown sand particles to surround and eventually bury the plants altogether. With other grasses of similar habitats, stabilized dune systems are rapidly produced. For this reason, *A. arenaria* is most valuable in coastal maintenance and reclamation.
GRAMINEAE, 2 species.

Amorphophallus a genus of herbaceous perennials native to tropical Africa and Asia, some species of which, although rarely purposely cultivated, are used as an emergency food in the Pacific Islands, Indonesia and other Asian countries.

A. campanulatus [WHITESPOT GIANT ARUM, TELINGO POTATO] is the most important edible species, although the swollen corms of *A. variabilis*, *A. prainii* and *A. rivieri* [DEVIL'S TONGUE, LEOPARD PALM] are also used as food in various parts of Indonesia, Malaya and China respectively. In all cases the corm is

The Tassel Flower (Amaranthus caudatus) exists in both red-flowered and, rarely (as here), green-flowered forms. ($\times \frac{1}{4}$)

peeled and then boiled. Harvesting can be delayed for more than two years after planting, when the corm may reach 10kg in weight, hence their value as an emergency food. The protein deficiency of the corm is compensated in the case of *A. campanulatus* by a high content of vitamin A and minerals such as calcium.
ARACEAE, about 100 species.

Ampelopsis a genus of deciduous climbers found in Central and northern Asia and North America. The leaves may be variously shaped, from simple to pinnate or palmate. *A. arborea* [PEPPER VINE], a bushy climber with finely divided bipinnate leaves, is sometimes cultivated as a wall climber. Other cultivated species are *A. brevipedunculata* and *A. japonica*. *Ampelopsis* of horticulture is *Parthenocissus*, for example *P. quinquefolia*.
VITACEAE, about 20 species.

The pyramidal spike of bright pink flowers with long, slender nectar-containing spurs of the Pyramidal Orchid (Anacamptis pyramidalis). (×4)

Anabaena a genus of blue-green algae closely related to *Nostoc* but the unbranched filaments are separate or united by their gelatinous sheaths into small groups, instead of forming large gelatinous colonies, and they can also move with a gliding motion. Some species such as *A. flosaquae* are sometimes very abundant in freshwater lakes and have filaments which contain gas bubbles causing them to float on the surface of the water, producing a "bloom". Other species such as *A. torulosa* are abundant on the surface of the mud of salt marshes. Other species again occur in soils, especially under moist conditions and are particularly common in rice fields.

It has been shown that many *Anabaena* species can "fix" atmospheric nitrogen, converting it into nitrogenous compounds available to other organisms, including crop plants such as rice. The ability of rice to produce good crops in the absence of fertilizer is thought to be due to the activities of *Anabaena* in the soil.
NOSTOCACEAE.

Anacamptis a genus represented by a single species, *A. pyramidalis* [PYRAMIDAL ORCHID], occurring in Europe, western Asia and North Africa. The densely crowded pyramidal flower-spike bears fragrant flowers, bright pink or purple in color.
ORCHIDACEAE, 1 species.

Anacardium a small genus of tropical South American trees which includes *A. occidentale* [CASHEW NUT TREE], a species native to Mexico, extending to Peru and Brazil, which was introduced into the Old World tropics by Portuguese and Spanish colonists in the 15th to 16th century, and is now widely cultivated and naturalized there.

After fertilization, the stalk and receptacle of the flower swell into a soft pear-shaped accessory fruit up to 20cm long which tastes rather like apple. Cashew apples are popularly eaten raw or made into jam, and the juice made into beverages and wine. For cashew nuts, the crop is processed by roasting the whole fruit until the outer parts crack and peel off, leaving the cashew nut kernel.

The wood of *A. occidentale* is a valuable timber. In addition, when wounded the tree produces a resinous secretion known as cardol which is used to tar boats and to preserve fish nets, to protect wooden structures from insect damage in the tropics, as a stain and varnish for woodwork, and as a gum sometimes used in bookbinding (*acajou gum).

The tree *A. rhinocarpus* [ESPAVE MAHOGANY] produces a yellowish-brown timber similar to mahogany.
ANACARDIACEAE, about 15 species.

Anacyclus a small genus of annual or perennial herbs native to the Mediterranean region. Most species are weeds but at least two are cultivated for their attractive daisy-like flower heads, including *A. radiatus* and *A. depressus* [MOUNT ATLAS DAISY] from Morocco. *A. pyrethrum* [PELLITORY] is cultivated in Algeria for its essential oil used in alcoholic drinks and mouthwashes.
COMPOSITAE, about 25 species.

Anagallis [PIMPERNELS] a widely distributed genus of small annual or perennial herbs. *A. arvensis* [SCARLET PIMPERNEL, POOR MAN'S WEATHER GLASS] is a common European annual species of gardens and arable land on light, calcareous soils. The closing of the scarlet flower with the onset of clouds and fall in temperature, often after fine weather, is said to foretell rain. There are some blue-flowered varieties of *A. arvensis*.

Other well-known species include *A. monelli*, a variable perennial species from southwest Europe with blue, red or white flowers, depending on the variety, and *A. tenella* [BOG PIMPERNEL], a native of damp, acidic heaths and bogs in Western and central Europe, Greece and Crete, bearing attractive pale pink flowers. *A. linifolia*, a perennial from the western Mediterranean, bears gentian-blue flowers.
PRIMULACEAE, about 28 species.

Ananas   see PINEAPPLE.
BROMELIACEAE, 5 species.

Anaphalis [PEARLY EVERLASTING FLOWER] a genus occurring mainly in the north temperate regions of the world through Europe, Asia and America. They are erect, perennial herbs characterized by their white woolly appearance, which is caused by a dense covering of hairs. Several species are cultivated either as rock-garden plants or as plants for sunny borders; when cut and dried all make excellent silvery-creamy everlasting flowers. The most popular species is probably *A. margaritacea* which is especially good as an everlasting flower. *A. triplinervis* is a similar, though somewhat smaller plant from the Himalayas with larger leaves.
COMPOSITAE, about 35 species.

The silvery appearance of Anaphalis lerentizii *(photographed here at 3 000m. in West Java) is due to a dense covering of hairs. (× $\frac{1}{10}$)*

*Flowers of the Scarlet Pimpernel (*Anagallis arvensis*) close up in response to a change to cloudy or cooler weather — hence the common name Poor Man's Weather Glass. (× $\frac{1}{3}$)*

Anchusa a genus of annual, biennial or perennial herbs found in Europe, western Asia and Africa. They bear racemes of yellowish, bluish or whitish funnel-shaped flowers. Several species are cultivated in gardens, including *A. azurea*, with several cultivars varying in color and size of flowers, the blue-flowered *A. officinalis* [BUGLOSS] and the yellow-flowered *A. ochroleuca*.
BORAGINACEAE, 30 species.

Andreaea a genus of mosses mainly north and south temperate as well as bipolar in distribution. *Andreaea* species mostly appear blackish when dry, dingy olive-green or dull reddish-brown when moist. They grow chiefly on hard rocks. The stems form carpets or sparse turfs rather than cushions, but are seldom a very conspicuous component in the vegetation.
ANDREAEACEAE, about 120 species.

Andromeda a genus represented by two species, *A. polifolia* [MARSH ANDROMEDA, BOG ROSEMARY], a dwarf evergreen shrub native to acidic heaths of mountains and moorlands of Europe, Asia and North America, and *A. glaucophylla*, a taller species from North America. *A. polifolia* is a dense-growing species which is cultivated in a number of varieties and has linear, dark-green leaves and terminal clusters of pale pink and white bell-shaped flowers.
ERICACEAE, 2 species.

Andropogon an important genus of tall tropical and subtropical grasses distributed in a wide range of habitats of both hemispheres. Many species are valued for their use as fodder crops and as soil or sand binders. Commercially, several species, eg *A. aciculatus*, are used for their fibers.

A. gayanus, an African tufted grass which grows to well over 2m in height, is an important fodder crop, and has been introduced to various parts of the Asian tropics for this purpose; its stems are used for matting. In the dry parts of India *A. pumilus* is relished by cattle as both fresh grass and hay.
GRAMINEAE, about 115 species.

Androsace a genus of hardy, small, tufted perennial herbs native to north temperate regions and characteristic of the tundra, alpine pastures, screes and rock clefts. Some species, such as *A. alpina* (= *A. glacialis*), form loosely tufted rosettes and bear pink flowers while others such as the white flowered *A. chamaejasme* [ROCK JASMINE] develop distinct rosettes. Both of these species and the European *A. carnea* and *A. imbricata* are cultivated best in alpine houses. On the other hand the Himalayan rose-pink-flowered *A. sarmentosa* is easier to grow in rock gardens. Other cultivated species include *A. albana* (white, pink or rose flowers),

*The pink bell-shaped flowers and narrow leaves of the Bog Rosemary (*Andromeda polifolia*). (× $\frac{1}{3}$)*

A. armeniaca (white flowers), *A. lanuginosa* (rose flowers) and *A. septentrionalis* (white or pink flowers).
PRIMULACEAE, about 100 species.

Anemia a genus of primitive, erect, terrestrial ferns widespread in Central America, Africa and southern India, where they colonize open clay banks, roadsides and disturbed areas. They have no economic value but *A. seemannii* has been quoted as being formerly used by the women of Panama as an abortifacient.
SCHIZAEACEAE, about 90 species.

Anemone [ANEMONES, WINDFLOWERS] a genus of attractive herbaceous perennials widespread in cooler parts of the Northern Hemisphere but with a few representatives in South Africa and South America. Many species are cultivated in gardens or commercially for their showy and colorful

Anemone coronaria 'de Caen', one of the two main strains of florists' anemone, produces large, long-stemmed single flowers. (× ½)

flowers borne either singly or in clusters.

The florists' anemones are the tuberous-rooted *A. coronaria* [POPPY ANEMONE, CROWN ANEMONE], *A. pavonina*, *A. hortensis* and *A. × fulgens* [SCARLET ANEMONE]. Two main strains (and numerous clones) of *A. coronaria* are grown – de Caen, with large single, long-stemmed flowers in a wide range of showy colors, and St. Brigid, with large semidouble flowers made up of lanceolate sepals in three rows, again in a wide range of colors. These are widely grown for forcing. *A. pavonina* includes the St. Bavo anemones with long-stemmed single flowers and an involucre of three bracts and has forms with single or double flowers which are red, violet or rose,

sometimes yellowish or whitish at the base of the sepals. *A. hortensis* is closely related and in fact *A. × fulgens* is a hybrid between *pavonina* and *hortensis* with scarlet or flame-colored flowers. Also widely grown is *A. hupehensis* [JAPANESE ANEMONE], the semi-double form of which was introduced to Japan around the beginning of the century and is the historic *A. japonica* (now called *A. hupehensis* var *japonica*). It was crossed with *A. vitifolia* to produce the commonly cultivated *A. × hybrida* (= *A. elegans*) of gardens.

Other attractive, widely grown species include *A. apennina* from southern Europe, a tuberous-rooted species with large blue flowers, *A. narcissiflora* from Europe, Asia and North America, with large white or cream flowers, often backed with pink or purple, and the white-flowered *A. nemorosa* [WOOD

ANEMONE, GROVE WINDFLOWER] of European woodlands.
RANUNCULACEAE, 70–80 species.

Anethum see DILL.
UMBELLIFERAE, 4 species.

Angelica a genus of perennial herbs native to north temperate regions and to New Zealand. The angelica used in confectionery is obtained from the green stems or leaf-stalks of *Archangelica officinalis* (formerly known as *Angelica officinalis*). *Angelica sylvestris* [WILD ANGELICA] occurs in fens, damp meadows and woods. The flowers are white or pink and borne in umbels on stout purplish stems. The stalks of the Japanese *A. edulis* and the North American *A. atropurpurea* [PURPLE ANGELICA] are used locally as food and for medicinal purposes.
UMBELLIFERAE, about 60 species.

Angiopteris a genus of huge ferns found in tropical forests from Madagascar across to Southeast Asia, Japan and Polynesia. Compound fronds, often reaching 5m long and 3m across, arise from trunk-like rootstocks up to 1m high.
MARATTIACEAE, about 10 species.

Angiosperms see p. 32.

Angostura the name given to the aromatic bitter tonic obtained from the bark of the South American tree *Cusparia febrifuga*.
RUTACEAE.

Wild Angelica (Angelica sylvestris), showing its characteristic many-rayed umbels borne on hollow purplish stems. (× 1/15)

Flower of Angraecum giryanae, one of several species of this genus of orchids that are cultivated in warm greenhouses. (× 1)

Angraecum a genus of epiphytic, lithophytic (rock-inhabiting) and semiterrestrial evergreen orchids found throughout tropical Africa and as far east as Madagascar, Réunion, Mauritius, the Comores, Seychelles, Sri Lanka and the Philippines. A few species such as *A. distichum* are widespread, but most are endemic to fairly small areas. Many species are very small with short narrow leaves and insignificant green flowers, but some of the large-flowered species such as the white-flowered *A. distichum* (= *Mystacidium distichum*) and *A. infundibulare* (both from Africa) and the pale-green flowered Madagascan *A. superbum* (= *A. eburneum*) are frequently cultivated in warm greenhouses. ORCHIDACEAE, about 220 species.

Anigozanthos a small genus of perennial herbs native to southwestern and western Australia. At least two of the species, *A. flavidus* and *A. manglesii* [KANGEROO PAW], are cultivated as ornamentals for their racemes of large tubular, showy, predominantly yellowish-green flowers, tinged with red or blue. HAEMODORACEAE, about 10 species.

Anise the common name for *Pimpinella anisum*, an annual umbelliferous herb whose hard, light-brown fruits have been used as a flavoring, spice and medicine since 2000 BC. It originated in the eastern Mediterranean or Middle East but it is now widely established in Europe, Asia and North America. It is cultivated in many Mediterranean countries and in southern Russia and India.

Oil of anise, which is obtained from the fruits by steam distillation, is used in cough mixtures and lozenges. The seeds (aniseed) are also widely used as a major flavoring in various alcoholic cordials and liqueurs, such as absinthe, pastis, anis, aguardiente, ouzo, raki etc and sweet liqueurs such as anisette.

The flavoring in these products is usually obtained from *FENNEL and STAR ANISE as much as from ANISE itself. STAR ANISE is a quite unrelated plant, *Illicium verum* (Illiciaceae), a small evergreen Chinese tree. Its star-shaped fruits are strongly flavored and extracts are used in anise-flavored drinks and other commercial products along with or as a substitute for aniseed.

Anise pepper is a spice obtained from the dried red berries of *Zanthoxylum piperitum* [JAPAN PEPPER, JAPANESE PRICKLY ASH], a small spiny tree of the family Rutaceae.

Annatto a vegetable coloring obtained from the seeds of *Bixa orellana*, a small tree native to the lowland tropics of the New World but now widespread throughout the tropics as an ornamental and source of a reddish pigment which consists mainly of a carotenoid, bixin. The main commercial use of bixin today is for coloring cheese, butter and margarine, and chocolate. Use of bixin-based colors in the food industry has increased recently because many synthetic dyes have been banned as potentially toxic. World consumption of annatto has consequently doubled since 1950. BIXACEAE.

Annona a commercially important genus of mainly tropical American small trees or shrubs, several of which yield edible fruits and are widely cultivated. The commonly cultivated species were introduced to most parts of the tropics after the discovery of the New World.

A. cherimola [CHERIMOYA] produces delicious, smooth, heart-shaped to globular green fruits with a scaly surface. The CHERIMOYA is widely cultivated in Asia and also on a smaller scale in Madeira, the Canary Islands and parts of southern Spain.

A. muricata [SOURSOP] is widely cultivated in tropical America and has now spread throughout the lowland tropics. The very large dark-green ovoid fruits weigh up to 4kg and are covered with soft recurved fleshy spines. The white fibrous flesh, containing many black seeds, is somewhat acid and is best strained and used for drinks and ice-cream.

A. squamosa [SWEETSOP, CUSTARD APPLE, SUGAR APPLE] produces yellowish-green heart-shaped fruits with white, sweet, granular pulp like custard surrounding the brown glossy seeds, which are used as a dessert. Other cultivated species include *A. diversifolia* [ILAMA, ANNONA BLANCA], *A. glabra* [POND APPLE], *A. montana* [MOUNTAIN SOURSOP, WILD SOURSOP], *A. senegalensis* [WILD CUSTARD APPLE] and *A. reticulata* [BULLOCKS' HEART, COMMON CUSTARD APPLE]. ANNONACEAE, about 120 species.

Anoectochilus [JEWEL ORCHIDS] a small genus of tropical, mainly Asiatic, terrestrial orchids. Their most attractive feature is the leaves which in most species as well as being strikingly veined are usually of a velvety metallic maroon or blue-green hue, giving rise to the popular name, JEWEL ORCHIDS, which is also applied to *Goodyera* species. ORCHIDACEAE, about 25 species.

Antennaria a genus of woolly perennial herbs or dwarf shrubs widely distributed in temperate and subarctic regions and frequently in mountain habitats. One of the commonest species in Europe is *A. dioica* [CAT'S FOOT, MOUNTAIN EVERLASTING], found in heaths, dry pastures and dry mountain slopes, and often cultivated. The small flower heads are invested with white or pink bracts which because they do not lose color or shape on drying are valued as "everlastings". Other cultivated species with similar habit are *A. carpatica*, *A. neglecta* and *A. chilensis*. COMPOSITAE, about 15–100 species. (continued on p. 36).

Anthemis cupaniana, from Italy, with its daisy-like flowers, is one of several Anthemis *grown in borders or rock gardens.* (× ¼)

Angiosperms

THE FLOWERING PLANTS

THE FLOWERING PLANTS OR ANGIOSPERMS are the dominant group of plants on the Earth today. The 250 000 or so species recognized show an almost bewildering diversity of habit, form and color, ranging from the tiny green specks, about 1mm in diameter, of the aquatic *Wolffia microscopica*, which is the smallest known flowering plant, to Australian gum trees (*Eucalyptus* spp) some of which grow to nearly 100m in height, and the *BANYAN (*Ficus benghalensis*), a celebrated specimen of which, in the Botanic Garden, Sibpur, Calcutta, is over 416m in circumference. Angiosperms occupy almost every kind of habitat, from sunbaked arid deserts and windswept mountain tops to tropical swamps and forests, and lakes, rivers and even marine habitats, salt water and sandy shores. As such they form the basic framework for most terrestrial animal life, including Man.

The flowering plants (class Angiospermae) are divided into two large natural groups, the subclasses Dicotyledonae (dicotyledons) and the Monocotyledonae (the monocotyledons). The dicotyledons are by far the larger group, with about 180 000 species, and contain nearly all the well-known trees and shrubs.

Although most flowering plants manufacture carbohydrates by photosynthesis and absorb the necessary water and minerals from the soil through their roots, some of them have become partially or wholly parasitic on other vascular plants: examples are the *dodders (*Cuscuta* spp) and *mistletoes (*Viscum* spp) and the total parasites such as the remarkable *Rafflesia* whose vegetative body is reduced to cellular filaments which parasitize the roots of members of the vine family and develop gigantic fleshy flowers measuring some 45cm across, weighing 7kg and stinking of putrid flesh. Other flowering plants are saprophytes which are symbiotic with mycorrhizal fungi in or on their roots, such as the YELLOW BIRD'S-NEST (*Monotropa hypopitys*). Then there are the so-called carnivorous flowering plants which obtain supplementary nutrition by capturing and

digesting insects and other small animals.

The Evolution of Angiosperms. The sudden appearance of the flowering plants in the fossil record was described by Charles Darwin, in a letter to a friend, as "that abominable mystery". Their exact time and place of origin are still uncertain and there are many theories as to their ancestry. The facts that can be obtained from the fossil record are that there was an abrupt change from the early Cretaceous, dominated by typically Jurassic ferns and gymnosperms, to the middle and upper Cretaceous, when the angiosperms became the characteristic and dominant plants of most parts of the world. Although there are some doubtful earlier records, the first fossils that can definitely be attributed to the angiosperms appear about 125 million years ago in the early Cretaceous.

The appearance of the flowering plants was one of the most important events in the history of our planet: it made possible the evolution and diversification of the more advanced forms of animal life and eventually the emergence of Man himself. Man's dependence on the flowering plant is one of the fundamental truths of biology. Early Man soon had to learn to distinguish between those plants that would feed him or poison him, and to recognize those that could be used for shelter, clothing and so on. Increased familiarity with plants and their way of growth led to the beginnings of primitive agriculture and the formation of sedentary communities, and set in train the evolution of human society and cultures.

What is a Flowering Plant? The flowering plants form a class representing one of the two major groups of the Spermatophyta (or

Two extremes of flowering plant nutrition. Above The orange, twining stem of the parasite Cuscuta grandiflora *(Dodder) which penetrates host stems by means of peg-like haustoria. (×6)* Right *A fly about to be trapped in the carnivorous* Dionaea muscipula *(Venus Fly Trap). (×3)*

seed plants), distinguished from the other group, the gymnosperms (see Conifers and their allies, p.98), by having the ovules (the future seeds) enclosed in carpels, which make up the ovary and protect the ovules. Because of this enclosure of the ovules the pollen grains have to gain access to them by means of special stigmatic surfaces on which they germinate to form a pollen tube which carries the male nuclei to the nucleus of the egg in the ovule.

Flower and Inflorescence. The flower is the principal characteristic that marks off the angiosperms from other groups of plants. Typically, it consists of a number of green sepals (forming the calyx), colored petals (forming the corolla), stamens (forming the androecium) and carpels (forming the gynoecium), but there are many deviations from the basic pattern, and there is wide variation in size, number, arrangement and degree of fusion of the various parts, and in form, symmetry, sex distribution, pollination mechanisms and biology. Flowers may be solitary or grouped together into structures known as inflorescences. A wide range of inflorescence types is found in the flowering plants. These are classified according to their manner of branching into two main types: racemose and cymose.

A number of trends of evolution in the flower have been observed. The result of all these trends operating under the guidance of natural selection has led to the colorful and

Above Top *Tropical rain forest in eastern Panama. Such vegetation contains the greatest diversity of flowering plants in the world.* Above Bottom *A section through the inflorescence (capitulum) of a cultivated chrysanthemum, comprising a central mass of individual yellow disc flowers clustered on a curved receptacle surrounded by a ring of large reddish ligulate flowers.* (× 6)

bizarre array of flowers found in contemporary flowering plants, varying from the reduced catkinate flowers of the willows (*Salix spp) to the complex structure of the passion flowers (*Passiflora spp); from the single-stamen male flowers in the inflorescences of the spurges (*Euphorbia spp) or even a single anther in the *duckweeds (Lemna spp), to the highly specialized structures found in the orchids, which are adapted for insect pollination.

Pollination. The evolution of the flowering plants and the flower itself is closely

34

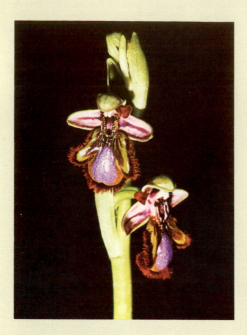

In species of the orchid genus Ophrys *the flowers of individual species resemble the females of their particular pollinator insects. Shown here is O. speculum*

Fruits. The array of fruit types found in the flowering plants is as diverse as that of the flowers, from which they are derived. Again, as with flowers, wind and various classes of animals are involved as their dispersal agents, and they have evolved numerous adaptations in response to this function. Wind-dispersed fruits or small seeds are often extremely light or have wings, hairs or parachute mechanisms. Animal-dispersed fruits show a range of adaptations according to the animal concerned, such as the colored fleshy pericarp of fruits like the cherry and grape, eaten by birds or mammals, the seeds being dispersed by passing unharmed through the digestive tract.

Habit. One of the most remarkable features of the flowering plants is their very great diversity in habit. The most generalized type (in

Raunkiaer's widely used classification of life forms) is called the phanerophytes, in which the buds and apical shoots are borne well above the ground and do not die back during unfavorable conditions of cold or drought. The phanerophytes comprise trees, woody lianas and shrubs.

Reduction in supporting tissue also leads, in the tropics, to phanerophytes with herb-

The flower of Passiflora *species is an example of an "advanced" flower with a complicated arrangement of parts. There are five purple sepals and five purple petals, above which is a "corona" comprising several rows of white and black filamatous structures. The stamens are fused at their bases with the style to form a structure known as a gynophore. Above the gynophore five free yellow anthers can be seen; these are overtopped by three stigmas with discoid tips.*

associated with the agent of pollination. The flower is not just a morphological structure but a biological unit whose prime function is the production of seed.

The flowering plants, unlike animals, cannot move from place to place, either to seek suitable growing conditions or to find breeding mates. This limitation is largely overcome, however, by the co-evolution of the flower and insect or other animal pollinators which in effect mean that the plant itself does not move but the pollinating agent does. In the earliest seed plants and "pre-angiosperms", with open leaf-like carpels bearing ovules on the surface, pollination was effected by wind and was a random process. The ovules exuded a sticky substance which trapped the pollen grains and attracted them to the micropyle of the ovule, thereby affecting pollination. Marauding insects, almost certainly beetles, would have come across these sticky secretions and the pollen grains (which are rich in proteins), and gradually started returning to this attractive food source, thus initiating a promiscuous form of insect pollination which supplemented the predominant wind pollination. Clearly those plants which evolved features that were more attractive to insects encouraged an increase in such visitors and greater seed production than those relying mainly on wind pollination and random visits. Thus by the forces of natural selection, a series of flower-insect adaptations evolved, leading eventually to the array of complex pollination mechanisms found in present-day flowering plants.

The damage caused by the visits of predatory insects to the open carpels and exposed ovules led to the closure of the carpel. This served to protect the ovules against insect attack and at the same time utilized the visits of the insects as pollinators.

Numerous stem modifications are found, as in the climbers which coil themselves around other plants or structures; modified branches called tendrils may be formed, as in the vines (*Vitis* spp). Other stems are modified for water storage, such as the fleshy, green swollen stems of various cacti and succulents. Thorns are modified branches which arise in leaf axils, as in *Crataegus* [*HAWTHORNS*] species. Other modified stems, such as rhizomes, tubers, corms and bulbs, act as food storage organs. Stems which are modified so as to resemble a leaf in form and function are referred to as cladodes or cladophylls, for example as in *Asparagus*.

Leaves. The leaf is one of the most variable of the organs of the flowering plants and certainly the most plastic. This can be shown not only by the enormous diversity of leaf shapes and sizes found in different species, genera and families, but also by the differences shown between the leaves of a single individual plant and between the leaves of different varieties of a single species. Only a few flowering plants are leafless, such as some kinds of cacti.

The basic parts of the leaf are the stalk or petiole, the blade or lamina, and the appendages at the base, the stipules (not always present). These all show variations and modifications. The arrangement of the leaves may be alternate (spiral), opposite (paired at the nodes of the stems), or whorled (three or more at a single point). The greatest range of variation is shown in leaf shape, especially in the dicotyledons, where there are three basic types: simple or undivided (although sometimes lobed); pinnately compound; and palmately or digitately compound. The leaf margins also show great diversity, ranging from entire to toothed or lobed; likewise the apex and base often show characteristic shapes and modifications. Then the pattern of the main veins (venation) in the lamina falls into three basic types – parallel, pinnate and palmate – although combinations of these may be found in a single leaf. There is also a wide range of features to be seen on the surfaces of leaves – color, hairs, scales, wax and other forms of covering (indumentum).

Leaves show enormous variations in size ranging from a few millimetres to many metres in length. Special leaf modifications include the spines found in several cacti, and tendrils.

Economic Importance. Today, flowering plants pervade all aspects of our life – they feed us, drug us, hallucinate us, intoxicate us, provide us with raw material for building, furniture and manufacturing; they provide pasture and grazing for our cattle and other animals and they afford us immense pleasure both in nature as the main part of the green landscape, and in parks and gardens, in the house and office as living ornaments. The myriad shapes and forms of plants, especially of the flowers themselves, have fascinated Man for centuries and played a major role in the development of his esthetic consciousness, as witness the flower in art, mythology and religion.

*Henequen (*Agave fourcroydes*) stacked in a plantation ready for processing. This is just one of the thousands of flowering plant species of use to Man.*

aceous stems, but the more normal herbaceous habit is found in life forms known as chamaephytes, hemicryptophytes and cryptophytes which are characterized by seasonal dying back as the plant becomes dormant during the unfavorable season. The buds that give rise to the new growth when favorable conditions return are situated progressively nearer the ground, culminating in the cryptophytes in which they are entirely underground. Most cryptophytes are also termed geophytes, of which four kinds can be recognized – those with rhizomes, stem tubers, bulbs or root tubers. It is generally accepted that within the herbaceous life forms, the evolutionary trend has been from perennial to biennial to the most specialized, the annual herb, which survives unfavorable conditions as seed only.

Stems. The stems of the angiosperms show considerable variation in their primary structure and in their production of secondary (woody) tissue. They differ in size and duration, features which are closely related to the life forms discussed. Herbaceous stems die down to the ground every year and the plant may be annual, biennal or perennial; other perennials may be somewhat woody (suffru-

tescent) or entirely woody (suffruticose), and may be shrubby or with a tree-like trunk (arborescent). The stem may be reduced, the leaves being all basal, borne on a very shortened stem base. Stems show a very wide range of surface features – hairs, spines, ridges etc – and colors.

*Capsules (fruits) of the Bluebell (*Hyacinthoides non-scripta*) split open to reveal the seeds. A vast diversity of fruit types are found in flowering plants.*

Anthemis a genus of more or less hairy, usually aromatic, annual or perennial herbs and dwarf shrubs, growing in Europe, the Mediterranean region and southwestern Asia. Several are cultivated as garden ornamentals.

Because of the complexity of classification, *CHAMOMILE, which was the common name for *A. nobilis*, is now ascribed to the closely related *Chamaemelum nobile*. Commonly cultivated border species include *A. cupaniana* with large white outer florets, *A. sancti-johannis* with bright orange outer florets and *A. tinctoria* [YELLOW CHAMOMILE, OX-EYE CHAMOMILE, GOLDEN MARGUERITE] with yellow outer florets (or white, lemon or cream in derived cultivars). The cushion-forming *A. montana*, with white or pinkish outer florets, is grown in rock gardens.

Little economic use is made today of *Anthemis* species. *A. tinctoria*, as the name suggests, has been used in the past as a source of a yellow dye, and *A. montana* has been employed in a poor quality herbal infusion as a substitute for chamomile tea.
COMPOSITAE, about 130 species.

Anthericum a large genus of perennial herbs, mostly found in tropical and South Africa, with a few in Europe, East Asia and America. The white or yellow flowers are borne in erect simple or branched inflorescences. The white-flowered *A. liliago* [ST. BERNARD'S LILY] from southern Europe is more commonly cultivated than *A. ramosum* from southwest Europe and the yellow-flowered *A. echeandioides* from Central America.
LILIACEAE, about 300 species.

Anthoceros a genus of liverwort-like plants native to Europe, tropical America, Asia and Australasia. They grow on mud or soil and several annual species are found in ephe-

Inflorescence of the St. Bernard's Lily (Anthericum liliago), which is the most commonly cultivated member of the genus. (× ⅓)

meral habitats like fallow fields or pond margins. *Anthoceros* has an irregularly lobed thallus which is a greasy translucent green and contains mucilage cavities in which a blue green alga, *Nostoc*, lives.
ANTHOCEROTACEAE, about 77 species.

Anthoxanthum a widespread genus of grasses occurring predominantly in Europe, but also in all north temperate countries and mountains in tropical Africa. They are very variable, either annuals such as the slender, branched, weed species *A. puelii* or perennials such as *A. odoratum* [SWEET VERNAL GRASS, SPRING GRASS], differing considerably in overall size, leafiness and hairiness.

All species are highly sweet scented, especially when dried as hay, owing to the presence of coumarin. Despite the pleasant smell, this compound gives the grasses a bitter taste, making them quite unpalatable to grazing cattle. Coumarin is particularly noticeable in SWEET VERNAL GRASS, a common grass of poor pastures, heaths and moors, hill grassland, meadows and open woodland.
GRAMINEAE, 6 species.

Anthriscus a small genus of annual or biennial herbs occurring naturally in Europe and temperate Asia. The flowers are white and are borne in compound umbels. *A. sylvestris* [COW PARSLEY, KECK, WILD CHERVIL, CICELY, OUR LADY'S LACE, QUEEN ANNE'S LACE] is the commonest early flowering umbellifer in southern England, the roadside verges often becoming banks of white in May. *A. cerefolium* [SALAD CHERVIL, GARDEN CHERVIL] is cultivated for its leaves, used for salads and for garnishing, especially in France.
UMBELLIFERAE, about 12 species.

Anthurium a large genus of perennial plants from tropical America, mostly from Colombia. They may be woody or herbaceous, erect or ascending, creeping or climbing by means of their roots. Some are epiphytes and many species have aerial roots at the base of the leaves. The flower stalk is usually long, bearing a spathe which subtends a cylindrical flower-bearing spadix. The spathe is often large, ovate-lanceolate in shape and either green, yellowish, red or purple.

Several species are grown as hothouse plants, one of the most robust being *A. scherzeranum*. It has a scarlet spathe subtending a spirally twisted, yellow-red spadix. The name *A.* × *cultorum* refers to complex hybrids involving *A. andraeanum* [FLAMINGO LILY], *A. lindenianum*, *A. nymphaeifolium* and *A. ornatum*. There are several cultivars.
ARACEAE, about 500 species.

Antirrhinum [SNAPDRAGON, LION'S MOUTH, CALF'S SNOUT] a herbaceous genus from the temperate Northern Hemisphere, consisting of the perennial European section *Antirrhinum* with 17 species, and the mostly annual section *Saerorrhinum* from North America. The genus is characterized by its

Anthoceros husnotii, one of the liverwort-like hornworts, showing the thalloid gametophyte and the characteristic horn-like capsules. (× 2)

racemes of large, showy, two-lipped, mouth-like or face-like flowers.

By far the commonest species in cultivation is *A. majus* [COMMON SNAPDRAGON], grown for cut-flowers or as a bedding plant in many varieties for its elongated racemes or spikes of flowers in a wide range of colors. Other members of the section *Antirrhinum* under cultivation are not very common and are mostly of the subshrubby rockery types such as *A. molle* and *A. hispanicum*, with mostly cream or pink flowers. Some of the annual section *Saerorrhinum*, such as *A. multiflorum* (= *A. glandulosum*) [WITHERED SNAPDRAGON] and *A. coulteranum* [CHAPARRAL SNAPDRAGON], are cultivated in North America, though rarely in Europe.
SCROPHULARIACEAE, about 28 species.

Aphelandra a genus of tropical and subtropical American forest shrubs, several of which are now well established as houseplants. The most popular species is *A. squarrosa* [ZEBRA PLANT], grown as several cultivars, the commonest being 'Louisae' and 'Leopoldii'. The former has attractive dark green leaves with ivory-white veining and produces a bright yellow, pineapple-shaped inflorescence. 'Leopoldii' is smaller with light green leaves with a silver-gray patch along the midrib and an inflorescence of pale lemon-yellow flowers almost entirely enclosed by red bracts. Other similar cultivars include 'Brockfield', 'Dania' and 'Louisae'.
ACANTHACEAE, about 80–200 species.

Apium a genus of marsh or aquatic herbs distributed in temperate and subtropical regions. One species, *A. graveolens*, has been used since earliest times as a food and medicinal plant (see CELERY). *A. nodiflorum* [FOOL'S WATERCRESS] is so called since it was occasionally eaten in Western Europe in mistake for watercress. Another species, *A. prostratum*, has been used as a vegetable, in vapor bottles in New Zealand and as an antiscorbutic.
UMBELLIFERAE, about 50 species.

Apocynum a small genus of perennial herbs mostly native to North America and Mexico. The latex extracted from the leaves and stems of *A. cannabinum* [HEMP DOGBANE] is used as a source of chewing gum by some North American Indians. Extracts from the root of this plant and of *A. androsaemifolium* [SPREADING DOGBANE] have medicinal properties as diuretics and emetics. The bark of *A. venetum* yields fibers used in the manufacture of string and fishing nets.
APOCYNACEAE, 7 species.

Aponogeton a genus of aquatic or amphibious perennials mostly native to warmer regions of the Old World but extending into Malaya and northeast Australia. *A. fenestralis* [LACE OR LATTICE LEAF] is often grown as an aquarium plant and has both submerged and floating leaves. On the other hand *A. distachyus* [CAPE PONDWEED, WATER HAWTHORN] has only floating leaves. The tubers of some species are eaten by Man and by livestock.
APONOGETONACEAE, about 45 species.

The Common Snapdragon (Antirrhinum majus) is popular for cut-flowers and as a bedding plant and has numerous cultivars. (× ¾)

Apple the name given to the cultivated, hardy, deciduous spring-flowering tree *Malus pumila*. The modern cultivated apple is of hybrid origin, the main ancestors being *M. sylvestris* and *M. pumila*, with contributions from various other species. Available evidence indicates that it originated in the upland regions between the Black Sea, Turkestan and India. Some of the forms spread westwards to be established elsewhere as varieties still capable of future variation.

Apple cultivars have arisen as chance seedlings, by deliberate selection from seedlings of unknown parentage or from naturally occurring bud sports. Irradiation techniques are also being increasingly used to induce desirable mutations. It is interesting to note, however, that 'Cox's Orange Pippin', 'Bramley's Seedling', 'Golden Delicious', 'Granny Smith', 'Ribston Pippin' and 'Discovery' all originated as chance seedlings.

Today, the apple has developed into one of the most important deciduous fruits. The worldwide interest in, and importance of, this crop is demonstrated by the existence of over 2 000 named cultivars.

Apples can be classified into four main groups: dessert, culinary, cider and ornamental. Worldwide emphasis has been on the development of dessert types, which are normally cultivars producing average sized fruits (6–7cm in diameter). They are mainly red and/or yellow (rarely green), with a high sugar content, the character being imparted by the amount of aromatics present. Culinary types are normally large-fruited cultivars (10cm or more in diameter), mainly green, with a high acid content. When cooked with sugar the flesh is reduced to a frothy consistency with a well-developed sharp flavor, eg 'Bramley's Seedling'.

Cider varieties are grown mainly in the UK, northern France and other northern European countries. They are subdivided into sweets, sharps, bitter sweets and bitter sharps depending on the proportions of sugars, acids and tannins in their expressed juices, all of which, along with certain organic and aromatic substances, have a profound effect on the "vintage" quality of the cider produced. Most commercial ciders are made by mixing juices from the different cultivars to produce the desired blend of sweetness, acidity or astringency. Present-day demands cannot be met by the juice from true cider varieties, which are normally supplemented by considerable quantities of downgraded dessert or culinary types.

Other apple products are unfermented apple juice, wine, liqueurs, vinegar, fillings for tarts, pies and sauces and pectin from the dried apple pulp left after juice extraction.

A number of *Malus* species are grown exclusively for their decorative character – the profusion of spring blossom and attractive leaves, shoots and bark. Many produce quantities of small fruits, of little use other than for making jelly, but they add distinctive splashes of color in parks and gardens in the autumn. Recently their potential as pollinators in commercial orchards and for pro-

Anthurium scherzeranum, one of the cultivated Flamingo or Tail Flowers, with its twisted spadix set in a brilliant scarlet spathe. (× 1¼)

viding useful characters for apple breeding has aroused considerable interest.

Apples can be grown only in regions where winter temperatures are sufficiently low to provide the chilling required to break bud dormancy. Distribution, therefore, is confined to the northern and southern temperate zones and the higher altitudes of

Cow Parsley (Anthriscus sylvestris) is common in Europe, particularly by hedges and roadside verges, and is naturalized in eastern North America. (× ⅕)

Aponogeton echinatus has thick, elliptic floating leaves and a single curved spike-like inflorescence borne on a long stalk. ($\times \frac{1}{2}$)

warmer regions where cooler winters occur.

Apple trees consist of two distinct parts: the underground portion made up of the rootstock, supporting the above-ground cultivar selected for cropping. Rootstocks exert profound effects on many aspects of tree performance, in particular on ultimate size. For centuries, and even up to the present time in some countries, the main sources of rootstocks were apple seedlings, but trees so produced showed uncontrolled variability and were usually vigorous to very vigorous in growth habit. In order to eliminate this disadvantage the East Malling Research Station in England embarked, some 60 years ago, on its important studies on selecting and breeding apple rootstocks, with the result that growers now have available a range of virus-tested, clonal rootstocks, which permits the production of trees of uniform behavior over a range of vigors suitable for the needs of modern orchard culture. ROSACEAE.

Apricot the name of the fruit of *Prunus armeniaca*, a hardy, deciduous, spring-flowering tree. The fruit is similar to the *peach, *cherry and *plum, and constitutes an important fruit crop. The trees are also widely grown as ornamentals. The species is thought to have originated in western China, with a secondary center in western Asia.

An apricot is smaller than a peach and is usually orange-yellow when ripe. Apricots are somewhat richer in vitamin A, proteins and carbohydrates than most common fruits. The USA is the largest single producer but Hungary, Turkey, Spain and France are also important producers and exporters. The leading all-purpose apricot cultivar is 'Royal' but breeding and selection to improve fruit size and climatic adaptability is important. ROSACEAE.

Aquilegia a genus of perennial herbs widespread over the north temperate zone. *A. vulgaris* [GRANNY'S BONNET, COLUMBINE] is

found over much of Europe and many long-spurred hybrids with large cream, yellow, red or blue flowers are widely cultivated as border plants.

A popular border species is the western American *A. caerulea* which bears white, long-spurred flowers with blue and yellow tints. *A. caerulea* with two other American species, the predominantly yellow-flowered *A. chrysantha* and the red-flowered *A. formosa* [SITKA COLUMBINE], played an important part, at the end of the 19th century, in the development of the long-spurred garden hybrids. Two other popular border species are *A. canadensis* [CANADIAN COLUMBINE] from Canada and the somewhat taller North American and Mexican *A. longissima*. There are several European alpine species, such as *A. alpina*, *A. bertolonii* and *A. pyrenaica*, all with blue or bluish-violet flowers. RANUNCULACEAE, about 100 species.

Arabis [ROCK CRESSES] a genus of annual, biennial and perennial herbs native to the north temperate zone. *A. caucasica* (= *A. albida*) is a commonly grown garden plant, forming a spreading mat with white flowers, borne in terminal racemes. It frequently escapes and becomes naturalized. Double forms are in cultivation. *A. muralis* is another vigorous spreading species with pinkish-purple flowers. Tight-rosette-forming species include *A. aubrietioides* with pink flowers and *A. androsacea* with white flowers and tiny silky, hairy leaves. Both are natives of Turkey. CRUCIFERAE, about 120 species.

Arachis a South American genus of usually perennial herbs, in which the typically pea-like flowers give rise to fruits that bury themselves and develop underground.

The cultivated apple comes in many varieties. Shown here are (Below) Cox's Orange Pippin; (Bottom) Jonathan McIntosh.

The Monkey Puzzle Tree (Araucaria araucana) is a common sight in many parks and gardens – a direct result of their popularity in the 19th century.

The main cultivated species is *A. hypogaea* [PEANUT, GROUND NUT, MONKEY NUT, EARTH NUT, GOOBER]. An annual or sometimes weakly perennial nonwoody plant up to 50cm high, it is grown throughout the tropics, subtropics and in many warmer countries of the temperate zones for the sake of its seeds (kernels), which contain up to 50% of a nondrying oil and about 25% protein. The oil is used domestically for food and cooking, and is also converted industrially into margarine and soap. The protein contributes about one-tenth of the estimated human dietary requirement in India and Nigeria, and (as peanut butter, salted peanuts and confectionery) about 5% of the average North American intake. The press-cake left after the oil has been extracted is a valuable part of poultry and cattle feeds, and the leaves and stems left after harvest are useful forage. At least one other species, the annual *A. prostrata*, is used as a green manure crop in parts of Brazil.

The peanut was first domesticated in South America, probably in the region of western and southwestern Brazil, Bolivia, northern Argentina, Paraguay and Peru east of the Andes. In and near this vast region there are perhaps 90 wild species of *Arachis*. LEGUMINOSAE, about 90 species.

Aralia a genus native to Indomalaysia, East Asia and North America, noted for the various medicinal uses of several species. They are mostly shrubs, occasionally herbs or trees, with pinnate leaves.

Among species used as vegetables are *A. chinensis* [CHINESE ARALIA, CHINESE AN-

GELICA TREE] and *A. cordata* in Japan. Two North American species are well known for their medicinal properties: *A. racemosa* [AMERICAN SPIKENARD] produces rhizomes which are a source of a purgative and can induce sweating, and *A. spinosa* [DEVIL'S WALKING STICK, PRICKLY ASH, PRICKLY ELDER, HERCULES' CLUB, ANGELICA TREE] yields a tonic. The main cultivated species is *A. elata* [JAPANESE ANGELICA] with large bipinnate leaves; there are several cultivars with variegated leaves.
ARALIACEAE, about 35 species.

Aramina the common name for *Urena lobata*, a shrub which reaches up to 5m in height under cultivation. It is perennial but is usually cultivated as an annual for its fibrous bark. It occurs throughout the tropics but is probably of Old World origin. Its use as a fiber crop is mainly local in Brazil, but it is exported from the Congo, where it is called Congo jute. The fiber is similar to jute, with which it is often mixed, and is used for hessian, ropes etc.
MALVACEAE.

Araucaria a genus of tall evergreen coniferous trees all confined to the Southern Hemisphere, notably in South America, Australasia and the islands of the South Pacific. The genus includes the well-known *A. araucana* (= *A. imbricata*) [MONKEY PUZZLE TREE, CHILE PINE], *A. heterophylla* [NORFOLK ISLAND PINE] and *A. angustifolia* [PARANA PINE, CANDELABRA TREE].

Members of *Araucaria* are imposing trees with branches in regular whorls. The leaves persist for many years and are flat and broad or awl-shaped and curved.

The wood of *Araucaria* is resinous, straight grained and easy to work. *A. angustifolia*, *A. araucana*, *A. bidwillii* [BUNYA BUNYA] and

A. cunninghamii [MORETON BAY PINE, HOOP PINE] are the most important timber trees. The timber is mainly used for general indoor joinery and carpentry and for boxes and masts as well as pulp for papermaking. For all practical purposes it can be used as a substitute for the SCOTS PINE.

The striking appearance of *A. araucana* has made it a popular subject for cultivation.
ARAUCARIACEAE, about 18 species.

Arbutus a small genus of shrubs and small trees native to North and Central America, southern Europe, the Canary Islands and western Asia. The distinctive fruit is reddish and resembles a strawberry, hence the common name of STRAWBERRY TREE, which is applied to the whole genus and specifically to *A. unedo*. The latter is cultivated for its attractive foliage, smooth red flaking bark, pendulous whitish flowers and the red or pinkish edible fruits which are used to make preserves and alcoholic drinks. The bark is used for tanning. A useful wood is yielded by the western USA species *A. menziesii* [MADRONA, MADRONO, MADRONA LAUREL], which is also cultivated as an ornamental.
ERICACEAE, about 14 species.

Archangelica a small genus of herbaceous perennials from north temperate regions. The crystallized petiole of *Archangelica officinalis* (=**Angelica archangelica*) provides the green angelica used in confectionery. The plant contains essential oils, which are distilled from the roots or seeds and used to flavor liqueurs such as chartreuse; distillations and infusions have been used medicinally and in perfumery. Today, ANGELICA is usually only seen as an ornamental in the flower garden. The small greenish flowers are borne in compound umbels on stout stems up to 2m tall.
UMBELLIFERAE, 12 species.

Arctostaphylos uva-ursi (*Bearberry*) is the only circumpolar species of this otherwise American genus. The berries for which this evergreen shrub is cultivated are edible but unpalatable – best left to the bears! ($\times \frac{3}{4}$)

Arctostaphylos a genus of shrubs, or rarely small trees, with a flaky bark, from North, Central and western America and including a north temperate and circumpolar species, *A. uva-ursi* [BEARBERRY], a prostrate and mat-forming shrub which flowers as soon as the snow melts.

Several species of *Arctostaphylos*, such as *A. glandulosa* [EASTWOOD MANZANITA], *A. glauca* [GREAT BERRIED MANZANITA], *A. pungens*, *A. nevadensis* [PINE-MAT MANZANITA] and *A. uva-ursi*, are commonly cultivated as ornamentals. The fruits of the genus formed part of the diet of early North American settlers, while the leaves of *A. uva-ursi* are used for tanning in Russia and Sweden and as tea (Kutai or Caucasian tea). The dried leaves of this species have some medicinal

The Hairy Rock Cress (Arabis hirsuta), *with its white flowers, erect leafy stem and basal rosette, here growing typically on limestone.* ($\times 1$)

Flowers and fruits of the Hybrid Strawberry Tree (Arbutus × andrachnoides), *the latter ripening 12 months after the flowers.* ($\times \frac{1}{2}$)

value when used as a tonic and diuretic. ERICACEAE, about 70 species.

Arctotis a genus of herbaceous annuals and perennials with a marked center of distribution in the Cape region of South Africa. A number of the species are cultivated as garden plants. *A. breviscapa*, for example, a perennial species growing to a height of about 30cm, produces flowers in many strikingly different shades of orange. *A. grandis* [AFRICAN DAISY] is a long-stemmed species which by contrast has distinctive white ray florets with a lavender blue reverse. *A. acaulis* is a stemless herb which has yellow, orange or reddish ray florets. COMPOSITAE, about 30 species.

Ardisia a large genus of tropical and subtropical evergreen trees or shrubs, many of which are cultivated as ornamentals, outside in warm countries or in the greenhouse in cooler temperate regions. Most cultivated species, such as *A. japonica* and *A. crenata* [CORALBERRY, SPICEBERRY], have attractive glossy leaves, sometimes variegated, and showy clusters of flowers followed by red or black berries. Various species such as *A. colorata* and *A. quinquegona* are used locally in Malaya and Vietnam respectively to treat various ailments such as colic and toothache. MYRSINACEAE, 300–400 species.

Areca see Betel nut. PALMAE, about 50 species.

The Sugar Palm (Arenga pinnata) is cultivated in the eastern tropics for its sap which is processed into a brown sugar (jaggery).

Flower of the South American Pelican Flower (Aristolochia grandiflora). Flies are attracted to the huge foul-smelling flowers. ($\times \frac{1}{4}$)

Arenaria [SANDWORT] a genus of annual and perennial herbs native to north temperate and subarctic zones but especially central and southern Europe. Many perennial species are cultivated as garden-border plants and a number, such as *A. purpurascens* (rose or white flowered) and *A. laricifolia* (white flowered), have a tufted habit, with erect or decumbent flowering stems. The Mediterranean *A. balearica* has a dwarf creeping habit (about 8cm high) so that it forms a dense mat in contrast to *A. grandiflora*, which is more erect (up to 25cm high) with a woody branching rootstock and more or less upright stems. CARYOPHYLLACEAE, about 200 species.

Arenga a genus of palms from Indomalaysia and northern Australia. The most important species), *A. pinnata* (= *A. saccharifera*) [SUGAR PALM], is cultivated in the tropics of both hemispheres for its sugary sap which produces palm sugar, known as jaggery. This and other species also produce a type of sago and edible seeds; the leaf sheaths yield fibers which are used for ropes, thatch and brushes. PALMAE, about 17 species.

Argemone a genus of annual and perennial herbs with prickly, deeply dissected leaves and poppy-like flowers. A number of species, including *A. platyceras* [CRESTED POPPY] with white or purple flowers, and *A. mexicana* [DEVIL'S FIG, PRICKLY MEXICAN POPPY] with orange or yellow flowers, are cultivated as garden ornamentals. The latter is also a pantropical weed. PAPAVERACEAE, 10 species.

Arisaema a genus of perennial, stemless, tuberous herbs native to East Africa, tropical and temperate Asia, North America and Mexico. The garden species are grown for their large, often lobed leaves, the attractive and showy clusters of reddish fruits and the variously colored, hooded spathe. The latter is whitish green in *A. dracontium* [GREEN

DRAGON, DRAGON ROOT], purple with pale stripes in *A. triphyllum* (= *A. atrorubens*) [JACK IN THE PULPIT, INDIAN TURNIP] and white with green and pink veins in *A. candidissimum*. ARACEAE, about 150 species.

Aristolochia a large genus of shrubs and some herbs, frequently climbing, which are widely distributed in tropical and temperate regions of Eurasia, Africa and the Americas. The plants are noteworthy for their pendent, bilaterally symmetrical flowers which exhibit a range of striking forms. Most species have a disagreeable odor that attracts the pollinating flies which are trapped inside the flower tube for a time. In *A. clematitis* [EUROPEAN BIRTHWORT] there are downward directed hairs in the greenish-yellow flower tube, which are covered with oil so that flies, once inside, cannot escape and pick up pollen and/or deposit it on the stigmas. After a day or two the hairs wither so that the flies can emerge and visit another flower.

Many species are cultivated for their ornamental flowers, whose diverse forms are reflected in names such as DUTCHMAN'S PIPE (*A. durior* = *A. macrophylla*), BIRD'S HEAD (*A. brasiliensis* = *A. ornithocephala*), SWAN or PELICAN FLOWER (*A. grandiflora*). Many species have local medicinal uses. The snakeroots, such as *A. serpentaria* [VIRGINIA SNAKEROOT] from North America, are reputed remedies for snakebite, while *A. clematitis* was formerly widely cultivated in Europe for its supposed efficacy in relieving birth pangs. ARISTOLOCHIACEAE, about 350 species.

Armeria a genus of perennial herbs mainly from the north temperate zone. The narrow simple leaves all grow from the base of the plant and the flowers are borne in spherical heads on slender stems. *A. maritima* [THRIFT, SEA-PINK] is abundant on coasts and mountain ranges of the Northern Hemisphere and reappears again in the Andes. It seems to be unknown as a lowland plant away from the coast but it is often cultivated. Many species such as the Spanish *A. juniperifolia* (= *A. caespitosa*) are alpines and are cultivated in rock gardens. PLUMBAGINACEAE, about 80 species.

Armillaria a widespread genus of basidiomycete fungi known mainly for one member, *A. mellea* [HONEY FUNGUS or BOOTLACE FUNGUS]. The fruiting body varies in color from yellow-green to light brown and the cap may have a light covering of scales. The members of this genus possess rhizomorphs (subterranean root-like structures by which means this parasitic fungus spreads its infection from tree to tree) which are creamy white to light brown at the growing tip but hard, black and shiny in the older portions, hence the bootlace description. TRICHOLOMATACEAE, about 40 species.

Arnebia a small genus of annual and herbaceous perennials native to the Mediterranean area, Africa and the Himalayas. The

only widely cultivated garden species is *A. echioides* (= *Echioides longiflorum, Macrotomia echioides*) [PROPHET FLOWER], a tufted plant bearing clusters of bright yellow flowers, normally in dry rock-garden situations. It is native to Armenia, the Caucasus and northern Iran.
BORAGINACEAE, about 25 species.

Arnica a genus of north temperate ánd arctic perennial herbs. The characteristically hairy and often aromatic leaves form a basal rosette, and the large, usually solitary, orange-yellow flower heads are borne on long stalks. The genus is the source of the medicinal compound arnicine, which is applied externally for bruises etc. *A. montana* [MOUNTAIN TOBACCO] is cultivated as a rock-garden plant.
COMPOSITAE, about 30 species.

Arracacia a tropical American genus of perennial herbs of which some, such as *A. xanthorrhiza* (= *A. esculenta*) [APIO, ARRACACHA, PERUVIAN PARSNIP], are cultivated for their starchy tuberous rhizomes, the flavor of which is similar to a combination of parsnips and potatoes. The ARRACACHA is cultivated in the Andes, in Central America and the West Indies.
UMBELLIFERAE, about 50 species.

Arrack a term (derived from an Arabic word meaning "juice") that can be applied to any spirituous liquor produced from the toddy of various palms or from rice. How-

*Thrift or Sea-pink (*Armeria maritima*), which is native to north temperate coasts and mountains and the Andes, is also a garden favorite. (× ½)*

ever, it most often refers to the alcoholic spirit obtained from the sap of *Cocos nucifera* [*COCONUT PALM]. After the sugary sap has been extracted from the young inflorescence it rapidly ferments to produce toddy (palm wine) with an alcohol content of 5%, which is then distilled to produce palm alcohol or arrack, with an alcohol content of 50%.

Oryza sativa [*RICE] also yields an arrack ("rice spirit") which is distilled from rice wine, after the introduction of fungi and yeast to convert starch to glucose and alcohol.

*The Honey or Bootlace Fungus (*Armillaria mellea*) produces black rhizomorphs ("bootlaces") which penetrate the wood which it parasitizes. (× 1)*

Arrowroot the name given to the starch products of a number of different plants but true arrowroot [WEST INDIAN, ST. VINCENT, BERMUDA ARROWROOT] is the starch derived from the rhizomes of the herbaceous perennial, *Maranta arundinacea*. This species grows wild in Brazil, northern South America and Central America. Although introduced to many other tropical countries, it is now cultivated on a large scale only in St. Vincent.

According to early records, the crushed or bruised rhizome of arrowroot was used for the external treatment of wounds caused by poisoned arrows, and this is said to be the origin of its name. It was adopted by Europeans as an antidote to other kinds of poisons and later for more general medicinal uses, as an easily digestible food for the sick, and even for starching clothes. The starch is now exported for use in arrowroot biscuits and dietetic foods, and as a thickener in soups, sauces, puddings etc; it is also used in face powders and in glues. It is only in Asia that the whole, young rhizome is boiled or roasted and used as a vegetable.

The name "arrowroot" is sometimes applied to other plants with starchy roots, including: TULEMA or TOUS-LES-MOIS ARROWROOT (*Canna species), QUEENSLAND or PURPLE ARROWROOT (*Canna edulis*), EAST INDIAN or BOMBAY ARROWROOT (*Curcuma species, especially *C. angustifolia*), BRAZILIAN PARA or RIO ARROWROOT (*Manihot esculenta*), TACCA or AFRICAN ARROWROOT (*Tacca pinnatifida*) and even PORTLAND ARROWROOT (*Arum maculatum*).
MARANTACEAE.

Artemisia a large genus of aromatic and bitter herbs and shrubs, containing several well-known medicinal plants and culinary herbs. Species are widely distributed in South Africa, South America and throughout temperate regions of the Northern Hemisphere. *A. tridentata* [COMMON SAGEBRUSH, BLACK SAGE] dominates a conspicuous type of vegetation in the southwestern USA; other species are common on arid soils in various parts of the world. The stems and leaves are often covered with a silvery, silky, down-like covering and several species are aromatic or strongly scented when crushed.

Economically important species include the potherb *A. dracunculus* [*TARRAGON], *A. abrotanum* [SOUTHERNWOOD, LAD'S LOVE, OLD MAN], *A. vulgaris* [MUGWORT] and *A. absinthium* [COMMON WORMWOOD, ABSINTHE] used for flavoring vermouth and absinthe; the alpine species *A. glacialis* [GENEPI], an aromatic, gray-glaucous, herbaceous perennial, is also used for flavoring vermouth and fortified liqueurs, and other species are used as medicinal herbs. Several species, such as *A. ludoviciana* (= *A. gnapholodes*) [SAGEBRUSH, WESTERN MUGWORT] and *A. lactiflora* [WHITE MUGWORT], are cultivated in gardens for their attractive scented foliage. COMPOSITAE, about 400 species.

The edible young flower head of the Globe Artichoke (*Cynara scolymus*). *If left to develop, a mass of blue flowers will appear.* ($\times \frac{1}{3}$)

Artichokes perennial herbs, the best-known of which is *Cynara scolymus* [GLOBE ARTICHOKE, FRENCH ARTICHOKE] (Compositae), native to the Mediterranean region and the Canary Islands. It was known to the Greeks and Romans and probably originated from *C. cardunculus* [*CARDOON]. Resembling the thistle in appearance, it is cultivated for the young flower heads (chokes) which are cooked and eaten as a delicacy. They are eaten before the blue flowers open. In some Mediterranean countries such as Spain and Italy, baby artichokes may be eaten whole, either as an appetizer or fried. The young, tender leaf-stalks can also be eaten, when blanched, like cardoons.

The tubers of *Helianthus tuberosus* [JERUSALEM ARTICHOKE] (Compositae), native to North America, and of *Stachys affinis* [CHINESE OR JAPANESE ARTICHOKE] (Labiatae) are also edible. Both globe and Jerusalem artichokes are suitable vegetables for diabetics, since the edible parts contain inulin rather than starch as a storage carbohydrate.

Artocarpus a genus of tropical Southeast Asian and Indomalaysian species containing several important plants widely cultivated in the tropics for their edible fruits. *A. altilis* [BREADFRUIT] is a handsome tree growing to 20m in height and native to Polynesia, where it is of ancient cultivation and an important food. It is now grown sparingly throughout the tropics and is used more as a vegetable than as a fruit. The seedless forms are eaten after cooking and may be boiled, baked, roasted, fried or made into soup. In the seeded form, known as the breadnut, the seeds are eaten after boiling or roasting. The fruit is globe-shaped, formed from the whole inflorescence, 10–30cm in diameter.

The timber of *A. anisophylla* and *A. dadak* is used locally in the building of houses and bridges.

Related to the BREADFRUIT is *A. heterophyllus* (= *A. integrifolius*) [JACKFRUIT] which is a large tree native to India and Malaysia. Its huge barrel-shaped fruits up to 90cm long and 30kg in weight are among the largest in the plant kingdom. Another related species is the CHAMPEDAK (*A. integer*). The fruits and seeds of both species are eaten raw or cooked. MORACEAE, about 50 species.

Arum a genus of tuberous herbs native from Western Europe to the Canary Islands and North Africa, eastwards to Afghanistan and Kashmir. They are notable for their conspicuous arrow- or spear-shaped leaves and curious sheathed flower-spikes; some are cultivated in gardens.

Some species have rather narrow ranges in the Mediterranean region, such as north-eastern Libya (*A. cyrenaicum*), Dalmatia and northern Greece (*A. nigrum*), Crete (*A. petteri* = *A. creticum*). The two most widely distributed are *A. maculatum* [CUCKOO PINT, WAKE ROBIN, LORDS AND LADIES] and *A. italicum*, both of which grow in woodlands from Britain and Western Europe to the Black Sea and are highly variable in shape and color of leaves, spathes and spadices.

Pollination is effected by flies, which are attracted by the fetid scent of the inflorescence. Fertilization of the female flowers results in scarlet berries, which remain after the purplish-greenish or white spathe has decayed.

The tubers of *A. maculatum* were the source of a very white starch esteemed in the Elizabethan era as a stiffener of the elaborately ruffed collars then in fashion. The tubers were also once used as food under the names Portland arrowroot and Portland sago.

The frequently used English name for *A. maculatum*, CUCKOO PINT, comes from the Anglo Saxon for "lively penis", originally a phallic reference to the enlarged apical appendage. ARACEAE, 12–15 species.

Arum lily the name given to a South African species of the genus **Zantedeschia* (*Z. aethiopica*) which is widely grown as an ornamental in greenhouses in temperate

The Arum Lily (*Zantedeschia aethiopica*) *has inconspicuous flowers massed on a yellow spadix surrounded by a large spathe.* ($\times \frac{1}{3}$)

zones and is much used by florists. It is a robust plant, bearing inconspicuous unisexual flowers on a yellow spadix which contrasts markedly with the large white enveloping spathe. ARACEAE.

Aruncus a small north temperate genus of tall perennial herbs, closely related to **Spiraea*.

Arundinaria is the largest and most widespread of the bamboo genera. Here stands of an Arundaria species are being cleared to make way for a road.

Most of the species bear attractive long finger-like branching flower heads, as in the case of *A. dioicus* [GOATSBEARD SPIRAEA], which is widely cultivated.
ROSACEAE, about 12 species.

Arundinaria in the wide sense, the largest and most widespread of the bamboo genera, consisting of shrubs and climbers distributed in a wide range of habitats throughout temperate and subtropical regions of the North and South Hemispheres.

Many species, such as *A. amabilis* [TONKIN BAMBOO] and *A. simonii* [SIMON BAMBOO], are economically important, particularly for use as poles or stakes or, in the case of *A. wightiana*, for making baskets and mats. Others are cultivated as ornamentals, such species as *A. tecta* [SMALL CANE] and *A. falcata* (= *Chimonobambusa falcata*) [HIMALAYAN BAMBOO] being relatively common in gardens.
GRAMINEAE, about 150 species.

Arundo a tropical and subtropical genus of REED GRASSES growing in wet places, with stout, somewhat woody stems, broad, flat leaves and large inflorescences. *A. donax* [GIANT REED, SPANISH REED], native to the Mediterranean region, is now widely cultivated for making sticks, fencing and fishing rods as well as being an ornamental. The variegated, striped-leaved variety *versicolor* is particularly attractive. *A. donax* is the source of the ''reeds'' used in the mouthpieces of wind instruments such as the clarinet, and the hollow stems were used for making the traditional panpipes.
GRAMINEAE, 12 species.

Asclepias a North, Central and South American and African genus of attractive semishrubby and perennial herbaceous plants with a milky latex. Many are cultivated for their umbels of bright showy flowers. *A. syriaca* [COMMON MILKWEED, SILKWEED] is naturalized in parts of Europe, where it was introduced from eastern North America for fiber and as a food plant for bees. The tropical American *A. curassavica* [BLOOD FLOWER] is commonly grown as a greenhouse plant because of its orange-red flowers, but is also a pantropical weed. *A. tuberosa* [BUTTERFLY WEED, PLEURISY ROOT] is a hardy outdoor species.

A number of species yield a stem fiber used locally (called ozone fiber in the case of the North American *A. incarnata* [SWAMP MILKWEED]).
ASCLEPIADACEAE, about 150–200 species.

Ascobolus a genus of fungi, most species of which grow on dung, some on soil. The fruiting bodies are in the typical form of the cup fungi, of which *Ascobolus* is a member, and vary according to species and environment between 1mm and 3cm in diameter.
ASCOBOLACEAE, about 50 species.

Ascophyllum a genus of brown algae, of which much the most important member is

*The Yellow Wrack (*Ascophyllum nodosum*), a brown alga common on rocky shores in temperate zones. In some areas it is harvested to produce alginates. ($\times \frac{1}{4}$)*

A. nodosum [YELLOW WRACK, KNOBBED WRACK, SEA WHISTLES], found at about mean sea level on rocky shores in temperate latitudes. This species is longer-lived than most seaweeds: a single plant may live for 12–13 years though the thallus becomes denuded in the winter. *A. nodosum* is sufficiently abundant around the coasts of northwestern Europe to be worth harvesting commercially for the production of alginates.
PHAEOPHYCEAE.

Ash one of the best-known tree names in temperate parts of the Northern Hemisphere. It is generally used to refer to species of the genus *Fraxinus* although many other genera have members given the name ash, often because of a superficial resemblance to *F. excelsior* [EUROPEAN ASH]. Thus species of *Eucalyptus* are called ashes in other parts of the globe, including *E. gigantea* [ALPINE ASH], *E. oreades* [BLUE MOUNTAIN ASH], *E. regnans* [MOUNTAIN ASH]. Among other ashes are the QUAKING ASH (*Populus tremula*), the CANARY ASH (*Beilschmiedia bancroftii*), and the PRICKLY ASH (*Aralia spinosa* and species of *Zanthoxylum*).

The EUROPEAN ASH is well known for its appearance, the handsome ''keys'' (bunches of winged fruits) being a characteristic sight in the autumn. Medicinally, decoctions of the bark were once used to cure jaundice and other complaints of the liver. However, the best use of the EUROPEAN ASH, as indeed for the North American *F. americana* [WHITE ASH], *F. quadrangulata* [BLUE ASH], *F. nigra* [BLACK ASH] and *F. pennsylvanica* [RED ASH], is for its valuable timber. It is hard, durable and can be used for many purposes, including furniture, wagon-building and tool handles. Other species, including *F. ornus* [FLOWERING

ASH, MANNA ASH] and *F. oxycarpa*, are grown as ornamentals in parks and gardens.
OLEACEAE, about 70 species.

Asparagus a genus of perennial herbs, shrubs and climbers including the culinary asparagus and other species cultivated for ornament or as vegetables. They mostly occur in dry habitats from the Mediterranean region to Central Asia, the East Indies, Australia and South Africa. The leaves are reduced to inconspicuous dry-looking scales, and are replaced as photosynthetic organs by narrow green modified branchlets (cladodes). *A. officinalis* [COMMON OR GARDEN ASPARAGUS] is grown on a commercial scale in North America and Europe for the protein-rich young shoots (spears). Numerous cultivars, including 'Martha Washington' and 'White Cap', have been selected for earliness or heavy cropping.

About 35 species are grown indoors or in the greenhouse as evergreen foliage plants with red or purplish berries. Some have woody stems climbing to a height of more

*The creamy-white panicles of the Flowering or Manna Ash (*Fraxinus ornus*) make this species a popular ornamental for parks and gardens. ($\times \frac{1}{6}$)*

than 6m. Compact variants are often preferred for indoor decoration. *A. plumosus* [ASPARAGUS FERN] is a tall climber from South Africa, in which the delicately compound branching of the stems presents an overall triangular shape, mimicking the fronds of *Dryopteris* ferns. It is the species most frequently used by florists as cut greenery for bouquets, and several varieties are grown indoors. *A. sprengeri* from Natal, with pendulous branches, is a popular plant for hanging baskets. The florist's SMILAX is *A. asparagoides*, also a native of South Africa. It has no prickles, and produces zigzag arranged stems.

A. acutifolius is collected from the wild and is widely eaten from Spain to Greece. The tubers of other species, such as *A. lucidus* from China and Japan, *A. pauliguilolmi* from tropical Africa and *A. sarmentosus* from India and Sri Lanka are eaten boiled as a vegetable or curried.
LILIACEAE, about 300 species.

Species of Asparagus *grown for foliage:* (Left) A. sprengeri: (Center) *Florist's Smilax* (A. asparagoides): (Right) *Asparagus Fern* (A. plumosus). ($\times \frac{1}{4}$)

Asparagus pea one of the common names for *Psophocarpus tetragonolobus*, probably native to tropical Asia, and cultivated in the Old and the New World, but chiefly in tropical Asia. It is a twining perennial herb but is usually cultivated as an annual, supported by stakes or trellis. It has tuberous roots, pale blue-white flowers, and the large four-angled pods have leafy fringes along each of the four ridges. Also known as GOA BEAN or ASPARAGUS BEAN, it is grown mainly for the immature pods, which are cooked and eaten like French beans. In Burma the roots are eaten raw or cooked; in Java the ripe seeds are eaten after parching, and the young leaves, shoots and even flowers are eaten as a vegetable. ASPARAGUS PEA or BEAN is a common name given also to both *Dolichos sesquipedalis* and *Vigna sesquipedalis*.
LEGUMINOSAE.

Aspen the name generally applied to five species of trees of the genus *Populus* found in the Northern Hemisphere and growing to heights of 20 or 30m. Aspens are characterized by the smooth, pale bark and the strongly flattened leaf-stalks which act as a pivot on which the leaf-blades tremble in response to even the mildest winds.

Although the light, soft wood is unsuitable for lumber, European and North American species are utilized in the manufacture of paper, matches, matchboxes and fiber. All aspen species are grown for ornament to varying extents in temperate regions

far removed from their native habitats.

P. tremula [EUROPEAN ASPEN] occurs almost throughout Eurasia, central China is the home of *P. adenopoda* [CHINESE ASPEN] and *P. sieboldii* [JAPANESE ASPEN] is native to the mountains of Japan. Two species occur in the western hemisphere. *P. tremuloides* [QUAKING, TREMBLING AMERICAN ASPEN] is mostly a component of coniferous forests in which various species of fir and spruce are dominant, ranging from Alaska to Newfoundland, south to northern Mexico. *P. grandidentata* [LARGE-TOOTHED ASPEN] grows in the spruce, fir and pine forests from Manitoba and Minnesota to Nova Scotia, south to Tennessee, North Carolina and Delaware; it is also planted extensively in Austria.
SALICACEAE.

Aspergillus a genus of mold fungi, many of which are exceedingly common as yellow, green, gray, brown or black molds on a wide variety of materials. Some aspergilli are very active fermenters, and several are of considerable commercial importance. The most important fermentation industrially is carried out by *A. niger*, resulting in the production of citric acid. In Japan, *A. oryzae* is used to break down the starch of rice to sugars; yeast is added to convert the sugars to alcohol, resulting in the drink known as saké.

Strains of the same species and of *A. soyae* are used in the manufacture of soy sauce.

Aspergilli are of significance in causing diseases of both plants and animals. Strains of *A. niger* cause "smut" of fig and date, and black mold of cotton and other tropical and subtropical crops. Aspergillosis is the general name used for various diseases of animals caused by aspergilli. Brooder pneumonia of chicks and sometimes of other birds in captivity, caused by *A. fumigatus*, is perhaps the most important: the air sacs become coated internally with the fungus. This disease is associated with moldy straw or compost, and can be controlled by avoiding the use of such material. In Man, an infection known as mycetoma, consisting of a mass of *A. fumigatus*, is commonly found in the diseased lungs of patients suffering from "farmer's lung". Recent work, however, has shown that the cause of the disease is the actinomycete *Thermopolyspora polyspora*, and that the *Aspergillus* is merely secondary.
HYPHOMYCETES, about 50 species.

Asperula [WOODRUFF] a genus of herbs native to Europe, Asia and eastern Australia. They are grown as border or rock-garden subjects, particularly in shady positions. Some, such as the blue-flowered *A. orientalis* are annuals, but most cultivated species are perennials. The best-known of these is *A. hexaphylla*, which produces dense clusters of pink flowers, and the white-flowered *A. tinctoria* [DYERS' WOODRUFF]. The SWEET WOODRUFF is now named *Galium odoratum*, but was formerly placed in *Asperula*.
RUBIACEAE, about 90 species.

Asphodel the common name for *Asphodelus*, a small genus of 12 species of annual and perennial herbs, with a distribution from the Canary Islands and the Mediterranean to the Himalayas. Asphodels grow on hillsides and plains and their presence is often a sign of

The handsome flower-spike of Asphodelus albus; *this species is frequent as a pasture plant in its native southern Europe, where it is rarely eaten by animals.* ($\times \frac{1}{4}$)

The Common Michaelmas Daisy (Aster novi-belgii), from which many of the garden varieties have been derived. ($\times \frac{1}{3}$)

overgrazed ground, as they are rarely eaten by animals. Among the cultivated ornamental species are *A. cerasiferus* [WHITE ASPHODEL], *A. liburnicus* and *A. microcarpus* [SILVER ROD]. The tubers of some species are rich in starch, from which a glue has been made for use in shoemaking. The tubers of *A. microcarpus* [ASPHODEL] are also believed to be edible.

The name BOG ASPHODEL is given to species of the genus *Narthecium*, GIANT ASPHODEL to *Eremurus* species, SCOTTISH ASPHODEL to *Tofieldia pusilla*, and YELLOW ASPHODEL to *Asphodeline lutea*.
LILIACEAE.

Aspidistra a genus of perennial herbs from East Asia with stemless evergreen leaves. The flowers are dull brown, mauve or purplish-green, but insignificant. Only *A. elatior* is commonly cultivated. It was introduced to Europe about 1884, and because it proved tolerant of neglect, coal-gas fumes and poor light it was given the popular name of "CAST IRON PLANT". It has handsome dark green leaves, striped with silver in var *variegata*.
LILIACEAE, 8 species.

Asplenium [SPLEENWORTS] a large genus of ferns commonly associated with cool temperate areas or mountain regions of the tropics, although many are tropical epiphytes. Species of the former group make excellent rock-garden plants. A number of New Zealand species, such as *A. falcatum* and *A. flabelliforme* [NECKLACE FERN], are grown in cool greenhouses. Other cultivated species include *A. nidus* [BIRD'S NEST FERN] from tropical Asia and Polynesia and *A. bulbiferum* [MOTHER SPLEENWORT] from India, New Zealand and Australia. The latter and a number of tropical species produce plantlets in the axils of the pinnae.
ASPLENIACEAE, about 700 species.

Assegai or **assagai wood** timber from both *Terminalia sericea* [YELLOW WOOD] and *Curtisia dentata* (= *C. faginea*) [CAPE LANCE TREE], trees native to Central and South Africa respectively. The wood of the former is used for furniture and the bark yields an extract employed in tanning. The wood of *C. dentata* is a deep red color and has a hard, durable nature particularly suitable for spear shafts, as in the assegai of the Zulus of southern Africa.

Aster a vast and varied temperate genus of American, Eurasian and African autumn- or summer-flowering herbs, living in a wide range of habitats from coastal salt marshes to high mountain pastures. They are nearly all leafy-stemmed perennials and only very rarely annuals. The small, daisy-like heads are usually in clustered inflorescences but are sometimes solitary.

About 12 species are frequently cultivated in gardens; *A. novi-belgii* is perhaps the most common (see MICHAELMAS DAISIES). Other important species include *A. amellus*, a bluish-lilac-colored autumn-flowering plant from the Mediterranean region and western Asia, *A. novae-angliae*, a very tall summer and mid-autumn-flowering species with violet-purple or pink flowers, the bushy, bright mauve summer-flowering border species, *A. sedifolius* (= *A. acris*) and the mauve- or rose-purple-flowered alpine dwarf *A. alpinus*.

Many popular plants grown as "asters" are in fact CHINA ASTERS belonging to the genus *Callistephus*.
COMPOSITAE, about 500 species.

Callistephus chinensis 'Pinocchio Mixed'; one of the many cultivars of China or Garden Aster that are grown in borders and as cut flowers under the common name "aster". ($\times \frac{1}{4}$)

The Bird's Nest Fern (Asplenium nidus) is a popular houseplant; in the wild it forms a tangled mass of roots which resemble a bird's nest. ($\times \frac{1}{8}$)

Asterionella a genus of *diatoms that has relatively few species although they are often extremely abundant in the plankton of the sea and freshwater lakes.
BACILLARIOPHYCEAE.

Astilbe [FALSE GOATSBEARD] a genus of small herbaceous perennials with very numerous small flowers in erect spike-like inflorescences found mainly in China and Japan, but with a few in the Himalayas, Southeast Asia and North America. They grow in damp or shady places and are useful garden plants for similar conditions. Nearly

all those seen in gardens today are hybrids, with flowers of varying shades of white, cream, pink, mauve and crimson. They are derived from crosses between the Chinese *A. chinensis* var *davidii*, with pinkish-purple flowers, and two white-flowered Japanese species, *A. japonica* (SPIRAEA of florists) and *A. thunbergii*.
SAXIFRAGACEAE, about 35 species.

Astragalus a very large genus of perennial shrubs or herbs with white, yellow or purple pea-like flowers. *A. gummifer* [TRAGACANTH] is a shrubby species from western Asia (especially Iran, Iraq and Turkey) which yields gum tragacanth, much used in the food, confectionery, medicine and cosmetics industries for the stabilization of emulsions and powder suspensions and as a fabric conditioner and glazer in the textile trades. Among the many other western Asian species of *Astragalus* that yield gums of good enough quality to be sold as gum tragacanth are *A. adscendens*, *A. echidnaeformis* and *A. microcephalus*. Tragacanth is now the most expensive of all natural gums and since it is collected from wild bushes which are becoming exhausted the source of supply is greatly at risk.

A. glycyphyllus [WILD LICORICE], of Europe and Siberia, has sweet-tasting leaves and is sometimes used as fodder. The roots of some species such as *A. canadensis* are cooked and eaten by North American Indians. The seeds of *A. edulis* are eaten in Iran and North Africa, while the seeds of *A. boeticus* have been used as a coffee substitute in Spain and Sicily. A soap substitute is derived from *A. garbancillo* [GARBANCILLO] in South America.

A few species are also cultivated as garden flowers: *A. alopecuroides*, from Siberia, and *A. ajubensis* (= *A. durhamii*), from the Balkans, have yellow flowers; *A. danicus*, from

Few Astragalus species are cultivated as ornamentals; one such is the Purple Milk Vetch (A. danicus) shown here. ($\times \frac{2}{3}$)

Europe, has blue-purple colored flowers.
LEGUMINOSAE, about 1500 species.

Astrantia [MASTERWORTS] a genus of hardy herbaceous perennials native to central and southern Europe, the Caucasus and Asia. The stems are erect, 30–100cm tall, branching and hairless. The roots are often dark colored and aromatic.

Two common European species are *A. major* [GREATER MASTERWORT] and *A. minor* [LESSER MASTERWORT]. A number of species are cultivated as garden-border plants, notably *A. carniolica*, with slender, pointed, much divided leaves and very small white flowers, tinged with pink, *A. maxima*, with three- to four-divided leaves and pink flowers, and *A. major* with leaves divided palmately into five lobes and greenish-pink flowers and bracts.
UMBELLIFERAE, 10 species.

Athyrium a genus of mainly tropical and subtropical ferns. The best-known species is *A. filix-femina* [LADY FERN] which is cultivated in a wide range of varieties. The fronds are bipinnate with the pinnae deeply cut or toothed. Especially popular are 'Plumosum' with finely divided golden green leaves and 'Victoria' which has deep green fronds crenate at the tips.
ATHYRIACEAE, about 180 species.

Atrichum a widespread genus of mosses which is related to *Polytrichum* but which differs most notably in the "filmy" leaf structure.
POLYTRICHACEAE, about 40 species.

Atriplex a genus of annual and perennial herbs and shrubs widespread in temperate and tropical regions, often in arid zones or on saline soils. A few species, including *A. canescens*, *A. halimus* and *A. portulacoides* [SEA PURSLANE], are grown for their attractive grayish foliage. *A. hortensis* [*ORACHE] is grown as a green vegetable and for ornament. (See also SALTBUSH.)
CHENOPODIACEAE, about 130 species.

Atropa a small European and Central Asian genus of perennial herbs of which the most celebrated species is *A. belladonna* [DEADLY NIGHTSHADE, BELLADONNA, DEVIL'S HERB]. It is still one of the main natural sources of the alkaloids atropine and hyoscine (scopolamine), which are obtained from the roots and leaves, and it is cultivated extensively in some European countries. Atropine is an important drug for the ophthalmologist because it induces prolonged dilation of the pupils, and it is also used as an antidote against various chemical poisons. The drug scopolamine may induce delusions of flight, which perhaps explains the anointed broomstick of witches. The black berries are toxic.
SOLANACEAE, 4 species.

Attar of roses [OTTO OF ROSES, ESSENCE OF ROSES] a fragrant, pale yellow liquid con-

Nearly all the False Goatsbeards (Astilbe spp) seen in gardens are hybrids of Far Eastern species. Shown here is Astilbe × arendsii 'Fanal'. ($\times \frac{1}{10}$)

taining essential oils, obtained by distillation of rose petals, especially those of *Rosa damascena* [DAMASK ROSE], *R. gallica* [FRENCH ROSE], *R. centifolia* [CABBAGE ROSE], and *R. moschata* [MUSK ROSE], which are grown in Bulgaria, Turkey and the south of France. The essence is used for its fragrance in the manufacture of perfumes, toiletries, cosmetics, soaps and lozenges.
ROSACEAE.

Aubergine a common name for the fruit of *Solanum melongena*, which is widely cultivated as a vegetable. It is also called CHINESE EGG-PLANT, EGG FRUIT, MELONGENE, GARDEN EGG, JEW'S APPLE, MAD APPLE and BRINJAL. The fruits are egg-shaped with a tough, glossy

The star-like whorl of pinkish bracts is characteristic of the European Greater Masterwort (Astrantia major). ($\times \frac{1}{3}$)

Flowers (Top) and fruits (Above) of the narcotic and very poisonous Deadly Nightshade (Atropa belladonna), which is cultivated in Europe for the alkaloids obtained from the roots and leaves. ($\times \frac{1}{6}$, $\frac{1}{2}$)

yellow skin which changes to deep purple as the fruit ripens. Numerous small brown seeds are embedded in the cream-colored flesh.

S. melongena is a perennial herb which probably originated in India but it is now widely cultivated in the tropics and sub-tropics, being particularly important in India and the Far East. The white-flowered *A. melongena* var *esculentum* is the true AUBER-GINE. Other cultivars are available with white, yellow or black fruits, or the fruit may be elongate, as in var *serpentinum* [SNAKE EGGPLANT].
SOLANACEAE.

Aubrieta [AUBRIETIA] a small genus of evergreen mat-forming perennials distributed over the mountain regions of Mediterranean countries. Most species grow in sunny crevices among rocks.

A. deltoidea, with a geographical range extending from Sicily to Asia Minor, is the best-known species, being commonly grown as an ornamental plant. The flowers are purple and about 2cm in diameter. Selective breeding and possibly crossing with other species has widened the color range and many named cultivars, including double forms (eg 'Barkers Double') exist. *A. erubescens*, a tall tufted species native to Greece, has whitish flowers that turn pink with age and is probably a parent of the garden forms.
CRUCIFERAE, about 15 species.

Aucuba a small genus of hardy East and Central Asian shrubs, the best-known being *A. japonica* [SPOTTED LAUREL, GOLD DUST TREE], which is frequently grown as a garden ornamental, the female plants bearing bright red berries in the autumn. Popular cultivars include 'Crotonifolia' and 'Salicifolia'.
CORNACEAE, 3 species.

Aulacomnium a genus of erect mosses in which several species produce gemmae (clearly modified leaves, scattered along the upper part of the stem). *A. androgynum* grows abundantly on rotting tree trunks and stumps and also on the spongy bark of trees like *Sambucus nigra* [ELDER] in woodland. *A. palustre* is a much larger moss forming dense tufts and cushions on bogs or wet moorland.
AULACOMNIACEAE.

Auricula the name given usually to the numerous varieties of *Primula auricula* [BEAR'S EAR], native to the European Alps, and to *P. × pubescens*, a group of hybrids between *P. auricula* and another alpine species *P. rubra*.

P. auricula is a mountain perennial growing to a height of about 15cm. The somewhat fleshy leaves of this variable species are borne in a rosette and are covered with a rather mealy waxy powder when young. The flowers are in many colors, borne in terminal, stalked, many-flowered umbels. This species has been developed in cultivation and by crossing with related species to produce two separate races, the garden or "show" auriculas and the "alpine" auriculas, both with a wide range of colors.
PRIMULACEAE.

Avena a relatively small but important genus of annual, self-pollinating grasses, one of which, *A. sativa* [CULTIVATED *OAT], is a major temperate cereal. The genus consists of wild, weedy and cultivated species, the cultivated and weedy races being widely distributed throughout the temperate regions but the truly wild being found only in the Mediterranean basin, southwestern Asia and Ethiopia. Important weed species include *A. strigosa* and *A. weistii*. *A. barbata* is a very aggressive weedy species introduced from Spain to California where it is now used as a range grass.

Agriculturally the most important species are *A. sativa* and *A. byzantina* (see OAT). The other two species in this group, *A. sterilis* [WILD RED OAT, ANIMATED OAT] and *A. fatua* [COMMON WILD OAT], are both noxious weeds of arable land, the common wild oat being particularly troublesome in Europe and

Lady Fern (Athyrium filix-femina) *seen here (Center) with Oak Fern* (Thelypteris dryopteris) *(Bottom Right) and Beech Fern* (T. phegopteris) *in the crevices. ($\times \frac{1}{10}$)*

North America (see WILD OAT). All members of this group are closely related and involved hybridization of *A. magna* or *A. murphyi* with another unknown species.
GRAMINEAE, about 50 species.

Avocado, alligator pear or **aguacate** the fruit of the Central American evergreen tree, *Persea americana* (= P. gratissima). It is one of the finest and most nutritious of all salad fruits with a fresh, smooth buttery pulp. The fruit is a large fleshy berry, pear-shaped or globose, 7–20cm long; the skin of the ripe fruit may be green, yellowish, maroon or purple; there is a large single seed in the center surrounded by the greenish edible mesocarp, which contains 3–30% oil de-

Aubrieta deltoidea 'Red Shades', which exists in several other cultivar forms, makes a colorful display on sunny walls and in rock gardens. ($\times \frac{1}{3}$)

*The ripe branched panicles of the Cultivated Oat (*Avena sativa*). Oats are a major cereal crop of temperate regions.* ($\times \frac{1}{2}$)

pending on the particular cultivar involved.

There are three groups of cultivars – West Indian, Guatemalan and Mexican. A common cultivar often sold in temperate countries is 'Fuerte', a Mexican × Guatemalan hybrid which provides about 75% of the Californian crop. Avocados are now found throughout the tropics, but commercial production is largely confined to California, Florida, Argentina, Brazil, South Africa, Australia, Hawaii, Israel and the Canary Islands. LAURACEAE.

Axonopus a distinctive genus of mainly perennial and a few annual grasses centered on the New World tropics. The most important species include *A. compressus* [SAVANNA, CARPET GRASS], which is widespread throughout the wet tropics of the world, and *A. affinis*, a dominant pasture grass of alluvial soils on the southern coastal plain of the USA. Several species are used for pasture, and *A. scoparius* is cultivated for forage in South America under the names GAMALOTE, CACHI and CARICACHI.
GRAMINEAE, about 35 species.

Azalea the name given to small deciduous shrubs which include tender greenhouse plants as well as hardy ones. The name persists in garden use only and almost all botanists include *Azalea* in the genus **Rhododendron*. The attempt is often made, however, to retain the *Azalea* species in one series on the basis of the type of hairs covering leaves and ovaries. *A. pontica* (= *R. luteum*) is a popular species from the Levant and Caucasus and should not be confused with *R. ponticum*.
ERICACEAE.

Azara a genus of evergreen shrubs native to Chile. The leaves possess unequal-sized stipules, one of which is enlarged to almost the same size as the leathery leaf. *A. microphylla*, the most frequently cultivated ornamental, may grow to about 4m high and produces clusters of bright yellow flowers and orange-red berries. *A. dentata* is somewhat similar but has rather larger flowers, and *A. lanceolata* differs from both the above in its undivided narrow leaves and its mustard-yellow flowers.
FLACOURTIACEAE, 12 species.

Azolla [MOSQUITO FERN] a genus of water

*Mostly natives of China or the Himalayas, Azaleas (*Rhododendron *spp) are splendid plants for lime-free soils.* ($\times \frac{1}{10}$)

*An Avocado Pear Tree (*Persea americana*) showing its glossy evergreen foliage. Avocados are the most nutritious of all salad fruits.*

ferns found in tropical, subtropical and warm temperate zones. They are very small bluish-green leafy plants which float on the surface of water, forming dense mats. The stem is fragile and delicate and much-branched. They live symbiotically with the blue-green alga **Anabaena* which occupies a cavity at the base of the leaf. The best-known species are *A. caroliniana*, *A. filiculoides* and *A. nilotica*.
AZOLLACEAE, 6 species.

Azorella a genus of small, tufted or spreading perennial herbs distributed from the North Andes to temperate South America, through New Zealand, the Falkland Islands and Antarctica. *A. caespitosa* [BALSAM-BOG] from the Falkland Islands and *A. peduncularis* from Ecuador are tufted species, the latter sometimes cultivated in alpine houses.
HYDROCOTYLACEAE, about 70 species.

Azotobacter a genus of rather large, oval-shaped bacteria often with a variety of cell shapes and sizes occurring within a single species. The most important characteristic of species of *Azotobacter* is that they can "fix" nitrogen, ie they convert atmospheric nitrogen to nitrogenous compounds such as ammonium salts. They are common in well aerated soils which are not too acid, but are unable to grow under waterlogged or acid conditions. Attempts have been made to increase the fertility of agricultural soils by increasing the fixation due to *Azotobacter*, but without much success.
AZOTOBACTERIACEAE.

B

Baccharis a large genus of New World shrubs and small trees which may be deciduous or evergreen. Several species, including *B. halimifolia* [GROUNDSEL TREE, BUSH GROUNDSEL], are cultivated for ornament or ground cover.
COMPOSITAE, about 400 species.

Bacillus a genus of rod-shaped bacteria which produce spores resistant to heat, desiccation, irradiation and low temperatures. They are aerobic (ie they require oxygen to live) and are common in soil and air, but are rare in polar regions and water. Some cause diseases of vertebrates (anthrax), plants (soft rots) and insects (foulbrood). Others cause problems in the preservation of food and other materials because of their resistance to high temperatures.
BACILLACEAE.

Bacopa a genus of tropical and subtropical terrestrial or aquatic perennial herbs, mostly native to America. A few species, including the blue-flowered *B. caroliniana* (= *B. amplexicaulis*) and *B. monnieri* [WATER HYSSOP], are cultivated in gardens, particularly as border specimens.
SCROPHULARIACEAE, about 100 species.

Bacteria see p. 50.

Bagasse the fibrous remains of sugarcane left after extraction of the juice. Being a by-product of sugar milling it was used chiefly to fuel mills in the past. Mixed with molasses, it has also been fed to livestock. It is a good source of cellulose, and is now recycled into plastics, papers and compressed fiberboard manufacture.
GRAMINEAE.

Bahia piassaba a stiff wiry fiber from the stately palm *Attalea funifera*, indigenous to dry forests of the Brazilian state of Bahia. It is obtained from dried leaf bases, which are exported for manufacture into stiff brooms and scrubbing brushes.
PALMAE.

Balata a nonelastic gum very similar to *gutta-percha. It is obtained from the latex of *Manilkara bidentata* (often incorrectly identified as *Mimusops balata*), a tall broad-leaved tree native to South America and Trinidad. Balata is used for the same purposes as *gutta-percha and as a substitute for *chicle (*M. zapota*).
SAPOTACEAE.

Ballota a genus of herbaceous perennials and deciduous subshrubs from Europe, the Mediterranean region and western Asia. *B. nigra* [BLACK HOREHOUND, FETID HOREHOUND], a coarse, weedy, evil-smelling hardy perennial, is sometimes grown in gardens and is the source of an essential oil, but the most popular ornamental species is *B. pseudodictamnus*, which has pale gray-green woolly leaves and small white flowers with purple spots.
LABIATAE, about 35 species.

Balm a term applied to several aromatic herbs (see BERGAMOT, *Commiphora*, *Melissa* and *Melittis*).

Balsa the lightest commercial wood in the world, with the young heartwood weighing only about half as much as cork. It comes from a lowland tropical American tree, *Ochroma pyramidale*, which grows from Mexico to Bolivia and in the West Indies.

Balsa's lightness is due to the tree growing so fast and continuously that it does not leave annual rings and consists only of large spongy cellulose cells. The lightness of dry balsa makes it particularly useful where strength without weight is required. It has particular value in ship and aeroplane partitions, survival equipment and model-making. It also provides a useful insulating material in acoustic and refrigeration equipment.
BOMBACACEAE.

Balsam an aromatic mixture of gums and oils exuded from certain plants, usually trees. They have medicinal and some other specialized uses and are generally traded as

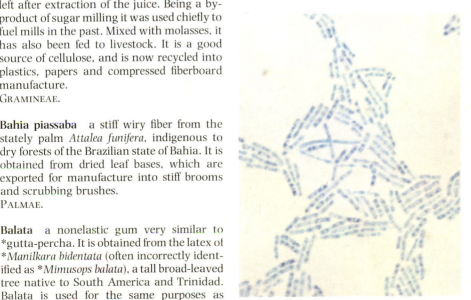

Photomicrograph of Bacillus cereus, *which comprises chains of rod-shaped cells; they have been stained to show the nuclear material (×2200).*

yellow or brown liquids with the consistency of treacle or honey. All balsams are insoluble in water but soluble in alcohol, benzene, ether or chloroform. They tend to have a bitter or burning taste – see *Abies* (Canada balsam), *Dipterocarpus* (Gurjun balsam) and *Myroxylon* (Peru and Tolu balsams).

Copaiba balsam is derived from a number of South and Central American trees of the genus *Copaifera*. It is used as a component in

*The Black Horehound (*Ballota nigra*); the whorls of flowers (verticillasters) are characteristic of members of the family Labiatae. Although foul-smelling, this species is sometimes grown in gardens. (×¼)*

medicinal preparations and as a fixative in porcelain painting, for sealing oil paintings and for making paper transparent.

The term "balsam" is also used popularly for species of the genus *Impatiens.

Bamboo a general term which refers to 70 or so genera and approximately 1000 species of perennial woody grasses. The characters which distinguish bamboos from all other grasses include the prominent development of branched, underground stems or rhizome systems, and thick, woody, segmented and often branched stems growing in ever-expanding thick clumps.

Bamboo stems range for the most part between 50cm and 45m in height, but one species, *Dinochloa andamanica*, has been known to grow to an incredible 90m. Geographically, bamboos are native throughout the tropics, although the greatest variety and abundance are to be found in southern and Southeast Asia, from India to Japan.

The bamboos are an important group of plants for a variety of different uses, chiefly as a result of their erect, fast growing, tree-like habit. Commercially important genera include *Bambusa, *Arundinaria, *Dendrocalamus, Melocanna, Ochlandra, *Phyllostachys and *Sasa, which are all used for (continued on p. 52)

Bacteria and Blue-green Algae

BACTERIA AND BLUE-GREEN ALGAE ARE treated together here since they are the oldest living organisms and have a unique cellular structure (prokaryotic), for which reason they are placed together in separate divisions within the kingdom Monera. Bacteria are notorious as agents of disease but the majority are harmless and indeed are vital components of the world we live in. Many of the characteristics of bacteria and blue-green algae are of a primitive kind and fossils which appear to have been blue-green algae have been found in Pre-Cambrian deposits about 3 000 million years old. At this stage oxygen was absent from the Earth's atmosphere and the development of an oxygen-producing type of photosynthesis by blue-green algae seems to have eventually brought about the major ecological transformation which made possible the evolution of the oxygen-requiring plants and animals. It has been suggested that the chloroplasts of higher plants originated as symbiotic blue-green algae.

Bacteria are the most abundant organisms in the world, being found in all habitats from the icy wastes of Antarctica to the boiling waters of hot springs, from the depths of the oceans to the hottest deserts, and from the soil to the guts of animals. Blue-green algae are usually to be found in moist or wet habitats where light is available and the reaction neutral or alkaline, their general resistance to adverse conditions such as extremes of temperature, desiccation and high salinity, enabling them to be successful in situations in which most other organisms except bacteria cannot survive.

On average the bacterial cell is only about one millionth of a metre ($1\,\mu m$) in diameter – there may be several million bacteria in just one cubic centimetre of milk. Basically bacteria come in three shapes: spheres (or cocci), rods (or bacilli), and spirals (or spirilla). Coccoid forms occurring in chains are termed streptococci. Many blue-green algae are unicellular and in some of these cases individual cells may be aggregated together into colonies. However, unlike bacteria, some are filamentous, usually unbranched, but sometimes branched. In the main, bacteria are heterotrophic, that is they are unable to synthesize the organic chemicals needed for life and hence must absorb them from their surrounding environment. Most of the heterotrophic bacteria are what is known as saprophytes, obtaining their nourishment from dead organic matter. At the other extreme, parasitic heterotrophic bacteria obtain their nourishment from living organisms. A few types of bacteria possess the green pigment chlorophyll that is found in green leaves of higher plants and the thalli of algae and are able to harness the energy of sunlight to produce the organic chemicals they need by the process of photosynthesis, but they do not evolve oxygen. Blue-green algae are photosynthetic and evolve oxygen.

The minute size of individual bacterial cells does little justice to their importance. Along with the fungi, bacteria are the main decomposers of organic waste material produced by plants and animals. They release back into the environment the materials of once-living organisms so that these materials are available once again to the living world. This is the most vital ecological role of bacteria and perhaps the one most often overlooked. A single gram of soil may contain 2.5 billion bacterial cells, all breaking down the remains of animals and plants. Man has exploited this aspect of bacterial life in his sewage works where the unacceptable detritus of human civilization is broken down by bacteria.

Nitrogen is one of the most important elements required for the growth of plants

Below Top The cells of Neisseria gonorrhoeae, *the causal agent of gonorrhea, occur in pairs (diplococci), a group of which are shown here (center) after being engulfed by a white blood cell. (×1600). Below Bottom* Treponema pallidum. *the causal agent of syphilis, is a spirochaete as can be seen from the single long twisted cell in the center of this smear (×1800).*

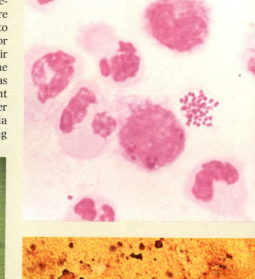

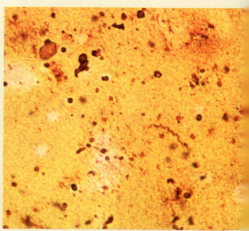

Water bloom formed by the unicellular blue-green alga Microcystis sp *on the waters of the Monkey Temple, Varanasi, India.*

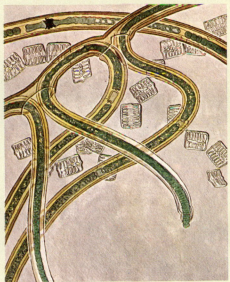

A dense mat of blue-green algae and flexuous bacteria in the hot spring region in the Yellowstone National Park USA, all busily fixing atmospheric nitrogen

A filamentous blue-green alga, Scytonema. *The empty-looking cells are heterocysts. In the background are cells of the fresh-water diatom,* Tabellaria. *(× 200)*

and this is absorbed in various forms from the soil. Bacteria and blue-green algae play a vital role in the cycling of nitrogen compounds in the soil. The bacteria genera *Azotobacter* and *Clostridium* are unusual in that they can absorb atmospheric nitrogen and fix it into ammonia or ammonium compounds. This ammonium nitrogen, and that produced by the decomposition by other bacteria of organic compounds, is oxidized by bacteria of the genus *Nitrosomonas* into nitrite compounds, which are in turn oxidized by the genus *Nitrobacter* into nitrate.

With leguminous plants such as beans, peas or clover this nitrogen fixation is achieved by bacteria living in nodules in the roots. These bacteria (of the genus *Rhizobium*) fix atmospheric nitrogen and release it into the host plant. This explains why leguminous crops actually enrich the soil they grow in. The ability to fix free atmospheric nitrogen is also of vital ecological importance in some of the blue-green algae; eg *Nostoc, *Anabaena, Rivularia*. Nitrogen fixation by blue-green algae makes a considerable contribution to the fertility of the habitats in which they occur. Thus, in certain types of paddy field they may fix up to 55–80kg/hectare of nitrogen per year and where fertilizer is not supplied make possible a reasonable crop of rice in fields which otherwise would produce little. It seems likely that many millions of people survive largely because of this activity.

Man also benefits from another group of bacteria – those that inhabit dairy products. These bacteria are vital in the production of cheese and yoghurt and the preparation of sauerkraut and pickles. On the other hand closely-related bacteria are involved in the spoilage of wine and beer.

This brings us to the detrimental side of bacteria – their involvement in disease. Some of the most serious diseases of Man are caused by air-borne bacteria, such as diphtheria (*Corynebacterium diphtheriae*), scarlet fever and rheumatic fever (*Streptococcus*), tuberculosis (*Mycobacterium tuberculosis*), bacterial pneumonia (*Diplococcus pneumoniae*), and whooping cough (*Bordetella pertussis*). Typhoid fever (*Salmonella*), dysentery (*Shigella dysenteriae*) and typhus (*Rickettsia prowazekii*) are bacterial diseases spread in food and water. As we have seen, bacteria are important agents of decay of once-living organisms and their products, and it is therefore not surprising to find that bacteria are immensely important as agents of decay of food, often causing very severe cases of food poisoning, as in botulism (*Clostridium*

Aerobic bacteria are active agents in the breakdown of sewage. The green scum is caused by the alga Euglena *which hampers bacterial activity.*

botulinum). Many diseases of plants are also caused by bacteria, particularly soft rots, blights and wilts.

Most bacterial diseases of Man and other animals can now be controlled by the use of antibiotics, but because of their rapid reproduction rate (on average, every half-hour) the production of resistant strains is not infrequent. However, bacteria are not without their natural enemies – a special type of virus called a bacteriophage specifically infects bacteria. As research tools, bacteriophages have been of importance in the study of bacterial breeding mechanisms and genetics. Bacteria themselves are useful subjects for work on genetics.

papermaking, building materials, fishing rods, handicrafts, medicine, musical instruments and gardening implements. Aqueducts, oars, masts, baskets, fish hooks, spearshafts, bows, arrows, knives, ladders, rafts, pails and churns all come from the same source. The joints of bamboo stems root as readily as willows and thus are invaluable in hedge-making. Bamboo also provides fiber for ropes and cordage, tiles for roofs, axles and springs for carts, frames for bird cages, and chopsticks. Bamboo shoots are eaten as a vegetable and the leaves used as fodder. GRAMINEAE.

Bambusa a commercially important genus of tropical and subtropical bamboos growing in continually expanding thickets to a height of about 16m. They are mainly Southeast Asian in distribution, although one species known only from cultivation, *B. vulgaris*, is pantropical.

Several species are grown as ornamentals and as hedge plants, such as *B. glaucescens* [HEDGE BAMBOO], a very variable Chinese species grown under various cultivars.

Several species, particularly *B. vulgaris* and *B. arundinacea* [SPINY BAMBOO], are utilized for building and other structural uses, as a result of their high stem strength and easy culture. They also provide a ready source of high-quality cellulose fiber for papermaking. The split stems of some bamboos, eg *B. textilis* and *B. tulda*, are used for weaving mats and the young tender shoots of *B. beecheyana*, *B. cornuta* and *B. spinosa* are consumed as vegetables. (See also BAMBOO.)
GRAMINEAE, about 100 species.

Banana one of the most important tropical fruit crops, a staple food as well as a commercial crop for export. Banana plants are giant herbs with huge leaves; the stems are made up of tightly packed leaf sheaths. The fruits form in large bunches. Cultivated bananas belong to *Musa acuminata* (= *M. cavendishii*) and *M. × paradisiaca* (= *M. sapientum*). There are still far more

banana plants in backyards and village gardens than there are in plantations, and bananas form an important starch-rich food throughout the tropics. Varieties cultivated for export are picked green and ripen to the characteristic yellow on the voyage.

A number of species are cultivated for their ornamental value, including *M. coccinea* [FLOWERING BANANA] and *M. basjoo* [JAPANESE BANANA]. (See also *Musa*.)
MUSACEAE.

Banksia a genus of evergreen trees and shrubs, rarely prostrate, endemic to Australia and Tasmania (mostly Western Australia), with one species (*B. dentata*) native to New Guinea and the islands of the Gulf of Carpentaria. The flowers are clustered in dense terminal globose or cylindrical heads of sometimes 1000, with conspicuous styles and abundant nectar [whence the once popular name, NATIVE HONEYSUCKLE]. The mature fruiting head has a cone-like appearance studded with the protruding, gaping "mouths" of the woody follicles.

B. grandis [BULL BANKSIA] seldom exceeds 8m tall and has golden flower spikes to 30cm long and ornamental serrate leaves. *B. ericifolia*, from New South Wales and Queensland, has orange flowers and small linear pale green leaves, while *B. serrata* has dark green serrate leaves, grayish-yellow flowers and purplish wood that is used for decorative furniture.

Banksias are not really hardy in cool temperate regions except in the very mildest areas. Exceptions are *B. grandis* and *B. serrata*, already mentioned, *B. littoralis* [SWAMP BANKSIA] and *B. integrifolia* [COAST BANKSIA].
PROTEACEAE, about 50 species.

Banyan the common name for *Ficus benghalensis*, a large tree native to the

Young Banana fruits ("hands"). The terminal part of the fruiting stalk consists of a series of red bracts subtending the male flowers.

Berberis darwinii, a hardy shrub cultivated as an ornamental for its evergreen foliage and orange flowers produced in the spring.

Himalayan foothills but now widespread throughout India, as for centuries it has been planted in villages for the excellent shade it provides. The Indians also regard it as sacred, for it is said that Buddha once meditated beneath a BANYAN tree (see *Ficus*).
MORACEAE.

Baptisia a genus of erect, herbaceous perennials native to North America, some of which are grown as ornamental garden-border plants. They produce attractive leaves and inflorescences of pea-like flowers. The cultivated species include *B. australis* [FALSE INDIGO], which bears wedge-shaped leaves and long spikes of rich indigo-blue flowers, and the shorter, softly hairy *B. bracteata*, with spikes of yellow-white flowers. The two other commonly cultivated species are *B. leucantha* and *B. tinctoria* [YELLOW INDIGO, RATTLE WEED].
LEGUMINOSAE, about 35 species.

Barberry the common name for the genus *Berberis*, which consists of deciduous and evergreen, often spiny shrubs widespread in temperate zones of Asia, Europe and the Americas. One of the commonest species is *B. vulgaris* [EUROPEAN BARBERRY] which grows 2–3m tall, has dense masses of arching three-spined branches, yellow racemes of flowers in spring, and brilliant succulent red berries in autumn. Despite their acid juice, the berries, such as those of the South American *B. buxifolia* [MAGELLAN BARBERRY] and the North American *B. canadensis* [ALLEGHANY BARBERRY], are made into preserves. A considerable number of species, including *B. floribunda* [INDIAN BARBERRY] (often confused in cultivation with *B. aristata*) from Nepal, *B. ruscifolia* from Brazil and *B. vulgaris*, are sources of a yellow dye. Extracts from the bark, stem or roots of some species have medicinal uses as tonics and aperients.

Many types of barberry are grown as ornamentals either for their evergreen glossy foliage, as in the case of *B. darwinii*, *B. julianae*, and *B. × stenophylla*, or for their brightly colored autumn berries, as in the case of the deciduous *B. thunbergii* and *B. × rubrostilla*.

B. vulgaris lost favor when it was confirmed to be the alternate host plant of the fungus causing wheat rust disease (see *Puccinia*).
BERBERIDACEAE, about 450 species.

Barleria a genus of shrubs and herbs, native to the tropics. Many of them have spiny stems and are adapted to dry conditions. The cultivated species bear conspicuous spikes of flowers, subtended by large spiny bracts. The most popular is the blue-flowered *B. cristata* [PHILIPPINE VIOLET] from India and Burma.
ACANTHACEAE, about 230 species.

Barley an important quick-growing, early maturing cereal crop of temperate regions. It shares with wheat the distinction of being one of the oldest of all domesticated plants. All cultivated forms are now grouped as one species, *Hordeum vulgare*, while all the wild and weedy races, from which these have been derived, are classified as *H. spontaneum*. Barley has its origins in southwestern Asia.

The world production of barley was 171 million tonnes in 1974. However, very little is consumed directly by humans. About 60% of the present production is used as animal feed, often as an alternative to maize. Most of the remaining tonnage is used for malt for beer and spirits. As animal feed, barley is considered primarily as a source of carbohydrate, the grain containing only 10–13% crude protein.
GRAMINEAE.

Basil a term loosely applied to several

The Salmon Barberry (Berberis aggregata) is one of the Berberis species grown particularly for its autumn fruits. Shown here is 'Barbarossa'. (× 3)

Fruiting heads of the Cultivated Barley (Hordeum vulgare) ready for harvesting. Most barley is used as an animal feed. (× ½)

species of *Ocimum* but normally referring to *O. basilicum* [SWEET BASIL], a fragrant aromatic annual herb of western tropical Asia that is widely cultivated especially in France, Hungary and the USA. The leaves are used for flavoring soups, sauces, ragouts and salads. It yields an essential oil (oil of sweet basil) used in various condiments, sauces and in perfumes and cosmetics.
LABIATAE.

Batrachospermum [FROG-SPAWN ALGA] a genus of red algae found in freshwater streams and lakes in many parts of the world. Plants growing in deep water or shade are dark violet or reddish in color but those growing in shallow water are olive-green.
RHODOPHYCEAE.

Bauhinia a large genus of a tropical and warm African, Asian and American shrubs, trees or climbers that includes several plants cultivated for their beautiful showy flowers. Among the cultivated species are the Asiatic *B. variegata* [ORCHID TREE, EBONY WOOD], a small tree with variegated purplish-white or yellow, orchid-like flowers and cleft leaves. The closely similar purple- or white-flowered *B. purpurea* [BUTTERFLY TREE, CAMEL'S FOOT TREE], another Asiatic species, is one of the showiest trees grown in India. Also widely cultivated are *B. tomentosa* [ST. THOMAS TREE] and *B. monandra* [BUTTERFLY-FLOWER, JERUSALEM DATE].
LEGUMINOSAE, about 300 species.

Bay rum a solution of some 1% of bay oil (distilled leaves and twigs of the West Indian *Pimenta racemosa* (=*P. acris*) [EVERGREEN BAYBERRY]) in alcohol, normally rum. The bay rum industry centers on Jamaica. Bay rum products include soothing lotions, soaps and perfumes.
MYRTACEAE.

Bean a term applied to a number of different species of the family Leguminosae. The pods (legumes) and seeds of beans form a valuable source of human and animal food as they are rich in protein, mineral salts and some vitamins (see Legumes, p. 194).
LEGUMINOSAE.

A flower of the Orchid Tree (Bauhinia variegata cv 'Candida'), which has five free petals, five stamens and an upturned style. (× ½)

Beech the common name for a closely-knit group of 8–10 species of deciduous trees belonging to the genus *Fagus*. They grow to a height of 30–45m and occur throughout the Northern Hemisphere, where they are frequently dominant or codominant in temperate forests. SOUTHERN BEECHES belong to the closely related genus *Nothofagus*.

Beeches have rounded, spreading canopies and smooth, gray bark. The leaves are usually thin and shining green and the slender, elongated winter buds are a very distinctive feature of the trees. Several beeches are widely cultivated for their handsome shape and foliage. *F. sylvatica* [EUROPEAN BEECH] is most frequently grown and many varieties [eg COPPER BEECH] are known. Beech wood deteriorates when exposed to the weather but is a hard and durable timber and that of *F. sylvatica*, *F. grandifolia* [AMERICAN BEECH] and *F. orientalis* [ORIENTAL BEECH] is much used for furniture and flooring. The nut is rich in oils and provides valuable food for stock such as pigs.

BLUE BEECH is an alternative common name given to *Carpinus caroliniana* [AMERICAN HORNBEAN] and BROWN BEECH is another name for *Cryptocarya glaucescens*.
FAGACEAE.

Beetroot or **garden beet** cultivated forms of the species *Beta vulgaris*. BEETROOT produces an edible swollen root that is round, flat or conical, and usually a deep purple color, though white and golden forms exist. Other forms of garden beet are the leaf beets. *SPINACH BEET is grown for its leaves, which are cooked and used as a green vegetable. SEAKALE BEET or SWISS *CHARD has white fleshy leaf midribs which are also used.
CHENOPODIACEAE.

Begonia a large and horticulturally important genus of subtropical and tropical herbs

Cultivated Beetroot (Beta vulgaris); cultivar 'Boltardy,' shown here, has round or globe-type swollen roots. (× ⅓)

There are many groups of cultivated begonias: shown here are male flowers of a tuberous-rooted begonia (Begonia 'Fireglow'). (× ⅓)

or subshrubs. The leaves are characteristically asymmetrical and many species are more or less succulent, making them popular ornamentals, as indoor or warm greenhouse plants, or as summer bedding specimens. The flowers are borne in axillary cymes and are usually showy – white, rose, scarlet or yellow being the predominant colors.

Horticulturally, begonias are divided into three groups – tuberous, rhizomatous or fibrous-rooted. Each of these is further subdivided according to the form of the stem. Thus the tuberous group consists of stemless species such as the bright-red-flowered *B. davisii* or species such as the pink *B. gracilis* var *martiana* [HOLLYHOCK BEGONIA], the yellowish-white-flowered *B. natalensis*, and the white-flowered *B. dregei* (= *B. caffra*) [GRAPE-LEAF or MAPLE-LEAF BEGONIA], which have stems from 0.3 to 1m high. The rhizomatous species have either a creeping or climbing habit, as in *B. × fuscomaculata*, or erect, as in the beautiful *B. rex* [PAINTED-LEAF BEGONIA], often with attractively marked foliage. The fibrous-rooted forms may be herbaceous, as in the hybrids *B. × carrieri* and *B. × erfordia* and the horticulturally important *B. × semperflorens-cultorum* which consists of many varieties grown as bedding plants. *B. glabra* (= *B. scandens*) has a trailing or creeping habit. However, most of the species and hybrids in this group are woody, with a shrubby habit. Among a number of popular forms are *B. × ascotiensis* (bright red flowers), *B. maculata* (= *B. argyrostigma*) (rose-white flowers) and *B. × kewensis* (yellow-white flowers). Among the taller species are the slender red-and-white flowered *B. acutifolia* (= *B. acuminata*) [HOLLY-LEAF BEGONIA] the red-flowered *B. coccinea* [ANGEL-WING BEGONIA] and the white, fragrant-flowered *B. venosa*.
BEGONIACEAE, about 350 species.

Bellflower the name for *Campanula*, a large genus which consists almost entirely of annual, biennial and perennial herbs, many of which are cultivated as ornamentals. They are widely distributed over the Northern Hemisphere including the Arctic, but with the main center in Europe.

Bellflowers are generally inhabitants of dry sunny habitats. Included in the numerous cultivated species are *C. medium* [CANTERBURY BELL], a biennial with blue or lilac flowers, but purple, pink or white in some forms, *C. latifolia* [GIANT BELLFLOWER], a perennial with blue or white flowers, *C. rapunculoides* [CREEPING BELLFLOWER], which like *C. isophylla* has trailing stems. A rich variety of shades of blue is seen in the flowers of *C. rotundifolia* [HAREBELL, BLUEBELL].

There are many other cultivated campanulas, including: *C. americana* [TALL BELLFLOWER], from North America, with blue or white flowers; *C. aparinoides* [MARSH BELLFLOWER], from eastern USA, with white or bluish flowers; *C. carpatica* [TUSSOCK BELLFLOWER], from the Carpathians, with several color forms; *C. elatines* [ADRIATIC BELLFLOWER], from the Adriatic region and eastward, with many color and form varieties; *C. glomerata* [CLUSTERED BELLFLOWER], from throughout Eurasia, with blue or white flowers; *C. persicifolia* [WILLOW BELLFLOWER, PEACH-BELLS], from Europe to northeastern Asia, with white to deep blue flowers and several cultivars; *C. pyramidalis* [CHIMNEY BELLFLOWER], from southern central Europe, with white to pale-blue flowers; and *C. rapunculus* [RAMPION], a biennial from eastern Europe and western Asia.
CAMPANULACEAE.

Bellis a small European genus of hardy perennial herbs. Its most important species is *B. perennis* [DAISY] a lawn weed which has been the source of many varieties cultivated as border plants. Another cultivated species is the North African *B. rotundifolia*.
COMPOSITAE, 15 species.

Berberis see BARBERRY.
BERBERIDACEAE, about 450 species.

Bergamot, sweet bergamot, bee balm or **oswego tea** the common names for *Monarda didyma*, a perennial aromatic herb native to North America. The leaves and flowers are used as a flavoring or as an infusion (Oswego tea). *M. fistulosa* and *M. punctata* [HORSEMINT] are also cultivated.

Oil of bergamot is an essential oil obtained from the peel of a variety of the *SEVILLE ORANGE*, *Citrus aurantium* subspecies *bergamia*. It is used in making perfumes.

Bergenia a genus of herbaceous perennials from temperate parts of Asia, with large, evergreen, glossy, leathery leaves and clusters of bell-shaped flowers. Several species are grown as garden ornamentals, including *B. ciliata* [WINTER BEGONIA], *B. cordifolia*, *B. crassifolia* and *B. purpurascens*, with pink to purple flowers.
SAXIFRAGACEAE, 10 species.

One of the most attractive bellflowers is the Harebell or Scottish Bluebell (Campanula rotundifolia), *which produces bright blue flowers.* (×1½)

Berkheya a genus of thistle-like perennial herbs and shrubs mostly native to southern Africa. Several species are cultivated in borders for their attractive flower heads – yellow in *B. adlamii*, *B. grandiflora* and *B. uniflora*, purple in *B. purpurea*.
COMPOSITAE, about 90 species.

Beta a genus of European and Mediterranean herbs, of which *B. vulgaris* [*BEET] is the most important. Subspecies *maritima* is the WILD SEA BEET, a perennial native to seashores of Europe, the Mediterranean, the Azores and North Africa. The cultivated forms of the same species include the leaf beets (see SPINACH), beetroot, mangel-

*The Wild Sea Beet (*Beta vulgaris *subspecies* maritima) *commonly grows on seashores and is very closely related to the Sugar Beet of commerce.*

wurzels, fodder beets and sugar beet.
CHENOPODIACEAE, 6 species.

Betel nut a nut containing a mild narcotic, which is chewed by the inhabitants of many tropical countries. It is derived from *Areca catechu* [BETEL NUT PALM], a feather palm probably native to the Malayan peninsula but widely cultivated in the tropics. It is always used in conjunction with the leaves of the BETEL PEPPER (see *Piper*). The semi-ripe nut is sliced and wrapped in a betel pepper leaf which has first been smeared with lime; clove and other spices may be added. Betel nut is chewed after meals to sweeten the breath and possibly to aid digestion, and at other times as a social and ceremonial custom. It is a mild stimulant and induces a feeling of well-being, but blackens the teeth and stains the saliva bright red.
PALMAE.

Betula see BIRCH.
BETULACEAE, about 50 species.

Beverage plants see p. 56.

Bidens [BUR MARIGOLD, STICK-TIGHTS, TICKSEED] a cosmopolitan but predominantly American genus of annual and perennial herbs, whose fruits have barbed bristles which catch in animals' fur as a means of dispersal. *B. pilosa* [WEED BLACKJACK] has somewhat irregular creamy flowers and var *radiata* is cultivated as a border plant as are a number of other species including the Mexican *B. ferulifolia* and the North American *B. laevis* (= *B. chrysanthemoides*).
COMPOSITAE, about 100 species.

Bignonia a genus from the southeastern USA represented by a single species, *B. capreolata* (= *Doxantha capreolata*) [TRUMPET-FLOWER, CROSS-VINE, QUARTER VINE], an evergreen, woody, tendril-climber.

A flower head of Bidens kilimandscharica, *a species from the mountains of East Africa.* (×1)

Beverage Plants

From earliest times Man has sought ways to make the liquids he drinks either more palatable or stimulating or both. Most of the popular beverages use some form of vegetable matter. They include simple unfermented fruit juices (now an important aspect of the soft drinks industry) such as blackcurrant, *orange, *lemon, *pineapple, granadilla, passion fruit, *coconut water, horchata de chufa (*Cyperus esculentus) etc, which are perhaps better regarded as nutrients and as a source of vitamins, minerals and carbohydrates. In the stricter sense, beverages can be divided into alcoholic, such as beer, *cider, wines, spirits, and alkaloidal such as *cocoa, *coffee, *tea, *kola, etc and alkaloid-substitutes such as *chicory, *dandelion, fig (coffee substitutes) and *lime, cape tea (tea substitutes).

Alkaloidal (non-alcoholic) beverages
Non-alcoholic beverages usually act as stimulants since they contain alkaloids which are extracted by hot water or other forms of processing. Most of the familiar stimulant non-alcoholic beverages are obtained from plants and are widespread throughout the world. The best-known are tea, coffee, cocoa and kola. They are not taxonomically related yet they all produce caffeine or related compounds: theobromine from the seeds of the CACAO TREE (*Theobroma cacao) and COLA TREE (*Cola nitida), and caffeine from the seeds of the COFFEE PLANT (Coffea arabica) and leaves of the TEA BUSH (*Camellia sinensis). The leaves of the COCA PLANT (*Erythroxylum coca), which contain the alkaloid *cocaine, are chewed to postpone fatigue.

Tea is derived from the leaf-tips of Camellia sinensis and is consumed by about half of the world's population. It has long been used in the Orient as an item of commerce and as a social custom in China dates from the 5th century AD; it then spread around the world. It was introduced to Western Europe at the end of the 16th century after the habit of coffee drinking had become established. The characteristic flavor and aroma of tea are provided by the essential oil theol. Tannins (quinones), along with pectins and dextrins, provide the color and astringency. The stimulant properties are due to the alkaloid theine (which is identical with caffeine) which occurs in the leaf at a concentration of 2–5%.

Coffee is one of the world's major international commodities. It is commercially more important than tea although consumed by fewer people. Only two species are of commercial importance: C. arabica, which produces about three-quarters of the world's good quality crop and C. canephora (= C. robusta), which is especially valuable for producing beans used in making instant coffee. The green beans are processed and roasted to develop their characteristic aroma (mainly caffeol and essential oils). The green beans contain up to 3% caffeine by weight.

The seeds of CACAO (Theobroma cacao) contain theobromine and traces of caffeine and are the source of cocoa and chocolate.

The seeds of KOLA (Cola nitida) are rich in caffeine and essential oils. Cola started off the fashion for cola drinks although the flavorings and stimulants are now obtained from other sources. Other plants containing alkaloids are used more locally to make beverages, although maté, which is restricted to South America, is drunk by a large part of the population there. Other stimulant beverage plants include *Ilex vomitoria [CASSINA, YAUPON], Paullinia cupana [GUARANA], P. yoco [YOCO] and Catha edulis [*KHAT].

Alcoholic beverages
Man has long taken advantage of the production of ethyl alcohol by the action of natural yeasts, and alcoholic drinks can be prepared by fermentation of any plant material that contains carbohydrate. Wine is universally the most important. The fruit of the GRAPE VINE (*Vitis vinifera) is the major source, but wines are also made on a smaller scale from other North American species. *Apples are used to make *cider, and a substrate more indirectly derived from plants, honey, is fermented to give mead.

Beers are produced from cereals which first have to be malted: the starch is converted to maltose by germination, the action of molds, or by saliva. Plants used for beer include WHEAT, *BARLEY, *RICE, *MAIZE, *MILLET, *RYE and palms.

Distilled liquors are made from many fermented plant sources, grape alcohol being the most important and providing the basis for brandy and many locally important spirits. Other carbohydrate sources include cereals (eg gin, whisky), *potatoes (eg vodka), *sugar cane (eg rum), sap (eg mezcal from *Agave species) and fruits (eg calvados from apples, slivovitz from *plums). Plants supply the predominant flavor in many non-alcoholic drinks, for example *ginger and *sarsaparilla. They also provide important additives to alcoholic beverages – eg *hops in beer, wormwood in absinthe, pine resin in retsina, *juniper berries in gin and various herbs in vermouths – and the making of liqueurs may require complex (and often secret) blending of herbs, essential oils and fruit extracts. Distilled liquors are often served diluted with other drinks of plant origin, and fruits are added to many drinks as a final garnish or flavoring. Spices such as *cardamom are used to give aroma to coffee, and flowers (eg jasmine) to scent tea, and for both these beverages there are many plant materials used as adulterants or substitutes. Several plant species are used as substitutes for alkaloid-containing beverages such as chicory and dandelion coffee and herbal teas such as lime.

(See also Alcoholic Beverages, p. 18.)

BEVERAGE PLANTS

Common name	Scientific name	Part used
Alkaloidal		
CASSINA	*Ilex vomitoria	Leaves, shoots
*COCOA	Theobroma cacao	Seeds
*COFFEE	Coffea arabica, C. canephora	Seeds
GUARANA	Paullinia cupana	Seeds
*KHAT (Qat)	Catha edulis	Leaves
*KOLA	Cola nitida	Seeds
MATÉ (*Yerba de maté, Paraguay tea)	Ilex paraguariensis	Leaves
*TEA	Camellia sinensis	Leaves
YOCO	Paullinia yoco	Bark
Alkaloid substitutes		
*CHICORY (coffee substitute)	Cichorium intybus	Taproot
*DANDELION (dandelion coffee)	Taraxacum species	Taproot
*BARLEY (coffee substitute)	Hordeum species	Fruits
FIG (fig coffee)	*Ficus carica	Fruits
*LIME (herbal tea)	Tilia species	Flowers
CAPE TEA (bush tea)	Cyclopia species	Leaves

(For **Alcoholic beverages** see p. 56.)

Agave cores ready to be roasted and distilled to produce mezcal, which is one of Mexico's traditional beverages.

is *B. nutans*, which is grown outdoors in warm climates, under glass or as a house-plant. Other cultivated species include *B. amoena*, *B. pyramidalis*, *B. saundersii* and hybrids between them.
BROMELIACEAE, about 50 species.

Bindweed the common name given to various weedy plants with climbing stems which twine around vegetation and other supports. Principal examples are species of the genera *Calystegia*, *Convolvulus* and *Polygonum*, such as *Calystegia sepium* [HEDGE BINDWEED], *Convolvulus arvensis* [FIELD BINDWEED] and *Polygonum convolvulus* [BLACK BINDWEED].

Birch the common name for members of the genus *Betula* which provides trees out-

The Silver Birch (Betula pendula) *is frequently the first tree species to colonize cleared woodland. It is shown here with Bracken.*

The bromeliad Billbergia reichartii *has pink bracts that subtend a flower with green petals with blue margins and yellow anthers. (×2)*

It is a popular ornamental in warm gardens and glasshouses, with its numerous large yellow-red flowers.
BIGNONIACEAE, 1 species.

Bilberry the common name for *Vaccinium myrtillus* (also known as BLAEBERRY, EURO-PEAN BLUEBERRY or WHORTLEBERRY]. It forms a low, branching, deciduous shrub which pro-duces small, sweet, blue-black berries. The species occurs widely over Europe, and ex-tends east into Asia. It is particularly com-mon in upland areas, and grows on acid heaths and moors and also in woodlands.

The berries have been traditionally used for food and are collected commercially in central Europe. Apart from being eaten as fresh fruit, they are stewed, preserved as compote or jam, or used in pies and pastries.
ERICACEAE.

Billardiera a genus of mostly climbing evergreen shrubs and subshrubs native to Australia. *B. scandens* [APPLEBERRY] and *B. longiflora* are cultivated for their creamy-white to purple flowers and blue, edible berries.
PITTOSPORACEAE, 9 species.

Billbergia a genus of mostly stemless epi-phytic plants native to the American tropics and grown mainly for their showy flowers. The leaves are somewhat scurfy, and usually have spiny margins. The bluish, greenish or reddish flowers are borne in a spike or panicle, often with showy colored bracts. The most commonly grown, and hardiest, species

The Bilberry (Vaccinium myrtillus) *grows on high moorland throughout Europe. The pink lantern-like flowers precede the blue-black berries. (×¾)*

standing for their beauty and usefulness. The birches are trees and shrubs native to the north temperate and Arctic regions. Most species are extremely hardy, the dwarf species *B. nana* reaching the tree limit in the Northern Hemisphere. The trees have an attractive and graceful appearance, possess-ing slender branches and twigs. The bark is often silvery or whitish and extremely hand-some, particularly in *B. pendula* [COMMON, WHITE or SILVER BIRCH], *B. pubescens* [DOWNY BIRCH] and *B. papyrifera* [CANOE or PAPER BIRCH], where it peels off in papery layers. Birches flourish on acid heathland among heather, spreading rapidly over clearings and wastelands by means of vast numbers of wind-dispersed seeds.

Useful timber is obtained from *B. lenta* [BLACK BIRCH, CHERRY BIRCH], *B. utilis*

[HIMALAYAN BIRCH, INDIAN PAPER BIRCH], *B. pendula* and *B. pubescens*. The flexible branches cut in winter are made into besom brushes, still much used by gardeners. The bark of *B. papyrifera* is used by the North American Indians in the construction of canoes. Birch twigs and bark also yield an oil which is used as a preservative and gives the fragrance to Russian leather.
BETULACEAE.

Bird's nest fungi fungi of several genera, including *Cyathus* and *Crucibulum*, which grow into small cup-shaped bodies about 1cm in diameter, each of which holds about 20 spore-containing reproductive bodies (peridioles). These look like eggs in a nest, hence the common name. Raindrops falling into the cup disperse the ripe peridioles by propelling them violently into the air, their filamentous "tails" twining around the surrounding vegetation. The spores contained in the peridiole are further dispersed when animals eat the vegetation.
NIDULARIACEAE.

Bischofia a genus represented by a single species, *B. javanica*, a tree growing in Indomalaysia, the southern Pacific Islands and northern Australia. It produces a dense, hard timber used in building construction,.
EUPHORBIACEAE, 1 species.

Bixa see Annatto.
BIXACEAE, 3–4 species.

Blackberries compound fruits produced by the prickly shrubs usually known as *BRAMBLES (*Rubus fruticosus* and other North American species, such as *R. laciniatus*, *R. ulmifolius* and *R. ursinus*). Blackberries are often collected from wild brambles but selected forms are cultivated and a few thornless varieties have been deliberately bred. The fruits may be eaten fresh or may be canned or frozen, and are used for jellies, jams or wine.
ROSACEAE.

Blackthorn or **sloe** the common names for *Prunus spinosa*, a thorny shrub or small tree frequently grown as an ornamental in a range of varieties. It produces a fine show of single or double white flowers and abundant bitter-tasting blue-black fruits used for making sloe gin.
ROSACEAE.

Blechnum a cosmopolitan genus of rather stiff ferns in which the spore-containing structures (sporangia) are arranged in two long uninterrupted chains on either side of the midvein of the leaf. One of the commonest species is *B. spicant* [HARD FERN, DEER FERN].
BLECHNACEAE, about 220 species.

Bletia a large genus of pseudobulbous terrestrial orchids with narrow grass-like leaves, native to Central America and the West Indies, but with some species now widely grown in temperate regions. The purple or whitish flowers are borne in terminal racemes. Most cultivated species, such as *B. campanulata* and *B. purpurea*, require to be grown under glass in temperate regions.
ORCHIDACEAE, about 45 species.

Bloodberry or **rouge plant** common names for *Rivina humilis*, a small perennial herb or subshrub native to tropical America. It bears small red berries from which a dye can be extracted. It is sometimes cultivated as an ornamental.
PHYTOLACCACEAE.

Blueberry the common name for several North American species of deciduous or evergreen shrubs of the genus *Vaccinium*. Some of the blueberries are commercially important. The so-called "low-bush" species have a low, sometimes straggling habit and are not normally cultivated, with the excep-

The Blackthorn or Sloe (Prunus spinosa) produces a mass of white blossom in early spring before the leaves expand. (×2)

Fruits of the Blackberry (Rubus fruticosus) are not berries in the strict botanical sense but are compound fruits made up of small drupes. (×½)

tion of *V. angustifolium* [LOW-BUSH BLUEBERRY, LOW SWEET BLUEBERRY, LATE SWEET BLUEBERRY] and *V. myrtilloides* (= *V. canadense*) [SOURTOP OR VELVET-LEAF BLUEBERRY].

The important sources of the more erect "highbush" cultivars are *V. ashei* [RABBIT-EYE BLUEBERRY] and *V. corymbosum* (= *V. australe*) [HIGHBUSH BLUEBERRY, SWAMP BLUEBERRY]. The fruits are small, 2cm in diameter, blue-black in color. They are commonly used in pies and pastries in the USA and are often sold canned or frozen.
ERICACEAE.

Blue-green algae see p. 50.

Blumea a genus of usually aromatic herbs native to tropical and southern Africa, Madagascar, Asia, Australia and the Pacific Islands. The most important species is *B. balsamifera* [NGAI CAMPHOR OR SAMBONG], a softly hairy, half-woody shrub smelling strongly of camphor. The camphor is distilled from the young leaves and is used medicinally.
COMPOSITAE, about 180 species.

Bocconia a genus of "TREE" POPPIES restricted to the West Indies and warm Americas. They are trees, shrubs or perennial herbs with large leaves. *B. arborea*, from Mexico and Central America is the source of a yellow dye used locally and the tropical American *B. frutescens* [TREE CELANDINE] is cultivated as an ornamental and for its latex which is used in the treatment of warts.
PAPAVERACEAE, about 9 species.

Boehmeria a Southeast Asian genus of herbs, shrubs and trees that includes *B. nivea* [RAMIE, CHINESE SILK PLANT, CHINA GRASS], whose bark is the source of the fiber ramie used for various fabrics (Chinese linen, Canton linen). Similar but rather inferior fibers are obtained from other species, including the tropical American *B. caudata* and the

Hawaiian *B. grandis* [HAWAIIAN FALSE-NETTLE, AKOLOA].
URTICACEAE, about 100 species.

Boletus a genus of large mushroom-like fungi which have the characteristic umbrella-shaped fleshy cap set on a stalk or stipe, but instead of having numerous gills, as in most mushrooms, the underside of the cap is permeated by a mass of small pores that lead to vertical tubes. The best-known species is *B. edulis* [EDIBLE CEP or CÈPE], much prized as a delicacy in many European countries. Many other species are edible. No bolete is known to be deadly poisonous, but *B. satanas* [SATAN'S BOLETE] causes vomiting.
BOLETACEAE, 50–60 species.

Bomarea a genus of perennial herbs native to Central and tropical South America, with slender twining stems. A few species, for example *B. caldasii* (red and yellow flowered), *B. racemosa* (red flowered) and *B. carderi* (bright pink flowered), are grown as ornamental climbers.
AMARYLLIDACEAE, about 100 species.

Bombax an important genus of tall, deciduous tropical Asian, African, Australian, Central and South American trees. The Asian *B. ceiba* (= *B. malabaricum*) [INDIAN SILK COTTON TREE, WHITE KAPOK TREE] has

Boletus bovinus is a member of the order Agaricales (gill fungi) but was once placed in the order Aphyllorales (polypores) because its fruiting body has pores rather than gills. (× 1)

seeds embedded in a mat of fine white fibers that are used as *kapok. A similar fiber is obtained from the tropical African *B. buonopozense* [SILK COTTON TREE]. An inferior floss is obtained from the Brazilian species *B. longiflorum* and *B. manguba*. The trees of several species often grow to heights of 45m or more and provide useful timber for light construction work, dugout canoes, balancer sheets in veneers, lightweight boxes, model-making and matchsticks.
BOMBACACEAE, 8 species.

Borago a small genus of European and Mediterranean annual or perennial herbs, one of which, *B. officinalis* [BORAGE], is a traditional garden herb used since the Middle Ages for its reputed medicinal and culinary value, but today grown more for its attractive, bright blue flowers.
BORAGINACEAE, 3 or 4 species.

Borassus a genus of palms from Africa, Madagascar, Asia and Malaysia. *B. flabellifer* [PALMYRA PALM] is cultivated in tropical Africa, India and Sri Lanka and provides a hard durable wood which is resistant to salt water, in addition to providing the raw materials for string, rope, brushes, thatching, matting, baskets and "Olas" paper.

The young male inflorescence of this palm is tapped for its palm juice (toddy), which is made into palm wine or vinegar, distilled into *arrack or evaporated to give jaggery, which is a form of sugar. Most of the sugar produced in Burma and southern India in fact comes from the PALMYRA PALM. The sweet fruity

The Hard-Fern or Deer Fern (Blechnum spicant) is a lime-hating fern typically found on grassy moorland. (× 1/10)

pulp around the nuts can be eaten, raw or roasted; it makes a good confection and is also used pickled.
PALMAE, about 7 species.

Boronia a genus of shrubs and subshrubs, rarely annuals, from Australia. *B. megastigma* [BROWN or SCENTED BORONIA] a very aromatic straggling shrub from Western Australia, with purple and yellow flowers, is grown commercially for its essential oil used in perfumery. Other species are also cultivated as ornamentals.
RUTACEAE, about 65 species.

Boswellia a genus of trees native to tropical Africa, Madagascar and Asia, including the

Boehmeria caudata yields a poor quality ramie-type fiber but, unlike B. nivea, it is not usually grown commercially. (× 1/6)

Borage (Borago officinalis) is an attractive garden flower also useful as a culinary herb; most parts are covered in rough hairs. (× 2)

frankincense trees of Arabia, Ethiopia and Somaliland, notably *B. carteri* [BIBLE FRANK-INCENSE]. The resin, gum olibanum or *frankincense, is used in the preparation of incense. *B. serrata* [INDIAN OLIBANUM TREE] is a characteristic tree of the dry hills of India, whose resin is also used for incense-making. BURSERACEAE, about 24 species.

Botrychium a genus of ferns, mostly north temperate but extending via New Guinea to Australia. The leaves (fronds) vary from pinnate in the European *B. lunaria* [MOON-WORT] to highly dissected in the North American species *B. dissectum*. The spore-bearing organs (sporangia) are clustered at the base of the leaf-stalk, having the appearance of bunches of grapes. OPHIOGLOSSACEAE, about 35 species.

Moonwort (Botrychium lunaria) showing a fertile leaf in the foreground, with a sterile leaf behind, the latter bearing halfmoon-shaped leaflets. (× 2)

Bougainvillea a genus of climbing shrubs and small trees native to South America. Bougainvilleas are often grown as defensive and decorative hedges in warmer climates, and as greenhouse plants farther north. They are valued for their brilliantly colored bracts. The two most commonly grown species are *B. glabra* and *B. spectabilis*. From these, and from *B. peruviana* and *B. × buttiana*, many hybrids have been produced. NYCTAGINACEAE, about 18 species.

Bouquet garni an aromatic mixture of various herbs, including bay, parsley, thyme and marjoram, usually tied together with string or placed in a muslin bag. Their principal use is for flavoring savory dishes such as soups, goulashes and stews.

Bouteloua a New World genus of grasses, many of which come from the drier plains and prairies of the USA and Canada where they make up a large part of the vegetation. They are valuable as fodder and forage grasses. The most important species are *B. repens* (= *B. filiformis*) [GRAMA GRASS], *B. gracilis* [BLUE GRAMA] and *B. eriopoda* [BLACK GRAMA]. These are important perennial pasture grasses, while *B. curtipendula* [SIDE-OATS GRAMA] is also used for the stabilization of sand dunes on the Great Plains. GRAMINEAE, about 40 species.

Bouvardia a genus of small tropical American evergreen shrubs with fragrant, showy, yellow, white, pink or red flowers. Although a number of species are cultivated as greenhouse ornamentals, most are probably hybrids. An example is *B. × domestica*, which grows to a height of 60cm, bearing clusters of pink or red flowers. Another popular type is *B. longiflora*, a rather taller shrub with glossy green leaves and clusters of white flowers. RUBIACEAE, about 50 species.

Box originally a name referring to the hard wood obtained from species of *Buxus*, a genus of shrubs and small trees mainly from temperate Eurasia, tropical and southern Africa, Central America and the West Indies. The best-known species is *B. sempervirens* [COMMON BOX], an evergreen tree which grows to 6m, with small, roundish, leathery leaves. If kept trimmed it forms very dense bushes and is used in topiary. The wood is very firm and close-grained, and is used for wood engraving and musical instruments; it has a yellowish color, very rarely splits, and has a good natural polish. In South Africa the local *B. macowanii* [CAPE BOX], is equally good.

As demand has increased, many substitutes for these true boxwoods have been exploited. These include *Cornus florida* [FLOWERING DOGWOOD] in North America, *Schaefferia frutescens* [FLORIDA BOXWOOD] in Central America, *Eucalyptus* species in Australia and *Gonioma kamassi* [KNYSNA BOXWOOD] in South Africa. They are not as good as true box for engraving work. Perhaps the best-known substitute boxwoods are WEST INDIAN BOXWOOD, a name which has been

As with most species of Bougainvillea, the flowers of B. spectabilis are inconspicuous and it is the bracts that provide the brilliant color. (× $\frac{1}{6}$)

applied to several diverse species with hard wood, but notably to *Casearia praecox* and SAN DOMINGAN or VENEZUELAN BOXWOOD (*Phyllostylon brasiliensis*).

Boykinia a small genus of herbs from Japan and North America. One or two species are cultivated in rock gardens or in wild plantings. The commonest is *B. jamesii* (sometimes placed in a separate genus as *Telesonix jamesii*), which has deep pink or purplish-pink flowers and kidney-shaped, toothed leaves. *B. tellimoides* is a Japanese species with wide-lobed leaves and greenish flowers. SAXIFRAGACEAE, 9 species.

Brachycome a genus of annual or perennial herbs mainly native to Australia and New Zealand. One species, *B. iberidifolia* [SWAN RIVER DAISY], is a branching annual used as a garden-border plant. The daisy-like inflorescences range in color from white to pink and bluish purple. COMPOSITAE, about 75 species.

Brachythecium a genus of mosses found in all continents but predominantly in cool temperate to subarctic regions. *B. rutabulum* is a very common and conspicuous woodland moss. Several other species are almost equally common in some habitats, eg *B. velutinum* on rocks and tree stumps, *B. rivulare* in wet places and *B. albicans* on dry, sandy or gravelly soils. HYPNACEAE, about 250 species.

Bracken the common name for *Pteridium aquilinum*, a fern found almost everywhere except for temperate South America and the polar regions. It is probably the most important weed in many countries. Under suitable conditions it can form a complete ground cover and, with the heavy litter layer during the winter, can exclude all competitors.

The plant grows from an underground stem (rhizome) about the size of a finger, and the fronds can reach about 2m in height in shrubby habitats where it can gain some support.

Bracken was once an extremely important source of potash, a crude form of potassium carbonate obtained from plant ash. It was much used in the production of glass and soap and in dyeing, bleaching and degreasing wool. Bracken was very suitable because of its high potassium content and also the large proportion of ash left. The fronds have been used as a floor-covering or bedding.

The use of bracken fronds and rhizomes as food in times of famine is well documented for both Man and animals, although it is poisonous when green and is carcinogenic in rats.
POLYPODIACEAE.

Brahea a small genus of palms native to the southern USA, Mexico and South America, characterized by their dense crowns of fan-shaped leaves. A few species are cultivated out of doors in subtropical gardens or in warm greenhouses in cooler regions. They include *B. nitida*, *B. dulcis* (= *B. calcarea*) and the blue-gray-leaved *B. armata* [BLUE FAN PALM] which grows up to 12m.
PALMAE, 7–12 species.

Bramble the common name for a number of species of prickly, straggling shrubs in the genus *Rubus*. The name is generally applied in Europe to *R. fruticosus* (but see also blackberry). Their purplish-black edible fruits are the familiar *blackberries. Brambles are distributed throughout the north temperate zone. Although very common in woods, they are now also abundant in scrub, hedgerows and neglected pastures, while some forms are found in wet heaths and marshes.

The stems are usually covered with hooked prickles, enabling the stems to climb over other vegetation. They are often arching, up to several metres long, frequently

rooting at the tips and bear inflorescences of white or pink flowers.
ROSACEAE.

Brassia a genus of high-altitude tropical American epiphytic orchids many of which are found in cultivation throughout the world. Their major distinguishing feature is the elongation of the sepals, and usually also the petals, of most species to give a spider-like appearance to each of the flowers on the many-flowered inflorescences.
ORCHIDACEAE, about 50 species.

Brassicas see p. 62.

Brazil nut the seed of *Bertholletia excelsa*, a large tree up to 40m tall, native to South America. The tree grows wild in the rain forests of the Amazon basin, and is also used for its timber. The trees are never cultivated, but the collection and export of the nuts is a major industry. About 50000 tonnes are shipped each year, mainly to Europe and North America.

Fruiting is never prolific, and the fruits take at least a year to ripen. As many as 24 seeds (nuts) are packed tightly together in a spherical fruit with a thick woody outer shell, the whole weighing up to 1.5kg. Each nut has a woody outer covering and white, creamy flesh, which has a high food value (60–70% oil and about 17% protein).
LECYTHIDACEAE.

Briar root the woody root of *Erica arborea*, a small tree native to the Mediterranean region, especially southern France, Italy and Algeria. The name is derived from the French "bruyère" (heath). Large woody swellings

Brachythecium rutabulum *is a very common moss in grassland and on rotting logs in woods. The red setae (stalks) bear pendulous capsules.* (× 3)

develop on the roots and the stem base, and these are used for the manufacture of tobacco pipe bowls. It is the only wood that is ideally suitable for making tobacco pipes and has a very hard, close grain.
ERICACEAE.

Briza a distinctive genus of annual and perennial grasses found in Eurasia and South America. The perennial *B. media* [COMMON QUAKING GRASS] is the most widespread Eur-
(continued on p. 64)

The typically arched stalks and tightly folded fronds of young Bracken (Pteridium aquilinum) emerging in springtime. (× ½)

The Swan River Daisy (Brachycome iberidifolia), from Australia, is a hardy annual with fragrant flowers. (× ¼)

Brassicas

obtained from the seeds of species which are particularly rich in glucosinolates, the substances that are precursors of the compounds that give the characteristic burning sensation. Many species can be used for mustard production, but until recently most types were mixtures of BLACK *(Brassica nigra)* and WHITE *(*Sinapis alba)* MUSTARDS. BROWN MUSTARD *(B. juncea)* has now largely replaced BLACK MUSTARD in cultivation.

The term brassica is applied to the six species of the genus *Brassica* (family Cruciferae) which are important as vegetables, animal feed and oilseed crops in many parts of the world, ranging from the tropics to the sub-arctic. Examples are *CABBAGE, *BRUSSELS SPROUTS, *KALE, *TURNIP, *SWEDE and *RAPE. By extension, the name covers also species of related genera such as *RADISH *(*Raphanus sativus)*, *CRESS *(*Lepidium sativum)* and *WATERCRESS *(*Nasturtium spp)*.

Brassicas as vegetables may have started as weeds of agricultural land that were used occasionally as garnishes and then later bred for a wide range of different types. The brassicas are in fact unique in the range of different parts of the plant that are exploited. Thus *Brassica oleracea*, which in the wild is a leafy perennial growing on seacliffs, has been selected to give forms with large compressed buds in the leaf-axils (BRUSSELS SPROUTS) or at the apex of the stem (CABBAGE), with precocious flowers (SPROUTING *BROCCOLI, *CAULIFLOWER), with disproportionately expanded leaf-margins (CURLED OR CURLY KALE) or with swollen stems (*KOHLRABI, MARROW STEM KALE), as well as leafy forms of kale (*COLLARDS in the USA).

A similar pattern is repeated in Asia with *B. campestris* (= *B. rapa*) which provides a range of vegetables resembling their Western equivalents as well as the more cosmopolitan *TURNIP.

Many of these brassica crops are of ancient origin – TURNIPS, LEAFY COLES, RADISH, and forms of CABBAGE being known from Greek and Roman times. Other forms evolved in the Middle Ages, such as CAULIFLOWER and SPROUTING BROCCOLI. After this, other CAULIFLOWER, CABBAGE and SPROUTING BROCCOLI varieties were bred to give a much wider range of maturity, so that today in the cool wet climate of northwest Europe these can be grown virtually all the year round. BRUSSELS SPROUTS, appeared in the 19th century, probably from a form of cabbage.

Brassicas notably SWEDE, RAPE, KALE (including the aptly named HUNGRY GAP KALE) and STUBBLE TURNIPS, helped to bridge the gap between the autumn harvest and the spring growth, when food was scarce. This extra fodder allowed more animals to be kept during the winter, and by the middle of the 19th century 800 000 hectares in the UK were occupied by turnips and swedes for fodder. In Asia, because of a different pattern of animal husbandry, brassicas occupied a more culinary role.

In recent times, 20–40% of the total consumption of vegetables in Eurasian countries has consisted of brassicas. Elsewhere they are less important. Small but significant areas of CAULIFLOWER, CABBAGE, *CALABRESE, SWEDES and BRUSSELS SPROUTS are grown in Australia and the USA where they represent about 10% of the total consumption of vegetables.

As oilseed crops, brassicas collectively now rank as the fifth most important oil crop, largely due to their ability to grow rapidly in the short hot summers of the USSR and Canada.

Brassicas are also used as green manure by exploiting the rapid growth of *MUSTARDS (WHITE, BROWN or BLACK) to form a dense sward which smothers weeds and which is ploughed in before flowering to improve soil texture and fertility.

Another use of brassicas is in the production of the condiment mustard which is

Cultivated "brassicas" 1 Main forms of Brassica oleracea *(a) Wild Cabbage; (b) Kale; (c) Curled Kale; (d) Round Cabbage; (e) Red Cabbage; (f) Savoy Cabbage; (g) Brussels Sprouts; (h) Kohlrabi; (i) Green Sprouting Broccoli or Calabrese; (j) Purple Sprouting Broccoli; (k) Cauliflower; (l) Flowering Cabbage.*
2 Other important crops of the genus Brassica *(a) Pak-choi; (b) Pe-tsai (c) Black Mustard; (d) Turnip; (e) Swede; (f) Oilseed Rape.*
3 Other members of the family Cruciferae to which the general term "brassica" is applied (a) Radish (Raphanus sativus); *(b) White Mustard* (Sinapis alba); *(c) Wild Watercress* (Nasturtium sp); *(d) Cress* (Lepidium sativum); *(e) Horseradish* (Amoracia rusticana); *(f) Sea Kale* (Crambe maritima) *– entire plant (lower) and blanched stalks (upper).*
1a, 2c, 3a, 3b, 3c, 3d ($\times\frac{1}{2}$); 2d, 2e, 2f ($\times\frac{1}{4}$); 1h, 1i ($\times\frac{1}{6}$); 1b, 1c, 1d, 1e, 1f, 1g, 3e, 3f ($\times\frac{1}{10}$). Details 1g ($\times\frac{1}{2}$); 3f ($\times\frac{1}{6}$).

IMPORTANT CULTIVATED MEMBERS OF THE GENUS *BRASSICA*

Group	Common name	Part used
B. oleracea		
Acephala (var *acephala*)	*KALE, SCOTCH (CURLED) KALE, FLOWERING KALE OR CABBAGE, *COLLARDS, COLE, COLEWORT, BORECOLE	Leaves
(var *medullosa*)	MARROW STEM KALE	Stem
Alboglobra (var *alboglobra*)	CHINESE KALE	Leaves
Botrytis (var *botrytis*)	*BROCCOLI, HEADING BROCCOLI, *CAULIFLOWER	Floral tissue (heads)
Capitata (vars *capitata, bullata, sabauda*)	*CABBAGE, SAVOY CABBAGE	Terminal bud
Gemmifera (var *gemmifera*)	SPROUTS, *BRUSSELS SPROUTS	Axillary buds
Gongylodes (var *gongylodes*)	*KOHLRABI	Stem-tuber
Italica (var *italica*)	SPROUTING OR ITALIAN BROCCOLI, *CALABRESE	Floral tissue
Tronchuda (var *tronchuda* = *costata*)	PORTUGUESE KALE OR CABBAGE	
B. campestris		
subspecies *oleifera*	TURNIP RAPE	Seed (oil)
subspecies *rapifera*	*TURNIP, DUTCH TURNIP	Roots, shoots
subspecies *chinensis*	PAK-CHOI, CHINESE MUSTARD, CELERY MUSTARD	Leaves (petioles)
subspecies *pekinensis*	PE-TSAI, CHINESE CABBAGE	Leaves (heads)
subspecies *narinosa*	T'A-KU-TS'AI, BROAD-LEAVED MUSTARD	Leaves
subspecies *nipposinica*	POTHERB MUSTARD, JAPANESE MUSTARD	Leaves
subspecies *dichotoma*	TORIA, INDIAN RAPE, BROWN SARSON	Seed (oil)
subspecies *trilocularis*	SARSON	Seed (oil)
subspecies *perviridis*	KOMATSU-NA	Leaves, roots
subspecies *ruvo*	RUVO KALE, TURNIP BROCCOLI, GREEN TOP TURNIP	Roots, leaves
subspecies *septiceps*	ITALIAN KALE	Leaves
B. napus	*RAPE, OILSEED RAPE, COLZA	Shoots, leaves, seed (oil)
Napobrassica Group	RUTABAGA, *SWEDE, SWEDISH TURNIP	Roots
Pabularia Group	EARLY KALE, SIBERIAN KALE, HANOVER SALAD, HUNGRY GAP KALE, RAGGED JACK KALE	Leaves, tops
B. nigra	BLACK *MUSTARD	Seed, seed oil
B. juncea	BROWN *MUSTARD, INDIAN MUSTARD, LEAF MUSTARD, MUSTARD GREENS	Seed, seed oil, leaves

1a
1b
1c
1d
1e
1f
1g
1h
1i
1j
1k
1l

2a
2b
2c
2d
2e
2f

3a
3b
3c
3d
3e
3f

asian species and perhaps the easiest to recognize by its loose panicles of flower spikelets which "quake" in the wind. Two other annual European species, *B. minor* [SMALL QUAKING GRASS] and *B. maxima* [PEARL GRASS] are also well-known ornamentals.
GRAMINEAE, about 30 species.

Broccoli a relative of *CABBAGE (*Brassica oleracea*)* in which the immature inflorescence is used as a vegetable. Similar in appearance, growth habit and utilization to cauliflower, broccoli is harvested during the winter and spring. The commonest form of broccoli (var *botrytis*) produces a dense, immature inflorescence (the curd) 7–20cm in diameter. The best varieties have a pure white curd which remains firm, with the inner leaves protecting the curd from frost. SPROUTING BROCCOLI (var *italica*) is an erect, branching plant which produces many small curds in the axils of the leaves.
CRUCIFERAE.

Brodiaea a genus of perennial herbs native to North and South America. Most of the cultivated species give rise to narrow, strap-shaped leaves and clusters of star-shaped, cup-shaped or tubular flowers on slender stalks. Popular garden species include *B. coronaria* (= *B. grandiflora*) [HARVEST BRODIAEA] with a bare flower stalk bearing a loose cluster of lilac-purple flowers, and *B. ida-maia* (= *Dichelostemma ida-maia*) [CALIFORNIA FIRECRACKER], with bright red, rarely yellow, flowers, tipped with green, which is also grown under glass. (See also *Triteleia*.)
LILIACEAE, about 40 species.

Bromelia a genus of herbs native to tropical America. Most have rosettes of stiff spiny leaves up to 2m long, growing from ground level. The inner leaves or bracts are often bright red, and the flowers are white or

Harvesting the immature curds of Broccoli (Brassica oleracea var botrytis). (×⅙)

Brown algae (Fucus and Pelvetia spp) in the mid zone of a rocky seashore. Also shown is, at the top, a band of the red alga Corallina and, at the bottom, the green alga Enteromorpha.

reddish purple, usually fleshy, borne on a central stalk. Various species are used locally in Latin America: *B. pinguin* for its edible, but very acid fruits; *B. karatas* for the young inflorescences, eaten as a vegetable; *B. fastuosa* and *B. serra* for their leaf fibers. These and other species, such as *B. balansae*, are also sometimes grown as ornamentals, despite their large size and vicious spines.
BROMELIACEAE, about 40 species.

Bromus a distinctive genus of annual, biennial and perennial grasses, the BROMES of the north temperate zone, South America and high mountains of the tropics. *B. inermis* [AWNLESS, SMOOTH or HUNGARIAN BROME], a native of Europe and Asia, has been cultivated for centuries in many countries as a forage grass and for hay. It is very drought-resistant and particularly important in temperate arid areas such as the steppes of Russia. The South American *B. unioloides* [RESCUE GRASS] fulfills a similar role in the New World. Other species used in agriculture include *B. arvensis* [FIELD BROME], a field weed sometimes used on poor light soils when all else fails, and the biennial weed *B. mollis* [SOFT BROME].
GRAMINEAE, about 50 species.

Broom a name applied to plants of three genera of the pea family (see *Cytisus*, *Genista* and *Spartium*).
LEGUMINOSAE.

Broomroot the common name for *Epicampes macroura* (= *Muhlenbergia macroura*), a Mexican grass, the tough roots of which yield strong fibers that are used to make brooms and brushes.
GRAMINEAE.

Brosimum a commercially valuable genus of mainly temperate and tropical South American trees. *B. guianense* [SNAKE WOOD], a medium-sized, slender tree from Brazil, produces an attractive wood which is one of the hardest, heaviest, most rot-proof woods known. It is used for measuring instruments, rulers and canes. *B. paraense* [BRAZIL REDWOOD] is another fine wood.

Several species, for example *B. galactodendron* [COW or MILK TREE], *B. utile* and *B. rotabile*, yield great quantities of a gray-white latex which is drunk like cows' milk. *B. alicastrum* [BREADNUT TREE], of tropical America and Jamaica, yields a fruit which is edible when cooked. The foliage of the same species is used as fodder.
MORACEAE, about 50 species.

Broussonetia a genus of trees and shrubs native to East Asia and Polynesia but sometimes cultivated in Europe for ornament. The male plant bears catkins; the female plant bears small round heads of white flowers. *B. papyrifera* [PAPER MULBERRY] is of considerable economic value; the fibers of the inner bark are used for paper (Japan), tapa or kapa cloth (Polynesia) and rope. The fruits are edible.
MORACEAE, 7–8 species.

Browallia a tropical American genus of shrubs and herbs, some of which make fine ornamentals, with blue or white flowers. Annual species are often grown in cool greenhouses as they have a very long flowering season, and will flower in winter provided that the temperature does not fall below 10°C. *B. speciosa* has showy, solitary, white, blue-violet or white-eyed flowers up to 5cm across. *B. americana* has several cultivars.
SOLANACEAE, 6 species.

Brown algae the general name for the seaweeds belonging to the algal division Phaeophyta. They are the most conspicuous plants on rocky shores of the temperate and cold oceans of the world and generally form a distinct zonation. The largest known algae, such as the GIANT KELPS, *Macrocystis* and *Nereocystis*, which may be up to 50m long, belong to this group. Brown algae are the source of the economically important alginates, which are used as emulsifying agents in food manufacture, especially of ice cream, and in cosmetic lotions and creams.

Brunfelsia a tropical genus widely cultivated as greenhouse shrubs in temperate

Greenish-yellow male flowers of the White Bryony (Bryonia cretica). The female flowers are borne on separate plants. (× 2)

climates. It consists of evergreen shrubs and small trees, originating from Central and South America and the West Indies. The large, showy, fragrant flowers are funnel-shaped and change color with age. In *B. pauciflora*, for example, the flowers open as violet-purple and fade to white, and in *B. americana* [LADY OF THE NIGHT] the white blooms change to yellow. *B. pauciflora* and *B. australis* (= *B. latifolia* of cultivation) are widely cultivated with several varieties.
SOLANACEAE, about 30 species.

Brunnera a genus of herbaceous perennials native to Europe. One species, *B. macrophylla*, is cultivated in gardens, producing heart-shaped leaves and racemes of small blue flowers.
BORAGINACEAE, 3 species.

Brussels sprouts cultivated forms of *Brassica* belonging to *B. oleracea* (*gemmifera* group), in which the swollen axillary buds are used as a vegetable in autumn and winter. The single erect stem can grow up to 1.5m tall but is usually 0.6–1m. In the axil of each leaf a shoot is initiated in the first year which does not elongate but develops into the sprout, a globe 2–5cm in diameter, consisting of the

tightly packed leaves on the reduced stem of the axillary shoot. If left, in the second year flowers will form. A good-quality sprout is dense and firm, with dark green outer leaves.
CRUCIFERAE.

Bryonia [BRYONY] a genus of perennial climbing herbs covered with bristles or short protuberances, native to Europe, Asia and Africa. *B. cretica* (= *B. dioica*) [WHITE BRYONY, REDBERRY BRYONY] is a European hedgerow plant. The roots of *B. cretica* and *B. alba* [WHITE BRYONY], cultivated in the Mediterranean, yield a purgative.
CUCURBITACEAE, 10 species.

Bryum a very large genus of mosses including *B. argenteum* which is cosmopolitan and *B. pallens* which occurs in all continents except Australasia. Many species tend to form loose cushions or compact turfs on walls, bare ground and rock ledges.
BRYACEAE, about 900 species.

Buckthorn the common name for species of *Rhamnus*, a widespread genus of small trees and shrubs native to both the Old and New Worlds. The EUROPEAN or COMMON BUCK-THORN is *R. catharticus*, a thorny shrub often used for hedging. Many *Rhamnus* species are popular ornamentals, varying from dwarf rock-garden shrubs to small trees, grown for their attractive foliage and berries, such as the North American *R. crocea* [REDBERRIED BUCKTHORN] and the southern European *R. alaternus*, which has varieties with variegated leaves.

The dye "sap-green" is derived from the berries of the EUROPEAN BUCKTHORN and green and yellow dyes are obtained from other species, such as *R. infectorius*. Buckthorns also have medicinal properties, notably as purgatives (PURGING BUCKTHORN is a popular

Buddleias are often found growing on waste ground. Buddleia davidii here has found a home in a disused railway station.

name), the most important species being the North American *R. purshiana* (see Cascara).
RHAMNACEAE.

Buckwheat the name usually given to *Fagopyrum esculentum* (Polygonaceae), a large-leaved annual herbaceous species grown as a cereal and fodder crop. It is not a member of the grass family, and hence not a true cereal, but the seeds are used in the same way as cereals.

BUCKWHEAT is a secondary agricultural crop in all countries where grain crops are cultivated. It is more important in mountain regions because of its short growing season and adaptability to poor soils. It is commonly grown for grain, green manure and animal fodder, and as a cover crop in orchards and vineyards. The grain, used for human food in various forms from pancake flour to buckwheat noodles, is also used as a feed for livestock and poultry.

F. tataricum [TARTARY, SIBERIAN or KANGRA BUCKWHEAT] is also grown, but is less robust; it is used for green manure and animal feed. The common name WILD BUCKWHEAT is sometimes given to the North American *Eriogonum microthecum*. The unrelated North American *Cliftonia monophylla* [BUSH BUCKWHEAT] (Cyrillaceae) is a source of honey.

Buddleia a genus of evergreen and deciduous trees and shrubs from tropical and temperate regions of America, southern Africa and Asia. The hardiest and commonest species in European and North American gardens is the Chinese *B. davidii* (= *B. variabilis*). This vigorous shrub has given rise to many colorful varieties producing long racemes of fragrant purple, blue, lavender, red, pink or white flowers. Also commonly cultivated is *B. globosa*.
LOGANIACEAE, about 70 species.

Bulbocodium a small genus of perennial herbs with corms, native to Europe.

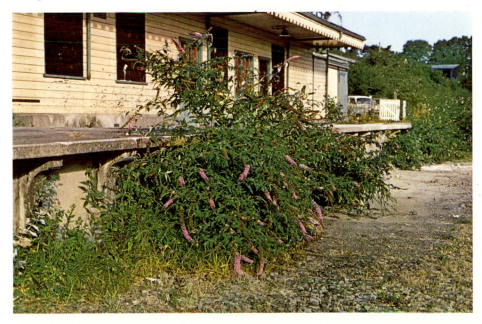

B. vernum is a very attractive crocus-like plant which grows well in sunny, well-drained rock gardens, producing flowers in March and early April.
LILIACEAE, 2 species.

Bulbophyllum a large pantropical genus of epiphytic orchids. The inflorescences are single to many-flowered, and every color is represented except pure blue; in many species several colors are contrasted or complemented in a single flower. Some species have developed bizarre fringes and gently waving hairy protuberances on their flowers, and are grown as ornamentals.
ORCHIDACEAE, about 1000 species.

Bulrush the common English name for the genus *Typha* which contains about 15 species of rush-like herbs growing in shallow fresh water in many tropical and temperate regions.

T. latifolia [REEDMACE BULRUSH, CATTAIL] is found throughout Europe in reed swamps, lakes, ponds, canals and slow-flowing rivers. It is a robust perennial with tall flower-spikes ripening to densely packed brown cylindrical fruiting spears that rupture when dry and ripe to release masses of wind-blown seeds. *T. angustifolia* [NARROW LEAF CATTAIL, LESSER REED MACE] is vigorous but less widespread.

BULRUSH is also a common American name for *Scirpus validus* [CLUB RUSH], a vigorous reed of similar habitat but with sedge-like flowers.

Buphthalmum [YELLOW OXEYES] a small Eurasian genus of perennial herbs, the common name of which is an allusion to the large, yellow, solitarily borne flower heads, which have a single row of overlapping ray florets. *B. salicifolium* and *B. speciosum* (=

Multicolored flowers of the orchid Bulbophyllum purpureorhachis. *Several Bulbophyllum species are grown as greenhouse subjects.* (× 2)

The Flowering Rush (Butomus umbellatus). (× $\frac{1}{10}$)

Telekia speciosa) are cultivated as ornamental border plants.
COMPOSITAE, 2–4 species.

Bupleurum [HARE'S EARS] a genus of annuals, herbaceous perennials or shrubs, native to Europe, Asia and Africa, with one species in North America. *B. fruticosum*, from southern Europe, is an evergreen shrub growing to 2.5m tall, frequently cultivated for its yellow flowers produced in late summer.
UMBELLIFERAE, 90–100 species.

Burning bush the usual common name for *Dictamnus albus* [also known as DITTANY, FRAXINELLA], a woody perennial herb of southern Europe and the East whose white or purple aromatic flowers exude an inflammable vapor during warm weather. The vapor can ignite with a momentary burst of flame that does not damage the plant. This is thought to be the "burning bush" seen by Moses in the Bible. The name is sometimes also given to the vigorous ornamental annual plant *Kochia trichophylla*, whose leaves turn brilliant scarlet and red in the autumn. Similarly, the North American deciduous shrub *Euonymus atropurpureus*, sometimes called the WAHOO TREE, is also given the common name BURNING BUSH.

Bursera a genus of small, shrubby aromatic trees native to tropical America. The most important species are *B. gummifera* [TURPENTINE TREE] and *B. simaruba* [GUMBO-LIMBO], which yield a balsam resin known as *elemi, and *B. delpechiana*, one of the sources of linaloe oil used in the production of soap and cosmetics.
BURSERACEAE, about 80 species.

Butea a genus of trees and shrubs from Indomalaysia and China. One of the most important is *B. monosperma* (= *B. frondosa*)

[BASTARD TEAK, PALAS, DHAK, BENGAL TEAK TREE], a dominant tree of the Indomalaysian monsoon forest. The large showy flowers yield an orange-red dye. The tree also yields a dark red resin (gum kino or Bengal kino). Oil is extracted from the seeds, the leaves are used for fodder and the timber is particularly water-resistant.
LEGUMINOSAE, about 30 species.

Butomus a genus consisting of a single aquatic perennial herb, *B. umbellatus* [FLOWERING RUSH]. It occurs in Europe, temperate Asia and has become naturalized in North America. It is cultivated for its handsome appearance and rose-colored flowers as an ornamental beside pools. The rhizomes are used as food in parts of the USSR.
BUTOMACEAE, 1 species.

Buttercup the name given to the species of *Ranunculus* commonly found growing in meadows: *R. acris* [MEADOW BUTTERCUP], *R. bulbosus* [BULBOUS BUTTERCUP] and *R. repens* [CREEPING BUTTERCUP] (see *Ranunculus*). The name has also been applied to other plants with similar-looking flowers, such as *Oxalis pes-caprae* [BERMUDA BUTTERCUP] and *Parnassia palustris* [WHITE BUTTERCUP].

Buxbaumia a peculiar genus of mosses of which the best-known, *B. aphylla*, is found in all continents except Africa. This and the other European species, *B. indusiata*, tend to occur sporadically and are rare over much of their range. The genus is remarkable for the underdeveloped leafy shoot system with very little chlorophyll, coupled with an outsize asymmetrical spore-producing capsule.
BUXBAUMIACEAE, about 10 species.

Buxus see BOX.
BUXACEAE, about 40 species.

C

Cabbage cultivated forms of *Brassica oleracea* in which the main stem is condensed. The plant rarely grows more than 30cm tall before flowering, and consists of a "head" of tightly packed leaves. Most cabbages are biennial. Cabbages are the most important vegetable crop of the brassicas. They are used as a vegetable, either cooked or raw, as a constituent of salads, and are also used for feeding cattle.

The most commonly grown are the white, smooth-leaved round or conical cabbages. Savoy cabbages have crinkled or blistered leaves and a similar range of shapes as the smooth-leaved cabbages. Red cabbages are smooth-leaved with round heads. There is a range of varieties of both these types which mature at different times of the year, though they are not usually grown for storage. (See also *Brassica*, and CHINESE CABBAGE.) CRUCIFERAE.

Cactus a general name referring to members of the family Cactaceae, a distinctive group of succulent plants characterized by

The Organ Pipe Cactus National Monument, Arizona, USA, with Organ-pipe Cactus (Lemairocereus thurberi), Barrel Cactus (Echinocactus acanthodes), Saguaro (Cereus giganteus), Cholla (Opuntia bigelovii) and Ocotillo (Fouquiera splendens).

thick, fleshy green stems, and sharp spines instead of leaves. There are some 87 genera of cacti altogether, and among the most important are *Carnegiea*, *Cereus*, *Echinocactus*, *Echinocereus*, *Echinopsis*, *Ferocactus*, *Lophophora*, *Mammillaria*, *Notocactus*, *Opuntia*, *Pereskia*, *Rhipsalis*, *Schlumbergera*.

Other succulent species from unrelated genera, such as *Euphorbia* and *Stapelia*, are sometimes also called cacti.

Caesalpinia a large genus of tropical and subtropical, chiefly New World shrubs, or trees, several of which are a source of wood dyes. Many are scrambling or climbing and may be armed with spines or prickles.

Among the several species grown as ornamentals, the West Indian *C. pulcherrima* [PRIDE OF BARBADOS, DWARF POINCIANA, PARADISE FLOWER, PEACOCK FLOWER] has colorful clusters of flame-red flowers (rarely rose or pure yellow) at the tips of the branches. It is widely planted in the tropics, *C. sappan* [SAPPANWOOD TREE, JAPAN WOOD TREE], from India and Southeast Asia, is the source of a red dye (brasil) obtained from the heartwood. A similar dye is found in *C. echinata* [BRAZIL WOOD, PEACH WOOD, ST. MARTHA WOOD] from tropical America. The pods of *C. coriaria* [*DIVI-DIVI], *C. spinosa* and *C. melanocarpa* are a source of commercial tannin. The dense heartwood of *C. coriaria*, *C. echinata* and *C. sappan* is used in cabinet-making. LEGUMINOSAE, about 200 species.

Caffeine or **theine** an alkaloid or organic base found in a wide variety of plants, many of which are used to make beverages. *Coffee is made from the beans of the COFFEE BUSH (*Coffea* species), tea from the young fermented leaves of the *TEA BUSH (*Camellia sinensis*) and cocoa from the beans of the *COCOA TREE (*Theobroma cacao*). Maté, cassine and *khat are infusions from the leaves of *Ilex paraguariensis*, *I. vomitoria* (see HOLLY) and *Catha edulis*, respectively. Guarana, which contains three times as much caffeine as

The dense head of leaves in a commercially grown Cabbage (Brassica oleracea) *'Mat Star'.* ($\times \frac{1}{8}$)

coffee, is made from the seeds of the Brazilian *Paullinia cupana*, and *kola, which is used in some soft drinks, comes from the nuts of *Cola acuminata*. Caffeine stimulates the circulation and the brain, producing a feeling of relief from fatigue.

Cajanus see PIGEON PEA. LEGUMINOSAE, 2 species.

Cakile a genus of annual herbs occurring on sandy seashores throughout Europe, North Africa, western Asia, North America and Australia. They have succulent leaves, and fruits with two unequal single-seeded segments, one of which breaks off, while the other remains on the plant and guarantees survival on the same site in the next generation. *C. maritima* [SEA ROCKET] is the most common species. The roots of the North American *C. edentula* [AMERICAN SEA ROCKET] can be dried, ground and mixed with flour. CRUCIFERAE, 4 species.

Calabash gourd the common name for two different types of gourd fruit produced by two unrelated genera. *Lagenaria siceraria* (= *L. vulgaris*) from the Old World tropics, a trailing vine of the cucumber family, is also called BOTTLE GOURD or WHITE-FLOWERED GOURD, while *Crescentia cujete* [CALABASH TREE, CALABOZO] is a tropical American member of the family Bignoniaceae. Both bear fleshy many-seeded fruits with hard rinds. Dried mature fruits are used as containers and are often elaborately decorated. Other uses include fish-net floats, penis sheaths, rattles and native musical instruments.

Calabrese a variety (var *italica*) of cabbage, *Brassica oleracea*. The enlarged green immature inflorescences are eaten as a vegetable. It is also called ITALIAN or SUMMER SPROUTING BROCCOLI.

Caladium a genus of perennial, tuberous, stemless herbs mainly native to tropical South America. Two species with numerous cultivars are grown, either as pot plants or for bedding in warm climates, for their attractive leaves: *C. humboldtii* with small variegated white and green leaves, the more popular *C. bicolor* with leaves in various shades of red, white and green, and *C. × hortulanum*, a hybrid with similarly variegated leaves.
ARACEAE, about 15 species.

Calamintha a genus of mainly herbs but including some shrubs and subshrubs from north temperate regions, particularly upland pastures and dry stony soils. The plants have long been associated with herbalists for their medicinal and curative properties. *C. grandiflora* and *C. nepeta* [LESSER CALAMINT] are grown in gardens as ornamentals.
LABIATAE, about 40 species.

Calamus [*RATTAN PALMS] the biggest and most widely ranging genus of the palm family, with nine species in Africa and the great majority in Indomalaysia. They are spiny palms and most are climbers. The inflorescences, which are complex panicles and often huge, are among the most complex in the plant kingdom. The most important commercial rattan canes are species of *Calamus* (see RATTAN).
PALMAE, about 370 species.

Calandrinia a large genus of annual and perennial herbs, mainly distributed in western North and South America. Some of the low-growing species, sometimes called ROCK PURSLANES, are cultivated in rock gardens and borders. They include the dwarf *C. umbellata*, with crimson-purple cup-shaped flowers, the taller *C. discolor*, with large pale purple

*The succulent Sea Rocket (*Cakile maritima*) seen growing on a sandy shore with a few shoots of Marram Grass (*Ammophila arenaria*). (×1/10)*

*The Bog Arum (*Calla palustris*) in flower, showing the greenish spadix and brilliant white spathe above masses of Duckweed (*Lemna sp*) and Water Milfoil (*Myriophyllum sp*). (×1/3)*

flowers about 5cm across, and *C. ciliata* with purple, white or crimson flowers.
PORTULACACEAE, about 150 species.

Calanthe a genus of epiphytic and terrestrial tropical and warm temperate orchids predominantly natives of mainland Asia, with a few species in Africa, Australia and the tropical Pacific Islands. Two deciduous terrestrial species, *C. masuca* and *C. vestita*, with tall spikes of brightly colored flowers, are among the most popular species cultivated in greenhouses. The predominant flower colors in the genus are purple, pink and white but there are also orange, yellow, red, and multicolored species in which the lobed lip (labellum) contrasts with the rest of the flower.
ORCHIDACEAE, about 120 species.

Calathea a genus of perennial herbs native to tropical America and the West Indies. Species cultivated as greenhouse or house pot plants for their attractive foliage (often variously colored in stripes or irregular blotches) include *C. ornata*, *C. micans* and *C. lindeniana*. Other species, including *C. allouia* [SWEET CORN ROOT, TOPEE-TAMBU], produce edible tubers which are the source of Guinea arrowroot.
MARANTACEAE, about 150 species.

Calceolaria [SLIPPER FLOWERS, SLIPPER-WORTS] a genus of herbs and shrubby plants mostly native along the Andes of Chile and Peru but extending south to Patagonia and north to Central America. Because of their colorful, often large, slipper-shaped corollas the flowers are prized greenhouse or garden subjects. Among the most popular species are the yellow-flowered *C. biflora*, and *C. darwinii* which has yellow flowers with large reddish spots on the lower lip. Another

Calanthe stolzii, a native of wet montane forest in East Africa. (×1)

popular species is the evergreen subshrub *C. integrifolia* which bears dense clusters of small reddish-yellow flowers.
SCROPHULARIACEAE, about 500 species.

Calendula a genus of yellow- or orange-flowered herbs and subshrubs widely distributed in the Mediterranean region and Macaronesia. *C. officinalis* [COMMON or POT MARIGOLD] is the popular cultivated species with a wide range of cultivars, from the double-flowered 'Apricot-Queen' and 'Chrysantha' to 'Radio' with deep orange or yellow petals.
COMPOSITAE, about 20 species.

Calla a north temperate and subarctic genus represented by a single species, *C. palustris* [BOG or WATER ARUM], an aquatic perennial of shallow ponds, swamps and lakes, with long-stalked, heart-shaped leaves that are carried above the water. The inflorescence is borne on a leafless stem, and is surrounded by an oval white spathe.
ARACEAE, 1 species.

Inflorescence of the Bottlebrush Tree (Callistemon citrinus). The protruding crimson stamens give the characteristic bottlebrush effect. (×1)

Calliandra a genus of trees or shrubs, some scrambling or twining, native mainly to tropical America with some species in Madagascar and tropical Asia.

A number of species are cultivated as ornamentals for their showy globose or brush-like inflorescences with usually long, white, pink or crimson stamens. They include *C. emarginata* (Mexico), *C. surinamensis* (Guiana to Brazil), and the beautiful tree from Ecuador and Bolivia, *C. haematocephala* (= *C. inaequilatera*) [RED POWDERPUFF]. Some species have local medicinal uses. The bark of the Mexican shrub *C. anomala* (= *Inga anomala*) is used in tanning and the close-grained timber of the tropical American tree *C. formosa* (= *Lysiloma formosa*) is used for carpentry.
LEGUMINOSAE, about 150 species.

Callicarpa a genus of tropical and subtropical deciduous shrubs and trees mostly native to Asia, Australia and North and Central America. Several species are grown for their ornamental flowers and large clusters of bright lilac, violet or purple berries in autumn. *C. bodinieri* (China) grows to about 3m and bears lilac flowers and violet berries. Most other cultivated species, such as the blue-flowered *C. americana* [FRENCH MULBERRY, BEAUTY BERRY] and the pink-flowered *C. japonica*, *C. rubella* and *C. dichotoma*, are smaller.
VERBENACEAE, about 140 species.

Callistemon [BOTTLEBRUSHES] a genus of shrubs and small evergreen trees native to Australia and Tasmania. The genus and common names derive from the very attractive, highly colored, long stiff stamens which protrude beyond the petals. *C. speciosus* from Western Australia, with crimson flowers and golden yellow anthers, is perhaps the most widely cultivated ornamental species.
MYRTACEAE, about 25 species.

Callistephus [CHINA ASTER], a herbaceous genus represented by one species, *C. chinensis* from China and Japan, an erect annual with coarsely toothed leaves and solitary daisy-like flower heads. The rays of the flowers are naturally dark purple but in cultivars they vary from violet to blue, red, pink and white; the inner florets are yellow. The cultivars are popular in garden borders and as cut flowers.
COMPOSITAE, 1 species.

Callitris [CYPRESS PINES] a small genus of Australian, Tasmanian and New Caledonian evergreen conifers. Several species, such as *C. endlicheri* (= *C. calcarata*) [BLACK CYPRESS PINE], *C. columellaris* [WHITE CYPRESS PINE] and *C. preissii* [ROTTNEST ISLAND PINE], are cultivated and the wood is valuable for its timber and as a source of resin.
CUPRESSACEAE, 16 species.

Calluna a genus comprising a single species, *C. vulgaris* [HEATHER, SCOTTISH HEATHER, LING], an evergreen shrub native to Western Europe, including Iceland, and extending throughout the Mediterranean region into Turkey. It is the dominant species on acid moorland soils. It has also been introduced to similar habitats in North America but is not fully naturalized. It is a woody evergreen shrub growing to about 1m high, bearing dense racemes of purple-pink flowers.

There are many named horticultural varieties, including some that display a very dwarf habit, such as 'Co. Wicklow' (pink double-flowered) and 'Joan Sparkes' (purple double-flowered). Other varieties which attain a height of 50cm include 'Alba Plena'

Heather (Calluna vulgaris) on the Brecon Beacons, South Wales. Acid moorland soils such as this are the natural habitat of Heather.

The China Aster (Callistephus chinensis) a half-hardy annual which exists in many cultivars that are used for bedding and as cut flowers. (×½)

(white single-flowered) and 'Golden Feather' (purple single-flowered and golden leaves). Some varieties, eg 'Gold Haze' (white single-flowered), grow to a height of 80cm.

Branches of *Calluna* are made into brooms, while the bark has been used for tanning and the flowers are a source of honey. The flowers yield extracts with various medicinal uses, as sedatives and diuretics, while an infusion of the leaves forms an ingredient of heath teas.
ERICACEAE, 1 species.

Calocephalus a small genus of annual and perennial herbs and shrubs native to temperate regions of Australia. *C. brownii* [CUSHION BUSH], a rigid woolly dwarf shrub, is cultivated as a garden plant for its silver-gray leaves and terminal yellow flower heads.
COMPOSITAE, 15 species.

Calochortus a genus of bulbous plants with grass-like leaves, native to North and Central America, a number of which are cultivated for their attractive, showy flowers under the following groupings: I. GLOBE TULIPS (eg *C. amabilis* = *C. pulchellus*), characterized by slender stems and globular, nodding flowers; II. SMALL STAR TULIPS (eg *C. elegans*), with more slender and flexuous stems and cup-shaped erect flowers and nodding capsules; III. LARGE STAR TULIPS (eg *C. apiculatus*) with sturdy stems and larger cup-shaped erect flowers and nodding capsules; IV. MARIPOSA TULIPS (eg *C. catalinae*), with erect, open flowers and erect capsules.
LILIACEAE, about 50 species.

Calophyllum a genus of valuable timber-producing trees, pantropical in distribution but not represented in Africa. One of the most important species is *C. inophyllum* [ALEXANDRIAN LAUREL], a seashore inhabitant of the Indian and Pacific Oceans which produces a red-brown water-resistant timber used in shipbuilding and sold in Europe as Borneo mahogany. *C. brasiliense* is widespread in tropical America and produces a very heavy, durable, pink-red timber used for furniture, floors and shipbuilding. Many species, including *C. calaba* [CEYLON BEAUTY LEAF], yield a kernel oil used in soap manufacture.
GUTTIFERAE, about 120 species.

Calotropis a genus of shrubs and small trees native to tropical Africa and Asia but

The Marsh Marigold (Caltha palustris), a common plant of temperate, marshy habitats. Double-flowered forms are often cultivated as ornamentals. (× ⅕)

also introduced into South America and the West Indies. *C. procera* [AURICULA TREE] and *C. gigantea* [TEMBEGA] are the most important species. *C. procera* tolerates water stress and stabilizes sand dunes in coastal regions. Both species produce inflated fruits packed with white seed hair (akund fiber, ak, madar or mudar), which has at times been used as a *kapok substitute. The leaves are used medicinally and for food and the bark produces a strong fiber.
ASCLEPIADACEAE, 6 species.

Caltha a genus of perennial herbs from moist areas of arctic and north temperate regions, with no petals and yellow, pink or white petal-like sepals. *C. palustris* [MARSH MARIGOLD, KINGCUP] is found on moist ground and the margins of ponds and streams, often in shallow water. It bears rich yellow flowers. Double forms exist and are often cultivated in gardens. The Caucasian *C. polypetala* is taller with larger flowers.
RANUNCULACEAE, about 20 species.

Calystegia a widespread genus of herbaceous perennials from temperate to tropical regions, most of which have vigorous twining stems and large, showy, pink or white, trumpet-shaped flowers. Cultivated species include *C. hederacea* [JAPANESE *BINDWEED], especially its double form [CALIFORNIA ROSE]. Some species, such as *C. sepium* [BINDWEED, HEDGE BINDWEED], are often serious weeds of cultivated ground. *C. soldanella* is a prostrate seashore species around the world.
CONVOLVULACEAE, about 25 species.

Camassia a genus of North American bulbous plants with blue or white flowers in erect racemes on stalks up to 1m high. The best-known cultivated species are *C. quamash* (= *C. esculenta*) [QUAMASH, CAMASS] and *C. scilloides* (= *C. fraserii*), sometimes called WILD HYACINTH.
LILIACEAE, about 5 species.

Camelina a European and Asian genus of annual or biennial herbs with all but one species, *C. sativa* [FALSE FLAX], occurring wild or as weeds. *C. sativa* is cultivated mostly in the USSR for its fiber and seeds, the latter yielding an oil similar to linseed oil. *C. alyssum* and *C. macrocarpa* are weeds of flax fields.
CRUCIFERAE, 10 species.

Camellia a genus of evergreen trees and shrubs native to East Asia, India, China and Japan, where their decorative value and economic importance as the source of tea and an oil has long been recognized. *C. sinensis* [TEA PLANT, COMMON TEA, JAPANESE TEA, CHINESE TEA] is the most widely cultivated member of the genus and the most important economically (see Tea). *C. oleifera* and *C. sasanqua* are cultivated in China for the valuable seed oil which is used for cooking and hair dressing. Camellia wood is often used as fuel and for carved utensils and souvenirs.

A flowering shoot of the Auricula Tree (Calotropis procera), showing the leaves in opposite pairs and the clusters of small purple flowers. (× 1)

For ornamental purposes the Japanese *C. japonica* [CAMELLIA] is outstanding for its diversity in habit, foliage, floral form and color in the 2000 named cultivars which have been produced. It is a spring-flowering shrub which bears blooms in every combination from white to red and from single forms to completely double. Another popular Japanese camellia, *C. sasanqua*, has scented flowers which appear in the autumn and winter. It is extensively planted in America and Australia, often to form hedges.

Cultivated forms of *C. reticulata*, from China, bear rose-red flowers 18cm in diameter. The white to pink-flowered *C. saluenensis*, from the same area, has

The brilliant white trumpet-shaped flowers of the Greater Bindweed (Calystegia sepium subspecies silvatica). This weedy species often grows in hedgerows. (× ½)

proved an excellent parent in the production of the modern hybrid race of camellias. THEACEAE, about 84 species.

Campanula see BELLFLOWER. CAMPANULACEAE, 300–500 species.

Camphor an organic compound produced by steam distillation of macerated wood of a large evergreen tree, *Cinnamomum camphora*, native to China, Japan and Taiwan and introduced to many tropical countries. Camphor has a penetrating, musty aroma, and has long been used medicinally and as a component of incense. It is used in insect repellents, medicines and chemical preparations. Much of the world supply of camphor is now produced synthetically from pinene, a turpentine derivative. Borneo or Sumatra camphor is obtained from the large tree *Dryobalanops aromatica*.

Campsis a small-genus of half-hardy, deciduous climbing shrubs. *C. grandiflora* (= *C. chinensis*) [CHINESE TRUMPET-FLOWER], from Chile, and *C. radicans* (= *Tecoma radicans*) [TRUMPET CREEPER, TRUMPET VINE], from the USA, are cultivated for their orange and scarlet trumpet-shaped flowers. BIGNONIACEAE, 2 species.

Camptothecium a small north temperate genus of mat-forming mosses. *C. lutescens* is the principal species to be found in abundance on the south-facing slopes of chalk hills and calcareous sand dunes. HYPNACEAE, about 15 species.

Cananga see YLANG-YLANG. ANNONACEAE, 2 species.

Canarina a small genus of attractive herbaceous perennials native to the Canary Islands and East Africa. All three species, *C. elmii*, *C. abyssinica* and *C. canariensis* (= *C. campanula*), are cultivated, usually under glass in temperate countries, for their attractive bell-shaped or trumpet-shaped showy flowers. The fruits are edible. CAMPANULACEAE, 3 species.

Canavalia a genus of woody or herbaceous perennial legumes which trail or climb by means of twining stems. The species are essentially subtropical and tropical in distribution, particularly in Africa and America. Some species are economically useful and others are grown for their attractive pea-like purple or violet flowers.

Four species have been domesticated: *C. ensiformis* [JACK BEAN, CHICKASAW LIMA] in tropical America; *C. plagiosperma* in the Andes; *C. regalis* in West Africa; *C. gladiata* [SWORD BEAN] in the Far East. They are used as minor food crops the young pods being eaten as a vegetable), forage plants and green manures. The fresh immature beans of *C. ensiformis* are, however, considered to be poisonous. In Hawaii, flowers of *C. cathartica* and other species are used in wreaths. LEGUMINOSAE, about 50 species.

Canarina canariensis, a famous Canary Island endemic of the laurel forests, now becoming much rarer than previously. ($\times \frac{1}{4}$)

Cane a general term for the commercially important stems of grass-like plants. These include the *BAMBOOS which are used for a variety of purposes, *MALACCA and *RATTAN canes for basket- and furniture-making, and, of course, *SUGARCANE. GRAMINEAE.

Canella a genus of small trees native to southern Florida and the West Indies. One of these, *C. alba* (= *C. winterana*) [WILD OR WHITE CINNAMON], is the source of canella bark which is strongly aromatic and is used as a stimulant and as a condiment or spice. CANELLACEAE, 2 species.

Canna a genus of tall erect, perennial herbs native to tropical and subtropical America, with a few species indigenous to Africa and Asia.

The best-known species is a common gar-

Camptothecium sericeum; the narrow leaves and closely pinnate branching give the plant a feathery and silky appearance. ($\times 2$)

den ornamental, *C. indica* [INDIAN SHOT], a tropical and subtropical plant with yellow and red flowers, used for bedding; it is naturalized in the tropics. Also widely cultivated is *C. × generalis* of garden origin. *C. bidentata* has starchy rhizomes and is a famine food in tropical Africa; *C. edulis* [PURPLE OR QUEENSLAND ARROWROOT], from South America (the Andes) and the West Indies, is cultivated commercially in Australia, South America, the West Indies, Africa and Asia for its starchy rhizomes. *C. speciosa* [AFRICAN TURMERIC PLANT] has tubers which produce coloring and spice-powders. CANNACEAE, about 55 species.

Cannabis a genus represented by a single herbaceous annual species, *C. sativa* [INDIAN HEMP, MARIJUANA]. An ancient crop plant once confined to Central Asia, it has been modified and spread over the earth by Man as a crop or weedy escape.

Next to alcohol, drugs from *Cannabis* are the world's most popular, although illegal, inebriant. The intoxicant ability lies in the resin (cannabinoids) secreted by all parts of

The attractive red flowers of the Purple or Queensland Arrowroot (Canna edulis), which is cultivated for its starch-rich rhizomes.

the plant, although a high yield of the drug is obtained only from the upper parts of female plants of certain cultivars grown in hot climates, as in India, the leading producer of the drug.

Cannabis drugs are generally smoked. The terminology used is variable, but strictly marijuana consists of coarsely pulverized leaves, flowering parts, and small twigs. The stronger intoxicant hashish consists of powdered plant material in a matrix of resin, prepared by scraping resinous particles from the plants and increasing the concentration of resin by filtering, heating and decanting, and extracting with solvents. *Cannabis* drugs are sometimes consumed orally as a slurry and in baked goods.

Cannabis is also the source of a textile fiber, hemp, found in the bark. Cannabis fruits, usually called "seeds", contain a drying oil like linseed oil, used in paints and varnishes. The fruits are edible and are often fed to cattle and birds.
MORACEAE, 1 species.

Caoutchouc   a substance occurring in the milky latex of many plants; it is better known by the name of india-rubber. Although caoutchouc is recognized as occurring in several genera, the greatest quantities come from the genus to which the term was originally applied, *Hevea* (= *Caoutchoua*).

The most important caoutchouc-producing species is *H. brasiliensis* [PARA RUBBER TREE] a lofty South American tree of the Amazon region, which is widely cultivated in many other tropical countries, especially in Asia. The milky juice is obtained by cutting deep vertical and linking slanting incisions in the trunk. After collection, the juice thickens to a paste which is then dried in the sun or on clay molds over slow fires. In the commercial market the caoutchouc of *H. brasiliensis* is also known as para rubber.

Some other plants from which caoutchouc is also obtained, of varying quality and quantity, include *Castilla elastica* [CAUCHO], *Manihot glaziovii* [CEARA RUBBER], many *Landolphia* species including *L. comorensis* and *L. dondeensis*, and many others. (See also RUBBER PLANTS.)

Capers   the pickled flower buds of the trailing Mediterranean shrub *Capparis spinosa*, used as a condiment.
CAPPARACEAE.

Capparis   a large genus of tropical and subtropical shrubs growing in both hemispheres but largely in Central and South America. Many are thorny climbing plants or scramblers. The flower buds of *C. spinosa* [CAPER BUSH], a trailing spiny southern European and Mediterranean shrub, are the capers of commerce which are pickled in vinegar or dry-salted and eaten as a condiment in sauces (sauce tartare, remoulade etc) and used in pickles and butters (Montpelier butter).
CAPPARACEAE, about 250 species.

Capsella   a genus of annual or biennial herbs, the best-known of which is *C. bursa-pastoris* [SHEPHERD'S PURSE], which is a cosmopolitan weed; the remaining species are native to southern Europe and southern Russia. SHEPHERD'S PURSE is an annual or biennial, preferring rich, loamy soils, but

Capparis aphylla, a thorny shrub native to the semidesert areas of southern Iran, Arabia and neighboring areas. (× ⅓)

One of the species of Chili Pepper (Capsicum frutescens), showing the long pendent fruits typical of cultivated varieties. (× ⅛)

growing in a wide range of soil types. It can occur as a weed of most crops and grows wild in many places. One plant can produce tens of thousands of seeds in a single year. It is self-pollinated, and many distinct forms have arisen, some of which have been named as distinct species.
CRUCIFERAE, 5 species.

Capsicum   a genus which includes the RED or *CHILI PEPPERS, *PAPRIKA, *CAYENNE and TABASCO PEPPERS as well as the non-pungent *GREEN or RED SWEET PEPPERS. The five cultivated species (*C. annuum*, *C. baccatum*, *C. chinense*, *C. frutescens* and *C. pubescens*) and the wild species, originated in the New World. Thereafter, *C. annuum* and *C. frutescens*, in particular, spread rapidly throughout tropical and warm temperate regions and liberal use of pungent chillies is now characteristic of much of the cooking of tropical Africa, the Indian subcontinent and the Far East as well as Latin America. Pungent peppers also have various medicinal uses. Sweet peppers [also known as PIMENTA, PIMENTO or PIMIENTO] are used, raw or cooked, as a vegetable and are good sources of vitamins A and C. Some small-fruited cultivars are grown as ornamentals. Sweet peppers are now widely grown in temperate areas.
SOLANACEAE, about 50 species.

Caragana [PEA SHRUBS]   a genus of mainly deciduous, spiny shrubs with pinnate leaves, native to temperate Central Asia. They produce yellow, pea-like flowers and bear straight pea-like pods. The young pods of *C. arborescens* [SIBERIAN PEA SHRUB] are eaten

as a vegetable in some parts of Siberia; the bark is also reputed to be used as a source of fiber. In China the flowers of *C. sinica* are sometimes eaten. Some species are grown as ornamentals, such as *C. frutex* [RUSSIAN PEA SHRUB], which has several cultivars.
LEGUMINOSAE, over 60 species.

Cardamine [BITTER CRESSES] a cosmopolitan genus (sometimes including *Dentaria*) of annual and perennial herbs chiefly confined to damp habitats in temperate regions, and in the tropics to mountains. The best-known European species is *C. pratensis* [CUCKOO FLOWER, LADY'S SMOCK], which is sometimes cultivated in rock gardens. The flowers are white or tinged lilac in terminal racemes.
CRUCIFERAE, about 160 species.

Cardamom the common name of a herbaceous perennial, **Elettaria cardamomum*, and of its dried fruits which are used as a spice, as a masticatory and medicinally. Cardamom grows wild in the evergreen rain forests of southern India and Sri Lanka, and it is here that it is mainly cultivated, although a considerable quantity is also produced in Guatemala.

As a spice, cardamom is used in curries, bread, cakes and confectionery, and also for flavoring coffee in Arab countries. In India it is often a constituent of the betel "quid", together with **BETEL NUT (Areca catechu)* and the leaf of BETEL PEPPER (**Piper betle*). The seeds yield a volatile oil that contains cineole, terpineol and limonene.

The name cardamom is applied to various substitutes and adulterants, including *Amomum* and *Aframomum* species, of the same family as cardamom.
ZINGIBERACEAE.

Cardiospermum a genus of annual or perennial climbers found throughout all warm countries, particularly in the American tropics. The fruit is a large, inflated capsule.

Spring Sedge (Carex caryophyllea). The larger yellow spikes contain male flowers, while the smaller spikes contain only female flowers. (× 1)

Lady's Smock (Cardamine pratensis), with its delicate pink inflorescences, is a common wild flower in damp meadows. (× ¾)

C. halicacabum [BALLOON VINE, HEARTSEED, HEART PEA, LOVE-IN-A-PUFF] is a climbing annual whose roots and leaves are used as an aperient and also as a hairwash. This and other species are occasionally grown for ornament.
SAPINDACEAE, 12 species.

Cardoon a thistle-like herbaceous perennial, **Cynara cardunculus*, native to southern Europe and North Africa. It is closely related to *C. scolymus* [GLOBE **ARTICHOKE*]. The blanched, crisp stalks of the inner leaves and the leaves themselves of both species are used as a vegetable. The thick main roots of the CARDOON are also used as food.
COMPOSITAE.

Carduus a genus of European, Mediterranean, southwest Asian, North and tropical East African stout annual, biennial or perennial, thistle-like herbs with spiny-winged stems and stalkless spiny-toothed or lobed leaves. The purple or white, rarely pink, globose flower heads are composed of long thin tubular florets on a densely bristly receptacle.

C. acanthoides [WELTED THISTLE] and *C. tenuiflorus* [SLENDER THISTLE] are typical members of the genus. *C. nutans* [MUSK or NODDING THISTLE] a biennial common on calcareous soils, is particularly attractive with its large nodding purple flowers.
COMPOSITAE, about 100 species.

Carex [SEDGES] a large cosmopolitan genus of grass-like perennial herbs most frequently found in wet places in cold and temperate regions. Although of little economic significance, they are very important ecologically, and some have a forage value. *C. arenaria* [SAND SEDGE], a useful sand binder, grows on loose sand dunes and can form a spectacular sight, with a series of aerial stems which appear to have been planted in long, straight

rows. Some of the larger, densely clumped species are used as ornamentals, such as *C. pendula* [PENDULOUS SEDGE], an attractive woodland species, and *C. paniculata* [TUSSOCK SEDGE], a striking subject for a water or bog garden.

Other well-known inhabitants of moist or wet habitats include *C. acutiformis* [LESSER POND SEDGE] and *C. rostrata* [BOTTLE SEDGE], often growing out into shallow ponds or lakes.

In North America *C. atherodes* is used as a hay grass; in Japan *C. dispalatha* is cultivated for its leaves, used in hat making; and in Europe the aromatic rhizome of *C. arenaria*, called German sarsaparilla, has uses in veterinary medicine.
CYPERACEAE, 1 500–2 000 species.

Carica an economically important tropical American tree genus which includes the valuable *C. papaya* [PAPAW, PAWPAW, PAPAYA, MELON TREE] (see PAPAW). Other species producing edible fruits include *C. chilensis*, *C. goudotiana*, *C. monoica* and *C. pubescens* (= *C. candamarcensis*) [MOUNTAIN PAWPAW].
CARICACEAE, about 22 species.

Carlina a small genus of European, Mediterranean and southwest Asian thistle-like herbs. *C. vulgaris* [CARLINE THISTLE], a stiff, spiny biennial with purple flower heads,

Male flowers of the Papaw (Carica papaya). (× ¼)

is abundant in most of Europe. Its roots are used in herbal remedies. *C. acaulis* [STEMLESS CARLINE THISTLE, WEATHER CLOCK] is a large, stemless, silvery leaved and white to purplish-brown flowered perennial, sometimes cultivated as an ornamental. This species is monocarpic, ie flowers once then dies.
COMPOSITAE, 28 species.

Carludovica a genus of tropical American short stemmed, palm-like plants, of which the most important species is *C. palmata* [PANAMA-HAT PLANT, PALMITA], whose young leaves form the basis of the important Panama-hat-making business of Ecuador. Of minor importance is *C. divergens* whose young leaves are a source of fiber in Peru. *C. palmata* is also often cultivated in tropical gardens as an ornamental.
CYCLANTHACEAE, 3 species.

Flower head of the Carline Thistle (Carlina vulgaris), with purple disk florets. The dried flowering heads have ornamental value. (×5)

Carmichaelia a genus of xerophytic shrubs which are usually leafless at maturity, the flat green branchlets acting as leaves. All but one of the species are endemic to New Zealand. The pea-like flowers range from white or yellowish to pink or purplish and are often borne in profuse masses. Several species are cultivated in park and gardens in frost-free regions, including *C. grandiflora* and *C. odorata*.
LEGUMINOSAE, 41 species.

Carnations highly popular cultivated flowers of the genus *Dianthus*. Modern carnations, which are the basis of a substantial cut-flower industry, are descended principally from *D. caryophyllus*. Two main groups are recognized by gardeners and florists: BORDER CARNATIONS and PERPETUAL CARNATIONS.

Border carnations are of more or less pure descent from wild species. They have a flat and circular flower 5cm–8cm in diameter. While completely hardy, they may be grown in a cool greenhouse for exhibition purposes. Examples of the groups grown commercially are: *clove-scented* – especially fragrant with a characteristic clove smell; *self colored* – only one color present in the flower; *fancy* – irregular markings in other colors, the main color being termed the ground, as in white ground fancy or yellow ground fancy; and *picotee* – narrow band of secondary color round each petal.

Because of their use in the cut-flower trade, PERPETUAL CARNATIONS are probably the best-known of all. Having been cultivated first in France, they were grown extensively in the USA during the 19th century. Hybridization with other species of *Dianthus* is regularly used to obtain new cultivars.

AMERICAN SPRAY CARNATIONS are true perpetuals, producing a very delicate spray of highly scented flowers, 6.5cm in diameter. Cultivars such as 'Exquisite' and 'Elegance' are very popular in America, and are now grown in Europe.
CARYOPHYLLACEAE.

Carnauba wax a product of the palm *Copernicia prunifera* (= *C. cerifera*), which grows wild in vast quantities along rivers in the arid regions of northeastern Brazil. The wax is beaten from the surface of dried leaves, sieved to remove debris, melted and formed into cakes for export. Carnauba is currently the most important plant wax, valued for its hardness and high melting point. It is used to make high-quality polishes and varnishes and in the manufacture of carbon paper, gramophone records and lipsticks.
PALMAE.

Carnegiea a genus represented by a single species, *C. gigantea* [GIANT CACTUS], native to Mexico and the southwestern USA. It is a large columnar cactus which can attain a height of 18m with a few erect branches.
CACTACEAE, 1 species.

Carob the common name for *Ceratonia siliqua*, a leguminous evergreen tree from the eastern Mediterranean. It has been cultivated since antiquity for its black pods (St. John's bread, *algarroba) and seeds. The former are fed to live-stock and also used in confectionery and food. The seeds yield a gum employed in papermaking and as a stabilizer in food products. The wood is hard and lustrous, and is used for marquetry.
LEGUMINOSAE.

Carolina allspice the common name for *Calycanthus floridus*, a shrub native to the eastern USA and often grown as an ornamental for its sweet-smelling reddish-purple flowers. The strongly aromatic bark is used as a spice by North American Indians.
CALYCANTHACEAE.

Carpinus see HORNBEAM.
BETULACEAE, about 60 species.

Flowers of the Perpetual or Border Carnation (Dianthus caryophyllus). (×1)

Carragheen or **Irish moss** the common name for the red alga *Chondrus crispus* that grows in the northern Atlantic Ocean. A polysaccharide, known as carrageenan, is extracted from it and is processed for use as a stabilizing emulsion in salad-dressing, ice cream, blancmanges, soups, paints, etc. RHODOPHYCEAE.

Carrot in the cultivated form, subspecies (variety) *sativus* of *Daucus carota*, a biennial herb native to Europe, the Orient and the Mediterranean.

The carrot is an important and popular root vegetable, both for human consumption and for feeding to cattle, widely grown in temperate regions and in the subtropics during the winter. Carrots are rich in sugars (4.5%) and in carotene, from which the body manufactures vitamin A. They are eaten raw or cooked and are often used to flavor soups and casseroles.

The Western carotene carrot has been derived from the Eastern anthocyanin carrot. The first carrots cultivated in Europe were purple in color (due to anthocyanin pigments) and then yellow and finally orange strains became more popular. All modern Western types are derived from three orange cultivars, differing in size and time of maturity: 'Late Half Long Horn' (the largest), 'Early Half Long Horn' and 'Early Scarlet Horn' (the smallest). Today's carrot types include the *shorthorn*, with short, conical roots; *stump-rooted*, with medium-length roots, blunt at the ends; *intermediate*, with middle-length, spindle-shaped roots; and *long-rooted*, with large, long and tapering rootstocks. Yield increases with root length, but roots longer than 30cm are difficult to harvest.
UMBELLIFERAE.

Carthamus a Mediterranean, central European and Central Asian genus of usually spiny herbaceous plants with yellow or orange flowers in ovoid-conical heads, including *C. tinctoria* [SAFFLOWER], an annual herb widely cultivated as an important oil-seed crop in the USA, India and Mexico and, on a smaller scale, in southern and central Europe. The oil is used in the manufacture of margarine, and as a salad and frying oil. The dried florets were once used as a source of a red and a yellow dye. Other species are occasionally grown in gardens for ornament.
COMPOSITAE, about 15 species.

Carum a genus of temperate and subtropical annual and perennial herbs. *C. carvi* [CARAWAY], an aromatic plant with fern-like leaves and clusters of small yellowish-white flowers, is cultivated throughout north temperate regions to the Sudan and northern India. Caraway seeds (actually the fruits), are used for flavoring bread, biscuits, cakes, potatoes, sausages and the liqueur, kümmel; they are also used in pickling spice, curries, goulashes and soups. The seeds yield an essential oil used in medicine. The seeds of *C. copticum* (= *C. ajowan*, *Trachyspermum*

copticum, T. ammi), a herb native to western Asia and cultivated from Iraq to western India, yield ajowan oil which contains the antiseptic thymol.
UMBELLIFERAE, about 30 species.

Carya a genus of large, fast-growing stately, deciduous trees, chiefly confined to the eastern part of North and Central America, with a few species in East Asia. The trees grow to over 30m and may be slenderly conical as in *C. cordiformis* [BITTERNUT] or broadly conical as in *C. glabra* [PIGNUT HICKORY, SMOOTH-BARK HICKORY] and *C. ovata* [SHAGBARK HICKORY, SHELLBARK HICKORY].

Species such as *C. ovata*, *C. tomentosa* [MOCKERNUT, BIG-BUD HICKORY] and *C. glabra* are cultivated principally for their wood. The timber produced is very tough and elastic and is used in the manufacture of tool handles. Other species are grown as ornamentals or for their edible fruits, such as *C. illinoinensis* (= *C. pecan*) (see Pecan nut). Apart from the pecan, the best hickory nuts are produced by *C. ovata*, and *C. laciniosa* [KING NUT, BIG SHELLBARK] but the trees are not generally cultivated specifically for their production.
JUGLANDACEAE, about 25 species.

Caryocar a genus of trees native to tropical America. The wood of *C. nuciferum* [BUTTER NUT, SOUARI NUT] is very durable and used in shipbuilding; an edible oil is extracted from the seeds (the butter nuts of commerce).
THEACEAE, about 20 species.

Caryopteris a genus of deciduous shrubs native to East Asia, some of which have been hybridized for use as garden ornamentals. Although both *C. incana* (= *C. mastacanthus*)

A Mockernut or Big-bud Hickory (Carya tomentosa) in full autumn color, which makes it a valued ornamental for parks and large gardens.

Cultivated Carrots (Daucus carota subspecies sativus), lifted, washed and bunched ready for marketing. Carrots are rich in sugar and vitamin A.

and *C. mongholica* are planted as ornamentals, it is the varieties of their vigorous hybrid, *C. × clandonensis*, that are more widely grown.
VERBENACEAE, 15 species.

Caryota [FISHTAIL PALM] a genus of tall palms native to tropical Asia and Australia. *C. mitis* is often grown in greenhouses for its attractive drooping leaves and is widely planted in the tropics. *C. urens* [KITUL PALM, WINE PALM, TODDY PALM, SAGO PALM, FISHTAIL PALM] has a wide range of uses, yielding palm sugar (jaggery) and sago. A black bristle fiber (kittul fiber) is made from the leaves and is used for making ropes, brushes and baskets, while the outer wood is used in cabinet-making.
PALMAE, about 12 species.

Cascara an extract with purgative and (in small doses) aperient properties made from the dried bark (cascara sagrada) of the North American shrubs *Rhamnus purshiana* [CASCARA *BUCKTHORN, BEARBERRY, BEARWOOD] and *R. californica* [COFFEEBERRY].
RHAMNACEAE.

Cassava the common name for *Manihot esculenta*, also known as MANIOC, BRAZILIAN *ARROWROOT or TAPIOCA, the roots of which constitute a major staple food crop in the tropical world, producing higher yields of carbohydrate than maize or rice. It is a particularly useful crop in dry areas where rainfall is too unreliable for grain crops. Roughly equal quantities are produced in South America, Africa, and Asia and various tropical islands.

Some of the adventitious roots of the shrubby cassava plant become swollen, the central white pith being rich in starch. Although cassava can be divided into

Cassava or Manioc (Manihot esculenta) on sale in a market in Bolivia. The swollen, tuberous roots can weigh several kilograms.

"sweet" and "bitter" varieties, they all contain a cyanogenetic glucoside, which produces prussic acid. This poisonous component is volatile and is removed by heating the extracted root material.

Tapioca can be prepared from the fine starch of cassava, while cassava flour is used for the production of macaroni and spaghetti in India. Cassava flour is mixed with wheat flour in Brazil to make "cassava bread". The dried, grated pulp is consumed as "gari" in Nigeria as an additive to meat or vegetable dishes. Cassava starch is also used industrially as a source of ethyl alcohol, acetone and glue. Cassava is now the subject of extensive research because of its excellent capacity to produce high yields of calories.
EUPHORBIACEAE.

Cassia [SENNA] a genus of trees, shrubs and herbs, a number of which are useful to Man. They are found throughout the tropics and warm temperate zone, excluding Europe.

The laxative senna (taken as a hot water infusion of either dried pods or leaves) is produced commercially chiefly from C. senna (= C. acutifolia) [ALEXANDRIAN SENNA], grown extensively on poor sandy soils of the Northern Province of the Sudan, and C. angustifolia [TINNEVELLY or INDIAN SENNA] cultivated in southern India.

Pods and leaves of C. holosericea [ADEN SENNA], C. obovata [ITALIAN or DOG SENNA] and C. marilandica) [AMERICAN SENNA] are also used as laxatives. Pulp from the pods of

the tropical American C. grandis and of C. fistula [PURGING CASSIA, INDIAN LABURNUM, GOLDEN SHOWER] is a powerful laxative.

Several species of Cassia are used to treat skin diseases. The leaves of C. alata [RINGWORM SENNA], found throughout the tropics, and of C. tora [SICKLE SENNA], from India, are used to treat ringworm. In Central America, leaves of C. emarginata are used to treat insect stings and leaves of C. sericea are used as wound dressings.

INDIAN LABURNUM, the most ornamental of all Cassia species, is found in gardens throughout the tropics for its racemes of yellow flowers up to 5cm in diameter. C. occidentalis [COFFEE SENNA] is also grown for ornament, and in Africa the seeds are roasted to give "mogdad" or "negro" coffee. C. laevigata [SMOOTH SENNA] and C. sericea are used as a coffee substitute in Guatemala and Brazil respectively. In Malaysia roots of C. nodosa are used for soap.

Cassia wood is hard and dark and is often used locally for tool handles and posts. C. siamea, from the Malay Peninsula and Indonesia, is cultivated for timber used in construction work, and the timber of C. sieberiana is much valued in tropical Africa for its resistance to termites. C. javanica has beautifully grained wood and in Indonesia the bark is used in tanning. The bark of C. auriculata [TANNER'S CASSIA] yields 18% tannin and is cultivated in its native India and Sri Lanka.
LEGUMINOSAE, 500–600 species.

Cassie the common name for *Acacia farnesiana [SWEET ACACIA], a small, thorny much-branched, tropical and subtropical American shrub, commonly grown throughout the tropics and subtropics. It is cultivated in the South of France for the flowers, which yield a valuable essential oil, often used in creating "violet" perfumes.
LEGUMINOSAE.

Cassiope a small genus of low, creeping evergreen, heather-like shrubs that have alpine–arctic distribution in the Northern Hemisphere. In cultivation, Cassiope is usually found in acidic peaty rock gardens or peat beds in semi-shade or in Alpine houses. C. lycopodiodes is minute and nearly prostrate, and C. tetragona is a small, neat shrub up to 30cm high, with white, pink-tinged, bell-like flowers. Other cultivated species include C. fastigiata, a much-branched shrub, and C. mertensiana [WHITE HEATHER], a small tufted shrub.
ERICACEAE, 8–12 species.

Castanea see CHESTNUT.
FAGACEAE, 12 species.

Castilla a tropical genus of large deciduous forest trees from Mexico, Central America and the northern areas of South America. The most important species is C. elastica [CAUCHO], the source of castilla or Central American rubber. The trees are tapped to obtain the latex which yields *caoutchouc

(Panama *rubber or caucho negro).
MORACEAE, 10 species.

Castor oil an oil obtained from the seeds of the shrub *Ricinus communis, an ancient and important oil crop which originated in India or Africa. The main countries of production are India, Brazil, the USSR, and China.

The seed contains approximately 50% oil and 18% protein. The use of the oil as a laxative is well known, but its utilization for numerous industrial purposes is far more important. It is used in the manufacture of enamels, paints, varnishes, soaps and other cosmetic products. Castor oil is the source of sebacic acid, some esters of which are used in the lubrication of jet engines.

Castor cake or pomace, the residue remaining after oil extraction, is rich in protein and is used as a fertilizer. In addition to utilization of the seed, the leaves have been a source of an insecticide and the stems a source of cellulose pulp.
EUPHORBIACEAE.

Casuarina a genus of hardwood trees and shrubs distributed in Malaysia, Australia, Polynesia and the Mascarene Islands, and introduced to some parts of East Africa. The wood of several species is extremely hard and valued for furniture manufacture. C. equisetifolia [HORSETAIL TREE, RED BEEFWOOD, SWAMP OAK, BULL OAK] is the most widespread cultivated species. Other valuable timbers include the Australian native C. stricta [SHE OAK] and the Australian cultivated C. cunninghamiana [RIVER OAK].
CASUARINACEAE, about 80 species.

Catalpa a genus of hardy deciduous trees native to America, the West Indies and East Asia, some of which are popular ornamen-

Cassia closiana in flower in the Chilean chapparal at Maipu, Santiago. A related species yields the laxative senna. (× ⅟₇)

*Fruits of the Castor Oil Plant (*Ricinus communis*).*
Each spiny fruit splits open to release several oil-rich
seeds. (× ½)

tals. The best-known species is *C. bignonioides* [COMMON CATALPA, INDIAN BEAN], a tree which attains a height of 20m and has a round, spreading head. It is quite extensively grown as a garden ornamental for its attractive leaves and large inflorescences of showy flowers. Another North American species, *C. speciosa* [WESTERN CATALPA, CATAWBA TREE], is somewhat taller (up to 30m). In their native North American habitats the coarse-grained durable timber of both these species is used for fence posts and rail ties.

The Chinese species *C. bungei*, a small tree, has flowers very similar to those of *C. bignonioides* and is sometimes grafted on to stocks of the INDIAN BEAN and cultivated under the name *C. bignonioides* var *nana*. Other attractive ornamental species are *C. ovata*, a small tree with yellowish flowers, and the HYBRID CATALPA, *C. × hybrida* (= *C. bignonioides × C. ovata*), which is a large tree showing characteristics intermediate between its two parents.

The only other commercially useful species is *C. longissima* whose bark is used for tanning in the Antilles and which is also planted in Florida.
BIGNONIACEAE, 11 species.

Catananche [CUPID'S DART] a small genus of annual and perennial herbs native to the Mediterranean region. The flowers are in typical composite heads with the florets all ligulate. *C. caerulea*, a perennial growing up to 1m high with slightly branched stems, is often cultivated. Typically it has blue flowers but white and bicolor forms are known. *C. lutea* is small annual with yellow flowers.
COMPOSITAE, 5 species.

Catasetum a genus of terrestrial or epiphytic pseudobulbous tropical American orchids.

The bizarre nature of the fleshy flowers and their often overpowering scents have made *Catasetum* species popular horticultural subjects. They are easy to grow, flower readily, and recently intergeneric hybrids with *Mormodes* (called *Catamodes*) and *Cynoches* (called *Catamoches*) have been reported.
ORCHIDACEAE, about 70 species.

Catharanthus a small genus of Old World tropical annual and perennial herbs. The best-known is *C. roseus* (= *Lochneria rosea*) [ROSE or MADAGASCAR PERIWINKLE], native to Madagascar and India but now naturalized throughout the tropics, where it is a noxious weed. It is an attractive plant, bearing shiny green leaves with a white midrib and conspicuous red and purple flowers, and an important source of medicinally valuable alkaloids.
APOCYNACEAE, 5 species.

Cattleya a genus of pseudobulbous epiphytic orchids native to tropical America. They are widely grown throughout the world either as species or as increasingly complex hybrids involving not only true cattleyas but many other genera such as *Laelia*, *Epidendrum* and *Brassavola*. With all species the flowers are relatively large, attractive and frequently vividly colored, generally brilliant mauve-pink or rose-purple.
ORCHIDACEAE, about 60 species.

Cauliflower one of the varieties (var *botrytis*) of *Brassica oleracea* [*CABBAGE]. The developing inflorescence forms a "curd", a dense mass of young flower buds born on thickened fleshy stalks, which is cooked and eaten. A good-quality curd is firm and dense, globe-shaped and pure white.
CRUCIFERAE.

Cayenne pepper the dried and ground fruits of several species of *CHILI PEPPER, mainly

Ceanothus 'Delight', a hardy hybrid between Ceanothus papillosus and C. rigidus. Many Ceanothus species are valued as ornamentals. (× ⅛)

*The Cauliflower (*Brassica oleracea var botrytis*) is one of the numerous cultivated varieties of Brassica.*

Capsicum annuum and *C. frutescens*. Small-fruited, extremely pungent peppers are usually used, but there are also cultivars with larger, somewhat milder fruits such as 'Long Red Cayenne'. Red fruits are preferred, since they produce a brighter powder. The fruits are dried, ground and sifted to produce a fine powder which is used mainly in cooking, but has also been used medicinally.
SOLANACEAE.

Ceanothus a North American genus including some of the best-known blue-flowered garden shrubs. They are small trees or shrubs, evergreen or deciduous. Although predominantly blue, the flowers are also in shades of lavender, purple, pink and white. Popular garden forms include *C. arboreus* [CATALINA or FELTLEAF CEANOTHUS] and *C. thyrsiflorus* [BLUE BLOSSOM], both of which are upright evergreen shrubs, *C. prostratus* [MAHALA-MAT, SQUAW CARPET], a prostrate evergreen shrub and the beautiful azure-blue-flowered *C. × veitchianus*. The latter may well be a natural hybrid, unlike many other garden forms which have been nursery-bred such as the well-known lilac-blue 'Gloire de Versailles', which is a

hybrid between *C. americanus* × *C. coeruleus*.

C. americanus [NEW JERSEY TEA] is so called because it was once used by North American Indians to prepare a beverage resembling tea. RHAMNACEAE, about 50 species.

Cecropia an interesting American genus of deciduous trees found in humid tropical lowlands and abandoned forest sites. *C. peltata* [TRUMPET TREE] has hollow stems (used as trumpets by American Indians). The lightweight timber is used for floats, rafts and paper pulp.
MORACEAE, about 100 species.

Cedar the common name for species of the genus *Cedrus* (Pinaceae). In commerce and especially the timber trade, however, the name cedar can refer to as many as 70 different types of trees and their timber. It applies especially to the *JUNIPER and several genera of the family Meliaceae, eg *Cedrela and *Chukrasia*. Cedar wood oil is obtained by distillation of the wood of species of *Juniperus*, especially *J. virginiana*.

*Bunches of Celery (*Apium graveolens *var* dulce*), cut, trimmed and ready for marketing; celery is a widely used vegetable and salad plant. ($\times \frac{1}{6}$)*

Cedrela a genus of tall trees native to Mexico and tropical South America. A few species, such as *C. sinensis* and *C. toona* [TOON OR RED CEDAR], are grown as avenue trees and some, including *C. mexicana* and *C. toona*, produce a fragrant, reddish timber used in making furniture. The most valuable species is *C. odorata* [WEST INDIAN CEDAR, CIGAR BOX TREE], whose fragrant wood is used for making cigar boxes.
MELIACEAE, about 9 species.

Cedrus a genus of stately evergreen trees, commonly regarded as comprising four species: *C. atlantica* [ATLAS OR ALGERIAN CEDAR], *C. brevifolia* [CYPRUS CEDAR], *C. deodara* [DEODAR] and *C. libani* [CEDAR-OF-LEBANON], although some authorities consider them all to be geographical variations of a single species. They have a wide but discontinuous distribution in the Old World.

All species have long and short shoots, the latter with clusters of needle-like leaves, 0.5–5cm long, according to species. Male cones are erect, ovoid or conic, up to 5cm long, opening in September to November. Female cones are erect, up to 1cm long, borne terminally on short shoots. The fruiting cones, which take two or three years to mature, are oval to oblong, rounded at the apex and about 5–10cm long. On average, trees do not bear cones until they are 40 or 50 years old.

Apart from *C. brevifolia*, which is little cultivated, all the species have numerous forms. Probably the best-known is *C. atlantica* var *glauca* [BLUE CEDAR], which is prized for specimen planting and is probably the best colored of all "blue" forms of conifer. The wood of *Cedrus* is soft but durable and is widely used in building construction and for furniture.
PINACEAE, about 4 species.

Ceiba a genus of large deciduous trees native to tropical America, one of which,

*Celeriac (*Apium graveolens *var* rapaceum*), the swollen roots of which have a celery-like flavor and can be eaten raw or cooked. ($\times \frac{1}{12}$)*

C. pentandra [SILK-COTTON TREE, KAPOK TREE], is extensively cultivated in the tropics for the cotton-like fiber (see Kapok) obtained from the seed pods. It is widely planted as a specimen tree in the tropics and many reach 50m, with a spiny buttressed trunk up to 3m or more in diameter.
BOMBACACEAE, about 10 species.

Celastrus a genus of shrubs, mostly deciduous climbers, widespread in the tropics and subtropics. The North American *C. scandens* [WAXWORK, STAFF VINE, CLIMBING BITTERSWEET] is showy in fruit. The bark of this species has purgative properties and the seeds of *C. paniculatus* yield an oil which is reputedly useful as a nerve stimulant.
CELASTRACEAE, about 30 species.

Celery the common name for *Apium graveolens* which grows wild in temperate

Europe in marshes, ditches and other damp habitats. The wild species is tough and virtually inedible but it has two cultivated varieties used as vegetables and flavorings.

CULTIVATED CELERY is *A. graveolens* var *dulce*, a biennial with fleshy, upright petioles (leaf-stalks) terminating in compound leaves. It is the petioles that constitute the vegetable. Celery may be cooked, or eaten raw as a salad vegetable. The seeds, which contain essential oils, are used as a spice, particularly to flavor soups, stews and pickles, have an aroma like the vegetable, and a slightly bitter flavor. They are also used in perfumery.

CELERIAC, also known as TURNIP-ROOTED CELERY, is *A. graveolens* var *rapaceum*. Its large, turnip-like root has a flavor like that of celery and is similarly eaten raw, or cooked in stews and soups.
UMBELLIFERAE.

Celmisia a genus of Australasian evergreen, herbaceous perennial shrubs or herbs with white daisy-like flowers. A number of species, including *C. coriacea*, *C. hieraciifolia* and *C. spectabilis*, are handsome garden ornamentals.
COMPOSITAE, about 65 species.

Celosia a genus of annual and perennial herbs distributed in the tropics and warm temperate zones of America and Africa. A number are cultivated for their showy flower heads, notably *C. cristata* (= *C. argentea* var *cristata*) [COCKSCOMB] which produces crested heads of red, orange, white or yellow flowers. Some forms produce variegated leaves and others varying degrees of aggregation of the inflorescences.
AMARANTHACEAE, about 60 species.

Celsia a genus of biennial or perennial herbs, differing from *Verbascum* in having four stamens not five. They are mostly native to the Mediterranean region but with some

*Inflorescence of the Greater Knapweed (*Centaurea scabiosa*); the conspicuous outer flowers are sterile, the inner ones fertile. ($\times 2$)*

representatives in Ethiopia and India, and one species in South Africa. Most species bear loose racemes of predominantly yellow flowers but sometimes with different colored markings as in *C. cretica* [CRETAN MULLEIN], which has two brownish spots, or *C. arcturus* [CRETAN BEAR'S TAIL] cultivar 'Linnaeana' with a purple throat. These and a number of other species make popular bedding plants. (See also MULLEINS.)
SCROPHULARIACEAE, 35–40 species.

Celtis a genus of mainly deciduous trees, sometimes shrubs, with decorative leaves, mainly native to southern Europe and the southern USA, with a few species in the tropics.

C. australis (= *C. lutea*), the EUROPEAN HACKBERRY or NETTLE TREE, grows to about 20m tall and produces purple, edible fruits, the hackberries. It yields yellow-brown to gray timber which is hard but easily bent and is used for manufacturing walking sticks and fishing rods. The North American *C. occidentalis* [COMMON HACKBERRY, SUGAR BERRY] is a much larger tree reaching 35m whose timber is used commercially. *C. laevigata* (= *C. mississippiensis*) [MISSISSIPPI HACKBERRY or MISSISSIPPI SUGARBERRY] is sometimes grown as a specimen ornamental tree.

Other species yielding usable timber of medium to low quality, which is used locally, are *C. brasiliensis* from Brazil, two African species, *C. africana* (= *C. kraussiana*) [STINKWOOD] and *C. mildbraedii* and, from New Guinea and the Philippines, *C. philippensis*. *C. iguanaea* [GRANJENO] from tropical Central America, and *C. sinensis* [CHINESE HACKBERRY] from eastern China, both have edible fruits. *C. cinnamomea* from India and Indonesia has strongly scented wood which is powdered and known as "kajoo lahi". This is used medicinally as a blood purifier and tonic.
ULMACEAE, about 80 species.

The crested flower head of the Cockscomb (Celosia cristata), a popular summer bedding plant which is native to the warm tropics. ($\times \frac{1}{2}$)

Centaurea a genus of annual and perennial herbs, rarely dwarf shrubs, of usually dry open habitats, mostly in the Mediterranean region and western Asia but extending to northern Europe, tropical Africa, North America and, with only a few species, to the central Andes.

Some of the species are grown as garden ornamentals. *C. cyanus* [CORNFLOWER, BATCHELOR'S BUTTON, BLUEBOTTLE], an annual showing a great variety of color forms, from white to blue and deep pink, is widely cultivated for ornament, as is the perennial *C. montana* [MOUNTAIN BLUET] which has white to blue, pink, or yellow flowers. *C. moschata* [SWEET SULTAN], an eastern Mediterranean annual with pink, purple, yellow or white, fragrant flowers, is also cultivated. *C. americana*, sometimes called BASKET FLOWER, is better-known as the AMERICAN CORNFLOWER. Species such as *C. cineraria* [DUSTY-MILLER] are grown for their white-tomentose foliage.

Several species, such as the perennial *C. nigra* [KNAPWEED] and *C. scabiosa* [GREATER KNAPWEED] are weeds of cultivation and have been widely introduced into temperate regions throughout the world.
COMPOSITAE, about 500 species.

Centranthus [VALERIANS] a genus of annual and perennial herbs distributed in the countries bordering the Mediterranean and extending to the Canary Islands and the Caucasus. *C. ruber* [RED VALERIAN, SPUR VALERIAN, JUPITER'S BEARD] is a perennial grown for ornament for its red or pink star-shaped flowers (white in cultivar 'Albus'). Its leaves and roots are used as food in some parts of Europe. *C. macrosiphon* is an annual species which is sometimes grown in rockeries and contains a range of deep-rose

or white-flowered varieties and cultivars.
VALERIANACEAE, 9 species.

Cephaelis see Ipecacuanha.
RUBIACEAE, about 180 species.

Cephalanthera [HELLEBORINES] a genus of terrestrial orchids distributed throughout the temperate regions of the Northern Hemisphere. The Old World species, such as the white flowered *C. damasonium* (= *C. grandiflora*) [WHITE HELLEBORINE] occur in most types of habitat except the very wettest and most acidic. Flower color varies from purest white to deep pink as in *C. rubra* [RED HELLEBORINE].
ORCHIDACEAE, about 14 species.

Cephalaria a genus of annual and perennial herbs mainly native to Central Asia, North and South Africa and the Mediterranean region. Some species, but most notably *C. gigantea* (= *C. tatarica* of horticulture), which has globose cream or yellow flower heads, are cultivated as ornamentals.
DIPSACACEAE, about 60 species.

Cephalotaxus [PLUM YEWS] a small Asian genus of evergreen shrubs and small trees. *C. fortuni* [CHINESE PLUM YEW] is a small tree commonly cultivated for its evergreen foliage and reddish-brown bark which peels away in flakes. *C. harringtonia* var *drupacea* [JAPANESE PLUM YEW] is also cultivated.
CEPHALOTAXACEAE, about 5 species.

Cephalotus a genus consisting of a single species, the carnivorous *C. follicularis* [AUSTRALIAN FLYCATCHER], which is also the only member of its family. A native of the drier parts of peaty swamps in Western Australia, it is a perennial herb with both normal foliage leaves and some modified to form a pitcher about 5cm long. This is decorated with fringed ribs up the sides, has a ribbed neck and is crowned by a lid. Secretory

The Sword-leaved Helleborine (Cephalanthera longifolia) is nowhere common but is widely distributed in Europe, North Africa and eastwards to Japan. ($\times \frac{1}{2}$)

glands lining the inner surface of the pitcher produce digestive juices. Nitrogenous materials provided by the digestion and bacterial decay of trapped insects supplement the normal nutrition of the plant. The AUSTRALIAN FLYCATCHER is occasionally grown in greenhouses as a botanical curiosity.
CEPHALOTACEAE, 1 species.

Ceramium a genus of red algae with a worldwide distribution. The plant has a cylindrical banded thallus. *C. hypnoides* is commercially important in Japan as a source of agar.
RHODOPHYCEAE.

Cerastium [MOUSE-EARS] a genus of hardy annuals and herbaceous perennials, cosmopolitan in distribution. Some are weeds of cultivation and a few are grown in gardens as ornamental plants. They are usually low-growing and mat-forming in habit.

C. vulgatum [MOUSE-EAR CHICKWEED] is a weed of cultivation the world over.

Of the cultivated garden species *C. biebersteinii* and *C. tomentosum* [SNOW-IN-SUMMER] are invasive rhizomatous plants suitable for covering rough areas with their silvery foliage. Some of the European alpine species, such as *C. alpinum* [ALPINE MOUSE-EAR] and *C. latifolium,* forming compact cushiony mats, are also grown in rock gardens.
CARYOPHYLLACEAE, about 60 species.

Ceratodon a genus of low-growing, cushion or turf-forming mosses best known through the cosmopolitan species *C. purpureus* which is abundant on heaths and in various other habitats. It takes its name from the shining purple-red capsule

Pitcher of the Australian Flycatcher (Cephalotus follicularis), a carnivorous plant which supplements its "diet" with trapped insects. (× 1½)

Snow-in-summer (Cerastium tomentosum) is frequently grown in gardens for its rapid growth and abundance of white flowers. (× ½)

stalks (setae) which make it a conspicuous moss when fertile.
DICRANACEAE, about 50 species.

Ceratophyllum a cosmopolitan genus of rootless, submerged, herbaceous water plants. The species form floating masses that can form very rapidly and choke waterways. Because *Ceratophyllum* forms mats just below the water surface they protect fish fry and can inhibit the development of mosquito larvae.
CERATOPHYLLACEAE, about 2 species.

Ceratostigma a small genus of perennial herbs and small shrubs from the Himalayas, China and Africa. Three blue-flowered species of ornamental small bushes are commonly cultivated: *C. plumbaginoides* and *C. willmottianum* (western China, Tibet), and *C. griffithii* (Tibet, Bhutan).
PLUMBAGINACEAE, about 8 species.

Cercidiphyllum a genus represented by a single species, *C. japonicum* [KATSURA TREE], a deciduous, ornamental tree native to China and Japan. It grows up to 30m in its native woodland habitat. Trees in cultivation develop a spreading graceful form, and in the autumn, under favorable conditions, the heart-shaped leaves turn gold, yellow and red. In Japan the light, fine-grained wood is used for cabinet-making and interior woodwork in houses.
CERCIDIPHYLLACEAE, 1 species.

Cercis a small north temperate genus of deciduous shrubs and trees, native to North America, southern Europe and China. The best-known is *C. siliquastrum* [JUDAS TREE], so called because Judas is said to have hanged himself on one (which explains the frequent occurrence of this low and irregular spread-

ing tree beside churches and graveyards).

C. canadensis [RED BUD], from North America, is another ornamental species, whose rose-purple pea-like flowers appear before the distinctive bright green leaves; a white-flowered variety is also known in cultivation. *C. racemosa*, from China, is also cultivated for its ornamental value, ultimately producing pendulous racemes of pink flowers.
LEGUMINOSAE, about 7 species.

Cereals see p. 81.

Cereus [TORCH THISTLES] a genus of tree cacti characterized by spiny ribs and a much-branched candelabra habit, from the West Indies through tropical South America to Argentina. Most of the cultivated *Cereus* species cannot be positively assigned to any one species and are probably of hybrid origin. They are among the toughest, most adaptable and long-lived of tree cacti and are widely planted in gardens in frost-free parts of the world. Their red, fleshy fruits are edible.
CACTACEAE, about 25 species.

Ceropegia a genus of upright or twining herbs and subshrubs from the Canary Islands, tropical and southern Africa, Madagascar and Arabia to New Guinea and northern Queensland. The species vary greatly in their habit (leaves or stems are often succulent) but have distinctive waxy tubular flowers.

One of the popular cultivated species is *C. radicans* (South Africa) with succulent leaves and stems, and large, showy flowers up to 10cm long, zoned in purple, white and green. Other cultivated species include *C. rendallii* (Transvaal) with dainty purple and whitish green flowers on slender stems and *C. elegans* (India) with slender twining stems and purple flowers edged with long hairs. Six species grow in the Canary Islands, notably *C. fusca* and *C. dichotoma.*
ASCLEPIADACEAE, about 160 species.

Cereals

The cereals are a group of annual grasses grown primarily for their large swollen grains. They provide the main concentrated carbohydrate food for the peoples of moist temperate areas and many parts of the tropics. Although cereals are basically "energy" foods, they also supply a large part of the protein needs of peoples in the poorer areas of the world. They also provide feed for livestock. Altogether they provide sustenance for more than 2 000 000 000 people and their production runs to well over 1 000 000 000 tonnes annually. Over half the cultivated land of the world is devoted to cereals.

The true cereals are all members of the grass family (Gramineae). The main cereals are WHEAT (*Triticum*), *MAIZE or CORN (*Zea*), *RICE (*Oryza*), *BARLEY (*Hordeum*), *OATS (*Avena*), *RYE (*Secale*), *SORGHUM (*Sorghum*) and *MILLETS (a group of diverse small-seeded grasses such as species of *Setaria, *Eleusine, *Pennisetum, *Panicum, Digitaria, etc). The first three are the most important and most extensively grown – corn (maize) in the New World, wheat in Europe and western Asia and rice in the tropics of the Orient. The other cereals are, however, of great importance and in Africa, where cereal production is low compared with other areas, many of the minor cereals are cultivated.

Man's evolution has been closely linked with the origin and cultivation of cereals. Early agriculture involved the gathering,

cultivation and domestication of cereal grains and from that followed the development of modern civilizations, all of which have been dependent on one or more grain crops as the staple food. The word cereal derives from Ceres – the goddess of grain of the Ancient Greeks.

All the cereals can be used for brewing although barley is probably the most important, as the basic constituent of most beer. They may also be used for the production of spirits such as whisky, gin and vodka.

Cereal grains constitute a concentrated source of carbohydrate plus some protein, oil and vitamins. They can moreover be easily handled, stored and stockpiled over a long period, unlike starchy staples with a higher water content such as *POTATOES, *YAMS, *CASSAVA (MANIOC) and other vegetables which are bulky, require special handling and can only be stored for limited periods.

Structurally, the grain is a fruit (caryopsis) developed from a single-ovuled ovary in which the fruit wall and the seed coats are completely fused. In addition to the store of carbohydrate in the form of starch, the embryo (germ) and seed coat are rich in protein, oil and vitamins, but in the case of wheat these are lost in the milling for white flour. The major change that accompanied domestication of each of the cereals was the appearance of types with tough non-shattering heads (ears). The immediate wild progenitors, while possessing fairly large seed, had heads that shattered on maturity. Although this facilitated seed dispersal in the wild it was a definite hindrance to domestication inasmuch as grain had to be harvested before it was quite ripe. Conversion to non-shattering involves a very simple genetic

change and mutants of this type would have been unconsciously selected for and would have become fixed quite quickly.

Current and future trends in breeding aim at further increase in yield linked to a more general base for disease resistance. Improvement in both the quality and quantity of protein, especially in maize and feeding barley, is also receiving high priority.

A group of crop plants known as the pseudocereals are sometimes listed along with the cereals. These include *BUCKWHEAT (*Fagopyrum*), *GRAIN AMARANTHS (*Amaranthus*) and some other seeds which somewhat resemble cereal grains although in fact with quite a different structure.

Important cereal crops (a) Bread Wheat; (b) Hard (Durum) Wheat; (c) Rye; (d) Oats; (e) Six-rowed Barley; (f) Maize; (g) Rice; (h) Sorghum; (i) Finger Millet; (j) Common Millet (k) Foxtail Millet. (a), (b), (c), (e)×1; (d), (h), (i), (k)×½; (g), (j)×⅓; (f)×¼.

THE MAIN CEREAL CROPS

Common name	Scientific name
WHEAT	*Triticum*
WILD EMMER	T. diococcoides
CULTIVATED EMMER	T. diococcum
EINKORNS	T. monococcum var monococcum, T. m. var boeoticum
HARD (DURUM)	T. durum
TURGIDUM	T. turgidum
BREAD	T. aestivum var aestivum
SPELT	T. spelta
CLUB	T. compactum
BARLEY	Hordeum vulgare
TWO-ROWED	H. distichum
SIX-ROWED	H. hexastichum
*RYE	Secale cereale
*MAIZE	Zea mays
*RICE	Oryza sativa
AFRICAN RICE	O. glaberrima
*OATS	Avena
HEXAPLOID	A. sativa, A. byzantina, A. nuda
TETRAPLOID	A. abyssinica
DIPLOID	A. strigosa, A. brevis
*SORGHUM	Sorghum bicolor
*MILLETS	
FINGER or AFRICAN	Eleusine coracana
BULRUSH, PEARL, BAJRA	Pennisetum americanum
COMMON, PROSO	Panicum miliaceum
JAPANESE BARNYARD, SANWA	Echinochloa frumentacea
FOXTAIL, GERMAN, ITALIAN	Setaria italica
TEFF	Eragrostis tef
FONIO, FUNDI	Digitaria species
KODA, KODO	Paspalum scrobiculatum

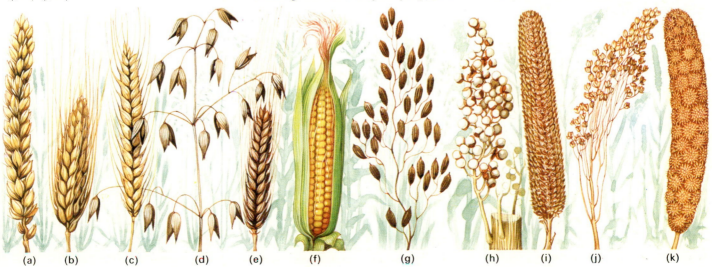

(a)   (b)   (c)   (d)   (e)   (f)   (g)   (h)   (i)   (j)   (k)

Half section of the flower of Ceropegia woodii, which is pollinated by insects trapped within the corolla tube by downward pointing hairs, which later wither to allow escape. (× 2)

Ceroxylon a genus of South American palms that includes the tallest palm species known. The Colombian *C. alpinum* (= *C. andicola* [WAX PALM] produces a wax on the trunk and leaves; this is peeled off and melted and used in the manufacture of candles and wax matches. A similar product is obtained from *C. klopstockiae*.
PALMAE, about 20 species.

Cestrum [BASTARD JASMINES] a genus of shrubs and small trees from tropical Central and South America which include some brilliant ornamentals. These have clusters of white, yellow or red trumpet-like flowers, eg *C. parqui* [WILLOW-LEAVED JASMINE] with yellowish flowers, *C. purpureum* and *C. fasci-culatum* with red flowers and *C. nocturnum* [LADY OF THE NIGHT, NIGHT-BLOOMING JESSAMINE] with creamy-white flowers which are very fragrant by night.
SOLANACEAE, about 150 species.

Cetraria a genus of leafy or shrub-like lichens, mostly arctic in distribution, grow-ing on the ground or on trees. *C. islandica* [ICELAND MOSS] is frequently eaten by rein-deer and caribou. *C. pinastri* [PINE LICHEN] and *C. nivalis* [SNOW LICHEN] have been used in some parts of Europe as sources of a wool dye.
PARMELIACEAE, about 40 Species.

Chaenomeles [FLOWERING QUINCE, JAPANESE QUINCE, JAPONICA] a small genus of spring-flowering shrubs native to China and Japan, several varieties of which are cultivated as ornamentals. The genus is closely related to *Cydonia* [COMMON QUINCE]. The white, yellow, pink or red flowers are typically

rosaceous and borne in clusters, often with very short stalks.

Two of the three species, *C. speciosa* (= *C. lagenaria*) [JAPANESE QUINCE, JAPONICA] and *C. japonica* [MAULE'S QUINCE], contain numerous varieties which are widely culti-vated for their attractive habit and flowers. The fruits of all species are edible and can be used for jams or jellies. *C. sinensis* is now placed in the genus *Pseudocydonia*.
ROSACEAE, 3 species.

Chaerophylum [CHERVILS] a north tem-perate genus of annual and biennial, glabrous or hairy herbs. *Chaerophyllum* is commonly called CHERVIL but should not be confused with another genus of the same family, *Anthriscus*, to which the SALAD CHERVIL belongs.

C. bulbosum [TURNIP-ROOTED CHERVIL or BULBOUS CHERVIL] from southern Europe, is cultivated on a small scale in Europe for its edible root. The base of the taproot swells like a small gray carrot; the flesh is whitish-yellow and tastes sweet when cooked.
UMBELLIFERAE, about 40 species.

Chaetomium a cosmopolitan genus of saprophytic fungi common on cellulose-rich substrates such as cotton textiles (*C. globosum*), sacking, straw (*C. elatum*) or

A Japanese Quince (Chaenomeles speciosa) in flower against a wall. Quinces are popular ornamentals, both for their flowers and edible fruits.

dung. *Chaetomium* species also cause soft rot in wood.
CHAETOMIACEAE, about 50 species.

Chamaecyparis [FALSE CYPRESSES, BASTARD CYPRESS, WHITE CEDAR] a small genus of valuable evergreen conifers of the Northern Hemisphere. They are found mainly in the western and southeastern coastal regions of North America and in Japan and Taiwan. They are hardy, mostly pyramidal trees with a habit very similar to that of the genus *Cupressus*.

The timber of most of the species of *Chamaecyparis* is of high quality, being generally light, durable, easily worked and resistant to fungus decay and insect attacks. The wood of most species has its own pleasant distinctive odor and color. *C. formos-ensis* is one of the most valued timber trees in Taiwan, while that of the North American *C. lawsoniana* [LAWSON CYPRESS] is no less useful and has an odor that can be described as "spicy". It is used for general building, for floors, furniture and fence posts, railway sleepers and in boat-building. The wood of *C. nootkatensis* [NOOTKA CYPRESS, YELLOW CYPRESS, ALASKA CEDAR] is also of excellent quality and is used in much the same way. It is known in the trade as "YELLOW CYPRESS", though this common name is also used for the SWAMP CYPRESS (*Taxodium distichum*). The excellent wood of *C. obtusa* [HINOKI CYPRESS] is much prized in its native Japan and is probably unsurpassed for the highest

Bursting capsules of the Rose Bay Willowherb (Chamaenerion angustifolium). (×4)

quality work in all kinds of construction. It is very straight, evenly grained and often beautifully marked.

Outside their native habitats, many species of *Chamaecyparis* are planted for ornamental purposes, especially *C. lawsoniana, C. obtusa* (numerous cultivars), *C. pisifera* [SAWARRA CYPRESS] (but mainly its cultivars), *C. noot-katensis* and *C. thyoides* [WHITE CEDAR, WHITE CYPRESS]. (See also *Cupressocyparis.*)
CUPRESSACEAE, 7 species.

Chamaedorea a genus of mostly small palms found mainly in the undergrowth of lowland tropical rain forest in Central and South America. Many form clumps by suckering and a few are weak scramblers. The fruits are fleshy and red, orange or black, often with irritant crystals but those of *C. elegans* are edible, as are the young flower clusters of *C. graminifolia* and *C. sartorii.* Being highly ornamental with bamboo-like stems, they are widely cultivated, especially as houseplants, eg *C. elegans* [PARLOR PALM]. PALMAE, about 100 species.

Chamaemelum see CHAMOMILE.
COMPOSITAE, 2–3 species.

Chamaenerion a genus of herbaceous perennials, native to the Arctic and north temperate regions. *C. angustifolium* [ROSEBAY WILLOWHERB, FIREWEED], sometimes known as *Epilobium angustifolium*, is abundant on newly burnt areas. It is a most attractive tall perennial herb, growing to 1.5m tall with numerous bright rose-purple flowers in a dense spike-like raceme.
ONAGRACEAE, about 10 species.

Chamaerops a genus represented by a single species, *C. humilis*, one of the two native palms of Europe. It is a fairly hardy, low, bushy fan palm, suckering from the base, and grows wild in coastal sandy or rocky areas of the western and central Mediterranean region. The leaves provide a fiber which is used for cordage and the young leaf buds are sometimes used as a vegetable.
PALMAE, 1 species.

Chamomile the common name for *Chamaemelum nobile* (= *Anthemis nobilis*), sometimes also called LAWN or SWEET CHAMOMILE, a prostrate or low-growing evergreen or semi-evergreen perennial herb native to Europe. The yellow and white daisy-like flowers of CHAMOMILE were at one time used as a tonic and medicinal herb regarded as a general restorative for all manner of ailments. Chamomile tea is still drunk as a tonic. Since medieval times, particularly in herb gardens, the plant has been used for making scented lawns. *Chamomilla recutita* (= *Matricaria chamomilla*) is known as WILD CHAMOMILE and *Anthemis cotula*, with a somewhat sickly smell, is commonly known as STINKING CHAMOMILE.
COMPOSITAE.

Chamomilla a genus of Eurasian annual herbs (formerly called *Matricaria*) with finely dissected leaves and daisy-like flower heads, as in *C. recutita* [WILD CHAMOMILE] or with flower heads in which all the flowers are tubular and greenish or yellow as in *C. aurea. C. recutita* (widely known as *M. chamomilla*) is cultivated in both the Old and New World as a medicinal herb which is used as a stimulant, antispasmodic and carminative. It produces oil of chamomile, used in shampoos and as a flavoring in liqueurs.
COMPOSITAE, 5 species.

Charlock (Sinapis arvensis) *is similar in appearance to the closely related* Brassica *species grown for oilseed and fodder. (×⅛)*

Fruits of the Chayote (Sechium edule), *a tropical vine originating in Central America. They are boiled and eaten as a vegetable. (×⅓)*

Chards a vegetable consisting of the stems or leaf midrib of a number of species. SWISS CHARDS, also known as SEAKALE BEET, are the large fleshy midrib and petiole of a form of *Beta vulgaris.* The blanched shoots of the GLOBE *ARTICHOKE and of *SALSIFY are also eaten as CHARDS.

Charlock the common name of *Sinapis arvensis* which before the development of chemical herbicides such as 2, 4-D was one of the most serious annual weeds of European arable land, particularly of spring-sown crops, and it was common to see wheat fields bright yellow with its flowers. It is native to north temperate regions but is now virtually cosmopolitan in distribution. Seeds are produced in abundance and retain their viability for at least 10 years, and ploughing-in will therefore not eradicate the weed. The seeds even resist the digestive enzymes of birds and animals, and they have been germinated from pigeon droppings and cow dung.
CRUCIFERAE.

Chaulmoogra or **hydnocarpus oil** an oil obtained from the seeds of *Hydnocarpus kurzii* (= *Taraktogenos kurzii*), a medium-sized evergreen tree of East Bengal, Burma and Thailand, which has been used in Hindu medicine for many centuries to treat leprosy, psoriasis and rheumatism. The seed residue after extraction of the oil is a valuable fertilizer. An oil with similar properties is extracted from seeds of *H. wightianus*, an evergreen tree of southwest India.
FLACOURTIACEAE.

Chayote, christophine or **choco** common names for the herbaceous perennial vine

Wallflowers (Cheiranthus cheiri) growing at the foot of a wall. In their native Greece and the Aegean islands they grow in rock crevices. (× ¼)

Sechium edule native to southern Mexico and Central America but now grown in many tropical countries. Almost all parts of the plant are edible: the greenish-white, pear-shaped fleshy fruits and the tuberous roots are boiled and eaten as a vegetable, the young shoots used somewhat like asparagus, and the seeds cooked in butter are considered a delicacy. In Central America, the fruits are also made into a kind of sweetmeat. The leaves are used as fodder, and the stems for plaiting.
CUCURBITACEAE.

Cheilanthes   a genus of tropical and temperate ferns, the majority of which grow in dry rocky places. They are often covered with woolly hairs; the sori occur at the ends of the veins and are protected by the reflexed leaf margins.
POLYPODIACEAE, about 180 species.

Cheiranthus   a small genus of perennial herbs and subshrubs native to the Mediterranean region. The most commonly cultivated and best-known species is *C. cheiri* [WALLFLOWER], but other ornamentals include *C. scoparius* [CANARY ISLANDS WALL-FLOWER] from high mountain regions and *C. × kewensis*, an attractive winter-flowering plant with brownish-orange flowers, turning purple. The latter is a hybrid derived from a cross between a red-flowered *C. cheiri* and another hybrid, *C. semperflorens × C. cheiri* which is yellow-flowered. The genus is closely related to *Erysimum*, some species of which are also commonly called WALLFLOWERS.
CRUCIFERAE, 10 species.

Chelone [TURTLE HEAD]   a small genus of North American herbaceous perennials.

Some species are cultivated as garden-border plants, including *C. lyonii* and *C. glabra* with terminal inflorescences of pink, rose or purple snapdragon-like flowers.
SCROPHULARIACEAE, 4 species.

Chenopodium   a genus of temperate herbs and small shrubs, rarely arborescent plants, which are often mealy or glandular. *C. album* [GOOSEFOOT, FAT HEN, LAMB'S QUARTERS] is the commonest species, occurring in Europe, Asia, America, Africa and Australia. An essential oil is obtained from *C. ambrosioides* [AMERICAN WORMSEED, MEXICAN TEA] and is used as a treatment for intestinal worms in America. *C. bonus-henricus* [ALL-GOOD, GOOD KING HENRY], a perennial herb native to Europe and naturalized in eastern North America, and *C. quinoa* [QUINOA] are among the edible species; the leaves of both are eaten like spinach and the seeds of the latter can be

A close-up of a variety of Goosefat or Fat Hen (Chenopodium album) that has distinctly striped stems and leaf-stalks. (× 5)

boiled like rice, and are the staple "cereal" in the Andean highlands. The seeds of the American *C. leptophyllum*, and *C. nuttaliae* are also eaten.
   Attractive ornamental species include the annuals *C. purpurascens* and *C. amaranticolor* with rose- or violet-purple-tinged leaves.
CHENOPODIACEAE, about 150 species.

Cherry   the commercially important fruit of various species of the genus *Prunus*. They are an important and long-established fruit crop. Cherry trees are also cultivated as ornamentals for their blossom.   The majority of cherry species evolved in Central Asia, spreading to China and Japan (where flowering cherry trees are especially prized for their attractive blossoms and other ornamental features) and to North America. The wild *P. avium* [MAZARD CHERRY, SWEET CHERRY] is the ancestor of the commercial sweet cherries. The noncommercial *P. fruticosa* [EUROPEAN DWARF CHERRY] also spread from Central Asia to Europe. Natural hybridization between *P. avium* and *P. fruticosa* produced *P. cerasus* [SOUR CHERRY], now native over most of Europe, from which have been derived the commercially important sour (acid) cherries.
   Fruiting cherries are often classified by type rather than by variety name because of the small differences between varieties. Cherries are used in pies, various other sweets, jams, confectionery and in various alcoholic drinks, most notably the Danish cherry brandy. Marascos [*MARASCHINO CHERRIES] yield a juice which is an ingredient of the Maraschino liqueur. Oil derived from cherry kernels is used in the cosmetics industry. The wood of *P. cerasus* is also valued for furniture and instrument making.
ROSACEAE.

Chestnut   the common name for species of *Castanea*, a genus of deciduous trees and shrubs from the Northern Hemisphere. *C. sativa* [SPANISH or SWEET CHESTNUT] is a native of southern Europe, North Africa and southwestern Asia. The trees are fast-growing, long-lived and can attain great size, branching low and spreading horizontally.
   In southern Europe chestnuts form an important article of food, and can also be variously processed to produce such diverse delicacies as marrons glacés and chestnut stuffing for turkey. The timber of young trees is used for hop-poles and hoops for barrels, and the bark is used in tanning.
   Other species of *Castanea* also supply timber which is often used for railway sleepers; many of these species also produce edible nuts. Best-known are *C. dentata* [AMERICAN CHESTNUT] from eastern North America, *C. crenata* [JAPANESE CHESTNUT], two species from China, *C. henryi* and *C. mollissima* [CHINESE CHESTNUT], and *C. pumila* [CHINQUAPIN, DWARF CHESTNUT] from the USA. Some of these species are also among the most popular ornamental trees of the genus.
   The well-known HORSE CHESTNUT belongs to the unrelated genus *Aesculus*.

Cherry blossom (Prunus sp). Many cherries, notably those of Chinese and Japanese origin, are grown as ornamentals for their blossom.

Chick pea, **grain pea** or **garbanzo** common names for *Cicer arietinum*, a small herbaceous annual plant cultivated for its seed, which is mainly used for food, but also as forage, principally in southern Europe, India, Pakistan and tropical America. The seeds are rich in protein and may be eaten raw or boiled, or they can be converted into flour, or used as a coffee substitute.
LEGUMINOSAE.

Chickweed the common name applied to many weeds of the Caryophyllaceae family, but principally to *Stellaria media* [COMMON CHICKWEED], a rapidly spreading annual herb abundant on cultivated land. The name also applies to several other genera of similar habit, notably *Cerastium*, eg *C. arvense* [FIELD

Sweet Chestnuts (Castanea sativa) *showing the protective spiny capsule split open to reveal the edible seeds inside.* ($\times 1\frac{1}{2}$)

MOUSE-EAR CHICKWEED], *Holosteum*, eg *H. umbellatum* [UMBELLATE CHICKWEED] and *Myosoton*, eg *M. aquaticum* [WATER CHICKWEED].
CARYOPHYLLACEAE.

Chicle the latex or gum derived from *Manilkara zapota* (= *Achras zapota*) [*SAPODILLA], which is native to Central America. Chicle was the basic ingredient in the manufacture of the original chewing gum and has always been harvested by tapping wild trees which are widely scattered throughout the dense forests of Latin America. Because of this irregularity of supply, alternatives were sought and nowadays most chewing gum bought in the shops contains synthetic chicle and other substitute natural gums.
SAPOTACEAE.

Chicory the common name for *Cichorium intybus*, a herbaceous perennial that is used both as a vegetable and as a coffee flavoring or substitute. Some confusion arises from one of its French names, ENDIVE, for in English this describes the related species *C. endivia*.

CHICORY is a native of Europe and is cultivated mainly in the Netherlands, Belgium, France, Germany, and in the USA, where it is now naturalized as a weed. The long fleshy taproot may be boiled and eaten as a vegetable, and it is dried, roasted and ground for use as a coffee substitute or as an additive, giving it color, body and a characteristic bitter flavor. The blanched leaves are sometimes cooked, but are more often used as a winter salad vegetable. Witloof is the name applied to the hearts of blanched leaves forced in winter or spring from stored roots.
COMPOSITAE.

Chilies or **Chili peppers** the fresh or dried fruits of pungent forms of *Capsicum annuum* and *C. frutescens*. South American types include the fiery little CHILE PEQUIN, the relatively mild CHILE ANCHO used in tamales, tacos and to make chilli powder, and the brown-fruited CHILE MULATO used, with chocolate, in mole sauce for poultry, etc.
SOLANACEAE.

Chimonanthus [WINTERSWEET] a small Chinese genus of deciduous and evergreen shrubs the best-known of which is *C. praecox*, (= *C. fragrans*), a deciduous twiggy shrub with shiny willow-like leaves, cultivated for its heavily scented fragrant yellow and purple flowers.
CALYCANTHACEAE, 3–4 species.

China jute a fiber obtained from an annual herbaceous plant, *Abutilon theophrasti* (= *A. avicennae*) [CHINA JUTE, INDIAN MALLOW], which is extensively cultivated in China, and in more recent years in parts of the USSR, for its stem fibers. It is native to tropical Asia but is now naturalized as a weed in the USA.

China jute is used in the same way as ordinary *jute (obtained from two species of *Corchorus*, *C. capsularis* and *C. olitorius*). There is every indication that China jute will

increase in importance, largely because it is hardier and more resistant to disease than *Corchorus*, while the fiber is closely similar in quality.

China jute is very tough and strong and readily takes up dyes. It is used for caulking boats, but the main use is for rough weaving such as sacks and cloth. In China it is used extensively in carpet and rug making, while the shorter fibers are used in the manufacture of paper.
MALVACEAE.

Chinese cabbage a name applied to a number of species of *Brassica* (*rapa* group), of which *B. pekinensis* [PE-TSAI], native to East Asia, is well known as a salad or vegetable plant and has been successfully introduced in recent years to Central America, West Africa, the USA and Europe. Another well-known CHINESE CABBAGE is *B. chinensis* [PAK-CHOI]. *B. perviridis* [TENDERGREEN] is cultivated in Japan and the southern parts of the USA.
CRUCIFERAE.

A Chicory plant (Cichorium intybus) *in flower at the edge of an arable field. Farmers sometimes sow this plant for cattle feed.* ($\times \frac{1}{8}$)

Chionodoxa [GLORY-OF-THE-SNOW] a genus of small, bulbous, spring-flowering perennial herbs found in the eastern Mediterranean region. Several species are grown as decorative plants in gardens and are frequently naturalized in grass and among deciduous shrubs. *C. luciliae* is the most commonly grown species; the normal form has blue flowers with white centers but white and pink forms are known.
LILIACEAE, 6 species.

Chives the common name for *Allium schoenoprasum*, a wild perennial herb, found throughout the north temperate zone, which has probably been domesticated on many separate occasions. CHIVES grow vigorously in spring and summer and their chopped

leaves are used for their onion-like flavor in salads and garnishes.
LILIACEAE.

Chlamydomonas a large genus of biflagellate single-celled green algae.
CHLOROPHYCEAE, about 350 species.

Chloris a genus of tropical grasses from Africa, America and Australia. Most are tufted with stems usually less than 1m tall. The most important species economically is *C. gayana* [RHODES GRASS]. Originating in Africa, it is now grown in many parts of the tropical world for pasture, hay and silage.
GRAMINEAE, about 50–70 species.

Chlorococcum a genus of unicellular green algae found *en masse* as thin green incrustations on damp soil, brickwork, stones, walls or perhaps, most familiarly, on the windward side of trees.
CHLOROCOCCACEAE.

This powdery green mass on the damp bark of a sycamore tree comprises thousands of unicellular green algae of the genus Chlorococcum. (× 3)

Chondrus a genus of red algae generally found growing on the lower part of rocky shores. It has a characteristic flattened frond which branches dichotomously. (See also Carragheen.)
RHODOPHYCEAE.

Chorizema an Australian genus of small evergreen shrubs with showy orange-red, scarlet or purplish, pea-like flowers. A few species are cultivated for ornament, especially *C. ilicifolium* and *C. cordatum* [FLAME PEA].
LEGUMINOSAE, 16 species.

Christensenia a tropical East Asian genus of ferns containing a single, very variable species, *C. aesculifolia*, found mainly on shady river banks in deep lowland or mid-mountain tropical forest. The leaves resemble those of *Aesculus* [HORSE CHESTNUTS], hence the specific name.
MARATTIACEAE, 1 species.

Christmas rose the common name for *Helleborus niger*. It is so called because it is a very early flowering perennial, its white or cream saucer-shaped flowers (rose-tinged in var *altifolius*) appearing in late December and January.
RANUNCULACEAE.

Christmas tree a name conventionally applied to any tree brought into the house at Christmas. There is no tradition as to a particular species, but in Europe the most widely esteemed is the NORWAY *SPRUCE (*Picea abies*). Other species of *Picea* and some PINES may be chosen, especially in North America. In New Zealand species of *Metrosideros* are also known as CHRISTMAS TREES.

Chrozophora a genus of branching and spreading herbs and undershrubs, especially characteristic of the Mediterranean region but also of tropical Africa and India. *C. tinctoria*, of Mediterranean coasts, is the source of the dye tournesol, or bezetta rubra, used in coloring wine, linen and cheeses.
EUPHORBIACEAE, 12 species.

Chrysalidocarpus a small genus of graceful feather palms mainly native to Madagascar but with some species widely cultivated for their ornamental value. They include *C. lucubensis*, *C. madagascariensis* and, perhaps the most attractive, *C. lutescens* [BUTTERFLY PALM]. The young terminal buds of *C. fibrosus* are used as a vegetable in Madagascar and the leaves provide a fiber used for fishing nets.
PALMAE, about 20 species.

Chrysanthemum a name often used in a wide sense to include some 200 species of

A heavy crop of cider apples at a cider mill in Taunton, Somerset, England. Before processing they are washed, here in the car park.

The term Chinese Cabbage is applied to several species of Brassica. This is B. chinensis (Pak-choi). (× ¼)

Chlorophytum a genus of evergreen grass-like plants widely distributed in the tropics and subtropics. The only cultivated species is *C. comosum* [SPIDER PLANT], often grown as a houseplant in temperate regions. It produces narrow, often white-striped, linear leaves.
LILIACEAE about 215 species.

Chloroxylon a genus native to southwest India and Sri Lanka represented by a single species, *C. swietenia* [SATIN WOOD], a large deciduous tree which yields a valuable timber, particularly useful in cabinet-making, furniture and paneling.
MELIACEAE, 1 species.

Choisya a genus of seven species native to Mexico and the southwest USA. The best-known species is *C. ternata* [MEXICAN ORANGE], an evergreen, aromatic shrub which has glossy leaves and clusters of white fragrant flowers in April and May. The plant is grown as an ornamental in temperate countries, particularly in sheltered sunny courtyards and against walls.
RUTACEAE, 7 species.

herbaceous, occasionally shrubby, plants from several parts of the world. However, the genus *Chrysanthemum* has now been divided up into numerous genera, viz. *Chrysanthemum*, *Tanacetum* (which includes *Pyrethrum*), *Leucanthemum*, *Balsamita*, *Dendranthema*, *Argyranthemum* and several smaller genera. Examples of well-known species are the annual garden chrysanthemums *C. carinatum* and *C. coronarium*, *T. coccineum* [GARDEN PYRETHRUM], *T. cinerariifolium*, which is the source of the insecticide pyrethrum, *L. vulgare* and *L. maximum* [the OXEYE and SHASTA daisies], and *A. frutescens* [PARIS DAISY].

By far the best-known are the autumn-flowering chrysanthemums derived from the Asian species *D. morifolium*, a complex hybrid of Japanese and Chinese parentage, including *D. sinense* and *D. indicum*. The numerous types of modification of the florets and types of branching habit of the plant have given rise to particular groupings. For instance broad-petaled ray florets of ligulate shape have given rise to the large ball-like flowers with incurved or recurved petals, which are still the main interest of amateurs growing prize specimens. Numerous other shapes and sizes of corolla tubes among central and ray florets give rise to a whole range of named types of chrysanthemum flowers, eg anemone-flowered, daisy forms, etc. Readily branching forms have been selected for pendant display and are described as "spray" or "charm" varieties.

The chrysanthemum has long been regarded as the funeral flower *par excellence*. Although the cut flower has a remarkably long vase-life, chrysanthemums as flowering pot plants are an equally important aspect of the flower trade. Usually such pot plants are treated by dwarfing chemicals to produce attractive short-stemmed bunches of flowers. COMPOSITAE.

Chrysobalanus a genus of shrubs and small trees native to tropical Africa and America but naturalized in Asia and the Pacific. *C. icaco* [COCO PLUM], of tropical America and the West Indies, has an edible fruit [SPANISH NECTARINE] which is reddish purple, plumlike and borne at the tips of the branches. It is used in jam, and in Colombia and Venezuela (as icaco, icacillo) is stewed, bottled in syrup and marketed. It has been used in the treatment of diarrhea, and can be used in tanning leather.
CHRYSOBALANACEAE, 4 species.

Cicer a genus of annual and perennial herbs distributed mainly in Central and western Asia. The most important species is *C. arietinum* (see CHICK PEA).
LEGUMINOSAE, 40 species.

Cichorium a genus of herbs from Europe and the Mediterranean region with usually blue flowering heads. *C. intybus* [*CHICORY, SUCCORY, WITLOOF] and *C. endivia* [*ENDIVE, ESCAROLLE] are cultivated as food plants.
COMPOSITAE, 9 species.

A pot-grown chrysanthemum coming into flower. Autumn-flowering chrysanthemums are now placed by many botanists in Dendranthema. (×2)

Cicuta a genus of perennial herbs native to north temperate regions and all highly poisonous. The European *C. virosa* [COWBANE, WATER HEMLOCK] grows in the shallow water of ditches and marshes as does the North American *C. maculata* [MUSQUASH ROOT].
UMBELLIFERAE, about 10 species.

Cider or **cyder** an alcoholic drink made from fermented *apple juice. The drink is of ancient origin but production is now mainly associated with the apple-growing areas of southwestern England and northwestern France. The apples used for cider are sour varieties that are unsuitable for culinary use. They are crushed in a cider press to extract all the juice which is then fermented by yeasts.

Cimicifuga a genus of perennial herbs from the north temperate region. They thrive in moist shady places, and their handsome foliage makes them attractive plants for shady garden borders, though some species have an unpleasant smell. *C. foetida* [BUGBANE] was once used as an insecticide, hence its name. A decoction of the roots of *C. racemosa* [BLACK COHOSH, BLACK SNAKEROOT] has been used against snake bites. It is favored as a garden plant as its pink-tinged flowers have no smell.
RANUNCULACEAE, 8–15 species.

Cinchona see QUININE.
RUBIACEAE, about 40 species.

Cineraria a mainly African genus closely related to *Senecio*, from which it differs in having flattened fruits. The "cinerarias" of cultivation (*Senecio × hybridus*) have been derived from *S. cruentus* and *S. heritieri*, of the Canary Islands. Commercially produced hybrid cinerarias have large flower heads 4–8cm in diameter, clustered together on a short stem. They vary from red to deep blue-violet. The deep purple and magenta shades, and the 'Feltham Beauty' strain with a startling white center to the ray florets, are especially popular.
COMPOSITAE, about 50 species.

Cinnamomum [CINNAMON] a genus of trees and shrubs native to East Asia and Indomalaysia. It includes the important economic species *C. zeylanicum* [TRUE CINNAMON, CEYLON CINNAMON], *C. cassia* [CASSIA, CHINESE CINNAMON], *C. camphora* [*CAMPHOR TREE], and others.

The bark and leaves of many of the species are aromatic, owing to the presence of essential oils. The spice cassia consists of bark

Cinerarias make attractive winter-flowering pot plants. The cinerarias of cultivation, as here, are varieties of Senecio × hybridus. (×½)

Spear Thistles (Cirsium vulgare) are handsome wild plants which can be troublesome weeds, particularly of grassland. ($\times \frac{1}{10}$)

stripped from trees which are felled when about 6–10 years old. The essential oil that gives cassia its aromatic qualities is extracted and used medicinally and as a flavoring substance. Cassia buds (immature fruits) contain the same essential oil.

C. zeylanicum is coppiced in cultivation. The young shoots that grow up are cut twice a year, and the bark removed in two semi-cylindrical sections. After fermentation and drying the pale brown bark is exported in the form of hand-rolled quills. The trimmings etc can be used for distillation of the oil, which is used as a flavoring and perfume.

Many species of *Cinnamomum* are of economic significance. Some are used as substitutes for true cinnamon, such as *C. massoia* [MASSOIA BARK] and *C. loureirii* [SAIGON CINNAMON, CASSIA-FLOWER TREE], while others yield timber, such as *C. burmanii* [BATAVIA CINNAMON, PADANG CASSIA].
LAURACEAE, about 250 species.

Circaea a small genus of perennial herbs inhabiting the north temperate and arctic zones. The best known species are *C. lutetiana* [ENCHANTER'S NIGHTSHADE] of woodland and shady places, and *C. alpina* [ALPINE ENCHANTER'S NIGHTSHADE], an inhabitant of mountain woods.
ONAGRACEAE, 12 species.

Cirsium a genus of hardy annuals, biennials and perennial herbs found throughout the Northern Hemisphere. Several of the species are well-known thistles which can become troublesome weeds of agriculture, while a few are grown as garden ornamentals.

Most are tall, erect plants with spiny margins to the leaves. The flower heads are large, purple and surrounded by spiny bracts. In many species, white-flowered plants occur sporadically among the purple-flowered ones, this being particularly characteristic of *C. palustre* [MARSH THISTLE]. Some are yellow flowered, for example *C. oleraceum* [CABBAGE THISTLE], a plant of fens in central Europe

sometimes eaten as a vegetable. *C. oleraceum* and *C. helenioides* [MELANCHOLY THISTLE], a large handsome species with leaves covered with white cottony hairs on their underside, are notable non-spiny species.

The biennial species are rarely important weeds, but the perennials can be serious pests of grassland and, in the case of *C. arvense* [CREEPING THISTLE] of arable land. The presence of species such as *C. spinosissimum* in grass or hay obviously lowers its palatability. Other common weeds are *C. acaule* [STEMLESS THISTLE, PICKNICKER'S THISTLE] and *C. vulgare* [BULL or SPEAR THISTLE]. Although not generally grown as garden plants, some species are grown as ornamentals in borders, including *C. kerneri*, *C. nutans* and the crimson-flowered *C. rivulare* 'Atropurpureum'.
COMPOSITAE, about 120 species.

Cissus a large genus of mostly tropical, but a few subtropical, species of deciduous or evergreen, mostly climbing shrubs, though some are erect shrubs and a few are perennial herbs. The stems and leaves are often succulent.

Several species are grown as indoor plants. *C. rhombifolia*, from Central and South America, is widely grown as a foliage plant. *C. antarctica* [KANGAROO VINE], from Australia, is a climber grown for its shiny, dark green, ovate leaves. *C. adenopoda*, from tropical Africa, is a very decorative climbing hothouse plant. It has trifoliolate leaves,

The Kangaroo Vine (Cissus antarctica) is one of the most robust of foliage house plants, tolerating smoke, poor light and wide temperature variations. ($\times \frac{1}{8}$)

Thistledown of the Creeping Thistle (Cirsium arvense) about to be dispersed by wind. The "parachute" of fine hairs attached to each fruit is called a pappus. ($\times 1\frac{1}{2}$)

green above and red with prominent veins beneath. Loose clusters of pale yellow flowers are followed in autumn by purple-black berries which ripen in spring. *C. discolor* from Java, which may be trained as a trailing plant, has red shoots and its leaves are rich green and mottled with white, red and purple above and purple beneath. *C. javalensis* from Nicaragua has purple-veined leaves and scarlet flowers.
VITACEAE, about 300 species.

Cistus [ROCK-ROSES] a genus of evergreen Mediterranean shrubs, many of which are cultivated in gardens although some are only half-hardy when grown in the Northern Hemisphere.

The best-known species are the white-flowered *C. salviifolius* and *C. ladanifer*, the pinkish-purple-flowered *C. albidus*, and the purplish-red-flowered *C. crispus*. A considerable number of hybrids are also cultivated as garden ornamental shrubs. These include *C. × aguilari* (a cross between *C. ladanifer* and *C. populifolius*) with large white flowers and *C. × cyprius* (a cross between *C. ladanifer* and *C. laurifolius*) with large crimson-blotched white flowers.

Ladanum or labdanum is a fragrant resinous substance used in soaps and perfumes, obtained from various species, especially *C. incanus* subspecies *creticus* and *C. ladanifer*, both commonly cultivated in southern Europe.
CISTACEAE, about 20 species.

Citron the large yellow fruit of the citrus tree, *Citrus medica*. The peel is very thick and fragrant and the pulp is small, dry and inedible. It is now grown mostly in Italy, Southeast Asia and the West Indies, for use in candied peel.
RUTACEAE.

Citrullus a small genus of climbing or trailing plants from southern and tropical Africa, Asia and the Mediterranean. *C. colocynth* [BITTER GOURD, BITTER APPLE, COLOCYNTH] and *C. lanatus* (= *Colocynthis citrullus*) (see WATER MELON) are cultivated for their globular fruits. The bitter gourd has a violent purging action due to alkaloids. The drug colocynth is obtained from the dried pulp of the unripe fully grown fruit. *C. lanatus* var *fistulosa*, with smaller round fruits, is grown in India where it is used as a vegetable and in preserves and pickles.
CUCURBITACEAE, 3 species.

A pink-flowered Rock-rose (Cistus × purpureus), one of several cultivated hybrids. (× ¼)

Citrus fruits a well-known and distinctive group of commercially important tropical and subtropical fruits originating in Southeast Asia. The important genera are *Citrus* (including most cultivated species, such as SWEET *ORANGE, *SEVILLE ORANGE, *MANDARIN, *GRAPEFRUIT, *SHADDOCK, *LEMON, *LIME and *CITRON), *Fortunella* [*KUMQUAT] and *Poncirus* (used as cold-resistant root stock and for breeding purposes). Proportions in the fruit of peel and pulp, size, juiciness, acidity and flavor vary widely according to species.

Total world cultivated area of citrus is about 2.2 million hectares (0.6 in the Mediterranean area). More than 14% of the citrus world crop is exported as fresh fruit because of their exotic flavor, fragrance and appearance, high vitamin C content and very good shipping properties.
RUTACEAE.

Cladonia a large genus of lichens with worldwide distribution. They are unusual among lichens in that the plant body is of two parts, a flat, usually small, leafy horizontal portion and an upright portion. The latter may be in the form of small cups on the end of a stalk with brown or colorful red fungal fruit bodies around the margin, as in *Cladonia pyxidata*. In others, the upright portion may be highly branched, as in the REINDEER MOSSES so common in the Arctic. Reindeer moss covers a mixture of species, the commonest being *C. alpestris* and *C. rangiferina*, which provide fodder for reindeer and caribou.
CLADONIACEAE, about 300 species.

Cladophora a genus of branched filamentous green algae found in both marine and freshwater habitats. *Cladophora* often grows abundantly as "blanket-weed".
CHLOROPHYCEAE, about 160 species.

Cladosporium a genus of fungi most of which are saprophytes occurring on all forms of moribund or decaying plant materials as olive-green molds. *C. herbarum* and *C. cladosporioides* frequently cause blackening of the ears of cereals. Other species are parasitic and of economic significance on greenhouse groups, eg *C. fulvum*, which causes leaf mold of tomatoes. *C. resinae* is common in soils but is best-known as an inhabitant of creosoted wood. The species is also frequently isolated from resins, aviation fuel and face creams.
DEMATIACEAE, about 50 species.

Clarkia a genus of annual herbs restricted to western North America and South America. They have slender spikes or racemes of usually showy purplish or violet to pink (rarely white) flowers. A number of species in a range of colors and in single and double forms are cultivated as garden-border plants, including *C. unguiculata* and *C. pulchella*.
ONAGRACEAE, about 36 species.

Clavaria (Clavulinopsis) formosa *belongs to a genus known as fairy clubs or coral fungi.* (× ½)

Clary, clary sage or **clary wort** the common names for *Salvia sclarea*, an erect biennial herbaceous plant whose aromatic leaves may be used fresh or dried as a flavoring in savory dishes. This European plant also yields an extract used in the perfumery industry, and is grown as a garden-border plant for its tubular whitish-blue flowers.
LABIATAE.

Clavaria, [FAIRY CLUBS, CORAL FUNGI] a cosmopolitan genus of basidiomycete fungi. They are predominantly terrestrial species, and the club-like appearance of their fruiting bodies varying from about 1cm to 30cm high, is reflected in the common names. A few species are edible but none is of any commercial importance.
CLAVARIACEAE, about 25 species.

Claytonia a genus of annual and perennial herbs from eastern Siberia and North America. The white to rose-colored flowers

Rosette and white flowers of Claytonia megarhiza *growing in a carpet of clover (*Trifolium nanum*).*

are borne in cymes. *C. virginica* [SPRING BEAUTY] is cultivated in rockeries. *C. parviflora*, from western North America, is occasionally eaten as a vegetable.
PORTULACACEAE, about 35 species.

Clematis a large genus of mostly woody climbers. The majority are deciduous, originating in the cool temperate region of the Northern Hemisphere. Some less hardy species come from warmer climates; these are usually evergreen with a few herbaceous non-climbing species.

Clematis may be seen blooming from April through October with the odd winter-flowering species. The early-flowering species, such as *C. alpina* and *C. macropetala*, generally have smaller flowers than those starting in May and June onward, such as *C. henryii* and *C. flammula*. Most cultivated forms of *Clematis* have been derived from three eastern species (*C. florida*, *C. patens* and *C. lanuginosa*) and two European species (*C. viticella* and *C. integrifolia*). The color range of the flowers of *Clematis* species and their hybrids extends from white, through cream to yellow, various shades of blue, purple-pink and red. Popular white-flowered types include *C. alpina* 'White Moth' and *C. × jackmanii* 'Alba'. Yellow-flowered species include *C. orientalis* and *C. tangutica*. There are many blue-flowered species and hybrids, such as *C. macropetala* 'Maidwell Hall' and *C. × jackmanii* 'Perle D'Azur'. There is a wide range of purple forms, including *C. × jackmanii* 'Superba' and *C. viticella* 'Gipsy Queen'. Popular red varieties include *C. montana* 'Rubens' and *C. × jackmanii* 'Rubra'.
RANUNCULACEAE, about 250 species.

Clementine a form of the small tree *Citrus reticulata* (see MANDARIN). The fruits are very similar to tangerines, both in shape and in the ease with which the coarse peel can

Cladonia fimbriata, a common lichen found on tree stumps and soil, showing its characteristic upright stalks forming green powdery cups. (× 3)

be removed from the fine-flavored flesh.
RUTACEAE.

Cleome a large mainly tropical and subtropical genus of herbs, shrubs and subshrubs, usually with a fetid smell. The pink, white, yellow or purple flowers are solitary or in racemes. *C. hasslerana* (often called *C. spinosa* in cultivation) [SPIDER FLOWER] is often grown as an ornamental. It is a sticky, strongly scented annual with white or pink flowers.
CAPPARACEAE, about 150–200 species.

Clerodendrum a very large, tropical, mostly African and Malaysian genus of shrubs, trees and climbers. Many are ornamental climbers, notably *C. thomsoniae* with white and crimson flowers, *C. splendens* with scarlet flowers fading yellow or white, and *C. umbellatum* with white flowers. Hardy ornamentals include the Japanese tree *C. trichotomum* with red and white flowers and blue fruits, and the Chinese shrubs *C. bungei* with dense purple red flowers and *C. fragrans* [GLORY BOWER] with rose-colored flowers.
VERBENACEAE, about 560 species.

Clethra a genus of shrubs and trees of mainly American and Chinese origin. Some species are cultivated for their floristic beauty, notably *C. arborea*, LILY-OF-THE-VALLEY TREE of Madeira, and *C. alnifolia* [SUMMER-SWEET, SWEET PEPPERBUSH] from the eastern USA and Mexico.
CLETHRACEAE, about 30 species.

Clianthus a genus consisting of two species of Australasian evergreen plants. Both are cultivated as ornamentals in warm temperate zones. *C. formosus* [STURT'S DESERT PEA, GLORY PEA] from Australia, is a silver-gray, hairy, prostrate subshrub with large striking claw-shaped scarlet-red flowers (the center or boss of each flower is black in the wild form, red or white in some cultivated forms). *C. puniceus* [PARROT'S BILL], from New Zealand, is a scrambling shrub with similar but larger pendulous flowers.
LEGUMINOSAE, 2 species.

Clintonia a genus of small bulbous plants native to North America, northern India and East Asia. White, pink or deep rose, funnel-shaped flowers are borne terminally in long-stalked umbels, or rarely solitarily. Several species are known in cultivation, for example the North American *C. uniflora*, with solitary white flowers, and *C. andrewsiana*, with a terminal umbel of rose-pink flowers.
LILIACEAE, 6 species.

Clitocybe a genus of gill fungi, almost cosmopolitan in distribution. Some are edible, but often may be dangerous if kept some days before being eaten, (eg *C. geotropa*). Small white species of the *C. dealbata* group can be deadly, there being a reported high content of muscarine. An antibiotic, clitocybin, has been extracted from *C. gigantea*

Clematis 'Percy Picton', one of the numerous large-flowered clematis cultivars. The "petals" are in fact sepals. (× ¼)

forma *candida*.
TRICHOLOMATACEAE, about 250 species.

Clitoria a genus of perennial herbs and shrubs, often climbers, from the warmer parts of the globe. They have pea-like flowers. *C. ternatea* [BUTTERFLY PEA OR BEAN, KORDOFAN PEA] is a twining herb originating in Asia but now pantropical and grown widely in drier parts of the Middle East and Australia as a fodder crop. This species and *C. heterophylla* are sometimes cultivated for their decorative blue flowers.
LEGUMINOSAE, about 30 species.

Clivia [KAFFIR LILIES] a small genus of leek-like plants native to South Africa. They are grown in greenhouses for their attractive, dark green, glossy, strap-shaped leaves and orange-red flowers. The most widely grown species are *C. miniata* and *C. nobilis*.
AMARYLLIDACEAE, 3 species.

Clover the common name for *Trifolium*, a genus of some 300 species of half-hardy and hardy, annual and perennial herbaceous plants. Some species of *Medicago* are also called CLOVERS. Deciduous and evergreen forms of *Trifolium* are found throughout the temperate and subtropical parts of the Northern Hemisphere, and many have been introduced to the Southern Hemisphere. The herbaceous plants bear red, white or yellow flowers, often in a spherical or globose head. One to four roundish seeds are borne in small pods.

While a few species are grown as decorative plants (eg *T. alpinum*, *T. repens*, *T. parnassii*), their most important use is in agriculture, as animal feed, forage, hay and silage, and as a green manure crop. As active fixers of nitrogen in their root nodules, they are beneficial to soil fertility, and are therefore included in many crop rotation schemes.

T. pratense [RED CLOVER] is a short-lived perennial, usually grown as an annual, with pink to mauve flowers in a globular head.

The dried flowers contain a volatile oil and are also a source of a yellow dye.

T. hybridum [ALSIKE or SWEDISH CLOVER] is a perennial grown in many parts of the world as it is less susceptible than RED CLOVER to clover sickness (caused by soil fungi or eelworms) and more tolerant of wet acid conditions.

T. repens [WHITE or DUTCH CLOVER] is a creeping perennial with a large root system which enables it to withstand drought. The white flowers grouped in a globose head are fertilized by hive bees, and are an important source of honey, which is light and mild in flavor. Among many cultivars the Kentish ones are particularly favored.

T. incarnatum [CRIMSON or ITALIAN CLOVER] is a hardy annual grown mainly in rather warmer areas, sometimes as a catch crop after cereals. It is said to be immune to clover sickness.

T. dubium [YELLOW SUCKLING CLOVER] is an annual which germinates well in dry conditions on poor soils.

T. subterraneum [SUBTERRANEAN CLOVER] is an annual but behaves like a perennial, burying its fruits and seeds. It has a spreading mat-forming habit which enables it to choke

Clerodendrum trichotomum *var* fargesii *has blue fruits subtended by the persistent red calyx, the lobes of which become fleshy.* (× ½)

out most weeds, and grow well on hilly ground with poor soil, as long as the ground is soft enough to allow the plant to bury its flowers. This species is particularly useful as a forage crop.

T. fragiferum [STRAWBERRY CLOVER] is a pinkish-white-flowered perennial that will grow on heavy, swampy land near coasts, and tolerate prolonged flooding by sea water. It is especially useful as a pasture plant on partly consolidated coastal sand.

T. alexandrinum [EGYPTIAN CLOVER, BEER-SEEM] is of minor importance as a cover or forage crop in drier tropical areas such as India, Egypt and Zimbabwe. *T. amabile* [AZTEC CLOVER] is eaten as a herb, mixed with cereals, in Argentina and Mexico.
LEGUMINOSAE.

Cloves the unopened, sun-dried flower buds of *Syzygium aromaticum* (= *Eugenia caryophyllata*). Indigenous only to some of the Moluccas, clove trees are now widely cultivated in the islands of Zanzibar, Madagascar and Pemba, and to a lesser extent in Indonesia. Madagascar is the only source of clove oil.

Once used to check tooth decay and counter halitosis, and as the sweet aromatic component of pomanders, cloves are now used in the East as a table spice, as a component of curry powder and to flavor tobacco. In the West, cloves are used as a flavorer in desserts, hams and herrings. Clove oil is an extremely important component of perfumes, and is used in flavoring, toothpastes and to remedy toothache.
MYRTACEAE.

Club mosses see p. 92.

Clytostoma a South American genus of climbing evergreen shrubs with large, showy, trumpet-like flowers. The principal cultivated species is *C. callistegioides* (= *Bignonia speciosa*), a native of Brazil and Argentina which is grown in greenhouses for its lavender-colored flowers. *C. binatum*, a similar species from Venezuala, Guiana, Brazil and Paraguay, is also cultivated.
BIGNONIACEAE, about 8 species.

Cnicus a genus now considered to consist of a single species, *C. benedictus* (= *Carduus benedictus* [BLESSED THISTLE], probably native to the Mediterranean region and the Near East but naturalized in several countries. It is cultivated in some countries partly for its medicinal value in herbal remedies for respiratory and other ailments and for its seeds which yield an oil. It is also sometimes cultivated for the ornamental appearance of its silvery-white, spiny leaves and pale yellow flower heads.
COMPOSITAE, 1 species.

Sturt's Desert Pea (Clianthus formosus) *growing in its native habitat in Western Australia. It is also cultivated in hanging baskets.* (× ⅙)

Cobaea a genus of tropical American climbers, the best-known being the Mexican *C. scandens* [CATHEDRAL BELL, CUP AND SAUCER PLANT], which is known for its ability to grow rapidly, covering large areas by means of hooked leaf-tendrils which if touched may be seen to respond over an interval of minutes. It has bell-shaped flowers, first green, later purplish in color.
POLEMONIACEAE, about 10 species.

Cocaine or **coca** a narcotic alkaloid obtained from the leaves of the economically important species *Erythroxylum coca* [COCA, COCAINE PLANT] and *E. novagranatense* [TRUXILLO COCA], native to South America, where they are extensively planted, the center of (continued on p. 94)

Heads of the Red Clover (Trifolium pratense) *growing in a field. As well as providing valuable forage for cattle, clover is an important bee plant.*

Club Mosses and their Allies

T**HE CLUB MOSSES AND THEIR ALLIES ARE A** small group of simple green plants. Indeed they represent the most primitive of the vascular plants (division Tracheophyta), ie those plants that possess a well-developed internal system for conducting nutrients around the plant. Unlike the *mosses and liverworts, the dominant form we see (sporophyte) produces spores and not the sexual gametes. The sexual phase (gametophyte) is insignificant.

This group comprises two subdivisions of vascular plants: the Lycophytina which contains five genera: *Lycopodium* and *Phylloglossum* (order Lycopodiales – the TRUE CLUB MOSSES), *Selaginella* (order Selaginales – the SMALL CLUB MOSSES) and *Isoetes* and *Stylites* (order Isoetales – the QUILLWORTS); and the Psilophytina (psilophytes) which contains just two living genera – *Psilotum* [WHISK FERNS] and *Tmesipteris*.

Club mosses and their allies have negligible economic or ornamental value. Their chief interest lies in the fact that they are the most primitive of vascular land plants and for this reason are sometimes grown as botanical curiosities. *Lycopodium* is of limited economic importance in that the spores of various species, including *L. clavatum*, *L. complanatum* and *L. annotinum*, are used medicinally as a dusting powder and as a preventative of adhesion of pills.

There are perhaps 400 species of *Lycopodium*, mostly found in the tropics and subtropics, but with some Arctic and tem-

Isoetes histrix (Sand Quillwort) is one of the few terrestrial species of this mainly aquatic genus. (×1)

perate representatives. Most tropical species are epiphytic and those of Arctic and temperate zones are terrestrial. All have relatively small herbaceous or shrubby sporophytes. Some of the terrestrial forms are erect, others are creeping, whereas the epiphytes are either pendent or erect.

The stems of all species are slender and invested with spirally arranged or whorled, small, lanceolate leaves (microphylls). They do not possess ligules (see later). There is no primary root system in mature club mosses, all roots being adventitious. In many of the tropical epiphytic species the roots become attached to the branches of trees and absorb nutrients and water from the humus in the crevices and cracks. Single compressed kidney-shaped sporangia occur on the adaxial surface or in the axil of each fertile leaf (sporophyll). The sporophylls in most species (eg *L. phlegmaria* and *L. clavatum*) are arranged in definite cones (strobili) at the apices of branches. In *L. selago*, however, the sporophylls are arranged along the stem in "fertile" zones alternating with "sterile" zones. The spores are liberated as a result of the longitudinal dehiscence of the sporangium. Although in some species (eg *L. cernuum* and *L. inundatum*) they may start to germinate within a few days, germination may be delayed for several years in others. The rate of development of the resultant gametophyte stage (prothallus) depends on the amount of green photosynthetic tissue that it possesses. It may take up to 15 years or more to mature in species such as *L. clavatum* which have subterranean prothalli more or less devoid of chlorophyll. In species such as *L. cernuum*, which has a photosynthetic prothallus growing on the surface of the soil, maturation takes only about one year. The development of the gametophyte of all species of *Lycopodium* is dependent on the presence of an endophytic fungus which has a symbiotic relationship with it.

Lycopodium produces only one type of spore (ie it is homosporous) and each prothallus bears both types of sex organ. The male organ (antheridium) gives rise to pear-shaped antherozoids (spermatozoids). The female organ (archegonium) is flask-shaped with a single egg cell. One of the antherozoids, which swims down the archegonial neck, fuses with the egg to produce a zygote. This gives rise to an embryo which is at first dependent on the prothallus for nutrition,

but ultimately becomes free-living when it develops its stem, leaves and roots and the prothallus decays.

Phylloglossum is monotypic, its single species (*P. drummondii*) occurring only in Australia, Tasmania and New Zealand. It has very small sporophyte, which grows from a small tubercle or protocorm. It produces several cylindrical sterile leaves and an elongated erect axis bearing a small cone at the apex. At the end of the season a new protocorm is formed. The spores develop a green gametophyte similar to that of *L. cernuum*.

Selaginella is a genus of between 400 and 700 species of small, creeping or climbing plants closely resembling, but smaller than, the true club mosses. They usually grow in moist habitats (a few inhabit deserts) and are mainly found in tropical climates. The stems are prostrate and creeping or climbing and bear four ranks for single-veined leaves (microphylls), which are normally of two or more kinds. Unlike *Lycopodium* each leaf has

The Club Moss Lycopodium selago *showing fertile shoots which are crowned by "bulbils" that are easily detached and grow into new plants. (×1)*

a small scale-like outgrowth called a ligule arising near the base on the upper surface. The sporangia develop on sporophylls that are borne in cones at the tip of the stems.

Selaginella exhibits heterospory, ie two types of sporangia develop (megasporangia and microsporangia), both of which occur in the same cone, the megasporangia at the base. The megasporangia produce large spores (megaspores), which germinate within the megasporangia to produce the female gametophyte. This often protrudes through a

The Club Moss Lycopodium alpinum *is a creeping species with its stems tightly clothed with very small leaves (microphylls)* (× 1½)

Selaginellas are easy to grow as foliage ornamentals. Shown here is Selaginella kraussiana *'Aurea', an attractive variegated leaf form.* (× ⅓)

tear in the wall, in which the egg-producing archegonia are formed. The microsporangia produce smaller spores (microspores), which germinate within the microsporangium to form the male gametophyte upon which the sperm-producing antheridia develop. Fertilization does not usually take place until the gametophytes are shed from the sporangia, when, in the presence of water, the sperms swim to and fertilize the egg produced within the female gametophyte. This marks the start of the new sporophyte generation.

Isoetes is a widespread temperate and tropical genus of about 75 species. Most are aquatic, eg *I. lacustris* and *I. macrospora*, but a

The Small Club Moss Selaginella pulcherrima *showing its leaves in two rows, each leaf with a surface ligule, and fertile shoot tips.* (× 3)

few are terrestrial, such as the European species *I. histrix* [SAND QUILLWORT]. The sporophylls of *Isoetes* are linear structures up to about 60cm long in some tropical species but more usually less than 20cm. The bases of the sporophylls overlap to form a "bulb" which is on top of an organ called a corm or caudex. The sporangia occur at the base of the sporophylls. These are protected by a flap (velum) and often have quite large membranous ligules, which arise early in the development of the sporangia. The outer sporophylls have the female megasporangia sunk into them and the inner ones the male microsporangia. There is no special dispersal apparatus, the spores simply remaining in the sporangia until the sporangia rot. The prothalli in both microsporangia and megasporangia develop inside the spore wall, which splits open. Fertilization may take place while the plant is decaying but before dispersal and the young sporophytes often develop around the parent corm, to be washed away later.

Many populations of *Isoetes* are extremely old but because the upward growth of the corm is so slow there is no accurate way of assessing the age of the plants.

In 1954 an apparent *Isoetes* was found on the margin of a high Andean lake at about 4 500m above sea level. It was described not only as a new species but also as a new genus – *Stylites*. It is obviously closely related to *Isoetes* but differs in its ability to branch dichotomously and adventitiously to form highly characteristic clumps or tufts. Furthermore, the roots are borne on one side of the corm only.

Psilophytes. *Psilotum* is a unique genus which is probably more closely related to the ferns than to the Devonian fossil psilophytes as formerly believed. There are two species, *Psilotum complanatum* and *P. nudum*, which are distributed widely in warm moist tropical and subtropical areas. They grow epiphytically, hanging from tree trunks or rocks, or

erect from the soil and are easily cultivated in greenhouses. The sporophytes have short, forked, rootless basal rhizomes from which arise the green dichotomously branched aerial fronds, 15–20cm long in *P. complanatum* and 75–100cm in *P. nudum*. These bear small simple and forked scales and globose three-chambered sporangia. The gametophytes are subterranean, colorless, saprophytes, relying for nutrition on a mycorrhizal relationship with fungi.

Tmesipteris is a genus closely related to *Psilotum* with about 10 species in tropical Asia, Australia, New Zealand and the Pacific islands. Some species, such as *T. tannensis*, grow as epiphytes hanging from tree trunks; others like *T. viellardi* grow erect from the soil. The sporophytes have a much-branched basal rhizome from which arise aerial fronds. In the epiphytes these bear two rows of flattened laterals; in the terrestrial species these are attached all around. The fertile laterals are forked and bear a large two-chambered sporangium at the point of branching.

The strobilus of Selaginella selaginoides *showing the cream-colored sporangia in the axils of each fertile leaf (sporophyll).* (× 4)

Left *Coconut* (Cocos nucifera) *harvest on the island of Zanzibar off the east coast of Africa. The whole fruits (left) are being split to separate the familiar "nuts" (foreground) from the fibrous husks (background).* ($\times \frac{1}{4}$)

Above *A broken "coconut" showing the hard inner fruit wall (endocarp) enclosing the oil-rich endosperm of the seed, called copra when dried.*

production being in the Andes of Peru and Bolivia. Coca is also grown elsewhere in the tropics, notably Indonesia (Java) and Sri Lanka, but under strict legal controls.

The leaves, which contain 0.25–2.25% alkaloids, principally cocaine, are harvested four times a year, dried rapidly and exported. For centuries, coca leaves have been widely used by South American Indians as a stimulating masticatory. The cocaine when released from the leaves acts as a stimulant to the cerebral cortex, producing restlessness, excitement and a greater capacity for muscular activity. It also allows the participant to resist physical and mental fatigue for long periods without food and drink, and constricts the blood vessels so that body heat is retained. The leaves are also rich in calcium, iron and vitamins.

In the late 19th century coca was widely used in stimulants and tonics, eg Coca-Cola. The coca now used as a flavoring is "de-alkaloidized". Cocaine is still an important drug in Western medicine. It provides local anesthesia for dental surgery and other minor surgical operations.
ERYTHROXYLACEAE.

Coccoloba a large genus of tropical and subtropical American and Caribbean trees and shrubs with often very large and showy leaves. The fruits are fleshy, berry-like and sometimes edible. *C. uvifera* [SEA or SEASIDE GRAPE] is a characteristic beach plant whose purple fruits resemble grapes and are used for making jelly. *C. diversifolia* (= *C. laurifolia*) [PIGEON PLUM] bears small edible pear-shaped fruits. The timber of both these species is used locally for general carpentry and furniture making.
POLYGONACEAE, about 125 species.

Cochlearia [SCURVY GRASS] a genus of annual and perennial herbs distributed in the cool temperate and arctic or alpine Northern Hemisphere, and occurring commonly around coasts or on inland mountains. The maritime species such as *C. danica* [EARLY SCURVY GRASS] were formerly eaten by sailors because of their antiscorbutic qualities arising from a high vitamin C content. The basal leaves of *C. officinalis* [COMMON SCURVY GRASS] are still occasionally used in salads for their slightly acrid and bitter taste.
CRUCIFERAE, about 30 species.

Cocoa the name usually given to the manufactured products of the small tree *Theobroma cacao* [CACAO] originally native to the eastern equatorial slopes of the Andes. The center of cultivation was, however, Central America, where it has been cultivated for more than 2000 years.

The flowers and fruits are borne on the trunk of the tree. The fruit is called a pod, which when ripe is about 30cm long and contains 20–30 seeds ("beans") surrounded by a mucilaginous pulp.

With the introduction of the crop into West Africa at the end of the 19th century, a tremendous expansion in production occurred to reach 1.5 million tonnes per year today. Over 70% of the crop now comes from Ghana, Nigeria, the Ivory Coast and Cameroon, most of the remainder from the Americas and the Caribbean.

Two types of cocoa are distinguished in world trade: fine or flavor cocoa and ordinary or bulk cocoa. Flavor cocoa is derived from the white or pale violet seeded variety (Criollo cocoa) and from natural hybrids (Trinitario cocoa, a fine cocoa) between it and the purple-seeded variety (Amazonian Forastero cocoa). Bulk cocoa is produced from the Forastero type and all the West African crop (called West African Amelonado) is of Forastero origin.

The freshly picked seeds are highly astringent and have no chocolate flavor. This only develops after a two-stage processing. The seeds are first fermented by placing in wooden boxes or in heaps on the ground. Naturally occurring yeasts and *Acetobacter* species flourish on the sugary pulp and lead to a rise in temperature which kills the seeds and allows various enzymatic reactions to occur. At the end of five to seven days the beans are dried. The next stage in manufacturing is the roasting of the bean which then develops its chocolate flavor. After grinding the beans are pressed to reduce the cocoa butter content from 55% to 22%, and what is left becomes cocoa powder.
STERCULIACEAE.

Coconut the common name for *Cocos nucifera*, a tall unbranched palm probably originating in Southeast Asia, although a South American origin has also been suggested. It is now common along tropical coasts throughout the world. It does not actually require saline conditions and will grow well inland at altitudes of up to 1 200m, but the structure of its roots enables it to survive salt water.

Coconuts provide the fat ration for about 7% of the world's population. The most important commercial product is the oil, which is extracted from the dried endosperm (copra). It is exceptionally low in unsaturated fatty acids, and is used to make margarine, soap and detergents. *Coir is another important product and comes from the coarse, highly lignified fibers of the mesocarp. Desiccated coconut consists of shredded and dried endosperm and is exported from Sri Lanka and the Philippines.

Other parts of the coconut are used locally.

Coconut water, from the unripe fruits, provides a refreshing drink and toddy (*arrack) is tapped from the young axillary inflorescence. The cotyledon (coconut apple) can be eaten and the terminal bud of the adult provides palm cabbage. Coconut leaves are used for thatching, making screens, baskets, mats etc, while the shells (endocarps) are used for fuel.
PALMAE.

Cocos see COCONUT.
PALMAE, 1 species.

Codiaeum a genus of evergreen trees and shrubs native to Malaysia, Polynesia and northern Australia. The many cultivars of the most important species, *C. variegatum* var *pictum*, are commonly referred to as "crotons", but should not be confused with the genus *Croton*. *C. variegatum* var *pictum* is a native of Malaysia which is cultivated throughout the world for its ornamental value as a foliage plant. The leaves vary in shape from linear to ovate and are variously colored in shades of green, spotted or blotched with yellow, pink red or crimson patches. The young leaves of yellow leaved races are used as a sweet flavoring agent.
EUPHORBIACEAE, 16 species.

Codium a genus of mainly tropical and subtropical green algae having a soft worm-like appearance. Some species such as *C. geppii* (Philippines) and *C. lindenbergii* (Japan) are used as food, either raw, dried or boiled.
CHLOROPHYCEAE, about 50 species.

Codonopsis a genus of annual and perennial herbs from Central Asia east into the Himalayas, China and Malaysia, all of them making attractive garden climbers, bearing bell-shaped flowers.
Cultivated species include the robust

Codonopsis convolvulacea twining up bushes and other vegetation at around 2450–3000m in the Himalayas, where it is native. (×1)

C. clematidea from western Asia, with white blue-tinged flowers; *C. meleagris* from Yunnan Province, western China, with flowers which are cream or bluish-veined with brown on the outside and purple-violet with a green base on the inside; and *C. ovata* from the Himalayas, with clear pale blue flowers.
CAMPANULACEAE, about 35 species.

Coeloglossum a genus represented by the single terrestrial species *C. viride* [FROG ORCHID] distributed in grassy places throughout temperate Eurasia and North America. The inflorescence bears up to 35 dull greenish and often reddish-tinted flowers. The North American race (var *bracteatum*) has each flower subtended by a large, often leaf-like bract.
ORCHIDACEAE, 1 species.

Coelogyne a large genus of tropical and subtropical epiphytic or rock-inhabiting orchids growing wild throughout tropical Asia, especially in the Chinese and Indo-malaysian regions. Most species are capable of prolonged survival in extremes of temperature and drought and as a consequence are widely cultivated throughout the world.
Although the flowers can be borne singly or in pairs (eg *C. fimbriata, C. speciosa*), they are usually in an erect or pendulous few-flowered spike (eg *C. lawrenceana, C. asperata*) or arranged regularly in two rows on a long drooping raceme as in *C. massangeana*. Flower colors vary from pure white through creams, buffs, dull salmon-pink and many browns to bright yellows and orange and, as in *C. virescens*, a unique translucent green.
ORCHIDACEAE, about 200 species.

Coffee the name given to the very important and valuable genus *Coffea* comprising about 90 species, of which two are grown commercially in quantity for their seeds (beans). Roasted and ground the beans are infused with hot water to make coffee.
The most widely cultivated species is *C. arabica* ('Typica'), which accounts for over 80% of world coffee production. The other species cultivated in quantity is *C. canephora* ('Robusta') grown mainly in the warmer lowland areas of central and West Africa. *C. arabica* originated in the Ethiopian highlands and it was probably from seeds or plants taken from Ethiopia to the Yemen in Arabia that coffee was first cultivated. Coffee is now grown commercially throughout the tropics, especially in Central and South America. Brazil now accounts for more than half of world production. Several other species, for example *C. liberica* [LIBERICA COFFEE], *C. dewevrei* var *excelsa* and *C. congensis*, make a small contribution to world coffee production.
After harvest, the fleshy fruits ("cherries") are either dried further in the sun and then hulled to release the two beans they contain or, if the "cherries" have been picked fresh, they are pulped wet, fermented to remove the fruit tissues, and the cleaned beans dried in the sun. For the beans to acquire their

A coffee (Coffea arabica) plantation near Antigua, Guatemala. Coffee is obtained from the roasted and ground seeds (beans).

characteristic flavor and quality it is essential that at some stage they are dried in the sun. The oil known as caffeol is the principal component which confers upon coffee beans their characteristic flavor when roasted. The beans also contain 1–2% *caffeine, which is responsible for the stimulatory properties of the beverage.
RUBIACEAE.

Coir the long stout fiber obtained by retting the husk of the fruit of *Cocos nucifera* [COCONUT PALM] in water for 15–20 days. Coir is used for ropes, mats, brushes, and yarn.
PALMAE.

Coix a small genus of tropical Asian grasses one of which, *C. lacryma-jobi* [JOB'S TEARS],

Job's Tears (Coix lachryma-jobi) growing in Papua New Guinea. The egg-shaped "tears" are the fruits enclosed by a hard covering. (×⅓)

has long been cultivated as an annual forage plant and has been introduced throughout the tropics, sometimes as an ornamental. The variety *ma-yuen* has soft-shelled grains used for food in East Asia. The grains are also used as beads for necklaces and rosaries.
GRAMINEAE, 5 species.

Cola   a genus of shrubs and trees native to tropical Africa. The seeds of a number of species, for example *C. anomala, C. cordifolia* and *C. verticillata,* are edible but only those of *C. nitida* and, to a lesser extent, *C. acuminata* are commercially important, almost exclusively in West Africa. Because the alkaloids they contain are stimulating and depress symptoms of thirst and hunger, the seeds are widely used for chewing amongst the indigenous population. (See also KOLA.)
STERCULIACEAE, about 50 species.

Colchicum   a genus of perennial herbs with a distribution range extending from the Mediterranean to Central Asia and northern India. The best-known species flower in the autumn before producing leaves which follow in the spring. They bear a superficial resemblance to *Crocus.*

C. autumnale [MEADOW SAFFRON, AUTUMN CROCUS], C. byzantinum and C. speciosum are a few of the many species in cultivation. All these species have pinkish purple flowers – the usual color of the genus. A distinct and unusual species from northern India and Afghanistan is *C. luteum,* with yellow flowers which appear in spring with the leaves.

The colchicums contain the highly poisonous alkaloid colchicine, which is used as a specific remedy for gout. It has also been tried in leukemia because of its ability to arrest cell division.
LILIACEAE, about 65 species.

Coleosporium   a genus of rust fungi parasitic on flowering plants, particularly on members of the family Compositae, such as *C. senecionis* on *Senecio* species.
COLEOSPORIACEAE about 80 species.

Colchicum speciosum growing in a garden in western Nepal. Colchicum species have a superficial resemblance to Crocus species, although they belong to different families. ($\times \frac{1}{4}$)

Coleus   a genus of African and Asian annual or perennial herbs (some subshrubs) most of which are blue- to violet-flowered and several of which are economically important. The popular bedding and houseplants are varieties of *C. blumei* or hybrids (*C. × hybridus*) involving this and other species. They are bushy in habit, with heart-shaped leaves attractively variegated in shades of red, green, purple, brown, white, yellow and gold.

In tropical Africa *Coleus* species are used as a root crop, especially *C. dazo* and *C. dysentericus.* The tuberous root of the latter is known as the HAUSA POTATO or SUDAN POTATO, because of the reputed similarity in texture and taste to the temperate potato (*Solanum tuberosum*).
LABIATAE, about 150 species.

Collards   a term used mostly in the USA for a form of *Brassica oleracea* (var *acephala*) intermediate between *KALE and *CABBAGE. It has loosely packed green leaves, used mostly in early spring as a vegetable.
CRUCIFERAE.

Colletia   a genus of temperate and subtropical South American shrubs and small trees. *C. cruciata* is quite commonly cultivated in temperate gardens for its urn-shaped white flowers. The branchlets are modified into flat triangular spines on old plants. The timber is used locally in house and wagon building.
RHAMNACEAE, 17 species.

Colletotrichum   a small cosmopolitan genus of fungi which cause diseases known as anthracnoses, a name which refers to the limited, shallow, coal-like lesions, eg *C. lindemuthianum* on runner beans and *C. coffeanum* on coffee. The diseases that they cause are more severe where heavy rain

causes some water-soaking of the host tissues.
MELANCONIACEAE, about 40 species.

Collinsia   a North American genus of annual herbs. The markedly two-lipped flowers are borne solitarily or in small clusters in the axils of the upper leaves. Several species, including *C. heterophylla* (= *C. bicolor*). *C. grandiflora* [BLUE LIPS] and *C. verna* [BLUE-EYED MARY] are cultivated for their ornamental value.
SCROPHULARIACEAE, about 20 species.

Columnea   a large genus of small and softly hairy, often epiphytic shrubs and climbers confined to the American tropics in woodland habitats. The paired leaves are often broad and sometimes attractively colored by hairs above, or pigment beneath. The ornamental value of these often interfertile species and hybrids, make columneas popular as pot or basket plants. Cultivated species include *C. gloriosa, C. microphylla* and *C. schiedeana,* all with pendulous stems carrying bright orange to scarlet tubular flowers.
GESNERIACEAE, about 250 species.

Colutea   [BLADDER SENNA] a genus of deciduous shrubs native to southern Europe and Central Asia. *Colutea* species have pealike flowers, yellow or brownish red in color, followed by attractive, inflated, bladder-like pods. Because they are hardy and the flowers are produced over a long season some species make useful garden shrubs.

C. arborescens [COMMON BLADDER SENNA], from the Mediterranean regions, is a vigorous shrub up to 4m high with yellow or orange flowers in racemes and very inflated pods, reddish or coppery in color. It has strong laxative properties (like true sennas, genus *Cassia*). C. × media (= C. × orientalis), one of the ornamental species, has copper-colored flowers and attractive grayish foliage.
LEGUMINOSAE, about 26 species.

Coleus blumei seen to advantage here in a bedding display in Hyde Park, London, England.

Colza the name given to the semi-drying oil derived from the seed of specially developed forms of the closely related species *Brassica campestris* and *B. napus*. The oil is incorporated into foods and soap.
CRUCIFERAE.

Commelina a genus of tropical and subtropical perennial herbs, including some popular ornamentals. The leaves in cultivated forms often have white stripes. Of the many species sufficiently hardy to be grown out of doors in temperate regions, *C. coelestis* [DAY FLOWER] is the most popular. It grows to 45cm tall and bears blue flowers in June; cultivar 'Alba' has white flowers while 'Variegata' has blue to white. *C. africana* (yellow flowers) and *C. elliptica* (white flowers) are greenhouse species.
COMMELINACEAE, about 180 species.

Commiphora a large genus of sometimes spiny resin-producing trees and shrubs from tropical and subtropical Africa, Madagascar, Arabia and India. *C. abyssinica* [ABYSSINIAN MYRRH TREE], *C. schimperi*, *C. molmol* and other species yield the famous biblical resin myrrh (an oleo-gum-resin), which is used to make incense. Other species, including *C. africana* [AFRICAN MYRRH TREE] and *C. dalzielii*, yield a similar gum resin, bdellium, which is used in varnishes and in medicinal preparations. The black seeds of *C. africana* are threaded as rosary beads. *C. opobalsamum* is one of the species said to yield balm of Gilead.
BURSERACEAE, about 200 species.

Conifers see p. 98.

Conium a small genus of tall annual and biennial herbs native to Europe, Asia and North Africa. The best-known species is *C. maculatum* (see HEMLOCK).
UMBELLIFERAE, about 4 species.

The Archer Plant (Colletia cruciata), native to Uruguay, is heavily armed, the smaller branches consisting of flattened, triangular spines with some resemblance to a drawn bow. ($\times\frac{1}{10}$)

Conocephalum a genus of thallose liverworts consisting of two species, *C. conicum*, which is a conspicuous liverwort in Europe, Asia and North America and *C. supradecompositum*, which is found in China and Japan. Terpenoid contents make the thallus of *C. conicum* strongly scented.
MARCHANTIACEAE, 2 species.

Conophytum a genus of dwarf succulent perennial herbs, all of which are native to South and Southwest Africa. The plants grow in clumps, each plant consisting of a pair of fused leaves, opening only by a small slit on the upper surface through which the daisy-like flowers appear. Popular cultivated species include the yellow-flowered *C. bilobum* (= *Mesembryanthemum bilobum*) and *C. calculus* (= *M. calculus*), the pink-flowered *C. globosum* and the white-flowered *C. pellucidum*, *C. multipunctatum*, *C. peersii* and *C. saxetanum*.
MESEMBRYANTHACEAE, about 260 species.

Conopodium a genus of European and Mediterranean herbs, none of which are of economic importance except as weeds that are resistant to most herbicides. *C. majus* (= *C. denudatum*) [PIGNUT, EARTH NUT] has much dissected, parsley-like leaves and white-flowered umbels. Its tuberous rootstock is edible either raw or roasted.
UMBELLIFERAE, about 20 species.

Convallaria see LILY OF THE VALLEY.
LILIACEAE, 1–3 species.

Convolvulus a widespread genus of annuals, erect or climbing herbaceous perennials and deciduous or evergreen shrubs most of which occur in warm temperate and subtropical regions of the Northern Hemisphere. The best-known species, the pernicious weed *C. arvensis* [FIELD *BINDWEED, PINK BINDWEED, LESSER BINDWEED], is found throughout most of the world. It bears pink or white, sweetly-scented, trumpet shaped flowers on long trailing or twining, herbaceous stems.

The Mediterranean *C. althaeoides* [RIVIERA BINDWEED], with deep pink flowers, is sometimes grown as a garden ornamental climber. *C. cneorum*, a shrubby Mediterranean species, has attractive silvery leaves and white flowers and, along with a few other shrubby species including *C. sabatius* (= *C. mauritanicus*), is cultivated for ornament. A few species form dwarf cushions on mountains in southern Europe, and are prized by alpine gardeners. A further group are short-lived annuals, often with attractive flowers which contribute to the colorful flower-meadows of Mediterranean regions. Of these, *C. tricolor* [DWARF MORNING GLORY] is grown as a bedding plant in a range of blue, rose and pink varieties.

Scammony, a resinous juice formerly used as a purgative, comes from the roots of *C. scammonia* [SCAMMONY] from southwest Asia. A fragrant powder known as "bois de rose" is obtained from the roots and stems of

The inflated pods of the Common Bladder Senna (Colutea arborescens). The pods burst with a loud pop if squeezed when almost ripe. ($\times\frac{3}{4}$)

C. floridus and *C. scoparius* from the Canary Isles.
CONVOLVULACEAE, 200–250 species.

Copals natural resins, some of which are collected as exudations from living trees while others are found buried in a semi-fossilized condition. The latter are rather similar to amber, but of much more recent origin. Congo (*Guibourtia demeusii*, *Copaifera* (continued on p. 100)

The Lesser Bindweed (Convolvulus arvensis). ($\times\frac{1}{10}$)

Conifers and their Allies

THE GYMNOSPERMS

ＴHE GYMNOSPERMS ARE A MAJOR GROUP OF vascular plants characterized by having seeds that are not enclosed in an ovary, unlike the flowering plants (class Angiospermae). The name in fact is derived from the Greek *gumnos* meaning "naked" and *sperma* meaning "seed". Present-day gymnosperms comprise the class Coniferae (conifers and the *MAIDENHAIR TREE – Ginkgo biloba*), class Cycadinae (the cycads) and the isolated class Gnetinae, which includes the three genera *Gnetum*, *Ephedra* and *Welwitschia*. The gymnosperms, however, are an ancient group with a long and rich fossil history extending back to the Devonian period of geological time (about 350 million years ago) and were much more numerous and varied in the past than today. The cycads and *Ginkgo* represent the last survivors of groups that were important and abundant during the Mesozoic era, declining only towards the end of this era as flowering plants advanced and replaced them. Even the conifers, although still quite numerous in genera and species, were more plentiful and varied in the past, several whole families having become extinct.

Today, the conifers still dominate certain vegetation types. Most are trees of cooler climates (either high latitude or high altitude). Few occur in tropical or subtropical lowland. Large areas in higher latitudes (boreal region) are still naturally dominated by conifer forest, though the virgin forests have largely been exploited for timber. The same is true of many mountain forests of temperate regions. Many conifers of particular value as timber trees have been planted extensively in many parts of the world. *Pinus radiata* [MONTEREY PINE], for example, is native to a limited area of California, but has been planted in many areas of Australia, New Zealand and South Africa and has become a very important timber tree. *Ginkgo biloba*, the only member of its order, is thought still to exist in the wild state in eastern China, although it was was once believed that it had only survived in cultivation.

Cycads, the second largest order of living gymnosperms, are easily distinguished by their palm-like habit. Most of the species occur in tropical or subtropical areas, with the majority in the Southern Hemisphere, where some species reach semi-dominance in the understory of forests. The only species of *Welwitschia* is found in a 800km coastal strip of southwestern Africa. The 40 or so species of *Ephedra* are mostly low, creeping shrubs found widely distributed in both hemispheres. The 30 species of *Gnetum* are mostly climbing shrubs inhabiting moist tropical forests.

Economic Importance. The "softwoods" of commerce are by definition the timbers produced by conifers. The term is not altogether appropriate, for the softest known woods are in fact derived from flowering plant trees (broadleaves) and some conifer woods, eg yew, are fairly hard. Softwoods are used not only as an important constructional material, but also in various manufacturing industries (paper, textiles, synthetic board and packing materials, chemicals etc). Resins produced by conifers are important as the source of turpentine and various other substances used in the paint, pharmaceutical and perfumery industries. As a source of food, conifers are of little value. The seed kernels of a number of pines, especially *Pinus pinea*, are marketed.

Another use of increasing importance is in horticulture. There are numerous horticul-

The cycad Encephalartos hildebrandtii, *from Africa, here showing a display of male cones.* ($\times \frac{1}{20}$)

tural varieties of many conifers which have been selected for their special ornamental value. Some members of the cypress and pine families are particularly rich in named horticultural varieties, many of them of dwarf habit and prized for "alpine" gardens. The only known intergeneric hybrids among conifers (*× Cupressocyparis*) have arisen in cultivation. The best-known hybrid is × *C. leylandii*, a cross between *Chamaecyparis*

The Cedar of Lebanon (Cedrus libani) is one of the most majestic conifers and is widely used for specimen planting.

nootkatensis and *Cupressus macrocarpa*. This hybrid species is remarkably vigorous and is now widely grown as a hedge plant and screen-forming tree.

Other gymnosperms are of little practical use. Cycads were cultivated as interesting "stove plants" by Victorian collectors and are still grown outdoors as ornamentals in the warm regions of the world. A kind of sago can be made from the soft, starch-rich pith tissues of certain cycads but the plants are far too slow-growing to be of any real commercial value as a food source. Seed kernels of *Ginkgo* are eaten in the Far East where the tree has been cultivated for many centuries as an ornamental.

Structure and Reproduction. The major groups of gymnosperms have certain features of structure and life history in common. Unlike the flowering plants, all gymnosperms are woody trees or shrubs in which the primary vascular (conducting) tissue soon becomes augmented by the production of secondary vascular tissue (wood).

The leaves of gymnosperms vary greatly in form and anatomy but most share certain features, such as a thick cuticle, sunken stomata and a certain amount of fiber or stony tissue in the internal parts. Such features are often associated with plants inhabiting dry regions and are called xeromorphic characters.

Most gymnosperms possess secretory canals. In all conifers, except the *yews (Taxus spp), resin canals occur in leaves and stems and in some, notably certain members of the pine family, resin canals are present in the wood itself. Resin may exude from the cut stems of conifers and may sometimes exude naturally, emerging as droplets on the surface of cones.

Reproduction. As in flowering plants, the ultimate result of sexual reproduction is the seed. The processes, as well as the morphology of the pollen- and seed-bearing parts, however, are significantly different in gymnosperms. There are also important differences among the various major gymnosperm groups. Pollen is produced in capsules (called pollen sacs or microsporangia), which are borne on microsporophylls. These vary in detailed morphology from one major group to another but in most living gymnosperms they are scale-like, with the pollen sacs on the underside. The pollen sacs are generally few in number, commonly two in conifers and *Ginkgo*, but much more numerous in cycads. The ovules (later the seeds) are borne on scales (macrosporophylls) usually in cones. In conifers, female cone morphology constitutes an important basis for classifi-

cation into families. Pollination is by wind and following fertilization the seed is produced.

In conifers the seed is shed when mature and contains a differentiated embryo that is embedded in endosperm (food reserves). It usually separates from the scale on which it was borne but sometimes, as in species of

Autumn foliage of the Maidenhair Tree (Ginkgo biloba), *a tree often grown for ornament, but once thought to be extinct in the wild.* ($\times \frac{3}{4}$)

Araucaria, for example, the whole scale, together with its embedded seed, is shed as a unit. In cycads and in *Ginkgo* the seeds are often shed before embryo formation, sometimes as early as about the stage of fertilization. This is a peculiar habit which has attracted some interest in these obviously primitive seed-plants.

Gymnosperm seeds do not show the range of specialized structural features associated with dispersal that flowering plant seeds present. Among conifers, the development of some kind of wing, thus increasing the surface area, may be regarded as a modification aiding wind-dispersal. In many members of the cypress family the wall is expanded laterally into two wing-like portions, while in the pine family the single seed-wing attached to the base of the seed is formed from the upper surface of the scale. In *Juniperus* the scales may become fleshy so that the cone forms a kind of berry at maturity and in the yew there occurs a cup-like growth (aril) around the seed, which becomes fleshy and bright red at maturity. Such modifications may be regarded as aiding dispersal by animals. In cycads the seeds tend to be hard and very large, up to several centimetres in diameter in some. Dispersal of such seeds is particularly inefficient.

Welwitschia bainesii, *from the coastal deserts of southwestern Africa, is quite unlike other gymnosperms. This specimen is reckoned to be 2000 years old.* ($\times \frac{1}{40}$)

demeusii), Sierra Leone or West African (*C. guibourtiana*), Zanzibar or West African (*Trachylobium verrucosum*) and South American and Central American (chiefly *Hymenaea courbaril*) copals all derive from leguminous species. *Kauri copal, from New Zealand, and Manilla copal are fossilized and semi-fossilized resins from species of *Agathis* (Araucariaceae). Copal used to be very important in the manufacture of varnishes, but is now being superseded by synthetic resins.

Copernicia a genus of American and West Indian palms, the most important of which is *C. prunifera* (= *C. cerifera*) (see Carnauba wax).
PALMAE, about 30 species.

Coprinus [INK CAPS] a genus of gill fungi characterized by the early deliquescence or breakdown of the gills into a black ink-like fluid, at about the time of spore maturation. Most species are found growing on dung, manured ground etc, but also on wood. A few species, especially *C. comatus* [SHAGGY

The Magpie Ink Cap (Coprinus picaceus) is an inedible species which grows in European Beech woodlands. Some species are edible. ($\times \frac{3}{4}$)

CAP, SHAGGY MANE, LAWYER'S WIG] are edible. On the other hand, *C. atramentarius* causes severe nausea and vomiting if consumed with alcohol.
COPRINACEAE, about 110 species.

Coptis a small genus of low, white-flowered, perennial herbs native to the Arctic and north temperate regions. No member is of commercial importance although extracts of a few species have local medicinal uses. *C. asplenifolia* and *C. orientalis* are two white-flowered species sometimes cultivated for ornament. The roots of *C. brachypetala*

The fruiting bodies of Cordyceps ophioglossoides, which is a parasite on the underground fruiting body of the False Truffle (Elaphomyces sp). ($\times 2$)

(Japan) yield a yellow dye used locally.
RANUNCULACEAE, about 15 species.

Corallina a genus of small red algae in which the thallus is heavily calcified (coralline). They are found on seashores and especially in rock pools. The internodal sections of the thallus become encrusted with calcium carbonate giving the plant a stiff, articulated appearance.
RHODOPHYCEAE.

Corchorus a genus of annuals native to the tropics. The stems of *C. capsularis* and *C. olitorius* are the main sources of the important fiber *jute. The leaves and young shoots of *C. olitorius* are used as potherbs in eastern Mediterranean countries.
TILIACEAE, about 100 species.

Cordia a genus of shrubs and trees native to the tropics and subtropics, some of which are cultivated as ornamentals. *C. sebestena* [GEIGER TREE], from the southeast USA, the West Indies and Venezuela, is an evergreen shrub or small tree with sizeable terminal clusters of funnel-shaped, orange-red flowers and edible fruits. *C. myxa* from India to Australia is a small deciduous tree with smaller white flowers and blackish edible fruits used in local medicine. A considerable number of species, including *C. dentata* (= *C. alba*) (tropical America), *C. dodecandra* (Central America), *C. irwingii* (West Africa) and *C. gerascanthus* [SPANISH ELM] (tropical America, West Indies), produce valuable timber used for carpentry, construction work or cabinet-making.
EHRETIACEAE, about 250 species.

Cordyceps a genus of fungal parasites of insects or false truffles (*Elaphomyces*). Forms parasitic on insects, such as *C. militaris* and *C. gracilis*, develop club-shaped, often brightly colored fruiting bodies on the

mummified bodies of the host insect larvae.
CLAVICIPITACEAE, about 100 species.

Cordyline a large genus of tufted or tree-like plants, some up to 80m high, with the leaves clustered at the ends of the branches. The species are mostly tropical in distribution (India, Australia, and one species in tropical America), growing in open forest, particularly if moist, swamp margins and scrub. The seeds are very rich in fatty acids, particularly the commercially valuable linoleic acid, but are not exploited on any scale.

Several species and hybrids are valuable horticultural subjects, both for landscaping and as tub plants, notably *C. terminalis* [GOOD LUCK PLANT] with its pinkish, purple, yellow and green leaves and yellowish, white or reddish flowers, and *C. australis* [CABBAGE TREE, PALM LILY] with its long panicles of fragrant white flowers.
AGAVACEAE, about 15 species.

Coreopsis [TICKSEED, TICKWEED] a large genus distributed throughout North and Central America, Hawaii and tropical Africa. They are mostly summer- and autumn-blooming annual or perennial herbs (or seldom shrubs) with showy clustered or solitary daisy-like flower heads. Several species, such as *C. grandiflora*, *C. lanceolata* (perennials) and *C. tinctoria* [CALLIOPSIS] (annual), are attractive garden plants cultivated in Europe and the USA.
COMPOSITAE, about 120 species.

Coriander the fruit of *Coriandrum sativum*, one of the most ancient spices, herbs and flavorings known. *C. sativum* is an annual

The Shaggy Cap, Shaggy Mane or Lawyer's Wig (Coprinus comatus), an edible ink cap fungus. The bases of the gills are already deliquescing into the black, ink-like fluid that gives the genus its popular name, ink cap. ($\times \frac{3}{4}$)

A Flowering Dogwood (Cornus florida) growing in its native eastern USA; it is also a popular ornamental, both for its flowers and its autumn foliage. (× 1/20)

herb, native to the Mediterranean region, in North Africa and western Asia and widely naturalized in other parts of the world, where it has probably escaped from cultivation. The leaves [GREEN CORIANDER] are still very commonly used in cooking in many subtropical and tropical countries.

The fruits are hard, ovoid or globose, and ribbed. They are used, either entire or ground, to flavor soups, stews, sausages, cakes, liqueurs, gins, and as an ingredient of curries and pickles. Coriander oil, extracted from the seeds, is used as a flavoring in confectionery. It is also used in soaps and perfumes. Medicinally, coriander fruits are employed as a carminative, diuretic and aphrodisiac.
UMBELLIFERAE.

Cork   a layer of dead cells produced on the outside of woody stems and roots. The early history of cork is of considerable significance, for it is in that tissue that Robert Hooke first observed in 1665 what he described as "cellula" or cells. These appeared to him as small cavities surrounded by walls.

Cork is a specialized tissue with waterproof properties, and its occurrence in plants is usually associated with protection. Cork layers of the outer bark replace the epidermis of the primary stem as secondary growth in thickness takes place. Cork also normally develops in perennial plants over scars left after fruit, leaves or branches have fallen. Wounds quite often become covered by cork layers as part of the healing process.

From antiquity the peculiar properties of cork have led to its commercial exploitation. The commercial source is *Quercus suber [CORK OAK], a tree native to western Mediterranean countries. Because cork cells are mostly air-filled, they have good heat- and sound-insulating properties which are put to use in lagging and table mats, amongst other things. The water- and oil-resisting properties find application in gaskets, for example. Because it also has a high coefficient of friction, cork makes successful bottle stoppers. It is both compressible and resilient, making it good for floor tiles and parts of

footwear. Cork in granulated or other forms is used in a very wide range of products.

Cornel   the common name for several members of the genus *Cornus [*DOGWOODS], notably *C. canadensis* [DWARF CORNEL], *C. officinalis* [JAPANESE CORNEL or JAPANESE CORNELIAN CHERRY]; *C. mas* is known as CORNELIAN CHERRY.
CORNACEAE.

Cornflour   the finely ground, starchy flour derived from the milled grains of non-glutinous varieties of *Oryza sativa [*RICE]. Cornflour is preferred to other milled grain products in cooking, particularly for thickening sauces and puddings, since it consists for the most part of very fine starch and does not impart the mealy taste of flours derived from most other cereals. However, a similar product derived from *Zea mays [*MAIZE] is also called cornflour (or cornstarch) and is used in cooking and as an industrial starch.
GRAMINEAE.

Cornus   see DOGWOOD.
CORNACEAE, about 50 species.

Corokia   a small genus of evergreen shrubs mainly native to New Zealand, all of which are cultivated for ornament. The yellow flowers are small and star-like, usually arranged in panicles as in *C. buddleioides*, in racemes as in *C. macrocarpa*, or in small clusters as in *C. virgata*. The fruits are an attractive feature, globose or ovoid in shape and normally red or orange in color.
ESCALLONIACEAE, 5–6 species.

Coronilla   a genus from the Mediterranean, western Asia and the Canary Islands. They are herbs or shrubs with clusters of pea-like yellow, purple or white flowers. Several species are cultivated for their ornamental value, including the shrubby *C. emerus* [SCORPION SENNA], *C. glauca* and *C. valentina*, and the herbaceous *C. varia* [CROWN VETCH].
LEGUMINOSAE, about 20 species.

Correa   a small genus of shrubs and small trees native to Australia. The commonly cultivated species are *C. alba* [BOTANY BAY TEA TREE] and *C. reflexa* (= *C. speciosa*), both of which are small shrubs with showy flowers.
MYRTACEAE, 11 species.

Cortaderia   a small genus of perennial grasses, native to South America and New Zealand. The best-known is the widely cultivated ornamental *C. selloana* (= *C. argentea*) [PAMPAS GRASS] from Brazil, Argentina and Chile. It is a tufted plant, forming large clumps with leaves up to 3m long, and beautiful erect silvery-white to pink inflorescences up to 60cm long.
GRAMINEAE, about 15 species.

Cortinarius   the largest genus of gill fungi, mainly native to Europe. They are predominantly mycorrhizal species found in temperate zones, and are rare in the tropics

Coreopsis stillmanii is one of the Tickseeds grown in gardens. It is a hardy annual suitable for lighter soils and a sunny position. (× 1/12)

except at high altitudes. These fungi are of major importance in forestry because of their mycorrhizal associations with trees. Many of the larger species such as *C. emodensis* are edible. However, *C. orellanus* and *C. speciosissimus*, among others, are poisonous.
CORTINARIACEAE, more than 500 species.

The Crown Vetch (Coronilla varia) showing the pea-like flowers and pinnate leaves so characteristic of many members of the pea family. (× 1/2)

Cortusa a small genus of perennial, often hairy herbs, native to mountainous regions from central Europe to Japan. Cultivated species have heart-shaped long-stalked leaves and tubular to bell-shaped flowers, yellow in *C. semenovii*, purple in *C. matthioli*. PRIMULACEAE, about 10 species.

Corydalis a large genus of mostly perennial herbs distributed over the north temperate region and South Africa. None are of any great economic importance but some are cultivated as ornamentals, particularly in rockeries and alpine houses. Two of the best-known cultivated species are the yellow-flowered *C. saxicola* (= *C. thalactrifolia*) and *C. lutea* [YELLOW FUMITORY]. The tubers of some species, notably *C. bulbosa* (= *C. solida*), are used as a vegetable.
FUMARIACEAE, about 300 species.

Corylopsis a genus of deciduous shrubs and

Corylopsis veitchiana, *a fragrant-flowered shrub, native to China, with leaves similar to those of the hazels* (Corylus spp). (× $\frac{1}{16}$)

small trees native to East Asia and the Himalayas. A number of species, such as *C. glabrescens*, *C. pauciflora* [BUTTERCUP WINTER HAZEL] and *C. spicata*, are cultivated for their attractive fragrant flowers, which are produced early in the season, before the leaves, in catkin-like pendulous spikes or racemes.
HAMAMELIDACEAE, about 20 species.

Corylus see HAZEL.
BETULACEAE, about 15 species.

Corypha a small genus, from tropical Asia to Australia, of huge solitary fan palms forming a gigantic terminal tree-like inflorescence, then dying. *C. elata* [GEBANG PALM] is the most widespread species, extending from Bengal and Burma to northern Australia. The most massive species is *C. umbraculifera* [TALIPOT PALM], from Sri Lanka and the Malabar coast of India, whose 6m-long inflorescence is reputed to be the longest in the plant kingdom. For centuries these two species were favored in India to produce the paper used for permanent documents.
PALMAE, about 8 species.

Coryphantha a genus of small globose or cylindrical cacti native to Mexico, Cuba and the southwestern USA. Many species are cultivated for ornament, including *C. andreae*, *C. bumamma* and *C. palmeri* all with yellow flowers, *C. asterias* (white to pink flowers), *C. macromeris* (= *Mammillaria macromeris*) (purple flowers) and *C. vivipara* var *arizonica* (= *M. arizonica*) (pink flowers).
CACTACEAE, about 64 species.

Cosmos a genus of annual and perennial herbs with usually pinnately divided, opposite leaves, native to tropical America and the southwestern USA. A number are cultivated as garden-border or pot plants for their showy capitulate inflorescences (white, crimson-rose and pink varieties from *C. bipinnatus* and predominantly yellow and orange varieties from *C. sulphureus*).
COMPOSITAE, about 25 species.

Costmary the common name for *Balsamita major* (= *Chrysanthemum balsamita*), an herbaceous perennial bearing sweet aromatic leaves which are used as a herb in cooking, and were once used to flavor beer.
COMPOSITAE.

Costus a large tropical genus of perennial rhizomatous herbs many of which are cultivated for their ornamental attractively marked leaves and showy flowers. Popular species include *C. malortieanus* (= *C. zebrinus*) [STEPLADDER PLANT] with yellow flowers, marked with red, and *C. pulverulentus* (= *C. sanguineus*) [SPIRAL FLAG] with fleshy greenish-blue leaves with a central silvery band and a greenish yellow zone at the margin. The leaves are deep red on the lower surface. Various species of *Costus* are used in native medicine, including the tropical African *C. afer* [GINGER LILY] and the Central American *C. spicatus*.
COSTACEAE, about 150 species.

Cotinus a small genus of deciduous shrubs and trees represented by species sometimes included in the genus *Rhus*. *C. obovatus* (= *C. americanus*), from America, and *C. coggygria* (= *Rhus cotinus*) [SMOKE TREE], from Europe and Asia, are cultivated for their brilliant autumn coloration, their leaves turning various shades of scarlet and purple, or orange in one cultivar.
ANACARDIACEAE, 3 species.

Cotoneaster a genus of deciduous and evergreen shrubs and small trees native to the Northern Hemisphere, with main centers of distribution in western China and the Himalayas; many species are cultivated as ornamentals.

The genus exhibits a wide range of habit, from prostrate, ground-loving plants (eg *C. dammeri*) to erect bushes and trees (eg *C. frigidus*).

Many species of *Cotoneaster* are cultivated in gardens for their attractive clusters of flowers, red or blackish fruits and foliage. Some of the deciduous species (eg *C. adpressus*) have leaves that change from green to rich golden red in autumn. Many of the evergreen species (eg *C. conspicuus*) are useful for hedges and screens, and the creeping forms (eg *C. dammeri*) are grown as ground cover or in rock gardens. Another popular garden species is the spreading ever-

Cotoneaster microphyllus *photographed in its natural habitat at 1 800m in central Nepal. It is also a popular garden ornamental.* (× $\frac{3}{4}$)

Defoliant being applied mechanically to a field of cotton in the USA before harvesting of the mature cotton fruits (bolls).

green shrub *C. conspicuus*, especially var *decorus*, which is more prostrate and therefore suitable for rock gardens.

Useful rock-garden or border-edge species include *C. adpressus*, *C. microphyllus* and *C. horizontalis*. The deciduous *C. bullatus* and *C. frigidus* and the evergreen *C. salicifolius* are commonly grown taller shrubs. Many hybrids are also cultivated, including cultivar 'Cornibus', a tall, vigorous, semievergreen shrub, probably the result of a cross between *C. frigidus* and *C. henryanus*.
ROSACEAE, about 50 species.

Cotton the most important fiber-producing plant of the world. It is the product obtained from the seeds of a number of species of the subtropical genus of subshrubs and shrubs, *Gossypium*. The crop is grown in some 70 countries, major producers being the USA and the USSR.

Although naturally a perennial shrub, the cotton plant is normally grown as an annual crop, eventually growing to a height of 1–1.2m. The flower matures into a fruit (boll) which splits to expose the fluffy white seed cotton for picking.

Commercially important cotton is grown as a large number of varieties belonging to one or other of the four species: *G. barbadense* (from the West Indies), *G. hirsutum* (from Central America), *G. arboreum* (from Africa, Arabia and India) and *G. herbaceum* (from Asia).

The two most valuable types of cotton, with the longest staple (seed hair length), are SEA ISLAND COTTON and EGYPTIAN COTTON, both cultivars of *G. barbadense*. The short-stapled Asiatic cottons belong to the two Old World species and that of *G. herbaceum* in particular is used for poorer quality fabrics, carpets etc and is often blended with wool.

Despite the extensive development of man-made fibers, cotton is still of great importance in the world economy. The current oil crisis has highlighted the dependence of synthetic fibers on the petrochemicals industry. With the likely future restraint on the economic production of such fibers, cotton would appear to have an assured future.

Cotton is harvested as seed cotton, ie with the fiber still attached to the seed, and this has to be separated in the ginning factory into lint, the raw cotton of commerce, and seed. The seed is also extremely valuable for its considerable oil content, which after extraction yields a high protein residue (cotton "cake"), the well-known cattle feed. Other parts of the plant, notably the stalks which contain a fiber used in papermaking, add to the high commercial value of cotton.
MALVACEAE.

Cotula a genus of annual to perennial creeping herbs from the Southern Hemisphere, mainly South Africa, Australia and New Zealand; a few species are naturalized in the Northern Hemisphere. Mat-forming in habit, they bear yellow or purple daisy-like flowers. *C. coronopifolia* [BRASS BUTTONS], *C. reptans* and *C. squalida* are being used increasingly in gardens as ground cover and carpeting plants.
COMPOSITAE, about 50 species.

Cotyledon a genus of tender succulent plants from South and Southwest Africa, with outliers in Ethiopia and southern Arabia. The flowers, often yellow, orange or red, are relatively large, tubular and pendent. *C. paniculata* and *C. wallichii* are shrubs with grotesque, thick trunks, peeling bark and fleshy but deciduous leaves – the whole forming a characteristic feature of the African landscape over wide areas of the Cape Province. Some species, including *C. paniculata*, *C. undulata* and *C. orbiculata* [PIG'S EAR] are cultivated in warm temperate gardens, or

under glass, only the latter being at all widely grown.
CRASSULACEAE, about 40 species.

Courgette a dwarf bush variety of the annual herb *Cucurbita pepo* [MARROW]. The green, cylindrical, ribbed fruit is eaten as a vegetable, either pickled, boiled or in stews.
CUCURBITACEAE.

Couroupita a small genus of trees from tropical America and the West Indies, one species of which, *C. guianensis* [CANNONBALL TREE], is widely cultivated for its large showy flowers and reddish-brown woody capsular fruits up to 20cm in diameter ("cannonballs").
LECYTHIDACEAE, about 20 species.

Cowpea an ancient and valuable annual herbaceous legume, *Vigna unguiculata*, grown widely throughout the tropics and subtropics. *V. unguiculata* is currently divided into five interfertile subspecies, two of which are wild and three of which are cultivated, viz. the ubiquitous subspecies *unguiculata* (= *V. sinensis*) [COMMON COWPEA], the Indian and Far Eastern subspecies *cylindrica* (= *V. catjung*) [HINDU PEA, JERUSALEM PEA, CAJANG] and subspecies *sesquipedalis* (= *V. sesquipedalis*) [YARDLONG BEAN, ASPARAGUS PEA, ASPARAGUS BEAN]. The many cultivars may be grouped according to floral, pod and seed characters as well as habit (erect, suberect, trailing or climbing).

In Sri Lanka, cowpeas are grown as a cover crop in rubber plantations. The dried seeds form an important pulse crop and may be used as either human or cattle food. They are an important source of protein (especially of lysine) in the staple diets of subsistence and peasant-farming communities of semi-arid Africa and Asia. The leaves and stems of immature plants are used as a potherb in

Succulent leaves and woody stem of Cotyledon orbiculata, *crowned by a cluster of red waxy flowers.* ($\times \frac{1}{2}$)

tropical Africa. The fresh seeds and green, immature pods may be eaten as a boiled vegetable, like French beans. The entire plant may be harvested if it is to be used as a forage or silage crop, or it may be ploughed in as a green manure.
LEGUMINOSAE.

Crab apple a name freely applied to all sour-fruited apple trees (*Malus* species). They may be cultivars and varieties of *M. baccata* (= *Pyrus baccata*) [SIBERIAN CRAB], *M. coronaria* [AMERICAN CRAB APPLE, WILD SWEET CRAB], *M. angustifolia* (= *Pyrus angustifolia* [SOUTHERN WILD CRAB APPLE], *M. floribunda* (= *M. pulcherrima*) [SHOWY CRAB APPLE], *M. prunifolia* (= *Pyrus prunifolia*) [PLUM-LEAVED OF CHINESE CRAB APPLE], *M. ioensis* [WILD OF PRAIRIE CRAB APPLE] and some other less well-known species and hybrids.

Many have fragrant, white or pink-tinged blossoms and some, such as cultivar 'Plena' of *M. ioensis* have double flowers. Many, such as *M. floribunda* and *M. prunifolia* cultivar 'Rinkii' (= *M. ringo*), are cultivated for their showy red or yellowish fruits. Cider is made from the acid fruit of *M. coronaria* – this and many other crab apples also make good preserves and sauces.
ROSACEAE.

Crambe [SEA KALE] a genus of vigorous annual or perennial herbs native to the Canary Islands, Europe, western and Central Asia and tropical Africa. *C. maritima* has long been cultivated in England as a forced vegetable, the blanched leaf stalks usually being boiled. *C. cordifolia,* from the Caucasus, is cultivated for its striking appearance – up to

Crassula comptonii is one of the miniature crassulas that are popular in cultivation. The flowers emit an intense, sweet fragrance. (×2)

2m high with basal leaves up to 60cm across with large flowers in a big terminal panicle.
CRUCIFERAE, about 20 species.

Cranberry the name given to a few of the species of *Vaccinium*, a genus widely distributed throughout the Northern Hemisphere. All cranberries are low, creeping, shrubby plants.

V. oxycoccos [EUROPEAN CRANBERRY, SMALL CRANBERRY] is native to bogs and swamps in temperate and arctic regions of Asia, Europe and North America while *V. vitis-idaea* [CRANBERRY OF COWBERRY], also occurs in Europe and northern Asia. *V. vitis-idaea* var *minus*, from North America is known as the MOUNTAIN OF ROCK CRANBERRY. The berries of all these species are edible and are sometimes used for jellies, but none is grown on a commercial scale. *V. macrocarpon* [AMERICAN CRANBERRY] is native to swamps with acid soils, and is cultivated commercially in a range of varieties in northeast and northwest USA and in Canada.

The fruit is canned or consumed cooked or as a jelly, or is made into a beverage. Cranberry sauce is the traditional accompaniment to roast turkey in the USA.
ERICACEAE.

Crassula a large, mainly South African, genus of shrubs and perennial herbs with a diverse vegetative habit, ranging from thick stemmed shrubs to dwarf rosette plants. Many species are cultivated in the open, or under glass, for their vegetative form, including their often succulent attractive leaves.

C. argentea (= *C. portulacea*) [JADE PLANT] forms a miniature tree with smooth trunk and thick glossy green leaves. A cultivar with variegated foliage is also in cultivation, and both are popular houseplants. *C. falcata* has long enjoyed popularity for its broad corymbs of rich pink flowers, but is now rivaled by

Sea Kale (Crambe maritima) is common around seashores, and is here seen establishing itself in crevices in an unstable rubble cliff. (× $\frac{1}{10}$)

recent hybrids. *C. arborescens* (= *C. cotyledon*) is a profusely branched shrubby plant with rounded grayish-green leaves, ringed red at the margin. In warmer temperate zones it produces white flowers which turn red with ageing.
CRASSULACEAE, about 300 species.

Crataegus see HAWTHORN.
ROSACEAE, about 200 species.

Cratoneuron a cosmopolitan genus of mosses with shoots often regularly pinnately branched. The best-known species are *C. filicinum* and *C. commutatum*.
HYPNACEAE, about 15 species.

Crepis [HAWKSBEARDS] a large genus of perennial, biennial and annual herbs, widely distributed in the Northern Hemisphere and tropical and South America. They bear a rosette of leaves and a tall dandelion-like inflorescence of yellow, orange, pink or white flower heads. As ornamentals, the red- and pink-flowered species *C. rubra* and *C. incana* are grown in gardens. The weeds *C. capillaris* [SMOOTH HAWKSBEARD] and *C. vesicaria* subspecies *taraxacifolia* [BEAKED HAWKSBEARD], together with the rare alpine *C. incana*, inhibit bacterial growth of *Staphylococcus aureus*.
COMPOSITAE, about 200 species.

Crescentia a genus consisting of a single small tree species native to tropical America, *C. cujete* (see CALABASH TREE). Other species formerly included in *Crescentia* are now referred to *Enallagma*.
BIGNONIACEAE, 1 species.

Cress the name given to a number of species whose young shoots are used as a green salad or as a garnish. They include *Lepidium sativum* [TRUE CRESS], *Barbarea praecox* [AMERICAN OF LAND CRESS] and *Nasturtium officinale* and *N. officinale* × *microphyllum* [WATER CRESS].
CRUCIFERAE.

Crinum a genus of bulbous herbs of the tropics and subtropics of both hemispheres, found notably in southern Africa and Asia, characteristically near the coast and sometimes in marshes. The bulbs of some species are poisonous to Man and animals. *C. asiaticum* [ASIATIC POISON BULB] was formerly used in the East as an emetic and for poulticing. *C. kirkii* [PYJAMA LILY], from Zanzibar, yields an efficient rat poison from the outer scales of its bulb. Many species are cultivated as pot plants and several good hybrids have been raised, such as *C.* × *powellii* (*C. bulbispermum* × *C. moorei*). AMARYLLIDACEAE, about 100 species.

Crithmum [SAMPHIRE] a genus consisting of a single species, *C. maritimum*, which occurs on sea cliffs, sand or shingle, around Mediterranean and western European coasts. It is a perennial herb, somewhat woody at the base. SAMPHIRE is collected and cultivated for its aromatic succulent pinnate leaves which are salted, boiled and pickled in vinegar. UMBELLIFERAE, 1 species.

Crocosmia a small genus of perennial herbs native to South Africa. A number of species and hybrids are cultivated in gardens, such as *C. aurea* (= *Tritonia aurea*) with its tall stems of double-ranked tubular orange or yellow flowers. IRIDACEAE, 6 species.

Crocus a small genus of cormous herbs with yellow, purple or white flowers, distributed mainly in Europe and the Mediterranean, but also reaching as far east as the Tienshan mountains of western China. *Crocus* has been cultivated for ornament at least since the 16th century when forms of *C. flavus*, *C. vernus* and *C. nudiflorus* were popular. Today an increasing number of spe-

The Pyjama Lily (Crinum kirkii) flowering after the first rains. The outer scales of the bulb make an efficient rat poison. (× ⅙)

Crocuses produce a welcome burst of color in early spring and are consequently often planted in parks and gardens.

cies, such as *C. speciosus* and *C. tomasinianus*, are being grown, together with a wide range of cultivars. These include forms of *C. biflorus*, F₁ hybrids between *C. biflorus* and *C. chrysanthus*, such as *C. chrysanthus* 'Advance' and the large polyploid forms of *C. vernus*, such as *C.* 'Pickwick'.

Saffron, the dried stigmata of *C. sativus*, has been of economic importance since very early times and was probably already valued as a dye, perfume and medicament in 1600 BC. The "saffron robes" of religious orders occur many times in history and in different parts of the Old World. The culinary uses of the herb are many, eg saffron rice, saffron cakes etc. The Romans scattered the flowers in theaters and public meeting places to scent the air, and saffron has long been used as a panacea for many ills and and as a general tonic. It is, in fact, the richest known source of vitamin B₂. *C. sativus* is a sterile triploid clone still cultivated in Turkey, the Middle East, France and Spain.

Recent research has shown that *Crocus* is a highly complex genus cytologically with a very wide range of chromosome numbers, karyotypes and variation, even within species. IRIDACEAE, about 90 species.

Crossandra a genus of mainly tropical herbs and shrubs, a number of which are cultivated in temperate countries as greenhouse ornamentals. Two of the most popular cultivated species are the tropical African *C. nilotica* (with red flowers), and *C. infundibuliformis* (with orange red flowers) from southern India and Sri Lanka. ACANTHACEAE, about 50 species.

Crotalaria [RATTLE BELL, RATTLE BOX] a large genus of tropical and subtropical shrubs and herbs, cosmopolitan in distribution but concentrated in Africa. *C. juncea* [SANN or SUNN HEMP] is an important green manure crop in India, is second only to jute in the production of twine and cord and is also used in making cigarette and tissue paper. *C. anagyroides*, and *C. mucronata* are grown as green manure crops in the Old World tropics and Florida (USA), respectively. Other species are similarly used, but many contain toxic alkaloids: *C. burkeana* causes crotalism (styfsiekte) in cattle in southern Africa, inflaming

Crocosmia masoniorum, with its tightly packed heads of flowers on arching stems, is the hardiest of the Crocosmia *species. (× ¼)*

Codiaeum variegatum, *a popular houseplant, is commonly known as Croton, but does not belong to the genus of that name.* (×1/10)

the horn-forming tissues of the hoof and *C. dura* causes pulmonary and liver diseases in cattle.

Many species are highly ornamental and have been introduced to Europe as greenhouse plants (eg *C. agatifolia* [CANARY BIRD BUSH]) while the seeds of others (eg *C. striata*) have been used as a coffee substitute.
LEGUMINOSAE, about 600 species.

Croton a very large genus of herbs, shrubs and trees from all the tropical and subtropical regions of the world, including 600 American species. *C. tiglium* [CROTON OIL PLANT, PURGING CROTON] a shrub or small tree of India and Sri Lanka is the source of croton oil, prepared from the dried ripe seeds, which has human and veterinary medicinal uses as a purgative. *C. laccifer*, another species of India and Sri Lanka, is a host plant to lac-producing insects (*Laccifer lacca*) and important in the varnish industry.

C. glabellus of Central America produces a hard dense fine-grained yellow-brown wood which is used in building construction work and for box and container manufacture. Several Brazilian species yield a "*dragon's blood" resin. *C. cascarilla* (Bahamas) is the source of cascarilla bark from which a tonic

is prepared. The genus *Croton* should not be confused with the widely cultivated foliage plants of horticulture, genus *Codiaeum, to which the common name CROTON is applied.
EUPHORBIACEAE, 700–1000 species.

Cryptanthus [EARTH STARS] a genus of epiphytic, stemless herbs native to Brazil. Some species are grown as greenhouse plants in pots or in moss on tree trunks for their attractive stiff leaves with prickly margins. In *C. acaulis* [STARFISH PLANT] these are up to 15cm long and wavy, green on the upper surface and whitish below. *C. zonatus* [ZEBRA PLANT] has rather longer leaves (up to 23cm), cross-banded with white or brown stripes on the upper surface.
BROMELIACEAE, about 22 species.

Cryptocarya a genus of trees from the tropics and subtropics of Africa, Asia, Australia, and South America. *C. erythroxylon* [ROSE MAPLE], *C. glaucescens* [SILVER SYCAMORE, BROWN BEECH, JACK WOOD] and *C. oblata* from Australia produce timber used for joinery. *C. latifolia* in Africa yields nitronga nuts which contain a vegetable fat. The fruits of *C. moschata* [BRAZILIAN NUTMEG] are used as a spice. Some species are planted as street trees.
LAURACEAE, about 200 species.

Cryptogramma a small genus of small

alpine or boreal ferns found in America, Europe and Asia. The leaves are much divided and the edges of the fertile segments are reflexed, protecting the sori. *C. crispa* [PARSLEY FERN] is often cultivated in rock gardens. POLKPODIACEAE, 4 species.

Cryptomeria a genus represented by a single species, *C. japonica* [JAPANESE RED CEDAR], an important evergreen ornamental and timber tree in its native Japan where about one-third of the area under forestation is devoted to *C. japonica*. It is also widely grown as an ornamental in mild temperate regions. It produces a durable wood that is easy to work and has the advantage of being resistant to insects. The wood is used in building construction and for furniture. The bark is used as a roofing material.
TAXODIACEAE, 1 species.

Cucumber the common name for the annual vine *Cucumis sativus* and its fruit which is used mainly as a salad vegetable. The plant is of Indian origin. Fruits vary greatly in size, shape and color, according to cultivar; commonly they are often cylindrical and shortly spiny. There are three major classes of cultivars: field or ridge cucumbers with black or white spines; English or forcing cucumbers with long, almost spineless, seedless fruits; and pickling cultivars with small fruits used in the production of gherkins. Cucumbers are grown from the tropics to temperate regions, in cool temperate countries mostly under glass.
CUCURBITACEAE.

Cucumis a genus of trailing or climbing annual or perennial vines with tendrils, native to the Old World. Most species are African, and a few are found in Asia and Australasia. Many of the wild species are poisonous and, like *C. myriocarpus* [WILD CUCUMBER, BITTER APPLE] in southern Africa, are known to have poisoned livestock. Two of the four species known to be cultivated are widespread important food crops: *C. sativus* [*CUCUMBER], probably originating in India, and *C. melo* [MUSK MELON, *MELON, *CANTALOUPE]. The fruits of *C. anguria* [WEST INDIAN *GHERKIN] are used mainly in pickles,

Foliage, mature cones (brown), immature cones (left), and male cones (right) of the Japanese Red Cedar (Cryptomeria japonica). (×1/2)

A variety of Squash (Cucurbita pepo), from tropical America. Like its close relative the Marrow, it is eaten as a vegetable. ($\times\frac{1}{6}$)

and grown also as a curiosity. *C. melo* is indigenous to Africa, south of the Sahara, the wild forms in India probably being escapes from cultivation.
CUCURBITACEAE, about 25 species.

Cucurbita a small but economically important genus native to the New World with the center of diversity near the Mexico-Guatemala border. They are annual or perennial herbaceous vines with tendrils and large, solitary, short-lived, yellow flowers. Several perennial species, such as *C. foetidissima*, are xerophytic and inhabit the arid regions of southern North America. Five species are known only in cultivation in temperate to tropical regions. They are *C. pepo* [VEGETABLE MARROW, COURGETTE, SUMMER and WINTER SQUASHES, PUMPKIN, ORNAMENTAL GOURD], *C. maxima* [WINTER SQUASH, PUMPKIN], which produces the largest known fruits, *C. moschata* [WINTER SQUASH, PUMPKIN], the species most suited to and widely grown in the tropics, *C. mixta* [WINTER SQUASH, PUMPKIN, CUSHAW] and the perennial *C. ficifolia* [FIG-LEAF GOURD], which is restricted to the highland tropical regions of Mexico, Central America, and part of South America.
CUCURBITACEAE, about 26 species.

Cumin seeds the fruits of *Cuminum cyminum*, an annual herb originally native to southwestern Asia and North Africa, now widely cultivated in the Mediterranean region. Frequently confused with *CARAWAY, which has largely replaced it, CUMIN is used in curries, and to flavor cheese, bread, cakes etc.
UMBELLIFERAE.

Cunninghamia [CHINESE or CHINA FIR] a genus of coniferous trees native to China and Taiwan. The CHINESE FIR (*C. lanceolata*) is infrequently found in cultivation but in its native China it is an important timber tree. They are evergreen trees with spreading branches. The leaves are stiff, decurrent,

linear-lanceolate with serrulate margins, white-banded beneath and spread in two ranks, but arise spirally. Two species have been positively identified: *C. lanceolata* (south and west China) and *C. konishii* (Taiwan). The doubtful species is *C. kawkamii*, also from Taiwan, which is intermediate in form between the two previous species. The timber of *C. lanceolata* is used for making coffins.
TAXODIACEAE, 2 or 3 species.

Cuphea a large genus of herbs and shrubs native to tropical America, a number of which are grown for their attractive and showy flowers, in the open in the subtropics and tropics or in greenhouses in temperate regions. The most popular are *C. ignea* [CIGAR FLOWER] with long, tubular, bright scarlet flowers, tipped purple, black and white at the end, and *C. cyanea*, a subshrub with tubular yellow flowers blotched with scarlet, and with tiny purple petals.
LYTHRACEAE, about 230 species.

× **Cupressocyparis** [HYBRID CYPRESSES] a bigeneric hybrid (× *Cupressocyparis leylandii*) between *Cupressus macrocarpa* [MONTEREY *CYPRESS] from California and *Chamaecyparis nootkatensis* [NOOTKA or YELLOW CYPRESS] from the Pacific Coast of northwestern America. Although it is thought to have arisen in cultivation in England in 1888 the hybrid cypress only really began to attract widespread attention in the 1950s for fast-growing shelterbelts and hedges.
Today × *C. leylandii* is widely planted as a very fast-growing, hardy and adaptable conifer. It exists now in about 10 or 11 clones which display some differences in growth habit, color and texture. The hybrid cypress seems to inherit its hardiness from the Nootka parent and its fast growth from the Monterey. It withstands clipping and can be made into a dense hedge from 2m in height.
Two other hybrids have also emerged: × *C. notabilis* (*Chamaecyparis nootkatensis* × *Cupressus glabra* [SMOOTH-BARKED ARIZONA CYPRESS]) and × *C. ovensii* (*Chamaecyparis nootkakatensis* × *Cupressus lusitanica* [MEXICAN CYPRESS]).
CUPRESSACEAE, 3 species.

Cupressus see CYPRESS.
CUPRESSACEAE, about 20 species.

Curare a South American arrow poison which has acquired great importance as a muscle-relaxant drug used mainly in anesthesia. It is prepared from *Chondodendron tomentosum* [MOONSEED VINE] and other species and from *Strychnos toxifera* [CURARE POISON NUT]. It contains several active ingredients which are extracted by soaking the bark in water and then concentrating by evaporation. The resultant thick black substance is packed in tins and sent to the drug companies for refinement.
Curare kills by causing paralysis: it causes relaxation of the skeletal muscles, blocking off the nerve endings adjacent to the muscles, finally causing breathing to stop.

Male and female flowers of the Vegetable Marrow (Cucurbita pepo), a species which encompasses several other fleshy vegetables. ($\times\frac{1}{2}$)

Curcuma an Indomalaysian and Chinese genus of rhizomatous herbs many species of which are useful to Man. The powdered rhizomes of *C. domestica* (= *C. longa*) [CURCUMA, *TURMERIC] yield an important curry spice and dye, turmeric, a substitute for saffron. *C. pallida* (= *C. zedoaria*) [ZEDOARY], *C. aromatica* [YELLOW ZEDOARY] and *C. caesia* [BLACK ZEDOARY] all have aromatic rootstocks which are used in the preparation of dyes, cosmetics, drugs and perfumes. Starch from the rhizomes of *C. pallida* and *C. angustifolia* [INDIAN or EAST INDIAN ARROWROOT] is extracted and used in the same way as *ARROWROOT (*Maranta arundinacea*).
ZINGIBERACEAE, about 70 species.

Currant the dried fruit of the *GRAPE, *Vitis vinifera* var *corinthiaca* or var *apyrena*, long cultivated in Greece and Cyprus, and widely

Mechanical harvesting of Black Currants (Ribes nigrum), which are highly prized for their rich vitamin C content.

The Sago Palm (Cycas revoluta) *is a palm-like gymnosperm often grown outdoors in the tropics, but also as a houseplant in temperate regions.* ($\times \frac{1}{15}$)

used in bakery items, cooking, etc in some countries. The term currant is also applied, especially in the USA to small-berried fruits of *Ribes nigrum* [BLACK CURRANT], *R. rubrum* [RED CURRANT] and *R. sativum* [COMMON or GARDEN CURRANT]. *R. sanguineum* [FLOWER-ING CURRANT] consists of many varieties cultivated for ornament.

Cuscuta   see DODDER.
CONVOLVULACEAE, about 170 species.

Cusparia   a genus of shrubs and trees native to South America. The Brazilian *C. febrifuga* (= *C. trifoliata*) is a tree whose bitter bark yields a product used locally for treating paralytic infections as well as diarrhea and dysentery. The commercial angostura bitters are produced from the bark of this tree.
RUTACEAE, about 30 species.

Cutch   a grayish resinous substance obtained by boiling chips of the heartwood of the Indian tree *Acacia catechu* in water, followed by evaporation. The extract is used as a brown dye and in tanning. Cutch is frequently included in the Asian narcotic *betel* "quid". The term "cutch" is sometimes used for mangrove bark extracts and for dye from *Uncaria gambier* [BENGAL *GAMBIER*].

Cutleria   a genus of subtropical marine brown algae.
PHAEOPHYCEAE.

Cyathea   a large cosmopolitan genus of tree-ferns occurring in subtropical tropical areas and often the dominant members of high altitude woodland. All have a tree-like habit and can attain 25m in height. Some

species, such as *C. arborea*, are widely grown in tropical America as a garden plant. *C. tricolor* (= *Alsophila tricolor*) and *C. medullaris* (= *Sphaeropteris medullaris*), both from New Zealand, produce edible sago from the stem pith.

The trunks of *Cyathea* are used in house-building by various native communities, being resistant to fungal decay. In New Zealand, Maoris produce "Pongo-ware" – carved jugs and vases from tree-fern stems – for sale to tourists.
CYATHEACEAE, 600–800 species.

Cycads   ten living genera, *Bowenia*, *Ceratozamia*, *Cycas*, *Dioon*, *Encephalartos*, *Lepidozamia*, *Macrozamia*, *Microcycas*, *Stangeria*, and *Zamia*, representing the largest order of living gymnosperms. They grow in the tropics, subtropics and warm temperate regions. Cycads are distinguished from the conifers by a number of important characters, most conspicuously by their palm-like habit, with an unbranched stem bearing a crown of large pinnately compound leaves, but also by quite numerous structural and reproductive characters.
(See also Conifers and their allies. p. 98)

Cycas   a genus of palm-like gymnosperms widely distributed from Madagascar to northern Australia, Polynesia and Japan. The pith of *C. circinalis* and *C. revoluta* [SAGO PALM] yields a type of *sago* which is now known to be carcinogenic. Both are widely cultivated outdoors in warm regions or in conservatories in cooler areas.
CYCADACEAE, about 20 species.

Cyclamen   a genus of small perennial herbs native from the Mediterranean to Iran, mostly on mountains and hills, and widely cultivated elsewhere. The rootstock is a tuberous corm and the leaves are all radical, heart- or kidney-shaped or almost round, sometimes with blunt lobes and often silver-mottled. The flowers are solitary on leafless stalks and are pendent, with the petals (corolla lobes) sharply reflexed.

C. hederifolium (= *C. neapolitanum*), from southern France to western Turkey, is an autumn-flowering species which has proved very amenable to cultivation with its pink or white flowers appearing just before the leaves. *C. coum* (= *C. orbiculatum*, *C. ibericum*) is a variable species and contains a number of closely related forms, or possibly hybrids, with rounded leaves and small dumpy flowers of pink, purple or white, often with a basal blotch of darker color. *C. purpurascens* (= *C. europaeum*), from Switzerland to Bulgaria, has scented flowers over a long period from summer to autumn or later. *C. persicum* (= *C. indicum*) from the eastern Mediterranean is the parent of the cyclamen extensively grown as a pot plant [FLORIST'S CYCLAMEN].
PRIMULACEAE, 15–17 species.

Cydonia   a genus represented by a single west Asian species, *C. oblonga* [COMMON

QUINCE], a deciduous thornless tree which has been in cultivation in Europe for many centuries and is thought to have originated in Central Asia. The flowers are pink or white and followed by light golden, aromatic, pear-shaped fruits in late summer. The fruit is too acid and astringent to be eaten raw but the ripe fruits can be used to flavor apples in cooking or made into preserves and jellies. There are a number of named varieties. THE COMMON QUINCE also has an important use as a rootstock for pears, several clones being used to confer degrees of vigor to the grafted trees. *C. sinensis* (= *Chaenomeles sinensis*) is now placed in a separate genus *Pseudocydonia*.
ROSACEAE, 1 species.

Cymbalaria   a genus of short-lived, small, creeping, perennial herbs native to Europe and Asia. *C. muralis* [KENILWORTH IVY, IVY-LEAVED TOADFLAX] commonly grows on walls and, with several other species, is cultivated in gardens.
SCROPHULARIACEAE, about 15 species.

Cymbidium   a genus of tropical, terrestrial and semi-epiphytic, pseudobulbous orchids found throughout Asia from the Himalayas to China, Japan and Australia. The flower-spikes arise from the base of the pseudobulb and can be erect or arching and up to 1m long. The flowers can be borne singly or up to 40 in the single, unbranched inflorescence and vary from 2cm to 15cm in diameter. Breeders have succeeded in producing an enormous range of flower color combinations and more *Cymbidium* hybrids are produced and grown throughout the world than all other popular orchid genera.
ORCHIDACEAE, about 45 species.

Cymbopogon   [OIL GRASS]   an Old World genus of tropical and subtropical, perennial, densely tufted, coarse grasses, common in dry places and grasslands throughout Asia

Cymbidium × *'Vieux Rose'*, *one of the numerous hybrids in this popular orchid genus.* ($\times \frac{1}{2}$)

and Africa. Several species yield essential oils which are important components of perfumes in the soap and talcum powder industries, including *C. flexuosus* and *C. citratus* [LEMONGRASS], which both yield lemongrass oil; the leaf buds of the latter are sold in the East as an additive for curries. *C. martinii* produces palma rosa oil and gingergrass oil and *C. nardus* [CITRONELLA GRASS] citronella oil used in perfumery.
GRAMINEAE, about 60 species.

Cynara a genus of large and spectacular perennial herbs from the Mediterranean region. It includes *C. cardunculus* [*CARDOON] and *C. scolymus* [GLOBE *ARTICHOKE].
COMPOSITAE, about 12 species.

Cynodon a genus of tropical and subtropical creeping grasses occurring mostly in South Africa and Australia. The cosmopolitan sand and dune species *C. dactylon* [BERMUDA GRASS, COUCH GRASS, DOUB] is widely cultivated for pastures and lawns.
GRAMINEAE, 10 species.

Cynoglossum a genus of biennial and perennial herbs found in temperate and subtropical regions. The long basal leaves are gray-green and downy. *C. grande*, *C. nervosum* and *C. amabile* [CHINESE FORGET-ME-NOT] are all blue-flowered species cultivated in garden borders. A decoction of the leaves and young roots of *C. officinale* [COMMON HOUND'S TONGUE] has been used as a sedative.
BORAGINACEAE, about 60 species.

Cynosurus a genus of grasses of temperate regions of the Old World. The most common is *C. cristatus* [CRESTED DOG'S TAIL], a very widely distributed annual grass found in Europe and Asia and also introduced to North America and Australasia. It is a tough, wiry species which has been used with other species in the formation of lawns and amenity turf on account of its resistance to drought and hard wear.
GRAMINEAE, 3–4 species.

Cypella a small genus of cormous plants from Central and South America with blue, yellow or orange iris-like flowers. Several species are cultivated, especially *C. herbertii*.
IRIDACEAE, 15 species.

Cyperus a genus of grass-like herbs including edible and ornamental species, and some plants used in making mats. They are annuals with fibrous roots or, more often, perennials with rhizomes, native to tropical and warm temperate regions.

The genus is economically important for *C. papyrus* [PAPYRUS, PAPER REED], a tall perennial herb originally native to riverside regions in North and tropical Africa and now widespread in the Mediterranean region and southwest Asia. The long (1–5m) stems, when split into strips and pressed while wet, made the papyrus paper of ancient times. The stems are also used locally in building rafts and boats. The rhizomes of this species, the tuberous rootstocks of the southern European *C. esculentus* var *sativus* [RUSH NUT, TIGER NUT, CHUFA, GROUND ALMOND] and the tuberous roots of the North American *C. aristatus* are edible when cooked, roasted, or made into beverages (the Spanish horchata de chufa). The stems of many species, such as *C. malacopsis* (tropical Asia), *C. cenus* (tropical America) and *C. tegetiformis* (China), are used for matting. The sweet-scented, fragrant roots of *C. articulatus* and *C. rotundus* [COCO GRASS, NUT GRASS] are used for perfuming clothing, while the rhizomes of *C. longus* [ENGLISH GALINGALE] yield a violet-scented extract sometimes added to lavender water. Medicinal tonics are obtained from *C. iria* (India) and *C. lenticularis* (East Asia).

A number of species are cultivated as ornamentals, either as pot plants under glass or in the open at the edges of pools and lakes in summer in temperate zones. These include *C. papyrus*, *C. alternifolius* [UMBRELLA GRASS], and *C. vegetus*.
CYPERACEAE, about 550 species.

Cyphomandra [TREE TOMATO] a little-studied South American (Andean) genus of perennial herbs, shrubs and short-lived trees closely related to *Solanum*. *C. betacea* (= *C. crassifolia*), native to Peru, is grown in the tropics and subtropics (especially in Ecuador) for its edible, egg-shaped, usually red fruits.
SOLANACEAE, about 30 species.

Cypress the common name for members of *Cupressus*, a genus of coniferous evergreen trees, rarely shrubs, widely distributed in the New and Old Worlds, from Oregon to Mexico in North America, in the Mediterranean

Papyrus (Cyperus papyrus) growing beside a lake in Uganda. In Lake Victoria, floating islands of Papyrus have seriously interfered with boats.

The cyclamens commonly grown as pot plants are all derived from Cyclamen persicum. They are difficult to grow indoors but remain extremely popular. ($\times \frac{1}{4}$)

area, western Asia, the western Himalayas and China. The limits of the modern genus have been reduced by the transfer of a number of species to *Chamaecyparis* [FALSE CYPRESSES]. The species of *Chamaecyparis* are much more hardy than those of *Cupressus*, which, with the exception of a few species such as *C. arizonica* [ROUGH-BARKED ARIZONA CYPRESS] need a milder climate. *C. arizonica*, *C. sempervirens* [ITALIAN, MEDITERRANEAN or FUNERAL CYPRESS], *C. lusitanica* [PORTUGUESE CYPRESS] and *C. macrocarpa* [MONTEREY CYPRESS] are the most widely planted species.

The wood of many cypresses is valuable, being durable and easily worked. It is used in general building construction, carpentry and for posts and poles of all sorts. The most commonly used timbers are those of *C. macrocarpa* and *C. sempervirens*. The timber of *C. torulosa* [BHUTAN or HIMALAYAN CYPRESS] in China and of *C. arizonica* in Arizona and New Mexico is used for general construction work, although both of these species and others such as *C. glabra* [SMOOTH ARIZONA CYPRESS] are grown mainly as decorative ornamental trees.
CUPRESSACEAE, about 20 species.

Cypripedium [LADY'S-SLIPPER ORCHIDS] a genus of terrestrial orchids occurring throughout the temperate and subtropical regions of the Northern Hemisphere in woods, grassy and marshy places. The most characteristic feature of the *Cypripedium* flower is that part of it is inflated to form a

Cyrtanthus obliquus, *one of the numerous bulbous species from southern Africa which are grown as decorative pot plants.* ($\times \frac{1}{2}$)

slipper, moccasin or sabot-like pouch, associated with the complex pollinating mechanism.

Because of their unusual and generally attractive flowers cypripediums are widely grown, especially the North American *C. reginae*, with its pure white and rich pink flowers, and the widespread *C. calceolus* which has mahogany and bright yellow flowers. The latter species is in such demand that plants of it are frequently removed from the wild for sale and transplanting, and it is consequently in great danger of extinction.
ORCHIDACEAE, about 50 species.

Cyrtanthus a genus of bulbous perennials from tropical and southern Africa. The genus is of little economic importance but some

species, such as *C. contractus* [FIRE LILY] and *C. mackenii* [IFAFA LILY] are grown as ornamental plants in gardens for their tubular or trumpet-shaped flowers.
AMARYLLIDACEAE, about 47 species.

Cyrtomium a small genus (sometimes united with the American genus *Phanerophlebia*) of subtropical and tropical ferns, one of which, *C. falcatum* [HOLLYFERN] with shiny leathery pinnate fronds with ovate leaflets, is widely cultivated as a houseplant. The rhizome of *C. fortunei* is used in China as a treatment for parasitic worms.
ASPIDIACEAE, about 20 species.

Cystopteris [BLADDER FERNS] a genus of temperate and subtropical rock ferns with delicate foliage. *C. fragilis* [BRITTLE BLADDER FERN] and *C. bulbifera* [BERRY BLADDER FERN] which has bulb-like bodies on the under surface of the fronds, are both commonly cultivated in rockeries.
ATHYRIACEAE, 18–20 species.

Cytisus a genus of mainly deciduous and a few evergreen shrubs mostly native to Europe, the Mediterranean region and the Atlantic islands. It is one of three genera commonly called BROOM (*Genista* and *Spartium* are the others).

The species are mostly trifoliolate, rarely one-foliolate (some almost always leafless), and range from only a few centimetres to some four metres in height. In some Mediterranean species, the leaves are reduced to scales and the leaf functions are performed by the stems. Most species flower in late spring and early summer; few flower towards the autumn. The pea-type flowers are borne in leafy terminal racemes. The best-known species is *C. scoparius* (= *Sarothamnus scoparius*) [COMMON BROOM], common throughout Europe as a vigorous colonizing shrub

The Common Broom (Cytisus scoparius)*, a common European shrub which often covers considerable areas of dry heathland, producing a blaze of color in late spring.* ($\times \frac{1}{3}$)

The Lady's Slipper Orchid (Cypripedium calceolus) *occurs virtually throughout the north temperate zone, but has been so heavily depleted by collectors that it is now legally protected in most countries.* ($\times \frac{1}{4}$)

on dry acidic heaths and sandy places.

Many species are cultivated, most varieties being developed from *C. scoparius* and having red and crimson coloring in the petals. Among these numerous hybrids and varieties are the popular 'Johnsons Crimson' with crimson flowers and 'Goldfinch' with purple flowers with red and yellow wings. Other popular species include *C. multiflorus* (= *C. albus*) [WHITE BROOM], *C. nigricans* (= *Lembotropis nigricans*) and the mat-forming golden-flowered *C. ardoini* only 15cm tall. Of the numerous hybrids, the dwarf shrub *C. × beanii* (*C. ardoini* × *C. purgans*) and the bushy *C. × praecox* [WARMINSTER BROOM], the result of a cross between *C. multiflorus* and *C. purgans*, are two which are widely cultivated. *C. battandieri*, from Morocco, is a spectacular plant up to 5m high with large silvery leaflets and erect racemes of golden-yellow flowers smelling of pineapple. It is now widely cultivated.
LEGUMINOSAE, about 30 species.

D

Daboecia a very small genus of low-growing, evergreen, lime-hating shrubs related to *Erica. D. cantabrica* [IRISH HEATH, ST. DABOEC'S HEATH], from western Europe, with several cultivars, and the rather smaller *D. azorica*, endemic to the Azores, are widely grown in gardens. The rose-purple bell-shaped flowers are borne in terminal racemes.
ERICACEAE, 2 species.

Dacrydium a genus of coniferous evergreen trees and a few shrubs native to New Zealand, Tasmania, Australia, New Caledonia, New Guinea, Malaya, the Philippines, and Fiji. The genus, while including many stately trees of ornamental value, such as *D. cupressinum* [RED PINE] and *D. franklinii* [HUON PINE], also includes smaller species such as *D. taxoides* from New Caledonia and *D. bidwillii* [MOUNTAIN PINE] of New Zealand. The RED PINE and the HUON PINE both produce a valuable timber used in building and for furniture.
PODOCARPACEAE, about 25 species.

The Common Spotted Orchid (Dactylorhiza fuchsii), a poor relation of the extravagant orchid blooms found in florists' shops. This is probably the commonest of the temperate orchids. (×1)

Stem and leaves of a rattan belonging to the genus Daemonorops, showing the whorls of interlocking spines that form galleries in which ants live.

Dactylis a small genus of temperate Eurasian, North African and North American tufted perennial grasses recognized easily by their persistent, dense, one-sided and wedge-shaped inflorescences. *D. glomerata* [COCK'S-FOOT, ORCHARD GRASS] is a coarse, densely tufted perennial of meadows, pastures, roadside and rough grasslands. It is introduced and widely cultivated in many temperate countries as an important pasture and hay fodder crop. Var *variegata* with striped silver and green leaves is grown as a border ornamental.
GRAMINEAE, 5 species.

Dactylorhiza a genus of terrestrial orchids with purple-spotted leaves and conspicuous spikes of yellow, red, pink, mauve, lilac or white flowers. Most species are native to Europe and North Africa and normally grow in open grassland, marsh, fen or bog. All have been reported as hybridizing with each other as well as with plants of other genera such as *Platanthera* and *Aceras*.
ORCHIDACEAE, about 30 species.

Daemonorops a genus of palms from Indomalaysia, most of which are very spiny climbers, but some are shrub-like, growing in clumps. The slender, flexible stems of some members of the genus such as *D. forbesii* and *D. longipes* are used as *rattan canes, while others produce edible fruits. Some species, notably *D. draco* [MALAYAN DRAGON'S BLOOD], yield a red resinous exudation (hence the common name) from the fruit scales; this substance is still used in varnishes.
PALMAE, about 100 species.

Daffodil the popular name for species of the horticulturally important genus *Narcissus*. The term tends to be restricted (particularly in the UK) to those species with long, trumpet-shaped coronas, the rest being called narcissuses. (See also *Narcissus*.)
AMARYLLIDACEAE.

Dahlia a small genus of herbaceous perennials with tuberous roots, native to Mexico, Central America and Colombia. The wild species are rarely cultivated. The garden dahlias with their large showy flower heads are all grouped under the name *D. pinnata* (= *D. variabilis*) and are all hybrids with an obscure history, probably involving *D. pinnata* and *D. coccinea*.

Because of the large numbers of forms, garden dahlias are classified into sections for show and exhibition purposes. Single-flowered dahlias have a single outer row of overlapping rays (ray florets) and a center of disk florets, as in 'Coltness Gem' and 'Mignon'. Anemone-flowered dahlias also have a single row of outer rays, but the disk florets are tubular, and usually of a different color from the rays. Decorative dahlias have

Garden Dahlias (Dahlia pinnata) come in a wide range of forms and colors. Above top: A pompom dahlia with its typical compact bloom. (×1) Above: A cactus dahlia with its quilled petals. (×⅓)

double flowers and no central disk. The rays are usually broad and bluntly pointed. Show dahlias, with double flowers over 10cm in diameter, and most of the globular Pompom dahlias are like the previous group but smaller in diameter, and Cactus dahlias have completely double flowers with the rays rolled backwards and usually pointed.

Other cultivated species include the shrubby *D. excelsa* [FLAT TREE DAHLIA], *D. merckii* [BEDDING DAHLIA], *D. imperialis* [BELL TREE DAHLIA, CANDELABRA DAHLIA] and *D. blagayana*, a mat-forming species. COMPOSITAE, about 15 species.

Daisy the common name given principally to members of the genus *Bellis*, such as *B. perennis* [COMMON DAISY] but which also applies to many other members of the Compositae. Examples include species of *Brachycome*, such as *B. scapigera* [ALPINE DAISY], *B. iberidifolia* [SWAN RIVER DAISY], *B. insignis* [ROCK TREE DAISY], species of *Aster* [MICHAELMAS DAISY], species of *Olearia*, such as *O. tomentosa* [BUSH DAISY], and species of *Leucanthemum*, such as *L. vulgare* [OXEYE DAISY] and *L. maximum* [SHASTA DAISY]. COMPOSITAE.

The characteristic plumed fruit of the Common Dandelion (Taraxacum officinale) caught in a cobweb. (×2)

Dalbergia an important large genus of trees, shrubs and woody climbers widespread in the tropics and subtropics. Many tropical species grow as scrambling shrubs, but many other species, especially those of India and Brazil, are valuable timber trees. Scrambling species include the purple-stemmed *D. ecastaphyllum* of coastal mangrove swamps, and *D. variabilis*, in open habitats, a scrambling shrub with pendulous twigs, but a climbing liana in forests. *D. saxatilis* climbs by spiral hooks and produces showy panicles of white flowers, worthy of cultivation for their ornamental value.

D. nigra [BRAZILIAN ROSEWOOD, RIO ROSEWOOD, JACARANDA] yields a purple-black wood and is one of the finest cabinet-making timbers. *D. latifolia* [EAST INDIAN ROSEWOOD, BLACK ROSEWOOD, MALABAR ROSEWOOD, BOMBAY ROSEWOOD, JAVANESE PALISSANDER], a tree native to the East Indies, is an important timber-producing species in India and Sri Lanka. The purple heartwood streaked with green, gray or black is used in cabinet-making. In Asia the timber is widely used for railway sleepers. The tree frequently grows in association with *D. sissoo* [INDIAN SISSOO

WOOD]. This species attains about 25m and grows wild only in the foothills of the Himalayas, but is extensively cultivated in India and Pakistan for its excellent timber, which is hard and strong and comparable to teak. *D. parviflora* is a Malaysian species which grows as a large liana; the roots and heartwood of the thick stems near the ground are scented and are supplied to China for the joss-stick trade.

One of the showiest, most exotic woods comes from *D. retusa* [COCO BOLO, NICARAGUA ROSEWOOD, PALO NEGRO] from Mexico and Panama. This tree produces a very hard, orange heartwood streaked with jet black, which is used for instrument handles and inlay work. *D. melanoxylon* [AFRICAN BLACKWOOD, AFRICAN GRENADILLE WOOD, SENEGAL EBONY], from tropical West Africa, is used as an ebony substitute. About 12 other species including *D. cochinchinensis* [THAILAND ROSEWOOD], *D. sissoides* [MALABAR ROSEWOOD], *D. cearensis* [BRAZILIAN KINGWOOD] and *D. stevensonii* [HONDURAS ROSEWOOD, HAGAED WOOD] produce luxury timbers with remarkable coloration. Today rosewoods are amongst the most highly esteemed and costly timbers in the world. LEGUMINOSAE, about 250 species.

Dammar an aromatic resin obtained by tapping the bark of many tropical trees, notably *Balanocarpus heimii* [CHENGAL], *Shorea robusta*, *S. eximea* [ALMOND DAMMAR] and *S. balangeran*. Dammar is used in East Asia for caulking boats and is burnt for incense. Dissolved in turpentine, it makes an excellent final varnish for oil paintings, which yellows only slowly with age. It is also used by picture-restorers, mixed with wax, as a reversible adhesive for backing damaged paintings with fresh canvas. DIPTEROCARPACEAE.

Damsons small sour purple fruits usually classified with the Bullace, Mirabelle and St. Julien as *Prunus insititia*. They are rarely eaten raw but they can be stewed with sugar or made into jellies and jams. The distinctive

Mezereon (Daphne mezereum). Like many Daphne species, it is often grown for its sweet and distinctive fragrance. (×1/20)

Flower heads of the Common Dandelion (Taraxacum officinale). (×1/3)

flavor varies very little between typical varieties such as 'Prune', and 'Frogmore'.

The common names BUTTER DAMSON, DAMSON PLUM and WEST AFRICAN DAMSON are given to the species of *Simarouba*, *Chrysophyllum* and *Sorindeia* respectively.

Dandelion the common name usually given to members of the genus *Taraxacum*, the best-known species being *T. officinale* [COMMON DANDELION], a widespread weed. COMPOSITAE.

Daphne a genus of usually small evergreen or deciduous shrubs many of which have sweetly scented flowers. They are distributed over Europe, North Africa, and temperate and subtropical Asia through the Pacific to Australia.

D. blagayana, which produces white flowers in the spring, and *D. cneorum* [GARLAND FLOWER], which bears pink flowers in the summer, are two of the cultivated evergreen European species. *D. laureola* [SPURGE LAUREL], an evergreen widespread in Europe and western Asia, is about 1m high and produces green flowers in early spring. Species from China include the deciduous *D. genkwa* (= *D. fortunei*) and the evergreen *D. retusa*, both with purplish flowers in early summer and about 1m high.

Most species of daphne are poisonous. *D. mezereum* [MEZEREON] provides mezereon bark, once used both internally and as a rubefacient. The bark of other species is used in the Mediterranean to stupefy fish. The inner bark of some species, including *D. papyracea* is used for papermaking in India and elsewhere in the Far East. THYMELAEACEAE, about 70 species.

Daphniphyllum a small genus of evergreen shrubs and trees native to China, Japan, Taiwan, Indomalaysia and South Australia. *D. humile*, a low, spreading, much-branched shrub, is cultivated as a ground-cover plant, and its leaves are used for smoking in Japan. *D. macropodum* 'Variegatum' is grown as a neat, rounded, red-wooded tree with

variegated leaves. The wood is used locally for turnery.
EUPHORBIACEAE, 25 species.

Darlingtonia a Californian genus represented by a single species, *D. californica* [CALIFORNIAN PITCHER PLANT, COBRA PLANT], an insectivorous plant of boggy meadows and stream banks. The leaves have tubular petioles which are enlarged upwards and terminated in a pitcher with a rounded hood and a circular orifice on one side, which secretes honey and attracts insects to the pitcher.
SARRACENIACEAE, 1 species.

Dasylirion [SOTOL] a genus of tree-like desert plants, all from Mexico and the USA. Some species are grown in greenhouses for the sake of their rather unusual appearance. The two most commonly grown are *D. acrotriche* and *D. graminifolium* which produce very short trunks to about 2.5m tall surmounted by a dense rosette of evergreen leaves. The leaves of several species are used for thatching and the fiber for cordage and the alcoholic beverage, sotol, is obtained from the trunks.
LILIACEAE, about 10 species.

Dates the fruits of the Old World subtropical palm *Phoenix dactylifera* [DATE PALM] which is closely related to *P. sylvestris* [WILD DATE] and considered by some authorities to be a cultivar derived from this species which is common in southwest Asia and the Indian subcontinent.

The fruit is fleshy and finger-like, about 3–4cm long and is a true berry. The food reserve is cellulose which is converted to sugar during germination. The date is also rich in vitamins A, B_1, B_2 and nicotinic acid.

Commercially, the date is an extremely important crop in the Arab countries of North Africa, being a principal source of carbohydrate. Outside North Africa and the Middle East, however, the date is considered

The carnivorous California Pitcher Plant (Darlingtonia californica) bears an uncanny resemblance to a cobra. ($\times\frac{1}{10}$)

A Date Palm (Phoenix dactylifera) in full fruit. The date is a staple food crop in North Africa and the Middle East.

to be a luxury product and the export varieties are rich in sugars but have a lower cellulose content. The cultivated dates fall into three main categories: *soft dates*, grown mainly in Iraq for consumption within the Arab countries and normally eaten raw; *semi-dry dates*, most commonly exported to Europe from Egypt and Iran, the variety 'Deglet Noor' being the most widely grown; and *dry dates*, which are very important in the countries of origin as they can be stored for considerable periods of time and can be ground into flour or soaked and softened.
PALMAE.

Datura a genus of annual or perennial herbs of warm temperate and tropical regions, mostly in America, some of which are violently narcotic and very poisonous. Several species are cultivated for their showy, funnel-shaped, white, orange or purple flowers, including *D. × candida* (which is correctly placed in the related *Brugmansia* as *B. × candida*), *D. inoxia* [DOWNY THORN-APPLE, INDIAN APPLE, ANGEL'S TRUMPET, SACRED DATURA] and *D. metel* (= *D. chlorantha*, *D. fastuosa*) [DOWNY THORN-APPLE, HORN-A-PLENTY]. *D. inoxia* bears pink or lavender flowers up to 20cm long. Subspecies *quinquecuspida* (mistakenly referred to as *D. meteloides*) from China is widely grown as an ornamental.

In addition to its value as an ornamental species, *D. metel* is widely cultivated in both hemispheres as a source of the alkaloid drug hyoscine (scopolamine). The North American *D. stramonium* (= *D. tatula*) [COMMON THORN-APPLE, JIMSONWEED, JAMESTOWN WEED] is widely naturalized throughout the world and is sometimes grown as a source of the alkaloidal drug hyoscyamine.
SOLANACEAE, about 8 species.

Daucus a genus of herbaceous plants mainly native to Europe and the Mediterranean, with a few species in the Americas and Australasia. *D. carota* subspecies *sativus* is the CULTIVATED *CARROT.
UMBELLIFERAE, 21 species.

Davallia a genus of terrestrial and epiphytic ferns widely distributed in temperate, tropical and subtropical regions. A number of species are cultivated as house or greenhouse plants. One of the most popular is *D. canariensis* [HARE'S FOOT FERN], native to the Canary Isles, which has leathery triangular quadripinnate fronds up to 45cm long and 30cm wide. The East Asian species *D. mariesii* [SQUIRREL'S-FOOT FERN, BALL FERN] has lighter green triangular fronds. The creeping surface rhizomes are used in Japan for making various ornaments and figures. Popular cultivated ornamental species include *D. trichomanoides* (= *D. dissecta*) [SQUIRREL'S FOOT FERN, BALL FERN] and

The well-named Thorn-apple (Datura stramonium), with its spiny seed-capsule, is a very poisonous member of the family Solanaceae. It is now widely naturalized throughout the world. ($\times\frac{1}{3}$)

D. fejeensis [RABBIT'S-FOOT FERN] grown as ground-cover or basket plants.
POLYPODIACEAE, about 35 species.

Davidia a genus consisting of a single species, *D. involucrata* [POCKET HANDKERCHIEF TREE, DOVE TREE, GHOST TREE], from central and western China, a deciduous tree distinctive because the globose heads of petal-less flowers are subtended by large pendulous bracts which give the impression of dainty white handkerchiefs hanging from the branches.
NYSSACEAE, 1 species.

Dawsonia a genus of mosses with a center of distribution in tropical Southeast Asia and extensions into Australia and New Zealand. It is remarkable for the great stature attained by some of the species (eg *D. superba*) and for the strong resemblance of the leafy shoots to those of the larger species of *Polytrichum*.
DAWSONIACEAE.

Decumaria a genus of two species of climbing shrubs, one from the southeastern USA, the other from China. They climb by attaching themselves to their supports by aerial rootlets. Neither the deciduous *D. barbara* [CLIMBING HYDRANGEA, WOOD-VAMP], the American species, nor the evergreen *D. sinensis*, the Chinese species, is commonly cultivated.
SAXIFRAGACEAE, 2 species.

Delonix a very small genus of beautiful ornamental trees. *D. regia* (= *Poinciana regia*) [FLAMBOYANT, PEACOCK FLOWER, ROYAL POIN-

Delphinium brunonianum, a common plant of alpine regions in the Himalayas. ($\times \frac{1}{2}$)

Flower of the Pocket Handkerchief Tree (Davidia involucrata) showing the white bracts that have given the tree its popular name. ($\times \frac{1}{3}$)

CIANA], from Madagascar, is one of the most spectacular trees in cultivation throughout the tropics and subtropics, with its racemes of vivid bright scarlet flowers borne at the ends of the branches, and flat pendulous woody pods about 30cm long, borne after flowering. *D. elata*, with less spectacular white flowers fading yellowish-orange, is native to India, Arabia and Ethiopia.

Related species are the yellow-flowered *Peltophorum pterocarpum* [YELLOW POINCIANA] and *P. roxburghii* [YELLOW FLAMBOYANT].
LEGUMINOSAE, 2 species.

Delphinium a large genus of annual and perennial erect herbs native to the north temperate zone. It contains some popular garden ornamentals. *D. ajacis* [LARKSPUR, ROCKET LARKSPUR] and allied species are frequently placed in a separate genus (*Consolida*). Annual LARKSPURS and a race of perennial hybrids [DELPHINIUMS] derived from *D. elatum* [ALPINE LARKSPUR] are widely grown as garden-border plants, mainly with blue, purple or white showy inflorescences.

Hardy annuals include hybrids derived from *D. ajacis* (eg the double-flowered "hyacinth" group) and from *D. consolida* [FORKING LARKSPUR] (eg the tall "giant imperial" and "stock-flowered" groups).

A wide range of perennials derived from crosses between *D. elatum* and *D. grandiflorum* (native to Siberia and China) are popular with gardeners. They include tall plants called Pacific hybrids with 1.5–2m stems, eg 'Black Night Violet' and 'Galahad White', as well as smaller cultivars such as 'Blue Tit' (indigo blue) and 'Blue Jade' (sky blue).

D. tatsienense (China) is a perennial that grows to a height of only 25–40cm and is cultivated in rock gardens. Two North American species, *D. cardinale* (California) and *D. nudicaule* (California and Oregon), are somewhat unusual in having reddish flowers, the former with scarlet sepals and yellow upper petals tinged with red, the latter

with orange-red or occasionally yellow sepals.
RANUNCULACEAE, about 250 species.

Dendrobium a very large genus of tropical and subtropical epiphytic orchids growing wild from the lower slopes of the Himalayas through India, Burma, the Far East, Malaysia and the Pacific Islands as far east as Tahiti and as far south as Australia and New Zealand. Those from the humid tropical forest regions such as New Guinea are mainly evergreen but the remainder are deciduous, usually flowering on the thin, leafless pseudobulbous stems.

The great variability of vegetative, floral and ecological characteristics has led to *Dendrobium* species being widely cultivated throughout the world; they are widely hybridized to produce a great number of striking plants of considerable commercial value. Popular cultivated species include the deciduous *D. aureum* (= *D. heterocarpum*) and *D. loddigesii* (= *D. pulchellum*), and the evergreen *D. bigibbum*, *D. kingianum*, *D. nobile*, and *D. phalaenopsis*, the latter two species containing many cultivars with varying flower colors.
ORCHIDACEAE, about 900–1200 species.

A cultivated delphinium. The modern perennial garden cultivars are derived from crosses between Delphinium elatum and D. grandiflorum. ($\times 2$)

Dendrocalamus a genus which includes the bulkiest of all *bamboos from China and Indomalaysia. The tallest is *D. giganteus* [GIANT BAMBOO] of Burma, which can grow at the rate of 45cm per day. The stems may attain 30m in height and 25cm in diameter, in clumps up to 15m across. Many species are employed in all the uses to which bamboos are put, but especially in building and the manufacture of large water buckets. *D. strictus* [CALCUTTA BAMBOO], from India and Java, is the most common, widespread and universally used bamboo in India, where it is the principal source of pulp for paper. It is also widely cultivated elsewhere. *D. asper* is cultivated by the Chinese in Malaysia and the Javanese in Indonesia for its excellent edible

young shoots, which are boiled and sometimes pickled in vinegar.
GRAMINEAE, about 20 species.

Dendromecon a genus consisting of a single variable species, *D. rigida* [BUSH POPPY], an evergreen much-branched shrub found only in California, Baja California and offshore islands. It differs from *Romneya* in having smaller yellow flowers and grayish-green, entire, leathery leaves. It is planted for ornament in warmer climates.
PAPAVERACEAE, 1 species.

Derris root the enlarged rootstock, also known as TUBA ROOT, of *Derris elliptica* (and less importantly *D. malaccensis* and *D. ferruginea*) which, when powdered, has been used for centuries as an insecticide and fish poison. The active principle is rotenone. The plants are small, weak, vine-like shrubs. They are native to the forests of the Old World tropics, but are cultivated in small plantations throughout the tropics, mainly in the Malay peninsula, Indonesia and the Philippines. The resin is extracted from mature roots and is widely used to control, without hazard to warm-blooded animals, such animal pests as ticks and fleas and such plant pests as bean beetles, apple and pear aphids.
LEGUMINOSAE.

Deschampsia a genus of tussock-forming grasses occurring in temperate and cool regions mainly in the Northern Hemisphere. *D. caespitosa* [TUFTED HAIR GRASS, TUSSOCK GRASS] is a hay and pasture species occurring throughout Europe and North America. Another important species is *D. flexuosa* which is probably best-known for its use in lawns on acid soils.
GRAMINEAE, about 60 species.

Desfontainea a genus comprising a single evergreen shrubby species with attractive

Sweet Williams (Dianthus barbatus) originated in eastern Europe but now many differently colored cultivars have been developed. (×⅛)

spinous holly-shaped leaves, native to the Andes of Peru and Chile. *D. spinosa* is widely cultivated for its shiny leaves and crimson and yellow funnel-shaped flowers.
LOGANIACEAE, 1 species.

Desmodium a genus of tropical or subtropical herbs or semi-woody shrubs. *D. motorium* (= *D. gyrans*) [TELEGRAPH PLANT], from Indo-malaysia, makes a fascinating houseplant because of the curious movement of the two small lateral leaflets, which move steadily in elliptical orbits during the daytime in temperatures over 22°C. Species such as the tropical American *D. tortuosum* [BEGGAR-WEED] are often grown in the West Indies and the southern USA for forage or manure because they have a high protein and calcium content, and the crop reseeds itself.
LEGUMINOSAE, about 300–400 species.

Deutzia a genus of usually deciduous shrubs of moderate height and vigor, mostly from China and Japan, with a few in Central America. They have opposite, finely-toothed leaves on reddish-brown stems and the mostly white, pink or purplish flowers are produced in attractive clusters. There are many cultivars and hybrids flowering in early to mid-summer, such as *D. discolor*, *D. scabra*, *D. gracilis*, *D.* × *rosea* and *D.* × *lemoinei*.
SAXIFRAGACEAE, about 60 species.

Dianthus a large genus of annual, biennial and perennial herbs or small shrubs native to Europe (especially the Mediterranean region), Asia and extending to South Africa. The narrow or linear leaves are borne oppositely in pairs, frequently united at the base where they are attached to the stem, which is swollen at this part. Most cultivated species have showy, rather tubular flowers with five petals, solitary, in panicles or clustered in tight heads. *Dianthus* species have long been cultivated, frequently in rock gardens, mainly for their ornamental value, including their leaf color, especially in winter, as well as for the strong characteristic fragrance and bright colors of their flowers.

D. caryophyllus [CARNATION, CLOVE PINK, DIVINE FLOWER], a bright grayish-green-leaved perennial native to southern Europe, but widely cultivated elsewhere, is the main ancestor of the border and perpetual-flowering carnations (see CARNATIONS). It has been hybridized with other species (eg *D.* × *allwoodii* – the result of a cross with *D. plumarius*) and is considered the ancestor of all modern pinks.

Another popular ornamental is *D. barbatus* [SWEET WILLIAM], an annual, biennial or perennial which although native to Europe is naturalized elsewhere, notably in China and North America. All cultivars are characterized by the flowers (sometimes double) being borne in dense, flattened heads. Cultivars exist in a range of colors, eg 'Albus' (white-flowered), 'Atrococcineus' (deep red-flowered), 'Nigricans' (violet-purple-flowered). Numerous species are commonly

Perpetual Carnations are descendants of Dianthus caryophyllus *and are widely used as cut-flowers.*

called pinks. These include the widely grown perennial *D. carthusianorum* (= *D. atrorubens*) [CLUSTER-HEAD PINK], with dense terminal clusters of bearded pink to purple flowers (rarely white), the annual, biennial or short-lived perennial *D. chinensis* [RAINBOW PINK], which bears loosely clustered rose-lilac flowers with a purple eye, the perennial rock-garden species, *D. deltoides* [MAIDEN PINK], with a range of cultivars including 'Albus' (white-flowered) and 'Coccineus' (scarlet-flowered), and *D. gratianopolitanus* (= *D. caesius*) [CHEDDAR PINK], with bearded, toothed, rose-pink flowers. Another excellent rock-garden pink with several cultivars and varieties, is the mat-forming perennial *D. plumarius* (= *D. blandus*) [COTTAGE or GRASS PINK], which bears fragrant flowers, solitary or in inflorescences of up to three. The petals are frequently fringed and may be white, pink, purple or multicolored.
CARYOPHYLLACEAE, about 300 species.

The Maiden Pink (Dianthus deltoides), from Europe, makes a good rock-garden plant. (×¼)

Diatoms a group of microscopic algae sometimes called "the grass of the sea", for they are the most important primary producers of organic material in the open sea and in many lakes. Their main habitat is in the plankton but they are also found abundantly as epiphytes on marginal vegetation and living on mud or sandy surfaces. They are even found in soil. Diatoms have been found in strata at least as old as the Jurassic period and in some parts of the world extensive fossil deposits, called diatomaceous earth, are worked for use in industry.
BACILLARIOPHYCEAE, about 5000 species.

Dicentra a genus of herbaceous rhizomatous perennials native to North America, China, Japan and other parts of Asia. A number are cultivated in gardens for their attractive fern-like foliage and pendulous inflorescences. *D. spectabilis* [BLEEDING HEART], with its pinkish-red, heart-shaped flowers, is one of the most popular garden species; others are the white-flowered *D. cucullaria* [DUTCHMAN'S BREECHES] and *D. canadensis* [SQUIRREL CORN], and the yellow-flowered *D. chrysantha* [GOLDEN EARDROPS].
FUMARIACEAE, 19 species.

Dicksonia a distinctive genus of tree-ferns in which the younger parts of the trunk, the unfurling fronds and the leaf-stalks are covered with coarse orange or red-brown bristles. The fronds are 2- to 3-pinnate, coarse in texture and up to 3m in length. Most species are native to South America and Australia, reaching north to Mexico and the Philippines. *D. antarctica* and *D. fibrosa* are two species grown frequently in warm American and European gardens and, in colder areas, in frost-free conservatories.
CYATHEACEAE, about 25 species.

Dicranum a genus of mosses of wide distribution, but found predominantly in cool temperate and cold climates. The best-known, and highly polymorphic, species,

Dicranum majus, a turf-forming moss, which often forms a substantial part of the ground cover in moist sheltered woods. (×1)

A tree fern of the genus Dicksonia *growing in a burned* Eucalyptus *forest east of Armidale in New South Wales, Australia.*

D. scoparium, occurs in Europe, North, Central and South America, North Africa, much of Asia and New Zealand. Its habitat range is also wide, including wood (living and dead), heaths, grassland and rock ledges. All species grow erect, forming cushions or turfs and bear narrowly lanceolate, often strongly curved, glossy leaves.
DICRANACEAE, about 100 species.

Dictamnus a group of woody perennial herbs from southern Europe to Asia, probably representing a single variable species, *D. albus* (= *D. fraxinella*) (see BURNING BUSH).
RUTACEAE, 1–2 species.

Dictyosperma a small genus of solitary palms, from the Mascarene Islands. *D. album* [PRINCESS PALM] of Mauritius is widely cultivated throughout the tropics for the edible shoot or "cabbage", and as an ornamental in gardens. It is nearly extinct where native.
PALMAE, 2–3 species.

Dictyostelium a genus of slime fungi which occur naturally in soil, where they feed on other microorganisms such as bacteria. Species such as *D. discoideum* and *D. mucoroides* can be easily cultured in the laboratory, where their simple life cycle, with the clear distinction between growth and developmental phases, has made them increasingly important organisms for research into cell and developmental biology.
DICTOYSTELIACEAE, about 7 species.

Dictyota a genus of brown algae found in warm seas. The thallus of this alga is dichotomously branched with the branches flattened like leaves. *D. acutiloba* and *D. apiculata* are used locally as food in certain Pacific islands.
PHAEOPHYCEAE, about 35 species.

Dieffenbachia a genus of poisonous, erect, evergreen perennials from tropical America. The stem is fleshy, the leaves usually green but sometimes blotched with yellow or white, and the flowers are borne on a spadix with a greenish spathe. Various species, including *D. maculata* and a plant incorrectly known as *D. amoena*, are cultivated as pot plants for their attractively blotched and veined leaves. *D. seguine* [DUMB CANE, DUMB PLANT, MOTHER-IN-LAW PLANT] has a gruesome history; it was used in the West Indies to torture slaves who were made to bite the plant. The excessively acrid juice caused the mouth to swell, rendering the person speechless for several days. Several cultivars of this species are in cultivation.
ARACEAE, about 30 species.

Dierama a genus of South African cormous perennials with graceful grass-like foliage and tall arching stems of drooping terminal panicles of funnel-shaped pink, white or purple flowers in June and July. The species most widely grown are the attractive *D. pulcherrimum* [ANGEL'S FISHING ROD] and *D. pendulum* [WANDFLOWER, GRASSY-BELL, ANGEL'S FISHING ROD] and *D. gracile*.
IRIDACEAE, about 20 species.

Diervilla a small genus of deciduous shrubs native to the eastern USA. *D. sessilifolia* is sometimes seen in cultivation as is a hybrid, *D.* × *splendens*, between it and *D. lonicera*. The Asiatic relatives of *Diervilla* are placed in the genus **Weigela*, which has a regular corolla (not two-lipped).
CAPRIFOLIACEAE, 3 species.

Digitalis see FOXGLOVE.
SCROPHULARIACEAE, about 30 species.

Dill the common name for *Anethum graveolens*, an easily grown aromatic herb native to India, southwest Asia and possibly North Africa. It is widely cultivated as a herb and often naturalized, especially in the Mediterranean. The seeds give an aniseed flavor to pickles, salads, sauces, meats, soups, etc, and if soaked in wine vinegar give dill vinegar, often used for pickling gherkins. The seeds are used dried as a substitute for caraway seeds in baking, and the oil (distilled also from leaves and stems) is used medicinally. The chopped leaves are used as a garnish. The commercial crop comes chiefly from Europe, India and Japan. INDIAN or JAPANESE DILL is *A. sowa*.
UMBELLIFERAE.

Dillwynia a genus of small heath-like pea-flowered legumes from Australia and Tasmania. The attractive bright red, yellow, or red and yellow flowers have led to the cultivation in mild climates of several species as ornamentals, such as *D. ericifolia* and

*Dill (*Anethum graveolens*). The seeds and chopped leaves are used as a garnish and have an aniseed flavor. (× ½)*

D. cinerascens, the former including var *glaberrima*.
LEGUMINOSAE, 12–15 species.

Dimorphotheca [CAPE MARIGOLD] a genus of annual or subshrubby perennial herbs native to southern Africa. The flower heads are very showy, the disk may be yellow, golden-brown or metallic blue, and the ray florets vary from orange to pink, purple, yellow or white, often white, often with an inner band of a contrasting color. Most of the species are cultivated as garden ornamentals, the most popular being *D. pluvialis* (= *D. annua*) and *D. sinuata* (= *D. aurantiaca* of horticulture) [STAR OF THE VELDT].
COMPOSITAE, about 7 species.

Dionaea a genus represented by single in-

Dieffenbachia maculata, native to tropical America, is popular as a houseplant. On a mature plant the leaves may be up to 60cm long. (× 1/12)

sectivorous species, *D. muscipula* [VENUS'S FLY TRAP], restricted to damp habitats on the pine barrens of North and South Carolina, USA. It has a short rhizome which bears a rosette of leaves with kidney-shaped to circular, two-lobed blades, lying near soil level. Each leaf, up to 12cm long, is hinged along the midrib. The margins of the lobes are fringed with tooth-like cilia, and the leaf resembles a shell clam in the half-open position.

If the three trigger hairs on the upper leaf surface are irritated, for example by an insect, the two halves of the leaf quickly and tightly close, the marginal teeth interlocking and trapping the insect inside. Digestive glands on the leaf surface secrete juices which break down the proteins of the animal into a soluble form which the leaf then absorbs. When digestion is complete the leaf opens again. This mechanism supplements the nitrogen requirements of these plants growing in soils with a low nitrogen content.
DROSERACEAE, 1 species.

Dionysia a genus of tufted, cushion-forming plants found in the mountains of Central Asia, northern Iraq, Iran and Afghanistan. Specialist growers of alpine plants attempt the cultivation of such species as *D. bryoides* and *D. curviflora*.
PRIMULACEAE, about 35 species.

Dioon a genus of *cycads containing three to five species found in Mexico and Central America, and sometimes grown as ornamentals. The seeds take at least a year to ripen in the female cones. The starchy endosperm of seeds of *D. edule* [PALMA DE MACETAS] can be ground into an edible meal or consumed after boiling or roasting.
ZAMIACEAE, 3–5 species.

Dioscorea see YAMS.
DIOSCOREACEAE, about 600 species.

Diospyros a largely subtropical and tropical genus of deciduous and evergreen shrubs and trees, several of which produce edible berry fruits (*persimmons). *D. lotus* [DATE PLUM], from East Asia is a tree which pro-

duces a round, yellow or purple fruit eaten as the date plum. However, this species is not widely cultivated for its fruit, but is more often used in the East as a rootstock for grafting *D. kaki* (see KAKI).

D. virginiana [AMERICAN PERSIMMON, COMMON PERSIMMON, POSSUM APPLE, POSSUM WOOD] (see PERSIMMON), a native of eastern and southern USA and *D. digyna* [BLACK SAPOTE] from Mexico and Central America bear edible fruits with a soft flesh, orange, in the former, chocolate-colored in the latter. The former species is also cultivated for the ornamental value of its autumn coloring and tasselled bark.

The black heartwood of *D. ebenum* (= *D. ebenaster*) [*EBONY, EAST INDIAN EBONY, MACASSAR EBONY] from India and Sri Lanka is the source of the best ebony of commerce. In other species the heartwood is merely streaked black (as in *D. kurzii*, the source of zebra wood) and may be present as only a small core. (See also EBONY.)
EBENACEAE, about 200 species.

Dipelta a small genus of deciduous shrubs allied to *Diervilla* and *Weigela*, native to central and western China. Three species are in cultivation, namely *D. floribunda*, *D. ventricosa* and *D. yunnanensis*, all of which have two-lipped pink flowers.
CAPRIFOLIACEAE, 4 species.

Diplophyllum a genus of leafy liverworts in which there are two ranks of leaves on the stem, and each leaf is divided almost to the base into two lobes. One lobe lies across the other thus creating the impression of a "double leaf", hence the generic name. The genus occurs in Europe, Asia and North America, with isolated representatives in South America and Hawaii. *D. albicans* is the commonest leafy liverwort of many acidic

A tuft of Dionysia tapetodes, *pictured here growing on a shaded limestone cliff – its usual habitat in its native Iran and Afghanistan. (× ½)*

habitats over wide areas of the north temperate zone of Europe and North America. SCAPANIACEAE, about 30 species.

Dipsacus a genus of thistle-like biennial or perennial herbs from Europe, Asia and Africa with spiny and prickly stems, notable for the highly distinctive fruiting heads of *D. sativus* [FULLER'S TEASEL], which are still used on a small scale for combing the nap on woollen cloth. Several species are grown in gardens for ornament or as dried flowers, such as *D. fullonum* [TEASEL], a weedy species which like *D. sativus* is widely naturalized. DIPSACACEAE, 15–20 species.

Dipterocarpus a genus of medium-sized to very tall trees distributed from India to the Philippines. Members of the genus are dominant in parts of the evergreen forests of Malaysia, and other, deciduous species are equally important in the monsoon forests of India and Burma.

Species of *Dipterocarpus*, together with other dipterocarp genera, notably *Hopea*, **Shorea* and *Vatica*, are the world's main source of hardwood timber. Much is processed into plywood locally or in Korea and Japan. Dipterocarp forests are destined to disappear in the next few decades due to over-exploitation and failure to replant on an adequate scale.

D. tuberculatus [INDIAN ENG TREE] yields eng oil and its wood is used for the construction of derricks, rolling stock, railway sleepers and parquet flooring. Among monsoon-forest species valued for their timber are the Burmese *D. turbinatus* and *D. alatus*; like other species these yield gurjun oil or balsam, used variously in boat-caulking, varnishes and medicine. Keruing is a commercial timber from some 20 Malaysian species, but this is used more for local purposes than for export. DIPTEROCARPACEAE, about 76 species.

Disa an orchid genus native to South Africa and Madagascar which, although spectacular when seen colonizing large areas of grassland and marsh, is rarely cultivated.

The wiry stems of a Dodder (Cuscuta campestris) form a solid mat as they twine around Lucerne (Medicago sativa) which it is parasitizing. (× ¾)

Male shoots of Diplophyllum albicans, a common leafy liverwort found on banks and damp rock faces in non-calcareous districts. (× 4)

The outstanding exception is *D. uniflora*, a native of South Africa which has large butterfly-like scarlet-red flowers borne singly on long, erect stems. ORCHIDACEAE, about 130 species.

Divi-divi a small tree, **Caesalpinia coriaria*, from tropical America and the West Indies. It is commercially valuable for its pods, the source of the tannin divi-divi, one of the most astringent of all vegetable substances and hence of great value for tanning leather, to which it imparts a yellow tone. *C. coriaria* also makes a useful shade tree in tropical gardens. LEGUMINOSAE.

Dizygotheca [FALSE ARALIAS] a small genus of deciduous shrubs or small trees native to New Caledonia and Polynesia. Many of the species cultivated as ornamentals have been described under the name of the related genus **Aralia*. Their ornamental value lies in the attractive palmately compound leaves of juvenile plants, as in *D. elegantissima* (= *A. elegantissima*, *A. laciniata*), *D. ternata* and *D. veitchii* (= *A. veitchii*). ARALIACEAE, about 17 species.

Dodder the common name for *Cuscuta*, a cosmopolitan genus of some tropical and temperate herbs, usually annuals, which parasitize a variety of other flowering herbs and shrubs.

Dodders are almost totally dependent upon other plants for their nutrient supply, with only their seedlings able to live for a short time in soil. Their stems are extremely thin and sensitive to the contact of other plants. They twine tightly around a branch or stem, rarely making more than three turns in the process. At the points of closest contact they produce haustoria, root-like suckers which penetrate the host plant. Because of their parasitic existence, all species of *Cuscuta* have lost the ability to photosynthesize. Their leaves have no green tissue and are reduced to minute brown scales along the stem.

Some species are particularly well-adapted for one host plant, while others infect a wide range of hosts. Considerable damage can be inflicted on the infected plants, especially agricultural crops. The cultivated species attacked by dodders are numerous, perhaps the most important being **Trifolium pratense* [RED **CLOVER*], which is affected by *C. campestris*, a pernicious weed of Europe and America. A widespread species common to all parts of Europe is *C. europaea* [GREATER DODDER] which usually affects **Humulus lupulus* [HOP], **Urtica dioica* [STINGING **NETTLE*] and many other hosts. *C. epithymum* [COMMON DODDER] is another widespread European species whose host plants include **Calluna vulgaris* [HEATHER], **Ulex europaea* [GORSE] and various **clovers* (**Trifolium* species). CONVOLVULACEAE, about 170 species.

Dodecatheon a genus from North America, with one species in northeast Asia. They are perennial herbs, sometimes with fleshy bulblets among the roots. The flowering stalks each bear an umbel of pendent, pink, purple or white flowers with reflexed corolla lobes, rather like those of a **cyclamen*. Some species are cultivated in shady places, wild or rock gardens, such as *D. meadia* [SHOOTING STAR], with its magenta, pink, lavender or white petals and yellow stamens, and *D. alpinum* with magenta to lavender flowers and stamens. PRIMULACEAE, about 50 species.

Dogwood the common name for *Cornus* (and its allies), a genus of mostly deciduous trees and shrubs of temperate regions of the Northern Hemisphere, especially USA and

Asia, with a sparse representation in South America and Africa. A number of species are cultivated as ornamentals, notably *C. alba* [RED BARKED DOGWOOD, TARTARIAN DOGWOOD], which is characterized by thickets of red stems on the younger growth during winter. There are several cultivars. Another attractively stemmed species is the shrub *C. sericea* (= *C. stolonifera*) [RED OSIER DOGWOOD] which produces yellow-green stems.

The most attractive flowering species include the North American *C. florida* [FLOWERING DOGWOOD], with numerous cultivars, and *C. nuttallii* [WESTERN FLOWERING DOGWOOD, MOUNTAIN DOGWOOD], both with large creamy-yellow bracts and the European and west Asian *C. mas* [CORNELIAN CHERRY] with greenish-yellow flowers and bright red cherry-like fruits, again with several cultivars. Some dogwoods produce particularly attractive foliage especially as the leaves turn color in the autumn. The best example is *C. controversa* [GIANT DOGWOOD], from China and Japan, with its purple-red autumn colors.

Some species, notably *C. nuttallii*, *C. mas* and the European *C. sanguinea* [BLOOD-TWIG DOGWOOD, DOGBERRY], provide valuable timber for tools, turnery and cabinet-making.
CORNACEAE, about 45 species.

Dolichos a genus of climbing beans mostly from Africa to India and East Asia, and closely related to the *MUNG and *KIDNEY BEANS and *COWPEAS of the genera *Phaseolus* and *Vigna*. Many are of agricultural importance, the seeds or beans being used for human food or the whole plant being used as animal food or green manure. *D. lablab* [LABLAB or HYACINTH BEAN] is the most important (see LABLAB). *D. uniflorus* [HORSE GRAM] is cultivated as a food and fodder crop chiefly in

Dorstenia barnimiana, a minute plant of the grasslands of Kenya. This species has an inflorescence consisting of a flattened receptacle surrounded by tendril-like projections.

Leopard Bane (Doronicum pardalianches), a reputedly poisonous species, with dandelion-like flowers and heart-shaped leaves. ($\times \frac{1}{8}$)

southern India. Some varieties are grown for their ornamental purple or white flowers, eg *D. lignosus* [AUSTRALIAN PEA].
LEGUMINOSAE, about 50 species.

Dombeya a genus of rounded shrubs and small trees native to tropical Africa, Madagascar and the Mascarene Islands. The bark of many species, including *D. cannabina* and *D. perrieri*, is widely used as a form of rough cordage and, occasionally for the manufacture of coarse sacks. A few species, such as *D. burgessiae* and *D. spectabilis*, are occasionally grown for their white, pink or red, mallow-like flowers which are borne in large, dense heads.
STERCULIACEAE, about 100 species.

Dorema a genus of hollow-stemmed perennials from the deserts of Central and southwestern Asia, the most important being *D. ammoniacum*, native to Iran and western India. It is the source of a gum resin, gum ammoniacum, a milky juice exuded by the stem, which is used medicinally and also in the perfume industry.
UMBELLIFERAE, 16 species.

Doronicum a genus of perennial herbs from Europe and temperate Asia. They are noted in gardens for their early colorful golden-yellow "daisy" flowers and soft-green, heart-shaped leaves. *D. plantagineum* and *D. cordatum* have produced some good garden varieties.
COMPOSITAE, about 25 species.

Dorstenia a genus of herbaceous plants, often stemless, with the leaves arising on long stalks directly from the perennial rhizome. Most species are found in tropical America and Africa, and some in the Middle East and India. The Central American and

Caribbean *D. contrajerva* [CONTRA HIERBA, TORUS HERB] is the source of contrajerva root, used as an antidote to snake-bites and also a stimulant and tonic. Other species are occasionally cultivated for their unusual inflorescences.
MORACEAE, about 170 species.

Doryanthes [SPEAR LILY] a small Australian genus of herbaceous plants, some of which are cultivated as greenhouse pot ornamentals. They produce scarlet and crimson flowers in globose heads with bright colored bracts, as in *D. excelsa* and *D. palmeri*.
AMARYLLIDACEAE, 2–3 species.

Double coconut or **coco de mer** the common name for the giant fan palm *Lodoicea maldivica* (= *L. callipyge*, *L. sechellarum*) endemic to the Seychelles. The stem reaches 30m in height and bears fruits each resembling a huge flattish double coconut weighing 12–20kg and containing one seed with a stony wall. The seed is the largest in the world and takes up to 10 years to develop. Locally the tree has reputed aphrodisiac properties.
PALMAE, 1 species.

Douglas fir the common name for members of the genus *Pseudotsuga*, native to western North America, China, Japan and Taiwan. They are evergreen coniferous trees, generally pyramidal in the early stages, finally somewhat widely spreading. The

The Double Coconut (Lodoicea maldivica) is restricted to a few valleys in the Seychelles.

leaves are flat and linear, bearing two white bands on the lower surface; the female cones are drooping.

The timber of *P. menziesii* (= *P. douglasii*, *P. taxifolia*) [GRAY OR OREGON FIR] is in great demand but, being very variable in strength and grain, it requires careful grading to ensure reasonable uniformity. It is used for practically every type of constructional work – houses, bridges and boats, for general carpentry, rolling stock and all kinds of poles and posts including commemorative flag-staffs. A mature DOUGLAS FIR is also a fine sight as an ornamental tree, especially when cleared of the lower and dead branches.

P. menziesii contains numerous cultivars, most of which are commercially valuable in North America and produce timber used extensively in building construction and carpentry. A number of species such as *P. japonica*, native to Japan, and *P. macrocarpa* [BIG-CONE SPRUCE], native to California, and *P. sinensis*, native to western China, are also cultivated as ornamental trees.
PINACEAE, 5–7 species.

Draba [WHITLOW GRASS] a genus of small tufted annual and perennial herbs mainly distributed in cold regions and mountains in the Northern Hemisphere. A number of species, including *D. aizoides*, *D. alpina* and *D. rigida*, are cultivated as rock-garden plants for their cushion-like habit and racemes of

Douglas Firs (Pseudotsuga menziesii) *growing in natural forest in the Cascade Mountains, Oregon, USA.*

Dragon Arum (Dracunculus vulgaris) *on a stony hillside in Crete, showing the segmented leaves and maroon spathe surrounding the black spadix.* ($\times \frac{1}{8}$)

yellow flowers. Other species have white, pink or purple flowers.
CRUCIFERAE, about 300 species.

Dracaena an unusual genus found mostly in the Old World tropics but extending westwards to the Atlantic Macaronesian islands, and with a single American species. Most species are shrubs with various large different-shaped leaves, ranging from long, sharp, sword-shaped to broad oval blades which are often marked in stripes and when mature are crowded to the top of the plants. The best-known species are *D. cinnabari* [DRAGON'S BLOOD TREE] of Socotra (and its close relatives, the DRAGON TREES of Somaliland, Ethiopia) and the Canary Island *D. draco* [DRAGON TREE], the source of *dragon's blood, a cinnabar-like resin that exudes from the trunks of tree dracaenas when they have passed their flowering phase. Perhaps the best-known DRAGON TREES are those of the Canary Islands, which grow to enormous sizes and live to great ages. The best living example today is the "Millennium Tree" at Icod, Tenerife, Canary Islands, which is reputed to be over 1 000 years old, but is probably much less. A much larger and older tree, reputed to be 6 000 years old was blown down in 1868.
AGAVACEAE, about 50 species.

Dracocephalum [DRAGON HEAD] a genus of mainly European and temperate Asiatic labiate herbs and dwarf shrubs, with a few species in North Africa and one species in North America. They are often grown in gardens. The common name refers to the shape of the showy flowers, which are two-lipped, blue, purple or (rarely) white and are borne in axillary clusters or in terminal leafy spikes. Some species, such as *D. peregrinum* and *D. nutans*, are cultivated for ornament, as honey plants or as medicinal plants.
LABIATAE, about 45 species.

Dracunculus a genus of two species of herbaceous perennials. *D. vulgaris*, which attains a height of up to 1m, bearing large divided leaves, flesh-colored stalks, flower stems mottled with black, and a large deep purple-red spathe sheathing a black-purple shiny spadix, is sometimes cultivated. *D. canariensis*, which grows in the Canary Islands, and Madeira, has a smaller green spathe.
ARACEAE, 2 species.

Dragon's blood a name applied to several resins, from various botanical sources, of a deep-red or ruby-red color. The first to be collected were those exuded from *Dracaena cinnabari* [DRAGON'S BLOOD TREE] of Socotra, and *D. draco* [DRAGON TREE] of the Canary Islands. The dragon's blood of old European medicine and modern varnish manufacture comes from resin-yielding *rattan canes

Three Dracaena *species growing as houseplants. From left to right:* D. marginata *'Tricolor';* D. deremensis *'Warneckii';* D. terminalis. ($\times \frac{1}{12}$)

which form a distinct group in the genus *Daemonorops*, the climbing jungle palms of the East Indies and Malaya. The resin is derived from the fruits of eight or more species.

Draparnaldia a genus of freshwater filamentous green algae, the species of which are characterized by producing tufts of delicate threads of cells arising from the main filaments. Species are often to be found on stones and rocks at the bottoms of pools or slow-running streams.
CHLOROPHYCEAE.

Drepanocladus a genus of mosses with representatives occurring in all five continents. They are found mainly in wet habitats, often growing submerged. In most species the leaves are strongly curved, sickle-shaped to coiled, and the habit is notoriously variable.
HYPNACEAE, about 25 species.

Drimys a small genus of aromatic evergreen trees or shrubs from Mexico south to Tierra del Fuego and the Juan Fernandez islands. The related genera *Bubbia, Belliolum, Pseudowintera, Tasmannia, Zygogynum* and *Exospermum* occur from New Zealand and Tasmania to Borneo and the Philippines, with one species in southern Brazil and Guiana and another in Madagascar.

D. winteri [WINTER'S BARK] is a South American tree-like species with lustrous polished leaves, ivory-white fragrant flowers and very bitter aromatic bark. This bark was used as a preventive for scurvy, and is still used locally as a tonic and for stomach complaints. *D. lanceolata* (= *D. aromatica*), from Southeast Australia and Tasmania is a highly aromatic pungent evergreen whose fruits are occasionally used for pepper.
WINTERACEAE, 5 species.

Drosera [SUNDEWS] a genus of perennial, rarely annual, insectivorous herbs occurring throughout the world but especially in the Southern Hemisphere. The plants generally occur in soils poor in nutrients, frequently in bogs. The leaves are usually in basal rosettes and are covered and edged with sensitive gland-tipped hairs which, when stimulated, change direction and trap and hold insects; subsequently, digestion is achieved by proteolytic enzymes, apparently often aided by bacterial activity. Some species are grown in greenhouses for their showy flowers and foliage, or as curiosities, the commonest being *D. capensis*.
DROSERACEAE, about 80 species.

Drosophyllum a genus consisting of a single species, *D. lusitanicum*, which grows in dry places in southern Spain, Portugal, and Morocco. About 25cm high, it has a short stem which is woody at the base in older plants, and numerous crowded linear leaves which are covered with stalked glands which secrete mucilage essentially for capturing insects, and sessile glands which digest and absorb the insect's proteins.
DROSERACEAE, 1 species.

Dryas a small genus of evergreen, prostrate subshrubs from alpine regions of the Northern Hemisphere. All three species are cultivated as rockery plants. *D. drummondii* has creamy yellow flowers and *D. integrifolia* has white flowers; both are from North America, and the white-flowered *D. octopetala* [MOUNTAIN AVENS] is from Eurasia and North America.
ROSACEAE, 3 species.

Dryopteris a large genus of mostly terrestrial ferns widespread in temperate and tropical regions. The leaves are once, twice or thrice pinnate-pinnatifid, with unequal-sided pinnae having their largest basal pinnules on the side away from the frond apex. They bear circular sori protected by large, kidney-shaped indusia.

Some of the species are cultivated in the garden or greenhouse and include the North

American *D. arguta* [COASTAL WOOD FERN], *D. cristata* (= *Aspidum cristata*) [CRESTED WOOD FERN, CRESTED BUCKLER FERN], native to North America, Europe and Asia, and *D. varia* (*Polystichum varium*) [JAPANESE HOLLYFERN], from China and Japan. *D. marginalis* (= *Aspidium marginale*) [MARGINAL SHIELD FERN, LEATHER WOOD FERN], from eastern North America, characteristically bears sori close to the leaf margin. *D. filix-mas* (= *Aspidium filix-mas*) [MALE FERN], native to North America and Europe, contains several variant forms with crisped, crested or forked leaves; its rhizome, frond-bases and apical bud are used in medicine for the expulsion of tapeworms. *D. fragrans* [FRAGRANT FERN, FRAGRANT CLIFF FERN], found naturally on mountain ledges of subtundra Canada and Eurasia, is sometimes cultivated for its highly glandular aromatic foliage.
ASPIDIACEAE, about 150 species.

Duckweed the common name for plants of *Lemna*, a genus represented by small floating or submerged aquatic herbs. Duckweed, often seen as a green floating carpet on stagnant water, is leafless, possessing instead

Duckweed (Lemna sp) growing as a green carpet on the still surface of a pond. Such dense mats can be a nuisance on ornamental ponds.

Leaves of the carnivorous Sundew (Drosera sp) clothed in gland-tipped hairs. (× 3)

green fronds or thalluses, usually with a single root, which perform the functions of leaves. They are among the smallest known angiosperms. *L. minor* [COMMON or LESSER DUCKWEED] is a cosmopolitan floating herb comprising mostly solitary plant bodies, while *L. trisulca* [STAR or IVYLEAFED DUCKWEED], also cosmopolitan, consists of submerged delicate plant bodies that remain interconnected. DUCKWEEDS can be valued feed for waterfowl.
LEMNACEAE, about 15 species.

Dudleya a genus of perennial succulent plants, mostly of Mexico and California, closely related to *Echeveria*. They produce clusters of mostly red to yellow flowers, sometimes white, borne in spring and summer on lateral stalks. They grow in rocky places and a few species are occasionally grown by rock gardeners.
CRASSULACEAE, about 40 species.

Duguetia a genus of shrubs and small trees from the West Indies and tropical South America. An economically valuable species is *D. quitarensis* [JAMAICAN, CUBAN or GUIANAN LANCEWOOD], a small tree which yields a tough elastic timber that is used for

Dyes and Tannins from Plants

Vegetable dyes have been used since earliest times in various parts of the world. They were, for example, used by the ancient Egyptians, Greeks, Romans and other early civilizations. *WOAD (*Isatis tinctoria*) was the imperial blue dye of Europe for many centuries and was used by the ancient Britons as a body paint. *INDIGO (*Indigofera anil, I. tinctoria* and other species) has been grown in India and Africa from time immemorial to provide another blue dye. *MADDER or TURKEY RED (*Rubia tinctorum*) was used by the ancient Egyptians and other Middle Eastern peoples to provide a red dye. A yellow dye and food colorant has been obtained from

SAFFRON (*Crocus sativus*) since the times of the ancient Greeks. Another yellow dye is obtained from DYER'S GREENWEED or WELD (*Reseda luteola*), one of the earliest dye plants known. Many hundreds of plants are still used for dyeing, especially in tribal societies in remoter regions of the world. They include lichens, club-mosses, mosses, ferns, gymnosperms and flowering plants and the parts used range from the bark, twigs, needles, flowers and fruits to the whole plant.

Vegetable dyes are substances which, when dissolved in water are able to color yarns, textiles, leather, wood and some foodstuffs. They are distinguished from animal

dyes and, in more recent times, from synthetic dyes. There is currently a resurgence of interest in natural plant dyes as part of home or cottage industries.

Water soluble dyes have to be made insoluble to prevent them running. This is achieved by pre-treating the material to be dyed with substances known as mordants, which make the dye fast and help attain an intensive coloration. Mordants include alum (used from very early times), common salt, cream of tartar, iron, chromium and tin with which the dyes form an insoluble compound.

Tannins and Tanning   Mention can also be made here of another group – the tannins. The term tannin describes a group of pale-yellow to light brown substances that are widely found in plants and which have been used for centuries for dyeing fabrics, making ink and in medical preparations. Their main use, however, lies in tanning leather – the process of converting animal skins into leather. This consists of steeping skins in infusions of vegetable materials such as bark, wood, leaves, nuts or galls which are rich in tannins. Putrefaction of untreated animal skins is promoted by enzymes secreted by various microorganisms. During tanning the tannins inactivate these enzymes and combine with the proteins of the skins, converting them into compounds which the enzymes cannot attack and binding together the protein fibers.

The tannins used in commerce are mainly obtained from the bark and wood of a few species which produce exceptionally high concentrations of tannins. Other materials such as leaves, galls and fruits are used less often. Few of these materials can be produced cheaply enough as a crop and most are either collected from the wild or are a by-product of some other activity.

OAK barks have been widely used for tanning for thousands of years. In Europe, *Quercus robur* and *Q. petraea* are mainly used, while in the New World *Q. montana, Q. alba, Q. prinus, Q. borealis, Q. dentata* and other species are important. Bark of HEMLOCK SPRUCES (*Tsuga canadensis* and *T. heterophylla*) is an important source of tannin in North America and is preferred for tanning heavy skins. WATTLE (*Acacia decurrens* and *A. dealbata*) bark has a tannin content of up to 40% and these fast-growing trees can be harvested at 5–10 years old. *EUCALYPTUS bark (from *Eucalyptus astringens* and *E. wandoo*) is widely used in Australia. Although *MANGROVE BARK is rich in tannins it is not so popular since the leather it produces is not of good quality. *BIRCH (*Betula*) and conifer barks are the main tannin sources in the USSR. *QUEBRACHO from South America (*Schinopsis balansae* and *S. lorentzii*) produce a wood rich in tannins. *CHESTNUT wood is also used to some extent, notably from *Castanea dentata* in North America and *C. sativa* in Europe. Tannins extracted from the autumn leaves of SUMACS (*Rhus typhina* and *R. glabra* in America and *R. coriaria* in Europe) produce fine leathers for bookbinding.

VEGETABLE DYES

Common name	Scientific name	Color
CHINESE TALLOW TREE	Sapium sebiferum	Black
GUAVA	*Psidium guajava	Black
*GAMBIER	Uncaria gambier	Black
*PERSIMMON	Diospyros kaki	Black
*WALNUT	Juglans nigra	Black, dark brown
KAVA	*Piper methysticum	Black
*LOGWOOD	Haematoxylon campechianum	Black, brown, gray
*INDIGO	Indigofera tinctoria, I. anil, I. arrecta	Blue
*WOAD	Isatis tinctoria	Blue
DYER'S KNOTWEED	*Polygonum tinctorium	Blue
FUSTIC	Chlorophora tinctoria	Yellow
LADIES' BEDSTRAW	*Galium verum	Yellow, red
DYER'S BROOM	*Genista tinctoria	Yellow
SAFFRON	*Crocus sativus	Yellow
SAFFLOWER	*Carthamus tinctorius	Yellow, red
GOLDEN ROD	*Solidago species	Yellow
YELLOW CHAMOMILE	*Anthemis tinctoria	Yellow
OSAGE ORANGE	*Maclura pomifera	Yellowish-tan, gold
WELD, DYER'S GREENWEED	Reseda luteola	Yellow, gold
YELLOW-BARKED OAK	*Quercus velutina	Yellow
*TURMERIC	Curcuma longa	Yellow
JACKFRUIT TREE	*Artocarpus integrifolius	Yellow
ZEDOARY	Curcuma zeodaria	Yellow, green
*GAMBOGE	Garcinia hanburyi	Yellow
SUMAC, VINEGAR TREE	*Rhus glabra	Yellow-tan, gray, brown
BUTTERNUT	Juglans cinerea	Tan
*BRACKEN	Pteridium aquilinum	Gray, yellow green
*BLACKBERRY	Rubus species	Gray
*BIRCH	Betula species	Gray
*PINEAPPLE	Ananas comosus	Green
MUNDU	*Garcinia dulcis	Green
MYROBALAN	*Terminalia chebula	Green
*COCONUT	Cocos nucifera	Green
BROOM	*Cytisus scoparius	Green
*CHERVIL	Anthriscus sylvestris	Yellow-green
TANSY	*Tanacetum vulgare	Yellow-green
*ONION	Allium cepa	Orange
*HENNA	Lawsonia inermis	Orange
COREOPSIS	*Coreopsis tinctoria	Orange, yellow
*JUNIPER	Juniperus communis	Brown
*CUTCH, CATECHU	Acacia catechu	Brown
*ASH	Fraxinus excelsior	Brown, green
*ALDER	Alnus glutinosa	Brown, gray, black, yellow
	A. incana	
*BLACKTHORN	Prunus spinosa	Reddish brown
RED STONE-LICHEN	*Parmelia omphalodes	Brown
REINDEER MOSS	*Cetraria species	Brown
	*Cladonia species	Yellow
WILLOW	*Salix pentandra	Brown
ORSEILLE	Roccella tinctoria	Red
*ALKANET	Alkanna tinctoria	Red
CROTTLE, CROTAL	*Parmelia saxatilis	Brownish red
BRAZIL WOOD	*Caesalpinia brasiliensis	Red, purple
BRAZIL WOOD, CAPPAN	*Caesalpinia sappan	Red
*ANNATTO	Bixa orellana	Red, yellow
BLOODROOT	*Sanguinaria canadensis	Red
*MADDER	Rubia tinctorum	Red

ornamental work and for fishing rods.
ANNONACEAE, about 7 species.

Durian the common name for *Durio zibethinus*, one of the most famous trees of the East, widely distributed and cultivated in Malaya and the East Indies, and to a small extent in the New World tropics, for its delectable fruit. The tree attains a height of about 30m, with buttresses, and bears massive spiny fruits, with leathery walls containing several large seeds, each embedded in a creamy yellow, or white, aromatic and strongly flavored pulpy aril. The fruits fall mainly at night and have to be collected very quickly, as they decay so rapidly that the pulp becomes rancid and inedible within four days.
BOMBACACEAE.

The rose-like flowers and feathery fruits of Dryas × suendermannii. (× ½)

Dyckia a genus of succulent plants from South America. Several species are grown for their foliage as indoor pot plants or greenhouse plants, or, in the tropics and subtropics, for bedding. The flowers are small and usually yellow, orange or red, borne on spikes, racemes or panicles. The commonly cultivated species include *D. altissima*, *D. brevifolia* and *D. frigida*.
BROMELIACEAE, about 70 species.

Dyes see p. 122.

Dysoxylum a genus of trees mainly of Indomalayasia and the Pacific. Many are locally important timber trees, such as *D. fraseranum* [AUSTRALIAN MAHOGANY, ROSEWOOD, PENCIL CEDAR], used in cabinet-making. Other commercially exploited species include *D. malabaricum* [WHITE CEDAR] of India, used in barrel-making, and the Australian *D. muelleri* [TURNIP WOOD, PENCIL CEDAR], used for furniture. A few are cultivated for ornament, such as *D. lessertianum* of Australia. The leaves of some species have been used as an onion substitute in Malaysian cooking.
MELIACEAE, about 60 species.

Earth balls the common name for certain gasteromycete fungi, such as *Scleroderma aurantium*, possessing tuberous fruit bodies which rupture irregularly to release masses of spores.
SCLERODERMATACEAE.

Earth stars the common name for the fungal genus *Geastrum*. The powdery spore mass of the mature fruit body is enclosed within a double wall. The thick tough outer wall breaks into lobes at maturity, bending backward in a characteristic star shape to expose the inner spore sac.
GEASTRACEAE, about 30 species.

Ebony the heavy, hard, dark heartwood derived chiefly from species of the genus *Diospyros such as *D. ebenum* [MACASSAR EBONY], the source of the best ebony, and *D. dendo* [GABOON EBONY, CALABAR EBONY]. It takes a high polish and is now used principally for small objects such as piano keys, knife handles and chessmen, hairbrushes and walking sticks. Although the word is synonymous with black, ebony wood can be other colors such as that of *D. hirsuta* [COROMANDEL EBONY], which has brown or gray mottling. Other genera, such as *Dalbergia, *Bauhinia, *Caesalpinia and *Jacaranda, have been used for ebony, and *HOLLY and *HORNBEAM have been dyed to make false ebony.

Ecballium a genus represented by a single species, *E. elaterium* [SQUIRTING CUCUMBER], a prostrate, perennial herb with stout, fleshy stems. It is native to the Azores, the Mediterranean, the Middle East and the southwestern USSR. The mature fruit detaches itself from its stalk at the slightest touch and the fruit contents are violently expelled through the resultant hole. The juice is exceedingly bitter and highly irritant to the eyes. The plant yields the drug, elaterium, a powerful purgative and cathartic used in local medicine and produced commercially in Malta.
CUCURBITACEAE, 1 species.

Eccremocarpus a small genus of evergreen climbers native to Peru and Chile. *E. scaber* [CHILEAN GLORY FLOWER], which bears tubular orange-scarlet flowers is cultivated to cover walls and pergolas. It is cultivated as an annual or perennial according to the climate.
BIGNONIACEAE, 4–5 species.

Echeveria [HEN-AND-CHICKENS] a genus of perennial leaf-succulents from Mexico extending into the southernmost USA and southwards to the Andes. Many species are cultivated for their attractively colored leaves arranged in flower-like rosettes at the ends of usually short stems, and for their graceful loose inflorescences of bell-shaped red to yellow flowers. Popular species, cultivated as house or greenhouse plants for their attractive leaf habit and color, include *E. agavoides* [MOULDED WAX], with large rosettes of thick, apple-green pointed leaves, *E. derenbergii* [PAINTED LADY], with broad concave pale green leaves with red margins, and its hybrid with *E. elegans*, *E.* 'Fallax', with bluish shell-like leaves with pallid margins.

Species which in addition to attractive leaves are grown for their flowers include

The colorful inflorescence of waxy flowers of Echeveria secunda *var* glauca, *one of the many ornamental succulents from Mexico.* (× 1½)

E. derenbergii, with orange flowers, *E. gibbiflora* (and several cultivars), with scarlet flowers, and *E. setosa* [MEXICAN FIRECRACKER], which enjoys considerable popularity for its showy red, yellow-tipped flowers.
CRASSULACEAE, about 150 species.

Echinacea [PURPLE CONE FLOWERS] a genus of three species of perennial herbs native to North America. *E. purpurea* is commonly cultivated, in a range of cultivars, for its daisy-like, purple flower heads about 10cm in diameter on stems up to 1m tall.
COMPOSITAE, 3 species.

Echinocactus a genus of large thick, ribbed, hemispherical to cylindrical cacti from the southwestern USA and Mexico. Among these are the largest of the aptly-named BARREL CACTI, which form a solitary main stem of great age and bulk up to 1.2m in diameter. One of these is *E. ingens* [MEXICAN GIANT BARREL]. By far the finest is *E. grusonii* [GOLDEN BALL CACTUS, GOLDEN BARREL], which

Echinocactus polycephalus, *one of the barrel cacti, forming a mound of several heads in the South Nevada Desert, USA. (×1/20)*

has bright glossy green stems, numerous ribs and radiating stout yellow spines. This is one of the easiest cacti to cultivate and one of the most popular. The name "MOTHER-IN-LAW'S ARMCHAIR" is also popularly applied to it.
CACTACEAE, about 10 species.

Echinocereus a genus of cacti from Mexico to the northern USA. They are all low-growing and decumbent, clump-forming or rarely solitary, with soft-fleshed, short, columnar or globose, ribbed stems. The flowers are mostly showy, up to 10–12cm in diameter and ranging in color from intense magenta or purple through pink, orange, yellow to brownish and lime-green. The fruits have a pleasantly acid taste. The frost-hardy *E. triglochidatus* (= *E. paucispinus*) [CLARET-CUP CACTUS] has brilliant claret-

Close-up of the Barrel Cactus (Echinocactus grusonii) showing the swollen stem, with numerous ribs covered with radiating yellow spines. (×1/3)

colored funnel-shaped blooms from stout but very soft ribbed stems which cluster and, in old age, form mounds. *E. pectinatus* var *rigidissimus* [RAINBOW CACTUS, COMB HEDGEHOG], from Arizona, is so-called from the successive rings of pink, white and straw-colored spines circling the stem. Many species withstand freezing, and a few such as *E. viridiflorus* [TORCH CACTUS] survive many years unprotected outdoors in milder areas. A few species, including *E. enneacanthus* [STRAWBERRY CACTUS] and its variety *stramineus* (= *E. stramineus*) [MEXICAN STRAWBERRY] produce edible fruits.
CACTACEAE, about 70 species.

Echinops [GLOBE THISTLE] a genus of biennial or perennial herbs occurring in open, often dry habitats from the Mediterranean east to Ethiopia and India. They are rather thistle-like, usually whitish-woolly plants with large globose heads of blue, pink or white flowers and an involucre of stiff, colored bracts. Several species are grown as border plants, especially *E. ritro* (= *E. ruthenicus*) and *E. sphaerocephalus*, with blue and silvery-gray flowers respectively.
COMPOSITAE, about 70 species.

Echinopsis a genus of South American globose to oblong, prominently ribbed cacti. Nearly all species bloom at night, with large white or pink, often sweetly scented flowers that last for a few hours only. The habit in cultivated forms ranges from dwarf solitary stems broader than wide, through large clumps as in *E. multiplex* [EASTER LILY CACTUS], to species such as *E. houttii* with cylindrical stems making offsets. The dwarf globular species include *E. eyriesii*, *E. oxygona* and *E. tubiflora*, which is often seen under glass. Globular *Echinopsis* species and hybrids are among the toughest and most trouble-free of cacti grown as houseplants.

Crosses with *Lobivia* species (× *Echinobivia*), notably the "Paramount Hybrids" from California, have large blooms in shades from pure white through yellow, orange, apricot and pink to red.

E. pachanoi contains hallucinogenic alkaloids used in religious ceremonies in Peru.
CACTACEAE, about 35 species.

Echium a genus of herbaceous annuals, biennials, perennials or shrubs found mainly in the Mediterranean region, North Africa and Macaronesia. The plants have tubular or bell-shaped flowers and a complex series of spines and hairs on the stems and leaves. Among several ornamental species are the biennial *E. plantagineum* (also a serious weed in many countries, particularly Australia) and *E. vulgare* [VIPER'S BUGLOSS, BLUE WEED, BLUE DEVIL] from Europe. Also cultivated in warmer climates such as California are some of the spectacular shrubby Canary Island species such as *E. candicans*, *E. simplex* and *E. wildpretii*, the latter with red flowers.
BORAGINACEAE, about 50 species.

Ectocarpus a genus of small branched

filamentous brown algae with a worldwide distribution. The filaments are often interwoven to form narrow spongy strands. Many of the species such as *E. tomentosus* are epiphytic on larger seaweeds.
PHAEOPHYCEAE.

Edelweiss the German/Swiss name for *Leontopodium alpinum*, a small alpine plant which has been much sought after by collectors. The flower stems and undersides of the leaves are covered in dense white woolly hairs and each flower head is surrounded by petal-like, white, woolly bracts.
COMPOSITAE.

Edraianthus [GRASSY-BELLS] a small genus of compact tufted perennial herbs native from Italy to the Caucasus and mainly centered in the Balkan peninsula. Some species are grown as rock-garden plants. Most cultivated species, including *E. pumilio*, *E. graminifolius* (= *E. caricinus*) and *E. tenuifolius* have bell-shaped or trumpet-shaped, blue to purple or violet flowers (white in *E. graminifolius* subspecies *niveus*), similar to those of *Campanula*.
CAMPANULACEAE, about 10 species.

Egg fruit the common name given to the fruits of two unrelated species: *Solanum melongena* (Solanaceae) (see AUBERGINE) and *Pouteria campechiana* (= *Lucuma nervosa*) [CANISTEL, SAPOTE AMARILLO, SAPOTE BORRACHO] (Sapotaceae), a slender tree which reaches a height of 16m in its native tropical America. The fruit is pear-shaped to globose in shape, with sweet, orange or yellow, mealy pulp with a musky flavor, eaten fresh. The plant is also the source of the spice canistel and is sometimes grown as a garden

The cactus Echinopsis macrogona, *showing the funnel-form flowers and ribbed stems that are typical of the genus. (×1/30)*

Water Hyacinth (Eichhornia crassipes) and Peacock Hyacinth (E. azurea) colonizing a stream in the Lower Amazon Basin.

tree in the Caribbean. A related species, also with edible fruits, is *P. sapota* (see SAPOTE).

Eichhornia a small genus of freshwater aquatic herbs. All species, except the African *E. natans*, occur naturally in tropical South America where they commonly form a conspicuous component of the aquatic vegetation of shallow pools, lakes and rivers. Depending on water depth, species such as *E. azurea* [PEACOCK HYACINTH] and *E. crassipes* (= *E. speciosa*) [WATER HYACINTH] can exhibit either an emergent or free-floating life form. The spike-like or paniculate inflorescence is composed of blue or lilac flowers. In *E. crassipes* and *E. azurea* these flowers are large and attractive and thus they are cultivated ornamentally in garden ponds and greenhouses.

A major problem is represented by the WATER HYACINTH which in the past century has spread throughout the world – often with the help of Man. It is now considered one of the world's most menacing aquatic weeds, and large sums of money are spent annually in attempts to eradicate its extensive mats from reservoirs, rivers and canals. The most important factor responsible for the success of *E. crassipes* as a weed is its spectacular powers of vegetative growth, coupled with the buoyancy provided by its inflated petioles. It has been estimated that in suitable growing conditions 10 plants can produce a solid acre of plants within eight months.
PONTEDERIACEAE, 5–7 species.

Elaeagnus a genus of hardy evergreen or deciduous shrubs and small trees native to Europe, Asia and Australia, with one species, *E. commutata* [SILVERBERRY] in North America. The densely hairy leaves are sometimes attractively variegated (eg *E. pungens* 'Ma-

culata'). Several species are cultivated as ornamental shrubs for their attractive foliage and inconspicuous fragrant flowers. One of the most widely grown is *E. angustifolia* [OLEASTER], a deciduous shrub with narrow silver-green leaves, silver-white flowers and yellow fruits. *E. umbellata* and *E. orientalis* are other deciduous shrubs grown as ornamentals, while *E. glabra*, *E. macrophylla* and *E. pungens* are among the evergreen cultivated species. The fruits of some species, especially *E. multiflora*, are consumed as preserves or as jelly.
ELAEAGNACEAE, about 45 species.

Elaeis a very small genus of palms, comprising the tropical American *E. oleifera* [AMERICAN OIL PALM] and the very important West African *E. guineensis* [OIL PALM]. These tall palms are grown in the wet tropics, principally West Africa and the Malay peninsula for their clusters of red fruits from which is extracted palm oil and palm kernel oil, which are extensively used as lubricants and for soap and margarine production. Palm toddy is extracted from the tender parts of the stem (see Arrack). They are also grown as ornamentals in the tropics.
PALMAE, 2 species.

Elaphomyces [FALSE TRUFFLES] a genus of widespread saprophytic fungi with subterranean, globose, nut-like fruiting bodies 1–3cm across, often associated with deciduous tree roots. The fruiting bodies of certain species, notaby *E. cervinis*, are used in veterinary medicine.
ELAPHOMYCETACEAE, about 20 species.

Elemi the name applied to a group of resinous substances and to a group of tropical forest trees from some of which the resin is obtained. The resin is soft and translucent,

Inflorescence of the Peacock Hyacinth (Eichhornia azurea) showing the conspicuous yellow "nectar guides" on the flowers. ($\times \frac{1}{2}$)

becoming yellow on exposure to the air and is used in the pharmaceutical and paint industries. The trees are mainly members of the Burseraceae such as *Canarium schweinfurthii* [AFRICAN ELEMI], *C. luzonicum* [MANILA ELEMI], *C. commune*, *Bursera gummifera* [WEST INDIAN or AMERICAN ELEMI], *B. jorullensis* [MEXICAN ELEMI] and species of *Protium*, *Amyris* and *Dacryodes*.

Eleocharis a genus of small perennials and tufted annuals, cosmopolitan but mainly found in wet places in the temperate zones, particularly the colder regions. *E. palustris* [SPIKE RUSH] is typical of many members of the genus with tufted stems and small inconspicuous flowers confined to a terminal congested spike. *E. dulcis* (= *E. tuberosa*) [MATAI, CHINESE WATER CHESTNUT], from tropical East Africa, the Pacific, Africa and Madagascar has starchy tubers which are eaten cooked in Chinese dishes. It is grown like rice in flooded fields and some of the crop is canned and exported. *E. acicularis* is widely grown in aquaria and several species, including *E. austro-caledonica*, are used in basketwork.
CYPERACEAE, about 200 species.

Elettaria see CARDAMOM.
ZINGIBERACEAE, 6–7 species.

Eleusine a small genus of annual or perennial grasses native to Africa, with one species in South America (*E. tristachya*). The annual *E. indica* [CROWFOOT, GOOSE GRASS, WIREGRASS, YARDGRASS] is now a pernicious pantropical weed. The most important species economically is *E. coracana* [FINGER MILLET, AFRICAN MILLET, RAGI, KORAKAN] (see MILLET).
GRAMINEAE, about 9 species.

Elm the common name for members of *Ulmus*, a genus of deciduous trees widespread throughout the north temperate zone but with three species extending south to the tropics. Elms are widely grown as shade and avenue trees, and as specimen trees for parks. They are mainly hardy and of easy cultivation. On the other hand, they are susceptible to insect attack as well as Dutch elm disease (see below).

Some of the better-known cultivated species are: *U. americana* [AMERICAN or WHITE ELM], with several cultivars; *U. minor* (= *U. carpinifolia*) [EUROPEAN FIELD ELM], with several cultivars and distinctive local populations, such as var *cornubiensis* [CORNISH ELM], which are sometimes treated as separate species; *U. glabra* (= *U. montana*) [SCOTCH or WYCH ELM]; *U. × hollandica* [DUTCH ELM], a hybrid of *U. glabra* and *U. minor*; *U. procera* [ENGLISH ELM]; *U. rubra* [SLIPPERY or RED ELM]; *U. serotina* [SEPTEMBER ELM]; and *U. × vegeta* (*U. minor* × *U. glabra*) [HUNTINGDON or CHICHESTER ELM].

Elm timber has some valuable characteristics, especially as it is cross-grained and resists splitting. Consequently it has been the timber of choice for wheelwrights for the hubs of spoked wheels. It has also been much

A flowering shoot of the Common Elm (Ulmus procera). *The projecting stamens and small perianth are adaptations to pollen-dispersal by wind.* (× 1)

used for furniture-making and for posts, barrels and handles for tools. Among the most important species for these purposes are *U. procera, U. americana, U. rubra, U. thomasii* [ROCK ELM] and *U. glabra*. The polished wood of elm shows a beautiful zigzag pattern, the so-called partridge-breast grain. The fact that the timber is resistant to decay under continuously waterlogged conditions accounts for its use for underwater piles.

The one known medicinal product of the elms is the mucilaginous bark of the SLIPPERY ELM, which is used as a soothing poultice for inflammations and internally as a treatment for diarrhea and other disorders. The foliage of the western Himalayan *U. wallichiana* is still important as animal fodder.

The fungus *Ceratocystis ulmi* is the causal agent of Dutch elm disease, which became prevalent in the 1920s and 1930s whence it was introduced into North America. In Europe, the disease attenuated after this outbreak, but another flare-up began after 1965, apparently through reintroduction of virulent strains from North America, where *U. americana* is particularly susceptible. The fungus kills by inducing blockage of the vessels. It is distributed by bark-boring beetles, mainly *Scolytus* species.
ULMACEAE, about 30 species.

Elodea a genus of submerged freshwater herbaceous plants. They are native to the New World but one species from temperate North America, *E. canadensis* [CANADIAN WATER-WEED], has been naturalized throughout Europe since 1836 in slow-flowing fresh water. A number of species, notably *E. canadensis, E. nuttalli* (North America), *E. densa* (South America) and *E. callitrichoides* (South America) are used in aquaria. In Europe usually only female plants of *E. canadensis* are found as reproduction is almost entirely vegetative.
HYDROCHARITACEAE, 10 species.

Elymus a widespread genus of mostly perennial Northern Hemisphere and particularly American grasses. *E. arenarius* [SEA LYME GRASS] is a widespread European species growing in sand dunes along the coasts of Britain and northwestern Europe. It is an effective sand-binder, spreading by its extensively creeping rhizomes. In North America, other species, such as *E. aralensis* and *E. racemosus*, both from the USSR, have been introduced for dune stabilization. *E. chinensis*, from China, is grown in the USA in wind-eroded areas.

In North America many species are good forage grasses: *E. glaucus* [SMOOTH BLUE WILD RYE] is one of the most valuable crops in the northwestern USA. *E. sibiricus* is cultivated in the Great Plains and intermountain district, *E. canadensis* and *E. virginicus* are common in the eastern states and *E. triticoides* [ALKALI RYE GRASS] occurs in the alkaline soils of the west.
GRAMINEAE, about 70 species.

Embothrium a small genus of evergreen shrubs or trees from the central and southern Andes of South America. The best-known is *E. coccineum* [CHILEAN FIRE TREE]. In its native habitat it forms a magnificent tree up to 15m high with dark glossy green leaves and striking, brilliant crimson-scarlet, axillary and terminal racemes of flowers.
PROTEACEAE, about 8 species.

Emilia a small genus of annual and perennial herbs widespread through tropical Asia, China, Polynesia and Africa, with a few naturalized in the New World tropics. The only species which appears to be cultivated is *E. javanica* (= *E. sagittata, E. flammea, Cacalia sagittata, C. flammea*) [TASSEL FLOWER]. It is an erect annual, bearing loose heads of scarlet flowers – golden-yellow in cultivar 'Lutea' (= *Cacalia aurea*).
COMPOSITAE, about 30 species.

Empetrum a small genus of dwarf evergreen procumbent or low growing heath-like shrubs widespread and often dominant in oceanic heathlands of cool temperate areas in the Northern Hemisphere, southern South America and Tristan da Cunha. They are regarded by some authorities as three species, *E. nigrum* (= *E. eamsii, E. scoticum*) [CROWBERRY, ROCKBERRY, CRAKEBERRY], *E. rubrum* and *E. hermaphroditum* [CROWBERRY], others as only one. The black or red fleshy fruits are used locally for making jellies and preserves, and are eaten by birds.
EMPETRACEAE, 1–3 species.

Encephalartos the second largest genus of *cycads distributed in tropical and southern Africa. Meal can be prepared from the starch-rich pith tissues of the stem of a number of species, including *E. caffer* [KAFFIR BREAD] and *E. altensteinii* [BREAD TREE]. Some are grown for their ornamental foliage.
ZAMIACEAE, about 20–30 species.

Endive the common name given to *Cichorium endivia*, an annual or biennial herb grown as a salad plant for its stocky head of pale green crisp leaves which when young and tender are used in salads and other dishes. A related species is *C. intybus* [WILD *CHICORY], which is grown as a salad vegetable [WITLOOF] and whose dried roots are roasted, ground and mixed with coffee.
COMPOSITAE.

Endothia an ascomycete fungal genus, the best-known species being *E. parasitica*, the cause of chestnut blight or canker. After its introduction into the USA from China in about 1900 it virtually destroyed the forests of *Castanea dentata* [AMERICAN *CHESTNUT]. It was introduced into Italy in 1938, and has caused great damage to *C. sativa* [SWEET CHESTNUT].
DIAPORTHACEAE, 10 species.

Endymion see *Hyacinthoides*.

Enhalus a marine genus represented by a single species, *E. acoroides*, an aquatic herb which grows at and below low-water level on the shores of Indomalaysia and Australia. The species forms underwater "meadows" which provide the chief source of food for the

The Chilean Fire Tree (Embothrium coccineum) *can be grown in temperate regions.* (× $\frac{1}{30}$)

dugong, a herbivorous sea mammal of Indian seas. *Enhalus* fibers are used to make fishing nets, and the fruits are edible.
HYDROCHARITACEAE, 1 species.

Enkianthus a genus of deciduous shrubs native to the Himalayas, China and Japan. Several species, including *E. campanulatus* and *E. cernuus*, are cultivated for the small white or cream, bell- or urn-shaped flowers which are produced in great profusion in spring, and for their colorful orange or red autumnal foliage.
ERICACEAE, 10 species.

Ensete a genus of large perennial herbs found in tropical Africa, Madagascar, southern China, Southeast Asia and Indomalaysia. They flower once and then die. *E. ventricosum* [ABYSSINIAN BANANA] is cultivated for the pulpy starch content of the swollen pseudostem and corms, which is eaten as bread, and for the young shoots which are eaten cooked. The seeds are also eaten. Fibers of the pseudostems are used for cordage and sacking.
MUSACEAE, 6–7 species.

Enteromorpha a genus of very common green algae found in marshes and watercourses near the sea. The genus can be a considerable nuisance as it tends to foul the bottoms of ships. Some species are consumed as a vegetable, including *E. prolifera* [locally called LIMU ELEELE] in the Hawaiian Islands.
CHLOROPHYCEAE.

Entomophthora a fungal genus (frequently known as *Empusa*) the members of which are all parasitic upon insects. *E. muscae* parasitizes flies, and dead flies attached to window panes, surrounded by a white halo of discharged fungal spores are a familiar sight. *E. grylli* is an important cause of death of grasshoppers in the USA.
ENTOMOPHTHORACEAE, about 40 species.

Ephedra a genus of gymnosperms widely distributed in warm temperate regions of the Northern Hemisphere and South America. Species tend to inhabit semi-arid or seasonally dry regions. They are mostly low creeping shrubs with slender green stems but a few are climbers and small trees. Some species, including *E. distachya*, from southern Europe and Asia, and *E. equisetina*, from Asia, are the source of vasoactive drug, ephedrine.
EPHEDRACEAE, about 40 species.

Epidendrum a large genus of American epiphytic and terrestrial orchids native to tropical and subtropical areas southwards from Florida but commonly cultivated and introduced into most parts of the Old World tropics. They exhibit a considerable range of plant form: many are pseudobulbous with a few apical, leathery leaves of varying shapes and sizes, but others have reed-like stems bearing many leaves.

Flower color varies with the predominant hues being whites (such as *E. nocturnum*),

The Abyssinian Banana (Ensete ventricosum) produces inedible fruits but the pseudostem and corms yield a starch-rich pulp.

yellows, and greens (such as *E. ciliare*), but several species bear brilliant magenta, purple, scarlet and crimson inflorescences, as in hybrids of *E. ibaguense* (= *E. radicans*). Often, many contrasting complementary colors coexist in a single flower as in *E. cochleatum* [COCKLE-SHELL or CLAM-SHELL ORCHID].

Epidendrum species enter into the pedigree of many inter-generic hybrids such as × *Epicattleya* (*Epidendrum* × *Cattleya*).
ORCHIDACEAE, about 1 000 species.

Epilobium a diverse genus mainly of erect or creeping, annual to perennial herbs in cool temperate areas of both hemispheres and on mountains in the tropics, usually in open habitats. Several species are aggressive weeds, while some of the creeping New Zealand species, such as *E. glabellum* and *E. pedunculare*, are grown for ornament in rock gardens. ROSEBAY WILLOWHERB or FIREWEED, also known as *E. angustifolium*, is here considered as a species of *Chamaenerion*.
ONAGRACEAE, about 200 species.

Epimedium a genus of rather woody perennial herbs from the north temperate Old World. They are hardy ornamental species of the semi-shade, grown for their heart-shaped leaflets (often beautifully veined and tinted) and the reds, yellows and whites of the flowers, and include *E. alpinum* [BARRENWORT, BISHOPS HEAD], *E. grandiflorum* (= *E. macranthum*) and their hybrid *E.* × *rubrum*, *E. pinnatum*, and *E.* × *versicolor* (*E. grandiflorum* × *E. pinnatum*).
BERBERIDACEAE, about 25 species.

Epipactis a genus of orchids most of which are north temperate, but with representatives in tropical Africa, Thailand and Mexico. Some species have been grown horticulturally, including two marsh plants with conspicuous open flowers. These are *E. gigantea* [GIANT HELLEBORINE, GIANT ORCHID, CHATTERBOX], from the western USA and Mexico, with yellowish and red flowers, and *E. palustris* [MARSH HELLEBORINE], a European species with rosy flowers.
ORCHIDACEAE, about 24 species.

The Cockle-shell Orchid (Epidendrum cochleatum) has twisted pale green sepals and an erect shell-like purple labellum. (× 2)

Epimedium × versicolor, *one of the Barrenworts, is often cultivated for its yellow flowers and attractively veined leaves.* (× $\frac{1}{5}$)

Epiphyllum a genus of mostly tropical American epiphytic cacti. The true species have long, jointed, three-winged or flattened, green leaf-like stems with aerial roots. *Epiphyllum* species have typically cactus-like flowers: usually large, white and nocturnal, and commonly intensely fragrant. There are numerous hybrids, many ascribed to *E. ackermannii* (= *Nopalxochia ackermannii*) but are probably hybrids of this species with *Heliocereus*. Commonly cultivated greenhouse species include *E. anguliger* [FISHBONE CACTUS] and *E. oxypetalum* [DUTCHMAN'S PIPE CACTUS]. In tropical countries, species of *Epiphyllum* are sometimes planted as hedges and make a magnificent sight when in bloom. It should be noted that the name *Epiphyllum* has been misused for the CHRISTMAS and EASTER CACTI now classified under *Schlumbergera*.
CACTACEAE, about 16 species.

Episcia a genus of showy-flowered tropical American herbs, often grown under glass, in hanging baskets and as houseplants. Commonly cultivated species include the scarlet-flowered *E. cupreata*, from Colombia and Venezuela, and the white or pale-lilac-flowered *E. lilacina* (= *E. chontalensis*), especially cultivar 'Panama'.
GESNERIACEAE, about 10 species.

Equisetum see Horsetails, p. 174.

Eragrostis [LOVE GRASSES] a large cosmopolitan genus of temperate and tropical, annual and perennial grasses. Species of *Eragrostis* occupy a wide range of different habitats. In South Africa, for example, there are many important veld species, such as *E. curvula* [WEEPING LOVE GRASS], while others occur in the desert, such as *E. cyperoides*, and a few even occur by water, such as *E. nebulosa*. In the USA the principal species include *E. pectinacea*, a pernicious perennial tumbleweed, which readily drops its panicles at maturity. *E. chloromelas*, *E. curvula* and *E. lehmanniana* (all from South Africa) have

been introduced into the USA for erosion control.

Few species have economic value. However, *E. tef* [TEFF] is grown as a millet in Ethiopia, where it is the most important crop, the grain being ground into a brownish flour called "ingera". Elsewhere in Africa, and in Australia, it is grown for hay. *E. curvula* has been introduced as a forage grass into North and South America. *E. amabilis*, *E. curvula*, *E. superba* and *E. obtusa*, are valuable ornamentals.
GRAMINEAE, about 300 species.

Eranthemum a genus of shrubs and perennial herbs, native to tropical Asia, some of which are cultivated as ornamentals. *E. pulchellum* (= *E. nervosum*), a small shrub with blue flowers is normally grown under glass in cooler temperate zones. *E. cinnabarinum* is a tropical garden flowering shrub with reddish flowers.
ACANTHACEAE, about 30 species.

Eranthis a genus of small, tuberous, herbaceous, spring-flowering perennials from the north temperate Old World. *E. hyemalis* [WINTER ACONITE], *E. cilicica*, from Asia Minor, and the hybrid *E. × tubergenii* are all commonly cultivated for their large solitary yellow flowers.
RANUNCULACEAE, 7 species.

Eremurus [FOX-TAIL LILY, DESERT CANDLE] a genus of perennial herbs with narrow leaves in rosettes or tufts, native to the dry steppes of western and Central Asia. Some, including *E. stenophyllus* (= *E. bungei*), *E. himalaicus* and *E. robustus* and many hybrids, are cultivated for their spectacular flowering racemes, up to 3m high, bearing white yellow or pale pink flowers depending on the species.
LILIACEAE, about 50 species.

Ergot a term applied to (i) a fungal disease of grasses and sedges caused by a species of *Claviceps*, (ii) the fungus causing such a disease and (iii) the sclerotia formed by *Claviceps* species, especially *C. purpurea*. Essentially, the sclerotium is a hard compact mass of fungal mycelium which forms after *Claviceps* parasitizes the ovary of grasses (notably RYE) and develops in place of the normal grain.

The sclerotia of *Claviceps* species contain toxic alkaloids which, if eaten by animals or man, can cause serious poisoning (ergotism) and even death. The arterioles are contracted with the result that the supply of blood to the extremities is impaired, and gangrene and loss of limbs may follow. They also cause convulsions and hallucinations. In the Middle Ages human ergotism was called "St. Anthony's Fire" and resulted from eating rye bread infected with *C. purpurea* (particularly toxic). With routine screening of cereal grain for impurities, ergotism in humans has become more or less unknown although serious outbreaks occur occasionally, the last being in 1953 in Belgium and France.

The same alkaloids from *C. purpurea*, purified and, in some cases chemically modified, are today used in obstetrics and in the treatment of migraine and high blood pressure. Ergot alkaloids required by the pharmaceutical trade were formerly obtained from rye crops naturally or artificially infected with *C. purpurea*, but these sources are being rapidly replaced by preparations produced in fermentation tanks using selected strains of *Claviceps* species.

Eria a genus of epiphytic orchids native to tropical Asia, with strikingly colored flowers. Unlike most orchids, the whole plant is covered with a mat of hairs.
ORCHIDACEAE, about 375 species.

Erianthus is a genus of large grasses found mainly in the warmer regions of the world. *E. ravennae* [PLUME OR RAVENNA GRASS] from southeast Europe, is cultivated for its ornamental large, silvery, plume-like inflorescences. The genus is now included within *Saccharum*.
GRAMINEAE, about 30 species.

'Maiden Erlegh' *is typical of the group of hybrids commonly called epiphyllums, although they are of polygeneric origin.* (× $\frac{1}{5}$)

Erica a genus of evergreen shrubs and occasionally dwarf trees which are commonly called HEATHS. However, the term "HEATHER" is also applied to a number of *Erica* species as well as the true HEATHER, *Calluna vulgaris*.

Ericas have an unusual distribution in two distinct regions of the world: most species are native to the Cape region of South Africa [CAPE HEATHS] and the remainder to Europe [HARDY HEATHS]. Their leaves are small and needle-like and the flowers are either solitary or in terminal umbels, panicles or racemes, with usually bell-shaped corollas, variously colored pink, white or red.

Once very popular as cool greenhouse and conservatory plants, very few species of CAPE HEATHS are now grown in cultivation. However, the HARDY HEATHS remain popular rock-garden subjects, often blooming in winter and early spring. The hardiest are *E. tetralix* and *E. cinerea*. *E. tetralix* [BOG HEATHER, CROSS LEAVED HEATH] is native to bogs and wet heaths in Western Europe. The stems, like those of *Calluna*, root at the base, but there are no short axillary shoots. A number of varieties are in cultivation, usually growing best on lime-free soils, eg 'Con Underwood' (red flowered) and 'Mollis' (white flowered). This species has hybridized with other species in the genus, eg with *E. ciliaris* [FRINGED HEATH] to give *E. × watsonii* and with *E. vagans* [CORNISH HEATH] to give *E. × williamsii*.

E. cinerea [BELL or PURPLE HEATHER] is another low-growing shrub, with profusely branched stems rooting at the base, and many short, leafy axillary shoots. This species is more usually found on dry moors in its native Europe. A number of varieties are in cultivation, including 'Atrorubens' (red flowered), 'Carnea' (pale lavender flowered) and 'Alba Minor' (white flowered). Other hardy species include *E. vagans* [CORNISH HEATHER or HEATH] and *E. carnea* [SPRING HEATH, SNOW HEATHER] which has many cultivars that vary in leaf and flower color.

E. arborea [TREE HEATH] from southern Europe is the source of the *briar root used in the manufacture of pipe bowls, while the stems of *E. tetralix* and *E. cinerea* as well as those of *Calluna* species, are the source of a yellow dye used for dyeing wool in the Scottish Highlands.
ERICACEAE, 500 species.

Erigeron [FLEABANE] a large genus of annual, biennial and perennial herbs, worldwide in distribution in temperate and mountainous regions, especially in North America. Most cultivated species flower in late spring or early summer.

Some of the perennial species are grown as garden-border plants, notably the perennial, fibrous-rooted *E. speciosus*, whose varieties, including *macranthus*, grow to a height of 60cm, and bear pink, blue, or purple ray florets. Another, taller (1.5m), popular garden species is the coarse-leaved annual or perennial *E. annuus* [DAISY FLEABANE, SWEET SCABIOUS, WHITE-TOP], with clusters of flower

The tall yellow flower-spikes of Eremurus stenophyllus, *one of the Fox-tail Lilies, a native of eastern Iran and Afghanistan.*

heads with normally white ray florets. A number of the lower-growing species are grown in rock gardens. These include the hairy perennial *E. alpinus* with large pink to violet flower heads, the attractive bright-orange-flowered *E. aurantiacus* [DOUBLE ORANGE DAISY], the white-, pink- or blue-flowered *E. flagellaris* [RUNNING FLEABANE], and cultivars of the somewhat succulent *E. glaucus* [BEACH ASTER, SEASIDE DAISY] and *E. thunbergii* (= *Aster japonicus*).

Medicinal extracts have been obtained from the roots of the Mexican *E. affinis* and from the leaves and stem tips of the North American *E. canadensis* [CONYZA].
COMPOSITAE, about 200 species.

Erinacea a genus represented by a single species, *E. anthyllis* (= *E. pungens*) [HEDGEHOG BROOM, BRANCH THORN], a tough, spiny shrub native to southwest Europe in dry, hilly areas. It bears racemes of rather large bluish-violet flowers and is occasionally cultivated for ornament.
LEGUMINOSAE, 1 species.

Erinus a genus consisting of one variable alpine and Pyrennean species *E. alpinus*, a tufted dwarf perennial herb cultivated in rock gardens with clusters of white (cultivar 'Albus'), carmine ('Carmineus'), lilac ('Lilacinus') or pink ('Roseus'), starry flowers.
SCROPHULARIACEAE, 1 species.

Eriobotrya a genus of evergreen shrubs and trees mainly distributed in East Asia. They have large leathery leaves and panicles of whitish flowers. *E. japonica* [LOQUAT, JAPANESE MEDLAR], from China and Japan, is a very important fruit tree in Japan and northwestern India and is widely cultivated for its

edible yellow plum-sized fruits. The wood of this species and of *E. bengalensis* is of economic value. *E. japonica* is also widely cultivated as a subtropical ornamental.
ROSACEAE, about 30 species.

Eriogonum [WILD BUCKWHEAT] a genus of annual and perennial herbs or shrubs native to western and southeastern USA and Mexico. Two perennial species, *E. umbellatum* [SULPHUR FLOWER] and *E. ovalifolium*, are cultivated as ornamental garden plants. Both bear clusters or umbels of yellow flowers, the latter with pinkish veins on the sepals. The leaves and stems of some species such as *E. longifolium* [INDIAN TURNIP] are used as food by some Indian tribes.
POLYGONACEAE, about 150 species.

Eriophorum [COTTON GRASS] a genus of mainly north temperate and Arctic, tufted or creeping, perennial herbs characteristic of wet peat moorland. The perianth of the flower is made up of bristles which elongate and become cottony after flowering and act as a means of seed dispersal. The "cotton hair" was once used as pillow stuffing.
CYPERACEAE, about 21 species.

Eriophyllum a small genus of herbs and subshrubs, all native to western North America. Their stems and leaves are covered in dense hairs which give them a woolly appearance. *E. lanatum* (= *E. caespitosum*, *Bahia lanata*), an erect perennial herb, is occasionally cultivated in garden borders for its attractive daisy-like yellow flower heads.
COMPOSITAE, about 11 species.

Erodium [STORKSBILL, HERON'S-BILL] a genus of hardy annual to perennial herbs distributed from Europe to southern Asia, as well as temperate Australia and tropical South America. They bear white, pink or purplish flowers with long-beaked carpels.

The beaks twist spirally with the carpels attached at the base separating from each other.

Many of the compact species make popular rock-garden or border plants. They include *E. corsicum* (pink flowers), *E. macradenum* (light purple flowers, but rose-pink in cultivar 'Roseum'), the densely-tufted, white-flowered *E. chamaedryoides* [ALPINE GERANIUM] and *E. petraeum* subspecies *lucidum* (white-flowered with red veins). A few species are valuable forage plants, notably the annual *E. cicutarium* [ALFILARIA, RED-STEMMED FILAREE, WILD MUSK, PIN CLOVER, PIN GRASS], which is native to the Mediterranean region but is now widely naturalized in North and South America, as is *E. moschatum* [WHITE-STEMMED FILAREE, MUSK CLOVER]. *E. hirtum* and *E. malacoides* are both used as food in North Africa, the former for its edible roots, the latter as a leaf-salad.
GERANIACEAE, about 90 species.

Eryngium a large genus of cosmopolitan perennial herbs with spiny-toothed leaves and small flowers arranged in dense hemispherical to cylindrical heads subtended by spiny bracts at the base. Hardy cultivated rock garden and border plants include cultivars 'Grandiflorum' and 'Superbum' of *E. alpinum*, with attractive blue or white flowers and deeply cut leaves, *E. giganteum*, a tall biennial or perennial Caucasian species with silvery-blue or greenish flowers, and the blue-flowered Eurasian *E. planum* (pink flowered in cultivar 'Roseum').

Another well-known species is *E. maritimum* [SEA HOLLY, SEA HOLM, SEA ERYNGIUM], with its glaucous-blue leaves and pale blue flowers. Although native to Europe this species is naturalized on the Atlantic coast of the USA. One of the better-known North American species is *E. yuccifolium* [RATTLESNAKE-MASTER, BUTTON SNAKEROOT], which grows to a height of about 1m and bears whitish flowers. Most material cultivated in the USA under the name of *E. aquaticum* is in fact *E. yuccifolium*. The

Fruiting heads of the Narrow-leaved Cotton Grass (Eriophorum angustifolium), a common colonizer of acid peaty bogland. (×1½)

roots of both the latter species and of the European *E. campestre* [SNAKEROOT] have medicinal uses.
UMBELLIFERAE, about 200 species.

Erysimum [FAIRY WALLFLOWER, BLISTER CRESS, TREACLE MUSTARD] a genus of annual to perennial herbs from the north temperate zone. Many of the species are cultivated in borders or rock gardens for their showy flowers in various shades of yellow to orange, purple or reddish, but the identity of many of the cultivated plants is not known. Most of them are probably hybrids between various *Erysimum* and *Cheiranthus* species.

Cultivated forms include the yellow-flowered European *E. helveticum* and *E. decumbens* (= *E. ochroleucum*) and the North American *E. suffrutescens* [BEACH WALLFLOWER], *E. torulosum* and *E. asperum* [WESTERN WALLFLOWER, PRAIRIE ROCKET]. Cultivated purple-flowered species include the rather dwarf Asian perennial *E. purpureum* and the European *E. linifolium* (pink and white flowers in cultivar 'Bicolor').
CRUCIFERAE, about 80 species.

Erysiphe a most important cosmopolitan genus of powdery mildews, including the devastating parasite of cereals and wild grasses, *E. graminis*, and *E. polygoni*, which has more than 500 hosts. In barley, *E. graminis* can cause crop losses as high as 40% in a season. The pathogen produces cottony white to brown mycelial mats on the surfaces of the leaves. Badly infected leaves turn brown and die, and the plant may be dwarfed with consequent failure to produce flower heads and grain.
ERISYPHACEAE, about 10 species.

Erythrina a genus of tropical and subtropical, quick growing, often spiny trees and shrubs, rarely herbs, bearing conspicuous

red, pea-like flowers. Some are grown as ornamentals in warm climates or under glass in cooler temperate zones. For example *E. crista-galli* [CORAL TREE], native to Brazil and neighboring countries, is renowned for its ornamental red flowers and seeds. The seeds of many species, including *E. herbacea* (= *E. arborea*) [CHEROKEE BEAN, RED CARDINAL], from Florida and Mexico, *E. tahitensis* (= *E. monosperma*) [HAWAIIAN CORAL TREE, WILWILLI] and *E. caffra* [KAFFIR BOOM] are used as beads and are often highly poisonous. The seeds of *E. herbacea* are used as rat poison in Mexico.

Several species are used as shade trees because of their large crowns, including *E. lithospermum* in Southeast Asia and Indonesia over tea and coffee, and *E. glauca* in Venezuela and Guyana over cacao. Other species called IMMORTELLE are grown in cacao and coffee plantations, including *E. mitis* (= *E. umbrosa*), *E. fusca* [SWAMP IMMORTELLE] and *E. poeppigiana* [MOUNTAIN IMMORTELLE]. In Asia and Australia *E. variegata* (= *E. indica*) provides support as well as shade for the weak-stemmed pepper plants.

The flowers and buds of some species such as the Central American *E. rubrinervia* are eaten, as are the leaves of the Indonesian *E. fuseda*. Several species yield extracts from the leaves, seeds and bark with various local medicinal uses such as febrifuges and diaphoretics. The wood of *Erythrina* species is soft and light: that of WILWILLI is used for surfboards and fishing net floats. The bark of *E. suberosa* is used as a cork substitute.
LEGUMINOSAE, 80–100 species.

Erythronium a genus of erect hairless perennial herbs from southern Europe and temperate Asia, but mainly from temperate North America. The nodding white, yellow, pink or purplish flowers are either solitary or

Erysimum suffruticosum is a commonly cultivated perennial used in rockeries or herbaceous borders. (×⅕)

The Coral Tree (Erythrina crista-galli), *from Brazil, is a shrubby species grown for its waxy brilliant red flowers.* ($\times \frac{1}{2}$)

in racemes. *E.dens-canis* [DOG-TOOTH VIOLET] is known in cultivation in a range of cultivars including the white-flowered 'Album' and the purple-violet flowered 'Purpureum'. Popular cultivated North American species include *E. albidum* [WHITE DOG-TOOTH VIOLET, BLONDE LILIAN], with long, solitary, bluish to purplish flowers, *E. grandiflorum* [AVALANCHE LILY] with flowers solitary or several, large and golden-yellow in cultivar 'Robustum', and white with yellow centers in cultivar 'Album', and *E. americanum* [YELLOW ADDER'S TONGUE, TROUT LILY, AMBERBELL], which bears solitary yellow flowers up to 6cm long. LILIACEAE, about 25 species.

Erythroxylum a genus of shrubs and trees occurring in subtropical and tropical countries, mainly in tropical America. The most important economic plants are *E. coca* and *E. novagranatense* (see COCAINE).
ERYTHROXYLACEAE, about 250 species.

Escallonia a South American (mainly Andean) genus of deciduous or evergreen shrubs, a number of species and hybrids of which are in cultivation as ornamentals in parks and gardens. The inflorescences of white, purplish or red flowers are sometimes showy and fragrant.

E. *rubra* (= E. *punctata*) is a widely cultivated and very variable species containing such varieties as *macrantha* (= E. *macrantha*), which is tolerant of salt-laden winds and forms a good hedge on the western coasts of Europe. It hybridizes in cultivation with most of the other species. One of the best-known hybrids is the generally low-growing shrub E. × *langleyensis* (= E. × *edinensis*), which is the result of cross between E. *rubra* and E. *virgata*. The latter, from Chile and Argentina, is the hardiest of all the cultivated species of this genus. The

extent of interspecific hybridization between garden species is such that many of the specimens in cultivation are hybrids which have replaced the original species.
SAXIFRAGACEAE, about 40 species.

Escherichia a non-spore-forming Gram-negative rod-shaped bacterium. Its major species, *E. coli*, is found in the intestinal tract of Man and other mammals. *E. coli* is extensively used in experimental work by biochemists and microbial geneticists.
ENTEROBACTERIACEAE.

Eschscholzia a genus of poppy-like annual or perennial herbs from western North America to northern Mexico, and widely naturalized in warm, dry areas elsewhere. The genus has provided gardeners with some of the most brilliant and easily grown hardy annuals. The species are of variable flower color: the widely cultivated *E. californica* [CALIFORNIAN POPPY], the state flower of California, has white and cream to yellow, orange, pink and scarlet forms. It is an annual or a perennial which flowers in its first year and, in mild areas, may overwinter after flowering.
PAPAVERACEAE, about 10 species.

Esparto a fiber used for the manufacture of cordage, shoes and paper, known in Arabic as "halfa", in French as "alfa", in German as "afriemengras" and in the USA as Spanish grass. It is derived from the leaves of two unrelated western Mediterranean grasses, *Stipa tenacissima* and *Lygeum spartum*. *S. tenacissima* [ALFA] can be divided into two commercial varieties – cordage alfa and papermaking alfa. *L. spartum* [FALSE ALFA] yields a long fiber that is useful only for making cordage.
GRAMINEAE.

Escallonia rubra var *macrantha, from South America, showing its glossy leaves and pinkish-red flowers.* ($\times 1$)

The Dog-tooth Violet (Erythronium dens-canis) *is not a true violet but a member of the lily family* (Liliaceae). ($\times \frac{1}{3}$)

Essential oils see p.132.

Eucalyptus a large, important genus of evergreen trees and shrubs, chiefly native to Australia and Tasmania with a few extending to New Guinea, eastern Indonesia, Timor and the Philippine Islands. They range in height from less than 1m to over 100m, as in *E. regnans* [MOUNTAIN ASH, SWAMP GUM, AUSTRALIAN OAK], the tallest of flowering plants. By far the greatest part of natural forest and woodland in the non-arid regions of Australia from the tropics to cool-temperate southern Tasmania is dominated by species of *Eucalyptus*, usually unaccompanied by any other genus in the upper story.

Eucalypts constitute the major hardwood timber resource of Australia: among the particularly valuable species for bridge, ship and house construction are *E. marginata* [JARRAH], *E. pilularis* [BLACKBUTT], *E. obliqua* [MESSMATE], *E. delegatensis* (= *E. gigantea*) [ALPINE ASH], *E. botryoides* [BLUE GUM, BASTARD MAHOGANY], *E. diversicolor* [KARRI GUM], *E. regnans*, *E. resinifera* [RED MAHOGANY] and *E. robusta* [SWAMP MAHOGANY]. Recently, emphasis has swung to pulp for paper and similar products, and especially for chipwood for hardboard production. Among minor products derived from the genus the best-known is a terpenoid essential oil (cineole or eucalyptol) used medicinally, as an antiseptic and to relieve colds, and in flavoring. Species which yield such an oil include *E. citriodora* [LEMON-SCENTED GUM], *E. dives* [BROAD-LEAVED PEPPERMINT] and *E. dumosa* [MALLEE, CONGOO MALLEE]. However, perhaps the major commercial source of eucalyptus oil is from the fresh leaves of *E. globulus* [TASMANIAN BLUE GUM]. Several species, including *E. leucoxylon* [WHITE IRONBARK] yield a kino, an astringent resin-like substance used in medicine and tanning. The bark of *E. astringens* [BROWN MALLEE] con-

Essential Oils

Essential oils are a class of vegetable oils which are made up of complex mixtures of volatile organic chemicals, usually responsible for pleasant odors or tastes. They are distinguished from fixed or fatty oils such as palm oil, groundnut oil and maize oil (see Oil Crops, p. 244).

Essential oils occur in a wide variety of plants – herbs, trees and shrubs – drawn from some 60 families of gymnosperms and flowering plants, notably the Myrtaceae, Labiatae, Umbelliferae, Lauraceae and Compositae. They frequently occur as droplets in the cells of glandular hairs or in secretory cavities or ducts or in canals which permeate the tissues. They can occur in most organs of the plant, especially the leaves, flowers, fruits and stems. Often, for a single species, production is confined to a single organ but in some cases oils are produced by different organs, as in citrus fruits where they are found in the sap hairs, the rind and seed.

The essential oils contain a great variety of terpenes and terpenoids which are hydrocarbons basically derived from a branched 5-carbon unit, isoprene. Their function in plants is still not clear and as by-products of secondary metabolism they have been regarded as waste-products or even demonstrations of the plant's biochemical virtuosity! There is accumulating evidence, however, that they play a significant part in attracting pollinating agents to flowers, in encouraging animals to eat their fruits and thereby aid in seed dispersal, and in defence mechanisms against predators.

Essential oils are used extensively in perfumes, flavorings and medicines. They can be conveniently divided into perfume, wood and flavoring oils, although some particular oils come under more than one of these categories.

Perfume oils

The use of essential oils as perfumes and scents dates from earliest times and they were known to have been used by the ancient Egyptian, Hebrew, Greek and Roman civilizations. Despite the increasing use of synthetic chemicals in perfumery today, there is still a flourishing trade in a wide array of naturally produced essential oils for use in scents, cosmetics, detergents, polishes etc. Today the perfume industry is centered around the town of Grasse in the south of France. There is extensive cultivation of species such as MIMOSA (*Acacia dealbata*), JASMINE (*Jasminum officinale*), *NARCISSUS, VIOLETS (*Viola* species), TUBEROSE (*Polianthes tuberosa*), and DAMASK ROSE from which essential oils are extracted for use in perfumes. Geranium oil is obtained from the leaves of several *Pelargonium* species, such as *P. graveolens* [ROSE GERANIUM] and *P. × asperum* (= *P. graveolens × P. radens*) grown in France, Spain and East Africa. *Attar or otto of Roses is obtained by distillation of the oil obtained from the flowers of the DAMASK ROSE (*Rosa damascena*) and to a lesser extent from the CABBAGE ROSE (*R. centifolia*) and the MUSK ROSE (*R. moschata*), which are grown for this purpose in Bulgaria, Italy, France, North Africa and India. Lavender oil is obtained from the flower tops of *Lavandula angustifolia* (= *L. officinalis*) and *L. latifolia*, grown in France, Spain and England.

Grass oils are obtained from the leaves of several species of *Cymbopogon*, such as *C. nardus*, grown in Java and Ceylon, which yields oil of citronella, and *C. citratus*, cultivated widely in India, tropical Africa and tropical America, which gives lemongrass oil. The roots of another grass, *Vetiveria zizanioides* [*KHUS-KHUS], are used in the production of oil of vetiver, widely used in perfumery.

Wood oils

Several well-known essential oils are obtained mainly from the wood of a wide range of plant genera. Examples include *camphor and camphor oil, obtained by steam distillation of the wood, twigs and leaves of *Cinnamomum camphora*. Cedar wood oil is obtained, not from cedars, but from *Juniperus virginiana*. Sandalwood oil, used in perfumery and medicine, comes from *Santalum album*, whose fragrant wood is also made into cabinets, while several species of *Eucalyptus*, notably E. dives and E. globulus, are the source of eucalyptus oil, which is used in medicine and in the refinement of mineral oils by flotation.

Flavoring oils

Numerous essential oils are used for flavoring in the food and confectionery industries, such as *mint oils from *Mentha* species, eg M. × piperita [*PEPPERMINT] and M. spicata [SPEARMINT], and umbelliferous oils such as caraway from *Carum carvi*, and aniseed from *Pimpinella anisum* and from *Illicium verum* [STAR ANISE]. *Clove oil, obtained from the CLOVE TREE (*Syzygium aromaticum*) and *nutmeg oil from the NUTMEG TREE (*Myristica fragrans*) are also widely used. (See also Flavorings from Plants, p. 142.)

A scene from the back of the throne of Tutankhamum depicts the Queen holding a salve-cup and spreading perfumed oil on her husband's collar. (Egypt: 18th dynasty – 1567–1320BC).

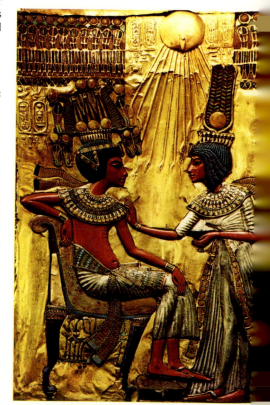

IMPORTANT ESSENTIAL OILS

Common name	Source
Perfume Oils	
GERANIUM	*Pelargonium* spp
*BERGAMOT	*Citrus aurantium* var bergamia
*ATTAR (OTTO) OF ROSES	Rosa damascena, R. centifolia R. moschata
*YLANG-YLANG	Cananga odorata, C. latifolia
TUBEROSE	*Polianthes tuberosa
CITRONELLA	*Cymbopogon nardus
LEMON GRASS	C. citratus
PATCHOULI	*Pogostemon patchouly
VETIVER, KHUS-KHUS	*Vetiveria zizanioides
SCENTED BORONIA	*Boronia megastigma
LAVENDER	*Lavandula latifolia, L. angustifolia
JASMINE	*Jasminum officinale, J. niloticum, J. odoratissimum
GARDENIA	*Gardenia florida
Wood Oils	
CADE	*Juniperus oxycedrus
*CAMPHOR	Cinnamomum camphora
CEDAR WOOD	*Juniperus virginiana
EUCALYPTUS	*Eucalyptus dives, E. globulus
*SANDALWOOD	Santalum album
SASSAFRAS	*Sassafras albidum
PINE NEEDLE	*Pinus sylvestris
Flavoring Oils	
*PEPPERMINT	Mentha × piperita
SPEARMINT	*Mentha spicata
STAR ANISE	*Illicium verum
*CELERY	Apium graveolens
*CLOVES	Syzygium aromaticum
*LEMON	Citrus limon
*ORANGE	Citrus aurantium, C. sinensis
*LIME	Citrus aurantifolia
CARAWAY	*Carum carvi
*NUTMEG	Myristica fragrans
*THYME	Thymus vulgaris, T. capitatus
ANISEED, *ANISE	Pimpinella anisum

*The Alpine Snow Gum (*Eucalyptus pauciflora *ssp* niphophila*) is one of the hardiest members of its genus. (×$\frac{1}{12}$)*

tains 40–50% tannin and was formerly much used in tanning processes.

Eucalypts are widely planted (eg in Brazil, North Africa, the Middle East, southern and tropical Africa, the Mediterranean, California, India and on the Black Sea Coast of the USSR) for timber, pulpwood, firewood, shelter, erosion control, essential oil production and ornament.

About 200 species have been introduced into cultivation but the most important species grown outside Australia number only 30 or 40 and are often different from those of greatest importance for timber in Australia. They include *E. camaldulensis* [MURRAY or RIVER RED GUM], *E. globulus*, the most widely cultivated species in the world, especially as the cultivar 'Compacta', *E. microtheca* [COOLABAH, FLOODED BOX], *E. sideroxylon* [RED IRONBARK], and *E. tereticornis* [FOREST RED GUM]. *E. viminalis* [MANNA GUM] is also widely cultivated for its timber, used mainly for house construction, while its bark produces a manna eaten by the natives. Species cultivated for their ornamental value include *E. polyanthemos* [SILVER-DOLLAR GUM] and *E. pulverulenta* [SILVER-LEAVED MOUNTAIN GUM, MONEY TREE]. The leaves of juvenile trees of both these species are used in decorative flower arrangements.
MYRTACEAE, about 550 species.

Eucharis a small genus of bulbous, highly fragrant perennials native to tropical Central and South America, mainly Colombia. *E. grandiflora* [AMAZON LILY], from the Andes of Colombia and Peru, with flowers 10–12cm across, in umbels of four to six flowers, has

been extensively cultivated in hothouses in Europe and elsewhere for its very beautiful and fragrant flowers.
AMARYLLIDACEAE, 10 species.

Euchlaena see MAIZE.
GRAMINEAE, 2 species.

Eucomis a tropical and southern African genus of bulbous perennials. All the species make good pot plants; the more commonly cultivated include *E. comosa* (= *E. punctata*) [PINEAPPLE FLOWER], with purple-spotted leaves, *E. autumnalis* (= *E. undulata*), with green flowers and undulate leaf margins, and *E. bicolor* with purple-edged flowers and crisped leaf margins.
LILIACEAE, about 14 species.

Eucryphia a genus of evergreen (rarely semievergreen or deciduous) trees and shrubs from Chile and Australasia, which bear numerous cup-shaped white flowers. They provide useful timber and are also cultivated for the ornamental value of their flowers and leaves. Among the most widely grown are the Chilean tree *E. cordifolia* and the shrub *E. glutinosa*, and the Tasmanian tree *E. lucida*. A number of hybrids are also valued as ornamentals, notably *E.* × *intermedia* (*E. glutinosa* × *E. lucida*).
EUCRYPHIACEAE, 5 species.

Eugenia a large tropical and warm temperate genus of evergreen trees and shrubs. Edible fruits are obtained from *E. uniflora* [FLUTED or RED SURINAM CHERRY], *E. brasiliensis* [BRAZIL CHERRY], *E. pitanga* [PITANGA] and other, chiefly Brazilian, species. Among species formerly ascribed to *Eugenia* and now transferred to other genera are the Old World species now placed in *Syzygium*, including *S. aromaticum* (= *E. aromatica*, *E. caryophyllata*) [CLOVE TREE], *S. jambos* (= *E. jambos*) [YELLOW ROSE APPLE] and *S. cuminii* (=

The spectacular midsummer display of Eucryphia × nymansensis, *a garden hybrid originating in England.* (×$\frac{1}{12}$)

E. jambolana) [JAVA PLUM, JAMBOLAN].
MYRTACEAE, about 1 000 species.

Euglena the most common algae genus; it is often called a "plant-animal" because of the presence of chloroplasts in an organism which has a marked animal-like movement. Members of this genus are mainly found in fresh water and are particularly abundant in richly nitrogenous effluents.
EUGLENOPHYCEAE, about 50 species.

Eulophia a predominantly African genus of mainly terrestrial orchids with a few tropical American and tropical Asian species. The majority have tuberous, underground stems. The flowers, which are borne on a long erect spike, vary in color but are usually bright with yellows and reds, although many of the species, especially those from Asia, have muted reds, browns, greens, and dingy creams. The pseudobulbous species from drier areas, such as *E. quartiniana* can be cultivated in a succulent house.
ORCHIDACEAE, about 200 species.

Euodia a genus of evergreen or deciduous trees and shrubs occurring in East Asia, Australasia and Madagascar, of which several, including *E. daniellii* and *E. velutina* are cultivated as ornamentals. An infusion of the flowers and leaves of some species, such as *E. lunuankenda*, is used as a tonic.
RUTACEAE, about 50 species.

Euonymus a genus of deciduous and evergreen trees and shrubs (rarely creeping) native to Europe, Asia, North and Central America and Australia, but with the greatest concentration of species occurring in the Himalayas and East Asia. Many species are cultivated for their foliage, especially for the autumn colors of leaves and fruits.

The evergreen *E. japonica* is a popular hedging plant often seen in towns and by the seaside. Golden- and variegated-leaf forms are known in cultivation, such as cultivar 'Albomarginata' and cultivar 'Aureovariegata'. The deciduous *E. europaea* [EUROPEAN SPINDLE TREE], which grows to a height of 6m, has several cultivars including 'Aldenhamensis' with bright pink fruits and 'Burtonii' with orange-red fruits. The Asian *E. hamiltoniana* (= *E. sieboldiana*) is another deciduous tree species with several cultivars. Most of the numerous cultivars such as 'Acuta' and 'Vegeta' of the evergreen *E. fortunei* have a low trailing or climbing habit. *E. alata* [WINGED SPINDLE TREE] is one of the most valuable and spectacular autumn-coloring shrubs.

Several species have particularly unusual fruits for which they are cultivated as in the case of *E. americana* [STRAWBERRY BUSH, BURSTING-HEART], which has red, spiny, warty fruits. *E. nanus*, a dwarf species, is cultivated in rock gardens. Apart from the economic value in cultivation, the wood from members of the genus has been used for making spindles, clothes pegs and skewers. The powdered leaves of *E. europaea* have been used to eradicate lice from children's hair. CELASTRACEAE, about 175 species.

Eupatorium a very large genus of mainly perennial herbs or shrubs, with a few annuals, mostly from Mexico, the West Indies and tropical South America, but a few from Europe, Asia and Africa. The white, pink or purple flowering heads are borne in terminal clusters.

Many species have local medicinal uses. *E. cannabinum* [HEMP AGRIMONY, WATER HEMP] contains the glucoside eupatorin and was once used to treat dropsy. *E. ayapana* from Brazil is cultivated for medicinal Ayapana tea, a stimulant and cure for dyspepsia. A few species, such as *E. coelestinum* [MIST FLOWER], and *E. purpureum*, are ornamentals grown in wild gardens, borders and in temperate climates for autumn flowers. *E. glandulosum* and *E. micranthum* are cultivated in cool greenhouses and *E. atrorubens* and *E. macrophyllum* in warm greenhouses. COMPOSITAE, about 1200 species.

Extreme habits of the genus Euphorbia. *Below* Poinsettia (E. pulcherrima). *Bottom The columnar* E. canariensis *and cushions of* E. aphylla. *(×$\frac{1}{50}$)*

Euphorbia a large and important genus distributed throughout the world, with heavier concentrations in certain areas, for example Mexico, the Mediterranean region, southwest Asia and South Africa. The genus includes prostrate annual herbs, biennial and perennial herbs, spiny cushion-plants, shrubs, trees and cactiform succulents ranging from small subglobose types to large candelabra-like arborescent forms. They all, however, share a common type of inflorescence known as a cyathium – a cup-shaped structure bearing glands (four or five) around the rim and containing numerous male flowers surrounding a single female flower. Furthermore, all species exude a poisonous milky latex when cut or damaged.

The most popular cultivated species is *E. pulcherrima* (= *Poinsettia pulcherrima*) [*POINSETTIA, CHRISTMAS FLOWER], a pot shrub with beautiful crimson floral bracts. Some of the smaller succulent euphorbias have become popular pot plants in recent years. *E. milii* [CROWN OF THORNS], native to Madagascar, with bright scarlet or cream-colored leaves subtending the cyathia, is widely cultivated in the tropics. *E. marginata* ["SNOW-ON-THE-MOUNTAIN"], native to the USA, is often cultivated in northern gardens for its green and white banded upper leaves. *E. lactea* [CANDELABRA CACTUS] is cultivated in the open in warmer climates. Other favorite garden species include the perennials *E. griffithii*, *E. sikkimensis*, *E. tirucalli* and *E. wulfenii*.

Although of no great economic importance some species yield locally useful substances. For example *E. antisyphilitica* [CANDELILLA], a shrub of Mexico and South western USA is the source of candelilla wax which when refined is used for polishes and varnishes and incorporated into an assortment of products which require waterproofing. Medicinally useful extracts are obtained from many species including *E. ipecacuanhae* [IPECACUANHA SPURGE] and

*The brilliant autumn foliage of the Winged Spindle Tree (*Euonymus alata*), a deciduous species from China and Japan. (×$\frac{1}{2}$)*

E. lathyris [CAPER SPURGE]. The latex of the Mexican *E. calyculata* [CHIPIRE] and *E. fulva* has been used as a source of rubber. Several species, including *E. lathyrus* and *E. esula* [LEAFY SPURGE], are troublesome weeds. EUPHORBIACEAE, about 2 000 species.

Euphrasia [EYEBRIGHTS] a taxonomically difficult group of semiparasitic annual herbs (rarely perennials), mainly in the temperate zone. The name "eyebright" may be derived from the former use of extracts of the leaves of *E. officinalis* as an eyewash. SCROPHULARIACEAE, 200 or more species.

Eurhynchium a genus of mosses of mat-forming or weft-forming habit. HYPNACEAE, about 45 species.

Euterpe a genus of tall, ornamental and graceful palms from tropical Central and South America. The fruits of *E. edulis* [ASSAI PALM] yield a thick, plum-colored liquid. It makes a very popular beverage and is used to flavor ices and sweetmeats. The terminal buds of *E. oleracea* [ASSAI PALM, CABBAGE PALM] and other species are used as a source of "palm hearts", which are eaten fresh or canned, especially in Brazil. PALMAE, about 20 species.

Exacum a subtropical and tropical genus of annual, biennial or perennial herbs and shrubs from Africa and Asia, some of which are grown as ornamentals under glass, especially *E. affine* [GERMAN VIOLET], with fragrant purple flowers, and *E. macranthum*, with showy, rich blue flowers. GENTIANACEAE, about 20–30 species.

Exochorda a genus of hardy deciduous flowering shrubs native to China, Korea, Manchuria and the southern USSR. They are cultivated for their showy white flowers, up to 5cm across, which are produced in great profusion in early spring, as exemplified by *E. korolkowii* [PEARL BUSH]. ROSACEAE, 5 species.

Fagopyrum [BUCKWHEATS] a genus of perennial and annual herbaceous species of leafy plants often with succulent stems, native to temperate regions of Eurasia. Best-known are *F. esculentum* (= *F. sagittatum*) [COMMON BUCKWHEAT] and *F. tataricum* [INDIAN WHEAT, TARTARY or SIBERIAN BUCKWHEAT], which are cultivated mainly for their grain in many regions of the world. POLYGONACEAE, about 6 species.

Fagus see BEECH.
FAGACEAE, 8–10 species.

× **Fatshedera** a bigeneric hybrid:*Fatsia × *Hedera.* × *F. lizei* is derived from a cross between *Fatsia japonica* 'Moseri' and *Hedera helix* var *hibernica.* A trailing evergreen shrub with deeply lobed palmate leaves, it originated in cultivation in France.
ARALIACEAE.

Fatsia a genus consisting of a single widely grown ornamental evergreen shrub species, *F. japonica* (= *Aralia sieboldii*) [FORMOSA RICE TREE, PAPER PLANT], native to Japan. It produces dark green shiny leaves and umbels of

Common Buckwheat (Fagopyrum esculentum) is a minor crop in many parts of the world. (×$\frac{1}{10}$)

white flowers in large terminal panicles.
ARALIACEAE, 1 species.

Faucaria [TIGER JAWS] a genus of dwarf succulent, almost stemless perennial herbs all native to South Africa. Each plant has several pairs of leaves united at the base, keeled underneath and interlocking when young by their toothed margins, hence the popular name. The daisy-like flowers are mainly yellow in the most commonly cultivated species, *F. felina,* *F. tigrina* and *F. tuberculosa,* but there are one or two white-flowered varieties. Normally cultivation is confined to the greenhouse in the cooler temperate regions.
AIZOACEAE, about 36 species.

Feijoa a genus comprising two species of shrubs or small trees native to subtropical South America. *F. sellowiana* [PINEAPPLE GUAVA] is cultivated as an ornamental and for its edible fruits. It has glossy green leaves (whitish beneath) and ovoid fruits whose fleshy pulp may be eaten raw, stewed or as a preserve.
MYRTACEAE, 2 species.

Felicia a genus of annual or perennial shrubby herbs from tropical and southern Africa. They are prized for their daisy-like flowers and are often grown in rock gardens. A favorite is *F. bergerana* [KINGFISHER DAISY] with its mass of small azure blue flower heads. Taller species, such as *F. amelloides* [BLUE DAISY, BLUE MARGUERITE] make good pot or bedding plants.
COMPOSITAE, about 60 species.

Fennel the common name for *Foeniculum vulgare,* a commonly cultivated herb. Both leaves and seeds have a pleasant aniseed flavor and are often added to fish, and to soups, poultry, salad and vegetable dishes. *F. vulgare* var *azoricum* [FINOCCHIO, FLORENCE FENNEL or SWEET FENNEL] is a shorter plant with swollen leaf-bases. These are eaten in salads, and the so-called bulbs used as a vegetable in their own right, or simply as a flavoring. *F. vulgare* var *dulce,* which does not have swollen leaf-bases, is cultivated for the essential oils in the fruits.

The name GIANT FENNEL is given to species of another genus of the carrot family, *Ferula.* In addition LOVE-IN-A-MIST (*Nigella,* family Ranunculaceae) is often called FENNEL FLOWER.
UMBELLIFERAE.

Ferns see p. 136.

Ferocactus a genus of cacti from the deserts of Mexico and the southern USA popularly known as BARREL or HEDGEHOG CACTI. They are large globose to cylindrical cacti with prominently ribbed stems covered in large, straight or hooked spines. They produce long bell-shaped yellow, pink or red flowers. Many species are grown as indoor or greenhouse pot plants, such as *F. acanthodes* (= *Echinocactus acanthodes*) with yellow flowers and

Giant Fennel (Ferula communis) growing on the Aegean Island of Patmos. (×$\frac{1}{10}$)

F. latispinus [DEVIL'S TONGUE] with red to purple flowers.
CACTACEAE, about 30 species.

Ferula a genus of stately herbaceous perennials, native to the Mediterranean, western and Central Asia. *Ferula* species, such as *F. communis* [GIANT FENNEL], make handsome garden plants, providing fine foliage in the larger herbaceous border, or specimen plants in isolation. The roots of *F. foetida* and *F. narthex* yield the resin asafoetida, used as an antispasmodic in medicine and as a condiment in Iran, where it is popularly known as "food of the gods". *F. galbaniflua* exudes galbanum, an oleo-gum-resin used medicinally as an expectorant and antispasmodic.
UMBELLIFERAE, about 130 species.

Fescue the common name for *Festuca,* a genus of annual or perennial grasses mostly from temperate regions of the Northern Hemisphere. The most interesting and horticulturally important species are those upland fine-textured species that have been selected for use in cultivated turf and lawn grass mixtures.

F. rubra [CREEPING OR RED FESCUE] is a fine-leaved, hard-wearing species with many new strains emerging for use in amenity turf. *F. ovina* [SHEEP'S FESCUE] is a delicate, tufted species for cold, acidic and dry soils. *F. glaucescens* is an exceptionally fine, dwarf species of blue-green color found in marsh turf in Cumbria, England. Other species, such as *F. ovina* var *glauca* [BLUE FESCUE], have considerable ornamental value and are used in contemporary landscape planting for ground cover. Several species, such as *F. elatior, F. ovina* and *F. rubra,* are cultivated as meadow and forage grasses.
GRAMINEAE, about 100 species.

Fibers see p. 140.

Ferns

THE FERNS ARE THE MOST ADVANCED OF THE spore-producing plants and today represent the largest class (Filicophytina) of the lower vascular plants (pteridophytes), which also includes the *club mosses and *horsetails. They are, however, unique among the pteridophytes in that they, mostly, bear large photosynthetic leaves (fronds) which more closely resemble those of the more advanced seed plants (*conifers and *angiosperms).

The fossil record of true ferns is generally accepted as beginning with the Carboniferous Period, although fragments of plant fossils with fern-like characteristics have been found as early as Devonian times. Today, for the most part, ferns are mesophytic plants, with thin leaf tissue, preferring moist shady habitats such as those found in woods and forest undergrowth, and beside rocky streams and in gorges. They are most abundant in the everwet climate of the tropical rain forest or the warm-temperate oceanic areas, eg New Zealand and the Canary Isles. In such climates they not only form the dominant herb layer in the forest but are also abundant as epiphytes on the trees or festooning rock walls. However, by no means all have a typical fern-like appearance and form. Some, such as *Cyathea and *Dicksonia reach the stature of small trees, while others, such as *Azolla, are minute and aquatic. So-called filmy ferns (family Hymenophyllaceae) have very flimsy, almost transparent leaves. They inhabit moist areas, such as rain forests and

Polypodium injectum is an epiphytic fern that grows in bark crevices. (× ¼)

cloud zones on mountains, and many tropical species are epiphytic. Some species of fern are specially adapted to drier ecological niches, eg exposed rockfaces, but these species require a deep crevice or seasonal wet periods for sexual reproduction to take place. So-called nest ferns, such as members of the genus *Platycerium [STAGHORN FERNS], produce a mantle of leaves which cling closely around tree branches. Humus and water collect in these structures, thus providing a substrate of nutrient and moisture for the roots of the plant.

Economic Uses. Ferns have limited uses in the modern world and even primitive peoples who rely on wild plants for food, medicines and building materials have few uses for them. Mature ferns are mostly poisonous or at least contain alkaloids that cause sickness and diarrhea; in their young stages, however, some species may be eaten. In New Guinea and other places in Southeast Asia a common species on river banks and open places is *Diplazium esculentum* (known in Malay as PAKU); this is commonly eaten boiled like spinach, which it resembles in taste. The same species – called NINGRO in Sikkim – is used there as a vegetable.

Young fronds of tree ferns (*Cyathea* spp) are eaten in New Guinea in a similar way and in the United States fronds of the OSTRICH FERN (*Matteuccia struthiopteris*) are sold deep-frozen or in cans. Very young unfurling croziers of *BRACKEN (*Pteridium aquilinum*) may be blanched and the cortical tissue eaten although mature bracken contains carcinogenic chemicals and for this reason this species should be avoided. The ripe sporocarps (nardoo) of *Marsilea* species are pounded and mixed with water by Australian aborigines to make a starchy "bread". The bulbous tubers of *Nephrolepis cordifolia* are similarly used by tribes in Sikkim.

Medicinal uses of species are no longer important. Both the MALE FERN (*Dryopteris filix-mas*) and the POLYPODY (*Polypodium vulgare*) were used by early apothecaries as kidney flushes and vermifuges. Primitive tribes in New Guinea and South America utilize the vascular strands of the SUN FERN (*Gleichenia* spp) and of the CLIMBING FERN (*Lygodium* spp) to weave armlets and neck ornaments and today such skills are redirected to make trays, place-mats and coasters for the tourist trade. The trunks of tree ferns (*Cyathea* spp) are prized as long-lasting

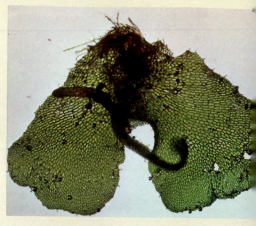

The prothallus of a fern. This is the sexual (or gametophytic) stage of the life cycle. (× 25)

house posts throughout New Guinea. Tree ferns in Central and South America are in danger of extinction through overexploitation for use as hanging baskets and as a source of fiber.

In many parts of the world there has been in recent years an upsurge of interest in ferns as desirable plants for cultivation. They fill the need for restful greens as a contrast to the bright colors of flowers and in the garden they usefully occupy shady places and can be grown in light woodland, on banks, along the margins of ponds and streams and in wall crevices.

Structure and Reproduction. Ferns have two distinct stages in their life-cycle: firstly there is the sporophyte (or spore-producing stage)

The nest fern Platycerium bifurcatum *(Staghorn Fern). (× ⅕)*

which is the typical fern that we recognize and secondly there is the gametophyte (or sexual stage), which is difficult to find by the untrained eye.

For the most part the stem of the sporophyte generation is confined to a rootstock or rhizome. This should not be confused with

Various forms of sori (clusters of sporangia) found in some common temperate ferns.
Left Top Dryopteris filix-mas: round, mixed sori covered by a kidney-shaped indusium. Left Bottom Asplenium sp: a marginal sorus covered by a flap-like indusium. Right Top Platycerium sp: this lacks true sori, the sporangia being widely distributed over the whole leaf. Right Middle Phyllitis scolopendrium: a linear sorus bordered by a long narrow indusium. Right Bottom Pteridium sp: a linear mixed sorus that lacks an indusium (ie is naked), but is protected by the inrolled edges of the leaflets. ($\times 5$, $\times 5$, $\times 1$, $\times 6$, $\times 5$)

form a vascular cambium. The traditional leaf shape of the fern, as seen in the LADY FERNS (*Athyrium* spp) or BUCKLER FERNS (*Dryopteris* spp), is compound, twice or three-times pinnately dissected. Such leaves, each with a central axis (rachis) and side leaflets, which themselves may be feather-like, are by

In leptosporangiate forms the sporangia are minute, stalked, globose structures, with walls only one cell thick. In intermediate forms the sporangia tend to be separate and thin-walled but sessile. Dehiscence of the sporangia also varies: in the eusporangiate and intermediate forms the wall cells simply dry and shrink, the capsule finally opening along a predetermined line of weaker cells (stomium). In the leptosporangiate forms a specialized row of cells (annulus) possess thickened walls, which on drying straighten out, thus breaking the cells at its weakest point. The inner cell walls of this "hinge" retain enough moisture to contract again and act as a spring, pulling the sporangium shut and projecting the spores into the surrounding air currents. In many ferns the sporangia arise usually in clusters (sori) from a pad of cells (receptacle) on the underside of the leaf. The structure positioning and arrangement of sori can vary greatly, but there are three basic types. In the simple type the sporangia are the same age and mature together. In the gradate sorus those at the apex of the receptacle mature first, while in the mixed sorus the receptacle continues to produce sporangia randomly over a period.

The dispersed spores germinate to produce the gametophyte. This varies in morphology but is always small and insignificant and most often heart-shaped. The gametophyte tends to be overlooked by the average person or thought to be a liverwort. The fern gametophyte (prothallus) produces motile sperm-cells within the male sex organs (antheridia), which are released and swim to fertilize the enveloped egg-cells which are produced in the female sex organs (archegonia). The fertilized egg germinates to give rise to the next sporophyte stage.

Ferns used as houseplants. From left to right Pteris cretica (Ribbon Fern), Adiantum capillus-veneris (Common Maiden-Hair Fern), Pteris sp, Asplenium nidus (Bird's Nest Fern), Adiantum sp, Platycerium bifurcatum (Staghorn Fern). ($\times \frac{1}{10}$)

the upright stalk-bearing compound leaves as seen in bracken or the central stalk (rachis) of other ferns; these are midribs of the compound leaf system, not stems. The creeping rhizome may be a thin, wiry, branched system as in the tropical filmy ferns or the EUROPEAN OAK FERN (*Gymnocarpium* spp) or it may be thicker as in *Polypodium vulgare* or the NARROW BUCKLER FERN (*Dryopteris carthusiana*). The rhizome is normally horizontal, but may be more upright. This upright form is seen fully developed in the tree ferns of the tropics and the warm temperate zones of the Southern Hemisphere. These belong mainly to the genus *Cyathea* or *Dicksonia*. Some species of *Blechnum* may also have short stems 1–2m high. Tree ferns may reach over 18m.

Ferns can usually be distinguished from seed plants in having no secondary growth and therefore no wood or bark. The single exception to this is the genus *Botrychium* [MOONWORT and GRAPE FERN] which may

no means the dominant kind, although the immediate ancestors of ferns may well have had this kind of leaf. Many species characteristically found festooning the trunks of tropical rain forest trees have simple leaves. Yet other fern leaves are palmate or dichotomously divided. Size of leaf may vary from a few millimetres in *Didymoglossum* to 2m or more in some tree ferns.

The spore-bearing structures (sporangia) may be borne on the underside of foliage leaves, on modified leaves or on separate stalks. Sporangia form is one of the major characters in classification. In what is known as eusporangiate forms the sporangia walls comprise two layers or more, often over 1mm across and are without stalks. In these cases the sporangia may be fused in clusters (synangia) or fused into a column as in the characteristic "adders tongue" of *Ophioglossum*. In *Botrychium* they are not fused but are globose and sessile on a branched organ (sporophore) and resemble a bunch of grapes.

Ficus a large pantropical genus of small shrubs, large trees up to 45m tall and woody, root-clinging climbers. They are frequently evergreen in tropical zones, but tend to become deciduous in temperate zones. Some species are glabrous, as in *F. elastica* [INDIA RUBBER FIG, CAOUTCHOUC], others hairy, or with stinging hairs (*F. minahassae*) or with silica bodies in the leaves (*F. scabra*). The presence of latex vessels is universal throughout the plant and, in fact, the latex of *F. elastica* was used extensively in rubber manufacture up to the mid-19th century.

Figs have a characteristic inflorescence, termed a syconium, a flask-shaped fleshy container with many minute flowers densely arranged on the inner walls. Fertilization is achieved by gall wasps (*Blastophaga* spp) which enter the syconium to lay their eggs, the newly-hatched adults carrying pollen to another syconium.

Several species of fig are classed as "stranglers": here the fig grows around a host plant, slowly enclosing it and often eventually killing it. This strange habit is common in tropical species, for example *F. pertusa* and *F. cordifolia*. The strangler fig begins from a seed dropped in the fork of a twig by a fig-eating mammal or bird. As the seed germinates it begins to grow downwards, the roots wrapping round the host. The effect is to crush the bark, thus ringing the tree and destroying its food-bearing vessels. Eventually the fig survives as a free-standing tree.

Other species are called *banyans, the best-known being *F. benghalensis* [BANYAN TREE, EAST INDIAN FIG, INDIAN BANYAN]. This species is a large tree which, though indigenous to the Himalayan foothills, is now widespread throughout India, as for centuries it has been planted in many villages for the excellent shade it provides. The Hindus also regard it as sacred, for it is said that Buddha once meditated beneath a banyan tree. Perhaps its most interesting botanical features are pillar-like aerial roots which grow vertically downwards from the branches and, once established in the ground below, quickly thicken, so that the tree assumes the unusual appearance of being supported by pillars. By this manner of growth the tree is able to spread outwards almost indefinitely, and many examples are of immense size and great antiquity. The timber is fairly hard and durable but light in weight. It is used for a variety of purposes in India, including furniture, door panels and cart shafts. A type of paper is made from the bark. The COMMON or EDIBLE FIG (*F. carica*) has a long history of cultivation, beginning in Syria probably before 4 000 BC. Since then it has held an important place in folklore and literature. The art of fig culture was first documented by the Greek poet Archilochus, around 700 BC, and there are many references to the fig in the Bible. It grows successfully in many tropical and some temperate habitats, usually on dry, higher ground. The main areas of cultivation are California, Turkey, Greece and Italy. The COMMON FIG is a small tree, less than 10m tall,

Fissidens serrulatus is a mainly western Mediterranean species with a distribution extending into southern England. (×4)

with large palmately lobed leaves 10–20cm long. There are two fruiting types, Adriatic and Smyrna. The more common ADRIATIC FIG does not have male flowers and the fig fruits develop from the female flowers without the need for pollination and fertilization. Its seeds are undeveloped and infertile. The SMYRNA FIG also has no male flowers, but differs in that the female flowers require pollination. To achieve this, branches of WILD FIGS with male flowers are attached to SMYRNA FIG trees at the time fig wasps are expected to emerge, thus allowing cross-pollination. The WILD FIG with male flowers is

*Inflorescence of the Meadowsweet (*Filipendula ulmaria*), which is common in damp meadows and is also cultivated.* (×1)

known as the CAPRIFIG, and the process described is known as caprification. All fig cultivars produce a main crop which develops on the current season's growth and matures in August to November. However, in some cultivars, syconia emerge on the previous autumn's shoots, producing a first or "breba" crop in June to July.

There are many other species of *Ficus* with edible fruits, for example *F. racemosa* [CLUSTER FIG], a common tree of East Asia. Although edible, the fruits are hardly palatable by Western standards, the figs being full of insects or hard seeds. *F. religiosa* [PEEPUL or PEEPAL TREE], venerated by Hindus and Buddhists, is often planted as a shade tree because of its dense, spreading crown. Several species are used as host trees for the lac insect, for example *F. semicordata* (= *F. cunia*) and *F. rumphii*. The insect secretes a resinous substance (*lac) which in its purified form (shellac) is used for several purposes, particularly in the manufacture of varnish and for electrical insulation. In temperate zones, some species are grown as houseplants. The commonest is the RUBBER PLANT, which is a sapling of *F. elastica* var *decora*. Also grown are *F. pumila* [CLIMBING FIG] and *F. benjamina* [WEEPING FIG or JAVA WILLOW].
MORACEAE, about 2 000 species.

Filipendula a genus of hardy herbaceous perennials of the north temperate region, often grown as garden ornamentals. *F. ulmaria* [MEADOWSWEET, QUEEN OF THE MEADOW], common in wet meadows and woods, is characterized by its crowded irregular clusters of creamy white flowers and sweet scent. Oil of meadowsweet is distilled from the flower buds and used in perfumery. The yellow cultivar 'Aurea' and the double-flowered cultivar 'Plena' are grown in gardens. Varieties of *F. vulgaris* (= *F. hexapetala*, *Spiraea filipendula*) [DROPWORT], *F. purpurea* (= *Spiraea palmata*) and *F. rubra* [QUEEN OF THE PRAIRIE] are also cultivated.
ROSACEAE, 10 species.

Fissidens one of the largest genera of mosses, cosmopolitan but with the chief concentration of species in the tropics. Most are small delicate plants and some are almost microscopic. The leafy shoots are instantly recognizable as *Fissidens* because in contrast with most mosses, the leaves are strictly two-ranked.
FISSIDENTACEAE, about 1 000 species.

Fittonia a small genus of creeping perennial herbs with attractive foliage, native to South America but sometimes cultivated in greenhouses. Both *F. gigantea* and *F. verschaffeltii* have broad, entire leaves with colored or white veins.
ACANTHACEAE, 2–3 species.

Fitzroya a genus represented by a single species, *F. cupressoides* [PATAGONIAN CYPRESS], a large evergreen tree with reddish bark, native to Chile. It yields a valuable timber, particularly useful for general carpentry and construction as well as for musical instruments and pencils.
CUPRESSACEAE, 1 species.

Flavorings from plants see p. 142.

Flax the common name for *Linum usitatissimum*, an annual plant which is grown widely in Canada, the USA, the USSR, Europe, Argentina, Uruguay and India for its stem fiber (fiber flax) which is spun to produce linen threads and yarns for weaving into pure linen cloth, or to be mixed with other fibers in production of other materials. Its particular value lies in its strength and durability compared to other natural fibers such as cotton or wool.

The plant is also grown for its seed oil [OIL FLAX, SEED FLAX, *LINSEED*], a drying oil used in paints, varnishes, putty and linoleum. The residue, linseed cake, is a valuable animal feed. Although flax may be grown for both fiber and oil (dual purpose flax) it is mainly grown for oil production in warmer drier regions. The best quality flax for fiber is grown in Belgium and Northern Ireland, as it is suited to moist soils and damp climates.

"Flax" is a name given to many other species of *Linum*, eg *L. narbonense* [BLUE FLAX], *L. catharticum* [FAIRY FLAX], *L. pubescens* [PINK FLAX], and to species of other genera, eg *Daphne gnidium* [SPURGE FLAX], *Phormium tenax* [NEW ZEALAND FLAX] and *Camelina sativa* [FALSE FLAX].
LINACEAE.

Flour is usually understood to be a very finely pounded form of WHEAT (*Triticum aestivum*), but other foodstuffs may be milled into flour, namely *potato, *rice, *maize, *rye, *barley, *oats, *soybean, *peas and *beans. When whole wheat grain is pounded, a brown whole wheat or 100% extraction flour is produced. If part of the germ (embryo in the grain) and bran (pericarp, testa and aleurone) are removed the flour formed is called pale brown wheat meal or 85–95% extraction. However, consumer preference is for white flour or 72–76% extraction, which has improved baking and keeping qualities because it is only ground from the starchy endosperm.

Bread flour has a protein content of 11–12%, with a high level of good-quality gluten proteins which are required to produce well-risen bread with a fine internal crumb structure. Wheat varieties inherently containing good breadmaking proteins are termed "strong", whereas "weak" wheats

Flax plants (Linum usitatissimum) *in flower. The flowers can be blue or white, the blue-flowered varieties producing the finest yarn. An oil is also extracted from the seed. (×1)*

contain poor breadmaking proteins. Bread flour must contain low levels of enzymes that degrade carbohydrates and proteins because these enzymes can ruin bread texture, making it coarse and sticky. Bread flour is milled so that it contains a high level of damaged starch grains, which are required to make non-sticky doughs with a high water content. Household flour, used for cakes and pastries, and biscuit flour are milled from weak low-protein wheats that have low enzyme activity. Soup flours must have good gelling characteristics in the starch and also low enzyme activity.

In several countries, all types of flour, except wholemeal, have to contain certain minimum quantities of iron, vitamin B and nicotinic acid. Calcium is added to all except self-raising flour and wholemeal flours.

Foeniculum a genus of tall biennial or perennial herbs native to Europe and the Mediterranean. The leaves are dissected into narrow segments which with the umbels of yellow flowers make them easily recognizable. (See also FENNEL.)
UMBELLIFERAE, 2–3 species.

Fodder crops see p. 145.

Fomes a fungal genus characterized by having tough, bracket-shaped fruiting bodies. Most species are actively wood-rotting, and several cause serious diseases and economic losses of trees and shrubs. One of the commonest is *F. annosus*, which attacks a wide variety of conifers. It enters stumps either through the freshly cut surface by means of air-borne spores or through roots already infected at the time of felling. Stumps then act as sources of infection to surrounding trees, the fungal hyphae passing from root to root.

There are several tropical representatives, such as *F. noxius*, which attacks the roots of important crops, including cocoa, coffee, tea and rubber.
HYMENOCHAETACEAE, about 100 species.

Reaping Flax (Linum usitatissimum). *The fiber-producing varieties are sown close together and have tall unbranched stems.*

Fibers from Plants

It consisted of strips of water reeds (*Cyperus papyrus*) laid in a network, soaked with water and beaten flat. Papyrus was the staple writing surface until it gradually lost its popularity to animal parchment after 200 BC.

A chinese eunuch named Ts'ai Lun has been accredited with the invention of paper in the year AD 105, but a further 700 years elapsed before the secrets of the technique

The use of plant fibers antedates civilization. Perhaps torn strips of bark first "held things together" for emerging *Homo sapiens*, and braided palm fronds might have sheltered him as he subsisted a million or more years ago on the inland African lakes. There is little archaeological evidence regarding the use of fibers as they do not preserve very well, but diggings of sites dating back to ten thousand years or even earlier, have revealed the use of many plant fibers, often in a very sophisticated manner. *Agave* fibers 8 000 years old are known from excavations in the Tehuacan Valley, Mexico and remains of woven palm-leaf fabric about 12 000 years old have also been found in Mexico.

Xylem (wood) of trees is the world's most prolific source of fiber. Wood fiber can be separated to make paper (see later) or be dissolved and restructured to make synthetic fibers such as rayon (viscose). Plant cells from seed, bark, and monocotyledon leaves, used for textiles, cordage and stuffing, are more "conventional" fibers. Like wood fibers they consist of cellulose cell walls with some lignin and other substances. Cellulose is fiber material par excellence. Lignin is more resistant to decay than cellulose, but is generally unwanted because it makes fibers harsh and prone to discoloration; it is normally dissolved away in the making of chemical paper pulps.

A wide assortment of plants belonging to 44 families yield useful vegetable fibers. They fall more or less naturally into three categories: (1) surface fibers, borne externally, the most important of which are *COTTON and *KAPOK; (2) "soft" or bast fibers, from dicotyledonous bark, of which *FLAX, RAMIE, *HEMP and *JUTE are the most important; and (3) "hard" or structural fibers, the fibrovascular bundles in foliage of monocotyledons, such as *ABACA and *SISAL. Surface fibers are used chiefly for textiles and as stuffing materials, soft fibers mostly for weaving, and hard fibers mostly for cordage (twines and ropes). A miscellaneous assortment of twigs, split stems, sectioned palm leaves and so on are used for brooms, brushes, baskets, and for coarse weaving into mats.

In recent years the production of synthetic fibers exceeded that of natural fibers, but following the dramatic increase in oil prices and conservationist policies, there is a renewed interest in natural fibers.

Paper. The biggest and most important use of vegetable fibers is in the manufacture of paper. PAPYRUS, in use as early as 3 500 BC, can be described as the forerunner of paper.

COMMERCIAL VEGETABLE FIBERS
(excluding wood fibers and those used for paper making)

Family and Scientific name	Popular name	Principal growing areas	Uses
AGAVACEAE			
Agave cantala	MAGUEY	SE Pacific, SE Asia	Cordage, sacking
Agave fourcroydes	HENEQUIN	Mexico	Cordage, sacking
Agave heteracantha	*ISTLI FIBER	Mexico	Brooms
Agave sisalana	*SISAL (HEMP)	E Africa	Cordage, sacking
Furcraea gigantea var *willemettiana*	MAURITIUS HEMP	Mauritius	Cordage, mats
Phormium tenax	NEW ZEALAND FLAX	New Zealand	Cordage
Sansevieria cylindrica	BOWSTRING HEMP	Africa, Asia, Florida	Cordage
Sansevieria trifasciata	SNAKE PLANT		Cordage
Sansevieria hyacinthoides (= *guineensis*)	AFRICAN BOWSTRING HEMP	Africa, Asia	Cordage
Sansevieria roxburghiana	INDIAN BOWSTRING HEMP	India	Cordage
Sansevieria zeylanica	CEYLON BOWSTRING HEMP	Africa, Asia	Cordage
APOCYNACEAE			
Apocynum cannabinum	INDIAN HEMP, HEMP DOGBANE	N America	String, cloth
Apocynum venetum	HEMP DOGBANE	N America	String, fishing nets
ASCLEPIADACEAE			
Asclepias incarnata	SWAMP MILKWEED, OZONE FIBER	N America	Cordage
Asclepias syriaca	MILKWEED	N America	Cordage
BOMBACACEAE			
Bombax ceiba	RED SILK COTTON TREE, WHITE KAPOK	India, SE Asia	Stuffing
Ceiba pentandra	*KAPOK, SILK COTTON TREE	Asia	Stuffing
Ochroma pyramidale	*BALSA, CORKWOOD	C America, W Indies	Packing, insulation
BROMELIACEAE			
Aechmea magdalenae	PITA	Colombia	Cordage, sacking
Ananas comosus	*PINEAPPLE	Philippines	Cordage, sacking
Neoglaziovia variegata	CAROA	Brazil	Cordage, basketry
Tillandsia usneoides	SPANISH MOSS	Tropical America	Packing
CYCLANTHACEAE			
Carludovica palmata	PANAMA HAT PALM	C America	Panama hats, mats
GRAMINEAE			
Bambusa and many other genera (eg *Arundinaria, *Dendrocalamus, Melocanna, Ochlandra, *Phyllostachys, *Sasa*)	*BAMBOO	Pantropical	Mats, baskets, cordage, paper
Lygeum spartum	*ESPARTO	N Africa	Cordage, paper
Stipa tenacissima	ESPARTO	N Africa	Cordage, paper
LEGUMINOSAE			
Crotalaria juncea	*SUN HEMP	Tropical Asia	Cordage
Sesbania aculeata		Africa, Asia	Fishing nets
Sesbania aegyptica		Tropical India	Cordage
LINACEAE			
Linum usitatissimum	*FLAX, LINEN	Europe	Textiles
MALVACEAE			
Abutilon avicennae	*CHINA JUTE, INDIAN MALLOW	China, USSR	Sacking, tough cloth
Gossypium spp	*COTTON	N America, Africa, Asia	Textiles
Hibiscus cannabinus	KENAF	India	Sacking, tough cloth
Hibiscus sabdariffa	ROSELLE	India	Sacking, tough cloth
Pavonia bojeri		Madagascar	Cloth
Pavonia shimperiana		Africa	Cloth
Sida cordifolia	QUEENSLAND HEMP	India, China	Ropes, cloth
Sida rhombifolia	QUEENSLAND HEMP	India, China	Ropes, cloth
Urena lobata	*ARAMINA, CONGO JUTE	S America, Africa	Ropes, sacking

left China and reached the Arabs. The Moors introduced paper making to Spain in the 12th century and it was a further 300 years before it was in use in the rest of Europe.

The basic principles of paper making have not been changed since their inception.

Paper, simply described, is a thin tissue composed of any fibrous vegetable material. The individual fibers are first separated by mechanical or chemical action and then reconstituted in a sheet form, by depositing the fibers on to a wire mesh using water.

Although any vegetable matter can be used for paper making, the principal raw materials fall into the following categories: seed hairs (COTTON, once used for high-grade writing and printing papers); bast fibers (FLAX, HEMP, JUTE and RAMIE which are particularly strong but resistant to bleaching); grass fibers (cereal straws and bagasse); leaf fibers (ESPARTO, SISAL and MANILA HEMP which have strength and are tear resistant, but hard to bleach); wood fibers (the major group, particularly the conifers of North America and Scandinavia, eg *SPRUCE, *Picea excelsa*, and PINE, *Pinus sylvestris*). The deciduous trees, the hardwoods, produce shorter fibers (1.5mm compared to 3.5mm of the conifers). *Eucalyptus*, *POPLAR*, *CHESTNUT and *BIRCH are particularly in demand.

MORACEAE			
Broussonetia papyrifera	TAPA (KAPA)	E Asia, Polynesia	Cloth, rope
Cannabis sativa	*HEMP	Asia	Rope, sacking
MUSACEAE			
Ensete ventricosa	ABYSSINIAN BANANA	African, SE Asia	Cordage, sacking
Musa textilis	*ABACA, MANILA HEMP	Philippines, C America	Rope, cordage
PALMAE			
Arenga saccharifera	SUGAR PALM		Rope, thatch, brushes
Attalea funifera	*BAHIA PIASSAVA	Brazil	Brooms, brushes
Borassus flabellifer	PALMYRA PALM	Africa	Brooms, baskets, fencing
Calamus spp	*RATTAN CANE, *MALACCA CANE	SE Asia	Baskets, mats, chair seats
Caryota urens	KITUL FIBER	Tropical Asia, Australia	Ropes, brushes, baskets
Chamaerops humilis	DWARF FAN PALM	NW Africa	Horsehair, cordage, baskets
Cocos nucifera	*COCONUT, *COIR	Polynesia, Tropics	Mats, brushes, cordage
Daemonorops spp	*RATTAN CANE	SE Asia	Baskets, mats, chair seats
Metroxylon sagu	*SAGO PALM	SE Asia	Thatch
Nypa fruticans	*NYPA PALM	SE Asia	Thatch, matting
Raphia spp	*RAFFIA	Africa	Twine, matting, baskets
Raphia hookeri	*RAFFIA *PIASSAVA FIBER	Africa	Brooms, baskets, fencing
Sabal palmetto		C America, W Indies	Thatch, baskets
Trachycarpus excelsus	CHINESE WINDMILL PALM	E Asia	Cordage, brushes
STERCULIACEAE			
Abroma augustum	DEVIL'S COTTON	Asia, Australia	Ropes, cordage
TILIACEAE			
Corchorus capsularis	*JUTE	India, Bangladesh	Sacking
Corchorus olitorius	JUTE	India, Africa	Sacking
URTICACEAE			
Boehmeria nivea	RAMIE, CHINA GRASS	Asia	Rope

Important fiber-producing plants. 1 Hemp—stem ($\times\frac{1}{4}$, $\times 1$); 2 Cotton—seed hairs ($\times\frac{1}{3}$); 3 Manila Hemp—"stem" of sheathing leaf bases (shown in cross-section) ($\times\frac{1}{100}$, $\times\frac{1}{20}$); 4 Jute—stem ($\times\frac{1}{6}$, $\times 1$); Flax—stem ($\times\frac{1}{3}$); 6 Raffia—leaves ($\times\frac{1}{300}$); 7 Kapok—seed hairs ($\times\frac{1}{5}$); 8 Dwarf Fan Palm—leaves ($\times\frac{1}{50}$); 9 Sisal—leaves ($\times\frac{1}{50}$); 10 Esparto Grass—leaves ($\times\frac{1}{4}$); 11 Ramie—stem ($\times\frac{1}{16}$); 12 Pineapple—leaves ($\times\frac{1}{20}$); 13 Coir—husk surrounding the coconut ($\times\frac{1}{300}$, $\times\frac{1}{12}$).

Flavorings from Plants

A wide range of plants are used to provide materials that are added in small quantities to food and beverages to give flavoring or improve their taste. Flavoring plants include all *herbs, spices and condiments and may be used fresh, dried or preserved in some way. They normally have no nutritive value. Many of them have been known since earliest times, especially in the Greek, Roman and other civilizations bordering the eastern Mediterranean and in India and the East Indies.

Spices have played an important role in the history of civilization, exploration and commerce. Most spices are tropical in origin and are mainly produced on islands. They were among the first products to be exported from the tropics to the temperate regions. They were usually very expensive. Early Arabs had the monopoly of the spice trade, obtaining the spices in India, some of which had been brought there from further east. It was largely a desire to take part in the lucrative spice trade which led to the early explorations from Spain and Portugal. Columbus went west in 1492 in the hope of reaching the spice islands of the east, but instead of discovering the East Indies he found the West Indies. In 1498, the Portuguese sent Vasco da Gama around Africa to reach India. Later the Portuguese discovered the Moluccas, the source of *CLOVES, *NUTMEGS and *MACE. The Portuguese retained the monopoly of the spice trade for 100 years, until this was taken over by the Dutch, who maintained it for 200 years.

The only spices of New World origin are CAPSICUM or CHILIES, *ALLSPICE and *PIMENTA, the latter being the only one exclusively produced there. Most spices are lowland crops of areas with relatively high rainfall of 2 000–2 500mm and with an optimum temperature of about 30°C. Most spices are derived from fruits or seeds and their characteristic pungent flavors are due to the presence of *essential oils.

Some of the best-known flavoring plants are listed in the table (for Herbs see p. 168). In addition to these, there are also some curious and unexpected cases of plants providing flavor. Carragheen, with its pungent flavor due to a high potassium iodide content, comes from a seaweed. The characteristic flavor of several famous French cheeses, such as Brie, Camembert and Roquefort, is due to the presence of different molds belonging to the genus *Penicillium. The Greek wine known as retsina is flavored with the resin from the pine casks it is stored in. The only other flavoring originating from conifers are *juniper berries, which are used in the production of gins and sometimes as a flavoring in meats, stews, roasts and in *sauerkraut. The berries are sometimes used ground as a component of mixed spices. Many fruits are used as flavorings, either eaten entire or sliced or as juices and extracts. Some are candied, notably citrus peel and angelica stems.

The scientific basis of flavoring is highly complex and largely not elucidated, involving assessments of both taste and smell. Human tastes are grouped into four main classes: salty, sweet, bitter and sour. Sweetness of flavor and aroma is generally attractive to the palate while bitterness, acidity and astringency tend to be repellent although often welcome in some degree to counteract insipid foods. Bitterness can be an acquired taste – in beer, for example, when the responsible constituents are the hupulones and lupulones derived from hops. Mouth feelings (sensations produced by foods that are important in taste) are pain (provoked by chilli pepper, mustard, horseradish, etc), anesthesia (provoked by cloves, vanilla, etc), coolness (induced by menthol, mints, etc) and astringency (as caused by sloes, cashew fruit, etc). Some flavors may be pleasing and attractive in small quantities, at least to some people, as in the case of sulfur-containing mustard oils in numerous crucifer crops, but they become unacceptable and repellent if taken in large quantities. Likewise onion and garlic, which contain other sulfur compounds, are highly distasteful to many people while greatly appreciated by others.

Some fruit flavors may be the result of a single chemical compound as in apple and peach; others are caused by several compounds; in some, for example apricot, as many as 10 compounds are involved. In some cases it has not yet been possible to isolate and identify the chemical principles responsible for flavor, as in blackcurrants and strawberries. In coffee and chocolate over 700 compounds have been isolated but it is still not known which, if any, are responsible. In other cases trace compounds are important and contribute more to the flavor than the major compounds. In the lemon, for example, limonene makes up 70% of the oil but it is the 5% of the oil citral that is responsible for the lemon flavor.

Flavor potentiators and modifiers are naturally occuring compounds which have little effect on their own but which enhance the flavor when combined with other taste and flavor molecules. A common example is sodium chloride (salt) and more recently monosodium glutamate. More striking is the glycoprotein miracularin. This is obtained from the *MIRACULOUS BERRY (Synsepalum dulcificum) and has the property of eliminating sourness or acidity temporarily. Sour lemons taste as sweet as oranges if eaten after chewing the berries of Synsepalum but the effect wears off in about an hour.

Flavor is greatly affected by temperature; chilling may reduce flavor while heating can both enhance and destroy it as in the case of *essential oils which may be released and then dispersed. The drying of herbs causes some flavor loss although this is minimized by quick drying methods. Some flavors are only released after mixing with water as in the case of mustard which activates enzymes to release the characteristic pungent mustard oils.

COMMON FLAVORING PLANTS

Popular name	Scientific name	Part of plant used
*ALLSPICE	Pimenta dioica	Fruit
*ANGELICA	Archangelica officinalis	Stem
*ANGOSTURA	Cusparia febrifuga	Bark
*ANISE	Pimpinella anisum	Fruit
*BALM, LEMON BALM	Melissa officinalis	Leaf
*BASIL	Ocimum basilicum	Leaf
BAY LAUREL	*Laurus nobilis	Leaf
*CAPERS	Capparis spinosa	Flower bud
CARAWAY	*Carum carvi	Seed
*CARDAMOM	Elettaria cardamomum	Fruit
CHILLI	*Capsicum spp	Fruit
*CHIVES	Allium schoenoprasum	Leaf
CINNAMON	*Cinnamomum zeylanicum	Bark
*CLOVES	Syzygium aromaticum	Flower bud
*CORIANDER	Coriandrum sativum	Fruit
*CUMIN	Cuminum cyminum	Fruit
FENUGREEK	*Trigonella foenum-graecum	Seed
*GARLIC	Allium sativum	Bulb
*GINGER	Zingiber officinale	Rhizome
*HOPS	Humulus lupulus	Fruit
*HORSERADISH	Armoracia rusticana	Root
JASMINE	*Jasminum spp	Flower
*JUNIPER	Juniperus communis	Fruit
*LEMON	Citrus limon	Fruit
*LICORICE	Glycyrrhiza glabra	Root
*MACE	Myristica fragrans	Aril
*MARJORAM	Origanum majorana	Leaf
*MUSTARD	Brassica juncea B. nigra Sinapis alba	Seed
*NUTMEG	Myristica fragrans	Seed
*PARSLEY	Petroselinum crispum	Leaf
*PEPPER	Piper nigrum	Fruit
*PEPPERMINT	Mentha × piperata	Leaf
*POPPY	Papaver somniferum Glaucium flavum Argemone glauca	Seed
ROSELLE	*Hibiscus sabdariffa	Calyx
*RUE	Ruta graveolens	Leaf
SAFFRON	*Crocus sativus	Stigma
SARSAPARILLA	*Smilax spp	Root
SASSAFRAS	*Sassafras albidum	Bark and wood
SESAME	*Sesamum indicum	Seed
*TAMARIND	Tamarindus indica	Flower
*TARRAGON	Artemisia dracunculus	Leaf
*THYME	Thymus vulgaris	Leaf
*TURMERIC	Curcuma domestica	Rhizome
VANILLA	*Vanilla planifolia	Fruit

Some common flavoring plants. 1 Caraway; 2 Anise; 3 Cumin; 4 Allspice; 5 Bay; 6 Capers; 7 Saffron Crocus; 8 Cloves; 9 Vanilla Orchid; 10 Turmeric; 11 Sorrel (Rumex acetosa); 12 Tarragon; 13 Nutmeg; 14 Mace; 15 Ginger; 16 Cardamom; 17 Pepper; 18 Cinnamon; 19 Fenugreek; 20 Chilli; 21 Juniper; 22 Licorice. 1, 2, 3, 9, 14, 19, 20, 21, 22 ($\times\frac{1}{2}$); 4, 6, 7, 11, 12, 18 ($\times\frac{1}{3}$); 5, 8, 15 ($\times\frac{1}{4}$); 17 ($\times\frac{1}{6}$); 10, 16 ($\times\frac{1}{8}$). Details 1, 2, 3 ($\times 2$); 4, 7, 15, 16, 18 ($\times\frac{1}{2}$); 8, 17 ($\times\frac{3}{4}$).

Fontanesia an Asiatic genus of deciduous shrubs, including *F. fortunei*, of Chinese origin, and *F. phillyraeoides*, from Syria. Both species bear panicles of creamy-white flowers; they resemble *PRIVET (Ligustrum) and are similarly used as hedging plants.
OLEACEAE, 2 species.

Fontinalis a genus of mainly aquatic mosses which are confined almost exclusively to north temperate regions. However, *F. antipyretica*, which is the best-known and one of the largest species, also occurs in South America. In rivers, *Fontinalis* species commonly form trailing shoot systems well over 1m in length.
FONTINALACEAE, about 40 species.

Forestiera an American and West Indian genus of mainly deciduous trees and shrubs. *F. acuminata* [SWAMP PRIVET], *F. ligustrina* and *F. neomexicana* are sometimes cultivated as ornamentals for their privet-like appearance, small yellowish flowers and black or purplish fruits.
OLEACEAE, about 15 species.

Forsythia a small genus of hardy deciduous shrubs from Asia, some of which are cultivated in European parks and gardens. The conspicuous bright yellow flowers are carried on lateral buds on the previous

Ocotillo (Fouquieria splendens) in bloom in the southern Californian desert, USA. This species is occasionally used for hedging. ($\times \frac{1}{50}$)

season's wood and emerge very early in the growing season. The most popular cultivated species are *F. ovata*, *F. suspensa* and the hybrid *F. × intermedia* (= *F. suspensa* × *F. viridissima*), particularly the free-flowering cultivar 'Lynwood'.
OLEACEAE, 6–7 species.

Fortunella see KUMQUAT.
RUTACEAE, 4–5 species.

Fothergilla a small genus of deciduous shrubs native to the southeastern USA and very popular as garden ornamentals. The three most popular species, *F. gardenii* (= *F. alnifolia*), *F. major* and *F. monticola*, all produce dense white or cream sweet-smelling inflorescences of bottle-brush type in spring and show a range of attractive red and orange autumn leaf colors.
HAMAMELIDACEAE, 4 species.

Fouquieria a small genus of spiny trees or shrubs native to Mexico, with one species extending into the southwestern USA. The best-known species is *F. splendens* [OCOTILLO] which, like other members of the genus, produces attractive inflorescences of showy red flowers. It produces long branching cane-like stems and is used as a hedging plant. A wax is obtained from the stem or bark of some species.
FOUQUIERIACEAE, 7–9 species.

Foxglove the common name for members of the small but important European, Mediterranean and Macaronesian genus *Digitalis* which includes several attractive garden plants and species of major medicinal importance, such as the purple-flowered *D. purpurea* [COMMON FOXGLOVE]. The plants are mainly biennial or perennial herbs, sometimes flowering in their first year, but a few species are small shrubs, such as *D. obscura* [SPANISH RUSTY FOXGLOVE] from Spain and North Africa, and the Macaronesian group of species, such as *D. canariensis*, sometimes treated as a separate genus, *Isoplexis*.

D. purpurea is a highly variable species widely cultivated both for ornament and for the extraction of cardiac glycosides such as digitalin. Another widespread species, the yellow-flowered *D. grandiflora* (= *D. ambigua*) [LARGE YELLOW FOXGLOVE], is grown as a garden border plant. The hybrid named *D. × mertonensis* is the result of a cross between *D. purpurea* and *D. grandiflora*, followed by chromosome doubling. It has large attractive, often strawberry-colored flowers and is widely cultivated in gardens. Also widely grown are *D. lutea* [SMALL YELLOW FOXGLOVE], which has pale yellow to whitish tubular-cylindrical flowers, *D. ferruginea*, with yellowish or reddish flowers, and *D. lanata*, with a similar-shaped corolla but white or yellowish-white with brown or violet veins. The latter species is widely grown as a source of digoxin and other alkaloids used in the manufacture of heart drugs.

The *Isoplexis* group of species from the

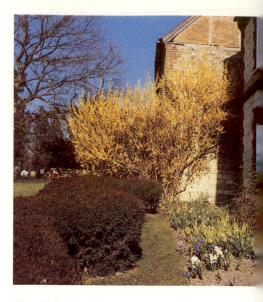

Forsythias (Forsythia spp) are one of the first shrubs to burst into bloom during early spring. The flowers are produced on the previous season's wood.

Canary Islands and Madeira are attractive small shrubs with dark green leathery leaves and racemes of striking reddish-orange or copper-colored flowers, sometimes cultivated in the open or in conservatories. The best-known species of this group is *D. canariensis* (*Isoplexis canariensis*).
SCROPHULARIACEAE, about 23 species.

Fragaria see STRAWBERRY.
ROSACEAE, about 15 species.

Francoa [BRIDAL OR MAIDEN'S WREATH] a very small genus of herbaceous perennials from the temperate regions of Chile. According to some botanists it consists of only a single variable species, *F. appendiculata*, which is cultivated as a border plant for its long, slender spikes of white or pale pink flowers. Others regard it as comprising five species.
SAXIFRAGACEAE, 1–5 species.

Frangula a genus of shrubs or small trees, mainly American, Mediterranean and Eurasian in distribution. The bark of some species has purgative properties. Dyes are obtained from the bark, leaves and berries of *F. alnus* (= *Rhamnus frangula*) [ALDER BUCKTHORN], whose branches are also used for making charcoal.
RHAMNACEAE, about 50 species.

Frankincense the aromatic resin (gum olibanum) obtained from trees of the genus *BOSWELLIA, especially *B. carteri*, by tapping the trunks. The milky resin hardens when exposed to the air and forms irregular lumps, the form in which it is usually marketed. Frankincense burns with a pleasant odor and is used today as a constituent of incense; it is also in fumigants, and for fixatives in perfume.
BURSERACEAE.

Fodder Crops

Fodder crops or forage crops are a diverse group of plants used directly or in a preserved form (such as hay or silage) for feeding ruminant livestock and thereby to transform plant carbohydrate and protein into meat and dairy products. They play an exceedingly important role in the farming economy of most parts of the world and represent an indirect way by which Man uses vegetation.

Much of the world's grazing land (which vastly exceeds that devoted to grain or other crops) is under-utilized. Natural pasture is being increasingly improved by planting it with high-yielding cultivated fodder crops. At present most of the important forage crops are grown in the temperate countries and all the important cultivated forage or fodder plants belong to four families – the Gramineae, Leguminosae, Cruciferae and Chenopodiaceae in that order, the first two greatly exceeding the other two in importance. Many wild, as well as cultivated species of grasses and legumes, are of value as fodder plants especially on a local scale. Many are being tried out in cultivation and, in the tropics in particular, the planting of fodder crops is still a novelty in most areas, yet tropical grassland represents a major resource and supports some 500 million head of cattle as well as other grazing animals.

The oldest fodder legume is probably *LUCERNE (Medicago sativa). It is native to western Asia and reached Western Europe in the mid-16th century and through the colonizing activities of the Spaniards reached Mexico in the 16th century. It did not reach the United Kingdom till 1650 and was quite unknown in the USA until the middle of the 17th century when it came in from Mexico.

The grasses (family Gramineae) are ideally suited for grazing because during vegetative growth the growing apex remains at about ground level and grazing animals only remove the leaves above the growing point.

Fodder plants used by domestic animals in the tropics embrace a vast range of species, from thorny shrubs browsed by camels and goats in the desert to the most luxuriant giant grasses and lush pastures of the humid tropics. In the natural grazings which are mainly of the *savanna (mixed tree-grass) type, dominant grasses differ from area to area. Typical genera are *Andropogon, Hyparrhenia and Themeda in Africa, *Cynodon and Dichanthium in India, and *Paspalum in Brazil. The feeding value is frequently low as a result of human interference (eg fire) or unenlightened management (eg heavy stocking).

Sheep grazing on Rape (Brassica napus) *which is a valued forage crop used for late autumn and early spring pasturing. The seed is often sown after a cereal crop.*

TEMPERATE FODDER CROPS

Common name	Scientific name
Grasses – Gramineae	
*RYEGRASS	*Lolium perenne*
	L. multiflorum
TIMOTHY GRASS	*Phleum pratense*
COCK'S-FOOT, ORCHARD GRASS	*Dactylis glomerata*
*MAIZE	*Zea mays*
*RYE	*Secale cereale*
WHEATGRASS	*Agropyron fragile* subspecies *sibiricum*
CRESTED WHEATGRASS	*A. cristatum*
SLENDER WHEATGRASS	*Elymus trachycaulum* (= *Agropyron trachycaulum*)
BENT-GRASS	*Agrostis stolonifera* (= *A. alba*)
	A. capillaris (= *A. tenuis*)
BLUESTEM	*Andropogon* species
OATGRASS	*Arrhenatherum elatius*
CARPET GRASS	*Axonopus affinis*
BLUE GRAMA	*Bouteloua gracilis*
SIDEOATS	*B. curtipendula*
SMOOTH BROME	*Bromus inermis*
RESCUE GRASS, PRAIRIE BROME	*B. willdenowii* (= *B. unioloides*)
BUFFALO GRASS	*Buchloe dactyloides*
RHODES GRASS	*Chloris gayana*
BERMUDA GRASS	*Cynodon dactylon*
CANADA WILD RYE	*Elymus canadensis*
SIBERIAN WILD RYE	*E. sibiricus*
BLUE WILD RYE	*E. glaucus*
LOVE GRASS	*Eragrostis curvula*
TEFF	*E. tef*
TALL, REED *FESCUE	*Festuca arundinacea*
SHEEP'S FESCUE	*F. ovina*
MEADOW FESCUE	*F. pratensis*
CANARY REED GRASS	*Phalaris arundinacea, P. aquatica* var *stenoptera* (= *P. tuberosa*)
TEXAS BLUEGRASS	*Poa arachnifera*
CANADA BLUEGRASS	*P. compressa*
KENTUCKY BLUEGRASS	*P. pratensis*
Sedges – Cyperaceae	
SEDGE	*Carex lyngbyei*
	C. nigra
	C. norvegica
	C. rariflora
Legumes – Leguminosae	
KIDNEY VETCH	*Anthyllis vulneraria*
CROWN VETCH	*Coronilla varia*
*LUCERNE, ALFALFA	*Medicago sativa*
BLACK MEDICK, YELLOW TREFOIL	*M. lupulina*
WHITE SWEET CLOVER	*Melilotus alba*
YELLOW SWEET CLOVER	*M. officinalis*
SAINFOIN, ESPARCET	*Onobrychis viciifolia*
SERRADELLA	*Ornithopus sativus*
RED CLOVER	*Trifolium pratense*
WHITE CLOVER	*T. repens*
STRAWBERRY CLOVER	*T. fragiferum*
ALSIKE CLOVER	*T. hybridum*
CRIMSON CLOVER	*T. incarnatum*
SUBTERRANEAN CLOVER	*T. subterraneum*
*BROAD BEAN	*Vicia faba*
SPRING VETCH	*V. sativa*
HAIRY VETCH	*V. villosa*
Kales – Cruciferae	
KALES, RAPE (see Brassicas p. 62.)	*Brassica oleracea*
	B. campestris
	B. rapa

TROPICAL and SUBTROPICAL FODDER CROPS

Common name	Scientific name
Grasses – Gramineae	
BLUESTEMS	*Andropogon* species
DIAZ BLUESTEM	*Dichanthium annulatum*
BRAHMAN BLUESTEM, ANGLETON BLUESTEM	*D. aristatum*
	Hyparrhenia hirta
	H. rufa
RHODES GRASS	*Chloris gayana*
HAIRY CRAB GRASS	*Digitaria sanguinalis*
PANGOLA GRASS	*D. decumbens*
	D. longifolia
GUINEA GRASS	*Panicum maximum*
BROOMCORN MILLET	*P. miliaceum*
PARA GRASS	*P. purpurascens*
DALLIS GRASS	*Paspalum dilatatum*
BAHIA GRASS	*P. notatum*
VASEY GRASS	*P. urvillei*
PEARL MILLET	*Pennisetum americanum*
BUFFEL GRASS	*P. ciliare*
NAPIER GRASS, ELEPHANT GRASS	*P. purpureum*
GUATEMALA GRASS	*Tripsacum laxum*
—	*Themeda triandra*
Legumes – Leguminosae	
BUTTERFLY PEA, CONCHITA	*Centrosema plumieri*
	C. pubescens C. virginianum
SESBAN	*Sesbania sesban* (= *S. aegyptiaca*) *S. cinerascens*
BEGGARWEED	*Desmodium tortuosum*
TICK CLOVER	*D. discolor D. heterophyllum*
	D. heterocarpon D. triflorum
WHITE POPINAC	*Leucaena glauca*
	L. leucocephala
VELVET BEAN, BENGAL BEAN	*Mucuna deeringiana*
*INDIGO	*Indigofera arrecta*
	I. pauciflora
MACROPTILIUM	*Macroptilium atropurpureum*
	M. geophilum
KUDZU VINE	*Pueraria lobata* (= *P. thunbergiana*)
KUDZU	*P. phaseoloides*
STYLO, WILD LUCERNE	*Stylosanthes guianensis*
	S. capitata S. humilis
	S. hamata S. scabra

and most of which are bird-pollinated. Most members of the genus are native to mountainous regions of Central and South America and to Tahiti and New Zealand.

Fuchsias are mainly of ornamental value. Amongst the popular shrubby species are *F. corymbiflora* (Ecuador, Peru) with long crimson flowers, *F. fulgens* (Mexico) with red stems and scarlet, green-tipped flowers, *F. magellanica* (= *F. macrostemma*) (Argentina, southern Chile) in a range of varieties including *alba*, the red- and purple-flowered *gracilis variegata*, *F. procumbens* (New Zealand), a prostrate, hanging-basket species and *F. triphylla* (Haiti, St. Domingo), a subshrub with brilliant orange-scarlet flowers. Most hybrids, referred to as *F. × hybrida*, have been derived from crosses between *F. fulgens* and *F. magellanica*, and include numerous named varieties, such as 'Mrs Popple' and 'Mission Bells' which can be grown outdoors, and the more tender greenhouse types such as 'Cascade'.
ONAGRACEAE, about 100 species.

Franklinia a genus consisting of a single species, *F. alatamaha*, originally discovered in Georgia in the southeastern USA. It is now only known in cultivation as an ornamental deciduous camellia-like tree or shrub with solitary, showy, attractively scented white flowers.
THEACEAE, 1 species.

Fraxinus see ASH.
OLEACEAE, about 60 species.

Freemontia see *Freemontodendron*.

Freemontodendron (formerly *Freemontia*) a small genus of flowering trees or shrubs, native to Mexico and California. *F. californicum* is cultivated as an ornamental shrub for its large cup-shaped golden-yellow flowers, the sepals being petal-like and the petals absent. Although *F. mexicanum* with its yellow-orange star-shaped flowers is less widely cultivated, a hybrid between these two species, 'Californian Glory', is a more popular shrub in frost-free areas.
STERCULIACEAE, 2 species.

Freesia a genus of cormous perennial herbs from South Africa. The leaves are two-ranked, forming a flat fan, and the richly scented flowers are borne in one-sided spikes on branched inflorescences with their flowers turned upwards. The best-known species are *F. refracta* with pale yellow flowers, and *F. armstrongii* with pink flowers. Modern hybrids, *F. × hybrida*, derived from these and possibly other species have a wide color range. Freesias are popular as cut flowers and are grown commercially for this purpose.
IRIDACEAE, up to 20 species.

Fritillaria a genus of bulbous plants from all the north temperate zones except eastern North America.
F. meleagris [SNAKE'S HEAD FRITILLARY] is

Franklinia alatamaha has camellia-like flowers and is known only in cultivation. (× ½)

commonly cultivated for its attractive nodding spring flowers, purple, pink or white, with checkering throughout of a darker shade.
F. imperialis [CROWN IMPERIAL], from southeast Turkey to the western Himalayas, has long been cultivated for its attractive, pendent, yellow, red or orange flowers on a stout stem often as much as 1.2m tall. One of the best garden species is the Siberian *F. pallidiflora* which is very hardy and flourishes in any well-drained soil. The easiest species to cultivate is *F. pontica*, from the Balkans to the Pontus mountains of Turkey, with apple-green flowers suffused with brown at the tips and margins.

The bulbs of a number of Asiatic species, notably *F. verticillata*, have various uses in Chinese medicine, particularly in treating fever and dysentery. The chocolate or blackish-purple bulbs of *F. camschatcensis* [KAMCHATKA LILY, BLACK SARANA], from East Asia, Japan to North America, are eaten by the natives of southeast Alaska.
LILIACEAE, about 100 species.

Fruits see p. 147.

Frullania a large genus of liverworts of essentially warm climates; most of its species are therefore restricted to the tropical areas of all continents. A few species, however, reach the temperate zones and one or two species are able to survive beyond the Arctic Circle. Most of the species are epiphytes that grow on the trunks of trees and more rarely on leaves in tropical rain forests; many others grow on rocks.
FRULLANIACEAE, about 1 000 species.

Fuchsia a genus of tender to hardy woody plants with showy pendulous tube-like flowers, all of which produce copious nectar

Fucus one of the most common genera of brown algae distributed on temperate and cold-water shores. The various species usually occur in a distinct zonation on the intertidal zone. The plants have tough, leathery, divided ribbon-like fronds, a short stem terminating in a disk-shaped holdfast, attached firmly to rocks. Various species, including *F. vesiculosus* [BLADDERWRACK, LADY WRACK, SEAWARE, BLACK TANG, BLADDER FUCUS], are used as a manure and as a forage for sheep and cattle. Because it is rich in iodine, it has been used as a treatment for thyroid deficiency.
PHAEOPHYCEAE.

The Crown Imperial (Fritillaria imperialis), a tall Himalayan species commonly grown in gardens, in spite of its unpleasant smell. (× 1/12)

Fruits

The practice of fruit gathering and growing in gardens and farms stretches back into antiquity. Woodlands and hedgerows are still a major source of such fruits as *cranberries and *blackberries, while garden production of fruits is important in many countries such as Germany. Some fruit crops are grown in mixed farming systems, as in China and Korea where vegetables and cereals are grown under fruit trees, but in the West the traditional orchard with grass providing grazing for animals has almost disappeared. Most fruit entering commerce today is grown on specialist plantations ranging from those for tropical *bananas and *pineapples to those for *dates, *citrus fruits, *apples, *raspberries, and *currants. All these crops are perennials (which increases the build-up of pests and diseases), are vegetatively propagated and are of high value if they can be produced to a high standard of visual appeal as well as good eating quality. They are also highly perishable, so need special methods of storage and transportation.

Fruit crops give high returns but are expensive to produce, so the grower cannot risk using poor sites. Good drainage combined with an adequate supply of rainfall or irrigation water is very important. Freedom from spring frost, or the ability to protect against it, is also essential for citrus fruits in Florida and *peaches in New Zealand as well as apples and *pears in Europe and America. Cold-temperate fruit plants require chilling in winter to stimulate normal growth and flowering in the following year. Those that require little chilling need a long growing season and are damaged by dormant season frosts; these are classed as warm-temperate fruits. By selection and improved cultural practices, cold-temperate fruits can be grown over a wide climatic range.

In the accompanying tables are listed the main temperate and tropical fruits. The term fruit is used here in the strictly horticultural sense of the fleshy edible part developed from the flower or flowers of a perennial plant.

TEMPERATE FRUITS

Common name	Scientific name	Family
*APPLE	*Malus pumila*	Rosaceae
*PEAR	*Pyrus communis*	Rosaceae
*QUINCE	*Cydonia vulgaris*	Rosaceae
*PEACH, NECTARINE	*Prunus persica*	Rosaceae
SWEET *CHERRY	*P. avium*	Rosaceae
SOUR CHERRY, COOKING CHERRY, MORELLO	*P. cerasus*	Rosaceae
*PLUM	*P. domestica*	Rosaceae
BULLACE, DAMSON	*P. insititia*	Rosaceae
GAGE, *GREENGAGE, MIRABELLE	*P. insititia*	Rosaceae
	var *italica*	Rosaceae
	var *syriaca*	Rosaceae
CHERRY PLUM	*P. cerasifera*	Rosaceae
JAPANESE PLUM	*P. salicina*	Rosaceae
AMERICAN PLUM	*P. americana*	Rosaceae
*APRICOT	*P. armeniaca*	Rosaceae
*MEDLAR	*Mespilus germanica*	Rosaceae
*RASPBERRY	*Rubus idaeus*	Rosaceae
AMERICAN RED RASPBERRY	*R. ideaus var strigosus*	Rosaceae
BLACK RASPBERRY	*R. occidentalis*	Rosaceae
*BLACKBERRY, BRAMBLE	*R. fruticosus*	Rosaceae
EVERGREEN BLACKBERRY	*R. laciniatus*	Rosaceae
CLOUDBERRY	*R. chamaemorus*	Rosaceae
PACIFIC DEWBERRY	*R. ursinus*	Rosaceae
*LOGANBERRY, BOYSENBERRY, VEITCHBERRY	*R. × loganobaccus*	Rosaceae
WINEBERRY	*R. phoenicolasius*	Rosaceae
*STRAWBERRY	*Fragaria × ananassa* (= *F. virginiana × F. chiloensis*)	Rosaceae
*GOOSEBERRY	*Ribes uva-crispa* (= *R. grossularia*)	Rosaceae
BLACK CURRANT	*R. nigrum*	Rosaceae
RED CURRANT	*R. rubrum*	Rosaceae
FIG	*Ficus carica*	Moraceae
*OLIVE	*Olea europaea*	Oleaceae
*MULBERRY, BLACK MULBERRY	*Morus nigra*	Moraceae
RED MULBERRY	*M. rubra*	Moraceae
*GRAPE	*Vitis vinifera*	Vitaceae
FROST GRAPE	*V. riparia, V. vulpina*	Vitaceae
BUSH OR SAND GRAPE	*V. rupestris*	Vitaceae
FOX OR SKUNK GRAPE	*V. labrusca*	Vitaceae
MUSCADINE, BULLACE GRAPE	*V. rotundifolia*	Vitaceae
*BILBERRY	*Vaccinium myrtillus*	Ericaceae
*CRANBERRY	*V. oxycoccus, V. macrocarpon*	Ericaceae
COWBERRY	*V vitis-idaea*	Ericaceae
LOWBUSH *BLUEBERRY	*V. angustifolium*	Ericaceae
HIGHBUSH BLUEBERRY	*V. corymbosum*	Ericaceae
STRAWBERRY TREE	*Arbutus unedo*	Ericaceae
CHINESE GOOSEBERRY KIWIBERRY	*Actinidia chinensis*	Actinidiaceae

See also:
Marrows, Squashes, Pumpkins and Gourds p. 216
Nuts p. 238
Vegetables p. 338

TROPICAL AND SUBTROPICAL FRUITS

Common name	Scientific name	Family
SWEET *ORANGE	*Citrus sinensis*	Rutaceae
SOUR, *SEVILLE OR BITTER ORANGE	*C. aurantium*	Rutaceae
*LIME	*C. aurantifolia*	Rutaceae
*LEMON	*C. limon*	Rutaceae
RANGPUR LIME, MANDARIN LIME	*C. × limonia*	Rutaceae
*SHADDOCK, PUMMELO	*C. maxima*	Rutaceae
*CITRON	*C. medica*	Rutaceae
KING ORANGE	*C. × nobilis*	Rutaceae
*GRAPEFRUIT	*C. × paradisi*	Rutaceae
*MANDARIN, SATSUMA, TANGERINE, CLEMENTINE	*C. reticulata*	Rutaceae
*KUMQUAT	*Fortunella japonica*	Rutaceae
LOQUAT, JAPANESE MEDLAR	*Eriobotrya japonica*	Rosaceae
BREADFRUIT	*Artocarpus altilis*	Moraceae
JACKFRUIT	*A. heterophyllus*	Moraceae
CHERIMOYA	*Annona cherimolia*	Annonaceae
CUSTARD APPLE, BULLOCK'S HEART	*A. reticulata*	Annonaceae
SOURSOP, GUANABANA	*A. muricata*	Annonaceae
SUGARAPPLE, SWEETSOP	*A. squamosa*	Annonaceae
*BANANA, EDIBLE PLANTAIN	*Musa acuminata, M. × paradisiaca*	Musaceae
FEHI BANANA	*M. fehi*	Musaceae
*AVOCADO, AGUACATE, ALLIGATOR PEAR	*Persaea americana* (= *P. gratissima*)	Lauraceae
*COCONUT	*Cocos nucifera*	Palmae
*DATE	*Phoenix dactylifera*	Palmae
*PINEAPPLE	*Ananas comosus*	Bromeliaceae
*MANGO	*Mangifera indica*	Anacardiaceae
CASHEW APPLE	*Anacardium occidentale*	Anacardiaceae
GRANADILLA, PASSION FRUIT	*Passiflora edulis*	Passifloraceae
SWEET GRANADILLA	*P. ligularis*	Passifloraceae
YELLOW GRANADILLA	*P. laurifolia*	Passifloraceae
SWEET CALABASH	*P. maliformis*	Passifloraceae
CURUBA	*P. mollissima*	Passifloraceae
GIANT GRANADILLA	*P. quadrangularis*	Passifloraceae
*PAPAW, PAWPAW	*Carica papaya*	Caricaceae
*DURIAN	*Durio zibethinus*	Bombacaceae
*MANGOSTEEN	*Garcinia mangostana*	Guttiferae
*RAMBUTAN	*Nephelium lappaceum*	Sapindaceae
LONGAN	*Euphoria longan*	Sapindaceae
*AKEE	*Blighia sapida*	Sapindaceae
GUAVA	*Psidium guajava*	Myrtaceae
CAPE GOOSEBERRY	*Physalis peruviana*	Solanaceae
TOMATILLO, JAMBERRY	*P. ixocarpa*	Solanaceae
*MAMMEY APPLE, MAMMEE	*Mammea americana*	Guttiferae
SAPODILLA	*Manilkara zapota*	Sapotaceae
*SAPOTE	*Pouteria sapota* (= *Calocarpum sapota*)	Sapotaceae
*TAMARIND	*Tamarindus indica*	Leguminosae
CARAMBOLA, CARAMBA, BLIMBING, BILIMBI	*Averrhoa carambola*	Oxalidaceae
*PERSIMMON	*Diospyros kaki*	Ebenaceae
*POMEGRANATE	*Punica granatum*	Punicaceae
LITCHI, LYCHEE	*Litchi chinensis*	Sapindaceae

Opposite *common temperate fruits:*
1 Blackcurrant; 2 Redcurrant; 3 Bilberry;
4 Highbush Blueberry; 5 Sweet Cherry; 6 Cranberry;
7 Black Mulberry; 8 Gooseberry; 9 Raspberry;
10 Plum; 11 Strawberry; 12 Fig; 13 Damson;
14 Greengage; 15 Medlar; 16 Quince; 17 Apricot;
18 Peach; 19 Apple; 20 Pear. (×1)

Below *tropical fruits: 1 Pineapple; 2 Durian;*
3 Carambola; 4 Mango; 5 Papaw; 6 Soursop;
7 Persimmon; 8 Mangosteen; 9 Pomegranate;
10 Litchi; 11 Akee; 12 Cherimoya; 13 Banana;
14 Guava; 15 Sapodilla; 16 Passion Fruit;
17 Loquat; 18 Cape Gooseberry; 19 Rambutan.
1 to 6, 12, 13 (×⅓); 7 to 11, 14 to 18 (×½).

*Male plant of Serrated Wrack (*Fucus serratus*) growing in the intertidal zone. (×1)*

Fumaria a genus of annual, often climbing herbs mainly native to Europe and the Mediterranean region but with a few species further east to Mongolia and south to East Africa. There are no horticulturally important members but *F. officinalis* [COMMON FUMITORY, EARTH SMOKE] was at one time the source of a yellow dye. Like some other species, including *F. parviflora*, it is used in herbal medicine as a tonic and laxative.
FUMARIACEAE, about 60 species.

Funaria a genus of mosses best-known for the common cosmopolitan species *F. hygrometrica* which tends to appear in abundance after fires.
FUNARIACEAE, about 220 species.

Fungi see p. 152.

Furcraea a genus of succulent perennials similar to *Agave and indigenous to tropical America. The best-known species is *F. foetida* (= *F. gigantea*) [GREEN ALOE], whose large fleshy leaves yield a strong fiber (Mauritius hemp) used for mats, cordage etc. It is cultivated widely in tropical America, and on a commercial scale in Mauritius, Madagascar and St. Helena. The cultivated plant is var *willemettiana*. The inflorescences are up to 8m long and contain many thousands of flowers.
AGAVACEAE, about 20 species.

Fusarium a large genus of imperfect fungi widely distributed in soils, especially cultivated ones. Various species are important as seedling pathogens of cereals, causing pre-emergence killing or damping-off. Others attack more mature plants causing root rots or ear blight. Growth of fusaria in feed grain may produce toxins which are harmful to horses, pigs and Man when eaten. Other important pathogens are those that cause severe wilt, such as *F. oxysporum*, the cause of Panama disease of bananas, or storage rots, such as *F. caeruleum*, the cause of dry rot of potato tubers.
HYPHOMYCETACEAE, about 70 species.

G

Gagea a genus of low-growing, Eurasian, bulbous monocotyledonous herbs with basal leaves and solitary or few-flowered umbels of yellowish-green flowers. *G. lutea* (= *G. sylvatica*) [YELLOW STAR OF BETHLEHEM] is sometimes cultivated for ornament. It is widespread in Europe and the Mediterranean. Some species have been used as diuretics.
LILIACEAE, about 70 species.

Gaillardia a genus of temperate American annual and perennial leafy-stemmed herbs. The showy radiate flower heads are red, purple, orange or yellow in color, and borne on long stalks. *G. pulchella* [BLANKET FLOWER], especially var *picta*, is widely cultivated. *G. aristata* [PRAIRIE FLOWER] was formerly cultivated but is now replaced by *G. × grandiflora* (*G. aristata* × *G. pulchella*), of garden origin but now naturalized in several parts of the western USA. There are numerous cultivars.
COMPOSITAE, about 28 species.

Galangal the common name for the herbaceous perennials *Alpinia officinale* (= *A. officinarum*) [LESSER GALANGAL] and *A. galanga* [GREATER GALANGAL]. Both of these Asian species yield from the rhizome a spice

*Inflorescences of the Blanket Flower (*Gaillardia pulchella*), which has many single- and double-flowered garden varieties. (×¾)*

used in flavoring curries and other savory dishes.
ZINGIBERACEAE.

Galanthus see SNOWDROP.
LILIACEAE, about 15 species.

Galega a small genus of busy, erect perennial herbs native to southern Europe, western Asia and tropical East Africa. *G. officinalis* [GOAT'S RUE], *G. orientalis* and *G. × hartlandii* (= *G. bicolor* var *hartlandii*), a natural garden hybrid, are grown for their attractive white or blue pea-like flowers.
LEGUMINOSAE, 6 to 8 species.

Galeopsis [HEMP NETTLES], a genus of annual herbs native almost throughout temperate Eurasia. Most species, such as *G. segetum* (= *G. dubia*) [DOWNY HEMP NETTLE] and *G. tetrahit* [COMMON HEMP NETTLE] are weeds of arable land. However, *G. segetum* has been used, in the form of a leaf-decoction, for lung, intestinal and spleen complaints.
LABIATAE, about 10 species.

Galinsoga a small genus of New World annual herbs, bearing small daisy-like flower heads. *G. parviflora* [GALLANT SOLDIER, JOEY HOOKER] is a widespread weed in the Northern Hemisphere. Young plants are eaten as a vegetable in Southeast Asia.
COMPOSITAE, 14 species.

Galium a large almost cosmopolitan genus of annual or perennial herbs. *G. aparine* [CATCHWEED, GOOSEGRASS, CLEAVERS, STICKY WILLIE] is a pernicious weed with seeds and stems which cling by means of small reflexed hooks. Attractive, profusely flowered species with filmy foliage are cultivated in rockeries, such as *G. verum* [YELLOW-FLOWERED BEDSTRAW, OUR LADY'S BEDSTRAW]. Dried plants were used to stuff mattresses (hence bedstraw), and plant juices were used to curdle milk in cheese making. In Germany and Austria, *G. odoratum* (= *Asperula odorata*) [SWEET WOODRUFF] yields an extract containing coumarin, used to flavor Maiwein ("May wine").
RUBIACEAE, about 350 species.

Galtonia [SPIRE LILY, SUMMER HYACINTH] a small genus of attractive bulbous plants native to South Africa. The fragrant *G. candicans* (= *Hyacinthus candicans*) with racemes of pure white flowers is the species most frequently cultivated in gardens.
LILIACEAE, 4 species.

Gambier a climbing shrub, *Uncaria gambir* [BENGAL GAMBIER], native to tropical Asia. The leaves and young stems yield a tannin extract used in dyeing, printing and leather preparation. Extracts of the leaves, known as catechu or pale catechu (not to be confused with black catechu from *Acacia catechu*), are still included in Western pharmacopeias as an astringent used in treating the symptoms of diarrhea.
RUBIACEAE.

The fruits and stems of Goosegrass or Cleavers (Galium aparine) are covered in tiny hooks, as an aid to animal dispersal. (× 3)

Gamboge a yellow dye obtained from the gum resin of species of *Garcinia*, particularly *G. cambogia* [GAMBOGE TREE] and *G. hanburyi* [SIAMESE GAMBOGE TREE]. The dye, being water-soluble is used by artists. It also imparts a golden tint to metal lacquers and varnishes. Gamboge is also used locally as a violent purgative and laxative.
GUTTIFERAE.

Ganoderma a genus of basidiomycete fungi with woody bracket-shaped fruiting bodies. The species are widely distributed in temperate and tropical regions on decayed wood. *G. adspersum* infects wounds on broadleaved trees such as beeches, causing extensive stem rot. *G. pseudoferreum* attacks roots of rubber, cacao and other tropical crops.
GANODERMATACEAE, about 50 species.

Garcinia a large genus of trees and shrubs confined to the Old World tropics, with the greatest diversity in Asia. Many species yield products useful to native populations, while some have special economic importance. *G. mangostana* [*MANGOSTEEN], *G. xanthochymus* (= *G. tinctoria*), *G. multiflora* and many other species bear delicious fruits. *G. cambogia*, *G. hanburyi* and other Asiatic species produce the gum resin, *gamboge. (See also GAMBOGE and MANGOSTEEN.)
GUTTIFERAE, about 400 species.

Gardenia a large genus of evergreen shrubs and small trees found in the Old World tropics and subtropics. The bark is often whitish and sometimes spiny-branched and the flowers are salver-shaped, white, greenish or yellow, solitary or in pairs. The most commonly cultivated species is *G. jasminoides* [CAPE JASMINE], which is grown for its showy white highly fragrant flowers.
RUBIACEAE, about 250 species.

Garlic the common name for *Allium sativum*, a small extremely pungent, onion-like plant which is widely used for flavoring in salads and meat and savory dishes, particularly in the Mediterranean countries, the Middle and Far East, South America etc. The garlic "bulb" consists of a cluster of swollen axillary buds ensheathed by dry foliage leaf-bases. Each swollen axillary bud or clove is a true bulb.
LILIACEAE.

Garrya a genus of evergreen shrubs native to southern and southwestern North America, Mexico and the West Indies. *G. elliptica* [SILK-TASSEL BUSH] with its attractive gray-green catkins appearing in winter, is the most widely cultivated. The bark and leaves of this and other species are used in local medicine to treat fevers.
GARRYACEAE, about 15 species.

Gasterias are popular houseplants and exhibit a great diversity of leaf form – some are biseriate, others form spiral rosettes. (× ⅓)

Wild or Wood Garlic (Allium ursinum) is abundant in damp, shady old woods of temperate regions of Europe and Asia. This species is also known as Ramsons, Buckrams, Gypsy Onion and Bear's or Hog's Garlic. (× 1)

Gasteria a genus of dwarf, more or less stemless, leaf-succulent rosette plants from the drier parts of southern and southwestern Africa. The leaves are thick and fleshy with often horny margins and usually a sharp white tip. The flowers are borne in tall, pendulous, lateral racemes or panicles; they are tubular with a markedly inflated base and are red or pinkish with usually greenish tips.

Unlike most succulents, species of *Gasteria* tolerate shade and are hence popular houseplants where full sun is not available. The most decorative are *G. batesiana*, *G. pulchra*, a miniature forming clumps, *G. armstrongii*, a miniature with a single rosette of blackish leaves in two series, and *G. obtusifolia* 'Variegata' with dramatically cream-striped leaves.
LILIACEAE, about 70 species.

Gaultheria a large genus of evergreen flowering shrubs, mainly in the Americas but also in India, East and Southeast Asia to Australasia. The urn-shaped flowers are usually pink or white. The low-growing *G. procumbens* [CREEPING or SPICY WINTER-GREEN, PARTRIDGE BERRY, CHECKERBERRY, TEA-BERRY], from northeast North America, with red berries and a spread of 1m or more, is the original source of wintergreen oil; it is also grown for ground cover. Several species are cultivated on lime-free soils for their edible berries, for example *G. antipoda*, from New Zealand and Tasmania, with white or red berries, and *G. shallon* [SALAL, SHALLON], (continued on p.156)

Fungi

The nutrition of fungi is typically absorptive – ingestive in slime molds – but never photosynthetic. Fungi are, therefore, dependent for their carbon on a great variety of organic compounds. Carbohydrates are particularly favorable carbon sources. Almost all fungi can assimilate glucose but the acceptability of other sugars and carbohydrates varies from one group (or even species) to another. Fungi cannot utilize elemental nitrogen but many can assimilate

One of the most distinctive of the cup fungi is the Orange-peel Peziza (Aleuria aurantia). ($\times \frac{1}{2}$)

Fungi form a kingdom of plants which lack the green pigment chlorophyll and reproduce by means of spores. Fungi are ubiquitous. They occur in air, water (both fresh and salt) and in the soil. As saprophytes, they are an important cause of the deterioration of stored organic products and they destroy structural timber. As parasites, they cause major diseases of plants, animals and Man. They also parasitize one another.

On the positive side, if some are poisonous, others are edible and fungi have been used since ancient times for the leavening of bread and in the preparation of a wide range of alcoholic beverages and other fermented foods. They find a use in medicine and the many metabolic fungal products commercially available include such valuable drugs as the antibiotics penicillin and the cephalosporins. As symbionts, fungi are the dominant partner in lichens and they form mycorrhizas in association with the roots of forest trees and many other plants. The most important role of fungi is as agents of decay of

A bracket fungus Inonotus hispidus (= Polyporus hispidus) *which causes decay on standing Common Ash* (Fraxinus excelsior) *trees and some other deciduous tree species.* ($\times \frac{1}{3}$)

organic residues, an activity that makes a major contribution to the maintenance of an environment suitable for life on this planet.

To define a fungus concisely is not easy because of the diversity of fungi and uncertainties regarding the circumscription of the group. Fungi lack chlorophyll (that is, their nutrition is heterotrophic) but they are eukaryotic and therefore distinct from the prokaryotic bacteria which were at one time commonly classified as "fission fungi" or Schizomycetes. Individual fungi vary from single cells a few microns in length, with a life span measured in minutes, to centuries-old "fairy rings" hundreds of metres in diameter. Although typically non-motile, creeping (plasmodial) forms exhibiting amoeboid movement occur and motile states are not infrequent in some aquatic fungi and their terrestrial relatives. Fungi may be asexual or sexual and exhibit simple or complex life cycles. One notable characteristic is the great variation in the form and function of the spores which are more diverse than in any other group of organsims.

Fungi are basically unicellular. Some are unicells, as is a typical yeast, but in most fungi the thallus (body) is derived from a unicellular spore (or from one cell of a multicellular spore) which germinates to give a short tube (germ tube) that elongates into a filament or hypha (plural: hyphae). These hyphae by branching and rejoining (anastomosis) give rise to a mycelium (plural: mycelia) which may undergo fusions with other mycelia.

The microscopic hyphae may become aggregated to form complex and quite massive structures. The simplest condition is the twisting together of hyphae to form the mycelial strands not infrequently shown by molds in laboratory culture and it is a similar aggregation which gives rise to the macroscopic rhizomorphs of which the most familiar example is the black "boot-laces" of the HONEY FUNGUS (*Armillaria mellea*) found associated with rotten wood. To ensure survival under adverse conditions, hyphae may become aggregated and consolidated as sclerotia which may range from 1mm or so in diameter to massive structures weighing several kilograms, such as the sclerotium of *Polyporus mylittae*, known as "blackfellows bread" in Australia. The fruiting bodies (so-called mushrooms and toadstools) of larger fungi are also constructed of hyphae.

nitrate nitrogen; for others, ammonium nitrogen or nitrogen in organic combination is needed.

Reproduction. Examples of the simplest methods of fungal reproduction are provided by the budding of a yeast cell to produce a new individual or the development of a hyphal fragment into a new mycelium. The usual method of reproduction among fungi is by spores which are produced in enormous numbers – it has been calculated that one fruiting body of the GIANT PUFFBALL (*Langermannia gigantea*) contains 7×10^{12} spores. Spore types form the basis of fungal classification. They are divided into two main categories (sexual spores and asexual spores) according to whether or not their formation is preceded by a process that can be interpreted as sexual.

Sexual spores in fungi are of four main types. Oospores result from the fusion of a female gamete with a smaller male gamete which may or may not be motile. Zygospores are naked, thick-walled spores resulting from the fusion of two morphologically undifferentiated gametes. Ascospores are produced (usually in eights) inside a microscopic, sac-like organ, the ascus. Basidiospores are borne externally (typically in fours) at the apex of a structure called a basidium.

Among asexual spores the development of the spores (sporangiospores) of *Mucor* within a rounded body (sporangium), the

wall of which finally ruptures to release the spores, is analagous to that of ascospores; but the development of conidia (a general term for all asexual spores other than non-motile sporangiospores and motile zoospores) is blastic, that is the spore is differentiated from part of the conidial mother cell in a manner analagous to basidiospore formation.

Spores, being the equivalent of seeds, have to fulfill several functions: to initiate a new individual in the appropriate ecological niche and to remain viable during periods unfavorable for growth. This last function is frequently effected by special thick-walled chlamydospores. For transfer to an appropriate site, passive dispersal by air currents, water and animals is usual. The first step in dispersal is frequently the active discharge of the spore. This may be to a distance of a few microns, in the case of the discharge of a basidiospore from its basidium in agarics and polypores by a "bubble mechanism", to a few centimetres in the violent discharge, due to turgor, of the spores from the asci of a discomycete (cup fungus). Air-borne spores are typically small and dry and often dark in color to offset the deleterious effects of light,

Above Coprinus micaceus (*Glistening Inkcap*) *which grows on old stumps in dense clumps and has distinctly grooved caps.* (× 1)

Left Young fruiting bodies of the puffball Lycoperdon perlatum. *Spores are produced inside the wall which ruptures to release them.* (× 1) *Right The evil-smelling Stinkhorn (*Phallus impudicus*). The spores mature while the fruiting body is still enclosed in a membrane and are only released when the stalk elongates rapidly to its characteristic form. The strong smell attracts insects which are involved in the dispersal of the spores.* (× ¾)

MEMBERS OF THE KINGDOM FUNGI

The following table gives details of the distribution, structure and reproduction of the main groups of fungi.

Abbreviations for taxonomic ranks: d = Division, sd = Subdivision, c = Class, sc = Subclass, o = Order

d MYXOMYCOTA [SLIME MOLDS]
Heterotrophic amoeboid organisms. Cells comprising naked (no cell wall), multinucleate plasmodia or pseudoplasmodia for most of the life cycle. Nutrition by ingestion.

c *Myxomycetes* [PLASMODIAL SLIME MOLDS]
Characteristic of wet wooded habitats, frequently bright orange in color. Comprising multinucleate plasmodia that creep by amoeboid movement. Multinucleate sporangia formed giving rise to many spores. About 400 species.

c *Plasmodiophoromycetes* [ENDOPARASITIC SLIME MOLDS]
Entirely parasitic slime fungi producing spores that give rise to biflagellate swarmers. About 50 species.

c *Acrasiomycetes* [CELLULAR SLIME MOLDS]
A group of slime fungi that inhabit soil with a high content of decaying organic remains. They do not produce flagellate cells or multinucleate plasmodia. Population comprises mostly uninucleate amoeboid cells. About 30 species.

c *Protosteliales*
A recently defined heterogeneous group of slime fungi, comprising a small number of species.

c *Hydromyxomycetes*
A very small group of marine parasitic slime fungi. The uninucleate cells form a network that moves in a gliding fashion. Number of species not yet defined.

d MYCOTA (EUMYCOTA) [TRUE FUNGI]
Unicellular and multicellular (basically filamentous) heterotrophic organisms. Cell wall present, cells often multinucleate. Nutrition by absorption. Both sexual and asexual phases in the life cycle.

sd Mastigomycotina
Zoospores motile and sexual spores typically motile.

c *Chytridiomycetes*
o Chytridiales [CHYTRIDS]
Zoospores with a single posterior whiplash flagellum. Mainly aquatic, some inhabit soil or organic remains, a few are parasites. Thallus either without a vegetative system (and entirely reproductive) or with specialized vegetative system bearing reproductive structures. About 750 species.

o Blastocladiales
Saprophytes in soil, water or organic remains. Thallus always with a well-defined vegetative system and having a conspicuous basal cell anchored by rhizoids. Thick-walled, resistant sporangia formed. About 50 species.

o Monoblepharidales
Soil- and water-inhabiting. Vegetative thallus always well-defined, but basal cell absent. No resistant sporangia formed. About 10 species.

c *Oomycetes* [EGG FUNGI]
Mainly aquatic (some parasites), with motile zoospores and oospores with two flagella.

o Saprolegniales [FISH MOLDS]
Mainly free-living in water, soil or plant debris (some parasitize fish and plant roots). About 150 species.

o Leptomiales
Saprophytic aquatic and soil fungi. About 20 species.

o Lagenidiales
Parasites of algae, protozoa and aquatic animals with one genus infecting roots of cereal crops. About 45 species.

o Peronosporales [DOWNY MILDEWS]
Mostly parasites of higher plants. About 350 species.

sd Zygomycotina
Terrestrial fungi with the hyphae septate only during reproduction; non-motile, non-flagellate asexual spores and sexual resting spores called zygospores.

c *Zygomycetes*
Comprises two orders.

o Mucorales [PIN MOLDS]
Widely found in soil and mostly saprophytic, with some parasites of other fungi. Asexual reproduction by sporangia containing one or more spores. About 300 species.

o Entomophthorales
Mainly parasites of insects and other animals. Asexual reproduction by conidia. About 100 species.

sd Ascomycotina [ASCOMYCETES]
Terrestrial or aquatic fungi with septate hyphae. Sexual reproduction involves production of an ascus containing ascospores. Asexual reproduction by formation of conidia.

c *Hemiascomycetes*
Ascocarp lacking. About 300 species.

o Endomycetales [YEASTS]
Saprophytes and parasites.

o Taphrinales
Plant parasites forming palisade-like layer of asci on surface of diseased tissue.

o Protomycetales
Plant parasites forming thick-walled spores.

c *Plectomycetes*
Ascocarps in the form of cleistothecia. About 500 species.

o Ascosphaerales
Some insect parasites asci in clusters in a spore cyst.

o Eurotiales
Mostly saprophytic, including many important agents of food spoilage and decay. Asci small and globose.

o Erysiphales [POWDERY MILDEWS]
Obligate parasites of flowering plants. Asci oval to club-shaped.

c *Pyrenomycetes*
Ascocarps in the form of flask-shaped perithecia. About 9 000 species.

o Sphaeriales
Saprophytes of soil and dung or weak parasites of woody hosts. Perithecia dark-colored.

o Hypocreales
Mainly parasites of flowering plants (particularly grasses) and insects, some saprophytic. Perithecia brightly-colored and fleshy. About 120 genera.

c *Discomycetes* [CUP FUNGI]
Ascocarps in the form of cup-shaped apothecia. About 3 000 species.

o Pezizales
Mostly saprophytes, many growing on dung. Ascocarps conspicuous. Asci opening by a lid.

o Tuberales [TRUFFLES]
Fungi with underground, strong-smelling ascocarps. Asci without a lid.

o Phacidales
Parasites of higher plants. Asci without a lid.

o Lecanorales
Forming symbiotic relationship with algae in lichens. Asci without a lid.

c *Loculoascomycetes*
Ascocarps in the form of pseudothecia or a stroma containing chambers with a distinct wall. About 500 genera, 2 000 species.

o Pleosporales
Saprophytes and higher plant parasites. Asci club-shaped, separated by sterile hairs.

o Myriangales
Asci globose.

sd Basidiomycotina [BASIDIOMYCETES]
Terrestrial fungi with septate hyphae. Sexual reproduction involves formation of a basidium containing basidiospores. Dikaryotic during most of life cycle.

c *Teliomycetes*
Parasites on vascular plants. Basidiocarps not formed and overwintering spores either grouped in clusters or pustules or dispersed in host tissue.

o Uredinales [RUSTS]
Spores forcibly discharged and borne on stalks (sterigmata). About 5 000 species.

o Ustilaginales [SMUTS]
Spores not forcibly discharged and not stalked. About 1 000 species.

c *Hymenomycetes*
Mainly saprophytic with some parasites. Basidiocarps well developed, typically with spores formed on a layer (hymenium) more or less exposed to the surroundings. Spores forcibly discharged.

sc *Phragmobasidiomycetidae*
Basidia variously septate.

o Tremellales [JELLY FUNGI]
Mainly saprophytes on wood, typically with gelatinous, yellow or orange basidiocarps. About 200 species.

o Auriculariales
Mainly saprophytes, with some parasites. About 100 species.

o Septobasidiales
Parasites on scale insects in the tropics and warm temperate regions. About 175 species.

sc *Holobasidiomycetidae*
Basidia not septate.

o Exobasidiales
Parasites on flowering plants. About 65 species.

o Brachybasidiales
Parasites causing leaf spot diseases; basidia protruding through the stomata or bursting the epidermis. Two species.

o Dacrymycetales
Saprophytes, often with yellow or orange gelatinous basidiocarps. Basidia of an unusual "tuning fork" type. About 45 species.

o Tulasnellales
Saprophytes and parasites with a wide range of hosts. About 50 species.

o Aphyllophorales (Polyporales) [BRACKET FUNGI]
Saprophytes and parasites with large basidiocarps with the hymenium mostly lining pores and not borne on gills. About 1 000 species.

o Agaricales [GILL FUNGI AND BOLETI]
Saprophytes and parasites, typically with large umbrella-shaped basidiocarps, with the hymenium spread over radiating gills (except in *Boletus*). About 3 000 species.

c *Gasteromycetes*
Basidiocarps mostly not exposed to the surroundings. Basidia not septate and spores not forcibly discharged. About 550 species.

o Lycoperdales [PUFFBALLS AND EARTHSTARS]
Hymenium present, collapsing at maturity to form a powdery mass surrounded by skin.

o Phallales [STINKHORNS]
Hymenium present, collapsing into a glutinous mass which is carried upwards as a stalk at maturity.

o Sclerodermatales [EARTH BALLS]
Hymenium not present and spores forming a powdery mass.

o Nidulariales [BIRD'S NEST FUNGI]
Hymenium not present and spore-mass in the form of hard seed-like pellets.

sd Deuteromycotina [IMPERFECT FUNGI]
Fungi in which the sexual (or perfect) phase is not known. Reproduction is by asexual spores (conidia). About 25 000 species.

but some spores have a moist surface when they are dispersed by rain splash or damp air. Spores dispersed by water may be motile (zoospores) or, as in certain aquatic hyphomycetes, much branched.

Life Cycles. In flowering plants the nuclear cycle is standardized. The plant cells are diploid. Meiosis immediately precedes the formation of pollen grains and egg, the fusion of which restores the diploid condition. In fungi there is much variation. For the assimilative thallus to be diploid (as in oomycetes) is rare. More frequently it is haploid with the diploid phase limited to the zygote, while in higher basidiomycetes during much of the assimilative phase each cell may contain two haploid nuclei of opposite mating type which undergo simultaneous (or "conjugate") division and only fuse in the basidium where a reduction division precedes basidiospore development. This last condition is called dikaryotic (each cell with its pair of comp-

Views of the underside of the fruiting bodies of (Below Top) a typical gill fungus (×5) showing the gills and (Below Bottom) a bracket fungus (×2) showing the pores. Spores are produced on the lining (hymenium) to the gills or pores, from which they are forcibly discharged.

lementary nuclei being a dikaryon).

Some fungi, particularly chytrids, oomycetes and certain ascomycetes, develop morphologically distinguishable sexual organs (antheridia, oogonia and ascogonia) but sometimes the individual fungus bearing both types of sex organs is self-sterile. More frequently there is no differentiation between the sexes but, as in the well-known case of the pin molds, there are two morphologically indistinguishable but compatible mating types, and when appropriate strains (+ and −) are brought together, in response to mutual stimulation by hormones, special branches develop which on fusion give rise to a zygospore. This condition is known as heterothallism, in contrast to homothallism, in which sexual fusions are possible between different branches of one mycelium.

Fungi frequently produce both sexual and asexual spores and there may be a more or less well defined alternation, as in many parasitic species where the sexual (or perfect) state constitutes the climax on dead host tissues. For others, the perfect state is of rare occurrence and for many only the asexual (or imperfect) state is known. Some of these last are certainly as yet unconnected with perfect states already described. For others, the perfect state will sooner or later be discovered, but there remains a large residue of forms that appear to have lost the ability to produce a perfect state.

The kingdom Fungi has two divisions, the Mycota (or Eumycota), in which the thallus is basically filamentous, and the Myxomycota (slime fungi) characterized by the possession of naked, multinucleate plasmodia or pseudoplasmodia. The main diagnostic characters by which the Mycota are divided are based on the types of the sexual spores. A summary of the main groups of fungi is given in the accompanying table.

The unmistakable cap of Amanita muscaria *(Fly Agaric) which is dangerous because of its poisonous and hallucinogenic properties. (×2)*

from western North America, with purple-black berries.
ERICACEAE, about 100 species.

Gaylussacia see HUCKLEBERRY.
ERICACEAE, about 40 species.

Gazania a genus native to South Africa, comprising low-growing, mostly perennial herbs or subshrubs, with basal leaf rosettes and large brightly colored daisy-like flower heads borne on long leafless stalks. They are easily hybridized and several cultivars, as well as a few of the parental species, notably *G. ringens* (= *G. splendens*) [TREASURE FLOWER] are popular half-hardy garden plants in north temperate regions.
COMPOSITAE, about 25 species.

Inflorescences of the Treasure Flower (Gazania ringens), a hybrid that is popular as a half-hardy garden ornamental. (×⅓)

Geastrum see EARTHSTARS.
GEASTRACEAE.

Gelidium a genus of red algae native mainly to the coasts of the North Pacific. It is economically important for the agar (used in the food, cosmetic, paper and textile industries) that is extracted from it by boiling. The chief producer is Japan, but it is also processed in many other countries.
GELIDIACEAE.

Genista a genus of shrubs or shrubby herbs, many of which are cultivated in gardens. They are native to Mediterranean regions with a few outliers in the Canary Islands and Madeira and a sparse representation in western Asia and northern Europe. The typically pea-like flowers are yellow, borne in terminal racemes or heads, or, in a few species, on short shoots in the leaf-axils. The genus as described here includes groups of species such as *Teline*, *Chamaespartium* and *Echinospartium* which are often treated as separate genera.

Species cultivated in gardens include spring shrubs such as *G. sylvestris* (= *G. dalmatica*) [DALMATIAN BROOM], a dwarf species of compact habit, *G. hispanica* [SPANISH GORSE], a much-branched, erect to decumbent, profusely flowering type, and *G. anglica* [PETTY WHIN, NEEDLE FURZE], which is erect to more or less prostrate with very spiny branches. Nonspiny cultivated species include *G. lydia* and *G. januensis* [GENOA BROOM], both of which are procumbent to erect, freely flowering forms, and *G. cinerea*, a sweetly scented erect shrub with flexuous branches. One of the most spectacular cultivated species is the tall, elegant *G. tenera* (= *G. virgata*) [MADEIRA BROOM], which bears deep yellow flowers in dense terminal racemes. *G. tinctoria* [DYER'S GREENWEED], the source of a yellow dye, is also widely cultivated under various varieties or cultivars.
LEGUMINOSAE, about 80 species.

Gentiana a large genus of perennial (rarely annual or biennial) herbs widely distributed in temperate regions and in mountains in the tropics. The genus is most evocative of the European Alps and is much favored by alpine gardeners. The flowers may be white, yellow, blue to purple, pinkish red, or mauve. The blue trumpet gentians of Europe include the widely cultivated *G. clusii* and *G. acaulis* (= *G. excisa*, *G. kochiana*) [STEMLESS GENTIAN]. The alpine species grow up to the snow level where few other flowering plants thrive. The North American cultivated species include the blue-flowered *G. andrewsii* [CLOSED OR BOTTLE GENTIAN] (white in var *albiflora*) and the very similar *G. clausa*, with the same common names. *G. catesbaei* (= *G. elliotti*) [CATESBY'S GENTIAN, SAMPSON'S SNAKEROOT] and *G. newberryi* [ALPINE GENTIAN] are two other North American blue-flowered species, while *G. autumnalis* (= *G. porphyrio*) [PINE-

Shoots of Genista lydia, a prostrate species native to southeastern Europe and Syria. It is cultivated both for its spineless gray-green shoots and bright yellow flowers that appear in the spring. (×1)

The Stemless Gentian (Gentiana acaulis), one of the best-loved wild flowers of the European Alps. (×1)

BARREN GENTIAN] has yellowish and blue flowers.

G. lutea [GREAT YELLOW GENTIAN], is the chief commercial source of the bitter tonic, gentian bitter, and, when fermented, of the aperitif drink Suze and the liqueur Enzian. Locally, other species have been used similarly and also with, or instead of, hops in beer. In addition to the many species in cultivation, many hybrids have been raised, eg *G.* × 'Inverleith' (*G. farreri* × *G. veitchiorum*).
GENTIANACEAE, 300–400 species.

Gentianella a genus of annual, biennial, and rarely perennial herbs from temperate regions of both hemispheres (not Africa). Although sometimes included in *Gentiana*, it differs from that genus in not having appendages between the corolla-lobes. The flowers are blue, mauve or white. It contains several taxonomically very complex species showing great variation and hybridization. Several species are occasionally cultivated, such as *G. amarella* and *G. tenella*. Some Northern Hemisphere species, such as *G. detonsa* [FRINGED GENTIAN], are sometimes separated out as a further genus, *Gentianopsis*.
GENTIANACEAE, about 125 species.

Geonoma a medium to large genus of slender palms of tropical America, where they ascend to cool mountain rain forests. Although of no economic importance, the leaves of several species such as *G. interrupta* (= *G. binervia*) and *G. dominica* are used locally for thatching, and the young inflorescences for food.
PALMAE, about 75–250 species.

Geranium [CRANESBILLS] a genus of annual, biennial or perennial herbs, often with a rhizome, tuber or rootstock. The genus is cosmopolitan but mainly found in temperate regions as far north as the Arctic Circle and south to Antarctica, while some species occur in mountains of tropical regions. The name 'Geranium' is widely applied to species and hybrids of *Pelargonium* from South Africa which are widely cultivated for ornament or for their essential oils.

The genus is not important economically and some such as *G. molle* [DOVE'S FOOT] are regarded as weeds. However, a number of the freer-flowering species are used horticulturally in herbaceous borders in rock gardens. These include the white-flowered *G. richardsonii* (with red veins on the petals), the pink-flowered *G. argenteum* [SILVER-LEAVED GERANIUM] and *G. cinereum*, the violet-blue-flowered *G. pratense* [MEADOW CRANESBILL], the purple-flowered *G. collinum* and *G. sylvaticum* (white in cultivar 'Album') and the red-to-magenta-flowered *G. robertianum* [HERB ROBERT, RED ROBIN] and *G. sanguineum* [BLOODY CRANESBILL] (white in cultivar 'Album'). The spectacular shrubby *G. maderense* with masses of carmine flowers is grown under glass.

The American species *G. maculatum* [WILD CRANESBILL, ALUM ROOT] provides a liquid extract from the roots which has astringent properties and is used to control bleeding. GERANIACEAE, about 300 species.

Gerbera a genus of daisy-like perennial herbs from Africa, Asia, Madagascar and Indonesia. *G. jamesonii* [BARBERTON or TRANSVAAL DAISY] is the most hardy species. Wild forms have yellow to orange, flame-colored flower heads and many color variants exist in cultivation. *G. asplenifolia* is also attractive, with purple flower heads and fern-like foliage.
COMPOSITAE, about 70 species.

Geum a genus of perennial herbs found in temperate regions of both the Northern and Southern Hemisphere. Many are grown as garden plants in herbaceous borders or rockeries. The species are generally readily separable except in a few groups; in Britain, for example, *G. urbanum* [WOOD AVENS] and *G. rivale* [WATER AVENS] show extensive hybridization, usually following human disruption of the habitat. Many species are grown as garden ornamentals for the color of their large flowers; double forms are frequent. The most commonly cultivated are *G. coccineum* and *G. quellyon* (= *G. chilense*, "*G. chiloense*"), both with scarlet flowers, and

The flowers of Water Avens (Geum rivale). (×1)

The stamens and stigmas of the Dove's Foot Cranesbill (Geranium molle) mature in close proximity, which assists self-pollination. (×2½)

G. montanum with yellow to orange flowers, together with hybrid derivatives of these species.

The rhizomes of *G. urbanum* have been used as a flavoring for wine liqueurs and beer, as a heart tonic and as a remedy for complaints of the digestive system.
ROSACEAE, about 50 species.

Gherkins the warty fruits of *Cucumis anguria*, a West Indian species similar to cucumber (*C. sativus*). The "gherkins" of today, commonly bought pickled, and often flavored with *DILL, are not true gherkins but small-fruited varieties of *CUCUMBER. Less warty, they are said to be of better flavor than the fruits of *C. anguria*.
CUCURBITACEAE.

Gigartina a genus of red algae often found in the intertidal zone on rocky shores. In the Pacific Ocean large parenchymatous species grow in the sublittoral zone (near the sea but not on the shore). Like *Chondrus*, which grows in similar situations, this alga is collected for carrageenan extraction (see CARRAGHEEN), notably from *G. mamillosa* and *G. stellata*.
RHODOPHYCEAE.

Gilia a genus of annual, biennial or perennial herbs mainly from western North America, many of which make attractive ornamental bedding plants with funnel-shaped or salverform flowers. Fine examples are the annuals *G. capitata*, with terminal heads of white, blue or violet flowers, and *G. latifolia*, with long inflorescences of pink-buff flowers. Many species previously placed in the genus have been transferred to *Ipomopsis*, *Eriastrum*, *Linanthus* and *Leptodactylon*.
POLEMONIACEAE, 20–30 species.

Gillenia a genus of only two species of perennial herbs from North America. Both *G. stipulata* [AMERICAN IPECAC] and *G. trifoliata* [INDIAN PHYSIC, BOWMAN'S ROOT] are cultivated as ornamentals for their showy white or pink flowers with narrow petals. The roots and bark were traditionally used by North American Indians as an emetic.
ROSACEAE, 2 species.

Gill fungi see Fungi, p. 152.

Ginger the common name for the plant and the product of *Zingiber officinale*, a monocotyledonous perennial herb, native to Southeast Asia but now cultivated in many tropical regions, particularly in the West Indies, Sierra Leone, China, Japan and India. The familiar ginger condiment is derived from the fleshy rhizomes of the plant. The common white ginger is obtained by washing, boiling, peeling and blanching the rhizomes, while the rarer black ginger is merely washed and boiled before drying. The dried rhizomes are frequently ground up into a powder as a means of efficient distribution and for its use as a flavoring. Ground ginger is used mainly in flavoring cakes and drinks such as ginger beer. Preserved ginger, produced mainly in China, is derived by boiling young peeled rhizomes or stem segments, and is packed in sugar syrup, candied or crystallized. In the preserved form it is frequently used in fruit cake and for the manufacture of ginger marmalade. Ginger is also the source of an essential oil, gingerol.
ZINGIBERACEAE.

Ginkgo see MAIDENHAIR TREE.
GINKGOACEAE, 1 species.

Gladiolus a large, predominantly African genus of cormous perennial herbs, especially well represented in tropical and South Africa, but also found to the north in Europe, the Mediterranean region and western and Central Asia, with one species occurring as far north as England. The dark green basal leaves are most commonly produced all in one plane and are fairly flat in section, although a number of species have narrow cylindrical or angular leaves. The inflorescence is a spike of stalkless flowers, each produced in the axil of a pair of bracts, the whole spike being either one-sided or less frequently two-sided. The color range in *Gladiolus* species is extremely wide, virtually all colors being represented. Many species are bicolored, with a blotch of different color on

The gladiolus is an important flower commercially. Many Gladiolus hybrids have been developed; cultivar 'Alderbaron' is shown here. (×⅓)

the lower segment or segments of the flower. Several species are sweetly scented.

In cultivation in the Northern Hemisphere the southwest Cape species are not hardy since they grow and flower during the most severe weather, therefore requiring greenhouse protection. The summer-rainfall east Cape species, however, are dormant during winter and some of them will survive out of doors. It is, therefore, largely the species of this group, such as *G. natalensis*, which have been used by plant breeders to produce the large and colorful hybrid cultivars of the *G. × hortulanus* [GARDEN GLADIOLUS]. One of the early cultivated forms of *Gladiolus* is the summer-flowering garden hybrid, *G. × gandavensis* (originating from either *G. cardinalis × G. natalensis* or *G. natalensis × G. oppositiflorus*), from which most modern gladioli have been derived.
IRIDACEAE, about 300 species.

Glaucium a genus of annual, biennial and perennial herbs found from Europe to Central Asia. The commonest, *G. flavum* (= *G. luteum*) [YELLOW HORNED or SEA POPPY] is maritime in distribution, growing on sand, shingle beaches and cliffs. It is naturalized in the eastern USA. The flowers are bright yellow with a darker blotch at their bases. The seeds yield an oil which is used in cooking and making soap. This species, the biennial *G. corniculatum* (= *G. rubrum*) [RED HORNED POPPY], with its showy crimson or orange flowers, and *G. grandiflorum* (which is often sold as *G. corniculatum*) are commonly cultivated as border plants.
PAPAVERACEAE, about 20 species.

Glaux a genus represented by a single species, *G. maritima* [SEA MILKWORT, BLACK SALTWORT], which is found on the coasts, salt marshes and inland saline areas of the north temperate zone. It is a small succulent herb with creeping stems and small pinkish flowers. The young shoots are edible and are sometimes used as an emergency food.
PRIMULACEAE, 1 species.

Glechoma a genus of perennial herbs native to Europe and Asia. The best-known species is *G. hederacea* (= *Nepeta hederacea*, *N. glechoma*), commonly known as GROUND IVY, ALEHOOF, HAYMAIDS, FIELD BALM or GILL-OVER-THE-ROAD. A bluish-purple-flowered, prostrate, trailing, perennial herb of thin woodland and shaded banks, it can become very invasive in cultivated borders and lawns. It is widely naturalized in North America. It is used as a ground cover and a number of cultivated varieties were at one time popular ornamentals, especially the variety *variegata*, used in hanging baskets and window boxes.
LABIATAE, about 6–10 species.

Gleditsia (or, incorrectly, *Gleditschia*) a genus of usually spiny, deciduous trees from eastern North America, China, Japan and Iran. Many species are cultivated for their attractive fern-like foliage, for example the

North American *G. triacanthos* [HONEY LOCUST, SWEET BEAN], and for shade. Cultivar 'Sunburst' has golden-yellow foliage in both spring and autumn, while 'Moraine' lacks the 30cm-long spines and is sterile, thus lacking the pods which rattle when dry. Also cultivated is the spineless hybrid *G. × texana*, which arose naturally where populations of *G. aquatica* [WATER LOCUST] and *G. triacanthos* grew together in Texas. Some species have domestic uses. The pulp from *G. triacanthos* is sweet and pods may be fermented to make "beer" or fed to stock. Pods of *G. caspica* [CASPIAN LOCUST] and also of *G. japonica* (Japan) and *G. macracantha* (China) are used in soapmaking. Pods of the latter are also used in tanning. The wood of many species is hard and durable, and that of *G. sinensis* (China) and *G. triacanthos* is used for general carpentry, fence posts, etc.
LEGUMINOSAE, about 10 species.

Gleichenia [SUN FERNS] a genus of ferns native to South Africa, Malaysia and Australasia. They are all terrestrial with long creeping rhizomes. The rhizomes of *G. linearis* (= *G. dichotoma*) [SAVANNAH FERN] are used as a source of edible starch by Australian aborigines, and its leaves are woven into matting in parts of Malaysia.
GLEICHENIACEAE, about 10 species.

Globularia [GLOBE DAISIES] a genus of mainly blue-flowered shrubs and perennial herbs largely found in the Mediterranean region, the Alps and Macaronesia. There are no species of economic importance though *G. alypum*, from the Mediterranean region, is poisonous and violently purgative and the leaves of *G. vulgaris* yield a yellow dye. Several species including *G. alypum*, *G. cordifolia*, and *G. meridionalis* (= *G. bellidifolia*) are cultivated as ornamentals, especially in rock gardens.
GLOBULARIACEAE, about 28 species.

Savannah Fern (Gleichenia linearis) here photographed in the foothills of the island of Oahu, Hawaii. These sun-loving ferns (hence the popular name "sun ferns") are not generally cultivated. (× ⅕)

The Glory Lily (Gloriosa superba). It climbs by means of tendrils extending from the end of each leaf. (× ½)

Gloriosa [GLORY LILY] a genus of climbing or creeping tuberous perennials native to tropical Africa and Asia. The showy flowers are pendulous with yellow, orange or red reflexed perianth segments. *G. rothschildiana* and *G. superba* are easily grown in a warm greenhouse in temperate regions.
LILIACEAE, 5–6 species.

Gloxinia the florists' name for the tuberous-rooted herbaceous species, **Sinningia speciosa* (= *Gloxinia speciosa*), a native of Brazil which is a popular houseplant in many parts of the world. It bears oblong-ovate leaves and large showy purple, pink or violet flowers (see *Sinningia*).

Gloxinia is also the generic name for about 6 species of erect perennial herbs, native to Central and South America, one of which, *G. maculata*, is widely cultivated in tropical countries for its large bluish-purple flowers (lilac and red in var *insignis*). *G. perennis* and *G. gymnostoma* are cultivated particularly in the USA.
GESNERIACEAE.

Glyceria a genus of tall, temperate and especially North American marsh or aquatic grasses, the sweet grasses of agriculture. All species provide excellent forage for cattle in swampy places. Important species include *G. fluitans* [MANNA GRASS, SUGAR GRASS, FLOATING SWEET GRASS] and *G. maxima* [REED SWEET GRASS].
GRAMINEAE, about 40 species.

Glycine a small genus of perennial (rarely annual) mostly climbing herbs native to Australia, Africa and Asia. The most important economically is the annual *G. max* [*SOYBEAN], which has never been found in the wild. The WILD SOYBEAN is *G. soja*, from eastern China, the USSR, Korea, Japan and Taiwan, which hybridizes with *G. max*.
LEGUMINOSAE, 9 species.

Glycyrrhiza a genus of temperate and subtropical American, Eurasian, North African and southeast Australian perennial herbs characterized by their spikes of blue or violet pea-like flowers. *Licorice is obtained from the dried rhizomes and roots of *G. glabra*. LEGUMINOSAE, about 18 species.

Gnaphalium a genus of yellow- or white-flowered hardy herbs or small shrubs, widely distributed throughout the world but with the main center in the mountains of Europe. *G. uliginosum* [CUDWEED, COTTONWEED] has been used locally as an astringent. They have limited horticultural value except for a few species such as *G. obtusifolium* which may be used as everlasting flowers. COMPOSITAE, about 135 species.

Gnetum a genus of tropical gymnosperms most of which are climbing shrubs, the remainder being small trees. *G. gnemon* (Malaysia) and certain other species are cultivated for their edible plum-like fruits. The bark of this species and of *G. scandens* (Indochina) is used as a source of fiber. GNETACEAE, about 30 species.

Godetias, such as this species – Godetia concinna – are a common garden favorite for herbaceous borders. ($\times \frac{1}{10}$)

Godetia a group of species once regarded as a distinct genus and now included in the genus *Clarkia* as a section of eight species. Many are grown as ornamental hardy annuals for their brightly colored single or double, funnel-shaped flowers, especially *C. amoena*, with pink to lavender flowers often with a central spot of bright red, and *C. concinna* (= *C. grandiflora*), with bright pink flowers; both are still often referred to under the old generic name *Godetia*. ONAGRACEAE.

Gomphrena a genus of annual and perennial herbs native to tropical America, Southeast Asia and Australia, with most species in the New World. The dense chaffy flower heads of the Old World species *G. globosa* [GLOBE AMARANTH] are useful for drying as everlasting flowers. They grow in a wide range of colors from white through yellow,

pink and red to purple, with yellow florets. *C. haageana*, from Texas and Mexico, with red heads and yellow florets is also cultivated. AMARANTHACEAE, about 100 species.

Gonyaulax a genus of dinoflagellates which is responsible in some areas for the phenomenon known as *red tide, in which a large concentration of the algae causes the sea to appear a reddish-brown color. *Gonyaulax* produces a toxin in the cells and when the plants grow to such high densities, filter-feeding animals such as shellfish concentrate the toxin in their tissues. This may affect humans who inadvertently eat such molluscs. DINOPHYCEAE.

Goodyera a cosmopolitan genus of terrestrial semi-epiphytic orchids. The terminal racemes or spikes of small white, pale green or dingy pink flowers and the usually tessellated or contrastingly veined and brittle leaves arranged in a rosette are characteristic of the great majority of species. *Goodyera* and its related genera are often called JEWEL ORCHIDS or RATTLESNAKE PLANTAINS because of the unusual leaves. ORCHIDACEAE, about 50 species.

Gooseberry the common name for *Ribes uva-crispa* (= *R. grossularia*) a much-branched deciduous bush with conspicuous nodal spines along the branches, native to Europe but widely cultivated in temperate regions. In North America cultivated gooseberries are either varieties of *R. uva-crispa* or of the closely related American species *R. hirtellum* and hybrids derived from it.

Gooseberries are often picked before they are ripe and stewed or made into jam. They

Gloxinias (Sinningia speciosa) are popular houseplants grown for their showy trumpet-shaped flowers. ($\times \frac{3}{4}$)

are also canned commercially and are an excellent fruit for domestic bottling. When ripe some varieties give excellent dessert fruits, although very few are grown commercially for this purpose.

Gooseberry varieties are usually classified by the color of the berries. Some, such as 'Whinham's Industry', are red, while others, such as 'Leveller', are yellow. 'Keepsake' has green berries, and 'Careless' and 'Whitesmith' bear whitish fruits when ripe. SAXIFRAGACEAE.

Gossypium an economically important genus of annual or perennial herbs, shrubs or small trees native to the tropics and subtropics (see COTTON). One or two species, eg *G. arboreum* [TREE COTTON], are sometimes cultivated for their large showy flowers. MALVACEAE, about 34 species.

Gourd the name given to the fruits of some members of the cucumber family, widely used in tropical countries for food, and as utensils. *Lagenaria siceraria* [BOTTLE GOURD, CALABASH] has been cultivated since prehistoric times, mainly for the hard fruit shells which are still used as containers, spoons, fishing floats, musical instruments etc; young fruits are sometimes eaten as a vegetable.

Other edible gourds include *Benincasa hispida* [WAX or WHITE GOURD], *Cucumis anguria* [GOOSEBERRY GOURD, WEST INDIAN GOURD], *Cucurbita foetidissima* [WILD GOURD], *Momordica charantia* [BITTER GOURD], *Trichosanthes anguina* (= *T. colubrina*) [SERPENT or SNAKE GOURD], *T. cucumeroides* [SNAKE GOURD], *Luffa aegyptiaca* (= *L. cylindrica*) [VEGETABLE SPONGE, SPONGE GOURD, DISHCLOTH SPONGE]. Some cultivars of *Cucurbita pepo* [MARROW] are grown as ornamental gourds. CUCURBITACEAE.

Grain amaranth a general name for the seeds of a number of annual herbaceous species of the genus *Amaranthus*. The seeds of three species are used by primitive peoples as food, viz. *A. caudatus* [INCA WHEAT, QUIHUICHA] and its subspecies *mantegazzianus* in the Andes, *A. hypochondriacus* (= *A. frumentaceus*, *A. leucocarpus*) in Mexico and *A. cruentus* (= *A. paniculatus*) in Mexico and Central America. The small seeds of all these cultivated species are normally milled for flour. Although they are all native to the New World, grain amaranth is now widely grown in India and parts of Asia. AMARANTHACEAE.

Grape the fruit of *Vitis vinifera* [GRAPEVINE]. They are eaten fresh or dried and numerous alcoholic beverages, such as wines and spirits, are derived from the grape. Dried grapes include *raisins, *sultanas and *currants. VITACEAE.

Grapefruit the large, yellow, roundish citrus fruit, borne mostly in clusters on the dense, dome-shaped evergreen trees *Citrus*

× *paradisi* in warm subtropical and tropical countries. It was first noticed in Barbados around 1750, and probably originated from a chance seedling or hybrid of imported *SHAD-DOCK with SWEET *ORANGE. It spread to nearby Florida, where most known varieties subsequently arose.

Total world grapefruit production was 3·7 million tonnes in 1975. The chief growing countries are USA (60% of the world production), Israel, Cyprus and South Africa. The leading cultivated variety is the yellow 'Marsh (or Marsh's) Seedless'. Pigmented varieties include the seedless 'Thompson' ('Pink Marsh') and the seedy 'Foster'.
RUTACEAE.

Grass pea the common name for *Lathyrus sativus*, an herbaceous annual native to southern Europe and western Asia. Known also as VETCHLING, GRASS PEAVINE, CHICKLING PEA or, in India, where it is chiefly cultivated, as KHESARI, it is a cold-weather crop and very tolerant of drought, waterlogging and poor soils. It produces the cheapest pulse available. Although it can be eaten by humans it is more usually grown for animal fodder. There is a suggestion that the consumption of large quantities of the raw peas may cause a paralytic disease of the lower limbs, lathyrism.
LEGUMINOSAE.

Green algae the common name for the algae placed in the division Chlorophyta where the pigment chlorophyll *a* is usually present in much larger amounts than the other chloroplast pigments. Green algae are mainly found in freshwater environments but a few inhabit the margins of the seas. (See also ALGAE, p. 20.)

Greengage a type of plum believed to be derived from hybrids between *Prunus domestica* [EUROPEAN PLUM] and *P. insititia* [DAMSON PLUM] or regarded as a variety (var *italica*) of *P. insititia*. The most commonly grown variety is 'Reine Claude'.
ROSACEAE.

Grapefruit (Citrus × paradisi) *growing in a small plantation in Mokwa, Nigeria.* ($\times \frac{1}{50}$)

The moss Grimmia pulvinata *showing the 16 teeth of the single peristome.* ($\times 2$)

Greenheart the common name for *Ocotea rodiaei* (= *Nectandra rodiaei*) (family Lauraceae), a large tropical forest tree, growing to 20m tall, which is exported in quantity from the Demarara River in Guiana. Its lustrous green to dark olive heartwood is very hard, dense and close-grained. It is thus immensely strong and durable and is highly resistant to attack by marine boring animals. Greenheart timber is used in the construction of marine piers, lock gates, boats and engineering work.

At least two other species are termed GREENHEART, viz. *Cylicodiscus gabunensis* [AFRICAN GREENHEART], of the family Leguminosae, and *Tecoma leucoxylon* [SURINAM GREENHEART], of the Bignoniaceae.

Greenovia a genus of tender leaf succulents from rocks and cliffs of the Canary Islands. All are in cultivation, and are hardy enough to thrive outdoors as far north as the Scilly Isles.
CRASSULACEAE, 4 species.

Green peppers the immature fruits of non-pungent cultivars of *Capsicum annuum*. Popular as salad or cooked vegetables, they are grown in most tropical and warm temperate countries and, under protection, in cool temperate regions.
SOLANACEAE.

Grevillea a large genus of evergreen trees and shrubs found mainly in Australia and Tasmania, extending to New Caledonia and Sulawesi. It has, however, been introduced into warm-temperate and subtropical countries throughout the world as a genus of ornamental value. In all the species the flowers are grouped together into long, usually extremely attractive racemes or panicles.

The best-known and most commonly cultivated species is *G. robusta* [SILK OAK, GOLDEN PINE], a large tree with fern-like pinnate leaves, highly prized for its golden inflorescences. It is also grown as an indoor foliage plant ("GREVILLEA") or as a summer

bedding plant in cooler climates. In addition, the usually red-flowered *G. banksii*, and *G. juniperina*, with pale yellow flowers, often tinged red, are sometimes cultivated.
PROTEACEAE, about 200 species.

Greyia a genus of South African shrubs and small trees. *G. sutherlandii*, a shrub with showy scarlet flowers, is grown as an ornamental in warm climates, sometimes under glass.
GREYIACEAE, 3 species.

Grimmia a large genus of mosses most of which form neat, dark green or gray-green cushions on rocks. They are found all over the world, although sparsely in the tropics where they occur mainly at high altitudes. Several species that are well known in north temperate regions also have very wide cosmopolitan distributions, such as *G. pulvinata*, and *G. trichophylla*.
GRIMMIACEAE, about 240 species.

Grindelia a genus of annual, biennial or perennial, sometimes shrubby herbs from North America and South America, south of the tropics. Several species are cultivated, including *G. chiloensis* (Argentina), *G. glutinosa* (Peru) and *G. squarrosa* [GUMWEED] (North America). The leaves and flower tops of the latter contain resin, saponin, tannin, robustic acid and grindol and have been used to relieve coughs and burns.
COMPOSITAE, about 6 species.

Griselinia a genus of half-hardy evergreen trees and shrubs from New Zealand and South America. The leaves are leathery and

The large rhubarb-like leaves of Gunnera manicata *from Colombia.* ($\times \frac{1}{25}$)

glossy and usually pale olive-green in color. *G. littoralis* and *G. lucida* are vigorous New Zealand evergreen shrubs that have proved very successful in southern Britain and Europe as attractive seaside hedge and screen plants. They are also grown outside in California. The timber of *G. lucida* is used in New Zealand for boat-building.
CORNACEAE, about 6 species.

Guaiacum a genus of tropical American and West Indian trees which have very hard, heavy wood. Two West Indian species (*G. officinale* and *G. sanctum*) are important as sources of *lignum vitae, the hardest of commercial timbers, and guaiacum resin, which has medicinal and other uses. They are planted in the tropics as ornamentals.
ZYGOPHYLLACEAE, about 6 species.

Guarea a genus of trees from tropical America and Africa. The African *G. cedrata* and *G. thompsonii* are trade timbers of West Africa and are much used in furniture-making. Many of the American species are small trees of the undergrowth of rain forest and exhibit the phenomenon of "evergrowing leaves". The leaves have some of the characteristics of shoots: the apex of the leaf is a naked meristem which periodically produces flushes of leaflets. In this way the leaf "bud" builds up a long pinnate leaf over several seasons. The purplish-red wood known as *acajou in the West Indies comes from *G. guara* (= *G. trichilioides*).
MELIACEAE, 35 species.

Guizotia a small genus of annual and perennial herbs native to tropical Africa. One species, *G. abyssinica* [NIGER SEED, NOOG], is a tall annual hairy herb, grown in Ethiopia and India as an oilseed crop. The fruits (cypselas) contain 30–50% of a pale yellow semidrying oil, mainly linoleic acid. Most niger seed oil is used locally for cooking and lighting and does not enter world trade. The protein-rich press-cake is fed to livestock. The seed is used as a food for cage birds.
COMPOSITAE, about 10 species.

Gunnera a largely Southern Hemisphere

A species of Gypsophila from the Mount Ararat region of Turkey, where it forms hard round cushions on the rocky slopes.

genus of small creeping or large, sometimes gigantic, erect herbs, with well-developed rhizomes or aerial stems and often round or kidney-shaped leaves.

The petioles of the large *Gunnera chilensis* (= *G. scabra*) [CHILE RHUBARB] are often eaten and are said to diminish fevers. The stems are used in tanning. The small mat-forming species, such as the South American *G. magellanica* and the New Zealand *G. arenaria*, are often cultivated as ground cover, while some of the large species, with enormous rhubarb-like leaves are grown around garden ponds for their striking appearance – the largest, *G. manicata* from Columbia, has leaves up to 3m across, borne on stalks up to 2.5m high.
HALORAGACEAE, about 47 species.

Gutta-percha a plastic gum obtained from the latex of the leaves or bark of trees mainly of the genus *Palaquium*, especially *P. gutta*, which is native to Indomalaysia. Gutta-percha has been used for insulating submarine and other electric cables, for machine belting and golf balls, and in dentistry, but is now largely superseded by plastics. *Payena leerii* (native to the same region and a member of the same family) produces a gum also called gutta-percha (gutta sundek).
SAPOTACEAE.

Gymnadenia a genus of tuberous, terrestrial, perennial orchids from Greenland, Europe and northern Asia. The flowers, which are usually fragrant, are borne in dense spikes. *Gymnadenia* hybridizes with other genera, eg *Anacamptis* and *Coeloglossum*; it is sometimes included in *Habenaria.
ORCHIDACEAE, about 10 species.

Gymnocalycium a South American genus of low growing globose cacti, many species of which are cultivated as ornamentals for their attractive ribbed, often spiny stems and large showy, often pink flowers.
CACTACEAE, about 60 species.

Gymnocladus a small genus of deciduous trees with one species in eastern North America and two or three species in China. *G. dioica* (= *G. canadensis*) [KENTUCKY COFFEE TREE] grows slowly to more than 30m producing greenish-white flowers. The flat pods contain seeds once used by settlers as a coffee substitute. Sometimes grown as an ornamental, the tree also yields a useful timber. The Chinese *G. chinensis* [SOAP TREE] has lilac-purple unisexual flowers, and smaller pods with lather-producing properties.
LEGUMINOSAE, 3–4 species.

Gymnodinium a genus of dinoflagellates. They have a simple form with a groove around the center of the cell and a short longitudinal furrow. *Gymnodinium* is found both in the sea and in fresh water. The disastrous Florida red tide is caused by *G. breve* (see Red tide).
DINOPHYCEAE.

A flowering shoot of Lignum Vitae (Guaiacum officinale), a species grown both for its attractive flowers and hard timber. (× ½)

Gymnosperms see Conifers and their allies, p. 98.

Gynura a genus of tropical and subtropical evergreen perennial herbs and subshrubs from Africa, Madagascar, Southeast Asia and Malaysia, some of which are cultivated under glass for their attractive foliage and stems. A popular species is *G. aurantiaca* [VELVET PLANT], with its leaves and stems covered with a dense mat of violet-purple hairs.
COMPOSITAE, about 50 species.

Gypsophila a genus of herbs or dwarf subshrubs from temperate Eurasia (a major center of distribution being the eastern Mediterranean), Egypt, Australia and New Zealand. The white or pink flowers, borne in cymose inflorescences, are often abundant and attract many insects. Several species and hybrids, including *G. repens* (= *G. dubia*), *G. paniculata* [BABY'S BREATH, MAIDEN'S BREATH] and *G. elegans* [BABY'S BREATH, CHALK PLANT] are cultivated in a wide range of cultivars for their ornamental value in rock gardens and borders. The latter two species are particularly valued by florists for bouquets.
CARYOPHYLLACEAE, about 125 species.

H

Habenaria a genus of widely distributed temperate and tropical, terrestrial and epiphytic orchids. The terminal inflorescence consists of dense or loose racemes of one to many flowers. The petals have a lip which may be long, lobed or fringed at the tip.

Many of the species are cultivated as ornamentals with predominantly yellowish or greenish white flowers. Among the better known temperate species are the Old World *H. bifolia* [LESSER BUTTERFLY ORCHID], with white or greenish white fragrant flowers, and the North American *H. ciliaris* (= *Blephariglottis ciliaris*) [YELLOW FRINGED ORCHID, ORANGE-PLUM, ORANGE-FRINGE], with racemes of bright yellow or orange flowers, and *H. clavellata* (= *H. tridenta*, *Gymnadeniopsis clarellata*) [GREEN WOODLAND ORCHID, GREEN REIN ORCHID, LITTLE CLUB-SPUR ORCHID], with greenish-white or yellowish-white flowers with a club-shaped spur, and *H. peramoena* [PURPLE FRINGELESS ORCHID, PURPLE-SPIRE ORCHID, FRET-LIP, PRIDE-OF-THE-PEAK], with showy rich purple flowers with a three-lobed lip and a club-shaped spur.
ORCHIDACEAE, about 100 species.

Hacquetia a genus consisting of a single species, *H. epipactis*, a small yellow-flowered herbaceous perennial of woodlands from the eastern European Alps to the Carpathians, which is sometimes cultivated in temperate

One of the Blood Lilies, Haemanthus multiflorus from tropical Africa, which bears a spherical head of red flowers. (× ¼)

rock gardens for its spring-time blooms.
UMBELLIFERAE, 1 species.

Haemanthus [BLOOD LILY] a genus of tropical and South African evergreen and deciduous bulbous plants. The dense umbellate heads of star-shaped, red or white flowers are borne on a solid stalk and enclosed by colorful fleshy bracts. The fruits are colorful berries. *H. albiflos*, *H. coccineus* and *H. katherinae*, all of South African origin, and other species, are cultivated for their ornamental value.
AMARYLLIDACEAE, about 90 species.

Haematoxylon a small genus of trees and shrubs ranging from Central America and the West Indies to southwestern Africa. Commercially the most important is *H. campechianum* [*LOGWOOD, PALO CAMPECHIO], a small spiny spreading shrub or tree valued for its very hard heartwood and the red dye haematoxylin which it yields. It is sometimes cultivated as an ornamental.
LEGUMINOSAE, 3 species.

Hakea a genus of Australian and Tasmanian erect evergreen shrubs and small trees. Several species are grown as ornamental shrubs in warmer areas. *H. laurina* [PINCUSHION FLOWER, SEA URCHIN], cultivated especially on the French and Italian Rivieras, has red or pink blossoms in globular heads. The leaves are broad and the young tips are a silky golden bronze, turning red in the autumn. *H. acicularis* (= *H. sericea*) [NEEDLE BUSH] is a tall silky shrub with clusters of white, pink-tinged or rich pink flowers. *H. subera* [CORK-BARK TREE] grows in arid central Australia where its contorted form is a conspicuous feature of the landscape. The large flowers are in torch-like spikes, either cream or yellow, sweetly scented and loaded with a honey which is much relished by the Aborigines.
PROTEACEAE, about 120 species.

Halesia a small genus of deciduous trees native to North America and China. The showy white drooping flowers hang like snowdrops in profuse clusters in May. Best-known are *H. carolina* (= *H. tetraptera*) [SNOWDROP TREE, SILVER BELL, OPPOSSUM-WOOD] from the southeastern USA, a beautiful small spreading tree which grows well in cultivation in sheltered, limefree sites, and the much taller pyramidal *H. monticola* [MOUNTAIN SILVER BELL, SNOWDROP TREE], from the mountains of the same area.
STYRACACEAE, 6 species.

Halimeda a genus of green algae found in tropical seas. The thallus is calcified, the breakdown of which contributes large quantities of calcareous debris to lagoons in coral atolls and along sandy tropical shores.
CHLOROPHYCEAE.

Halimione a small genus of plants adapted to saline habitats. One of the three species, *H. pedunculata* [STALKED ORACHE] is annual,

while *H. verrucifera* and *H. portulacoides* [SEA PURSLANE] are small shrubs. They are all found in salt marshes and estuaries mainly around Europe, but also in Central and southwest Asia.
CHENOPODIACEAE, 3 species.

Halosphaera a genus of green algae best known as the large (up to 1mm or 2mm) spherical cysts which form one phase of the planktonic organism. It is often found in abundance floating on the surface of the sea.
PRASINOPHYCEAE.

Hamamelis [WITCH HAZELS] a genus of deciduous shrubs or small trees which are native to East Asia and eastern North America and are popular with gardeners for their yellow or rusty-red flowers produced from October to March. The resemblance of the leaves to *HAZELS (*Corylus* species) led early settlers in North America to use the twigs for water divining, and because of the branches' pliant properties, the popular name arose. *H. virginiana* [COMMON WITCH HAZEL] produces yellow flowers in the autumn that are particularly resistant to cold weather. The bark, leaves and twigs yield the witch hazel of pharmacy which is widely used as an astringent and coolant, and can be applied to cuts and bruises. Other cultivated species and hybrids are *H. japonica* [JAPANESE WITCH HAZEL], flowering in March, *H. mollis* [CHINESE WITCH HAZEL], *H. vernalis* [OZARK WITCH HAZEL] and the hybrid *H. × intermedia*, the last three flowering in January and February.

Witch Hazel is a name used also for the WITCH *ELM (*Ulmus glabra*) and WITCH *HORNBEAM (*Carpinus betulus*), the term "witch" or "wych" being an old English word used to denote any tree with particularly pliant branches.
HAMAMELIDACEAE, 4–6 species.

Hamelia a small genus of evergreen shrubs from tropical and subtropical America and the West Indies. The cultivated species include *H. patens* (= *H. erecta*) [SCARLET BUSH], with small tubular orange to scarlet flowers and *H. ventricosa* with tubular yellow flowers.
RUBIACEAE, about 25 species.

Like most members of this genus, Hakea teretifolia has narrow leaves and axillary clusters of scented white to cream flowers. (× ¼)

Hancornia a Brazilian genus represented by a single species, *H. speciosa*, [MANGABEIRA RUBBER]. This tree, which may grow to 6m, bears yellow and red fruits, the size of a plum, which are eaten fresh or processed to make conserves and wine. The rubber is resinous and of secondary commercial quality on the commercial market. The wood is used locally for building.
APOCYNACEAE, 1 species.

Haplopappus a large American genus of annual and perennial herbs and shrubs widely distributed in North and South America. Most species possess either tufted or solitary inflorescences of yellow (rarely purple or saffron) daisy-like heads. Several shrubby species, including *H. croceus*, are cultivated as ornamentals in the USA.
COMPOSITAE, about 160 species.

Hardenbergia a small Australian and Tasmanian genus of evergreen twining shrubs. *H. comptoniana* and *H. violacea* are cultivated as ornamentals for their racemes of predominantly violet-blue or violet-purple to rose or white pea-like flowers.
LEGUMINOSAE, 3 species.

Haricot bean one of the common names given to the important legume *Phaseolus vulgaris*. It is a low, erect or climbing annual, native to Central and South America but now cultivated throughout temperate and tropical regions.

The varieties can be grouped by form (bush or climbers), seed colors (white, black, red, ocher or brown), or according to the use to which the fruit or bean is put and to age of harvesting. Thus FRENCH, DWARF, SNAP, GREEN or WAX, STRING and RUNNER BEANS are harvested for the entire young and tender pods which are eaten whole, sliced or snapped; FLAGEOLETS, GREEN SHELL BEANS or KIDNEY BEANS are picked for the seeds (beans) while they are still immature and therefore tender; and HARICOT or DRY SHELL BEANS are harvested when mature, and subsequently dried, when they can be stored over a long period. The latter are the basic ingredient of canned BAKED BEANS. Clearly, a single plant can yield products at all three stages, but over the years numerous cultivars have been developed which provide the best product for each need. The confusion in naming is compounded by differing national interpretations of the same names.
LEGUMINOSAE.

Haworthia a genus of more or less stemless leaf-succulent rosette plants from the drier parts of South Africa and Namibia. Many species are in cultivation for their attractive leaves, the surface of which has an almost infinite variety of textures and patterning, including pearl-like tubercles and incrustations, serrulate or finely haired margins and translucent "windows" which may enlarge to occupy the whole area of the truncate leaf tip. By contrast, the flowers are uniform, white or whitish, tubular and unattractive.

Haworthia is found naturally in the shade of shrubs and rocks and is hence suitable for shadier parts of the greenhouse. It can even be grown on sunless windowsills, provided that frost is excluded. The most commonly cultivated species are *H. attenuata*, *H. fasciata* [ZEBRA HAWORTHIA] and *H. margaritifera* [PEARL PLANT] each of which has several varieties and cultivars.

The most bizarre species are *H. maughanii* and *H. truncata*, with truncate windowed leaves in a spiral or two rows, and *H. graminifolia*, a rarity with linear grass-like leaves up to 50cm long arising from a small bulb.
LILIACEAE, about 68 species.

Flowers of the May Hawthorn (Crataegus laevigata). Shown here are those of the cultivar 'Paulii'. Hawthorns are popular ornamentals since they provide fine displays of color both when in bloom and when fruiting. (×3)

The Chinese Witch Hazel (Hamamelis mollis) is the most popular of the cultivated witch hazels and gives a welcome splash of color in winter.

Hawthorn the common name for members of *Crataegus*, a genus of the Rosaceae comprising 200 species of deciduous, usually thorny trees also known as MAYS or QUICKTHORNS. They are distributed throughout the north temperate region. Hawthorns have fragrant blossom and are often grown as ornamentals or, being thorny, make an effective hedging material. Some species produce useful timber although this is exploited less often now than formerly. The flowers are usually white, sometimes pink or red, and the fruits (haws) are blue-black, red or yellow.

Identification of individuals of the genus is difficult, partly due to extensive hybridization and partly because the constituent species do not fall into readily distinguishable groups. However, the color of the fruits is a particularly important character, the main divisions being bluish or blackish (such as *C. chlorosarca*), red (such as *C. mollis*) or yellow (such as *C. flava*; SUMMER HAW). The commonest European species are *C. laevigata* (= *C. oxyacanthoides*, *C. oxyacantha*) [MAY HAWTHORN] and *C. monogyna* [COMMON HAWTHORN].

The wood of *Crataegus* species is very hard and has been used for engraving. *C. crus-galli* [COCKSPUR THORN], of the USA, produces a heavy wood used for tool handles and that of the European *C. laevigata* is used for a variety of articles from wheels to walking sticks. Many species are popular ornamental trees, grown for their blossom and their fruit. Best-known is the PINK MAY, a variety of *C. laevigata*. The fruits of several species are made into jellies or preserves and those of *C. cuneata* are used in China for the treatment of stomach complaints. *C. laevigata* provides a leaf-infusion tea which reduces blood pressure, a coffee substitute from the seeds and a tobacco substitute from the leaves.

Another member of the same family is

Photinia serrulata [CHINESE HAWTHORN] while the unrelated *Aponogeton distachyus* is known as the WATER HAWTHORN.

Hazel the common name for members of *Corylus*, a genus of hardy decorative and economically valuable deciduous trees and shrubs from the temperate Northern Hemisphere, including Europe, Asia and North America.

The European *C. avellana* [COMMON HAZEL, COBNUT, FILBERT] is a vigorous shrub or small tree up to 7m tall. A formerly important economic species grown in coppices, it supplied wood for making hurdles and walking sticks, and also nuts for autumn cropping. The attractive male catkins appear in early

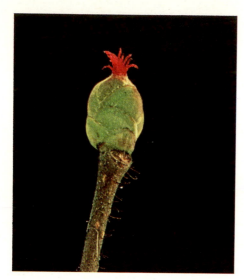

Bud-like inflorescence of the Common Hazel (Corylus avellana*) showing the feathery red styles which catch the pollen. (×10)*

spring. It is still grown commercially, but the true filbert is *C. maxima* [FILBERT], which is grown in four main areas: Turkey, which produces 65% of the commercial crop, Italy (20%), Spain (10%), and Oregon, USA (5%). It is also grown in southeast England, where the chief commercial variety is known as 'Kentish Cob'.

In North America the principal native cultivated species are *C. americana* [COMMON HAZEL] and *C. cornuta* (= *C. rostrata*) [BEAKED HAZEL]; *C. avellana* has also been introduced from Europe.

Also cultivated, both for their nuts and their ornamental value, are *C. chinensis* [CHINESE HAZEL], *C. colurna* [TURKISH HAZEL] and *C. tibetica* [TIBETAN HAZEL]. Ornamental cultivars include *C. avellana* 'Aurea', with soft yellow leaves, *C. avellana* 'Contorta' [CORKSCREW HAZEL], with strangely twisted branches, and *C. maxima* 'Purpurea', with purple leaves.
BETULACEAE.

Heath the common name given to many species of evergreen shrubs and occasionally dwarf trees which belong to the genus *Erica*. ERICACEAE.

Heather a term which in the strictest sense should apply only to the genus *Calluna*, eg *C. vulgaris* [HEATHER, SCOTTISH HEATHER, LING], while the term "HEATH" is applicable only to members of the genus *Erica*. However, at least eight species of the genus *Erica* are commonly known as HEATHERS: *E. cinerea* [BELL or PURPLE HEATHER], *E. tetralix* [BOG HEATHER or CROSS-LEAVED HEATH], *E. vagans* [CORNISH HEATH or HEATHER], *E. carnea* [SNOW HEATHER], *E. canaliculata* [CHRISTMAS HEATHER], *E. doliiformis* [EVER-BLOOMING FRENCH HEATHER], *E. hyemalis* [WHITE WINTER HEATHER] and *E. mediterranea* [MEDITERRANEAN HEATHER]. (See also *Cassiope*.)
ERICACEAE.

Hebe a genus of evergreen shrubs and small trees with opposite, four-ranked, often imbricate leaves, leathery in texture, mainly native to New Zealand, but with a few species in South America and New Guinea. Many species are cultivated in the open in warmer climates, otherwise under glass, for their attractive habit and many-flowered racemes of shortly tubular four-lobed, white, pink, reddish or bluish flowers. Members of this genus were at one time included in *Veronica* and it is also related to *Parahebe*.

Cultivated species include *H. lewisii* (= *Veronica lewisii*), an erect shrub with racemes of pale blue flowers, *H. menziesii* (= *V. menziesii*), a larger compact shrub (up to 4m), with racemes of white or pale lilac flowers, and the popular *H. speciosa* (= *V. speciosa*), a stout shrub growing to 1.5m with long racemes of reddish or bluish-purple flowers. The latter contains many varieties and cultivars (eg 'Imperialis') and hybridizes readily with other species. SCROPHULARIACEAE, about 80 species.

Hedera see IVY.
ARALIACEAE, about 15 species.

Hebe species are widely grown, particularly in mild climates, for their attractive flowers and evergreen foliage.

Most Erica *species are known as heaths, but E. cinerea, a native of dry moors in Europe, is known as Bell Heather. (×2)*

Hedychium a genus of erect rhizomatous perennial herbs native to tropical Asia and the Himalayas but widely cultivated in tropical countries and under glass in temperate regions. The cultivated species normally bear attractive terminal inflorescences of fragrant showy flowers with reflexed petals, petal-like lateral staminodes and a long stamen filament.

Popular species include *H. coronarium* [GARLAND FLOWER, BUTTERFLY GINGER, WHITE GINGER, BUTTERFLY LILY, GINGER LILY, CINNAMON JASMINE], a tropical Asian species with large fragrant white flowers, which is widely naturalized in tropical America, and the Indian *H. gardneranum* [KAHILI GINGER] which bears long spikes of light yellow flowers with protruding red filaments. The

rhizome of the Indian *H. spicatum* is the source of the perfume ingredient *abir.
ZINGIBERACEAE, about 50 species.

Hedysarum a genus of biennial or perennial herbs or subshrubs with pea-like flowers, from the Northern Hemisphere. Several species are cultivated in rock gardens, including *H. glomeratum* (= *H. capitatum*), *H. mackenzii* [LICORICE ROOT] and *H. hedysaroides* (= *H. obscurum*), all with pink to reddish-purple (rarely white) flowers. *H. coronarium* [FRENCH HONEYSUCKLE] is taller, with fragrant flowers, and is grown as a border ornamental. It is also grown agriculturally as green fodder [SULLA CLOVER].
LEGUMINOSAE, about 150 species.

Helenium a genus of annual or perennial herbs with usually daisy-like flower heads, from North America and Mexico. Species such as the North American *H. amarum* [BITTERWEED] are a pest on poor pasture as they are toxic to sheep, horses and cattle, imparting a bitter flavor to the milk. *H. autumnale*, also from North America, is a popular garden perennial, as it blooms for many weeks in late summer. Many varieties exist, including 'Bishop', which has orange-yellow semidouble flower heads with a dark brown center.
COMPOSITAE, about 40 species.

Heliamphora a tropical genus of Venezuelan and Guianan pitcher plants growing in very wet soil and air conditions. The funnel-shaped leaf 'pitcher' with a minute lid contains, in two apical, narrow wings, nectar-secreting glands to attract insects. All species bear white, rose-tinged flowers in racemes.
SARRACENIACEAE, 6 species.

Helianthemum a genus of usually ever-green shrubs, rarely herbs and sometimes annuals. They are native to North and South America as well as the Old World, notably the Mediterranean region and Iran. The American species are sometimes placed in a separate genus, *Crocanthemum*.

Although the dried leaves of the North American *H. canadense* (= *C. canadense*) [FROSTWEED] are used as a tonic and to stimulate the appetite, the main use of the genus is horticultural, and several species are popular rock-garden plants. The best-known of these is *H. nummularium* (= *H. chamaecistus*, *H. variabile*) [ROCK-ROSE, SUN-ROSE], which has yellow flowers but numerous subspecies (eg *grandiflorum*) and cultivars (eg 'Aureum', 'Roseum' and 'Rubro-plenum') exist with white, pink and scarlet flowers. Two other species with yellow flowers are *H. lunulatum* and *H. oelandicum*, while *H. apenninum* (= *H. polifolium*, *H. pulverulentum*) [WHITE ROCK-ROSE] has white flowers and gray leaves.
CISTACEAE, about 100 species.

Helianthus see ARTICHOKE and SUNFLOWER.
COMPOSITAE, about 67 species.

The Strawflower (Helichrysum bracteatum); the outer rays surrounding the disk of bisexual florets are female. (× 1)

Helichrysum a genus of annual to perennial herbs or shrubs ranging from southern Europe, Africa and Madagascar to southwest Asia, including southern India, Sri Lanka, Australia and New Zealand, with the center of distribution in South Africa. Many species are cultivated as ornamentals for their attractive, often gray or white, woolly foliage and for their strong colorful daisy-like flower heads, sometimes used as everlastings. Many cultivated species are half-hardy or green-house plants but some are quite hardy in temperate gardens.

The Australian annual *H. bracteatum* [STRAWFLOWER] and its various horticultural varieties have red, orange, yellow, pink or white petal-like bracts. Among the yellow-bracted perennials are European species such as *H. arenarium*, *H. angustifolium* [WHITE-LEAF], *H. orientale* and *H. stoechas* and South African species such as *H. splendidum*. White-bracted species suitable for rock gardens include the Corsican *H. frigidum* with a silvery mat of leaves.

Several species have aromatic foliage, some smelling strongly of curry or fenugreek,

One of the many garden varieties of Helianthemum nummularium, *the best-known of the Sun-rose or Rock-rose species.* (× ½)

eg *H. angustifolium* and *H. italicum*. A few species have culinary uses, including *H. cochinchinense* from China, which provides tasty young leaves for flavoring rice dishes, and *H. serpyllifolium*, the leaves of which are used for tea in South Africa. The European species *H. arenarium* and *H. stoechas* have been used to combat intestinal worms and skin diseases, and as diuretics.
COMPOSITAE, about 500 species.

Heliconia a genus of perennial banana-like herbs mainly in tropical America, with a few species in the islands of the western Pacific. They are sometimes placed in the Musaceae or the Strelitzeaceae. Like *BANANA, *Heliconia* possesses a many-flowered terminal inflorescence with colored bracts. Several species, including *H. bihai* [PARROT'S CLAWS, WILD PLANTAIN, BALISIER] and *H. psittacorum* [PARROT'S FLOWER], are cultivated in the tropics for their ornamental foliage, less frequently in greenhouses in temperate countries.
HELICONIACEAE, about 50 species.

Helictotrichon [OAT GRASS] a genus of mainly perennial grasses forming dense tufts. They are found mainly in northern and central Europe, Asia and North America, generally in chalky grassland. *H. sempervirens* (= *Avena candida*) is an attractive ornamental species with stiff, steely-blue foliage and graceful arching flowering stems up to 1m tall.
GRAMINEAE, about 95 species.

Heliophila a genus of mainly annual or rarely subshrubby perennial herbs native to South Africa. Some species are cultivated as perennial or half-hardy annual border plants, including *H. longifolia*, which bears racemes of blue flowers with a yellow or white claw to the petal, and *H. leptophylla* with much narrower leaves and racemes of blue flowers with a yellow center. CRUCIFERAE, about 85 species.

Heliopsis an American genus of annual and perennial sunflower-like herbs, some of which attain almost 4m in height. The large yellow, orange or golden flower heads are borne terminally on long erect stalks. Many varieties of *H. helianthoides* are known in cultivation as border plants, including cultivars with "double" ray flowers. COMPOSITAE, about 12 species.

Heliotropium [HELIOTROPES] a large genus of annual herbs, subshrubs and shrubs from tropical and temperate regions. The blue, purple, pink or white flowers are clustered in one-sided, sometimes axillary spikes. Many varieties of *H. arborescens* (= *H. peruvianum*) [COMMON HELIOTROPE, CHERRY PIE], from Peru, *H. amplexicaule* (= *H. anchusifolium*) [SUMMER HELIOTROPE], and *H. corymbosum* are grown for their clusters of fragrant, often forget-me-not-like flowers. *H. arborescens* is used in perfumery in southern Europe. BORAGINACEAE, about 250 species.

Helipterum a genus of mostly xerophytic herbs and subshrubs from South Africa and Australia whose daisy-like flower heads, like those of the related *Helichrysum*, often have petal-like involucral bracts which retain their color when dried and are sold as everlastings for flower arrangements. Popular cultivated species include *H. humboldtianum*, with a dense inflorescence of yellow flower heads, and *H. manglesii* (= *Rhodanthe manglesii*), with pink to reddish flower heads and a yellow center. COMPOSITAE, about 60 species.

Helleborine a name that is applied generally to several species of some genera of European terrestrial orchids. It is more com-

*The Green Hellebore (*Helleborus viridis*) produces a mass of greenish-yellow flowers during the early spring. (× ⅓)*

Inflorescence of Heliconia pendula *from Guatemala to Peru. Several* Heliconia *species are cultivated for their attractive flowers. (× 2)*

monly associated with species of *Epipactis* and *Cephalanthera*. This is due to the superficial resemblance of their flowers to those of the genus *Helleborus* (Ranunculaceae). ORCHIDACEAE.

Helleborus a very complex genus of perennial herbs, many of which are cultivated as garden-border plants, native from Europe to western Asia, with two outlying species in Tibet and China.

The flowers of the best-known species *H. niger* [*CHRISTMAS ROSE] are large and very showy, white or cream in color. In common with other species it flowers in early spring or in winter. *H. foetidus* [STINKING HELLEBORE, BEAR'S FOOT] has many small, pale green flowers, becoming edged with purple-red on aging. *H. orientalis* [LENTEN ROSE] has several cultivars with a range of flower color from creamy white and green to pink or purple and often spotted. The dried roots and rhizomes of *H. niger* yield a cardiac glycoside, helleborein. RANUNCULACEAE, about 20 species.

Helvella [SADDLE FUNGI] a genus of mainly north temperate saprophytic terrestrial fungi with fleshy fruiting bodies consisting of a hollow stipe and a saddle-shaped, reflexed cap. The best-known species is *H. crispa*, which forms white fruiting bodies in deciduous woods. HELVELLACEAE, about 25 species.

Helwingia a small genus of deciduous shrubs from the Himalayas eastwards to Taiwan and Japan. *H. japonica* is occasionally cultivated as a curiosity, but the flowers are not attractive. The leaves of this species have been eaten as a vegetable in Japan. CORNACEAE, 3 species.

Hemerocallis a genus of rhizomatous perennial lily-like herbs distributed in Europe and temperate Asia from Italy to the Far East, although the majority of species are from China and Japan. The flowers, which usually last only a day, are funnel- or trumpet-shaped, yellow, orange, red or purple in color. The modern cultivated hybrids have been produced as a result of crosses between several different species, notably *H. citrina*, *H. fulva*, [ORANGE DAYLILY, TAWNY DAYLILY, FULVOUS DAYLILY], *H. lilioasphodelus* (= *H. flava*) [YELLOW DAYLILY, LEMON DAYLILY] and *H. minor* [DWARF YELLOW DAYLILY]. The fragrant flower buds of *H. middendorffii* are eaten as a vegetable in Japan. LILIACEAE, about 15 species.

Hemileia a genus of rust fungi which occur on various tropical and subtropical hosts. The most important is *H. vastatrix*, the cause of coffee rust, which was probably native to East Africa but spread to Sri Lanka in the 1870s, where it wiped out the coffee industry. It has also spread to all coffee-growing parts of Africa and since 1965 has become widespread in Brazil and Central America. PUCCINIACEAE, about 35 species.

Hemlock the common name for the poisonous umbelliferous herb *Conium maculatum*, a tall biennial native to temperate Eurasia, though now widely naturalized in North and South America and in New Zealand. It is unique in its family in containing alkaloids that are extremely toxic to both Man and livestock, and was used by the ancient Greeks

*Hemlock (*Conium maculatum*) is notoriously poisonous and dangerous to both livestock and to Man. (× $\frac{1}{25}$)*

Like other hellebores, the Lenten Lily or Lenten Rose (Helleborus orientalis) produces its flowers in the early spring. (× ½)

to put criminals to death. The unripe fruit has been used in medicine as an anodyne, sedative and antispasmodic.

A species of *YEW, *Taxus canadensis*, is known as GROUND HEMLOCK.

Hemlock spruce or **hemlock** the common names for *Tsuga*, a genus of evergreen conifers native to North America, Japan, China, Taiwan and the Himalayas. In habit they are broadly pyramidal, the branches being horizontal to somewhat pendulous. Hemlocks produce valuable timber and pulpwood as well as ornamental trees.

The wood of *T. canadensis* [COMMON or CANADA HEMLOCK] and of the western North American species *T. heterophylla* [WESTERN HEMLOCK] is used for building construction generally and the bark for tanning. The resin of the former is also useful and is known commercially as Canada pitch. In China the soft wood of *T. chinensis* [CHINESE HEMLOCK] is used for shingles as is that of *T. brunoniana* [INDIAN HEMLOCK SPRUCE] in the Himalayas. The timber of *T. sieboldii* [JAPANESE HEMLOCK] is fine-grained and has been used in cabinet-making.

A number of species are cultivated as ornamentals or as hedging plants, notably COMMON HEMLOCK (including many garden forms and cultivars), WESTERN HEMLOCK, *T. mertensiana* [MOUNTAIN HEMLOCK] and *T. caroliniana* [CAROLINA HEMLOCK].
PINACEAE, about 12 species.

Hemp a tall-growing slender annual herb, *Cannabis sativa*, widely cultivated for its fibers throughout the warm temperate and subtropical countries of the world, but principally in the USSR (the leading producer), southern Europe and the eastern USA. The long, stout, strongly lignified bast-fibers are used in the manufacture of coarse fabrics. The term hemp is also used for fiber-

producing species of other genera, including *Musa textilis* [*ABACA FIBER, MANILA HEMP], *Crotalaria juncea* [SUNN HEMP], *Sansevieria* species [BOW-STRING HEMP], *Hibiscus cannabinus* [DECCAN HEMP], *Furcraea foetida* var *willemettiana* [MAURITIUS HEMP] and *Agave sisalana* [SISAL HEMP].

Henna the dye obtained from the shrub *Lawsonia inermis* (= *L. alba*) [MIGNONETTE TREE] native to the Old World tropics, cultivated since ancient times, and naturalized in tropical America. The dye is derived from the dried, ground leaves and young shoots and is used in many countries to dye various parts of the body, including the hair, an orange-brown color. It is a very fast dye for fabrics and leather. Henna is also used in local medicine for many disorders, and the fragrant flowers have been used in perfumery.
LYTHRACEAE.

Hepatica [LIVERLEAF] a genus of spring-flowering, low-growing, perennial herbs from the north temperate zone with long, petiolate, three- to five-lobed leaves. They are closely related to *Anemone in which they were formerly included. *H. nobilis* (= *Anemone hepatica*), with bluish-violet, purple, pinkish or white flowers, is often cultivated in rock gardens or shaded areas. Double forms are also known. *H. americana* and *H. acutiloba* (= *Anemone acutiloba*), from North America, are closely related species.
RANUNCULACEAE, about 10 species.

Heracleum a genus of coarse biennial or perennial umbelliferous herbs from north temperate regions, mainly in the Old World, with a few species in North America and tropical African mountains. *H. sphondylium* [COMMON COW PARSNIP, HOGWEED, KECK] is common in grassy places and roadsides in Europe, Asia and North America. *H. mantegazzianum* [GIANT HOGWEED], from the Caucasus, grows up to 5m tall, with a hollow stem which is purple spotted and up to 10cm in diameter. It is sometimes culti-

vated for effect and is locally naturalized. The huge inflorescences can be up to 1.2m in diameter.
UMBELLIFERAE, about 70 species.

Herbs see p. 168.

Herniaria a small genus of prostrate or mat-forming annual or perennial herbs from Europe and the Mediterranean region to Afghanistan and South Africa. Selected strains of *H. glabra* [RUPTUREWORT, HERNIARY BREASTWORT] have been used since the 19th century as the green carpeting component of formal bedding schemes in parks and gardens. It is also cultivated in rock gardens. RUPTUREWORT has been used medicinally in southern Europe as a diuretic and an antispasmodic. The only other cultivated species is *H. alpina*.
CARYOPHYLLACEAE, about 15 species.

Hesperis a genus of tall biennial or perennial herbs from Europe and the Mediterranean to Asia. Tall stems carry spikes of white, lilac or purple flowers with a delicate fragrance, especially in the evening. *H. matronalis* [DAME'S VIOLET, DAMASK VIOLET, SWEET ROCKET] is a perennial plant (sometimes biennial) long cultivated in cottage gardens. *H. tristis* and *H. violacea* are two cultivated biennial species. Oil from the seeds (damask oil) is used in perfumery.
CRUCIFERAE, about 30 species.

Heuchera a North American and Mexican genus of smallish perennial herbs with small pendent flowers in lax, delicate panicles or racemes. *H. sanguinea* [CORAL FLOWER, CORAL BELLS, ALUM ROOT], with red or white flowers, is the most widely cultivated species in borders. Other cultivated species or hybrids include *H. americana* [ROCK GERANIUM, ALUM ROOT], *H. cylindrica*, *H. micrantha*, *H. villosa* and *H. × brizoides*.
SAXIFRAGACEAE, about 40–50 species.

Various cultivars of Day Lilies are grown throughout the world. This is an orange cultivar of Hemerocallis lilioasphodelus. (× 1)

Herbs

The term herb refers, strictly speaking, to plants that die down after flowering and do not, therefore, have persistent aboveground woody parts. In a more restricted sense it is also applied to plants used for the aromatic, savory or medicinal properties of their stems, leaves or occasionally flowers. They may be used fresh or dried. In this sense it covers potherbs, culinary herbs, condiment herbs and medicinal herbs.

Herbs and spices have been used since earliest times to add variety and flavor to foodstuffs. Plants used in this way are normally pungent or aromatic and are used sparingly since they are often unpalatable if consumed directly or used in large quantities for flavoring.

The majority of the widely used herbs grow in warm regions with dry summers, especially in the Mediterranean area where aromatic herbs such as *THYME (*Thymus* species), ROSEMARY (*Rosmarinus officinalis*), *SAGE (*Salvia officinalis*) and SAVORY (*Satureja* species) are common components of the characteristic scrub vegetation that covers large areas of the Mediterranean where the forest cover has been removed. The cooks in the ancient Mediterranean civilizations of Rome and Greece must have learned by trial and error which plants possessed desirable properties for adding good flavor to their

dishes. Most of the herbs used were found growing wild locally. These Mediterranean herbs were in due course handed on to other cultures as the Greek and Roman empires extended. Many of the herbs grown in the kitchen and herb gardens of monasteries, and later in the early medicinal or botanic gardens were introduced by the Romans and are still grown today in our herb gardens. Even today, many of the commonly used herbs are grown commercially in Mediterranean countries such as France, Spain and Italy, where they are dried and exported to other parts of Europe and elsewhere.

In a similar way, herbs were introduced from Europe to North America and other parts of the world by colonizers and these continue to be cultivated today, sometimes in preference to local species with similar properties which otherwise would have replaced them. Other parts of the world have contributed some herbs, such as *GARLIC (*Allium sativum*), from Central Asia, *TARRAGON (*Artemisia dracunculus*) and LEMON GRASS (*Cymbopogon citratus*), from Southeast Asia and now widely grown in the United States, South America and Africa, LEMON *VERBENA (*Aloysia triphylla*), from the Argentine and Chile and now widely cultivated in the tropics,

and *BASIL (*Ocimum basilicum*), from the Old World tropics and widely grown for seasoning soups and sauces such as the classic Provençal "pistou".

Most of the culinary herbs belong to the families Labiatae and Umbelliferae. Labiate herbs include *BASIL (*Ocimum basilicum*), *MARJORAM (*Origanum* species), *MINT (*Mentha* species), SAGE, ROSEMARY and THYME. Their flavor and aroma are caused by the volatile essential oils present in the leaves and other parts of the plant. THYME, derived from several species of *Thymus*, notably *T. vulgaris*, is perhaps the most widely used of all culinary herbs and is used to flavor meats, stews, soups, sauces and stuffings. It is a component of "bouquet garni" and "mixed herbs". The mints (*Mentha* species) are amongst the oldest of European herbs and are native throughout Europe and widely cultivated in other temperate zones.

Umbelliferous herbs are aromatic due to the presence, throughout the plant and fruits, of secretory canals containing aromatic essential oils. The best known is *PARSLEY (*Petroselinum crispum*) which is widely used in the fresh state as a garnish rather than actually consumed, although it is extensively used in cooking as a mild flavoring. *FENNEL (*Foeniculum vulgare*), especially its finely dissected leaves, is used as a flavoring especially in fish dishes and is widely employed in Scandinavia. In

Some of the commonly cultivated herbs. 1 Garden Thyme; 2 Sweet Marjoram; 3 Parsley; 4 Coriander; 5 Rosemary; 6 Chervil; 7 Sage; 8 Summer Savory; 9 Fennel; 10 Tarragon; 11 Peppermint; 12 Sweet Basil; 13 Dill; 14 Tansy; 15 Mint, Spearmint; 16 Lemon Balm; 17 Lovage; 18 Costmary 1 to 11, 14 ($\times \frac{1}{2}$); 12, 13, 15 to 18 ($\times \frac{1}{3}$).

COMMONLY CULTIVATED HERBS

Common name	Scientific name	Family	Common name	Scientific name	Family
ALEXANDERS	*Smyrnium olusatrum	Umbelliferae	*MARJORAM; WILD OREGANO	Origanum vulgare	Labiatae
*ANGELICA	Angelica archangelica	Umbelliferae	*MARJORAM, SWEET	O. majorana	Labiatae
WILD ANGELICA	A. sylvestris	Umbelliferae	*MARJORAM, POT	O. onites	Labiatae
*ANISE	Pimpinella anisum	Umbelliferae	MELILOT, SWEET CLOVER	*Melilotus officinalis	Leguminosae
*BALM	Melissa officinalis	Labiatae	*MINT, LEMON, EAU DE COLOGNE OR BERGAMOT	Mentha × piperita var citrata	Labiatae
*BASIL, SWEET	Ocimum basilicum	Labiatae	*MINT, GINGER OR SCOTCH	M. × gentilis	Labiatae
BAY, SWEET BAY	*Laurus nobilis	Lauraceae	*MINT, HORSE	M. longifolia	Labiatae
*BERGAMOT, WILD	Monarda fistulosa	Labiatae	*MINT, APPLE	M. suaveolens	Labiatae
BERGAMOT, RED; OSWEGO TEA; BEE BALM	*M. didyma	Labiatae	*MINT, WATER	M. aquatica	Labiatae
BERGAMOT, LEMON	M. citriodora	Labiatae	*PARSLEY	Petroselinum crispum	Umbelliferae
BETONY	Betonica (*Stachys) officinalis	Labiatae	PENNYROYAL	*Mentha pulegium	Labiatae
BORAGE	*Borago officinalis	Boraginaceae	*PEPPERMINT	Mentha × piperita	Labiatae
BURNET, SALAD	*Sanguisorba minor	Rosaceae	*ROCKET	Eruca sativa	Cruciferae
CALAMINT	*Calamintha sylvatica	Labiatae	ROCKET, WALL	Diplotaxis muralis	Cruciferae
CAMOMILE; *CHAMOMILE	Chamaemelum nobile	Compositae	ROSEMARY	*Rosmarinus officinalis	Labiatae
CARAWAY	*Carum carvi	Umbelliferae	*RUE	Ruta graveolens	Rutaceae
CATMINT	*Nepeta cataria	Labiatae	*SAGE	Salvia officinalis	Labiatae
*CELERY	Apium graveolens	Umbelliferae	SAVORY, SUMMER	*Satureja hortensis	Labiatae
CHERVIL	*Anthriscus cerefolium	Umbelliferae	SAVORY, WINTER	S. montana	Labiatae
*CHIVES	Allium schoenoprasum	Liliaceae	SORREL, COMMON	*Rumex acetosa	Polygonaceae
*CORIANDER	Coriandrum sativum	Umbelliferae	SORREL, GARDEN	R. scutatus	Polygonaceae
*COSTMARY; ALECOST	Balsamita major	Compositae	SORREL, SHEEP	R. acetosella	Polygonaceae
*DILL	Anethum graveolens	Umbelliferae	SPEARMINT	*Mentha spicata	Labiatae
*FENNEL	Foeniculum vulgare	Umbelliferae	SWEET CICELY; *MYRRH	Myrrhis odorata	Umbelliferae
GARLIC MUSTARD	Alliaria petiolata	Cruciferae	SWEET WOODRUFF	*Galium odoratum	Rubiaceae
*HYSSOP	Hyssopus officinalis	Labiatae	TANSY	*Tanacetum vulgare	Compositae
LEMON GRASS	*Cymbopogon citratus	Gramineae	*TARRAGON	Artemisia dracunculus	Compositae
LEMON BALM	*Melissa officinalis	Labiatae	*THYME, GARDEN	Thymus vulgaris	Labiatae
LEMON VERBENA	Aloysia triphylla	Verbenaceae	*THYME, CARAWAY	T. herba-barona	Labiatae
*LOVAGE	Levisticum officinale	Umbelliferae	*THYME, LEMON	T. × citriodorus	Labiatae
MARIGOLD, POT	*Calendula officinalis	Compositae	VERVAIN	*Verbena officinalis	Verbenaceae

Provençal cuisine, the dried stems are used to form a bed on which fish such as sea bass is flamed with brandy (eg loup au fenouil). *DILL *(Anethum graveolens)* is grown both for its fruits, used in pickling cucumbers and for making dill vinegar, and for its young leaves (dill weed) used fresh or dried in flavoring soups and sauces as well as in dill pickles (in addition to the seeds) especially in Scandinavia, central and Eastern Europe. The leaves of CHERVIL (*Anthriscus cerefolium*) and *CORIANDER (*Coriandrum sativum*) are also used as herbs but more important are their seeds which along with FENNEL, PARSLEY, *CUMIN (*Cuminum cyminum*) and CARAWAY (*Carum carvi*) are very extensively used as culinary spices, in curries etc.

Few members of the Compositae are used as herbs. The best-known are TARRAGON, one of the great culinary herbs, whose leaves have a very characteristic flavor, used in tarragon vinegar, WORMWOOD (*Artemisia absinthum*), a bitter aromatic herb mainly used today to flavor herb wines and aperitifs, TANSY (*Tanacetum vulgare*), a traditional culinary and medicinal herb now little used, and *CHAMOMILE (*Chamaemelum nobile*) whose flower heads are used to make an infusion – chamomile tea.

The leaves of several members of the genus *Allium (Liliaceae), such as *CHIVES (*Allium schoenoprasum*) and CHINESE CHIVES or CUCHAY (*A. tuberosum*), are used as flavorings.

SORREL (*Rumex acetosa*) a member of the Polygonaceae, has very acid leaves which are used for flavoring. In India and the West Indies it is replaced by a malvaceous plant, *Hibiscus sabdariffa*, known as ROSELLE. From the laurel family comes the aromatic leaf of the BAY or SWEET BAY (*Laurus nobilis*).

Herb gardens have been grown for centuries and are still popular today. In earlier times they would include not only culinary herbs, but a selection of medicinal herbs used in tonics, lotions, potions, cough mixtures and so on. Despite the increasing tendency to return to herbal remedies, most of the herbs grown today are for culinary purposes.

Most culinary herbs are perennial and are either propagated from seed or by division. Some, such as CARAWAY, CLARY, DILL, FENNEL and SWEET MARJORAM are biennial or short-lived perennials, while a few, for example ANISE, CORIANDER, SUMMER SAVORY and SWEET BASIL, are annuals that are sown directly in the herb garden each year.

Herbs are normally distinguished from spices which are dried parts, such as roots, bark, leaves, berries or seeds, of aromatic plants found mainly in the Old World tropics. The fruits of some of the herbs, such as the umbelliferous CARAWAY, CUMIN and CORIANDER, mentioned above, are used as culinary spices.

Herbs and spices, together with other kinds of condiments, constitute the group of substances known as flavorings which are considered separately (see p. 142).

Hevea a genus of tropical American trees of which *H. brasiliensis* [PARA RUBBER] is of great economic importance as the world's major source of natural rubber (see RUBBER PLANTS). Two other related species, *H. benthamiana* and *H. guianensis*, also yield commercially acceptable latex. *Hevea* is native to the Amazon basin, and *H. brasiliensis* is grown on a commercial scale in plantations in the Old World tropics, mainly in Southeast Asia. The latex is obtained by tapping the bark. Cooked seeds of *Hevea* species are eaten by aborigines in the northwest part of the Amazon basin. EUPHORBIACEAE, about 9 species.

Hibbertia a large genus of mainly Australian heath-like or climbing shrubs also occurring in Madagascar, New Guinea, New Caledonia and Fiji. The leaves are small and often needle-like. The flowers are yellow or white, solitary or in few-flowered raceme-like inflorescences. Some species are grown out of doors in warm temperate regions, such as *H. dentata*, and *H. scandens*.
DILLENIACEAE, about 100 species.

Hibiscus a large tropical, subtropical and warm temperate genus of annual and peren-nial herbs, evergreen and deciduous shrubs and small trees. It contains numerous ornamentals and some plants of economic value such as *H. esculentus* [GUMBO, OKRA, LADY'S FINGER] (sometimes placed in the separate genus *Abelmoschus*) (see OKRA), *H. cannabinus* [INDIAN HEMP, DECCAN HEMP, KENAF, BASTARD JUTE], which yields a fiber like jute and an oil from the seed, and *H. sabdariffa* [ROSELLE, JAMAICA or INDIAN SORREL], whose calyces make a pleasing cordial, sauce or jelly. ROSELLE also produces a tough fiber used for cordage.

The most common tropical ornamental is *H.* × *rosa-sinensis* [SHOE FLOWER, ROSE OF CHINA, HAWAIIAN or CHINESE HIBISCUS], a name loosely applied to more than 1 000 cultivated varieties of uncertain origin, but probably involving hybridization and mutation. Participating species probably include *H. schizopetalus* from East Africa, itself most attractive, *H. kokio* and *H. arnottianus*, both from Hawaii. Some of the more familiar cultivars have scarlet, pink, white or yellow flowers, pendulous or erect, up to about 15cm in diameter. The habit of these bushes is normally rounded and to about 5m tall, although they also make a good dense hedge.

Para Rubber (Hevea brasiliensis) *in a west coast Malaysian rubber plantation showing the sloping tapping cuts and collecting cups.*

Other tropical ornamentals include the African *H. trionum* [FLOWER-OF-AN-HOUR], *H. mutabilis* [COTTON ROSE] from southern China, so called because the flowers change color to darker shades with age, and *H. elatus* [BLUE MAHOE], the national tree of Jamaica, a useful forestry species growing to 25m in the mountains and whose wood is used for cabinet-making etc. There are about 40 cultivars of *H. syriacus* (= *Althaea frutex*) [ROSE OF SHARON], a shrub native to warm temperate East Asia, with white, purple or mauve flowers borne on an erect bush. Another popular ornamental is the tall perennial herb *H. moscheutos* [COMMON or SWAMP ROSE MALLOW, WILD COTTON], a native of the eastern and southern USA which includes the subspecies *moscheutos* (= *H. oculiroseus*) and *palustris* (= *H. palustris*) [MARSH MALLOW, SEA HOLLYHOCK]. There are numerous cultivars of progeny from hybrids of this species and *H. coccineus* and *H. militaris*.
MALVACEAE, about 250 species.

Hieracium [HAWKWEEDS] a large genus of very variable, often hairy, perennial herbs distributed throughout temperate regions with the exception of Australasia. Only a few species are cultivated as they soon become rampant. Among the more popular are *H. pilosella* [MOUSE EAR HAWKWEED], with hairy gray leaves and lemon-yellow flower heads, *H. aurantiacum* [DEVIL'S PAINTBRUSH], with flame-colored flower heads, and *H. villosum* [SHAGGY HAWKWEED], which is covered with silky hairs and has large bright yellow flower heads.

Many thousands of apomictic forms have been described as species.
COMPOSITAE, 700–1 000 or more species.

Himantoglossum a small genus of robust terrestrial orchids found across Europe and in North Africa and the Near East. The commonest species is *H. hircinum* [LIZARD ORCHID] which, like the other members of the genus, possesses foul-smelling flowers basically greenish purple in color.
ORCHIDACEAE, about 4 species.

Hippeastrum a genus of bulbous plants widely distributed in the New world, mostly in tropical America, with one species in West Africa. The large showy funnel-shaped flowers have made these BARBADOS LILIES or AMARYLLIS, as they are commonly known, popular as greenhouse plants in the north for winter flowering, or as garden plants in the south.

Most of the cultivated species are hybrids or variants of *H. aulicum* (= *Amaryllis aulica*) [LILY-OF-THE-PALACE], which has red flowers with a green throat, *H. elegans* (= *A. elegans*), which has long greenish-white flowers, *H. puniceum* (= *A. belladonna*) which has red flowers, green at the base, *H. reginae*

Mare's Tail (Hippuris vulgaris) is a widely distributed hydrophyte that grows in mud in pools and slow-moving water. ($\times \frac{1}{5}$)

(= *A. reginae*), which has red flowers with a greenish-white throat, *H. reticulatum* (= *A. reticulata*), with mauve-red flowers, and *H. striatum* (= *A. rutila*), with bright crimson flowers with a green keel.
AMARYLLIDACEAE, about 75 species.

Hippophae a small temperate Eurasian genus of thorny deciduous shrubs or small trees covered with silvery scales, and with small inconspicuous flowers which appear before the leaves. *H. rhamnoides* [SEA BUCKTHORN] has an unusually wide distribution extending from Great Britain across Europe to Kamchatka and Japan. The bushes sucker freely and so are able to grow on loose soil, helping to stabilize it. The bright orange "berries" can be used to make a sharp-tasting jam or jelly.
ELAEAGNACEAE, 2–3 species.

Hippuris a genus comprising a single species, *H. vulgaris* [MARE'S TAIL], found in temperate and cold regions of the Northern Hemisphere. It is a perennial herb that normally grows in shallow water, developing erect flowering shoots from a creeping rhizome. The submerged shoots form an important winter food for many animals and for Eskimos.
HIPPURIDACEAE, 1 species.

Hoffmannia a genus of mainly shrubs, rarely herbs, native to Central and South America from Mexico to Argentina. A few species are grown under glass, more for their foliage than for their small tubular flowers. *H. discolor* has leaves which are purple beneath and red flowers. *H. refulgens* has leaves wine-red beneath and pale red flowers, and the leaves of *H. ghiesbreghtii* are reddish purple beneath and the flowers are yellow with a red blotch.
RUBIACEAE, about 100 species.

Hoheria a small New Zealand genus of shrubs or small trees cultivated as ornamentals for their attractive clusters of white flowers. A popular species is *H. populnea* [LACE BARK] an evergreen shrub which may reach a height of 10m, the wood of which is used in cabinet-making.
MALVACEAE, 2–5 species.

Holcus a small genus of perennial grasses occurring in the Canary Islands, North Africa, Europe, Asia Minor and the Caucasus, with one species in South Africa. The common tufted perennial *H. lanatus* [WOODY HOLCUS, YORKSHIRE FOG, VELVET GRASS] and the closely related rhizomatous *H. mollis* [CREEPING SOFT GRASS] are often troublesome weeds in north temperate regions, and are only of minor value for grazing.
GRAMINEAE, 8 species.

Holly the common name for the widespread woody genus *Ilex*, found mainly in temperate and tropical regions of Asia and North and South America. They are sometimes deciduous, usually evergreen trees and

Leaves and fruit of the Sea Buckthorn (Hippophae rhamnoides), a suckering shrub used to stabilize soil and sand. ($\times \frac{1}{3}$)

shrubs, with shoots often angled, the greenish or white unisexual flowers axillary and usually borne on separate male and female plants.

I. aquifolium [ENGLISH, EUROPEAN or OREGON HOLLY], from Europe, North Africa and western Asia, is used as a decoration during the Christmas season, as it was by the Romans during their Saturnalia. It is a tree to about 25m tall. Its dark green spiny foliage and red winter berries have made it one of the best-known and most popular of plants and there are over 100 different cultivated varieties, some referred to *I.* × *altaclerensis* (*I. aquifolium* × *I. perado*). Some are variegated, while others have crisped or puckered leaves of diverse shape. *I. opaca* [AMERICAN HOLLY] is the best-known evergreen American species, growing to about 15m and containing more than 110 cultivars. *I. verticillata* [BLACK ALDER, WINTERBERRY] is a large deciduous North American shrub or small tree, with several cultivars which bear copious red (rarely yellow) fruit in winter; its purple leaves turn yellow in autumn.

Asian evergreen hollies include two species each with several cultivars, *I. cornuta* [HORNED HOLLY] and the dwarf *I. crenata* [JAPANESE HOLLY] with black berries. The Chinese species *I. pernyii*, and *I. latifolia* [TARAJO HOLLY], from Japan, bear orange-red berries. *I. perado* var *perado* [MADEIRA HOLLY] and var *platyphylla* [CANARY ISLAND HOLLY] are attractive small evergreen trees native to Madeira and the Canary Islands, with dark green or yellowish-green leaves and dark red berries.

A bitter drink, cassine or the "black drink"

Female flowers of the Hop (Humulus lupulus) which after fertilization develop into the typical hop cone.

of North American Indians, is prepared from the dried leaves of the North American shrub or small tree, *I. vomitoria* [CASSINA, YAUPON, CAROLINA TEA HOLLY]. Birdlime is partly made from the bark of *I. aquifolium* and *I. integra* [MOCHI TREE]. The leaves of *I. paraguariensis* [YERBA MATE, PARAGUAY TEA] are the source of maté, the popular South American drink.

Hollyhock the common name for some of the tall leafy stemmed herbs belonging to the genus *Alcea*, particularly *A. rosea* (= *Althaea rosea*) [HOLLYHOCK] and *A. ficifolia* (= *Althaea ficifolia*) [FIGLEAF HOLLYHOCK]. Many varieties of the first-named species are cultivated in temperate gardens for their large showy flowers in a wide range of colors; they are also naturalized in many parts of the world. MALVACEAE.

Holodiscus a North American genus of deciduous flowering shrubs, two of which, *H. discolor* [CREAMBUSH] and *H. dumosus*, are

The fleshy leaves and round fruit of Sea Sandwort (Honkenya peploides). common on north temperate sandy and pebbly seashores. (× 1½)

A Silvery Moth (Plusia gamma) feeding at night on a Honeysuckle flower (Lonicera periclymenum), which brings about pollination. (× 3)

grown as ornamentals bearing dense inflorescences of tiny creamy-white flowers. ROSACEAE, 8 species.

Honeydew melons a name applied to a class of *melons more properly known as WINTER MELONS, fruits of cultivars of *Cucumis melo*. They are yellow or green, hard skinned, ellipsoid in shape with length greater than the diameter, smooth or shallowly corrugated. They ripen late, and can be stored for a month or more; in consequence, they are popular with growers who export to distant markets. CUCURBITACEAE.

Honeysuckle the common name for members of the genus *Lonicera*, consisting of evergreen and deciduous flowering shrubs and woody climbers, often with fragrant, long-tubed flowers containing nectar (see *Lonicera*). In countries where *Lonicera* is not native, other species have become known locally as honeysuckles. Thus *Halleria lucida* is the AFRICAN HONEYSUCKLE, *Tecomaria capensis* the CAPE HONEYSUCKLE and *Lambertia multiflora* the AUSTRALIAN HONEYSUCKLE.

Honkenya a small genus of perennial herbs of north temperate and circumpolar regions and southern Patagonia. *H. peploides* [SEA SANDWORT], a common species of sandy and pebbly shores, has a long creeping underground stem with scale-leaves and fleshy green leaves and can endure short periods of immersion in salt water. CARYOPHYLLACEAE, 2 species.

Hop the common name for the climbing perennial herb *Humulus lupulus*, which is cultivated for its use in brewing, giving to beer its characteristic bitter taste and hop

aroma. The plant has a perennial rootstock but the aerial twining stems (bines) die down to ground level each winter. The fresh roots which emerge in the spring climb by twining clockwise (without tendrils), and commercially they are grown up strings or wires which are attached each year to a permanent framework of poles and wire.

Hop plants are either male or female, and it is the female inflorescences or "cones" which are the commercial product. Usually the hops are grown seedless by eliminating all male plants from the district. The constituents of brewing value are the soft resins which provide the bitterness and the essential oils which contribute the aroma characteristic to beer. Both of these are produced in lupulin glands, which constitute up to a quarter of the dry weight of the cones.

Successful plant breeding by hybridizing European and wild American hops has resulted in cultivars with an increased soft resin content in their cones. The bittering power of hops depends upon the acid fraction of the soft resins, and the acid content of the traditional cultivars is from 4–5%. The most recently produced cultivars contain as much as 10–12% acid, so that the yield of brewing material has been more than doubled without increasing the land area grown. MORACEAE.

Hordeum see Barley. GRAMINEAE, about 20 species.

Horminum [DRAGON MOUTH] a genus represented by a single species, *H. pyrenaicum*, which occurs naturally in mountains from the Pyrenees to the Alps. A hardy herbaceous plant, sometimes cultivated in borders or rock gardens, it has a flowering stem up to 25cm long bearing numerous bluish-purple flowers. There are varieties with rose-purple or white flowers. LABIATAE, 1 species.

Hormoseira an important genus of brown

algae found in Australasia. The thallus appears rather like a string of hollow beads. PHAEOPHYCEAE.

Hornbeam   the common name for members of the genus *Carpinus*, deciduous trees distributed throughout the temperate regions of the Northern Hemisphere. The flowers are borne in unisexual catkins, with both sexes occurring on the same tree.

The spring-flowering *C. betulus* [COMMON HORNBEAM] is regularly cultivated and is one of nature's most handsome trees, ranging in height from 1.5–25m in its mature state. It coppices well, retains its leaves after clipping, and because it branches profusely, like beech, it makes an excellent hedge. The wood, which is heavy and strong, is used for turning, for tool handles etc. There are several well-known cultivars, varying in color, leaf shape, tree shape and branching habit. Hornbeams are extremely hardy trees and very handsome, especially when in flower and fruit.

C. caroliniana [AMERICAN HORNBEAM or MUSCLE TREE, BLUE BEECH], a native of the

Hornbeams (Carpinus betulus) were formerly much coppiced, both for their hard timber and for making charcoal.

eastern USA, is somewhat similar to the COMMON HORNBEAM, but is usually a smaller tree differing by its whitish hair-covered leaves, which turn orange yellow or scarlet in the autumn. Another well-known species, *C. japonica* [JAPANESE HORNBEAM], is a widely cultivated sturdy, pyramidal tree which grows to a height of 12–15m. It is particularly valued for its large, handsome leaves and its pendulous female catkins. Less frequently cultivated, but nonetheless beautiful trees include *C. orientalis* [ORIENTAL HORNBEAM] *C. henryana* [CHINESE HENRY'S HORNBEAM] and *C. cordata* [CORDATE HORNBEAM]. BETULACEAE, about 35–40 species.

Horseradish   the common name for *Armoracia rusticana* (= *Armoracia lapathifolia*), a vigorous rapidly spreading plant with a swollen, branching taproot and large simple leaves. The root is harvested, peeled and grated. It contains a very pungent mustard oil. The grated root can be used as a relish, but is most commonly used to make horseradish sauce, a traditional British piquant accompaniment to roast beef.

The traditional practice of growing a few aged plants in a garden to harvest periodically yields tough stringy roots. The best and most tender roots are obtained commercially

Venus' Necklace (Hormoseira banksii) is a brown alga, seen here growing on South Island, New Zealand. The conspicuous "beads" contain the reproductive organs. (×1)

by growing horseradish as an annual crop. CRUCIFERAE.

Horsetails   see p.174.

Hosta (= *Funkia*) [PLANTAIN LILIES]   a genus of perennial herbs with ornamental foliage and funnel-shaped white to dark violet flowers borne in erect racemes, mainly native to Japan, with a few species in China and Korea. Among the cultivated species (many of which occur in a range of cultivars) are: *H. plantaginea* [FRAGRANT PLANTAIN LILY] with a short dense raceme of upright, fragrant, white flowers subtended by large bracts; *H. ventricosa* [BLUE PLANTAIN LILY], with dark violet flowers with a narrow tube suddenly expanded into a bell-shaped mouth; *H. fortunei*, with pale lilac flowers; *H. lancifolia* (= *H. japonica*) [NARROW-LEAFED PLANTAIN LILY], with loose racemes of dark violet flowers; and *H. undulata* (= *H. lancifolia* var *undulata*), with pale lavender flowers in many-flowered racemes. LILIACEAE, about 40 species.

Hottentot bean   the edible seed, eaten by native peoples, of some of the 18 species of the southern African genus *Schotia*. Most are trees, such as *S. brachypetala* [TREE FUCHSIA], producing seeds with a large, rich, fatty aril. LEGUMINOSAE.

Hottonia   a genus comprising two species of hardy perennial, floating aquatic herbs from Europe, Asia and North America. The lilac-flowered *H. palustris* [WATER VIOLET], with its finely divided submerged leaves, is a valuable oxygenating species for ponds and aquaria. The other species, the white-flowered *H. inflata*, is North American in origin. PRIMULACEAE, 2 species.

Horsetails or Scouring Rushes

THE GENUS *Equisetum* [HORSETAILS, or SCOURING RUSHES] comprises 16 species and about the same number of hybrids. They are primitive plants and are the only members of the group known as the Sphenophytina, a subdivision of the vascular plants. They are intermediate in form between the *club mosses and *ferns. Horsetails are most frequently found in moist habitats such as marshes, flushes, dune slacks, lakesides and river banks but they also colonize drier sites such as derelict ground and railway embankments. They are often the dominant plants with colonies covering considerable areas: indeed, an unbroken stand of *E. hyemale* stretches for over 300km along the Mississippi river. The gametophytes (sexual generation) may be found on bare mud by lakes and rivers. Although spread is predominantly by rhizomes, the high incidence of hybrid horsetails indicates that sexual reproduction is probably more frequent than is generally assumed.

Although horsetails are found throughout the world (except Australia and New Zealand) they show greatest diversity in the Northern Hemisphere. Over half the species have distributions covering the entire range of the genus. Three species are restricted to central South America and *E. diffusum* is a Himalayan endemic. The occurrence of *E. telmateia* along the western edges of Eurasia and North America and its absence on the eastern sides of these continents suggests large-scale regional extinction.

Horsetails are perhaps best known as garden weeds whose deeply buried rhizomes make them difficult to eradicate. This same property, however, renders them important guides for the siting of wells above subterranean water supplies on the American prairies. Horsetails are poisonous to herbivores in pastures but, in Japan, the young cones are regarded as a delicacy. In the past, they were widely employed for scouring pans but today their abrasive powers are restricted to polishing tools and the reeds of wind instruments. Most remarkable of all, horsetails are said to have found favor with prospectors since they apparently accumulate unusual elements including gold.

The most conspicuous features of these plants are their jointed and ridged photosynthetic stems which bear whorls of branches and fused microphyllous leaves. The stem ridges, together with the leaves lying above them, show a regular alternation from one internode to the next. Their sandpapery texture, resulting from the presence of silica bodies in the epidermis, is especially interesting since silicon is an essential element for these plants. Horsetails grow prolifically by means of subterranean rhizomes which give rise at the nodes to erect shoots, roots and, in some species, starch-filled tubers.

With stems less than 15cm tall, *E. scirpoides* is the smallest species while *E. giganteum* reaches a height of 13m. The stems of the latter, however, are only 2cm in diameter

and have to rely on the surrounding vegetation for support. Some species are unbranched whilst others, such as *E. sylvaticum*, with two orders of lateral branches, have an almost lacy appearance.

All living horsetails are herbaceous, produce just one type of spore (homosporous) and lack secondary thickening, but in the past they displayed far greater morphological diversity. Tree-like and sometimes heterosporous plants, belonging to the extinct Calamitales formed a major component of Carboniferous coal-measure swamp forest. These existed side by side with herbaceous plants which so resemble extant horsetails that they are placed in the genus *Equisetites*.

The spore-forming bodies (sporangia) are borne on peltate stalks (sporangiophores) grouped together in cones. In some species they occur terminally on special unbranched stems that lack chlorophyll while in others the fertile stems are green and identical with the sterile ones. Material deposited around the developing spores forms into bodies (elaters) with spoon-shaped bands. These are hygroscopic (moisture-absorbing) and coil with changes in humidity, at the same time flicking spores from the sporangia. Ripe spores contain chlorophyll and are spherical.

Above *A mass of fertile and young vegetative shoots of* Equisetum fluviatile (Water Horsetail). ($\times \frac{1}{2}$)
Left *Cone of* E. arvense (Field Horsetail) *comprising a mass of brown stalks with white sporangia below. A typical vegetative shoot is shown on the right.* ($\times 2$)

The spores germinate to form the sexual (or gametophyte) generation. These are photosynthetic and surface-living. They are up to 3cm in diameter, and consist of cushions of soft tissue bearing unicellular root-like rhizoids on the underside and green tissues above. The female sex organs (archegonia) have projecting necks and each contain an egg. The male sex organs (antheridia) are borne on upright branches of the cushion and contain coiled male gametes (spermatozoids) each with 80–120 flagella. The gametophytes may be male or bisexual with

Above *The sponge-like prothallus of Equisetum fluviatile (Water Horsetail) on which the sex organs form and from which, after fertilization, the next sporophytes grow – three can be seen here.* (× 3) Below *The same species growing in shallow water.*

sexuality depending on environmental conditions. The next generation of sporophytes is readily produced by either self- or cross-fertilization and over 35 new plants have been recorded from a single gametophyte.

Equisetum arvense (Field Horsetail) can be a problematical weed of damp fields and gardens, and, because of its deep-seated rhizomes, it is difficult to eradicate. It is seen here in a newly-cultivated vegetable garden. (× ¼)

Houseleek the common name for *Sempervivum*, a genus of dwarf, rosette-forming, hardy leaf-succulents. They are typical of rocky mountainous habitats in Europe, North Africa, and Asia. Each plant forms a cluster of almost stemless rosettes of tightly packed, spirally arranged fleshy leaves of green, often overlaid with purplish-red. When flowering, the stem elongates and expands above into many-flowered clusters of small starry blooms. Even after the flowers have passed, the heads of the dried fruits have a decorative value.

From early times *S. tectorum* [COMMON or ROOF HOUSELEEK] has been credited with beneficial properties, perhaps because of its apparent coolness and longevity in hot dry situations lethal to other plants. It is often grown as a clump on the roof (hence the common name) in some rural areas of Europe, as a supposed protection from lightning. The sliced leaves were recommended as a poultice for burns or stings, and were said to cure warts and corns.

Sempervivums make ideal trouble-free rock-garden plants. Among the most popular garden species are the purplish-red-flowered *S. tectorum* with numerous cultivars, *S. arachnoideum* [COBWEB HOUSELEEK, SPIDER-WEB HOUSELEEK], whose leaves are connected by cobwebby strands, *S. montanum* with purple flowers (yellowish-white in var *braunii*), and the greenish-yellow-flowered *S. soboliferum* [HEN-AND-CHICKENS]. Numerous hybrids are in cultivation, including *S. × barbulatum* (*S. arachnoideum × S. montanum*) and *S. × fauconnettii* (*S. arachnoideum × S. tectorum*).
CRASSULACEAE, about 40 species.

Houstonia a genus of low-growing tufted perennial herbs native to the southern and western USA and Mexico. The genus is sometimes included in *Hedyotis*. The plants

Male flowers of the Hop (Humulus lupulus). *These occur on separate plants from the females and since only seedless female cones are required for brewing, great efforts are made to eliminate all male plants from a growing district.* (×2)

The Wild Hyacinth or English Bluebell (Hyacinthoides non-scripta) *flowers from April to June, usually forming a dense carpet in woodlands.*

have blue, purple or white flowers, and species such as *H. serpyllifolia* (= *Hedyotis michauxii*) [CREEPING BLUETS] and *H. caerulea* (= *Hedyotis caerulea*) [BLUETS] are frequently cultivated in rock gardens.
RUBIACEAE, about 50 species.

Houttuynia a genus represented by a single species, *H. cordata* (= *Gymnotheca chinensis*), a perennial creeping rhizomatous herb native to the Himalayas, China and Japan. It is cultivated for ground cover in moist places. The flowers are borne in a terminal spike subtended by a collar of four white bracts, the whole resembling a single flower.
SAURURACEAE, 1 species.

Howea (= *Kentia*) a genus comprising two species of ornamental palms from Lord Howe Island in the southwest Pacific. *H. belmoreana* [CURLY PALM] and *H. forsterana* [FLAT or THATCH-LEAF PALM, SENTRY], both with stout, erect stems and long feathery leaves, are grown as ornamentals indoors, or outside in the tropics.
PALMAE, 2 species.

Hoya a genus of evergreen, chiefly climbing or twining plants from China, southeast Asia, Indomalaysia and Australia, with fleshy, opposite leaves and axillary umbel-like clusters of waxy flowers. *H. carnosa* [WAX FLOWER, WAX PLANT] is a vigorous climber and has pendulous umbels of fragrant white flowers with pink centers. In temperate climates it can be grown on a wall in a greenhouse. In the tropics it does best on tree trunks. *H. bella* is more slender; it is grown in

hanging baskets for its white flowers with rose, crimson or violet centers.
ASCLEPIADACEAE, 70–200 species.

Huckleberry the common name normally given to several species of bushy shrubs of the North and South American genus *Gaylussacia*. They usually flower in spring, producing white or pinkish flowers in axillary racemes. Many species, including *G. baccata* [BLACK HUCKLEBERRY], *G. brachycera* [BOX HUCKLEBERRY], *G. ursina* [BEAR HUCKLEBERRY] and *G. dumosa* [DWARF HUCKLEBERRY, BUSHY WHORTLEBERRY], bear edible fruits similar to blueberries or blackberries.

Some species of *Vaccinium* are also called HUCKLEBERRY. Examples include *V. vacillans* [BLUE or SUGAR HUCKLEBERRY] and *V. hirsutum* [HAIRY HUCKLEBERRY].
ERICACEAE.

Humulus a small genus of perennial or annual climbing herbs widespread in the temperate Northern Hemisphere. Female plants of the perennial *H. lupulus* [*HOP] are widely cultivated for their catkins containing lupulin resins which during brewing are converted to isohumulones, the bitter principles of beer. The tender young shoots are sometimes eaten as a vegetable and the hops used in hop pillows. *H. japonicus* [JAPANESE HOP] is sometimes cultivated as a garden-twiner for its screening effect when in leaf in the summer. Although this is a perennial species it is usually grown as an annual from seed.
MORACEAE, 2 species.

Hyacinthoides a genus of bulbous monocotyledons native to Western Europe and northwest Africa. The genus is better known by the incorrect name *Endymion*. The flowers are bell-shaped or rotate in shape and white,

The Spanish Bluebell (Hyacinthoides hispanica) is often cultivated in gardens and is often naturalized in southern and western Europe. ($\times\frac{1}{10}$)

pink or blue in color, arranged in a raceme.

Cultivated species include the popular *H. non-scripta* (*Scilla non-scripta*) [WILD HYACINTH, ENGLISH BLUEBELL, HAREBELL] with various white (eg 'Alba'), blue (eg 'Caerulea') and pink (eg 'Rosea') cultivars, *H. hispanica* (= *S. campanulata*, *S. hispanica*) [SPANISH BLUEBELL, SPANISH JACINTH, BELL-FLOWERED SQUILL], again with several cultivars, and the lilac-flowered *H. italica* (= *S. italica*) [ITALIAN SQUILL].
LILIACEAE, 3–4 species.

Hyacinthus [HYACINTHS] a small genus of bulbous herbs native to the Mediterranean region, southwest Asia and the Middle East. The plants have a cluster of strap-shaped or linear basal leaves and a leafless stem bearing flowers in a raceme.

H. orientalis [COMMON HYACINTH, DUTCH HYACINTH] is a sweetly scented plant from

A white cultivar of the Common Hyacinth (Hyacinthus orientalis, together with Grape Hyacinths (Muscari sp). ($\times\frac{1}{12}$)

which all the colorful, large-flowered florists' hyacinths have been raised by selection. In the wild, it is a fairly small-flowered plant with only a few blue or lilac flowers in the raceme, whereas the named cultivars have densely flowered racemes with individual flowers about three or four times the size, in a wide range of colors from white through pink and blue to deep purple and red. They are widely used, both in outdoor bedding and for forcing in pots for indoor decoration. The "ROMAN HYACINTHS" resemble more closely the original wild species with fewer, smaller flowers.

Other species, previously included in *Hyacinthus*, are now placed in *Brimeura*, *Bellevalia* and *Hyacinthella*.
LILIACEAE, 1 species.

Hydrangea a genus of deciduous and evergreen shrubs or woody climbers, mainly from East Asia, the Himalayas, the Malaysian region and North and South America. The star-shaped flowers are small and numerous, arranged in corymbs or panicles, and in most species some of them are sterile, with greatly enlarged petal-like sepals and the other floral parts abortive.

The most popular hydrangeas, especially for pot-cultures, are those usually known as "hortensias", which are all sterile-flowered. They originated in Japanese gardens where the parent species, *H. macrophylla* [FRENCH HYDRANGEA, HORTENSIA] is native. In the "lace-cap" hydrangeas the inflorescence is a corymb with only the outermost flowers sterile. Hortensias and lace-caps have flowers of various shades of red, pink, mauve or blue; this is determined partly by the genetic constitution on the plant (in some varieties the flowers are always red or pink), but partly by the soil, all varieties which have blue flowers on acid soils changing to purple or pink on neutral or alkaline soils.

Among the popular cultivated hydrangeas are the climbers, *H. anomala* (subspecies *anomala* and subspecies *petiolaris*), both with white flowers, the small shrubby white-flowered *H. arborescens* [HILLS OF SNOW, SEVENBARK], and the medium-sized to taller shrubs, *H. paniculata* (flowers white, turning pink), *H. quercifolia* (white, turning pink), *H. aspera* subspecies *sargentiana* (= *H. sargentiana*) (white and lilac), *H. aspera* subspecies *aspera* (= *H. villosa*) (violet) and *H. heteromalla* (white).

The wood of *H. paniculata*, which is fine-grained and hard, is used in Japan for making umbrella handles and pipes.
SAXIFRAGACEAE, about 25 species.

Hydrocharis a small aquatic genus, widespread in temperate and tropical regions of the Old world and Australia. They are free-floating herbs with long stolons bearing leaves in groups at the nodes. The best-known species, *H. morsus-ranae* [FROG-BIT], has become naturalized in North America. This species is also widely grown in large freshwater aquaria.
HYDROCHARITACEAE, 3–6 species.

Hymenaea a Central and tropical American genus of trees, the most important of which is *H. courbaril* [LOCUST TREE, COURBARIL] from Brazil. It attains a height of 30m, frequently with buttress roots. This species is a major source of a yellow or red resin, *copal, which is exuded from every part of the plant. It is used in high quality varnishes. The hard, durable timber is of economic value. *H. verrucosa* [EAST AFRICAN COPAL] is often placed in the genus *Trachylobium*.
LEGUMINOSAE, about 25 species.

Hymenocallis a genus of bulbous perennials with shortly tubular or funnel-shaped flowers, native mainly to North and South America. Easily cultivated outside in warm regions, in cooler temperate countries they require to be kept in either hothouse or cool greenhouse conditions. The cool greenhouse species include the white-flowered *H. narcissiflora* (*H. calathina*) [BASKET FLOWER, PERUVIAN DAFFODIL] and the yellow-flowered *H. amancaes* (= *Ismene amancaes*). Hothouse species include *H. expansa* (white flowers tinged pale green) and the fragrant *H. latifolia*.

H. littoralis (= *H. americana*), which is native to tropical America, and *H. pedalis*, native to eastern South America, are naturalized in Africa where they have been incorrectly described as *H. senegambica*.
AMARYLLIDACEAE, 25–30 species.

'Pencil'd flower of sickly scent' – George Crabbe's poem well describes the poisonous Henbane (Hyoscyamus niger). ($\times1$)

Hyoscyamus [HENBANES] a genus of herbaceous, often poisonous plants widely distributed throughout the Western Hemisphere. *H. niger* [BLACK HENBANE] is a highly poisonous, fetid, annual or biennial, sticky-hairy plant from Europe and Asia. It is grown as a source of the sedative drug hyoscyamine.
SOLANACEAE, about 10 species.

Hypericum [ST. JOHN'S WORTS] a large and important genus of herbs or shrubs, rarely trees, native to temperate and tropical

The feathery shoots of the mat-forming moss
Hypnum cupressiforme *var* lacunosum, *a locally
common species found on chalky grassland.* (× 3)

regions. Best-known perhaps is the near-cosmopolitan *H. perforatum* [PERFORATED ST. JOHN'S WORT], which is poisonous to animals and contains photosensitizing compounds which can cause dermatitis.

Many species are cultivated for their attractive yellow flowers, including *H. calycinum* (= *H. grandiflorum*) [ROSE OF SHARON, AARON'S BEARD, CREEPING ST. JOHN'S WORT], an evergreen shrub, the Canary Islands *H. canariense* (= *H. floribundum*), a taller semievergreen shrub, and some smaller species, including the Asian evergreen *H. patulum*, the evergreen *H.* × *moserianum* [GOLD FLOWER], a hybrid between *H. calycinum* and *H. patulum*, the North American *H. stans* [ST. PETER'S WORT] and the Eurasian *H. androsaemum* [TUTSAN]. Useful rockery species are the prostrate shrub *H. reptans* and the low spreading perennial *H. concinnum* [GOLDWIRE].
HYPERICACEAE, about 300 species.

Hypnum a genus of mat-forming mosses, the best-known of which is the highly variable, cosmoplitan *H. cupressiforme*.
HYPNACEAE, about 40 species.

Hypochoeris [CAT'S EARS] a genus of perennial herbs, cosmopolitan in distribution. The sinuate or ovate leaves are arranged in a basal rosette and the dandelion-like, usually yellow flower heads are borne on leafless stalks. *H. scorzonerae* is used in Chile as a diuretic. The young leaves of *H. maculata* [EUROPEAN CAT'S EAR] are sometimes eaten in salads. *H. uniflora* from the Alps is sometimes cultivated in rock gardens. *H. radicata* [SPOTTED CAT'S EAR] is a weed that is naturalized in various parts of the world. COMPOSITAE, about 80 species.

Hypocrea a widespread genus of ascomycete fungi. The numerous fruiting bodies (ascocarps) are enclosed in a perithecium. Characteristic habitats include rotting wood and bark, bracket fungi and soil.
HYPOCREACEAE, about 100 species.

Hypoxis [STAR GRASSES] a genus of perennial herbs native to America, Africa, Indomalaysia and Australia. *H. hirsuta*, from North America, is one of several species cultivated as a garden ornamental. The shortly tubular yellow flowers are borne in a few-flowered terminal cluster at the end of a leafless stalk. *H. aurea*, from Asia, yields extracts used in local medicine as a tonic.
AMARYLLIDACEAE, about 100 species.

Hyssop the common name given to the genus *Hyssopus*, of which *H. officinalis*, an aromatic dwarf shrub native to Central Asia, is sometimes grown as an ornamental border plant or as a culinary herb. The leaves have a somewhat minty flavor and an oil distilled from them is used in perfumery and as a flavoring for liqueurs. In North America, the related *Agastache urticifolia* is called NETTLE-LEAVED GIANT HYSSOP, and was used by Indians medicinally and as a flavoring. *A. foeniculum* is ANISE HYSSOP or FRAGRANT GIANT HYSSOP, the dried leaves of which are used for seasoning and to make a tea.
LABIATAE.

*Hyssop (*Hyssopus officinalis*) is very attractive to insects such as the hoverfly shown here. The aromatic leaves are sometimes used as a culinary herb.* (× 3)

Iberis a genus of annual or perennial herbs, rarely dwarf shrubs, native to the Mediterranean region, some of which are cultivated in gardens for their umbel-like clusters of white, pink, red or purple flowers. *I. amara* and *I. umbellata* are the commonly grown CANDYTUFTS of gardens. They are variable, showy annuals with several cultivars. The perennial subshrubby *I. saxatilis*, *I. sempervirens* and *I. gibraltarica* are good rock-garden plants, sometimes used for edging.
CRUCIFERAE, about 30 species.

Incarvillea mairei *is a herbaceous perennial from East Asia. The pink flowers are typically tubular and two-lipped.* (× ½)

Ice plant the popular name originally given to *Mesembryanthemum crystallinum*, so called from the glittering crystaline papillae covering the whole plant. The name is now loosely used for other garden ornamental species of *Mesembryanthemum* and unrelated leaf succulents of fleshy habit such as *sedums.
AIZOACEAE.

Idesia a genus from China and Japan represented by a single species, *I. polycarpa* [IIGIRI TREE], a deciduous tree up to 15m tall with more or less horizontal branches. *I. polycarpa* is grown in the southern USA as a shade tree for parks or street planting and is hardy in parts of Europe, where it is occasionally cultivated in arboreta and botanic gardens for its fragrant flowers and clusters or orange-red berries.
FLACOURTIACEAE, 1 species.

Golden Samphire (Inula crithmoides) is usually found near the sea, growing on shingle, cliffs and rocks or on salt marshes. ($\times \frac{1}{10}$)

Ilex see HOLLY.
AQUIFOLIACEAE, about 400 species.

Illicium [ANISE TREES] a genus of evergreen trees and shrubs native to Asia, North America and the West Indies. They have red, purple, pink, white or pale yellow fragrant flowers and starlike fruits (clusters of many one-seeded follicles). Oil of anise, used for flavoring drinks and confectionery and in medicines, is obtained from *I. verum* [STAR ANISE], a native of China, and also from the American *I. parvifolium* [YELLOW STAR ANISE]. The unripe fruits of the former are used as a spice. The East Asian *I. anisatum* [JAPANESE STAR ANISE], with yellow flowers, and the North American *I. floridanum* [PURPLE ANISE, ANISEED TREE], with crimson to purplish flowers, are among the few species grown as ornamentals in warm regions.
ILLICIACEAE, about 40 species.

Immortelle the name given to a number of species of tropical trees used as shade trees in cacao and coffee plantations (see *Erythrina*). The name is also given to a wide variety of plants whose flowers retain their color after drying, such as *Helichrysum* species and *Xeranthemum annuum*.

Impatiens a large genus of annual and perennial herbs distributed through tropical and temperate Eurasia and Africa, especially Madagascar and the mountains of India and Sri Lanka. There are also a few species in North and Central America.
  I. wallerana (= *I. holstii, I. sultani*), from Tanzania to Mozambique, and many cultivars derived from it, are the widely cultivated BUSY LIZZIES with pink, red, white, orange, purple or variegated flowers. *I. balsamina* [GARDEN BALSAM], an Asian species, is also widely grown, especially as double-flowered forms. *I. roylei* (= *I. glandulifera*) [HIMALAYAN or INDIAN BALSAM] is cultivated in Europe and North America and has become natural-ized by rivers, lakes and other damp habitats.
BALSAMINACEAE, about 600 species.

Imperfect fungi see Fungi, p. 152.

Incarvillea a genus of annual or perennial herbs with woody or tuberous roots, native to Central and East Asia and the Himalayan region. Several species, such as *I. delavayi* and *I. mairei* var *grandiflora* (= *I. grandiflora*), are cultivated as garden-border plants, in rock gardens or under glass, for their large, showy, predominantly pink flowers.
BIGNONIACEAE, about 14 species.

Indigo a deep-blue and very fast natural dye obtained from leaves of the INDIGO plant, mainly from the Asian *Indigofera tinctoria* and the tropical American *I. suffruticosa*. The leaves contain the colorless glucoside, indican, which, when the water in which they are steeped is stirred, becomes oxidized and forms the permanent dye indigo. This settles out as a dark blue sediment which is then dried. Other leguminous plants are called INDIGO, including species of *Amorpha*, *Baptisia*, *Lonchocarpus* and *Swainsonia*.
LEGUMINOSAE.

Inula a large genus of mostly perennial herbs from Europe, Asia, Africa and Madagascar. Several species are commonly cultivated for ornament, including *I. helenium* [ELECAMPANE, SCABWORT, HORSEHEAL, YELLOW STARWORT], with daisy-like yellow flower heads, the shorter *I. ensifolia*, also with yellow flower heads and hairless leaves, and *I. orientalis* (= *I. glandulosa*), with orange flower heads and hairy leaves. The leaves of *I. crithmoides* [GOLDEN SAMPHIRE] are sometimes used as a potherb in England. The root of ELECAMPANE is used for flavoring absinthe and extracts are used medicinally to treat intestinal worm infestations. It is also sometimes used as a potherb.
COMPOSITAE, about 100 species.

Iochroma a genus of tropical Central and South American shrubs or small trees with large, showy, tubular or bell-shaped flowers. A number of species, including *I. coccineum* (scarlet flowers), *I. fuchsioides* (orange-scarlet flowers) and *I. cyaneum* (= *I. lanceolatum*) (blue to purplish-blue flowers) may be grown under glass or cultivated in the open in sheltered warm or subtropical areas.
SOLANACEAE, about 25 species.

Himalayan Balsam (Impatiens roylei) is now naturalized in many parts of Europe, especially along river banks.

Ionopsidium a small genus of annual Mediterranean herbs with the leaves in rosettes. *I. acaule* [VIOLET CRESS], from Portugal, is widely cultivated especially as an edging plant or in rock gardens. The whole plant is 5–10cm tall, and the solitary, white lilac or purple flowers are borne on long stalks.
CRUCIFERAE, 5 species.

Ipecacuanha the dried root or rhizome, sometimes called "ipecac", obtained from the tropical American shrub *Cephaelis ipecacuanha* (= *Psychotria ipecacuanha*) [IPECACUANHA, IPECAC]. Its active principles are emetine and cephaeline which are alkaloids used as emetics, expectorants and in the treatment of amebic dysentery. Commercial production still continues in India and Malaya. *Psychotria emetica* [FALSE IPECAC] is a source of ipecac although of lower quality than that obtained from *Cephaelis*.
RUBIACEAE.

Ipheion a small genus of bulbous, onion-scented perennials native to Central and South America. The best-known cultivated species is *I. uniflorum* (= *Brodiaea uniflora*), which usually bears solitary, star-like, whitish, or pale blue to deep blue, fragrant flowers.
LILIACEAE, about 20 species.

Ipomoea a large genus of herbaceous annuals and perennials, or less often woody shrubs or vines, including several ornamentals and the important tropical root crop *I. batatas* [SWEET POTATO]. Some species are the source of extracts with hallucinogenic properties. For example, the seeds of the herbaceous climber *I. tricolor* were used as a hallucinogen in religious ceremonies by the Aztecs. Species occur in tropical and warm-temperate regions of the world. The usually large flowers are borne singly or in clusters,

Flowers of the Morning Glory (Ipomoea tricolor) *showing their distinctly trumpet-shaped purple corollas.* (× ⅓)

Aerial view of cultivated and fallow Sweet Potato (Ipomoea batatas) *gardens in the New Guinea forest.*

and have white or variously colored trumpet-shaped corollas.

Many of the species with more attractive flowers are cultivated for ornament, either as annuals such as *I. hederacea* and *I. purpurea* [COMMON MORNING GLORY] in temperate regions or as perennials or woody twiners such as *I. acuminata* [BLUE DAWN FLOWER], *I. nil* (which includes the IMPERIAL JAPANESE MORNING GLORIES, ascribed to the horticultural species *I. imperialis*), *I. horsfalliae* and *I. tricolor* with its numerous cultivars. Some species have flowers which remain open all day, whereas in others they open around dawn and wither before mid-day [MORNING GLORY], or open only in the evening as in *I. alba* [MOONFLOWER].

Apart from their value as ornamentals and their innumerable uses in native medicine and animal foodstuffs, some species of *Ipomoea* constitute valuable human foodstuffs. *I. batatas* [SWEET POTATO] has edible subterranean tubers. Probably originally from South America, it is now cultivated in all warm regions of the world, and in some places it is the main starch food. The young shoots and leaves of *I. aquatica* are used as a vegetable in East Asia, and in some areas *I. batatas* is treated similarly, or used as a salad. *I. purga* [*JALAP] from Central America, produces subterranean tubers which are used to prepare a purgative. The seeds of *I. muricata* have similar properties. *I. pes-caprae* [BEACH MORNING GLORY, RAILROAD ROAD] is a useful sand binder.
CONVOLVULACEAE, about 500 species.

Iresine [BLOODLEAF] a genus of herbs and subshrubs native to South America, the Galapagos Islands and Australia. A number of the tropical South American species are grown as bedding plants, including *I. herbstii*

(= *Achryanthes verschaffeltii*), with its attractive purplish-red, crimson, green, greenish-red or bronze-colored leaves, depending on the variety.
AMARANTHACEAE, about 80 species.

Iris a large and very ornamental genus distributed throughout the northern temperate regions of the world. The genus can be divided into two major groups: subgenus *Iris* consists of the rhizomatous species and subgenera *Xiphium* and *Scorpiris* those that are bulbous. *I. nepalensis* is distinct from either of the above groups in possessing a very small rhizome terminating in a cluster of fleshy roots. It thus constitutes a third group or division – *Nepalenses*.

The stem in *Iris* varies from being more or less absent, as in *I. reticulata*, to extremely well-developed, as in *I. confusa*, where it grows to the stature of a small bamboo cane with a cluster of leaves at its apex. Leaves mostly arise from the stem base, are two-ranked and linear to sword-shaped.

Iris flowers are well known in their overall structure, and practically all of the species have this same basic make-up, differing from species to species in color and in size and shape of the individual perianth parts. There are three large and showy outer segments known as the "falls" because they are normally reflexed, and three inner segments which are usually smaller and erect, referred to as the "standards". The falls may be furnished with a raised crest or have a cluster of hairs on the middle and lower portion, known as the "beard". There are divergent groups which do not agree exactly with the basic flower form, notably the *Juno* group in which the standards are very small and held out horizontally or deflexed.

The rhizomatous division is further subdivided into bearded irises (*Pogon*), the beardless irises (*Apogon*), which have smooth falls without hair or crest, and the crested irises, with crests instead of beards on the falls.

These are further subdivided into many smaller groups. Although the genus as a whole has a very wide distribution, some of the sections are confined to distinct areas.

Obviously the main importance of the genus lies in its decorative value, although the dried rhizomes of *I. germanica* var *florentina* [*ORRIS ROOT] are used in perfumery to produce a violet-like scent. Most of the wild *Iris* species are extremely beautiful, but on the whole they are less popular for general garden display than the host of showy hybrids which have been raised. There is nowadays a large range of colors available in the tall bearded *Iris* group which do not occur in the original wild plants. Cultivated bearded irises include *I. × albicans, I. aphylla, I. kashmeriana* and the popular hybrid *I. × germanica* [FLAG, FLEUR-DE-LIS].

Among species of beardless irisis, several are widely cultivated, including cultivars 'Kermesina' and 'Rosea' of *I. versicolor* [WILD IRIS, BLUE FLAG, POISON FLAG], *I. spuria* [BUTTERFLY IRIS], *I. brevicaulis* [LAMANCE IRIS] with cultivars such as 'Brevipes' and 'Flexicaulis', *I. missouriensis* [WESTERN BLUE FLAG] and *I. pseudacorus* [YELLOW IRIS, YELLOW FLAG, WATER FLAG]. Important species of the crested forms include *I. cristata* [DWARF CRESTED IRIS] and *I. tectorum* [WALL IRIS, ROOF IRIS].

Cultivated bulbous species in the subgenus *Xiphium* include *I. reticulata, I. latifolia* (= *I. xiphioides, I. pyrenaica*) [ENGLISH IRIS] and *I. xiphium* (= *I. hispanica, I. lusitanica*) [SPANISH IRIS]. *I. persica* [PERSIAN IRIS] and *I. magnifica* are two of the much smaller number of cultivated species of the subgenus *Scorpiris*.
IRIDACEAE, about 250 species.

Iroko one of the common names for *Chlorophora excelsa*, a tropical African tree which

Iris songarica is a widespread species of the dry windswept lowland steppes of Iran, Afghanistan and southern USSR.

produces a valuable timber suitable for furniture and building construction. It has sometimes been called FUSTIC or AFRICAN TEAK and is often used as a substitute for true teak.
MORACEAE.

Ironbarks members of the Australian genus *Eucalyptus* named for their heavily furrowed, usually dark barks. The durable hardwood of such trees as *E. fergusoni* [BLOODWOOD IRONBARK], *E. crebra* [NARROW LEAFED IRONBARK], *E. fibrosa* (= *E. siderophloia*) [BROAD LEAFED or RED IRONBARK] and *E. nanglei* [PINK IRONBARK] is much used for heavy construction purposes, railway sleepers etc.
MYRTACEAE.

Ironwood a common name used locally, or more widely, particularly in commerce, for at least 24 species, from several genera, of shrubs and trees, and for their tough, hard timber, which is sometimes heavier than water. The woods are used for a variety of purposes including furniture, fine cabinet work and veneer, turnery, tool handles, floors, telegraph poles and railway sleepers. In North America, *Ostrya virginiana* [HOP HORNBEAM] and species of *Carpinus* [*HORNBEAM] are known as IRONWOOD. Central and South American IRONWOODS include *Olneya tesota* [SONORA IRONWOOD], *Cyrilla racemiflora, Reynosia septentrionalis* and *Rhamnidium ferreum*. Most IRONWOODS are from the Old World tropics and Australasia. From the latter come IRONWOODS of the myrtle family: *Choricarpia subargentea* [GIANT IRONWOOD], *Austromyrtus acmenoides* [SCRUB IRONWOOD], *Backhousia myrtifolia* and species of *Eugenia*. At least five tropical genera of the family Leguminosae yield timber called IRONWOOD: *Acacia, *Afzelia, Copaifera, Cynometra* and *Intsia*. Other tropical Old World IRONWOODS are *Lophira alata* [DWARF IRONWOOD], *Eusideroxylon zwageri* [BORNEO IRONWOOD], and species of *Casuarina* [SOUTH-SEA IRONWOOD], *Hopea, Sideroxylon* and *Toddalia* [WHITE IRONWOOD].

Isatis a genus of annual, biennial or perennial herbs, native to Europe, the Mediterranean and western and Central Asia. The small yellow flowers are borne in great profusion in graceful loose panicles. Only *I. tinctoria* is commonly cultivated in gardens. This is more usually known by its common name of WOAD, or DYER'S WOAD, and was formerly used as a deep-blue dye before it was replaced by *indigo.
CRUCIFERAE, about 45 species.

Isoetes see Club mosses, p. 92.

Istle or **istli fiber** the name given to fibers obtained from the leaves of several species of the genus *Agave, notably *A. lecheguilla* [TULA ISTLE] a large succulent stemless plant which is native to Texas and Mexico, and *A. funkiana* [JAMAUVE ISTLE]. The fiber is sometimes also called ixtle or ixtli.
AGAVACEAE.

The pure white flowers of Iris laevigata 'Alba' make it one of the most attractive ornamental varieties of the genus. ($\times \frac{1}{15}$)

Itea a genus of deciduous or evergreen trees and shrubs mainly from tropical and temperate Asia with one species in the eastern USA. The most popular cultivated species are the North American *I. virginica* [SWEET SPIRE, VIRGINIA-WILLOW], a handsome deciduous shrub with fragrant, creamy-white flowers in erect, cylindrical racemes. *I. yunnanensis* and *I. ilicifolia* are evergreen shrubs with spiny-toothed leaves, originally native to China.
SAXIFRAGACEAE, about 10–15 species.

Ivy the common name for *Hedera*, a genus of mostly evergreen climbers or occasionally woody shrubs native to the northern parts of the Old World. Most cultivated ivies are

Iris afghanica was discovered only in the 1960s in the Central Hindu Kush of Afghanistan where it has limited distribution.

grown for their attractive heart-shaped or lobed, leathery, glossy and often variegated leaves.

H. helix [COMMON IVY] is found throughout Europe, western Asia, North Africa and is naturalized in the USA. It is a very hardy, adaptable species with very many cultivars of great ornamental value. *H. helix* var *hibernica* (= *H. hibernica* of horticulture) [IRISH IVY], with cultivars 'Hibernica' and 'Scotica', is a lustrous, larger-leaved, more vigorous variety making magnificent ground cover beneath trees. *H. canariensis* (= *H. maderensis*) is the strong-growing CANARY ISLAND or MADEIRA IVY. *H. colchica* [PERSIAN IVY] has an attractive variegated form, cultivar 'Dentata-variegata'.
ARALIACEAE, 6–15 species.

Ixia   a genus of cormous herbs native to South Africa (Cape Province), with grass-like leaves and stems 30–40cm high, usually unbranched, bearing spikes of flowers with slender tubes spreading out into six lobes. *I. viridiflora* has a blue-green perianth with a black throat. In *I. monadelpha* the colors range from yellow through orange to red and also blue, lilac and purple.
IRIDACEAE, about 30–40 species.

The orange-flowered Ixora chinensis *is an evergreen shrub from the Malay Peninsula and China. There are also varieties with white or red flowers.* ($\times \frac{1}{10}$)

Ixora   a large genus of evergreen shrubs and small trees originally native to the tropics and now widely cultivated in greenhouses and in the open in warm climates, giving rise to a profusion of cultivars. They have showy tubular flowers usually borne in dense terminal clusters. Frequently cultivated shrubs include *I. coccinea* [FLAME-OF-THE-WOODS], with bright red flowers, *I. odorata*, with very fragrant, white, pink-tinged flowers, *I. chinensis*, with red to white flowers (dark orange in cultivar 'Dixiana'), and *I. williamsii*, with deep red flowers.
RUBIACEAE, about 400 species.

Jacaranda   a tropical American genus of trees and shrubs with feathery leaves and beautiful panicles of blue to violet, rarely white or pink flowers. Among the most widely cultivated species used as street trees or garden ornamentals in the tropics and subtropics are *J. mimosifolia* (= *J. ovalifolia*), from northwest Argentina, with blue flowers, the blue-violet-flowered *J. cuspidifolia*, from Brazil and Argentina, *J. obtusifolia*, from Venezuela and Guiana, with blue-mauve to lilac flowers borne on older leafless branchlets, and its variety *rhombifolia* (= *J. filicifolia*).

The timber of some species, including *J. copaia*, *J. micrantha* and *J. mimosifolia*, is used for general carpentry and house building.
BIGNONIACEAE, about 50 species.

Jalap   a purgative drug obtained as a resin from the tuberous roots of *Ipomoea purga*, a climbing plant of the Mexican Andes, and related species. The roots of the unrelated Peruvian *Mirabilis jalapa* yield a false jalap with similar properties. The name also applies to *Podophyllum peltatum* [WILD JALAP, MAY APPLE], from North America.
CONVOLVULACEAE.

Jasione   a small genus of annual, biennial or perennial herbs native to Europe, the Mediterranean, and Asia Minor. They are usually under 30cm high. The leaves are mostly in a basal rosette and the blue flowers are in a dense terminal head. *J. humilis* and *J. montana* [SHEEP'S BIT SCABIOUS] are sometimes cultivated as rockery or border-edging plants.
CAMPANULACEAE, about 10–20 species.

Jasminum [JASMINES]   a large genus from temperate and tropical regions, excluding North America. Jasmines are evergreen or deciduous climbers or scrambling shrubs of spreading habit, with tubular white, yellow or red, often fragrant flowers usually in terminal or axillary cymes. Jasmines are easily cultivated species ranging from hardy outdoor to tropical greenhouse plants.

J. fruticans is a hardy, semi-evergreen shrub, with clusters of yellow flowers in early summer. One of the best winter-flowering shrubs is the deciduous *J. nudiflorum* [WINTER JASMINE], which produces solitary yellow flowers from November to February. *J. officinale* [COMMON WHITE JASMINE], from the

Himalayas, is a climber, producing fragrant white flowers which contain an essential oil used in perfumery. *J. polyanthum* and *J. mesnyi* (= *J. primulinum*) [PRIMROSE JASMINE], both from China, are first-class climbing shrubs for very mild localities or cool greenhouses. *J. sambac* [ARABIAN JASMINE], long cultivated and probably of Asian origin, is used for making jasmine-scented tea, and the flowers of *J. grandiflorum* [CATALONIAN JASMINE], possibly from Arabia, yield an essential oil used in perfumery. The large fragrant flowered *J. humile* and its cultivar 'Revolutum' [ITALIAN JASMINE] have been cultivated for more than 100 years.
OLEACEAE, about 250 species.

Jatropha   a genus of subtropical and tropical species of herbs, shrubs or small trees with milky or watery juice. *J. curcas* [FRENCH

*Winter Jasmine (*Jasminum nudiflorum*) is a great favorite with gardeners and flowers from mid autumn to early spring.* ($\times \frac{1}{2}$)

PHYSIC NUT, BARBADOS NUT, PURGING NUT], from tropical America, with yellow, purple or red flowers, is used as a hedge plant, and others such as *J. multifida* [CORAL PLANT, PHYSIC NUT], from tropical America, and *J. integerrima* [PEREGRINA], from Cuba, with scarlet to rose flowers, are grown as ornamentals for their bright flowers and fruits and variously divided or lobed leaves. A strong purgative oil, also used for candles and soapmaking, is obtained from the seeds of *J. curcas*.
EUPHORBIACEAE, about 170 species.

Jelly fungi   see Fungi, p. 152.

Jew's ear fungus   a jelly fungus known for a long time as *Auricularia auricula-judae*, but correctly named *A. auricula*. The earlier name (*judae* = of Judas, *auricula* = ear: Judas' ear) derived from the supposed belief that Judas hanged himself on an elder tree, *Sambucus*, the commonest host for the fungus. The fruiting body is a well-known delicacy in the East.
AURICULARIACEAE.

Joshua tree   the common name for *Yucca brevifolia*, which occurs in the desert and arid hills of the southwestern USA. The plant

forms a tree 5–15m high with a stout stem. It has fleshy, stiletto-like fibrous leaves clustered near the ends of the branches and long branching inflorescences of greenish-white flowers.
AGAVACEAE.

Jubaea a genus comprising a single species, *J. chilensis* (= *J. spectabilis*) [COQUITO or CHILEAN WINE PALM]. It is native only to the coast of Central Chile but because of its massive, elegant, shiny gray trunk, up to 10m high and 1m in diameter, it has been extensively planted for ornament in parks

The complexly-branched, orange-stalked inflorescence of small flowers of the Purging Nut (Jatropha curcas).

and along avenues in many areas with a Mediterranean climate. On evaporation the sap yields a sweet palm honey (miel de palma).
PALMAE, 1 species.

Juglans see WALNUT.
JUGLANDACEAE, 15 species.

Jumping beans the small, half-round capsular fruits of a Mexican shrub, *Sebastiana pavoniana*. A small butterfly (*Carpocapsa saltitans*) lays its eggs in the young fruit. Ultimately the eggs hatch out into larvae or maggots. The rolling or jumping of the fruits ("beans") is caused by keeping them in warm conditions which makes the larvae active within them. In Mediterranean regions the beetle *Nanodea tamarisci* parasitizes the fruit of *Tamarix gallica* with similar results.

Juncus [RUSHES, BOG RUSHES] a genus of small, clumped herbaceous plants with underground stems producing a single leafy shoot each year, and leaves which are sometimes reduced to basal sheaths. They occur mainly in wet places of most temperate countries and rarely in the tropics. The stems of *J. acutus* [SHARP RUSH], *J. effusus* (= *J. polyanthemus*) [SOFT RUSH, JAPANESE-MAT RUSH], and *J. inflexus* (= *J. glaucus*) [HARD RUSH] are still used in many parts of the world in mats, chair bottoms and basket work. *J. effusus* is particularly valued in southwestern Japan

Jacaranda mimosifolia from Argentina is widely cultivated in the tropics and subtropics where it grows into a gracefully branched tree.

The Jew's Ear Fungus (Auricularia auricula) is one of the larger jelly fungi, having a gelatinous fruit body. (× 2)

where it is used for weaving tatami, the floor covering used in many Japanese houses. Rushes sometimes serve as winter grazing on hillsides when grass becomes scarce.

Two species are hardy evergreen waterside ornamentals: *J. effusus* cultivar 'Spiralis' [CORKSCREW RUSH], with curious twisted stems, and *J. inflexus*, with needle-like leaves bending over when mature.
JUNCACEAE, about 300 species.

Juniper the common name for *Juniperus*, a genus of evergreen trees and shrubs widely distributed throughout the Northern Hemisphere, from the mountains of the tropics as far south as the equator and ranging as far north as the Arctic. Junipers have leaves of two kinds: the normal adult leaves are scale-like and linear; the juvenile leaves are larger and awl-shaped. Some species retain their juvenile foliage but generally only one type of leaf is present on the shoots at any given time.

Juniper wood is generally durable and easy to work; the presence of oils is probably responsible for the juniper's resistance to many insect attacks. The timber is used in general building, for roof shingles, furniture, posts and fences. In Burma, *J. recurva* var *coxii* [COFFIN JUNIPER] is the favored wood for coffins. *J. virginiana* [PENCIL CEDAR], is extensively used in the manufacture of pencils.

Cedar wood oil is obtained from a distillation of the sawdust, shavings etc and, until recently, was the main immersion oil used in high-power light microscopy. *J. oxycedrus* [PRICKLY JUNIPER] yields oil of cade or juniper tar by distillation of the wood. This oil has been used as a treatment for skin diseases, especially psoriasis, but is now largely replaced by coal-tar products, which are more effective. It is also used in the perfumery industry. Oil of juniper is distilled from the fully grown but unripe berries of *J. communis* [COMMON JUNIPER] and is responsible for the characteristic flavor of gin. Oil of savin, from *J. sabina* [SAVIN], is obtained by distilling fresh leaves and shoots. It is a powerful diuretic and has been used as an abortifacient.

Junipers are important ornamentals, with species and varieties suitable for many situations. Particularly rich in named cultivars and varieties are *J. chinensis* [CHINESE JUNIPER], *J. horizontalis* [COMMON or CREEPING JUNIPER], *J. scopulorum* [COLORADO RED CEDAR] and *J. virginiana* [RED CEDAR]. As well as containing tree and shrub forms these species also have representatives suitable for rock gardens and ground cover, such as the

To produce jute fiber, fresh stalks of Corchorus *spp are first softened by retting (prolonged soaking in water). In Bangladesh a local method makes use of the floating mats of Water Hyacinths (Eichhornia sp), to keep the stalks submerged.*

J. chinensis var *sargentii* [SARGENT JUNIPER], *J. chinensis* var *procumbens*, *J. communis* var *depressa* [GROUND JUNIPER] and *J. virginiana*

*The Common Soft Rush (*Juncus effusus*).*

'Prostrata', to name but a few. Similarly "color" forms are widely available; within *J. chinensis* var *chinensis* there are many such forms, for example 'Alba' with creamy-white twig tips, 'Aureo-globosa', with golden-yellow leaves and 'Variegata' with cream-colored branchlets. Other junipers cultivated as ornamentals include *J. drupacea* [SYRIAN or PLUM JUNIPER], *J. excelsa* [GREEK JUNIPER], *J. occidentalis* [SIERRA or CALIFORNIAN JUNIPER], *J. rigida* [NEEDLE JUNIPER] and *J. silicicola* [SOUTHERN RED CEDAR].
CUPRESSACEAE, about 70 species.

Jurinea a genus of thistle-like, erect herbs or subshrubs native to central and southwest Europe, southwest and Central Asia. The inflorescence is a single flower head made up of purple disk florets. A few species are cultivated in gardens.
COMPOSITAE, about 250 species.

Justicia a large genus of shrubs or perennial herbs, mainly native to the tropics and subtropics, but extending into temperate North America. A number of species are cultivated as ornamentals outdoors in warm countries, and under glass in cooler temperate zones. Differences between groups of species have been the basis for some authorities to segregate this genus into a number of distinct genera, including *Beloperone* and *Jacobinia*. The following, however, refers to *Justicia* in the wide sense.

*The white flower and pinkish-brown bracts of the Shrimp Plant (*Justicia brandegeana = Beloperone guttata*), from Mexico. (×1)*

One of the most widely cultivated ornamental species is *J. brandegeana* (= *Beloperone guttata*) [SHRIMP PLANT, MEXICAN SHRIMP PLANT, FALSE HOP], a rather weak-stemmed evergreen shrub with long pendent terminal flowering spikes bearing large attractive brownish-red to yellowish-green bracts.

Another popular greenhouse species is *J. carnea* (= *Jacobinia carnea*) [BRAZILIAN PLUME, PLUME FLOWER, FLAMINGO PLANT, PARADISE PLANT, KING'S CROWN] which has four-angled stems growing to 2m, bearing large prominently-veined leaves and dense terminal panicles of long pink-purple flowers. The Mexican *J. spicigera* (= *J. atramentaria, Jacobinia mohintli*) [MOHINTLI] is a shrub of approximately the same size but bears smaller leaves and axillary or terminal one-sided racemes of several orange-red elongated flowers. The leaves yield a dye as well as an extract with medicinal uses.
ACANTHACEAE, about 300 species.

Jute perhaps the most important textile fiber of the world next to cotton, jute is derived from two annual species of the genus *Corchorus, C. capsularis* [WHITE JUTE] and *C. olitorius* [UPLAND or TOSSA JUTE]. *C. olitorius* is pantropical but is often an escape rather than wild and it has been regarded as primarily African. *C. capsularis* is probably native to Indo-Burma although some authorities prefer China. Although *C. olitorius* has been cultivated as a minor vegetable in Africa and the Middle East for a very long time, the domestication of it and *C. capsularis* in India is relatively recent. The domestication of both species was the result of deliberate research to find new fiber crops to replace *hemp. Jute became important in the mid 19th century when Dundee spinners learned how to spin it.

Although there are about 100 species in the genus and many that yield useful bast fibers, it is only these two species which yield commercial quantities of the fiber predominantly used for the manufacture of hessian cloth (burlap) and twine. Less widespread uses include carpet yarns, cloth backings for heavy rugs, carpets and linoleum, webbing and cable covers. About 90% of the world's jute supply comes from India and Pakistan. The most suitable soils for plantation are those which are inundated near the river banks – the largest production by far is in the Ganges-Brahmaputra delta. *C. capsularis* is the most widely cultivated species because it has a better tolerance of these conditions. There are numerous different strains which are grown to suit varying ecological conditions and fiber qualities required.

Bark fibers are obtained by retting, ie separation from the woody stems by a combined action of softening by water, microorganisms and enzymes. After retting, the bast fibers are removed from the long woody stalks by hand and then washed, dried and finally bleached in the sun before being baled up.
TILIACEAE.

Kaempferia (= *Kaempfera*) a genus of more or less stemless perennial herbs native to tropical Africa and Asia. The tuberous rhizomes usually have a gingery smell and those of *K. galanga* [GALANGA] and *K. aethiopica* are sometimes used as a spice. Some species, such as *K. rotunda* from Southeast Asia [RESURRECTION LILY] and *K. atrovirens* [PEACOCK PLANT], from Borneo, are occasionally grown as ornamentals for their attractive trumpet-shaped white flowers with a lilac, pink or violet lip.
ZINGIBERACEAE, about 55 species.

Kaempferia rosea is a common grassland ginger of East Africa. The flowers appear at ground level at the base of the stem. (×¾)

Kaki or **kakee** the widely used Japanese vernacular name for *Diospyros kaki*, otherwise known as the CHINESE or JAPANESE DATE PLUM or KAKI PERSIMMON. Native to East Asia, the KAKI is a small tree with deciduous, simple leaves, extensively cultivated in warm climates for its orange- to red-skinned apple-like fruits. It is the favorite fruit of much of China and Japan.
EBENACEAE.

Kalanchoe a widespread genus of shrubby, tender leaf-succulents centered in tropical Africa and Madagascar. The genus is subdivided into three sections accorded separate generic status by some authorities, viz. *Bryophyllum, Kitchingia* and *Kalanchoe*. The flowers are erect in *Bryophyllum, Kitchingia* and pendent in *Kalanchoe*. The corolla in members of *Bryophyllum* is frequently constricted against the pistils, a feature not seen in members of the other sections.

The Felt Bush (Kalanchoe beharensis) cv 'Nuda'. Like the wild species it has large heart-shaped leaves but without the shaggy hairs.

Several species are popular houseplants, notably *K. blossfeldiana* and *K. flammea* and their various hybrids. *K. pumila* combines dark purple leaves covered in white powder, with dainty, deep pink flowers. This species, *K. manginii* and *K. uniflora* [KITCHINGIA] are suitable as basket plants. Many species that are grown outdoors in warmer climates are suitable for greenhouse cultivation in cooler temperate zones. These range in habit from the woody, scrambling *K. beauverdii* and the felty shrub *K. beharensis* [FELT BUSH, VELVET-

The Mountain Laurel (Kalmia latifolia) is a handsome shrub from North America that grows best in partial shade on lime-free soil. (× ½)

LEAF, VELVET ELEPHANT EAR] to the smaller shrubs, *K. integra* [FLAME KALANCHOE], with its many red-flowered varieties, and the whitish- or yellow-flowered *K. marmorata* [PEN-WIPER KALANCHOE]. Other popular species include *K. pinnata* (= *Bryophyllum pinnatum*) [AIR PLANT, LIFE PLANT, FLOPPERS, MOTHER-IN-LAW, MIRACLE LEAF, SPROUTING LEAF], the densely felty *K. tomentosa* [PUSSY-EARS, PLUSH PLANT, PANDA PLANT] and *K. tubiflora* [CHANDELIER PLANT].
CRASSULACEAE, about 125 species.

Kale a name referring to two groups of plants allied to the common cabbage: *Brassica oleracea* var *acephala* (or Acephala group) and *B. napus* (Pabularia group). There are many types in both groups, most of which, because of their extreme hardiness, have been cultivated for thousands of years, as livestock feed for cattle during winter or as a winter vegetable. The vegetable forms [often known as CURLY, CRIMPED or COTTAGER'S KALE, or BORECOLE] are cultivars of *B. napus*, growing up to 1m high, commonly with green curly leaves, although smooth-leaved and purple-leaved varieties are also grown.

Kales for stock feeding are mainly derived from *B. oleracea* var *acephala* and have two main types: THOUSAND HEADED KALES, with many leafy shoots and MARROW STEM KALES with fewer shoots but with larger, fleshier leaves and a thickened stem. HUNGRY GAP KALE, SIBERIAN KALE and RAGGED JACK KALE are forms of *B. napus*.
CRUCIFERAE.

Kalmia a genus of shrubs native to North America and Cuba, with opposite or whorled, entire, leaves. A number are cultivated as garden ornamentals. *K. angustifolia* [SHEEP LAUREL] reaches a maximum of 1m in height and produces clusters of purple or crimson saucer-shaped flowers, while *K. latifolia* [MOUNTAIN LAUREL, CALICO BUSH] often grows to a height of over 3m and produces dense clusters of saucer-shaped white-rose colored flowers. Other cultivated species are *K. microphylla* [WESTERN LAUREL] and *K. polliifolia* [BOG KALMIA].
ERICACEAE, about 6 species.

Kapok a fiber made from the usually white lustrous seed hairs inside the fruits of *Ceiba pentandra* [KAPOK TREE, SILK COTTON TREE], which grows to 30m. The species is bicontinental in distribution, possibly originating in Central or South America with fruits or seeds being transported in pre-Columbian times via sea currents to Africa. It probably reached Southeast Asia via India from Africa before the 6th century.

One tree bears annually up to 400 fruits, yielding about 2kg of kapok which can be used for stuffing quilts, pillows, life-jackets or for insulation. Kapok is lighter than cotton, water-repellent, elastic, buoyant, durable, and has good thermal and acoustic insulation properties. Although kapok has been substantially replaced by synthetic fibers, it is still produced on a commercial

Numerous adventitious leaf buds are produced by Kalanchoe daigremontiana, earning it the popular name of Mother of Thousands. (× 3)

scale, particularly in Thailand and Kampuchea which are the main exporters.
BOMBACACEAE.

Kariba weed the common name for the free-floating water-fern *Salvinia auriculata*. This particular species grew extensively after the construction in 1955–59 of the Kariba dam on the Zambezi River, but is now less prevalent and occurs mainly in the more sheltered parts.
SALVINIACEAE.

Kauri or **kauri pine** the Maori names (now universally used) for species of the genus *Agathis* is *A. robusta* [QUEENSLAND KAURI]. handsome coniferous tree that grows to a huge size. It is restricted to the northwestern peninsula of North Island, New Zealand. Very little of the virgin forest still remains. It

rarely cultivated and most cultivated *Agathis* are *A. robusta* [QUEENSLAND KAURI]. Besides being a useful source of timber, most kauri pines yield a resin (kauri gum). In addition, large quantities of an amber-like fossil resin (*copal*), are found preserved in peat bogs where kauri pines formerly grew.
ARAUCARIACEAE.

Kelp a common term for seaweeds, usually the brown algae, found in intertidal and subtidal waters and washed up on the shore. The name refers especially to the genera *Macrocystis* [GIANT KELPS], *Laminaria* [OARWEEDS] and *Nereocystis* [BULL KELPS].

Kerria a genus represented by a single species, *K. japonica* [JEW'S MALLOW]. It is a deciduous, spring-flowering shrub growing to a height of up to 3m. It is much cultivated outside its native China and Japan, especially the cultivar 'Pleniflora', which has double, orange-yellow flowers.
ROSACEAE, 1 species.

Khat, kat, chat, catta, qat or **Arabian tea** common names for *Catha edulis*, a shrub native to Ethiopia and formerly Arabia. It is cultivated in Ethiopia, Somalia, Yemen and Zambia. The leaves and young twigs contain a stimulant and are either

chewed fresh or made into a tea-like beverage (Arabian tea).
CELASTRACEAE.

Khaya a genus of trees native to tropical Africa and Madagascar. They constitute the true African mahoganies, valued for their hard, reddish, insect-resistant wood. *K. ivorensis* [RED MAHOGANY], *K. senegalensis* [AFRICAN MAHOGANY] and *K. nyasica* [NYASALAND MAHOGANY] are among the most important species.
MELIACEAE, 8 species.

Khus khus or **khas-khas** the common name for *Vetiveria zizanioides*, a tropical Indian grass cultivated in Asia for its fragrant roots which are woven into fans, baskets and mats. The roots and sweet scented rhizomes also yield an oil on distillation (vetiver oil) which is used in the manufacture of perfumes and soaps.
GRAMINEAE.

Kino the name given to a number of astringent gum-resins, soluble in water and used in medicines and locally in tanning. The name is sometimes also used for the trees from whose bark the resins are derived (usually by tapping), for example the East Indian *Pterocarpus marsupium* [MALABAR KINO], *P. erinaceus* [WEST AFRICAN KINO, SENEGAL ROSEWOOD], *Eucalyptus* species, notably *E. resinifera* (Australia), *Coccoloba uvifera* (southern Mexico to South America) and *Butea monosperma* (India, Malaysia).

Kirengeshoma a genus represented by a single species, *K. palmata*, a robust perennial from Japan and Korea with handsome, yellow cup-shaped flowers, sometimes cultivated in moist herbaceous borders.
SAXIFRAGACEAE, 1 species.

A Kapok Tree (Ceiba pentandra) growing in northern Sonora, Mexico, showing the down which is harvested as kapok.

Kleinia semperviva, from Tanzania and Ethiopia, showing its succulent leaves and stems and the caudex (the above-ground portion of rootstock). ($\times \frac{1}{3}$)

Kiri wood the name given to wood from the Japanese tree *Paulownia tomentosa* [KARRI TREE, PRINCESS TREE]. The wood is reddish brown in color, very light in weight and is used in Japan mainly for cabinet-making and musical instruments.
BIGNONIACEAE.

Kleinia a genus of succulent herbs and shrubs, mainly native to southern Africa and Arabia, some of which are commonly cultivated in bowl gardens or as houseplants on account of their novel and cactus-like appearance. The flower heads consist mostly of disk florets, white or red, rarely yellow or orange, which appear in autumn and winter. Most modern authorities now include this genus within *Senecio*.

The main cultivated species can be divided into three types: (1) Subshrubs such as *K. galpinii* (orange flower heads) and *K. fulgens* (orange-red flower heads). (2) Dwarf shrublets or creeping, rooting perennials such as *K. rowleyana* [STRING-OF-BEADS PLANT], with relatively large fragrant, predominantly white flower heads. (3) Stem succulents with deciduous leaves, such as *K. anteuphorbium*, a shrub up to 1.5m tall, named for its reputed value as an antidote to the poison of the cactus-like *Euphorbia* species among which it grows.

Other species in this group include *K. articulata* [CANDLE PLANT] and *K. pendula* [INCHWORM], so called because the short, fat, jointed stems arch back to the soil and progress in serpentine fashion.
COMPOSITAE, about 50 species.

Knautia a genus of annual, biennial and perennial herbs which, although exhibiting very showy flower heads, unlike the closely related *Scabiosa*, is rarely cultivated. It is mainly native to Europe and the Mediter-

ranean region. The flowers show a range of colors from white or yellow to pink, purple or violet. *K. arvensis* [BLUE-BUTTONS] and *K. dipsacifolia* (= *K. sylvatica*), both extremely variable species, are sometimes cultivated, as is the yellow-flowered *K. tatarica*. DIPSACACEAE, about 50 species.

Kniphofia [TORCH LILY, RED HOT POKER] a genus of tufted, rhizomatous, perennial herbs from South and East Africa and Madagascar. A number are hardy in cultivation in Europe and North America but others need cool or heated greenhouse conditions. The erect many-flowered spikes of cylindrical flowers range in color from yellow to orange or scarlet. *K. uvaria*, with several cultivated forms, and the more compact *K. galpinii* are two of the most widely cultivated species. LILIACEAE, about 65 species.

Kochia a genus of annual and perennial herbs and subshrubs, one of which, *K. scoparia* [SUMMER CYPRESS], is grown as a bedding plant for its attractive foliage which turns reddish purple in the autumn. The Australian *K. aphylla* [SALTBUSH] is used for livestock feed in times of drought. CHENOPODIACEAE, about 90 species.

Koelreuteria a genus of deciduous trees native to East Asia and Fiji. The best-known is *K. paniculata* [GOLDEN RAIN TREE, PRIDE OF INDIA, VARNISH TREE], growing to about 20m high, native to China but long cultivated and naturalized on seashores in Japan; it is also planted around temples. It is prized for its impressive panicles of yellow flowers, which are used in medicine in China. The fruits are bladder-like capsules which separate into papery, colored segments, and the seeds are used for necklaces in China. *K. elegans* [FLAMEGOLD], from Taiwan and Fiji, is also cultivated. SAPINDACEAE.

Kohlrabi (Brassica oleracea var gongylodes). (× ¼)

Kola nuts are seeds from species of Cola, a genus of smallish forest trees from West Africa. They are consumed for their stimulatory effects. (× ¾)

Kohleria [TREE GLOXINIAS] a genus of rhizomatous terrestrial herbs or shrubs from Mexico to northern South America. Several species, notably *K. amabilis*, *K. bogotensis*, *K. eriantha*, *K. digitaliflora* and *K. warszewiczii*, and hybrids derived from them, are cultivated as greenhouse ornamentals for their attractive inflorescences of tubular foxglove-like flowers. GESNERIACEAE, about 50 species.

Kohlrabi one of the cultivated forms of *Brassica oleracea* (var *gongylodes*), with a swollen green or purple stem which is eaten like a turnip. It is popular in some continental European countries and is occasionally grown elsewhere. It is also used as a stock feed in some European countries. CRUCIFERAE.

Kola the common name given to the seeds, and to certain species, of the tree genus *Cola*. The main cultivated species is *C. nitida*, which is native to the rain forests of West Africa, while *C. acuminata* is a minor crop in the rain forests and savannas of eastern West Africa. The fleshy seed coat is removed to leave the embryos, called kola "nuts", which are creamy-white, through pink, to dark purple. There is a strong social preference for white nuts. Each nut is up to 5 × 3cm and contains 2% caffeine, traces of theobromine and a glucoside, colanin, all of which contribute to its stimulatory effects. Kola nuts are chewed to stave off hunger, thirst or fatigue. In times of scarcity, a substitute is found in the seeds of the unrelated *Garcinia kola* [BITTER KOLA].

The total production of kola nuts is an estimated 175 000 tonnes per year, of which the major part comes from the plantations of Nigeria. Both species were introduced to South America during the Slave Trade but production there (Caribbean Islands and Brazil) is low. There is little demand from soft drink manufacturers in America and Europe, for the "cola" ingredients commonly used by them are not extracted from kola nuts but are synthetic chemicals resembling the natural compounds. STERCULIACEAE.

Kolkwitzia a genus consisting of a single species, *K. amabilis*, a bushy shrub native to China. The flowers are pink with yellow throats, borne in pairs in corymbs, and the species makes an attractive bush in cultivation. CAPRIFOLIACEAE, 1 species.

Kumquat a name given to the four or five East Asian species of shrubs and trees of the genus *Fortunella*. They bear small edible orange-colored fruits with a thick pungent rind surrounding a pulpy center. Of the two species commonly cultivated, *F. japonica* [ROUND KUMQUAT] has round fruits and *F. margarita* [OVAL KUMQUAT] has oval fruits. Hybrids between *Fortunella* and several species of *Citrus* and with *Poncirus trifoliata* have been produced. RUTACEAE.

L

Lablab a common name for *Dolichos lablab* [also known as BONAVIST or HYACINTH BEAN], an important legume, thought to be of Asian origin and long cultivated in India. It has also been introduced to Africa and other tropical countries. A strong-growing herbaceous perennial climber, it is usually cultivated as an annual. The seeds and pods are normally eaten when young and tender, boiled or in curries. The seeds contain a poisonous glucoside which can only be destroyed by cooking. The whole plant is used as fodder, or it can be plowed into the ground as green manure. Some varieties are grown as ornamentals bearing pink, purple or white flowers. LEGUMINOSAE.

+ Laburnocytisus a graft chimera resulting from the grafting of a scion of *Cytisus* on the stem of *Laburnum*. + *Laburnocytisus adamii* is a small tree produced by the grafting of the purple-flowered *Cytisus purpureus* on to the yellow-flowered *Laburnum anagyroides*. *L. adamii* has purple-yellow flowers, but occasional branches with yellow or purple flowers of the parents also occur. LEGUMINOSAE.

Laburnum a small central and south-

Lactarius subdulcis is a common species of gill fungus often found in beech woods. ($\times \frac{3}{4}$)

eastern European genus of two species of small trees and shrubs with attractive foliage, pendulous racemes of bright yellow flowers, and poisonous leaves and seeds.

Both species, *L. alpinum* [SCOTCH LABURNUM] and *L. anagyroides* (= *L. vulgare*) [COMMON LABURNUM, GOLDEN-CHAIN, BEAN-TREE, GOLDEN RAIN TREE] are much-planted ornamentals, but their seeds (and those of their hybrid, *L. × watereri*) are freely formed and can be fatally poisonous to humans and cattle. The hybrid, although with equally poisonous seeds, forms far fewer seed pods and is therefore the most widely planted laburnum. The poisonous principles are the alkaloids cytisine and laburnine. Laburnum heartwood is used in cabinetwork and inlay. LEGUMINOSAE, 2 species.

Lac or **shellac** a resinous substance secreted on the twigs of many trees of India, Burma and Thailand, including *Ficus religiosa*, *Schleichera oleosa* [LAC TREE], *Ziziphus mauritiana*, *Butea monosperma*, by the lac insect (*Tachardia lacca* = *Laccifer lacca*). Shellac is used for several industrial purposes, particularly in the manufacture of varnish and for electrical insulation.

Lace bark the common name for *Lagetta lintearia* (= *L. lagetto*), a small tree native to the West Indies. The inner bark is made up of fine interlacing fibers which resemble lace and are used for ornament. The name is also applied to other species, such as the New Zealand evergreen shrub *Hoheria sexstylosa* and *Pinus bungeana* [LACE BARK PINE].

Lachenalia a genus of bulbous perennials native to South Africa. The tubular or bell-shaped flowers are borne on spikes or racemes. Ornamental species include *L. aloides* (= *L. tricolor*) [CAPE COWSLIP], with spikes of yellow to orange or red flowers, often tinged or tipped with green or red, and *L. glaucina*, with white flowers, flushed yellow or red. LILIACEAE, about 55 species.

A "loose leaf" or "salad bowl" variety of the Common Lettuce (Lactuca sativa). *Several other* Lactuca *species are grown as ornamentals.*

Lacquer an exudate obtained from *Rhus verniciflua* [LACQUER or VARNISH TREE], from East Asia, employed in the famous decorative techniques of China and Japan. It is cultivated in Japan. When freshly tapped from the tree, lacquer is a thick brownish fluid but when applied to the required object, usually of wood, in many layers, it becomes dry and hard. Lacquered articles have traditionally been decorated with gold dust and gold leaf, mother-of-pearl and precious stones, and the lacquer itself sometimes colored by an admixture of pigments. Burmese lacquer is derived from *Melanorrhoea usitata*. ANACARDIACEAE.

Lactarius a genus of gill fungi. Most species occur in temperate zones and although some species are poisonous, *L. deliciosus* and *L. sanguifluus* are eaten in Europe, Asia and North Africa. *L. resinus* and *L. scrobiculatus* are esteemed in the USSR after salting down. RUSSULACEAE, about 120 species.

Lactuca a genus consisting of Eurasian and North American herbaceous biennials, perennials and shrubs, with some species in tropical and temperate southern Africa. The yellow, blue, purple or white florets are all strap-shaped and arranged in cylindrical heads. Species cultivated as ornamentals include *L. macrantha* (= *Mulgedium macranthum*) and *L. racemosa* (= *M. albanum*), both with blue flower heads. *L. sativa* is the COMMON *LETTUCE*. *L. serriola* [PRICKLY LETTUCE] is a common weed in many parts of the world. COMPOSITAE, about 100 species.

Laelia a genus of pseudobulbous, epiphytic orchids, widely distributed from Mexico to Brazil. They vary very much in overall size and in the shape and size of their leaves and flowers. The inflorescences are erect and bear up to five flowers which are always strikingly showy or otherwise attractive. *Laelia* species are in great demand for horticultural pur-

A Larch (Larix sp) shoot in spring, showing the emerging leaves, previous year's cones (top) and new season's female cones (bottom); the latter are often called "larch noses". (×2)

poses, being grown either as "pure" species, such as *L. anceps*, with many cultivars, *L. autumnalis* and *L. purpurata*, or in the form of intrageneric or intergeneric hybrids such as × *Brassolaeliocattleya*. The great popularity of certain species has put them in considerable danger of extinction, but *L. jongheana* from Brazil, for example, is now one of many orchid species protected by an international convention.
ORCHIDACEAE, about 30 species.

× **Laeliocattleya** a group of bigeneric hybrids resulting from crosses between the orchid genera *Laelia* and *Cattleya*. Examples include × *L.-c. albanensis* (*C. warneri* × *L. grandis*), × *L.-c. amanda* (*C. intermedia* × *L. lobata*) and × *L.-c. verelii* (*C. forbesii* × *L. lobata*).

Popular cultivated greenhouse forms include 'Aconagua' (predominantly white and purple flowers) and 'Edgard van Belle' (bright yellow flowers with red and white markings). ORCHIDACEAE.

Lagenaria see CALABASH GOURD.
CUCURBITACEAE, 6 species.

Lagerstroemia a genus of deciduous or evergreen trees and shrubs from south and East Asia, south to New Guinea and Australia. The best-known species is *L. indica* [CRAPE MYRTLE], a very showy shrub widely cultivated in warm temperate or subtropical gardens and municipal plantings, with crinkled red, scarlet or pink flowers. The timber of a number of species, including the Asian *L. speciosa* (= *L. flos-reginae*) [QUEEN CRAPE MYRTLE] and the Indian *L. microcarpa* (= *L. lanceolata*) [BENTEAK, NANAN WOOD], is close-grained and used for general construction work such as bridge and house building.
LYTHRACEAE, about 30 species.

Lagurus a genus represented by a single species, *L. ovatus* [HARE'S TAIL] an annual grass native to the Mediterranean region but now widely grown as an ornamental and for use by florists. The beautiful inflorescences are very persistent and are often dried and dyed with bright colors for use in winter decorations, bouquets and floral displays.
GRAMINEAE, 1 species.

Laminaria [OARWEED] a genus of large brown algae normally found growing in the sublittoral fringe; they are only rarely exposed to the air. The genus is common in the colder waters of the Northern Hemisphere. They are large plants up to 3m long and consist of an attaching organ (holdfast), a narrow stipe and a flattened and variously-shaped lamina or blade. Plants of this genus

The White Dead-nettle (Lamium album) has been used in folk remedies, and the leaves and young shoots are occasionally used as a vegetable. (×½)

Brilliant in full sun, these Lampranthus flowers are closing in the late afternoon light, but still produce a bright display of color. (×1/10)

are consumed for food in Asia and are also widely collected for fertilizer and for the production of alginates.
PHAEOPHYCEAE.

Lamium [DEAD-NETTLES] a genus of herbaceous species from Europe, Asia and Africa. The flowers usually have a hooded upper lip, and the two lateral corolla lobes on the lower lip are reduced to small teeth. *L. album* [WHITE DEAD-NETTLE] and *L. purpureum* [RED DEAD-NETTLE] are common weeds, especially in Europe. *L. galeobdolon* [YELLOW ARCHANGEL], a tall garden ornamental with yellow flowers, is now placed in its own genus and called *Lamiastrum galeobdolon* (= *Galeobdolon luteum*). Cultivars of *L. maculatum* [SPOTTED DEAD-NETTLE], particularly the white-flowered 'Album' and pink flowered 'Roseum', and the species *L. garganicum* and *L. orvala* [GIANT DEAD-NETTLE] are cultivated in shady sites in gardens.
LABIATAE, about 50 species.

Lampranthus a genus of South African leaf-succulents forming small, much-branched shrublets. Cultivated species are hardy only in exceptionally sheltered areas but are much propagated for summer bedding as they produce brilliant displays of glossy, daisy-like flowers in all colors except blue.
AIZOACEAE, possibly 100 species.

Landolphia a tropical African genus of woody shrubs and climbers, which are the source of the *rubber known commercially as AFRICAN or *MADAGASCAR RUBBER. The latex is chiefly obtained from *L. owariensis*, *L. heudelotii* and *L. kirkii*. *Landolphia* fruits are globose or pea-shaped and about the size of an orange; they are edible raw and are also made into an alcoholic beverage. A blue dye is obtained from the leaves and flowers of

L. comorensis. A few species, including *L. florida* and *L. owariensis* are cultivated in warm greenhouses for their large white or yellowish salver-shaped flowers.
APOCYNACEAE, about 50 species.

Lantana a genus of evergreen shrubs or herbs mainly native to tropical and subtropical America and the West Indies, with a few species in the Old World. A number of species are cultivated as greenhouse ornamentals for the attractive red, golden or white flowers. They include *L. camara* (= *L. aculeata*) [YELLOW SAGE] a prickly-stemmed shrub with many varieties. The aromatic leaves of this species are used in tropical America to make a tea-like beverage.
VERBENACEAE, about 150 species.

Lapageria a genus consisting of a single species, *L. rosea* [COPIHUE, CHILEAN BELL-FLOWER], an evergreen creeper confined to the cool temperate forests of southern central Chile. This handsome plant, with red or occasionally white flowers up to 15cm long, is the national flower of Chile.
LILIACEAE, 1 species.

Lapeirousia a genus of cormous herbs from tropical and southern Africa. The leaves are arranged in opposite ranks to form a fan, and the long tubular flowers spreading out into six lobes are borne in spikes or racemes opening to one side. Of several cultivated species, *L. laxa* (= *L. cruenta*) has red flowers with darker blotches on the three lower lobes, and *L. anceps* and *L. corymbosa* both have deep violet or purple flowers.
IRIDACEAE, about 60 species.

Lapsana a genus of annual and perennial temperate Eurasian herbs. The inflorescence comprises numerous small yellow flower

The Tamarack (Larix laricina) flourishes under forestry conditions in a wide range of soils and climates.

Beds of Kelp (Laminaria spp), such as those shown here from Plymouth, England, only become visible at low tide and very rarely become completely exposed to the air.

heads borne in loose clusters. *L. communis* [NIPPLEWORT] has been used as a salad plant.
COMPOSITAE, 9 species.

Larch the generally accepted popular name of the coniferous genus *Larix*, which consists of graceful deciduous trees native to north temperate regions. A number are cultivated as ornamentals, both for their attractive shape and for their foliage which turns golden-brown in late autumn.

Widely cultivated species include *L. decidua* (= *L. europaea*) [EUROPEAN LARCH], *L. laricina* [TAMARACK, EASTERN LARCH] and one of the tallest species (to 60m), the North American *L. occidentalis* [WESTERN LARCH]. Two other widely grown species are *L. kaempferi* (= *L. leptolepis*) [JAPANESE LARCH] and its hybrid with *L. decidua*, *L. × eurolepis*. One of the smaller cultivated larches is the East Asian *L. gmelinii* [DAHURIAN LARCH].

The timber of many species, notably EUROPEAN, EASTERN and WESTERN LARCH is valued for its durability and strength and is therefore widely used in constructional work as well as for interior furnishings. The bark of some species has medicinal uses and that of EUROPEAN LARCH is the source of Venice or Venetian turpentine.

Larix see LARCH.
PINACEAE, about 12 species.

Larrea [CREOSOTE BUSH] a small genus of evergreen shrubs native to the southwestern USA, Mexico and South America. Characteristically, the stems are jointed and swollen at the nodes, the leaves are pinnate and the yellow flowers are solitary and terminal. All species are resinous, as in the case of

L. tridentata (= *L. mexicana*), a strongly-scented shrub which grows to a height of 3m. The flower buds are pickled in vinegar and eaten as *capers. Extracts of this species are used medicinally in parts of Mexico for treating rheumatism, and leaf decoctions are used for fomentations.
ZYGOPHYLLACEAE, 3–4 species.

Lasthenia a small herbaceous genus native to western North America and central Chile, all but one species of which are spring-flowering annuals. *L. chrysostoma* [GOLD-FIELDS] and, less often, *L. glabrata* (= *L. californica*) and *L. coronaria* are cultivated for their showy yellow flower heads.
COMPOSITAE, 16 species.

Latania a genus of large solitary fan palms originally from the Mascarenes but now (especially *L. loddigesii* from Mauritius) cultivated throughout the tropics and subtropics.
PALMAE, 3 species.

Latex the name given to the juice, usually white but sometimes colored, which exudes when certain plants are cut or wounded. Poppy and lettuce are familiar examples. The ramifying system which contains it is made up of living cells or tubes called laticifers or laticiferous ducts. Commercially, latex is very important, since both medicinal drugs like the morphine alkaloids and structural material like rubber are provided by it.

A wide range of different substances may be present in the fluid depending on the parent plant. In true solution are sugars, organic acids, mineral ions, alkaloids (eg morphine) and other small molecules; in colloidal solution are proteins, enzymes (eg papain), mucilages and, perhaps most characteristically, the hydrophobic (lacking affinity for water) particles which impart the usual opaque milky appearance. Among materials composing the colloids, the most important are hydrocarbons belonging to the family of the isoprenoids, which includes such substances as rubber, resins, steroids and essential oils.

Commercially, latex reaches its greatest importance in *rubber, the particles of which may comprise about 35% of the latex. Important families are Euphorbiaceae (*Hevea brasiliensis* is the source of rubber), Papaveraceae (*Papaver somniferum* yields the opium alkaloids), Apocynaceae, Araceae, Asclepiadaceae, Compositae, Moraceae, and Sapotaceae.

Lathraea a genus of wholly parasitic herbaceous perennials with leaves reduced to scales and lacking chlorophyll, occurring naturally through temperate Europe and Asia.

L. squamaria [TOOTHWORT] is a fleshy, white to purple colored plant bearing violet to dark purple flowers, parasitic mainly on the roots of species of *Corylus, *Fagus and *Alnus. *L. clandestina* [PURPLE TOOTHWORT], which parasitizes the roots of species of *Salix, *Populus and Alnus (rarely of other trees) is a most attractive spring-flowering plant with bright mauve flowers up to 5cm long, arising in the axils of the scales on the rhizomes, at or just below the soil surface. Both species are cultivated for their showy flowers.
OROBANCHACEAE, 5 species.

Lathyrus a genus of annual and perennial, frequently climbing herbs with branched tendrils, native to temperate regions of Eurasia and the New World. Economically the most important genus is *L. sativus* [CHICKLING or GRASS PEA], which is cultivated, chiefly in India, mainly for animal fodder [see GRASS PEA]. Experimental breeding of *L. odoratus*

The Everlasting or Perennial Pea (Lathyrus latifolius) is native to southern Europe, but is naturalized elsewhere as a garden escape. It is widely cultivated in gardens where it will produce a magnificent show of flowers each year. Several cultivars are in cultivation. ($\times \frac{1}{10}$)

[SWEET PEA] has produced every flower color except yellow, and a new shape (the 'Spencer' form) in which the standard is wavy-margined. Other ornamentals, some used locally as food or fodder, are *L. latifolius* [EVERLASTING PEA], *L. tingitanus* [TANGIER PEA], *L. tuberosus* [GROUNDNUT PEA VINE, EARTHNUT PEA] and *L. aureus*, a bushy plant without tendrils.

A few perennials are widespread in north temperate regions; examples are *L. pratensis* [MEADOW VETCHLING] which inhabits meadows and scrub, and *L. japonicus* [SEA PEA] which colonizes maritime sands, shingle and lake shores in Asia, Europe and North America.
LEGUMINOSAE, about 130 species.

Laurel a general name applied to various shrubs or small trees, particularly those of the family Lauraceae. The most notable examples are *Laurus nobilis* [BAY LAUREL, SWEET BAY] and *Umbellularia californica* [CALIFORNIA LAUREL]. Some species of other families are also known as laurels. Examples are *Calophyllum inophyllum* [ALEXANDRIAN LAUREL], *Kalmia angustifolia* [SHEEP LAUREL], *Prunus laurocerasus* [CHERRY LAUREL], *Aucuba japonica* [JAPANESE LAUREL or SPOTTED LAUREL], *Daphne laureola* [SPURGE LAUREL] and *Magnolia virginiana* [SWAMP LAUREL].

Laurelia a genus consisting of two species of aromatic trees. *L. novae-zelandiae*, from New Zealand, provides a commercial timber, and *L. aromatica* [PERUVIAN NUTMEG], from Peru and Chile, produces seeds which are used locally as a spice. *L. novae-zelandiae* is sometimes grown as an ornamental.
MONIMIACEAE, 2 species.

Laurus a genus consisting of two species of aromatic evergreen trees, the better-known being the evergreen *L. nobilis* [BAY LAUREL, SWEET BAY], native to the Mediterranean region and widely cultivated. In ancient times it was associated with victory and nobility. The dried leaves are used as a condiment to flavor meat and fish dishes, especially in France. The leaf oils and an oily extract from the seeds (bay fat or laurel berry fat) are used in perfumery and medicine. The other species, *L. canariensis* [CANARY ISLAND LAUREL], is restricted to the forests of the Canary Islands and Madeira.
LAURACEAE, 2 species.

Lavandula a genus of perennial herbs, shrubs and subshrubs mainly native to warm temperate regions from the Canary Islands to India. Several species are cultivated for ornament, for their pleasant aromatic scent, as honey plants and for the extraction of oil from the flowers.

Some species of *Lavandula* are used in the preparation of lavender oil, which is obtained by steam distillation of the flower heads. The commonest is *L. angustifolia* [COMMON LAVENDER], a mainly Mediterranean species which is widely cultivated, especially in France, Italy and England for perfumery and as an

Young leaves of the Bay Laurel (Laurus nobilis), which when dried are used as a flavoring for meat and fish dishes. ($\times 1$)

ornamental. The names FRENCH or SPANISH LAVENDER are applied to another species, *L. stoechas*, which has been used since classical times as a medicinal plant and as a toilet preparation. Lavender oil is also obtained from *L. latifolia* [SPIKE LAVENDER] (formerly called, incorrectly, *L. spica*, as was *L. angustifolia*).

Lavender oil is often obtained from wild plants and distilled *in situ* but the quality is inferior to that of *L. angustifolia*. French lavender is also sold as bunched flowers and the dried flowers are powdered for sachets and used in potpourris. Various medicinal properties have been attributed to lavender since classical times. Poor-quality oils are used in the manufacture of lacquers and varnishes and in cheap perfumery.
LABIATAE, about 28 species.

Lavatera a genus of herbs and shrubs mostly from the Mediterranean region, but some also from the Canary Islands, Asia, Australia and California. Several species are cultivated, including *L. olbia* [TREE LAVATERA], a western Mediterranean shrub with solitary axillary red-purple flowers 5cm across, and *L. arborea* [TREE MALLOW], a European biennial with clusters of pale purple flowers 5cm across. Large white- and pink-flowered varieties of the annual *L. trimestris* (= *L. rosea*) flower throughout the summer.
MALVACEAE, about 25 species.

Lavender the common name for a number of species of the genus *Lavandula*, particularly for *L. angustifolia* [COMMON LAVENDER]. However, species of other genera are also called LAVENDER, eg *Limonium vulgare* [SEA LAVENDER] and *Heliotropium curassavicum* [WILD LAVENDER].

Laver bread a food made from boiled, pulped and pressed algae of the genus *Porphyra*. It is eaten in various parts of Europe, particularly in South Wales. Laver bread has a jelly-like consistency and is usually mixed or coated with oatmeal and fried.
RHODOPHYCEAE.

Lawsonia see HENNA.
LYTHRACEAE, 1 species.

Layia a small genus of hairy annual herbs mainly from western North America (California). Their daisy-like flower heads have yellow disk florets and ray florets which may be white, golden-yellow or yellow with white tips. *L. platyglossa* (= *L. elegans*) [TIDY-TIPS], a bushy plant, is commonly cultivated as an ornamental.
COMPOSITAE, about 12 species.

Leek the common name for *Allium porrum*, a popular winter vegetable, especially in northern Europe. The edible portion consists of a false stem of concentric leaf-bases. It is a cultivated form of *A. ampeloprasum* along with the very similar KURRAT (*A. kurrat*). It was known as far back as 3200 BC in ancient civilizations of the Middle East and was popular as a vegetable in the Middle Ages. It is also used as a flavoring.
LILIACEAE.

Legumes and pulses see p. 194.

Leiophyllum [SAND MYRTLES] a small genus of evergreen shrubs native to eastern North

The Purple Toothwort (Lathraea clandestina) is a parasite of trees and shrubs, particularly willows (Salix spp). As a total parasite, it produces no green leaves, just short stems enclosed in white scales, gaining its nourishment from the host. (×½)

A grove of lemons (Citrus limon) in Greece, along the shore of the Corinthian Gulf. This scene is in great contrast to the vast and highly mechanized plantations of California, which is second only to the Mediterranean as a center of production.

America, two of which are cultivated as spring-flowering ornamentals. Both the prostrate, short, upright *L. buxifolium* and the prostrate *L. lyonii* (= *L. buxifolium* var *prostratum*), sometimes treated as a single variable species, produce attractive white or pink terminal inflorescences.
CYPERACEAE, 1 or 3 species.

Lemna see DUCKWEED.
LEMNACEAE, about 9 species.

Lemon the common name for the fruit and tree of *Citrus limon*. The species is considered to be of Himalayan or Southeast Asian origin, but is cultivated in areas of Mediterranean climate such as Spain, Italy, Cyprus and California. The pulp of the yellow fruits is juicy (minimum 28%) and acid (5–7%), mainly citric but also malic and other acids.

Lemons are the most important acid fruit. Acid and non-acid cultivars are cultivated, the rind producing lemon oil. In 1971, together with limes, they accounted for 10% of the total world citrus production, ranking second to oranges. Mediterranean countries accounted for 37% of production, the USA and Argentina being the other major producers.
RUTACEAE.

Lentil the common name given to the genus *Lens*, which contains five species of climbing herbs native to the Mediterranean region and southwestern Asia. The wild species are unimportant apart from *L. orientalis*, which is probably the progenitor of the valuable pulse *L. culinaris* (= *L. esculenta*). This is a small pea-like annual herb with white, blue-tinged flowers developing into short flattened pods containing two seeds (lentils).

Two main forms exist: lentils 6–9mm in diameter belong to a group known as *macrosperma*, while smaller forms (3–6mm in diameter) belong to the *microsperma* group. These types have existed for several thousand years. There is also great variation in color, from pale straw or greenish through light brown to dark brown. Lentils are mainly grown in the Mediterranean Area, Ethiopia, southwestern Asia, the Indian subcontinent, Chile, Argentina and the USA. World production in 1975 was estimated at over 1.2 million tonnes.

Lentils have a high food value and contain about 25% protein, 50% carbohydrate and 2% vegetable oil. They are usually cooked in boiling water as an ingredient of soups, stews or, in India, as "dal" or "dhal", a thick lentil sauce used with curry dishes. The young pods of some varieties may also be cooked as a vegetable dish. After harvesting, the pods, stems and leaves may be used as fodder. Lentil meal, prepared from milled lentils, may also be used as animal fodder and for making a type of bread.
LEGUMINOSAE.

Mature Leek plants (Allium porrum) ready for harvesting. (×⅙)

Legumes and Pulses

The family Leguminosae is second only to the grass family, the Gramineae, in its importance to Man. The term legume refers to the characteristic fruits of the family – basically a dehiscent pod that develops from a single carpel and splits into two valves, although there are many deviations from this general structure, some fruits in the family being indehiscent and drupe-like, others transversely divided, and they may be dry or fleshy, winged or not. The name legume is also applied to those members of the family which are edible – either the pods themselves or the seeds (when they are called pulses, grain legumes, or beans), or both. The leaves are rarely eaten by Man, as in the case of species of *Ptero-carpus* grown in parts of Nigeria, but several species are important fodder crops

Common temperate legumes and pulses: 1 Scarlet Runner Bean; 2 Kidney Bean; 3 Haricot Bean; 4 French; 5 Lentil; 6 Broad Bean; 7 Garden Pea; 8 Asparagus Pea; 1 to 8 (× ½); whole plant (× 1/10)

(see p. 145). Members of the Leguminosae also provide important timber trees, sources of dyes and tannins, gums and resins, oil seed crops, medicinal and insecticidal species as well as numerous well-known ornamental species. They are also an important component of the vegetation in many parts of the world. Here, however, we are only concerned with edible legumes and pulses.

Nutritionally, legumes are very important, second only to cereals as a source of human food. They are two or three times richer than cereals in protein, some are rich in oil, such as *soybeans (Glycine max)*, and ground nuts (*Arachis hypogaea*), and in terms of their amino-acid composition they complement the cereals, so that a mixed diet of pulses and cereals is nutritionally well balanced and traditional in several civilizations.

Legume grains or pulses are still major components of the diet in the Indian subcontinent (especially *lentil, Lens*

culinaris; *pigeon pea, Cajanus cajan*; and *chickpea, *Cicer arietinum*), the Far East (particularly *soybean), and Latin America (particularly the bean, *Phaseolus vulgaris*).

Only about twenty species of legumes (out of the thousands that are known to science) are widely used for food, such as *peas (Pisum species), *beans (Phaseolus species), lentil, ground nut, *cowpea (Vigna unguiculata), grams (*Vigna species), *mung bean (Phaseolus aureus) and *pigeon pea (*Cajanus cajan). Many tropical species have great potential, such as the bambara groundnut (Voandzeia subterranea) and the *lablab or hyacinth bean (*Dolichos lablab), and efforts are being made to exploit them more fully as human food. In temperate climates, peas, beans, lentils and lupins (Lupinus species) are the main edible legumes. The common or garden *pea (Pisum sativum) probably originated in the Near East and is now cultivated in most temperate regions and at high altitudes in the tropics. It is one of the four most important grain legumes and in the dried state was once a staple food of Western Europe, as pea meal or split peas. The immature pods are traditionally harvested for the young seeds which are used as the vegetable, fresh peas. They are also grown for canning, but a very large part of the crop is now harvested immature for freezing, garden peas being one of the frozen foods most in demand.

Beans derived from species of the genus *Phaseolus* are cultivated in both the Old and New Worlds. In tropical countries, the seeds are used largely as dry beans, whereas in temperate and Mediterranean countries, although there is some consumption of dry beans, cultivars have been developed for use as green vegetables such as the immature pods of *Phaseolus vulgaris* [french or snap bean]. The dried seeds of this species are the haricot beans of commerce used in stews and in sauce as canned baked beans. *P. coccineus* [scarlet runner bean], a Middle American species, is also grown in Europe for its fleshy immature pods. The tropical species *P. acutifolius* [tepary bean] is a drought-resistant crop which is grown for its dry beans which have a high protein content, but it is hardly ever grown outside its native America.

The field or broad bean (*Vicia faba*) is an important legume in many parts of the north temperate zone and in some subtropical areas at higher altitudes. The seeds are large and rich in protein and are consumed green and immature or ripe and dried.

The most important grain legume in terms of world trade and production is the soybean. The leading producer is the United States, where the rise to prominence of the soybean crop in the last 50–60 years has been spectacular, to the extent that it is now the most important cash crop in the United States and a major export. The main use of

LEGUMES AND PULSES

I. Cool Temperate and Warm Temperate

Common name	Scientific name	Part consumed
GARDEN *PEA	Pisum sativum	Seeds, young pods
FIELD PEA	P. arvense	Seeds
ASPARAGUS PEA, WINGED PEA	Tetragonolobus purpureus	
FRENCH, KIDNEY, *HARICOT, GREEN, RUNNER, STRING, SALAD, WAX BEAN	*Phaseolus vulgaris	Young pods, seeds
RUNNER, SCARLET RUNNER	P. coccineus	Young pods
BUTTER, SIEVA, CIVET, MADAGASCAR, CAROLINA SEWEE BEAN	P. lunatus	Seeds
LIMA BEAN	P. limensis	Seeds
*SOYBEAN	Glycine max (G. soja)	Seeds, sprouts, oil
*LENTIL	Lens culinaris	Seeds
BROAD BEAN	*Vicia faba	Seeds
LUPIN	*Lupinus albus L. pilosus, L. luteus, L. mutabilis	Seeds
*CAROB BEAN, LOCUST BEAN, ST JOHN'S BREAD	Ceratonia siliqua	Pods

II Tropical

Common name	Scientific name	Part consumed
TEPARY BEAN	*Phaseolus acutifolius var latifolius	Seeds
CLUSTER BEAN, GUAR	Cyamopsis tetragonolobus	Young pods, seeds
GOA BEAN, *ASPARAGUS PEA, WINGED PEA	Psophocarpus tetragonolobus	Young pods
	P. palmett-orum	Young pods
*YAM BEAN, CHOPSUI POTATO	Pachyrhizus erosus, P. tuberosus	Young pods, roots
*LABLAB, HYACINTH BEAN	*Dolichos lablab	Pods, seeds
MADRAS GRAM, HORSE GRAM	D. biflorus	Seeds
CHICK PEA	Cicer arietinum	Seeds
BAMBARA GROUNDNUT, KAFFIR PEA	Voandzeia subterranea	Seeds
KERSTING'S GROUNDNUT	Kerstingiella geocarpa	Seeds
*TAMARIND	Tamarindus indica	Pulp from pods, seeds
MOTH BEAN	*Vigna aconitifolia	Seeds
ADZUKI BEAN	V. angularis	Seeds
*COWPEA	Vigna unguiculata	Seeds
BLACK-EYED PEA	subspecies unguiculata	Seeds
YARD LONG BEAN	subspecies sesquipedalis	Pods
BLACK GRAM	Vigna mungo (Phaseolus mungo)	Seeds, young pods
GREEN GRAM, *MUNG BEAN	V. radiata (Phaseolus aureus)	Seeds Pods, sprouts
RICE BEAN	V. umbellata	Seeds
JACK BEAN	*Canavalia ensiformis	Young pods, seeds
SWORD BEAN	C. gladiata	Young pods, seeds
GROUNDNUT	*Arachis hypogaea	Seeds, oil
*PIGEON PEA, CAJAN, CONGO PEA, RED GRAM	Cajanus cajan	Seeds
AFRICAN LOCUST BEAN	*Parkia filicoidea, P. biglobosa	Seeds, pulp of pod
*YAM BEAN	Sphenostylis stenocarpa	Seeds

Common warm temperate and tropical legumes and pulses: 1 Cowpea; 2 Lablab; 3 Soybean; 4 Chick-pea; 5 Jack Bean; 6 Butter Bean; 7 Groundnut; 8 Pigeon Pea. 1, 2 ($\times \frac{2}{3}$); 3, 6, 7, 8 ($\times \frac{1}{2}$); 4, 5 ($\times \frac{1}{3}$); whole plant ($\times \frac{1}{10}$).

the beans is the production of protein-rich meal and oil. They are also consumed fresh, as bean sprouts and in liquid or curd form, especially in the Far East.

LENTILS are one of the oldest legume pulse crops of the New World and were involved in the origins of agriculture in the Near East along with wheat and barley. The seeds contain a high percentage of protein and are widely consumed in the Indian subcontinent, the Middle East and the Mediterranean. It has also been introduced into the New World, in Argentina, Chile and parts of the United States.

Tropical legumes used for human food are many and various, but few are cultivated on a major scale. The most widely cultivated crop species are the COWPEA, grown as a vegetable or as a pulse throughout the tropics and subtropics, the GROUND NUT, grown in warm temperate and tropical regions around the world, for vegetable oil or as an appetizer, the PIGEON PEA, a pulse crop grown by small farmers mainly in India, but with some production in Southeast Asia and equatorial Africa, and the SOYBEAN already mentioned. Minor tropical crop species are listed in the table.

It is often not realized that lupins have been in cultivation as agricultural crops, particularly in South America, since earliest times, being used for animal forage and as a source of grain. L. mutabilis was once a major source of protein in the Andes. The seeds, however, have a high alkaloid content and as agricultural practices improved and better legumes became available, they dropped out of favor. Interest in them as a protein source has now revived and research is being carried out into alkaloid-free varieties.

Legumes are additionally important in cultivation because of the association in their root nodules with nitrogen-fixing bacteria which are able to convert free atmospheric nitrogen into nitrates. Their value as green manure which can be ploughed in to enhance the nitrogen levels in the soils is especially important in shifting cultivation in the tropics.

Leontodon a genus of annual, more usually perennial herbs, common in Europe and southwest Asia. Although a number of species bear attractive yellow, orange, pink or purple flower heads, none appears to be cultivated.
COMPOSITAE, 50–60 species.

Leontopodium a genus of tufted, downy-woolly, perennial herbs found in the mountains of Europe and Asia, with two species in Andean South America. The best-known species is *L. alpinum* [EDELWEISS], which is widely grown as an alpine rock-garden plant.
COMPOSITAE, about 30–40 species.

Lepidium a cosmopolitan genus of annual to perennial herbs. The most important species is *L. sativum* [COMMON or GARDEN CRESS], a native of western Asia and Egypt, which is used as a garnish or a salad. It is the long, succulent hypocotyls of the seedlings which are used.
CRUCIFERAE, about 130 species.

Leptospermum a genus of evergreen shrubs and trees mainly from Australia but

Leontodon autumnale, *photographed in a dry meadow in September. Each strap-shaped "petal" is in fact a separate flower (floret). (×1)*

with some in New Zealand, Malaysia and the Caroline Islands. The New Zealand species *L. scoparium* [MANUKA, TEA TREE, BROWN TEA TREE] is an ornamental tree to 7m in height with small white flowers. Pink or reddish-flowered cultivars are also grown. The leaves of *L. scoparium* and *L. thea* have been used locally as a tea substitute. The hard durable timber of *L. ericoides* [HEATH TEA TREE] is used locally in New Zealand for spokes and wheels. *L. petersonii* (= *L. citratum*), native to Australia, is grown commercially in Kenya and Guatemala for the lemon-scented essential oil obtained from its leaves.
MYRTACEAE, about 50 species.

Lespedeza [BUSH CLOVERS] a genus of annual and perennial herbs and shrubs from the Himalayas, East Asia, Australia and North America. *L. bicolor* is cultivated for the

ornamental value of its rosy-purple flowers. The annuals *L. striata* [JAPANESE CLOVER] and *L. stipulacea* [KOREAN CLOVER] and the perennial *L. cuneata* (= *L. sericea*) are Asiatic species grown also in North America for fodder, hay and green manure.
LEGUMINOSAE, about 100 species.

Lettuce the common name for many of the species of the herbaceous genus *Lactuca* including the cultivated *L. sativa* and the closely related *L. serriola* [PRICKLY LETTUCE], *L. saligna* [WILLOW-LEAVED LETTUCE] and *L. virosa* [POISON, OPIUM or BITTER LETTUCE]. The three latter species are weeds of roadsides and waste places and originated around the Mediterranean, from where they have spread to most parts of the world. *L. sativa* does not occur in the wild except as an escape from cultivation and probably originated by

Commercially cultivated Lettuce (Lactuca sativa) growing under glass ready for out-of-season marketing.

Leptospermum flavescens, photographed after a rainstorm in the usually dry Pilliga scrub area in New South Wales, Australia.

Man's selection from its close relative *L. serriola*.

Lettuce is cultivated almost exclusively for fresh consumption as a salad vegetable and occurs in a number of forms, of which the most important are the cabbage lettuces and the cos or Romaine. In the cabbage lettuces, which include both the crisp-head and the butterhead, the relatively broad, succulent leaves overlap to form, when mature, a roughly spherical heart or head. The crisp-head type has leaves which are more crinkled and frilled at the edges, and of a more brittle texture than those of the somewhat limp-leaved butterhead type. Together these main types include several hundred cultivars.

Other types of lettuce include the "loose-leaf" in which relatively small leaves are copiously produced in a completely open rosette, and the asparagus or stem lettuce grown in and around China for consumption of the young fleshy stems.
COMPOSITAE.

Leucanthemum a genus of mainly perennial herbs with conspicuous flower heads of yellow disk and white ray flowers, mostly native to Europe and northern Asia. The commonest and most widespread species is the extremely variable *L. vulgare* (= *Chrysanthemum leucanthemum*) [OX-EYE DAISY, MOON DAISY, MARGUERITE]. Cultivated forms are grown in gardens in borders and used as cut flowers. The OX-EYE DAISY has traditionally been used as a medicinal herb and has properties similar to those of chamomile.

L. maximum, from the Pyrenees, is also widely cultivated in gardens for ornament. Several very large-headed forms, such as the

SHASTA DAISY, have been derived from it. These sometimes have double white centers or serrated or fringed strap-shaped flowers. Another related species, *L. atratum*, is also cultivated in borders. Like *L. vulgare* it is highly variable in the wild and contains several distinct subspecies.
COMPOSITAE, 20–50 species.

Leucobryum a genus of mosses, mainly tropical in distribution. The plants mostly form dense cushions which hold water like a sponge. The leaves, unlike those of most mosses, are several cell layers thick and when the moss becomes dry the water-holding layers become air-filled, thus imparting a characteristic whitish appearance.
DICRANACEAE, about 120 species.

Leucojum [SNOWFLAKES] a genus of bulbous perennial herbs from southern and central Europe and Morocco, closely related to the SNOWDROPS. *L. vernum* [SPRING SNOWFLAKE], from central Europe, has solitary flowers. *L. aestivum* [SUMMER SNOWFLAKE], from central and southern Europe, is more robust with stems 50cm high. The flowers are borne in umbels of two to nine flowers in April. *L. autumnale*, a Mediterranean species, has very narrow leaves, and pale pink flowers, one to three (usually two) per stem, in the autumn just before the leaves appear.
AMARYLLIDACEAE, about 12 species.

Leucothoe a genus of evergreen and deciduous shrubs from North and South America, East Asia and Madagascar. They bear crowded racemes of white, pink or greenish-white, urn-shaped flowers. Several species are known in cultivation, such as the deciduous shrub *L. racemosa* (= *Andromeda racemosa*) [SWEETBELLS] and *L. fontanesiana* (often offered as *L. catesbaei* in cultivation)

Leucobryum glaucum growing in a wood. The gray-green cushions of this moss are often collected for decoration.

[DOG-HOBBLE], a spreading evergreen shrub up to 2m in height, both from North America.
ERICACEAE, about 44 species.

Lewisia a genus of highly xerophytic, perennial, leaf-succulent rosette plants from the western USA through Mexico to Bolivia. The leaves are flat or cylindric and sometimes deciduous in the resting period and the flowers are often showy. Most of the plants in cultivation are of hybrid origin. *L. rediviva* [BITTER ROOT] is the State flower of Idaho.
PORTULACACEAE, about 16 species.

Leycesteria a small genus of Himalayan shrubs of which *L. formosa* is the most widely cultivated. It is a vigorous, deciduous, hollow-stemmed shrub with arches or drooping spikes of flowers, used in woodland coverts and as background planting.
CAPRIFOLIACEAE, 6 species.

Liatris [BLAZING STAR, GAY FEATHER, BUTTON SNAKEROOT] a genus of perennial, cormous or rhizomatous herbs native to North America. *L. spicata* is frequently cultivated in the herbaceous border for its dense spikes of bright purple, reddish-purple or white, thistle-like flower heads on stout stems up to 90cm tall. Other cultivated species are *L. graminifolia* and *L. scariosa*, both with purple or white spikes.
COMPOSITAE, about 40 species.

Libertia a genus of tufted perennial herbs native to New Guinea, Australia, New Zealand and Chile. The leaves are arranged in fans and the flowers are white or pale blue, borne in dense clusters in the axis of sheathing bracts, as in *L. grandiflora*. Although they are fine garden plants, they are not commonly cultivated.
IRIDACEAE, about 12 species.

Libocedrus a small genus of evergreen coniferous trees with scale-like leaves, native to

A columnar form of Libocedrus decurrens (= Calocedrus decurrens), a native of western North America where it normally has a conical form.

New Zealand and New Caledonia. These trees, especially *L. decurrens* [CALIFORNIA INCENSE CEDAR], sometimes placed in a separate genus, *Calocedrus*, *L. chilensis* [CHILEAN INCENSE TREE], sometimes placed in *Austrocedrus*, and *L. plumosa* [KAWAKA] are valuable sources of close-grained timber, which is soft but durable and used in general construction.
CUPRESSACEAE, about 5 species.

Lichens see p. 198.

Licorice or **liquorice** a medicinal and flavoring material extracted from the roots of *Glycyrrhiza glabra*, a perennial herb from southern Europe and Asia. Licorice has been valued since ancient times for medicinal purposes, not least for disguising the taste of nauseous prescriptions, but also as a demulcent and expectorant. It is also used to color and flavor confectionery, tobacco and beer, in shoe polish and in metallurgy. It contains glycyrrhizin, many times sweeter than sugar, and the waste fibers of the plant are used in wallboard. Today, it is mainly obtained from the USSR, Spain and the Middle East.

G. lepidota, from North America, is known as WILD LICORICE. Other leguminous plants known as LICORICE include *Abrus precatorius* [INDIAN or WILD LICORICE], *Astragalus glycyphyllus* [WILD LICORICE] and *Hedysarum mackenzii* [LICORICE ROOT].
LEGUMINOSAE.

Lichens

LICHENS ARE SLOW-GROWING PLANTS THAT consist of an intimate association between a fungus and an alga. In most lichens the bulk of the plant body is composed of the fungus, with the algae restricted to a thin layer near the surface. The algae are green or blue-green and are either unicellular or form simple short filaments. In nature, lichen fungi never occur free-living. Approximately 18 000 species of fungi occur in lichens, so they are a large group comprising about 25% of the known species of fungi. The commonest alga, found in about 70% of species, is the green unicellular *Trebouxia*. This alga is not found outside lichens, although all the other 26 genera of algae found in lichens may also be found free-living.

Most lichens occur as crusts either closely adhering to or actually within the surface of their substrate – typically rock or wood, but occasionally very stable soil surfaces. Much more prominent as components of the vegetation are the minority of lichens that are less closely attached to the substrate and are leafy, shrub-like or filamentous. Shrub-like forms such as REINDEER MOSS and ICELAND MOSS cover large areas of ground in the tundra regions.

Lichens have a worldwide distribution. They extend farther towards the poles and higher up mountains (up to 7 500m) than any other plants of comparable size, but may also feature prominently on trees in tropical and equatorial forests. Very few species live permanently submerged and these grow mostly on shallow rocks in freshwater streams. On rocky seashores, they occur almost down to the low water mark, but rarely grow below it.

Lichens are very slow-growing plants. Those existing as surface crusts rarely have a radial growth exceeding 1mm per year (and often much less than this), while even the leafy or shrubby kinds rarely grow more than 1cm per year. Because lichens grow so slowly, and because of the difficulty of continuously observing them over long periods of time in nature, it is not easy to observe how they reproduce. Many lichens have small powdery areas on their surface, each individual grain of powder consisting of an algal cell or short filament surrounded by fungal hyphae. In the few cases studied, these grains appear capable of slowly developing into new plants. Other lichens produce numerous fruiting bodies of the fungus. How-

ever, the germinating fungal spore must encounter the appropriate alga in nature before a new lichen can be formed, but such syntheses have only rarely been observed.

Very few lichens are found in urban areas because they are particularly sensitive to atmospheric pollution, especially sulfur dioxide. Indeed, they are totally absent from the centers of the more heavily polluted areas. It is only at relatively long distances from large conurbations that the full complement of lichen species can be found. One reason why lichens may be so sensitive to atmospheric pollution is that they have developed very efficient mechanisms for absorbing nutrients from the liquids passing over their surface – usually rainwater. Such an adaptation is

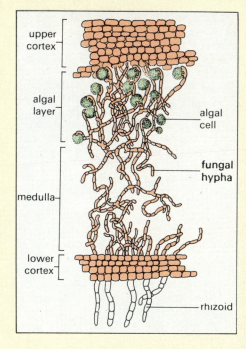

The bulk of the body is fungal (orange), while the algal cells (green) are confined to a layer below the upper surface.

The hanging lichen, Usnea longissima, *growing from a tree, near Mount Everest, at an altitude of nearly 2 500m. (× $\frac{1}{15}$).*

Below A mosaic of gray lichens growing on a rock in Australia. (× $\frac{1}{2}$)
Above Right Cross section of the thallus of a lichen.

Above *A mosaic of orange and yellow crustose lichens growing on a bare rock surface in California. Lichens are often the first colonizers of bare surfaces, slowly building up an organic-based surface which other plants can later colonize.* (×1)
Left *The fruticose (shrubby) lichen* Ramelina siliquosa *growing on rocks by the sea. This species thrives within the salt spray zone.* (×2)

antarctic regions, lichens of the reindeer moss type (*Cladonia* and *Cetraria* species) provide fodder for animals such as reindeer and caribou. However, most other animals (including human beings) gain little nutritive value from eating lichens since they do not have the digestive enzymes necessary for breaking down the unusual carbohydrates they contain. Additionally, many lichens contain substances generally called "lichen acids", some of which irritate the stomach unless the lichen is first boiled. Nevertheless ICELAND MOSS (*Cetraria islandica*) has been used as a substitute for flour during hard times in Scandinavia and Iceland.

Lichens have in the past had extensive use in dyeing. Litmus was originally produced from the lichen *Roccella*. They also have a long history in providing ingredients in certain kinds of perfumes, and the ability of powdered lichens to absorb and retain perfumes made them valuable as constituents of hair powders in the times when wigs were worn widely and bodily hygiene was less adequate than in modern times. More recently, lichens have been explored as a source of antibiotics. Many lichen antibiotics proved unsuitable for clinical use but a few (especially the common usnic acid) have had some use in combating tuberculosis and certain skin diseases.

stantial supply of nutrients enables lichens to live in their characteristically barren habitats removed from the supplies of external organic food normally needed for fungal growth. Many researchers also believe that the lichen alga must receive some "benefit" in return for supplying the fungus with an abundance of carbohydrate, although this has never been proved experimentally.

Lichens cover more of the Earth's surface than all other fungi, but their economic value to Man is negligible. In arctic and

Below Parmelia perlata. *a foliose (leafy) lichen found on trees and rocks.* (×1) Right *The crustose lichen* Rhizocarpon geographicum *is so-named because of the resemblance of its uneven outline to that of a map.* (×2)

presumably important for their existence in barren and nutrient-poor habitats. But this adaptation has its Achilles heel in that absorption of pollutants cannot be avoided.

Although very sensitive to atmospheric pollution, lichens are very tolerant of other environmental extremes, withstanding weeks and months of drought. They also tolerate extremes of temperature, as indicated not only by their luxuriant growth in arctic regions, but also by the fact that those in hotter regions can tolerate long periods of quite high temperatures when dry – over 60°C for several hours.

Both the fungal and algal components of lichens can be grown free-living in laboratory culture but neither bear any resemblance to the parent lichen. Clearly, there must be a remarkable interaction between the two components to produce the distinctive appearance and structure of the lichen plant. One aspect of the interaction is that the alga continuously supplies to the fungus 70–80% of the carbohydrates it manufactures by photosynthesis. This rapid and sub-

Lignum vitae the common name given to *Guaiacum officinale* and *G. sanctum*, attractive evergreen trees from tropical and subtropical America. They are grown for ornament and bear attractive blue or purple flowers. The wood yields a hard resin with medicinal properties called guaiacum, once used as a stimulant and purgative, and, in conjunction with mercury, as a treatment for syphilis. The very heavy durable wood (the hardest commerical timber) is pale brown outside, chocolate colored inside. It is used for fine carving and for making objects such as pulley blocks, where extreme toughness is required.
ZYGOPHYLLACEAE.

Ligularia a widespread temperate Eurasian genus of herbaceous perennials. Cultivated species include *L. dentata* (= *L. clivorum*, **Senecio clivorum*), with kidney-shaped leaves and orange-yellow flower heads, and *L. macrophylla* (= *S. ledebourii*), a massive species with 2–3m-stems, leaves up to 1m long and bright yellow flower heads.
COMPOSITAE, about 120 species.

Ligusticum a genus of perennial herbs occurring naturally all over the Northern Hemisphere. *L. scoticum* [SCOTCH LOVAGE] was once cultivated as a potherb, its roots, leaves and seeds used for medicinal purposes and the young stems candied and used like angelica.
UMBELLIFERAE, about 60 species.

Ligustrum see PRIVET.
OLEACEAE, about 50 species.

Lilac the common name for *Syringa*, a genus of deciduous shrubs or small trees from Asia and Eastern Europe with showy panicles of waxy, fragrant flowers ranging from white to deep purple in color. Cultivated species include *S. × persica* (*S. afghanica × S. laciniata*) [PERSIAN LILAC] and *S. vulgaris* [COMMON LILAC], of which there are over 500 named cultivars, many originating in France. These hardy, large, easily grown shrubs thrive in

Lilaea scilloides showing the stalked inflorescences bearing male and bisexual flowers, with the long styles of the female flowers projecting from the axils. (× ½)

towns and industrial areas. There are a number of excellent species from China and central Europe, such as *S. × chinensis, S. reflexa, S. villosa* and *S. josikaea* [HUNGARIAN LILAC], which have been used as parents for a great many hybrids.
OLEACEAE.

Lilaea a genus consisting of a single species, *L. scilloides*, a tufted, grass-like annual living in shallow water in western North and South America and in Victoria, Australia (where it was probably introduced). It has no known economic use.
LILAEACEAE, 1 species.

Lilium [LILIES] a horticulturally well-known genus of bulbous perennials spread throughout the north temperate region, although there are a few outlying species, such as *L. neilgherrense* which inhabits the mountains of southern India.

The inflorescence is a raceme with the flowers on long stalks, but in some species the flower is solitary and in others the axis of the inflorescence is condensed so much that the flowers appear to arise all at the same point, so giving an umbellate appearance. The flower consists of six, usually equal segments, free from each other but arranged in various ways to give a wide range of shapes to the different species. For example, in *L. martagon* [TURK'S CAP LILY, MARTAGON LILY] the flowers are pendulous with the segments curled right back, in *L. umbellatum* they are erect with the segments forming an upright cup, while in *L. regale* they are horizontal with the segments held together to form a long tube. Size varies from 2.5cm to as much as 25cm across.

Lilies are highly prized ornamental flowers, and are intensively cultivated and hybridized for indoor and garden display.

Ligustrum lucidum is an evergreen privet from China, where a wax secreted in response to insect damage is used commercially.

Enthusiasts and growers recognize nine divisions for exhibition purposes: Asiatic hybrids; martagon hybrids; candidum hybrids; American hybrids; longiflorum hybrids; trumpet and aurelian hybrids; oriental hybrids; all other hybrids; true species and their botanical forms and varieties.

Among the most important species, both in their own right and as a source of hybrids, are *L. auratum* [GOLDEN-RAYED LILY], *L. × aurelianense* (= *L. henryi × L. sargentiae*), *L. bulbiferum* [ORANGE LILY], *L. canadense* [CANADA LILY], *L. candidum* [MADONNA LILY], *L. chalcedonicum* [SCARLET TURK'S CAP LILY], *L. formosanum* [FORMOSA LILY], *L. hansonii, L. henryi, L. × imperiale* (= *L. regale × L. sargentiae*), *L. japonicum, L. lancifolium* (= *L. tigrinum*) [TIGER LILY], *L. longiflorum* [EASTER LILY], *L. × maculatum, L. martagon* [TURK'S CAP LILY], *L. pardalinum* [PANTHER LILY], *L. pyrenaicum, L. regale, L. superbum* and *L. × testaceum* (= *L. candidum × L. chalcedonicum*) [NANKEEN LILY].
LILIACEAE, about 90 species.

Lily of the valley the common name for *Convallaria*, a north temperate genus represented by a single species, *C. majalis*, a perennial plant with a creeping rhizome which grows in dry woods on chalky soils. Leafless stalks arise from the axils of the paired leaves and bear inflorescences of fragrant, bell-shaped flowers. Cultivated varieties include the white flowered 'Fortin's Giant' and the pink flowered 'Rosea'. The plant yields convallatoxin, a cardiac glycoside. Var *keiskei* from Japan is sometimes treated as a separate species.
LILIACEAE, 1 species.

Lima bean one of the common names [BUTTER BEAN is another] for the annual climbing herb *Phaseolus lunatus*, a number of varieties of which are commonly cultivated in the tropics. A native of tropical and subtropical Central and South America, it requires a higher humidity for growth than most beans. The varieties are divided into the larger 'Lima' group containing both climbing (pole) and dwarf varieties, and the smaller 'Sieva' group. While the short, flat pods may be eaten as a green vegetable, it is the usually pale green seeds which are eaten fresh, canned or frozen. Variety *limenanus* [DWARF SIEVA BEAN] is a dwarf bush form.
LEGUMINOSAE.

Flower of the Golden-rayed Lily (Lilium auratum), an outstanding species from Japan. This is a typical lily flower, having six perianth segments, six stamens and a single style. (×½)

Lime a very acid yellow citrus fruit, resembling the *lemon except for its greenish-yellow flesh. *Citrus aurantiifolia* [LIME TREE] is native to northeastern India and Malaysia but is widely cultivated throughout the tropics, especially in the West Indies, Mexico and India, for juice and for the oil which can be expressed from the rind. Limes may contain an acid or a sweet pulp.

The WEST INDIAN or KEY LIME fruits all the year round. It is highly acid, usually oblong or ovoid with a small nipple, while the PERSIAN or TAHITI LIME (= *C. latifolia*) is sweeter, larger, broadly ovoid with a broad nipple, and fruits only in autumn and winter. Historically, acid limes were the first citrus fruits to be used by sailors against scurvy.
RUTACEAE.

Limnanthes a genus of annual herbs from the west coast of North America, one of which, *L. douglasii* [POACHED EGG FLOWER], is grown as a border plant or as a greenhouse pot plant for its 2.5cm-wide flowers which have white-tipped petals sometimes yellow at their base.
LIMNANTHACEAE, 7 species.

Limonium [SEA LAVENDER, STATICE] a large genus of annual and perennial herbs and subshrubs widely distributed in coastal habitats throughout all the continents of the world. The flowers are small, carried in many branched panicles, pink to lavender in color, and persist in the dried form. Some species are used for decoration as everlasting flowers. Cultivated ornamental species include the annual yellow-flowered *L. bonduellii*, the rose-pink flowered *L. suworowii* (= *Psylliostachys suworowii*) and the perennial, lavender-blue flowered *L. latifolium*. *L. carolinianum* is a powerful astringent, formerly used to treat dysentery.
PLUMBAGINACEAE, about 300 species.

Linaria [TOADFLAX] a genus of annual and perennial herbs and subshrubs occurring mainly in Europe and the Mediterranean region. Many are found in gardens, perhaps the best-known being the yellow-flowered European *L. vulgaris* [TOADFLAX] or the purple *L. repens* [PALE or STRIPED TOADFLAX], both notorious as weeds. Similar to the latter, *L. purpurea* [PURPLE TOADFLAX] is grown for its dense purple or pink spikes. The annuals frequently cultivated for their variously purple, red, yellow or white flowers under the name *L. maroccana* are North African and perhaps of hybrid origin. *L. triornithophora* from Spain and Portugal is a tall perennial with large, showy, bluish-purple and yellow flowers, often found in cultivation. The commonest cultivated alpine species are the pale yellow *L. supina*, the yellow and brown *L. tristis* and the purple and orange *L. alpina* [ALPINE TOADFLAX]. All are short-lived perennials with masses of snapdragon-like flowers. *L. vulgaris* is now widely naturalized in North America.
SCROPHULARIACEAE, about 150 species.

Inflorescences of Sea Lavender or Statice (Limonium vulgare), a common inhabitant of muddy salt marshes, where it forms a carpet of color. (×1)

Lilium nepalense, a large-flowered lily native to the humid Himalayas at altitudes around 2 000–2 750m. It is grown in cultivation, but is slightly frost-tender. (×⅓)

Lindera a genus of aromatic trees or shrubs mainly native to Asia and North America. The most widely cultivated decorative species is the eastern North American *L. benzoin* [SPICEBUSH], a shrub which may grow to a height of 5m and whose aromatic bark also has medicinal uses. Other cultivated species include *L. mellissifolia* [JOVE'S FRUIT], *L. obtusiloba* and *L. umbellata*, the former North American, the latter two from China and Japan.
LAURACEAE, about 100 species.

Linnaea [TWIN FLOWER] a small genus of evergreen prostrate or trailing subshrubs, circumpolar in distribution. The common name derives from the arrangement of the flowers in pairs at the ends of long stalks. *L. borealis* is cultivated as a rock-garden or ground-cover plant bearing fragrant bell-shaped rose or white flowers from late spring.
CAPRIFOLIACEAE, 1–3 species.

Linseed the seed of *Linum usitatissimum* [*FLAX]. Linseed oil is the most important vegetable drying oil (ie when exposed to air it slowly oxidizes and becomes hard). It is used as an ingredient of emulsions, varnishes, putty (a mixture of chalk and linseed oil) and linoleum, which consists of the oil with resins and powdered cork on a jute backing. The residual oil-cake is a valuable cattle food. When extracted cold, it is important as an artists' medium for grinding oil colors. The seed is employed in medicine as a demulcent.
LINACEAE.

Linum a large genus of annual, biennial and perennial herbs and subshrubs found in temperate and subtropical regions of all continents. The annual *L. usitatissimum* [*FLAX], is the source of both *linseed oil and linen fiber. A number of other species are grown as ornamental border or rockery plants, including the yellow-flowered, eastern Mediterranean *L. arboreum* and the pale blue-flowered, European *L. austriacum*, both rather shrubby perennials, and the attractive annual North African *L. grandiflorum*, with large saucer-shaped, reddish-colored flowers (white in cultivar 'Bright Eyes', crimson in cultivar 'Rubrum' and purplish-blue in cultivar 'Caeruleum').
LINACEAE, about 230 species.

Lippia a large genus comprising perennial herbs and shrubs, mainly from the tropics of the New World, but with a few species in Africa and Asia. The genus is not often seen in cultivation, except for *L. citriodora* (= *Aloysia triphylla*) [LEMON or SWEET-SCENTED VERBENA], a shrub from Argentina and Chile up to 1.5m tall, which is grown for its fragrant lemon-scented foliage.
VERBENACEAE, about 200 species.

Liquidambar [SWEET GUMS] a small genus of trees from North America, southwestern Asia and southeastern China and Taiwan. SWEET GUMS are grown mostly as ornamen-

Perennial Flax (Linum perenne spp anglicum), a rare wild flower photographed in limestone grassland in northeastern England. ($\times \frac{1}{2}$)

tals for their autumn colors, and for their wood, but also for the fragrant gum known as American storax (or styrax) from *L. styraciflua* and liquid storax from *L. orientalis*, used in perfumery, primarily to scent soap, as an expectorant in cough pastilles, and as a fumigant in the treatment of skin diseases such as scabies. The Chinese *L. formosana* is another species cultivated for ornament. A reddish heartwood (satin walnut or red gum) is obtained from *L. styraciflua*; the white sapwood is sold as hazel pine.
HAMAMELIDACEAE, 3 species.

Liriodendron see TULIP TREE.
MAGNOLIACEAE. 2 species.

Liriope an East Asian genus of stemless perennial herbs with linear, tufted leaves and racemes of purple or white flowers. *L. muscari* (= *Ophiopogon muscari*) and *L. spicata* are grown in gardens for ground cover.
LILIACEAE, 6 species.

Listera a small genus of orchids native to the north temperate zone. They are characterized by having a pair of broad flat opposite leaves, and consequently are often known as TWAYBLADES, particularly *L. ovata*. This, like most other species of the genus, has a slender spike of rather insignificant greenish or purplish flowers.
ORCHIDACEAE, about 30 species.

Litchi a small genus of trees and shrubs from southern China, Southeast Asia and western Malaysia to India. The most important species is *L. chinensis* (= *Nephelium litchi*, *Dimocarpus litchi*) [LITCHI, LYCHEE, LEECHEE], an evergreen tree which yields an edible fruit. It is a native of southern China and has been widely introduced throughout the tropics, although it only flourishes well at

high altitudes. The red-brown globose fruits (drupes) are borne in clusters. The white edible juicy aril of the fruit is eaten fresh or canned in syrup. The aril is also dried and is known then as litchi nuts. The tree is also grown for its ornamental value and its timber.
SAPINDACEAE, about 12 species.

Lithops [LIVING STONES, PEBBLE PLANTS] a remarkable South African genus of leaf-succulents in which the entire body is reduced to one pair of opposite, pebble-like leaves forming a top-shaped growth which lies buried in the soil with only the flat leaf tips exposed. Each leaf pair is renewed annually, and some species form a cluster of heads with age. Daisy-like white or yellow flowers arise from the fissure separating the two leaves and are large for the size of the plant.

Lithops leaves are never plain green, but shades of gray-green and brown and variously mottled or striped. The colors and

The flower of the Tulip Tree (Liriodendron tulipifera), showing the cone-like mass of narrow carpels; it is closely allied to the magnolias. ($\times \frac{1}{3}$)

Livistona rotundifolia, a handsome palm tree with hanging branches of fruits, is widely cultivated throughout the tropics.

patterns of each species closely resemble those of the natural rock background of the habitat and have the effect of disruptive camouflage. *Lithops* species, especially *L. optica* and *L. fulleri*, are among the most popular of greenhouse succulents. AIZOACEAE, about 37 species.

Lithospermum a genus of perennial herbs or small shrubs with hairy leaves and usually blue flowers, native to Europe and Asia. Several are cultivated as perennials in flower borders and rock gardens and have been extensively hybridized. *L. diffusum* (= *Lithodora diffusum*) and *L. oleifolium* (= *L. oleifolia*), from Europe, bear blue flowers, while *L. canescens* [PUCCOON, RED ROOT] from North America, has yellow flowers. The European *L. purpuro-caeruleum* (= *Buglossoides purpurocaerulea*) [BLUE GROMWELL] has flowers which are at first red and then turn blue. BORAGINACEAE, about 50 species.

Littorella [SHOREWEEDS] a genus of creeping, perennial aquatic herbs with two species, *L. uniflora* in northern and central Europe and *L. australis* in southern Patagonia and Tierra del Fuego. The plants often form extensive turf in freshwater lakes and ponds down to depths of about 4m. PLANTAGINACEAE, 2 species.

The upper slopes of Mount Kenya support certain unique species, eg the spiky Lobelia plants (Lobelia deckenii ssp keniensis), and the giant tree-like Groundsels (Senecio sp) with rosettes of large leaves.

A Living Stone (Lithops olivacea) in flower. The pebble-like double leaves blend remarkably with the stony ground on which they grow. ($\times \frac{1}{2}$)

Liverworts see Mosses and Liverworts, p. 228.

Livistona a genus of tall, elegant solitary fan palms occurring from Assam and southern China to the Solomon Islands, New Guinea and Australia. Several species, especially *L. chinensis* and *L. rotundifolia* are widely cultivated as ornamentals throughout the tropics. The leaves of several species, for example *L. australis*, *L. saribus* and *L. jenkinsiana*, are used locally for thatching and the young leaves and buds of *L. australis*

are eaten as a vegetable but only locally. PALMAE, about 24 species.

Lobelia a large cosmopolitan genus of annual and perennial herbs, subshrubs and sometimes trees, some of which are familiar in temperate gardens. There are the fascinating giant lobelias, native to the East African highlands, as well as more conventional garden plants such as the North American *L. purpuro-caeruleum* (= *Buglossoides purpuro-* Chilean *L. tupa* growing to 2m high with deep red downy flowers, and the tender but graceful *L. laxiflora*, from Arizona to Mexico and Colombia, with red and yellow flowers.

The most commonly cultivated species is the dwarf annual or perennial *L. erinus* [EDGING LOBELIA], with blue flowers with a white or yellowish throat. There are several varieties differing in flower color, as well as double-flowered forms. *L. inflata* [INDIAN TOBACCO] yields the alkaloid lobeline, used medicinally. LOBELIACEAE, about 350 species.

Locust or **locust bean** a name given to the species of a number of different genera, including *Parkia filicoidea* [WEST AFRICAN LOCUST BEAN], *Robinia pseudacacia* [BLACK LOCUST], *Hymenaea courbaril* [LOCUST TREE] and *Gleditsia triacanthos* [HONEY LOCUST], but most commonly to the *CAROB TREE, *Ceratonia siliqua*, extensively cultivated in the Mediterranean area for its large sweet pods.

Lodoicea see DOUBLE COCONUT. PALMAE, 1 species.

Loganberries large, edible, purplish-red fruits produced by the widely cultivated shrub *Rubus × loganobaccus* (= *R. ursinus* var *loganobaccus*). The plant is generally considered to be a natural fertile hybrid of the AMERICAN BLACKBERRY (*R. vitifolius*) and an unreduced gamete of the EUROPEAN RASPBERRY (*R. idaeus*).
ROSACEAE.

Logwood or **haematoxylin** a dye produced from the red-brown heartwood of a large tree [also known as LOGWOOD], *Haematoxylon campechianum*, found growing wild in Central America and sometimes cultivated elsewhere in the tropics. Imported first into Elizabethan England, it was used for dyeing wool and silk. The heartwood provides the basis of several useful stains used in microscopy, such as haematoxylin.
LEGUMINOSAE.

Lolium see RYEGRASS.
GRAMINEAE, 10–12 species.

Lonas a genus represented by a single species, *L. annua*, an annual herb with yellow discoid flower heads, found in southern Europe and North Africa.
COMPOSITAE, 1 species.

Lonchocarpus a large tropical genus of climbers, shrubs and trees, some of which (eg *L. sericeus* and *L. latifolius*) yield useful timber. The Peruvian *L. utilis* is one of the major sources of rotenone, a powerful and commercially important insecticide extracted from the roots (see also Derris). Leaves of the West African vine *L. cyanescens* yield a blue dye, Yoruba indigo.
LEGUMINOSAE, about 150 species.

London pride the common name for *Saxifraga umbrosa* and *S. × urbium* (= *S. spathularis* × *S. umbrosa*), perennial herbs widely cultivated in gardens for their attractive fleshy leaves (in basal rosettes) and loose

*The Common Honeysuckle (*Lonicera periclymenum*). frequently found growing wild in woods and hedgerows. Below Left The fragrant flowers are*

inflorescences of small pink or white flowers.
SAXIFRAGACEAE.

Lonicera a large genus of erect or twining shrubs with opposite, simple leaves, widespread in the Northern Hemisphere, extending southwards to Mexico in the New World, and into North Africa from Europe; from Asia it extends southwards as far as southwestern Malaysia and into the Philippine Islands, and it is well represented in the Himalayan region.

The genus is popular in gardens where many shrubs and climbers are exploited as ornamentals. Most species and hybrids are hardy and deciduous although some are more or less evergreen. The genus can be divided into three sections, each of which provides a different group of garden plants. The first, including such well-known species

often pollinated at night by moths. (×½) Right The brightly colored berries extend the period of ornamental display. (×2)

Lophophora williamsii is one of the few completely spineless cacti. This is the fabulous Peyote which yields the hallucinogenic alkaloid mescaline, used by Mexican Indians for centuries. (×1)

as the European, North African and west Asian *L. periclymenum* [COMMON HONEYSUCKLE, WOODBINE], the yellow-flowered European and west Asian *L. caprifolium* [GOAT-LEAF HONEYSUCKLE, PERFOLIATE WOODBINE, ITALIAN WOODBINE] and the Mediterranean *L. etrusca*, with yellow and purple flowers, are climbers and often deciduous.

The second section, with *L. xylosteum*, a Eurasian species, *L. pyrenaica*, from the Pyrenees, and others, comprise upright shrubs ["BUSH HONEYSUCKLES"]. This section includes the evergreen Chinese *L. nitida*, a common hedging plant, and the numerous cultivars of the widely grown *L. tatarica* [TATARIAN HONEYSUCKLE], native from Russia to Turkestan.

The third section comprises mostly evergreen climbers, exemplified by the fragrant East Asian *L. japonica* [JAPANESE HONEYSUCKLE, GOLD AND SILVER FLOWER], with red and white flowers, the hairy *L. giraldii*, from China, and *L. hildebrandiana* [GIANT HONEYSUCKLE], from Burma, growing to 18–25m, with flowers more than 10cm long and yellow turning orange. Another popular species in this group is *L. sempervirens* [TRUMPET HONEYSUCKLE, CORAL HONEYSUCKLE], native to North America, cultivars of which include the yellow-flowered 'Sulphurea' and the scarlet-flowered 'Superba'.
CAPRIFOLIACEAE, about 200 species.

Lopezia a genus of erect herbs and shrubs native to Mexico and Central America. Although not extensively cultivated some species, especially *L. hirsuta* (= *L. lineata*), with pink to rose flowers, make attractive ornamentals in subtropical and warm-

A bee visiting the flower of the Bird's Foot Trefoil (Lotus corniculatus). It is a common plant in European meadows and is grown as a forage crop in Europe and North America. (×2)

temperate gardens or grown under glass in temperate regions.
ONAGRACEAE, about 17 species.

Lophira a genus represented by a single variable species, *L. lanceolata* (= *L. alata*) [AFRICAN OAK, RED IRON, DWARF IRONWOOD], a tropical West African tree which provides a commercial heavy red hardwood, mainly used locally for construction purposes. The seeds when pressed yield an edible oil (meni oil or niam fat).
OCHNACEAE, 1 species.

Lophocolea a large genus of leafy liverworts most of which occur in the tropics and subtropics, particularly in the Southern Hemisphere, although a few extend into the boreal climates of the Northern Hemisphere. They grow in moist habitats on soil, stones, tree trunks and decaying wood. The large dorsal "leaves" are commonly two-lobed, but in some species may also be entire or fringed.
HARPANTHACEAE, about 200 species.

Lophophora [PEYOTE, DUMPLING CACTUS, MESCAL BUTTON] a genus of dwarf, top-shaped, soft-bodied, gray, spineless cacti perhaps consisting of a single species, *L. williamsii*, from South Texas to North Mexico, although a second species, *L. diffusa*, from central Mexico, is recognized by some authorities. *L. williamsii* is the sacred cactus of Mexico and has achieved fame and notoriety for its hallucinogenic alkaloids (notably mescaline), used by Mexican Indians for over 2 000 years (as peyote or peyotl) in rituals. A ban was imposed on its cultivation in certain states in the USA and elsewhere when the scare over LSD plants was at its height.
CACTACEAE, 1–2 species.

Loranthus a large genus of woody parasites of the Old World tropics with a few in Eurasia and Australia. They usually parasitize trees by means of suckers, which are modified adventitious roots. One species, *L. europaeus*, is sometimes deliberately cultivated by growing it on *Castanea* or *Quercus*. It produces terminal racemes of yellowish-green flowers.
LORANTHACEAE, about 600 species.

Lotus a term embracing two entirely separate groups of plants. The genus *Lotus* belongs to the family Leguminosae, while LOTUS is also the popular name for the genus *Nelumbo* (Nymphaeaceae) (see *Nelumbo*). *Lotus* [BIRD'S FOOT TREFOILS] is a genus of annual and perennial herbs and some subshrubs, native mainly to North America, Australia, temperate Asia and Africa, and the Mediterranean region. Perhaps the best-known species is the Eurasian *L. corniculatus* [BIRD'S FOOT TREFOIL], a more or less spreading hairless perennial with yellow pea-like flowers streaked with red. It is widely found in European meadows and pastures and is grown there and in North America as a forage crop. *L. tenuis* [SLENDER BIRD'S FOOT TREFOIL] is similar but the stems are more wiry, slender and more profusely branched than *L. corniculatus*.

Species cultivated as ornamentals include *L. berthelotii* [PARROT'S BEAK, WINGED PEA, CORAL GEM], a silvery, scarlet-flowered shrub from the Cape Verde and Canary Islands, and *L. creticus* [SOUTHERN BIRD'S FOOT TREFOIL], a yellow-flowered perennial from the Mediterranean region. The European WINGED PEA formerly known as *L. tetragonolobus* is now placed in *Tetragonolobus* as *T. purpureus*.
LEGUMINOSAE, about 60 species.

A field of irrigated Lucerne or Alfafa (Medicago sativa) being mown on a new agricultural development in Saudi Arabia. It is used as a forage crop.

Lovage or **lovage angelica** the common names for *Levisticum officinale*, a tall perennial herb native to southern Europe and cultivated for its aromatic fruits and nuts which are the source of an oil used in flavoring. The stems may be eaten after blanching, like celery, and were once candied, like those of *Angelica*. SCOTCH LOVAGE is *Ligusticum scoticum*.
UMBELLIFERAE.

Lucerne [ALFALFA] the longest cultivated and the world's most important forage crop, with an ancient origin in southwestern Asia, but now grown from Norway to New Zealand. *Medicago sativa* [COMMON LUCERNE, COMMON ALFALFA] is the most widely grown species. It is a strong-rooted perennial usually persisting for three to five years depending on conditions. The flowers are violet in color and the pods are coiled in tight spirals. *M. falcata* [YELLOW-FLOWERED ALFALFA, SICKLE MEDICK] is a perennial which is very common in many countries, although it only produces one cut per year. It bears yellow flowers and sickle-shaped pods. Some of its forms are drought and cold resistant, and so it has been useful in breeding improved lucernes. *M. arabica* and *M. orbicularia*, both annuals, are sometimes grown as alfalfa.

Both *M. sativa* and *M. falcata* are related to the wild species *M. coerulea*. Hybrids between *M. sativa* and *M. falcata* show intermediate features between the two parents and were at one time ascribed to a specific rank as *M. media* or *M. varia*. Some forms of cultivated alfalfa may well be hybrids of either *M. sativa* or *M. falcata* with *M. glutinosa* or *M. glomerata*. Lucerne produces a good yield of nutritious fodder. This can be grazed, but is often cut for hay, producing a very valuable feed containing some 90% dry matter, about

half of which is digestible dry matter with 12–18% protein. Most of the protein is in the leaves, and haymaking methods should aim to lose as few leaves as possible. A feed with higher protein content can be produced by cutting the young foliage, which is then dried and usually ground into meal.
LEGUMINOSAE.

Ludwigia a genus of slender herbs, often floating or creeping, or large shrubs largely restricted to the Old and New World tropics but extending to eastern North America, temperate Eurasia and Hawaii in aquatic or moist habitats. Most species, such as the pantropical weed *L. hyssopifolia*, have been widely distributed as a consequence of human activity. A few species are grown as ornamentals as marsh plants or in aquaria. The stems of *L. repens* are eaten as a vegetable in parts of China.
ONAGRACEAE, about 75 species.

Luffa a genus of tropical herbaceous tendril-bearing climbers containing the economically important annual species, *L. aegyptiaca* (= *L. cylindrica*) [SPONGE GOURD, LOOFA], whose large fruit (20–60cm long) yields a fibrous skeleton on retting, used for bath sponges. The young fruits of *L. acutangula* [TOWEL or DISH-CLOTH GOURD] are used as a curry vegetable.
CUCURBITACEAE, about 6 species.

Lunaria a small herbaceous genus of central and southeast European origin. Two of the species, *L. rediviva* [PERENNIAL HONESTY] and *L. telekiana*, are perennial, and *L. annua* (= *L. biennis*) [HONESTY, BOLBONAC, SILVER DOLLAR, PENNY FLOWER] is a biennial. The white- or purple-flowered *L. annua*, as well as the cultivar 'Variegata', with red flowers and variegated leaves, and the white-flowered *L. rediviva*, are popular garden-border plants. *L. annua* is also grown for its nearly orbicular fruits with a flat, silvery-white, papery septum, which are used as a winter decoration.
CRUCIFERAE, 3 species.

The cucumber-like fruits of Loofa (Luffa aegyptiaca). The familiar loofas used as sponges are the "skeletons" left behind when the fleshy part has been retted away. (×⅛)

Lunularia a genus of thallose liverworts consisting of a single species, *L. cruciata*. Although restricted naturally to warmer and drier climates, especially in the subtropics, this species has been remarkably successful in extending its geographic range by becoming a conspicuous weed in moist greenhouses.
MARCHANTIACEAE, 1 species.

Lupinus [LUPINS] a large genus of annual or perennial herbs and some subshrubs mostly concentrated in western North America, but also distributed in other temperate regions, including the Andes and the Mediterranean. Several including *L. polyphyllus*, *L. perennis* [SUNDIAL LUPIN], *L. mutabilis* [PEARL LUPIN], are valuable herbaceous ornamentals. The best-known of the subshrubs is *L. arboreus* [TREE LUPIN]. The renowned lupin hybrids, the RUSSELL LUPINS, are amongst the most spectacular and colorful of all garden plants. They probably arose as hybrids between *L. hartwegii* and *L. polyphyllus*.

The five species cultivated for forage and as a green manure are of Mediterranean origin, with the exception of the South American *L. mutabilis*. *L. albus* [WHITE LUPIN] and *L. mutabilis* have been in cultivation for at least 3 000 years. Other commonly cultivated species are *L. angustifolius* [BLUE LUPIN], and *L. luteus* [YELLOW LUPIN]. Although normally poisonous, low-alkaloid forms are now generally available in cultivation, and improved varieties, especially of WHITE LUPINS, are increasingly being recognized as a rich source of oil (12–17%) and protein which

Ragged Robin or Cuckoo Flower (Lychnis floscuculi) is a common and attractive inhabitant of damp places. (×2)

may be used both as animal and human food.
LEGUMINOSAE, about 300 species.

Luzula [WOODRUSHES] a cosmopolitan genus of tufted grass-like perennials occurring chiefly in cold temperate regions in the Northern Hemisphere. A few species, including *L. albida*, *L. campestris*, *L. sylvatica* and *L. lutea* are cultivated in woodland situations.
JUNCACEAE, about 50 species.

Lycaste a genus of tropical American and West Indian pseudobulbous epiphytic or terrestrial orchids. Their appealing flower colors and often strong scents (such as *L. aromatica*) have led to a number of species being culti-

The green thallus of the liverwort Lunularia cruciata bearing disk-shaped gemmae within crescent-shaped gemmae cups. (×3)

Almost mature fruiting bodies of the puffball Lycoperdon perlatum, *showing the warty scars that are the remains of the spines.* $(\times \frac{1}{3})$

vated. These include *L. cruenta*, with bright yellow and rich apricot-colored flowers tinged with olive green and flecked with red. The wild white-flowered form of *L. virginalis* [NUN ORCHID] is the national flower of Guatemala; its cultivars have flowers that are white suffused with yellow, rose-white, purple, crimson or maroon.
ORCHIDACEAE, 40–50 species.

Lychnis a small genus of erect perennial herbs from north temperate Eurasia. Several species are cultivated, including *L. flos-cuculi* [RAGGED ROBIN, CUCKOO FLOWER] which grows in damp areas. Its cultivars include 'Pleniflora' with its double rose-colored flowers. *L. chalcedonica* [JERUSALEM CROSS], with dense inflorescences of scarlet flowers, has many varieties. Two attractive plants are *L. coronaria* [ROSE CAMPION], which has purplish or occasionally pale or white flowers, and *L. flos-jovis* with purplish or scarlet (rarely white) flowers.
CARYOPHYLLACEAE, 8–12 species.

Lycium a genus of temperate and subtropical shrubs from both hemispheres but mainly in America. Useful species include *L. afrum* [KAFFIR THORN], which is grown as a hedge plant in South Africa, and *L. chinense* [CHINESE WOLFBERRY, MATRIMONY VINE or TEA TREE], whose leaves are used as a vegetable. The fruits of a number of species, including those of *L. arabicum* [ARABIAN WOLFBERRY], *L. andersonii* [ANDERSON WOLFBERRY] and *L. pallidum* [RABBIT THORN] are eaten, either raw or cooked, locally in Arabia, Arizona, and Mexico respectively.
SOLANACEAE, about 100 species.

Lycoperdon a cosmopolitan genus of gasteromycete fungi which comprises the true puffballs. The powdery spore mass is contained in a sac which opens at maturity by an apical mouth, so that when the elastic

sides of the sac are struck by a raindrop a puff of spores emerges. The fruiting bodies of *L. fuligineum* and *L. gemmatum* are eaten in tropical Asia and the USA respectively.
LYCOPERDACEAE, about 50 species.

Lycopersicon see TOMATO.
SOLANACEAE, about 7 species.

Lycopodium [CLUB MOSSES] a genus of herbaceous prostrate creeping or erect perennial fern-allies widespread in temperate and tropical regions. The spores of *L. clavatum* [STAG'S-HORN MOSS, COMMON CLUB MOSS] are used as a dusting powder and for coating pills. (See Club mosses and their allies, p. 92.)
LYCOPODIACEAE, about 450 species.

Lygeum a genus comprising a single species, *L. spartum* (see ESPARTO).
GRAMINEAE, 1 species.

Lygodium a small tropical and subtropical genus of ferns which possess leaves with a twining habit. The sporangia lie singly in double rows on fertile leaflets and are not organized into sori. The stems of the tropical *L. scandens* are used for making hats and the stems of *L. circinatum*, from the Malayan peninsula, for basketwork. *L. japonicum* is cultivated for ornament.
SCHIZAEACEAE, about 40 species.

Lyonia a small genus of evergreen or deciduous shrubs native to North America and Asia, three of which are cultivated as ornamental shrubs, with axillary or terminal clusters of pink or white flowers. The early-flowering evergreen *L. lucida* [TETTERBUSH] and the deciduous *L. mariana* [STAGGERBUSH] grow to a height of about 2m, while the deciduous *L. ligustrina* [MALE BERRY] may reach 4m.
ERICACEAE, about 30 species.

The Yellow Loosestrife or Garden Loosestrife (Lysimachia vulgaris), *an attractive waterside perennial, native to Europe but now also naturalized in North America. It is also cultivated as an ornamental in gardens.* $(\times \frac{1}{2})$

The tall purple inflorescences of the Purple Loosestrife or Spiked Loosestrife (Lythrum salicaria) *are a familiar sight along river banks and the edges of lakes. A number of ornamental cultivars and varieties are now in cultivation.* $(\times \frac{1}{10})$

Lysichiton a small genus of two robust perennial stemless herbs which are cultivated in wet soil for their large ovate leaves and arum-like inflorescences. *L. americanum* [SKUNK CABBAGE] is native to western North America and bears leaves up to 1.75m long and an inflorescence enclosed by a bright yellow spathe. The other species, *L. camtschatcense* from East Asia, is similar but somewhat smaller with a white spathe. The latter is sometimes included in *L. americanum*.
ARACEAE, 2 species.

Lysimachia a large genus of erect or creeping herbaceous perennials, widespread in temperate and subtropical regions, especially East Asia and North America. *L. nummularia* [CREEPING JENNY, MONEYWORT] is an excellent trailing ground cover plant with cup-shaped yellow flowers. *L. vulgaris* [YELLOW LOOSESTRIFE] is a tall, 70–80cm, waterside perennial. Cultivated species include *L. clethroides* [GOOSENECK], with white-flowered spikes, and *L. ephemerum*, with white- or purple-flowered spikes. Various species, including *L. nemorum* [YELLOW or WOOD PIMPERNEL] were formerly used in healing wounds.
PRIMULACEAE, about 200 species.

Lythrum a small genus of annual and perennial herbs and small shrubs, usually found in damp places. *Lythrum* is centered round the Mediterranean and in western Asia, but extends into North America and East Asia, and is found in Australia and New Zealand. *L. salicaria* [PURPLE LOOSESTRIFE] is a common plant of reed-swamps with tall spikes of crinkly-petaled purple flowers. Although in nature living by water, it grows well in gardens and there are several cultivars, such as 'Roseum Superbum' with pink flowers. *L. virgatum* is a more slender species and also contains several cultivars with purple, pink or rose-pink flowers. *L. flexuosum* is a small trailing pink-flowered species sometimes grown in baskets.
LYTHRACEAE, 30–35 species.

M

Macadamia or **Australia nut** the seeds of two small ornamental evergreen trees, *Macadamia integrifolia* (= *M. ternifolia* of some authors) ["SMOOTH SHELL MACADAMIA"] and *M. tetraphylla* ["ROUGH SHELL MACADAMIA"]. The oily seeds which are expensive and highly prized are grown commercially mainly in Australia (New South Wales, Queensland), where they are endemic, and in California (USA) and Hawaii. They are eaten raw, roasted or fried.

A third species, *M. ternifolia* [MAROOCHY NUT], bears small bitter fruits which are inedible. Naturally occurring interspecific hybrids occur and several high-yielding cultivars are characterized also by their large kernels.
PROTEACEAE.

Macassar oil an oil obtained from the seeds of *Schleichera oleosa* (= *S. trijuga*, *S. trijugata*) [LAC TREE], a tree native to Southeast Asia. The oil is used in making ointments and

Many white-flowered magnolias are in cultivation. Shown here is the magnificent Magnolia kobus, *from Japan.* ($\times \frac{1}{40}$)

candles, as an illuminant and as a hairdressing.
SAPINDACEAE.

Mace the dried orange to scarlet aril of *Myristica fragrans* [*NUTMEG], from the Moluccas and widely cultivated in the tropics. This spice is used ground or whole to flavor meat, fish, cheese and vegetable dishes.
MYRISTICACEAE.

Macleaya a genus of two large, glaucous, stately perennials up to 2.5m tall, native to China and Japan. *M. cordata* [PLUME POPPY, TREE CELANDINE] has palmately lobed leaves up to 3cm wide and large showy panicles, up to 30cm long, of creamy or pink petal-less feathery flowers. Both it and *M. microcarpa*, with yellowish flowers, are cultivated in temperate gardens.
PAPAVERACEAE, 2 species.

Maclura see OSAGE ORANGE.
MORACEAE, 1 species.

Macrocystis a genus of brown algae commonly known as GIANT KELPS and often growing to a length of 30–50m. It is found mainly in the Southern Hemisphere but extends into the northern Pacific Ocean along the west coast of America. A sturdy holdfast attaches the plant to the bottom of the sea in water of up to 30m in depth. *Macrocystis* is important for production of alginates, plants being cut and collected from the kelp beds by special harvesting boats.
PHAEOPHYCEAE.

Macrozamia a genus of cycads inhabiting temperate regions of Australia. Some species, including *M. plumosa*, *M. miquelii*, *M. comm-*

The outstanding pink flowers of Magnolia campbellii *spp* mollicomata *produce a magnificent display in early spring.* ($\times \frac{1}{3}$)

unis and *M. lucida* are cultivated as greenhouse ornamentals. *M. hopei*, a native of Queensland, and sometimes placed in the genus *Lepidozamia*, is reputed to be the tallest of the cycads growing to about 20m.
ZAMIACEAE, about 14 species.

Madagascar rubber a type of *rubber obtained from the latex of a number of species of trees, shrubs and vines native to Madagascar. They include species of the genera *Landolphia*, *Marsdenia* and *Cryptostegia*. The latex is inferior in quality and yield to that of *Hevea brasiliensis*, and Madagascar rubber is no longer commercially important. (See also RUBBER PLANTS.)

Madder the common name for *Rubia tinctoria* and also for the red dye, alizarin, once obtained from its roots. The name is also given to other members of this genus, including *R. peregrina* [WILD MADDER, LEVANT MADDER], *R. cordifolia* [INDIAN MADDER], and to *Sherardia arvensis* [FIELD MADDER].
RUBIACEAE.

Madia a genus of annual, biennial or perennial Pacific American herbs which are usually glandular, strongly scented and hairy. The best-known is *M. sativa* [TARWEED], native to Chile and the west coast of North America but cultivated in many countries for the seeds which yield the sweet, edible madia oil.
COMPOSITAE, about 20 species.

Magnolia a genus of evergreen and deciduous trees or shrubs, native to Asia from the Himalayas to Japan and Java, and to North and Central America and Venezuela. They provide some of the most popular ornamentals, being unsurpassed in beauty when in full bloom. The solitary, showy flowers are usually large and star- or bowl-shaped, in a range of colors, with white, cream and rose predominant. The flower parts are arranged on the central axis with an outer perianth of two or more whorls of petaloid tepals, subtending numerous free, spirally arranged stamens with stout filaments. There are numerous spirally arranged carpels which are fused into a cone. The red or orange seeds are large, suspended when mature by a single thread-like attachment.

The ever-increasing popularity of magnolias as garden subjects is due mainly to the striking beauty of such deciduous precocious-flowering species as *M. acuminata* [CUCUMBER TREE], whose numerous cultivars include 'Cordata' (= *M. cordata*) [YELLOW CUCUMBER TREE] and 'Aurea', *M. campbellii* with such cultivars as the rose-purple-flowered 'Darjeeling' and the white-flowered 'Maharanee', *M. dawsoniana*, *M. heptapeta* (= *M. conspicua*, *M. denudata*) [YULAN], the white-to-pink-flowered *M. kobus*, the large-leaved, white-flowered, fragrant *M. macrophylla* [LARGE-LEAVED CUCUMBER TREE, GREAT-LEAVED MAGNOLIA], *M. quinquepeta* (= *M. discolor*, *M. liliiflora*, *M. purpurea*), *M. salicifolia*, *M. sargentiana*, with very large showy rose-purple flowers in the shrubby var *robusta*, *M. sieboldii* (= *M. parviflora*), *M. stellata* (= *M. halleana*) [STAR MAGNOLIA], of compact habit with many showy flowers as in cultivars 'Rosea' and 'Rubra', *M. tripetala* [UMBRELLA MAGNOLIA] and *M. virginiana* (= *M. glauca*) [SWEET BAY] which is evergreen in mild areas. Other

Macrozamia riedleri, one of some 14 species of this Australian genus of cycads, showing the typical palm-like form of the class.

evergreens include *M. delavayi* and *M. grandiflora* [BULL BAY, SOUTHERN MAGNOLIA].

Hybrids are now being cultivated in increasing numbers. Some have arisen spontaneously by accident, others by deliberate cross-fertilization. The most frequently grown hybrid is *M. × soulangiana* (= *M. heptapeta × M. quinquepeta*) [CHINESE MAGNOLIA, SAUCER MAGNOLIA], a small tree with numerous cultivars including 'Alba Superba' with large white flowers, 'Candolleana' with flowers tinged purple at the base and 'Lennei' with very large saucer-shaped flowers, white inside and purple on the outside.

Many authorities consider that the floral, vegetative and anatomical features of the genus are relatively unspecialized, and accordingly species of *Magnolia* could be regarded as some of the most primitive living examples of flowering plants.
MAGNOLIACEAE, about 85 species.

Mahogany a commercially important timber which is valued for its reddish-brown color, luster, strength and figuring. Over 200 types of wood are traded under the name, but true mahoganies are restricted to the tropical American and West Indian genus *Swietenia* and the African genus **Khaya*. *S. mahagoni* [TRUE, CUBAN or WEST INDIES MAHOGANY] was the original source of commercial mahogany but has now been largely replaced by *S. macrophylla* [HONDURAS, MEXICAN or BIG

The glossy spiny, divided leaves and clusters of bright yellow flowers make mahonias attractive ornamentals. Shown here is Mahonia 'Charity'. (× $\frac{1}{15}$)

LEAVED MAHOGANY], although *S. candollea* [VENEZUELAN MAHOGANY] is also a source of timber.

Of the AFRICAN MAHOGANIES, the best-known are *Khaya nyasica* [RED or NYASALAND MAHOGANY], *K. senegalensis* [SENEGAL MAHOGANY], *K. ivorensis* [IVORY COAST MAHOGANY] and *K. grandiflora*. The timbers vary in weight, color and figuring. The color of the wood varies with the species and also changes with age. The timber of *K. grandiflora* is pale at first and then darkens to a deep brown, while *S. macrophylla* becomes paler.

Other genera containing species whose wood often goes under the name of mahogany include **Cedrela*, **Dysoxylum*, **Guarea*, **Melia*, *Entandrophragma* and **Ptaeroxylon obliquum* [CAPE MAHOGANY].

Mahonia a genus of evergreen shrubs and trees native to Asia, from the Himalayas to Japan and Sumatra, and to North and Central America. The leaves are odd-pinnate, usually with spiny-toothed leaflets. Ornamental species include *M. bealei* and *M. japonica* (often confused in cultivation), and *M. aquifolium* [OREGON GRAPE], with bright yellow flowers in crowded racemes in spring, standing out against the shining dark foliage, followed by black berries. *M. lomariifolia*, with fragrant deep-yellow flowers in erect dense racemes up to 30cm long, is sometimes grown.
BERBERIDACEAE, about 100 species.

Maianthemum a small genus of low perennial herbs, native to north temperate regions, bearing very small white flowers followed by

the change of sugars to starch in the kernel is reduced, so keeping the grain sweeter and more tender. Other cultivars of maize belong to such varieties as *indurata* [FLINT CORN], in which the grain is hard and smooth, and *indentata* [DENT CORN], the principal commercial corn cultivated for grain fodder and silage.

Recent developments include the northward spread of maize for grain production into northwestern Europe, even as far as southeast England (51°N) and also into southern Ontario in Canada. This trend has followed the breeding of early hybrids that not only mature in short growing seasons but are tolerant of cold spring conditions and resistant to stalk-rot caused by *Fusarium species in the autumn. Breeding can also improve the protein content in maize grain, thus producing a more balanced food for livestock and Man. There is also a modern tendency to breed from plants which produce more numerous small cobs with an overall increase in yield.
GRAMINEAE.

Malacca cane the stems of some species of *Calamus [*RATTAN PALMS], notably *C. scipionum* and *C. bacularis*. Some species of *Licuala* are additional sources. The stems of these plants are used for walking sticks and for making baskets.
PALMAE.

Mallow the common name for several species, mainly of the genera *Malva and *Abutilon, such as *M. sylvestris* [COMMON EUROPEAN MALLOW], *M. crispa* [CURLED MALLOW] and *A. indicum* [INDIAN MALLOW]. The name is also used for species of other genera including *Althaea officinalis [MARSH MALLOW], *Lavatera maritima [SEA MALLOW] and others.
MALVACEAE.

Pounding maize (Zea mays) into flour in an African village. Maize is a staple food in Africa and widely used throughout the world as animal feed.

The Common European Mallow (Malva sylvestris) brings its bright color to hedgerows and waste places in early summer. (× ½)

brown or red berries in a terminal raceme. *M. bifolium* [MAY LILY] occurs extensively in Europe and parts of Asia in shady woodland habitats and, like the North American *M. canadense* [FALSE LILY OF THE VALLEY], is sometimes cultivated for ornament.
LILIACEAE, 2–3 species.

Maidenhair tree the common name for *Ginkgo biloba*, a tall deciduous tree sacred to the Buddhist religion and cultivated for many centuries in both China and Japan, especially in the grounds of temples. It also probably occurs in the truly wild state in eastern China. Within the last 200 years the tree has been planted widely and has been grown successfully under many different conditions of soil and climate. It is also remarkably free of disease and resistant to pests. The mature seed has a soft outer fleshy layer which has the unpleasant odor of rancid butter. For this reason male trees are usually preferred for avenue planting.
GINKGOACEAE, 1 species.

Maize or **corn** the common names for *Zea mays*, one of the most important cereal crops

in the world. It is the only cereal of American origin and it formed the staple diet of the American Indians [INDIAN CORN]. The USA produces nearly 50% of the world's maize but the crop is now grown widely in southeast Europe, Brazil, Argentina, Mexico, Africa and Indonesia. In some areas of Asia and Africa it forms the major part of the diet. There is evidence from pollen samples that maize and the closely related annual *Euchlaena mexicana* (= *Z. mexicana*) [TEOSINTE], or their possible common ancestor, were in existence 60 000 to 80 000 years ago in Mexico.

The grain is an important animal feed, particularly for poultry and pigs. For industrial purposes, maize is used in starch manufacture and for whisky distilling. Maize is also an important forage crop and is favored as a silage crop particularly in dry areas where grass growth may be poor. SWEET CORN, used for eating as "CORN ON THE COB", is a form of maize, var *rugosa*, in which

Young maize crop (Zea mays) in Wisconsin, USA. Half of the world's maize is grown in the United States and most of this is used as animal feed or as a forage crop.

Malope a genus of colorful annual herbs native to the Mediterranean region. Horticultural forms of *M. malacoides* and *M. trifida* are cultivated for their showy, mallow-like rose, pink, purple or white flowers.
MALVACEAE, 3–4 species.

Malpighia a tropical American and West Indian genus of evergreen trees and shrubs. Cultivated ornamental species include shrubs such as *M. coccigera* [SINGAPORE HOLLY], a small evergreen shrub with holly-like leaves and red fruits, and *M. glabra*, which has small pink or purple flowers and edible fruits with a high vitamin C content, resembling cherries but of poor flavor in comparison. The fruits of *M. punicifolia* [WEST INDIAN CHERRY] are of better quality and are used in preserves and sauces. The BARBADOS CHERRY, often referred to *M. glabra*, is probably a hybrid between this species and *M. punicifolia*.
MALPIGHACEAE, about 35 species.

Malus a genus of deciduous fruit-bearing trees and shrubs, some of which are known as crab apples. The genus also includes the many varieties of edible apples. (See APPLE and CRAB APPLE.)
ROSACEAE, about 35 species.

Malva [MALLOWS] a genus of annual, biennial or perennial herbs native to southern Europe, temperate Asia and North Africa. Most species are weedy and not cultivated. *M. sylvestris* [COMMON MALLOW] is common on roadsides and waste places; *M. moschata* [MUSK MALLOW], with attractive pink flowers and deeply divided leaves, and *M. alcea* are two of the species cultivated in herbaceous borders.
MALVACEAE, about 40 species.

Mammee the common name for *Mammea americana* [also known as MAMMEE APPLE, MAMEY, ST. DOMINGO APRICOT], a small tree native to the West Indies and widely cultivated in the Caribbean and Central and South America. It bears large, leathery, showy leaves, white, scented flowers and a globose, russet-colored fruit about the size of an orange. The tough rind encloses a sweet, scented, yellow, edible flesh. An aromatic liqueur (eau de creole) is distilled from the flowers in parts of the West Indies, while the fine-grained timber is valued for its use in cabinet-making.
GUTTIFERAE.

Mammillaria a large genus of cacti from Mexico and the southwestern USA, with a few species in the West Indies and northern South America. Most are dwarf, round or cylindrical stem succulents with spirally arranged tubercles tipped with the spine-bearing areole. The flowers are small but profusely borne in rings around the tops of the stems. They are followed by elongated red berries, often persistent.

Mammillaria is the most popular cactus genus in cultivation. The compact plants are ideal subjects for a specialist collection, taking up little space. Thus, *M. magnimamma* (= *M. centricirrha*), from Mexico, is almost hardy, pest-free and some would say indestructible, demanding a larger pot each year and eventually forming massive clumps. At the other extreme, *M. goldii* and *M. saboae* are at home in a 5cm pot although their blooms are larger than those of *M. magnimamma*.

M. hahniana [FEATHER CACTUS] and *M. plumosa* [OLD LADY CACTUS] look as if covered in white wool; in *M. candida* [SNOWBALL PINCUSHION] the whiter-than-white effect is due to the presence of multitudinous densely packed white spines. *M. spinosissima* is also heavily armed with spines which range in color from white through yellow to ruby red. The flowers are bright carmine. Hooked central spines are a feature of *M. magnifica*, where they are long and pale brown, and of *M. bombycina*, where the dark central spines contrast with the white radial spines and axillary wool. *M. deherdtiana*, *M. dodsonii* and *M. theresae* can be recommended for collectors who require tiny plants with comparatively large flowers.

The small but plentiful red fruits of *Mammillaria* are edible and are called chilitos.
CACTACEAE, about 220 species.

Mandarin or **mandarin orange** the collective name for the small, loose-skinned,

Mammillaria prolifera (= M. pusilla) produces abundant yellow flowers and red berries. (× ½)

orange-colored citrus fruits which are steadily gaining in popularity because of their easy-peeling properties. All mandarin oranges are generally regarded as belonging to a single species, *Citrus reticulata*, but there are many types and hybrids. Important varieties include SATSUMAS, TANGERINES and CLEMENTINES. Mandarin hybrids include *TANGELO and *UGLI (both crosses with *GRAPEFRUIT) and ORTANIQUE, a cross with SWEET *ORANGE. *C. × nobilis* (*C. reticulata* × *C. sinensis*) cultivar 'King' is the KING MANDARIN.
RUTACEAE.

Mandevilla a Central and tropical South American genus of tall climbing shrubs with racemes of large, showy, white, yellow or violet, funnel-shaped flowers expanding into five lobes. They are often cultivated in warm greenhouses in temperate regions. Popular species and hybrids include *M. × amabilis* (= *Dipladenia × amabilis*) (rose-colored flowers), *M. boliviensis* (= *D. boliviensis*) (white flowers with yellow throat), *M. sanderi* (= *D. sanderi*) (rose-pink flowers), and *M. laxa* (= *M. suavolens*) [CHILEAN JASMINE].
APOCYNACEAE, about 100 species.

Mandrake the common name for members of the genus *Mandragora*. All are herbs with narcotic or other medicinal properties and are native to the Mediterranean region. They have large fleshy roots and large bell-shaped flowers which are pale blue-violet, white or purple. Mandrakes have long had mythical associations with witchcraft and sorcery. Under the "doctrine of signatures" their root system was held to resemble the human form, and legend relates that anyone pulling up a mandrake will be driven mad by a terrible shriek coming from the root. *M. officinarum* and *M. autumnalis* both contain alkaloids that are pharmacologically useful.

SOLANACEAE, about 6 species.

Manettia a genus of evergreen twining shrubs and herbs, native to Central and tropical South America and the West Indies. Some of the species are grown as decorative ornamental plants in the greenhouse or in sheltered areas in warm temperate regions. One of the most widely cultivated species is the small twining shrub *M. inflata* (= *M. bicolor*), from Uruguay and Paraguay, which has tubular, reddish flowers, yellow at the tip where the lobes are expanded. *M. cordifolia* [FIRECRACKER VINE], a herbaceous vine, is also cultivated.

RUBIACEAE, about 130 species.

Mangel-wurzel, mangold or **field beet** the common names for a variety of *Beta vulgaris* which is cultivated for its swollen roots used as an important animal feed.

CHENOPODIACEAE.

Mangifera is a genus of tall evergreen trees from Southeast Asia, with the greatest concentration in the Malayan peninsula. Several species yield edible fruit but the most important by far is *M. indica* [MANGO] which is cultivated throughout the tropics, especially in India, where it is known wild in the northeastern part.

The celebrated fruit is a very variable fleshy, ovoid drupe 8–30cm long, sometimes with a beak at the proximal end, usually yellow or green at first and becoming reddish when ripe. The outer skin is thick and glandular, enclosing the orange or yellow flesh which varies from soft, sweet and juicy to somewhat fibrous and coarse. The ripe fruits are eaten raw, or canned or used in the manufacture of juices, squashes and jellies. Other mangoes are pickled for use in chutneys or curries. The fruits are also sliced, dried and seasoned with turmeric to produce "amchur", which is ground to a powder and

A Mangrove (Avicennia marina), *colonizing the mud of a rocky shore on the coast of southeastern Australia, showing the aerial roots.*

added to soups, chutneys and curries. The kernels are removed from the stones and dried, roasted and eaten, or ground to produce a flour.

The leaves of *M. indica* provide a yellow dye and are also used as cattle fodder. The bark yields a tannin and the timber is used in shipbuilding.

ANACARDIACEAE, about 40 species.

Mangosteen the highly esteemed fruit of *Garcinia mangostana*, a Southeast Asian village tree fruit rarely grown elsewhere in the tropics. The fruit is depressed-globose in shape and about 5cm across. Inside the thick, leathery, purple-brown pericarp lie a number of edible fleshy white segments, each enclosing a small seed.

GUTTIFERAE.

Mangrove bark the source of an important tanning material. All species of the mangrove genus *Rhizophora* contain tannin, and *R. mangle* [RED MANGROVE] and *R. mucronata*, both native to the coastal zones of the tropics, are the best sources. Most commercial supplies come from East Africa and Central and South America, yielding up to 40% tannin from the dried bark. Other mangrove genera whose bark yields quantities of tannin include *Avicennia*, especially *A. officinalis* [WHITE MANGROVE], *Ceriops*, especially *C. can-*

The thallose liverwort Marchantia polymorpha *is a common weed, particularly in soil disturbed by Man and by fire.* (×3)

dollaeanum, and *Kandelia*, especially K. rheedei.
RHIZOPHORACEAE.

Manihot a genus of shrubs and trees, all of which are native to subtropical and tropical areas, distributed from the southern USA through Central and South America as far as central Argentina. A number of shrubs and trees, including *M. glaziovii* and *M. dichotoma*, which are endemic to Brazil, contain latex in the stems and large fleshy roots which yield a high-quality *rubber (ceara and jequie rubber, respectively). Ceara rubber is grown commercially in Sri Lanka, India and other tropical countries. However, the trees are not as resilient as *Hevea brasiliensis* to continual tapping of the bark.

M. carthaginensis is a shrub, found from Mexico to Venezuela, with edible roots, and seeds which contain an oil with emetic and purgative properties. *M. dulcis* has sweet roots which are used as a vegetable in Brazil but the most important species commercially is *M. esculenta* (= *M. utilissima*) [see CASSAVA], which can produce more starch per hectare in a year than almost any other crop. The roots form an important staple food in many parts of Africa, India and South America, as well as providing a source of starch for many commercial purposes, for example in the paper and textile industries.

Plant breeders have successfully crossed *M. saxicola*, *M. dichotoma*, *M. melanobasis* and *M. glazovii* with *M. esculenta*, with the aim of producing cassava hybrids resistant to various virus, bacterial and fungal pathogens.
EUPHORBIACEAE, 98 or 128 species.

Manilkara (= *Achras*) a genus of evergreen trees native to tropical America and the West Indies. *M. zapota* [SAPODILLA, MARMALADE PLUM, BEEF APPLE, NASEBERRY, NISPERO, CHIKU] is cultivated throughout the tropics for its edible fruit which is pear-shaped or spherical, 5–10cm in diameter, with a thin, rough, rusty-brown skin and translucent brownish pulp containing several hard black seeds. Fully mature sapodillas are extremely sweet and are considered by some to be among the finest of dessert fruits. The trunk of the tree yields a latex, *chicle, the original main ingredient of chewing gum. *M. bidentata* is the main source of nonelastic rubber, *balata. The timber is hard, durable and commercially valuable.
SAPOTACEAE, about 85 species.

Manna a general term referring to plant exudates, usually sugary, which harden when dry and can be collected and eaten. Three types of manna are mentioned in the Bible. One has been tentatively identified as a sweet, gummy exudate from species of the genera *Tamarix* or *Alhagi*, while it has been suggested that the types that appeared on the damp ground and were sent from heaven may have been species of algae (*Nostoc) or lichen (*Lecanora*). Manna from *Fraxinus ornus* [MANNA ASH] is commercially collected in Sicily.

Manzanilla a name applied in Spain to a number of aromatic composite herbs and to the infusion made from their flower heads, especially *Matricaria aurea* (= *Chamomilla aurea*), *M. chamomilla* (= *Chamomilla recutita*) and *Chamaemelum nobile* [ROMAN CHAMOMILE]. Highly prized is *Artemisia granatensis* [MANZANILLA REAL], from the Sierra Nevada of Spain, which is greatly endangered through overcollection.

MANZANILLA is also the common name of several Central and North American *Crataegus* species such as *C. stipulosa* and *C. mexicana*, whose fruits are used as preserves or jellies. The name MANZANILLA means little apple and refers to the fruits.

Maranta a genus of herbaceous perennials native to tropical America. Some species such as *M. arundinacea* [ARROWROOT] are rhizomatous, while others such as *M. bicolor* are tuberous. The former produces erect branching leafy stems, while the latter and some other species, eg *M. leuconeura* [PRAYER PLANT] and varieties such as *kerchoviana*, grown as ornamentals for their attractive marked foliage, are nearly stemless.

M. arundinacea is found wild in Brazil, northern South America and Central America. It is cultivated in the tropics of the Old and New Worlds (with the main center of production in the Caribbean, especially St. Vincent) for the starch contained in its large fleshy rhizome (see ARROWROOT).
MARANTACEAE, about 25 species.

Maraschino cherry the cultivated fruit of and common name for *Prunus cerasus* cultivar 'Marasca'. It is used in canning and for desserts and jam as well as being the base of cherry brandy and the Italian liqueur maraschino.
ROSACEAE.

Marattia a genus of mainly tropical ferns with stout stems and pinnate leaves. Species cultivated for ornament include *M. alata* and *M. salicifolia*. The stem pith of *M. fraxinea* [KING FERN] was a source of starch for the New Zealand Maoris.
MARATTIACEAE, about 60 species.

Marchantia a very widely distributed genus of liverworts. The most common, *M. polymorpha*, is a cosmopolitan weed that extends from the tropics to the arctic regions. It is especially prevalent in areas that have been disturbed by the activities of Man, particularly by fire. It can be a pernicious weed in greenhouses because of the rapidity with which it covers soil.
MARCHANTIACEAE, about 50 species.

Marigold the name given to *Calendula officinalis* (also called COMMON or POT MARIGOLD), a hardy erect annual which is cultivated in many varieties. AFRICAN and FRENCH MARIGOLDS belong to the genus *Tagetes; the BUR MARIGOLD is *Bidens tripartita*; *Chrysanthemum segetum* is the CORN MARIGOLD; *Dimorphotheca* species are the CAPE MARIGOLDS; and *Baileya multiradiata* is the DESERT MARIGOLD. All of the above mentioned belong to

*Manioc or Cassava (*Manihot esculenta*) under cultivation in a recently cleared plot at Rio Casiquiare, southern Venezuela.*

the Compositae. However, *Caltha palustris (MARSH MARIGOLD) is a member of the Ranunculaceae.

Pot Marjoram (Origanum vulgare) used in flavoring meats, stews and stuffings and for bouquets garnis. (× ⅓)

Marjoram the name given to some species of culinary herbs in the genus *Origanum. O. majorana (= Majorana hortensis) [SWEET MARJORAM], from southern Europe, is cultivated for the leaves which are used to flavor meat dishes. Oil of marjoram, obtained by steam distillation, is used for the same purposes. O. vulgare [POT MARJORAM] yields a similar but not so sweet flavoring.
LABIATAE.

Marram grass the common name for *Ammophila arenaria (= A. arundinacea), a widespread perennial grass, abundant and often dominant along the coasts of Western Europe, and introduced to North America and Australia as a sand binder.
GRAMINEAE.

Marrows and squashes see p. 216.

Marrubium a genus of herbaceous perennials native to Eurasia and the Mediterranean region. M. vulgare [HOREHOUND] was once popular for home remedies. The leaves and young shoots were used in syrups and teas to counteract sore throats and colds. Essential oils from the leaves are used in liqueurs. M. incanum, with white-woolly leaves and stems and whitish flowers, is sometimes cultivated.
LABIATAE, about 40 species.

Marsdenia a small genus of tropical and subtropical African or Asian shrubs with small flowers usually in panicles or umbellate inflorescences. M. roylei, with yellow flowers, is one of several species sometimes grown for ornament, usually under glass. Some species, especially M. tenacissima, from the lower Himalayas and Bengal are a source of fiber and latex.
ASCLEPIADACEAE, about 70–100 species.

Marsilea [WATER CLOVERS] a genus of tropical and temperate aquatic or marsh ferns. The leaves and branches arise from a slender creeping rhizome. Individuals of a single species may be either aquatic or amphibious. In the former, the leaf-stalks are long, flexible and weak with the four leaf-lobes floating on the surface of the water. In the amphibious forms the leaf-stalks are shorter and thicker and stand erect. Some species vegetate during the dry season but others (eg M. vestita), which grow in areas with a distinct dry season, die down and persist in the form of bean-shaped sporocarps. Species cultivated in pools, aquaria etc include M. drummondii and M. quadrifolia.
MARSILEACEAE, about 60 species.

Martynia a genus represented by a single species, M. annua, a herb native to Mexico. The best-known "martynia", M. proboscidea, from southern North America to tropical America, is now referred to the genus Proboscidea as P. louisianica, commonly called the RAM'S HORN or UNICORN PLANT on account of its horned fruit. It is sometimes cultivated as an oddity, or for the young fruits which are used as pickles.
MARTYNIACEAE, 1 species.

Masdevallia a large genus of tropical American epiphytic or terrestrial orchids found growing mainly at high altitudes. The plants are without pseudobulbs, and the thick, sometimes succulent leaves grow in tufts usually evenly spaced along a creeping rhizome. The inflorescences are erect and bear up to 12 flowers. They are frequently grown for the bizarre somewhat funnel-shaped flowers, some of which have a "shot-silk" look. M. chimaera and M. coccinea, with various cultivars, are among the most popular species.
ORCHIDACEAE, about 300 species.

Marsilea, the Water Clovers, are in fact ferns. Shown here is M. mutica with the four leaf lobes floating on the surface; the leaf stalks are attached to a slender creeping rhizome. (× ¾)

Mastic a resin which has two principal uses: as a varnish and as a chewing gum. American mastic is tapped from *Schinus molle [PEPPER TREE], Bombay mastic from *Pistacia mutica, and Chios mastic, an ancient resin, obtained from Pistacia lentiscus, from the eastern Mediterranean. The Turks chew mastic to sweeten the breath, to aid digestion and to preserve the gums. Indeed, mastic has been used as a temporary stopping for teeth. In Europe it is used as a varnish for coating paint and in lithography.

Maté or **yerba de maté** the leaves of *Ilex paraguariensis, an evergreen tree, cultivated and wild, from South America (Paraguay, Argentina, Brazil), used as a tea in many parts of that continent. It was originally a native drink which was taken up by settlers and has since become widespread and second only to coffee, tea or cocoa. The oven-dried and threshed product is called, in diminishing size of the greenish leaf fragments, maté grosso, maté entrefino and maté fino.
AQUIFOLIACEAE.

Matricaria the name formerly used for species of the genus *Chamomilla.
COMPOSITAE.

Matteuccia a genus of large ferns with stout rootstocks, native to temperate regions of the Northern Hemisphere. The leaves are of two forms: the fertile leaves are smaller and less dissected than the sterile bipinnatifid leaves. The most commonly cultivated species is M. struthiopteris (= *Onoclea germanica, Struthiopteris germanica) [OSTRICH FERN], with sterile leaves up to 1.75m long.
ASPIDIACEAE, 3–5 species.

Matthiola a genus of gray-pubescent herbs and subshrubs from Europe, southwest Asia and North Africa. The flowers, which may be purple, brown-purple, reddish, bluish, yellow or white, are borne in cylindrical racemes

Melaleuca preissiana *growing at the edge of a dried-out lake in Western Australia.*

and are often scented. The best-known is *M. incana* [GARDEN STOCK, GILLYFLOWER], variants of which are grown as short-lived perennials, biennials [BROMPTON or IMPERIAL STOCKS] or as annuals [TEN WEEK STOCKS]. An intermediate form [WINTER STOCK] is sown at the same time as TEN WEEK STOCKS but is hardier and blooms later even during cold weather. *M. longipetala* subspecies *bicornis* [NIGHT-SCENTED STOCK] is less showy than the previous species, but its lilac or purple flowers open in the evening and are sweetly scented throughout the night.
CRUCIFERAE, 30–50 species.

Maxillaria a genus of pseudobulbous epiphytic and terrestrial orchids occurring from southern Florida throughout Central and tropical South America and the West Indies to Argentina. Many of them are found at high altitudes and can be grown easily in a cool greenhouse. The inflorescences arise from among the leaves and bear one to a few flowers. Varying greatly in size, up to 15cm in diameter, the flowers can be either of one color or variously spotted, speckled or blotched in contrasting colors. The tepals are usually long and narrow and the overall effect can be very spider-like, as in *M. arachnites.*
ORCHIDACEAE, about 300–400 species.

Mealies a common name for *MAIZE (Zea mays).* The term is used mainly in South Africa.
GRAMINEAE.

Meconopsis a genus of annual, biennial or perennial herbs, one of which, *M. cambrica* [WELSH POPPY], is native to Western Europe, the rest to the Himalayas and the region across to western China. They are popularly known as HIMALAYAN or TIBETAN POPPIES.

They are handsome garden plants when given rich soil and cool shady conditions. The yellowish or orange-flowered *M. cambrica* is not nearly so demanding. *M. baileyi* [BLUE POPPY] has flowers up to 10cm in diameter, sky-blue in color if the conditions are right, otherwise a washy mauve. *M. grandis* bears one or more spectacular dark blue or purple flowers, 13cm in diameter, on stout stalks from the axils of the upper leaves. *M. regia* and *M. integrifolia* [YELLOW CHINESE POPPY] produce yellow flowers up to 12cm in diameter. The winter basal rosette of *M. regia*, often 1m in diameter with silver or golden-haired leaves, is extremely attractive.
PAPAVERACEAE, about 43 species.

Medicago a genus of mostly weedy annual or perennial herbs, from Europe, the Mediterranean and South Africa. *M. sativa* [ALFALFA, LUCERNE] is the world's most important forage crop, growing in both temperate and subtropical climates (see LUCERNE).

M. lupulina [BLACK MEDICK] is also a highly nutritious fodder plant but in gardens it is regarded as a weed. However, a few species are cultivated in gardens, eg *M. echinus* [CALVARY CLOVER], in which the coiled pods are covered with long interlocking spines like a crown of thorns. It has clusters of small yellow pea-type flowers and leaflets with a reddish spot in the center which in var *variegata* is more prominent.
LEGUMINOSAE, about 50 species.

Garden Stock (Matthiola incana *) has dense racemes of heavily scented flowers. (* × 1 *)*

Meconopsis horridula, *one of the brilliant blue Himalayan Poppies; this species is covered with distinctive straw-colored spines. (* × $\frac{1}{6}$ *)*

Medicinal and narcotic plants
see p. 220.

Medlar the common name for *Mespilus germanica*, a small spreading tree that is found in southeastern Europe extending eastwards to Central Asia. It is the only species in the genus *Mespilus*. Medlars often persist in cultivation as old, gnarled, but attractive specimens. The brownish apple-shaped fruits are traditionally eaten with wine after frosting or bletting (rotting) has softened the hard fruit tissues. Jellies and preserves can also be made from the fruits.
ROSACEAE.

Melaleuca a tropical genus of shrubs and medium-sized evergreen trees from Australia and the Pacific Islands, with one species extending to Indomalaysia. The latter, *M. leucadendron* [CAJUPUT TREE, RIVER TEA TREE], has a thick spongy bark, and its leaves are a source of a green medicinal aromatic oil, cajuput oil, used (mainly locally) as a stimulant and tonic, and in soothing ointments. The tree also produces a very durable red-violet timber, suitable for posts, piles, roofing and shipbuilding. *M. minor* also yields cajuput oil.

Ornamental species include *M. quinquenervia* [PAPERBARK TREE] (often incorrectly called *M. leucadendron* in cultivation) as well as such attractive and graceful shrubs as *M. incana*, with spikes of yellowish-white flowers, and *M. huegelii* [CHENILLE HONEYMYRTLE], with spikes of creamy-white flowers.
MYRTACEAE, about 100 species.

Melampsora a mainly north temperate genus of rust fungi parasitic on angiosperms. The most important species is *M. lini* [FLAX RUST] which has all the stages of its life cycle on *FLAX.
MELAMPSORACEAE, about 80 species.

Marrows, Squashes, Pumpkins and Gourds

Marrows, *squashes and *pumpkins are the edible fruits of *Cucurbita pepo, C. mixta, C. moschata and C. maxima. The application of the common names of these species is very confusing and varies from region to region, and country to country. Cultivars known as pumpkins and others known as winter squashes occur in all four species. In the United States, the cultivars used for pies, stock feed and lanterns are commonly called pumpkins. The name marrow is normally restricted to cultivars of Cucurbita pepo although C. maxima is sometimes known as ORANGE MARROW.

The term *gourd is applied to fruits, usually with hard durable rinds, related to the pumpkins, squashes, *cucumbers and *melons. The yellow-flowered gourds of North America are Curcurbita pepo var ovifera and the MALABAR GOURD is C. ficifolia but the white-flowered and other kinds of gourd belong to *Lagenaria, Trichosanthes and other genera.

They are all trailing or climbing herbs, with tendrils, and with large, alternate, often palmately lobed leaves, and large, usually yellow, unisexual flowers. The fruit is a large berry, known as a pepo, sometimes with a tough rind or "shell". The edible species are all annuals. Marrows, squashes and gourds probably all originated in the

A wide range of hard-shelled ornamental gourds belonging to Cucurbita pepo var ovifera [YELLOW-FLOWERED GOURDS]. *(× ¼)*

New World, mainly around the Mexico–Guatemala border area. They were domesticated in pre-Columbian times by American Indians and were an important part of their diet along with maize and beans, but their detailed history is not known. C. maxima, C. moschata, C. pepo, C. mixta and C. ficifolia are of cultivated origin and do not occur in the wild state. Marrows, squashes and pumpkins are cultivated on a commercial scale in the United States, Soviet Russia and parts of Europe, and they are widely grown on a local scale especially in the tropics. C. moschata and C. mixta are more adapted to tropical conditions than C. maxima and C. pepo, which grow best in temperate areas. The young shoots and flowers are eaten (as in marrow flower soup) as well as the fruits, and the seeds (known as pepitos) are highly appreciated when fried and salted. Marrows or vegetable marrows (C. pepo) are divided into "bush" or "trailing" cultivars and are green, whitish-green or striped. Squashes can be divided into summer or winter: summer squashes are fruits of C. pepo which are eaten when immature (from the day of flowering until the rind becomes hard). They occur in a wide range of shapes and types, including 'Summer Crookneck' with bright yellow or orange warty club-like fruits, scallop or pattypan squashes, which have white disc-shaped and ribbed joints, and the cylindrical or globular *courgettes or

zucchini which are solid green or striped. Vegetable marrows also come under the heading of summer squashes. Winter squashes are fruits of C. pepo, C. maxima, C. mixta and C. moschata which are eaten when mature or are stored for winter consumption as table vegetables, or for making pies, jam and as feed for livestock. They include the Turban and Chioggia squashes. Winter squashes usually have darker flesh, are less fibrous, contain less water, are higher in sugar and dry matter,

MARROWS, SQUASHES, PUMPKINS AND GOURDS	
Scientific name	**Common name**
*Cucurbita pepo	VEGETABLE MARROW; SUMMER AND AUTUMN PUMPKIN OR SQUASH; BRAZILIAN OR AMERICAN PUMPKIN; CUSTARD MARROW; SCALLOP GOURD; SUMMER CROOKNECK; COURGETTE; ZUCCHINI; VEGETABLE SPAGHETTI; ACORN SQUASH; PATTYPAN SQUASH; COCOZELLE SQUASH; BUSH SQUASH.
var ovifera	YELLOW-FLOWERED GOURDS
*Cucurbita maxima	AUTUMN AND WINTER SQUASH OR PUMPKIN; TURBAN SQUASH; ORANGE MARROW; BANANA SQUASH; HUBBARD SQUASH; SEA SQUASH; CHIOGGIA SQUASH; OHIO SQUASH.
*Cucurbita mixta	WINTER SQUASH; WINTER PUMPKIN; CUSHAW PUMPKIN; SQUASH; SILVER SEED GOURD; TENNESSEE SWEET POTATO.
*Cucurbita moschata	PUMPKIN; CANADA PUMPKIN; CROOKNECK; BUTTERNUT OR WINTER SQUASH; NAPLES SQUASH.
*Cucurbita ficifolia	MALABAR GOURD, FIG-LEAF GOURD; MALABAR MELON; SIAMESE GOURD.
*Cucurbita foetidissima	MISSOURI GOURD; CALABAZILLA; FETID WILD PUMPKIN.
*Citrullus lanatus (Citrullus vulgaris, Colocynthis citrullus)	WATER MELON; CITRON; PRESERVING MELON.
*Citrullus colocynthis	COLOCYNTH
*Cucumis anguria	WEST INDIAN GHERKIN; BUR GHERKIN; GOOSEBERRY GOURD; GOAREBERRY GOURD.
*Cucumis dipsaceus	HEDGEHOG GOURD; TEASEL GOURD.
*Cucumis melo	
var cantalupensis	*CANTALOUPE; ROCK MELON
var chito	MANGO MELON; ORANGE MELON; GARDEN LEMON; MELON APPLE; VEGETABLE ORANGE; VINE PEACH.
var conomon	ORIENTAL PICKLING MELON.
var dudaim	DUDAIM MELON; POMEGRANATE MELON; QUEEN ANNE'S POCKET MELON; STINK MELON.
var flexuosus	SNAKE MELON; SERPENT MELON.
var inodorus	WINTER MELON; CASABA MELON.
var reticulatus (var scandens)	MUSK MELON; NETTED MELON; NUTMEG MELON; PERSIAN MELON.
var saccharinus	HONEYDEW MELON.
*Cucumis metuliferus	AFRICAN HORNED MELON.
*Cucumis sativus	*CUCUMBER, "GHERKIN".
Lagenaria siceraria	WHITE-FLOWERED GOURD; *CALABASH GOURD; DIPPER GOURD; SUGAR-TROUGH GOURD; HERCULES'-CLUB GOURD; BOTTLE GOURD; KNOB-KERRIE GOURD; TRUMPET GOURD.

Trichosanthes anguina (*T. cumeraria*)	SERPENT GOURD; SERPENT CUCUMBER; SNAKE GOURD; CLUB GOURD; VIPER'S GOURD.
Trichosanthes cucumeroides	SNAKE GOURD.
Benincasa hispida (*B. cerifera*)	WHITE GOURD; WAX GOURD; ASH GOURD; TUNKA; CHINESE WATER MELON.
Sicana odorifera	CURUBÁ; CASSABANANA.
Sechium edule	*CHAYOTE; SOU-SOU; CHRISTOPHINE.
Ecballium elaterium	SQUIRTING CUCUMBER.
Luffa aegyptiaca (*L. cylindrica*)	TOWEL GOURD; SPONGE GOURD; DISHCLOTH GOURD; LUFFA; LOOFA.
Luffa acutangula	ANGLED LOOFA.
Coccinea grandis	IVY GOURD.

(See also Melons, Gourds)

richer in protein, fats and vitamin A than summer squashes. They are usually baked rather than boiled.

Pumpkins and squashes are inexpensive and highly digestible vegetables. They do not have a pronounced flavor and are low in nutritive value, being 90–95% water and 3–6% carbohydrate. The seeds on the other hand are rich in oil and protein. Some of the gourds are grown for food, such as the MALABAR or FIG GOURD (*Cucurbita ficifolia*), which is cultivated both in the Old and New Worlds. Its flesh is eaten candied and its ripe seeds are edible. It differs from other edible species of *Cucurbita* in being perennial. The SNAKE or SERPENT GOURD (*Trichosanthes anguina*), an Indomalaysian species with fruits up to 2 metres in length, cultivated in tropical Asia, and the WAX GOURD (*Benicasa hispida*), from Southeast Asia, with large oblong to cylindrical fruits, are also eaten when immature. Other species such as *Lagenaria siceraria* [WHITE-FLOWERED CALABASH, BOTTLE GOURD], probably native to tropical Africa and long cultivated in both hemispheres, and *Coccinea grandis* [IVY GOURD], an Old World species naturalized in South America, are grown for use as containers, musical instruments or ornaments.

Although primarily grown for their edible flesh or as ornamentals, the seeds of some species of *Cucurbita* yield a high quality oil which is utilized, particularly in Europe. The seeds of one mutant of *C. pepo* contain up to 50% oil.

Common marrows, squashes and other cucurbits: 1 Watermelon; 2 Honeydew Melon; 3 Netted or Nutmeg Melon, Musk Melon; 4 Gherkin; 5 Cucumber; 6 Courgette; 7 Summer Squash 'Summer Crookneck'; 8 Pumpkin; 9 Winter Squash 'Hubbard'; 10 Custard Marrow; 11 Vegetable Marrow. 1, 2, 3, 7, 9, 11 ($\times\frac{1}{4}$); 4, 5, 6 ($\times\frac{1}{2}$); 8, 10 ($\times\frac{1}{10}$).

Melastoma a genus of evergreen shrubs native to southern China, Indomalaysia and the Pacific. A number of species, notably *M. denticulatum*, with showy white flowers, and *M. malabathricum*, with large violet-pink or purple flowers, are cultivated as ornamental flowering shrubs in tropical gardens or in greenhouses in temperate climates. MELASTOMACEAE, about 70 species.

Melia a small genus of deciduous trees native to Australia and the East Indies, cultivated for their fragrant inflorescences. The hard seeds of the purplish-flowered Asiatic species *M. azedarach* [CHINA BERRY TREE, UMBRELLA TREE, BEAD TREE, PRIDE OF INDIA, PERSIAN LILAC] are sometimes used as beads for rosaries. The bark, leaves and fruits of this species have medicinal uses and the fruits are a source of insecticide. A number of species, including *M. azedarach*, *M. dubia* and *M. excelsa*, produce valuable timber; that of the latter resembles mahogany and is put to a wide range of uses, including general carpentry work, furniture and veneer making.

   M. azedarach is widely naturalized in tropical America, and widely planted in other tropical and warm temperate regions for its commercial and ornamental value. The best-known cultivars include 'Floribunda' (= *M. floribunda*) and 'Umbraculiformis' [TEXAS UMBRELLA TREE].
MELIACEAE, about 10 species.

Melica a genus of perennial grasses attractive spreading inflorescences native to temperate regions, except Australasia, of

*The leaves of the Lemon Balm or Sweet Balm (*Melissa officinalis*) are strongly lemon-scented and are used for seasoning. (× ⅓)*

*The Field or Corn Mint (*Mentha arvensis*) is closely related to the cultivated Spearmint and is a common weed in wheat fields. (× 1)*

which *M. altissima* and *M. ciliata* are cultivated as ornamentals.
GRAMINEAE, about 50–70 species.

Melicoccus a genus of only two species, both from tropical America and the Caribbean. One of these *M. bijugatus* (= *M. bijuga*) [SPANISH LIME, HONEY BERRY, GENIP, MAMONCILLO], forms a sizeable tree, cultivated for its green globular edible fruits. The single seed is surrounded by a yellow, sweet, juicy pulp.
SAPINDACEAE, 2 species.

Melilotus [MELILOTS, SWEET CLOVERS] a genus of annual or biennial herbs with pea-like flowers, mainly native to the Mediterranean and Eurasian regions. Unlike the true *CLOVERS (Trifolium)*, they grow to three metres tall, have mostly pinnate leaves with finely toothed leaflets, and their flowers are borne in long spikes or racemes. The pod is usually short, containing only one or two seeds.

   A number of species are grown as fodder crops or green manure. Most contain coumarin and produce a smell of new mown hay when cut or grazed. Their copious nectar is the source of a good quality honey. Useful fodder crops are grown from *M. officinalis* [COMMON MELILOT, SWEET CLOVER, BOKHARA or BUCKHARA CLOVER], *M. alba* [WHITE MELILOT, WHITE SWEET CLOVER, BOKHARA CLOVER] and *M. indica* (= *M. parviflora*) [SMALL FLOWERED MELILOT, SOUR CLOVER]. *M. officinalis* has additional uses in flavoring cheeses and as an insect repellent.
LEGUMINOSAE, about 20 species.

Melissa a small genus of bushy herbaceous perennials from southern Europe and Asia. The leaves are aromatic and hairy and the

pale yellow to white or pinkish flowers are borne in whorls. *M. officinalis* [BALM, LEMON BALM] is cultivated for its lemon-scented foliage used in seasoning and in medicine. There are cultivars with attractive golden or variegated leaves.
LABIATAE, 4–5 species.

Melittis a genus consisting of a single species, *M. melissophyllum* [BASTARD BALM]. It is a perennial herb growing up to 70cm tall, with ovate leaves bearing the large creamy-white, pink or purple flowers in their axils. It grows in western, central and southern Europe, extending to the Ukraine, and is cultivated as a border plant, particularly as the cultivars 'Album' and 'Grandiflorum', which are white- and creamy-white flowered respectively.
LABIATAE, 1 species.

Melon a name applied to two quite different species of widely cultivated plants of the cucumber family (Cucurbitaceae), *Cucumis melo* [SWEET MELON, MUSK MELON] and *Citrullus lanatus* (= *C. vulgaris*, *Colocynthis*

*Water Melons (*Citrullus lanatus*) ready for harvest in western Texas. Most USA production is in the southern states.*

citrullus) [WATER MELON].

   C. melo is an annual, trailing vine. Its angular stems bear large, alternate, kidney-shaped leaves, the unbranched tendrils and yellow flowers being produced in the leaf axils. There is a great variety of fruits. NETTED, NUTMEG or MUSK MELONS, usually named cultivars of *C. melo* var *reticulatus*, are smaller than most melons, with smooth skins often covered with net-like markings. The CANTALOUPE (var *cantalupensis*) is a little larger and has thicker, rough to warty skin. They are widely grown in Europe and a few hardy cultivars are grown under glass in north temperate regions. The OGEN MELON exported from Israel is a member of the cantaloupe group.

   The CASABA or WINTER MELON (var *ino-*

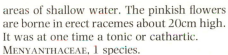

areas of shallow water. The pinkish flowers are borne in erect racemes about 20cm high. It was at one time a tonic or cathartic. MENYANTHACEAE, 1 species.

Menziesia a small genus of north temperate Asian and American species of deciduous, slow-growing shrubs. The alternate, entire leaves are variously pubescent and bristly, and the nodding white, pink, purple or deep red flowers are produced in terminal clusters on the previous year's growth. The Japanese group includes several cultivated species, eg *M. ciliicalyx* and *M. pilosa* [MINNIEBUSH]. ERICACEAE, 7 species.

Mercurialis a genus of annual or perennial herbs largely confined to the Mediterranean and Europe. *M. annua* [ANNUAL MERCURY] a weed of cultivated ground and waste places, is the most widespread, and *M. perennis* [DOG'S MERCURY] is a common woodland plant, forming pure stands on a wide variety

dorus) and the HONEYDEW (var *saccharinus*) are grown commercially but to a smaller extent. All these varieties have fruits with succulent flesh that is sweet and often scented, making a delicious breakfast or a dessert dish when served cold with a little sugar or limejuice. The ORIENTAL PICKLING MELON (var *conomon*) and the MANGO MELON (var *chito*) are used for both pickling and preserves. The SNAKE or SERPENT MELON (var *flexuosus*) has fruits up to 1m long and 7.5cm in diameter. They may be contorted into serpent-like shapes and are mainly grown for curiosity value. The DUDAIM MELON (var *dudaim*) has large flowers which produce small highly scented fruits about the size of an orange. (See also WATER MELON.) CUCURBITACEAE.

Mentha see MINT. LABIATAE, about 25 species.

The green-flowered Dog's Mercury (Mercurialis perennis) is common and often abundant in woodlands during early spring. ($\times\frac{1}{15}$)

The Bogbean (Menyanthes trifoliata) showing its trifoliolate leaves, erect pink-white racemes, and wet marshy habitat. ($\times\frac{1}{10}$)

Mentzelia a genus of annual, biennial and perennial herbs native in tropical and sub-tropical North and South America and the West Indies. *Mentzelia* does not exhibit the barbed, stinging hairs so characteristic of other members of the family. The showy orange, yellow and white flowers are often particularly fragrant in the evening. *M. laevicaulis* [BLAZING STAR, PRAIRIE LILY], a bushy annual or biennial, is widely cultivated for its large golden-yellow flowers, 6cm in diameter, produced two or three together and opening in the evening. Other cultivated ornamental species include *M. albescens* and *M. lindleyi* (= *Bartonia aurea*). LOASACEAE, about 70 species.

Menyanthes a genus consisting of a single species, *M. trifoliata* [BUCKBEAN or BOGBEAN], a north temperate perennial aquatic herb usually growing in bogs, ditches or other

A peasant crop of melons in Afghanistan, showing a wide range of form and quality — such crops contain great genetic diversity and are of importance in the future breeding of the crop. ($\times\frac{1}{10}$)

of soils. The latter species was at one time used as a dye. EUPHORBIACEAE, 15 species.

Merendera a small genus of crocus-like perennial herbs, native to the Mediterranean region, Afghanistan, the USSR and Ethiopia. They produce star-shaped flowers with narrow pink or rose petals. The best-known species is *M. pyrenaica* (= *M. montana*, *M. bulbocodium*), from the Pyrenees, Spain and Portugal. *M. filifolia*, from the western Mediterranean, is sometimes cultivated in rough grass, and the corms of *M. robusta* (= *M. persica*), from northern Afghanistan and adjacent USSR, are used as a treatment for rheumatism. LILIACEAE, about 10 species.

Mertensia a genus of north temperate perennial herbs distributed as far south as Mexico and Afghanistan. The leaves are entire, often with pellucid glands. The bluish, purplish or white, bell-shaped flowers are (continued on p. 222)

Medicinal and Narcotic Plants

For most of Man's history the fields of botany and medicine were, for all practical purposes, synonymous fields of knowledge, as the leading ethnobotanist R. E. Schultes has pointed out; the witch-doctor – usually an accomplished botanist – is probably the oldest of the professions in Man's culture. The close relationship between plants and Man can be traced back to earliest times when primitive man seeking to survive in this planet had to learn to distinguish and recognize plants that were useful as food or drugs from those that were ineffectual or poisonous. It has been noted that the number of plants that have been used in folk medicine greatly exceeds those used or cultivated for food or other purposes. The first botanic gardens were medical gardens or gardens of simples (ie herbs) attached to medical faculties.

There are references to drugs of plant origin in the early Greek literature and Hippocrates listed plants according to their uses, but the first important classification of drugs plants according to their use was compiled in the first century AD by Dioscorides. His compilation of plant lore formed the basis of materia medica for centuries until the publication of the more complete works, such as the *Dispensorium* of Cordus in 1515 and the first official pharmacopeia in Augsburg in 1564, soon followed by others.

The rapid development of the field of organic chemistry in the 19th century led to a wider understanding of the chemical structure of plant drugs and to their gradual replacement by drugs of synthetic origin. While the early official pharmacopeias comprised largely drugs of plant origin, today the percentage of drugs of natural plant origin in say the British and American Pharmacopeias of the 1960s is five or six per cent.

Nonetheless, in recent years there has been a resurgence of interest in the plant kingdom as a source of drugs and there is today a deep interest in understanding plants, their physiology and chemistry and their ethnobotanical significance. No longer does the pharmaceutical industry neglect plants in favor of "coal tar" drugs, largely due to the discovery in the last 10–20 years of the so-called "wonder drugs" such as reserpine from *Rauvolfia*, cortisone precursors from *Strophanthus* and *Dioscorea*, hypertensive agents from *Veratrum*, and the use of steroids of plant origin especially in oral contraceptives, the discoveries of antibiotics, and the more

recent interest in hallucinogens. As a consequence it is estimated that of the 300 000 000 new prescriptions filed in the United States each year, half contain at least one ingredient of natural plant origin. The spectacular development of phytochemical research in the past quarter of a century has only covered a small portion of the potential offered by the plant kingdom and the more extensive exploration of the tropical flora, coupled with an increasing interest in and knowledge of the plants used in folk medicines, suggests that the surface has only just been scratched.

Medicinal plants, or crude drugs as they are known commercially, are a diverse and economically important group of species belonging to many families, genera and species throughout the plant kingdom. They are used in the pharmaceutical industry for the production of drugs in the form of tablets, capsules and injections or for the preparation of tinctures, infusions and mixtures. The specific drugs contained by medicinal plants are known as "active principles". They may be divided chemically into a number of groups, principally the alkaloids (see p. 26), glycosides, volatile essential oils, resins and oleo-resins and steroids. Such drugs are isolated from poisonous plants – medicinal plants are useful as drugs because they are

toxic. Toxicity does not imply fatal results – it is a matter of dosage. Indeed the difference between a poison, a medicine and a narcotic is just one of dosage. Some crude drugs with low toxicity are used in medicine in the dried form or as infusions or tinctures. An example is valerian (from *Valeriana officinalis*) used as a sedative in medicinal preparations as a tincture. In modern pharmaceutics, the active principles in poisonous plants which comprise the group of specific drugs are isolated and a standard dosage is formulated.

The alkaloids are the major group of specific drugs. They were amongst the earliest plant poisons to be isolated – morphine was isolated in a pure form in 1806 from the *OPIUM POPPY (*Papaver somniferum*) – and research into their nature and structure led to the development of the sciences of organic chemistry and pharmacology. Examples include drugs used as local anesthetics, analgesics, muscle stimulants and relaxants, hallucinogens and tranquilizers. Glycosides are widespread in the plant kingdom and many have a pronounced pharmacological effect on mammals. There are several groups of glycosides, the most widely used in medicine being the cardiac or cardioactive glycosides which have a specific effect on the heart muscle. Glycosides isolated from *FOXGLOVES* (*Digitalis purpurea* and *D. lanata*) are widely used in the treatment of heart conditions. The FOXGLOVE was first introduced into medicine by William Withering in 1785 but the nature of the active principles was not discovered until 1933 by Stoll in Switzerland. Likewise, the purgative drug *cascara sagrada obtained from *Rhammus purshiana*, was introduced into medicine in 1870, based on its use by North American Indians, but it was not until 1975 that the

NARCOTIC or HALLUCINOGENIC PLANTS

Note: Some narcotic plants are mentioned under the other sections of this table.

Common name	Scientific name	Family
FLAG ROOT, SWEET FLAG	*Acorus calamus	Iridaceae
FLY AGARIC	*Amanita muscaria	Amanitaceae
YOPO, VILEA, SEBIL	Anadananthera peregrina	Leguminosae
BELLADONNA	*Atropa belladonna	Solanaceae
CAAPI	Banisteriopsis caapi,	Malpighiaceae
	B. inebrians,	
BOLETUS	Boletus manicus	Boletaceae
ANGEL'S TRUMPET	Brugsimania aurea	Solanaceae
ZACATECHICHI (BITTER GRASS)	Calea zacatechichi	Compositae
MARIJUANA	*Cannabis sativa	Cannabaceae
*ERGOT	Claviceps purpurea	Hypocreaceae
THORN APPLE	*Datura stramonium	Solanaceae
DOWNY THORN-APPLE	D. inoxia	Solanaceae
HORN-A-PLENTY	D. metel	Solanaceae
HENBANE	*Hyoscyamus niger	Solanaceae
MORNING GLORY	*Ipomoea tricolor (= I. violacea)	Convolvulaceae
OLOLOLIUQUI	Turbina corymbosa	Convolvulaceae
MESCAL OR PEYOTE BUTTON	*Lophophora williamsii	Cactaceae
*MANDRAKE	Mandragora officinarum	Solanaceae
CULEBRA BORRACHERO	Methysticodendron amnesiacum	Solanaceae
*NUTMEG, *MACE	Myristica fragans	Myristicaceae
T-HA-NA-SA, SHE-TO	Panaeolus sphinctrinus	Agaricaceae
SYRIAN RUE	Peganum harmala	Zygophyllaceae
INDIAN POKE	*Phytolacca acinosa	Phytolaccaceae
SACRED MUSHROOM	*Psilocybe mexicana	Agaricaceae
SWEET SCENTED MARIGOLD	*Tagetes lucida	Compositae

MEDICINAL PLANTS

I Alkaloid-containing Plants

Common name	Scientific name	Family	Alkaloid
*BETEL-NUT PALM	Areca catechu	Palmae	Arecoline
*HEMLOCK	Conium maculatum	Umbelliferae	Coniine
INDIAN TOBACCO	*Lobelia inflata	Lobeliaceae	Lobeline
*PEPPER	Piper nigrum	Piperaceae	Piperine
*TOBACCO	Nicotiana tabacum	Solanaceae	Nicotine
	N. rustica		
COCAINE	*Erythroxylon coca	Sterculiaceae	Cocaine
BELLADONNA	*Atropa belladonna	Solanaceae	Atropine
THORN APPLE	*Datura stramonium	Solanaceae	Hyoscine
HENBANE	*Hyoscyamus niger	Solanaceae	Hyoscyamine
*MANDRAKE	Mandragora officinarum	Solanaceae	Scopolamine
*CURARE	Chondrodendron tomentosum	Menispermaceae	Tubocurarine
*OPIUM POPPY	Papaver somniferum	Papaveraceae	Morphine, codeine papaverine
IPECAC, *IPECACUANHA	Cephalaeis ipecacuanha	Rubiaceae	Emetine
CALISAYA	*Cinchona calisaya	Rubiaceae	Quinine, quinidine
HUANACO	C. micrantha	Rubiaceae	Quinine
QUININE	C. officinalis	Rubiaceae	Quinine
*ERGOT	Claviceps purpurea	Hypocreaceae	Ergotamine
MADAGASCAR PERIWINKLE	*Catharanthus roseus	Apocynaceae	Vinblastine, vincristine
INDIAN SNAKEROOT	*Rauvolfia serpentina	Apocynaceae	Reserpine
STRYCHNINE	*Strychnos nux-vomica	Loganiaceae	Strychnine
HELIOTROPE	*Heliotropium europaeum	Boraginaceae	Heliotrine, lasiocarpine
BROOM	*Cytisus scoparius	Leguminosae	Sparteine
LABURNUM	*Laburnum anagyroides	Leguminosae	Cytisine
LUPIN	*Lupinus species	Leguminosae	Lupinine
ACONITE	*Aconitum species	Ranunculaceae	Aconitine
*COFFEE	Coffea arabica, C. canephora	Rubiaceae	Caffeine
*TEA	Camellia sinensis	Theaceae	Caffeine
*KOLA	Cola nitida	Sterculiaceae	Caffeine
CHOCOLATE	*Theobroma cacao	Sterculiaceae	Theobromine
EPHEDRA	*Ephedra sinica, E. distachya, E. equisitina	Ephedraceae	Ephedrine

II Glycoside-containing Plants

Common name	Scientific name	Family	Type of glycoside
*CASCARA	Rhamnus purshiana	Rhamnaceae	Anthraquinone
SENNA	*Cassia angustifolia	Leguminosae	Anthraquinone
ALEXANDRIAN SENNA	C. acutifolia	Leguminosae	Anthraquinone
*RHUBARB	Rheum officinale, R. palmatum	Polygonaceae	Anthraquinone
BUCKTHORN, ALDER BUCKTHORN	*Rhamnus frangula (= Frangula alnus)	Rhamnaceae	Anthraquinone
*ALOE	Aloe barbadensis (= A. vera)	Liliaceae	Anthraquinone
FOXGLOVE	*Digitalis purpurea, D. lanata	Scrophulariaceae	Cardiac
STROPHANTHUS	*Strophanthus sarmentosus, S. gratus, S. hispidus, S. kombe	Apocynaceae	Cardiac
OLEANDER	*Nerium oleander	Apocynaceae	Cardiac
QUEEN OF THE NIGHT	*Selenicereus grandiflorus	Cactaceae	Cardiac
LILY OF THE VALLEY	*Convallaria majalis	Liliaceae	Cardiac
STAR OF BETHLEHEM	*Ornithogalum umbellatum	Liliaceae	Cardiac
SQUILL	*Urginea maritima	Liliaceae	Cardiac
PHEASANT'S-EYE	*Adonis vernalis	Ranunculaceae	Cardiac
CHRISTMAS ROSE	*Helleborus niger	Ranunculaceae	Cardiac
YAM	*Dioscorea elephantipes	Dioscoreaceae	Saponin
*LICORICE	Glycyrrhiza glabra	Leguminosae	Saponin
GINSENG	*Panax pseudoginseng P. quinquefolius	Araliaceae	Saponin
WORMWOOD	*Artemisia absinthium	Compositae	Coumarin
MEZEREON	*Daphne mezereum	Thymelaeaceae	Coumarin
BLACK HAW, SWEET HAW	*Viburnum prunifolium	Caprifoliaceae	Coumarin

III Volatile (Essential) Oils

Common name	Scientific name	Family	Primary component
ANISEED	*Carum carvi	Umbelliferae	Limonene
*PEPPERMINT	Mentha piperita	Labiatae	Carvone
*SPEARMINT	M. spicata	Labiatae	Carvone
*CAMPHOR	Cinnamomum camphora	Lauraceae	Camphor
*CLOVES	Syzygium aromaticum	Myrtaceae	Eugenol
*SASSAFRAS	Sassafras albidum	Lauraceae	Safrole
WINTERGREEN	*Gaultheria procumbens	Ericaceae	
*GINGER	Zingiber officinale	Zingiberaceae	
*MUSTARD	Brassica and Sinapis spp	Cruciferae	
CAJUPUT	*Melaleuca leucadendron	Myrtaceae	Cineole
EUCALYPTUS	*Eucalyptus globulus	Myrtaceae	Cineole

IV Resins and Oleoresins

Common name	Scientific name	Family	Primary component
PEPPER TREE	*Schinus molle	Anacardiaceae	Urushiol
ASAFOETIDA	*Ferula foetida F. rubricaulis	Umbelliferae	Asafoetida
*STORAX	Styrax benzoin	Styracaceae	Benzoin
BALSAM OF PERU	*Myroxylon balsamum	Leguminosae	Cinnamein
ORIENTAL SWEET GUM	*Liquidambar orientalis	Hammamelidaceae	Storesin
SWEET GUM	Liquidambar styraciflua	Hammamelidaceae	Storesin

detailed structures of the active principles, cascarosides A and B, could be finally elucidated. Examples of these and of other groups of drug plants are given in the accompanying table.

Narcotic or hallucinogenic plants have been defined as those which contain "chemicals which in nontoxic doses, produce changes in perception, in thought and in mood, but which seldom produce mental confusion, memory loss or disorientation for person, place and time". The term narcotic comes from the Greek *narkoun*, to benumb, and strictly speaking refers to substances which terminate their action with a depressive effect on the central nervous system. In this sense both alcohol and tobacco are narcotics. Stimulants, on the other hand, such as *coffee, are not regarded as narcotics since they do not in normal dosage lead to a terminal depression. They are, however, psychoactive, a term often used in a broad sense to cover both narcotics and stimulants. The term narcotic is also used popularly to refer to substances which are dangerously addictive, such as opium and its derivatives. It is evident that there is no one term that adequately covers all psychoactive plants, but the term hallucinogen is widely used and understood. Hofmann, following an earlier classification of Lewin, divides psychoactive drugs into five groups: analgesics and euphorics (eg opium and *coca), sedatives and tranquilizers (eg reserpine), hypnotics (eg kavakava) and hallucinogens or psychotomimetics (eg peyote, marijuana).

Most hallucinogens derive from plants. Their activity is due to a limited number of chemical substances which act on a particular part of the central nervous system in a specific way. Most of them belong to the wide class of compound known as alkaloids. Hallucinogens include well-known plants such as *MORNING GLORY (*Ipomoea tricolor) and BELLADONNA (*Atropa belladonna). *Ergot derives from the pathogenic fungus *Claviceps purpurea* which grows on rye and other grasses, and was responsible for the horrifying disease known as St. Antony's fire, induced by eating bread made from rye contaminated with ergot. Lysergic acid diethylamide (LSD), a highly potent hallucinogen, is a synthetic derivative of ergot. It is believed now that the only deliberate use of ergot as a hallucinogen was in ancient Greece when it is said to have played a role in the Eleusinian mysteries.

It is only in the past 30 years or so that interdisciplinary research has led to a further appreciation of the potential role of hallucinogenic plants in modern medicine and society. As Schultes and Hofmann have written: "Plants that alter the normal functions of the mind and body have always been considered by people in non-industrial societies as sacred, and the hallucinogens have been 'plants of the gods' *par excellence*".

One of the group known as mesembryanthemums,
Lampranthus spectabilis *on the shores of the*
Monterey Peninsula, California.

borne in drooping racemose or cymose inflorescences. A number of species are grown
as garden plants in borders or rockeries, such
as *M. virginica* [VIRGINIAN COWSLIP], the
dwarf *M. primuloides* and *M. echioides*, whose
creeping habit makes it suitable for ground
cover. The rhizome of *M. maritima* [OYSTER
PLANT] is used as a food by Alaskan eskimos.
BORAGINACEAE, about 50 species.

Merulius a cosmopolitan genus of basidiomycete fungi usually found growing on
wood, with the fruiting body flat on the
substrate and the fertile layer (hymenium)
uppermost. Some species have recently been
transferred to *Serpula*, most notably the dry
rot fungus *Serpula lacrymans* (formerly
M. lacrymans).
CORTICIACEAE, about 40 species.

Mesembryanthemum [ICE PLANTS] a genus
of annual or biennial herbs with very soft,
succulent and brittle stems, cylindrical or
flat, expanded leaves and mostly inconspicuous pallid flowers borne singly or in
inflorescences. The most obvious characteristic is the overall presence of large glossy
papillae which make the whole plant sparkle
in the sun as if covered in hoar frost, hence
the common name. In recent years many
members of the original genus have been
transferred to 125 smaller genera of which
Mesembryanthemum is but one.
Centered in the Cape region of South
Africa, *Mesembryanthemum* has two widely
distributed species, *M. crystallinum* [ICE

PLANT], naturalized in Mediterranean
regions, the Canary Isles and California, and
M. nodiflorum found in Africa, southern
Europe, the Near East, Atlantic islands,
California and Mexico. *M. crystallinum* is
sometimes cultivated as an ornamental for
its glittering foliage. It can also be grown as
a substitute for spinach where the ground is
too dry for that crop, and the soft fleshy
shoots have been included raw in salad. The
seeds also have been used to make bread.
AIZOACEAE, 40–50 species.

Mespilus see MEDLAR.
ROSACEAE, 1 species.

Mesua a small genus of evergreen tropical
Asian shrubs or trees. *M. ferrea* [IRONWOOD]
is a handsome tree bearing fragrant large
solitary flowers. The flower buds and flowers
are used locally in cosmetic preparations and
perfumery, and the very hard timber is used
for walking sticks and cabinetwork.
GUTTIFERAE, 3–6 species.

Metasequoia an unusual genus in that it
was known as a fossil before being discovered
in 1945 as a living tree. It is a coniferous
genus consisting of a single living species of
deciduous tree *M. glyptostroboides* [DAWN
REDWOOD]. The fossil record of *Metasequoia*
extends back to the Cretaceous period (about
100 million years) and at one time the genus
was widespread in the Northern Hemisphere.
Although now geographically restricted as a
native tree to a small area in central China,

The evergreen foliage and crimson flowers make the
*New Zealand Christmas Tree (*Metrosideros
excelsus*) a common ornamental in the*
subtropics.

the DAWN REDWOOD will grow successfully in
a variety of climates and soils and has been
planted widely as an ornamental in Europe
and America.
TAXODIACEAE, 1 species.

Metrosideros a Pacific genus centered in
Polynesia, with some representatives in Australasia. Species are either evergreen trees,
shrubs or aerial-rooted climbers. They form
an extremely attractive group of plants as the
stiff, long stamens protrude beyond the
flowers and are frequently of a vivid red or
yellow. The hard wood of trees such as
M. robustus (confused in cultivation with
M. umbellatus), *M. collinus* (= *M. diffusus*)
and *M. umbellatus* (= *M. lucidus*) (the generic
name means "iron core") is used locally for
boatbuilding and carving. *M. excelsus* (=
M. tomentosus) [NEW ZEALAND CHRISTMAS
TREE, POHUTUKAWA], with dark crimson
flowers, is a common subtropical
ornamental.
MYRTACEAE, about 60 species.

Metroxylon a genus of stout, solitary or
clump-forming tree palms native from Thailand and Malaya to Fiji. *M. rumphii* and
M. sagus are the *SAGO PALMS widely cultivated in villages throughout the Asian
tropics for their starch (marketed as sago).
M. sagus is also a prime palm for thatching.
The Pacific Islands species are known as
IVORY NUT PALMS, the hard horny endosperm
being the vegetable ivory that was formerly
used for buttons.
PALMAE, 15 species.

Metzgeria a genus of tropical and subtropical thallose liverworts growing on rocks, tree
trunks and, more rarely, on living leaves. The

Common Millet (Panicum miliaceum) is widely cultivated as a food crop for Man or livestock, shown here in Hidalgo State, Mexico. ($\times \frac{1}{15}$)

thallus consists of a conspicuous cylindrical midrib with a thin wing, one layer of cells thick, on each side of it. The sex organs are produced in highly specialized branches. METZGERIACEAE, about 50 species.

Michaelmas daisies short-lived perennial herbs or shrubs belonging to the large genus *Aster. They are characterized by having tall, erect stems with alternate leaves and solitary or clustered inflorescences of small, daisy-like heads which have one or two series of white, blue, pink or red narrow strap-like ray florets and a central button of yellow disk florets.

Although there are several South African blue-flowered shrubby species of *Aster* cultivated in greenhouses, most MICHAELMAS DAISIES are tall, hardy, leafy perennial herbs of American origin. These are chiefly forms of *A. amellus, A. cordifolius, A. ericoides, A. laevis, A. lateriflorus, A. novae-angliae, A. novi-belgii, A. thomsonii, A. × versicolor (= A. laevis × A. novi-belgii)* and *A. vimineus.* Varieties of *A. novi-belgii* [NEW YORK ASTER] and *A. novae-angliae* [NEW ENGLAND ASTER] form by far the biggest group, although these are now rapidly being superseded by many new hybrid forms which are produced every year.

In addition to border plants there are a number of colorful dwarf species from the mountains of Europe, America and Asia well adapted for use in rock gardens. Important species include *A. alpinus, A. diplostephioides, A. falconeri, A. flaccidus* and *A. tongolensis (= A. subcoeruleus).*
COMPOSITAE.

Michelia a genus of Asian shrubs and trees very similar to *Magnolia. A number of species, notably *M. champaca* [SAPU] and *M. doltsopa (= M. excelsa),* produce a rather variegated timber used extensively in India, Sri Lanka and China for joinery and house building. These and other species are cultivated for their showy, often fragrant flowers, yellow or orange in *M. champaca,* yellow in *M. compressa,* yellowish green in *M. figo (= Magnolia fuscata)* and yellowish white, tinged green in *M. doltsopa.* Champaca oil, obtained from the flowers of *M. champaca,* is used in perfumery in Asia.
MAGNOLIACEAE, about 50 species.

Miconia a large genus of tropical American shrubs and trees with one species occurring in West Africa. Some species, are grown in greenhouses for their large attractive leaves. The flowers, which are white, yellow, pink or purple, are not very conspicuous; the fruits are berry-like. Cultivated species include *M. flammea* and *M. ovata* but perhaps the most striking is *M. calvescens (= M. magnifica),* from Mexico, which has very large, broadly ovate leaves, up to 70cm long, reddish-orange beneath with pale green or whitish veins.
MELASTOMATACEAE, about 600 species.

Micromeria a genus of herbs and sub-shrubs closely related to other herbs such as *THYME, *MINT and *SAGE. Some species are grown for their ornamental appearance and were formerly grown as potherbs. *M. croatica,* from Croatia, is a small herb producing pale rose-violet flowers in spring. *M. marginata (= M. piperella),* from the Maritime Alps has reddish-purple or violet flowers in the summer. All of these and a few other species are suitable rock-garden plants.
LABIATAE, about 100 species.

Mignonette the common name for certain

Metzgeria furcata is a small thalloid liverwort, here showing the hairy globular sheaths surrounding the male sex organs (antheridia). ($\times 4$)

annual to perennial herbs of the genus *Reseda,* native to Europe and the Mediterranean. The terminal spike-like racemes of small flowers are usually greenish-yellow, orange or whitish. *R. odorata* [COMMON MIGNONETTE] is grown as a border plant largely for its scent, although some cultivars produce spikes of attractive red flowers. This species yields a perfume oil. A reddish-yellow dye used to be obtained from *R. luteola* [WILD MIGNONETTE, DYERS' ROCKET, WELD]. *R. alba* is known as the WHITE UPRIGHT MIGNONETTE. RESEDACEAE.

Mikania a large genus of evergreen twining or creeping herbs and shrubs, mostly native to tropical America and the West Indies, but also found in Africa and tropical Asia. *M. scandens* [CLIMBING HEMPWEED] a quick-growing herb with ivy-like, bronze-olive-green leaves with brownish veins, is endemic to eastern North America, with related species pantropical in distribution. *M. apiifolia,* with softly hairy, lobed leaves is cultivated out of doors in warm climates or otherwise in heated greenhouses. The Brazilian *M. ternata,* a half-woody vine with purple hairy stems and yellowish flowers, is grown in hanging baskets.
COMPOSITAE, about 150 species.

Mildew see Fungi, p. 152.

Milium a genus of annual or perennial grasses native to north temperate regions. They have spikelets consisting of a single floret. *M. effusum* var *aureum* [GOLDEN MILLET] is cultivated as an ornamental.
GRAMINEAE, about 3–4 species.

Millet a loose term for a large number of cultivated grasses with small edible seeds. Most millets are believed to have originated in Africa or in Central and East Asia. Although largely replaced by other cereals, millets are still grown extensively in Asia, particularly in China, and in parts of Africa. They can produce grain under conditions of intense heat, scanty rainfall, relatively infertile soil, and a short growing season. Thus they tend to be "poor man's cereals", mainly grown for local consumption, and are therefore useful in dry areas of India, Africa and Australia.

Eleusine coracana [FINGER or AFRICAN MILLET, RAGI] is the most important species and constitutes a staple food in areas of East and central Africa and in the Mysore plains of India. The small brown (sometimes white) grains are ground into a flour used for porridge, or fermented to produce beer.

Pennisetum americanum (= *P. typhoides*. *P. glaucum*) [BULRUSH, PEARL, SPIKED or CATTAIL MILLET] is widely cultivated in Africa (particularly the Sudan) and India on rather sandy dry soils which cannot support other cereal species. In addition to the grain which is eaten unground or as porridge, this species makes a useful fodder crop.

Setaria italica [ITALIAN or FOXTAIL MILLET] is an important grain crop cultivated widely in India and Japan, and in parts of North Africa and southeastern Europe. It is also grown in the USA as a fodder crop, and in parts of the USSR where other grains are unprofitable. The grain is used for food and for brewing beer, and is the millet commonly seen in bird cages.

Echinochloa frumentacea [JAPANESE BARNYARD MILLET] is the quickest growing millet, producing a crop in only six weeks. It is used as a substitute for *RICE in China and Japan when the paddy fails. Other millets include *Panicum miliaceum* [COMMON or PROSO MILLET], *Digitaria exilis* [FUNDI], *Eragrostis tef* [TEFF] and *Paspalum scrobiculatum* [KODA or KODO MILLET].
GRAMINEAE.

A cultivated Sensitive Plant (Mimosa pudica *) showing the leaves semi-closed – a position typical of an hour after sunset. (×* $\frac{1}{10}$ *)*

Miltonia a largely tropical American genus of epiphytic and pseudobulbous orchids which are cultivated usually in the form of spectacular multispecific hybrids. In most species, the entire flower has a pansy-like appearance (hence the common name PANSY ORCHIDS), the basic flower colors being mauve, pink and white. In many of the species and hybrids, however, there are large overlapping and regularly spaced blotches of

Most species of Mimulus *are native to North Africa, but some are naturalized elsewhere, such as the Monkey Flower (*M. guttatus*), seen here growing in Britain. (×* $\frac{1}{5}$ *)*

contrasting maroon, purple and yellow. Miltonias have been widely hybridized with species from other genera such as *Oncidium (to give × *Miltonidium*), *Brassia (to give × *Miltassia*).
ORCHIDACEAE, 25 species.

Mimosa a large genus of annual and perennial herbs, shrubs, subshrubs, trees and climbers. Many are thorny or prickly and bear flowers borne in tight compact heads. The best-known is *M. pudica* [SENSITIVE PLANT, HUMBLE PLANT, SHAME PLANT], a tender subshrub native to Brazil, whose light green leaflets fold together when touched. *M. argentea* is a climber having attractive foliage with a silvery white midrib. The tropical American herb *M. invisa* [GIANT SENSITIVE PLANT] has been used as a green manure crop. The bark of the Mexican shrub *M. purpurascens* is used for tanning leather.

Florists' mimosa is *Acacia dealbata*.
LEGUMINOSAE, about 450 species.

Mimulus a cosmopolitan genus of annual and perennial herbs or rarely shrubs, especially frequent in North America but naturalized elsewhere. Many species are cultivated, notably *M. luteus* [MONKEY MUSK, MONKEY-FLOWER] from Chile, and the North American *M. moschatus* [MUSK PLANT], formerly grown for its musky scent which seems to have disappeared in cultivation. Many of the cultivated species, including the North American *M. ringens* [LAVENDER WATER MUSK, ALLEGHENY MONKEY-FLOWER], *M. cupreus*, from southern Chile, *M. variegatus*, from Chile, and their varieties, are grown in rock gardens, borders or beside pools, for their attractive showy snapdragon-like flowers, predominantly yellow, orange or crimson in color, variously blotched or spotted.
SCROPHULARIACEAE, about 150 species.

Mimusops a genus of mostly tropical Old World evergreen shrubs and trees with thick leaves and axillary clusters of whitish flowers. *M. balata* [BALATA] is a large tree, native to Madagascar (BALATA of tropical America is a different species, *Manilkara bidentata*, the source of a nonelastic latex). *M. elengi* [SPANISH CHERRY, MEDLAR] is a small tree native to India and Malaysia which bears edible, ovoid, yellow fruits.
SAPOTACEAE, about 20 species.

Mint the common name for *Mentha*, a well-known and widely grown genus of aromatic perennial herbs mostly native to north and south temperate regions. • It includes *M. spicata* [SPEARMINT, COMMON MINT], *M. × piperita* (= *M. aquatica* × *M. spicata*) [PEPPERMINT], *M. citrata* [ORANGE or BERGAMOT MINT], *M. arvensis* [FIELD MINT], *M. × rotundifolia* (= *M. longifolia* × *M. sauveolens*) [APPLE or ROUND-LEAFED MINT] and *M. arvensis* var *piperascens* [JAPANESE MINT].

Mints are characterized by the small, almost regular, rather unspecialized flowers, arranged in dense many-flowered verticillasters, spikes or heads. Hybrids between species are frequent and these have resulted in the description of a very large number of variants or "species". In many areas, hybrids are more frequent than the parent species.

Mints have a long history of cultivation and they are now grown throughout the world. The much-used essential oil, menthol, is present in most species, but menthol of commerce is mainly derived from JAPANESE MINT, an important crop in Japan, China and Brazil. SPEARMINT, the culinary herb, and PEPPERMINT, widely used in pharmacy and as a flavoring, are also widely grown. Several others are grown commercially and a few, including *M. × rotundifolia* and *M. × gentilis* (= *M. arvensis* × *M. spicata*) [AMERICAN APPLEMINT] make decorative garden plants.
LABIATAE.

Mirabilis a genus of American annual or perennial herbs (with one Himalayan species). They have opposite leaves, the lower stalked and the upper stalkless. The inflorescence is one-to-several-flowered but normally only the large, fragrant, central flower develops within a five-lobed petaloid calyx.

M. jalapa [FOUR O'CLOCK PLANT, BEAUTY-OF-THE-NIGHT, MARVEL OF PERU] has white, yellow, pink, red or striped flowers which open in the late afternoon. The tuberous roots of this species are the source of a purgative drug used as a substitute for *jalap. *M. jalapa* and *M. longiflora* (white to pink or violet flowers) are among several species cultivated as tender annuals in cooler temperate regions, for their ornamental value.
NYCTAGINACEAE, over 60 species.

Miraculous berry or **fruit** the common name for *Synsepalum dulcificum*, a West African shrub often planted near native dwellings. The fruit is fleshy and the pulp around the seeds has a sweet taste with a curious lingering after-effect which causes acid substances to taste sweet when taken up to three hours later. It is an ingredient of palm wine and is used to improve the taste of stale food. The active sweetener (a protein) is of considerable economic interest in the search for non-sucrose sweeteners.
SAPOTACEAE.

Miscanthus a genus of tall graceful perennial Asian grasses, some of which, especially *M. sacchariflorus* and *M. sinensis*, are commonly cultivated in a range of varieties as ornamentals in garden borders. The fan-shaped, feathery, terminal inflorescences are also useful when dried for use in floral bouquets.
GRAMINEAE, 15–18 species.

Mistletoe (Viscum album) infects trees at points where the seeds have been deposited by fruit-eating birds that eat the white berries but not the seeds. This species is harvested in Britain at Christmas time for decorations.

The widespread north temperate moss Mnium punctatum, *showing the large oval leaves and pendent spore capsules on orange stalks.* (× 3)

Mistletoe the name given to several members of the family Loranthaceae. They are semiparasites and the best-known is the Old World *Viscum album*. Other mistletoe genera include *Phoradendron* in North America, *Arceuthobium* in North America, Europe and Asia, and *Loranthus* in the Old World tropics and Australasia.

Seeds of parasitic species germinate on the host and absorb nutrient fluids from the host by means of haustoria. The commercial MISTLETOE of North America is *P. serotinum*. The much smaller *A. pusillum*, which is parasitic on conifers, is known as the DWARF MISTLETOE. The most attractive parasitic species is the tropical American *Psittacanthus elytranthe* [RED MISTLETOE], which bears conspicuous red flowers.

In Britain at Christmas the MISTLETOE is traditionally associated with the other evergreens HOLLY and IVY, their colored berries contrasting with the white mistletoe. The custom of kissing under the MISTLETOE reflects its supposed aphrodisiac powers. Extracts of MISTLETOE have been used to lower blood pressure.
LORANTHACEAE.

Mitella [MITERWORT, BISHOP'S CAP] a genus of perennial herbs native to North America and Japan. Most have basal heart-shaped leaves and an erect leafless flowering stalk. *M. diphylla* [COOLWORT], *M. breweri* and *M. nuda* are sometimes cultivated as border or rock-garden plants, producing numerous small, nodding flowers on an erect spike-like raceme.
SAXIFRAGACEAE, about 15 species.

Mnium a genus of mosses mainly distributed in north temperate regions. Several are notable for their large leaves which in texture and superficial appearance recall the fronds of filmy ferns. Damp woodland habitats are favored and several species, such as *M. hornum*, *M. punctatum* and *M. undulatum*, occur more or less throughout the north temperate zone.
MNIACEAE, about 50 species.

Molasses an important by-product (along with *bagasse) arising from the extraction of sugar from *Saccharum officinarum* [*SUGAR CANE]. It is a very viscous, blackish-brown, strong-smelling fluid which is separated from the crystalline sugar during the later stages of extraction in the refinement process. It is a complex mixture of substances but consists predominantly (about 30%) of sucrose which cannot be extracted economically from the remaining soluble invert sugar, gum, starch and carbonated ash.

Fermentation and distillation of molasses produces a variety of products, including a number of alcohols, acetone, citric acid, glycerol and, by fermentation, edible yeasts. Perhaps the most famous distillate is dark rum, produced mainly in the cane producing countries of the Caribbean, including Venezuela (and also in the Canary Islands). Other fermentation products based on raw molasses include ethyl, amyl and butyl acetate, vinegar and carbon dioxide.

Molasses is used in animal foodstuffs, particularly as an additive to grass for improving the palatability of silage. It is also extensively used as fertilizer and soil improver, especially on sandy soils, since it is a valuable source of potash and organic matter.

Molasses is also known as treacle and in a refined form it is sold for culinary use as black treacle. Cane syrup is a very different product from the true molasses since it is a very pure, clarified sugar product, manufactured for direct human consumption.

Mold see Fungi, p. 152.

Molinia a genus of perennial grasses of temperate Eurasian origin, the most common of which is *M. caerulea* [MOOR GRASS]. This is the characteristic species of a wet, acid, grass moor, producing valuable spring pasture and surviving fires because of its dense, tussock-forming leaf-bases. The variegated form is used as a bedding plant, having leaves with white stripes and long, erect panicles, bluish greenish or white.
GRAMINEAE, 2–3 species.

Moluccella a genus of annual and perennial herbs native in the Mediterranean region and extending to northwest India. The calyx is characteristically bell-shaped with relatively insignificant teeth. *M. laevis* [SHELL FLOWER, MOLUCCA BALM, BELLS OF IRELAND], with greenish calyces, is a hardy annual which dries well and is popular in flower arrangements. *M. spinosa* is a perennial, with the corolla longer than the calyx.
LABIATAE, 3–4 species.

Momordica a genus of annual or perennial, tendril-bearing, climbing herbs native to tropical Asia and Africa, but widely cultivated throughout the tropics for their white or yellow flowers, but principally for their round, oblong or cylindrical, bitter but edible fruits. Popular species include *M. balsamina* [BALSAM APPLE] and *M. charantia* [BALSAM PEAR, BITTER GOURD], whose fruits are used in pickles, curries and salads.
CUCURBITACEAE, about 40 species.

Monanthes a genus of mostly perennial, but a few annual, dwarf tender leaf-succulents from the Canary Isles, Salvage Isles and North Africa. Some of the perennials, such as *M. polyphylla*, are grown in succulent collections for their dainty miniature habit, colorful and often flossy, papillate leaves and starry flowers with conspicuous nectar scales. In addition to the purple-flowered *M. polyphylla*, cultivated species include the yellowish-flowered *M. pallens* and *M. anagensis*.
CRASSULACEAE, about 16 species.

Monarda a genus of annual and perennial aromatic herbs native to North America and Mexico. The flowers are in heads, usually with an involucre of bracts, which in some cases are very showy. *M. didyma* [SWEET BERGAMOT, BEE or FRAGRANT BALM, OSWEGO TEA] is widely cultivated in gardens and there are several cultivars with flower colors ranging from white and pink to scarlet and violet.
LABIATAE, 12 species.

Monardella a genus of annual and perennial aromatic herbs native to western North America. The terminal heads of flowers are rose-purple, lavender or white and resemble those of *Monarda*. Some species are grown as ornamentals, particularly in rock gardens, including *M. candicans* and *M. nana*, with white, rose-tinged flowers, and the red- to yellowish-flowered *M. macrantha*.
LABIATAE, about 20 species.

*The Yellow Bird's-nest (*Monotropa hypopitys*) is one of the few saprophytic herbs.* ($\times \frac{1}{8}$)

Monostroma a genus of parenchymatous green algae found on rocky seashores attached to the substrate by rhizoids. The plant has a ruffled thallus one cell thick, and the pale green fronds finally spread out flat, reaching a length of about 15cm. *Monostroma* is cultivated as a food plant in Japan.
CHLOROPHYCEAE.

Monotropa a genus of saprophytic fleshy herbs distributed over the Northern Hemisphere. They are devoid of chlorophyll and the stems and scaly leaves are yellow throughout. Their roots function in symbiosis with fungi which form mycorrhiza and thus enable the plant to absorb complex food materials. *M. hypopitys* [YELLOW BIRD'S-NEST] and *M. uniflora* [INDIAN PIPE] are the best-known species, both of them producing whitish or pinkish flowers.
MONOTROPACEAE, 3–5 species.

Monstera a genus of evergreen lianas from tropical America and the West Indies. They begin as climbers and have woody stems but become epiphytes with aerial roots. The leaves are thick and entire when young, but often the tissue between the veins ceases to grow and, becoming dry, tears away, leaving holes between the ribs. The most commonly grown species is *M. deliciosa* [MEXICAN BREAD-FRUIT, WINDOW PLANT, CERIMAN, SWISS-CHEESE PLANT], from Mexico and Central America, often grown indoors for its large perforated leaves. In the tropics, the inflorescence develops into a long succulent cone-like fruit smelling of pineapples.
ARACEAE, 30–50 species.

Moraea a genus of cormous or rhizomatous iris-like herbs native to tropical and southern Africa. The spreading flowers are usually fragrant and showy, either red, lilac, white or yellow in color, and usually borne in branched inflorescences. Cultivated species include *M. neopavonia* (= *M. pavonia*) [PEACOCK IRIS] with predominantly orange-red flowers, and the yellow-flowered *M. ramosissima* and *M. spathulata*.
IRIDACEAE, about 100 species.

Morels saprophytic terrestrial fungi of the mainly temperate genus *Morchella*, with fleshy fruiting bodies consisting of a hollow cylindrical stipe and a grayish to dark brown convoluted cap with interconnecting ridges and depressions lined by asci. The best-known species is the edible *M. esculenta* which grows in calcareous woodland, pasture or sand dunes in spring, and is much sought after and collected from the wild in Europe and the USA. The cap of FALSE MORELS (*Gyromitra*) is by contrast saddle-shaped or convoluted ("brain-like").
MORCHELLACEAE, about 15 species.

Morina a genus of thistle-like perennial herbs which make good garden-border plants, although they are not widely grown. They are found native from southeast Europe to the Himalayas and southwest China. The leaves are opposite or whorled, usually spiny. The showy flowers are borne in axillary whorls and are white, changing to crimson in *M. longifolia* [WHORL FLOWER], rose-red in *M. betonicoides* and white or purple in *M. bulleyana*.
DIPSACACEAE, about 10–15 species.

Morinda a genus of erect or climbing tropical and subtropical trees and shrubs. The funnel- or salver-shaped flowers are small, white or red, in heads, and the fruit is a conglomeration of berry-like carpels. *M. citrifolia* [INDIAN MULBERRY], whose flowers and roots produce a red and a yellow dye, respectively, is a small tree native to Southeast Asia and Australia. The red dye, known as "al" is used to color wood and linen. Several other species also yield dyestuffs from their flowers or roots. *M. citrifolia*, *M. bracteata* and the shrub *M. jasminoides* are also cultivated as ornamentals.
RUBIACEAE, about 80 species.

Moringa a genus of deciduous trees from the Mediterranean, Africa and India. *M. pterygosperma* (= *M. oleifera*) [HORSE-RADISH TREE], from India, is cultivated for an oil (ben oil) obtained from the seeds. Locally in India, the root is scraped and used in curries and pickles, and the flowers and bark

*The massive leaves of the Swiss-cheese Plant (*Monstera deliciosa*), growing here in its natural habitat in Trinidad, West Indies.* ($\times \frac{1}{30}$)

A morel (Morchella elata), showing the cylindrical stipe and "honeycomb" cap typical of the genus. (× 1)

have medicinal value. This species and *M. aptera* (which yields an oil from the seeds used in perfumery) are also grown for ornament.

MORINGACEAE, 5–12 species.

Morning glory the common name for several decorative species of the genus *Ipomoea* which bear flowers which open soon after dawn and wither by about noon. The name is also applied to species of *Calystegia*, *Convolvulus* and *Merremia*.

CONVOLVULACEAE.

Morus see MULBERRY.

MORACEAE, about 12 species.

Mosses see p. 228.

Mountain ash the common name given to some species of trees and shrubs of the genus *Sorbus*. *Eucalyptus regnans* is also called MOUNTAIN ASH.

Mucor one of the largest genera of the pin molds, usually saprophytic and fast-growing. In nature, *Mucor* species are of worldwide distribution, commonly occurring on decomposing vegetable matter, dung, soil and leaf litter. *M. hiemalis* and *M. circinelloides* can be obtained regularly from soil and soil-contaminated substrates. *M. pusillus*, a species which can grow at temperatures up to 55°C, has been reported as a pathogen of domestic animals, with feeding stuffs as the probable source of infection. It has also been reported from man. *Mucor* species are used in fermentations in China and Indonesia to make *soybeans a more palatable and digestible food.

MUCORACEAE, about 50 species.

Mucuna a genus of tropical and subtropical twining and climbing herbs from both hemispheres. *M. pruriens* [COWAGE, COWITCH] has intensely irritant hairs over the

pods. Cultivars of this and other species, including *M. aterrima* [BENGAL BEAN] and *M. deeringiana* [VELVET BEAN], also with irritant hairs on the pods and spectacular flowers, are used for fodder and silage. Spectacular tropical ornamental species include *M. bennettii* [NEW GUINEA CREEPER], with flame-colored pea-like flowers, *M. rostrata*, from tropical America, with orange flowers, and *M. imbricata*, from India, with purple flowers.

LEGUMINOSAE, about 120 species.

Muehlenbeckia a genus of climbing or prostrate shrubs or subshrubs native to Australasia and western South America.

Several species are cultivated in greenhouses, hanging baskets and outdoors, mostly for their curiosity value. For example, the shiny black nutlets of *M. complexa* [WIRE VINE], a deciduous climber up to 30m, from New Zealand, are enclosed in an expanded glistening white perianth.

POLYGONACEAE, about 15 species.

Mulberry the common name for members of the deciduous, woody, mainly tropical genus *Morus* and its fruits. The stems of *Morus* species contain a milky latex. The flowers are small and individually inconspicuous but are clustered into green, pendulous, male or female catkins. The mulberry fruits form around a central core in blackberry-like clusters in which the colored juicy parts have developed from the perianth segments of the individual flowers. Cultivated mulberry fruits may be about 2cm long, but those from the wild are usually less than 1cm long.

M. alba is the WHITE MULBERRY of China where its main use was to provide silkworm fodder from the soft, tender leaves and also to

The English or Black Mulberry (Morus nigra) is the most common mulberry and has the largest and juiciest fruits.

provide timber. The WHITE MULBERRY used to be considered one of the two most important Chinese timber trees and most Chinese homesteads had one or two planted nearby. It forms a wide-spreading tree up to 15m tall, with gray bark, small leaves and red fruits. *M. alba* was probably the parent of the original American "DOWNING" MULBERRY, selected for its fruit yield. *M. alba* var *tatarica* [RUSSIAN MULBERRY] is a very hardy variety not grown for its fruit but as an ornamental low-growing shrub in cold northerly climates.

M. rubra [RED MULBERRY, AMERICAN MULBERRY] is a native of eastern North America and a common tree of American woodland. The fruits are red to purple, and drop from the branches when ripe. The fallen fruit is juicy and very sweet, and is collected to make pies, jams and jellies. The tree also provides useful timber for boat building and fences.

M. nigra [BLACK MULBERRY, PERSIAN MULBERRY, ENGLISH MULBERRY], from Iran and Southwest Asia, is the most common species and has been cultivated in the Mediterranean area for centuries. It attains a height of no more than 10m and bears purple to black, juicy fruits.

M. mesozygia is a central African species up to 30m tall. It is utilized as a shade tree; the top branches are removed and the lateral branches are weighted to produce an umbrella-shaped crown. It yields edible mulberries, but they are small and not usually harvested.

Broussonetia papyrifera, a member of the same family, is called the PAPER MULBERRY.

MORACEAE, about 12 species.

Mullein the common name for some species of the genus *Verbascum* such as *V. thapsus* [GREAT MULLEIN, AARON'S ROD, HAG TAPER, ADAM'S FLANNEL, JACOB'S STAFF, SHEPHERD'S CLUB, BLANKET LEAF], *V. lychnitis* [WHITE MULLEIN] and *V. nigrum* [DARK or (continued on p. 230)

Mosses and Liverworts

THE BRYOPHYTES

A tussock of the moss Polytrichum commune, showing the upright capsules still with their hood-like coverings (calyptras). (×1)

MOSSES ARE POPULARLY SEEN AS PLANTS OF soft texture, vaguely defined growth form and strictly limited stature. They are clearly distinct from all other types of plants except for the liverworts which are placed along with them in the same division of the plant kingdom – the Bryophyta, the mosses in class Musci, the liverworts in class Hepaticae. All bryophytes are relatively small plants, most less than 20cm long, many below 2cm. They occur most commonly in warm and temperate climates and prefer moist conditions. All bryophytes have a common, and unique, life history which entails a regular alternation between a flattened or leafy sexual phase or gametophyte (the liverwort or moss plant as we know it) and a spore-bearing plant (sporophyte) which follows fertilization and takes the form of a "fruiting body" or capsule borne aloft on the sexual plant.

Liverworts and mosses may be regarded as the amphibians of the plant world. Although they are land plants, water is essential for the male gametes (spermatozoids) to swim through to reach and fertilize the eggs that are retained within the female sex organs. In many instances a thin film of dew or rain water covering the plants is sufficient for this purpose. Bryophytes do not have true roots although they do possess root-like append-

The semi-aquatic moss Drepanocladus lycopodioides growing with Salix repens and Equisetum variegatum. (×5)

ages (rhizoids) that help anchor the plants but do not absorb water or nutrients from the soil.

Having established that liverworts and mosses have many features in common, what then are the differences between them? About 15% of liverworts are thallose, that is they comprise a flattened green body that lacks leaves. These thallose liverworts are quite distinct from the erect, leafy moss plants and are thus not easily confused. The remainder of liverworts are "leafy" and may at first glance be confused with mosses. However, the leaves of these liverworts are generally arranged in two or three ranks, and are characteristically frail, often very small, and invariably without a true midrib, whereas those of mosses are spirally arranged, are of firmer texture and usually larger.

One small group of bryophytes which closely resemble thallose liverworts, and indeed were once thought to be liverworts, is now considered to be a separate class (Anthocerotae) within the bryophytes. These hornworts as they are called occur throughout the world, in moist shaded habitats. There most distinctive feature is the long cylindrical capsule which splits into two horn-like segments to release its spores. (See also Anthoceros.)

Although bryophytes mainly favor damp habitats, many mosses are able to survive in dry conditions. They have the power to enter a state of apparently total desiccation, re-

maining in it for weeks at a time only to recover with speed whenever rain comes. Bryophytes have no means of penetrating the soil beyond a depth of a few centimetres and mainly depend for the small amounts of mineral nutrients they require on water dripping from trees or even on the minute amounts conveyed by direct precipitation.

All bryophytes, but particularly mosses, are important primary colonizers of new habitats, preparing the ground for subsequent colonization by other plants. This is seen to the best advantage in the primary colonization of bare rock surfaces by species of *Grimmia, *Andreaea and *Camptothecium. They are seen as helping to collect a sufficient nidus of soil particles and organic debris to make possible the eventual germination of seeds.

Many mosses and liverworts are reliable indicators of the nature of the surface rock and soil conditions. Thus species such as Tortella tortuosa, *Neckera crispa and Scapania aspera indicate alkaline conditions while most species of *Rhacomitrium, all species of *Polytrichum and Scapania gracilis indicate acid conditions. It has even been suggested that certain mosses, such as Merceya species, which in some parts of the world have been found to be constantly associated with high concentrations of copper, might be of more than passing interest to ore prospectors.

Mosses occur in most types of plant community, but their importance varies greatly. They tend to be very sparse in grassland and savanna, and even in tropical forests, if there is a dry season, their role is limited. On the whole, the higher the latitudes the larger the part mosses will play within the vegetation.

A thick cover of bryophytes and lichens on the gnarled trunks of English Oak trees (Quercus robur) and boulders.

Throughout the vast coniferous forests of the boreal zone of Eurasia and North America, the moss component in the ground flora is important; farther North, through the taiga into arctic tundra, the bryophytes, chiefly mosses, but with liverworts freely represented, assume dominance along with lichens. In lower latitudes much the same

Pellia epiphylla, a thallose liverwort that grows in moist, rich soil, particularly on banks of streams. The plants have a flattened green thallus (the gametophyte or sexual generation), from which the sporophyte (spore-producing generation) emerges, the latter consisting of a stalk (seta) surmounted by a black spore-containing capsule. After dispersal the spores germinate to form new gametophytes. (×2)

situation is found in mountaintop vegetation. There is a single example of a genus of mosses achieving sole dominance in a particular habitat. This is the genus *Sphagnum* (the bog mosses) whose members achieve dominance in bog communities throughout the world. More often mosses form small locally concentrated communities, as in flushes on mountain slopes or on decaying logs in woodland.

Liverworts do not approach mosses in ecological importance. They never make up a major feature of the vegetation nor do they assume dominance over great tracts of country. In general, regions with a constantly high relative humidity and offering an abundance of "temporary" habitats for colonization are those that suit liverworts best. In temperate regions, liverworts are most prolific on western seaboards with their associated high rainfall. On boulders, rotting logs and sheltered banks in highland woods, leafy liverworts form a conspicuous part of the vegetation. In the dry tropics, liverworts only do well where there is a constant supply of surface water or where extreme shade and adequate shelter are provided. In the humid tropics, however, liverworts are found in a unique habitat – on the surfaces of large leaves, particularly those that are evergreen. Such epiphyllous liverworts share this habitat with lichens and other lower plants. They obtain no direct nourishment from the leaves and rely for their water on the saturated atmosphere.

Mosses and liverworts have remarkable powers of regeneration even from very small fragments. A fair degree of vegetative spread must be by this means, but many species also produce special vegetative propagules called gemmae.

Economically, mosses and liverworts are of limited importance. The economic role of mosses has largely been restricted to a limited range of uses in horticulture. The exception is *Sphagnum* moss, whose pronounced water-holding capacity and slight antiseptic properties have led to several uses, notably as a material for dressing wounds in World War I. There are also signs that certain mosses and liverworts may prove to be of increasing value as indicators of the extent of certain types of pollution, since many species are very sensitive to pollution, particularly sulfur dioxide in the atmosphere.

BLACK MULLEIN], once of importance in witchcraft, and the source of a dye. These species are grown as garden ornamentals, as well as others such as *V. phoeniceum* [PURPLE MULLEIN].
SCROPHULARIACEAE.

Mummy wheat or **miracle wheat** the usual names applied to the commonest branched form of the mealy-grained emmer wheat variety *Triticum turgidum* var *mirabile*. MUMMY WHEAT is a variety which differs from the normal *T. turgidum* by having long branches emerging from the lower spikelets instead of the normal three-to-five-grained spikelets. The name MUMMY WHEAT is also applied to various emmer wheat grains occasionally taken from sealed Egyptian tombs of the 18th dynasty. Dyed imitations are often sold in Middle Eastern markets to unsuspecting tourists.
GRAMINEAE.

Mung bean one of the common names given to *Phaseolus aureus*, an erect annual bean plant widely cultivated in the tropics and subtropics for its edible pods and green or yellow seeds. The germinated seeds are the chief source of Chinese bean sprouts.
LEGUMINOSAE.

Musa a Southeast Asian genus of large to gigantic perennial herbs, the source of the *BANANA and of *abaca fiber (Manila hemp). The basal corms are surmounted by "pseudostems" of leaf sheaths. The terminal inflorescences are thrust up through the center of the pseudostem. Most species fall into two major groups: (1) *Musa* (= *Eumusa*) and *Rhodochlamys*; (2) *Australimusa* and *Callimusa*. The former is most diverse in the region Malaysia–Indochina, the latter in the Borneo–New Guinea area. *M. acuminata* (= *M. cavendishii*) is of great importance as

Inflorescence of a species of Muscari. *Such Grape Hyacinths are popular subjects for cultivation in temperate gardens.* (× 3)

the primary source of the edible bananas to which the inedible *M. balbisiana* contributed by hybridity. *M. textilis* (*Australimusa*), from the Philippine Islands and Borneo is the source of abaca fiber or Manila hemp.

Both *M. acuminata* and *M. × paradisiaca* (= *M. × sapientum* (*M. acuminata × M. balbisiana*)) are called BANANA or EDIBLE PLANTAIN, each containing numerous cultivated forms of edible seedless bananas. *M. acuminata* contains 5 subspecies and includes such cultivars as 'Dwarf Cavendish' and 'Giant Cavendish'. Cultivars of *M. × paradisiaca* are of hybrid rather than of autopolyploid origin and include both dessert and cooking bananas. Another edible group of cultivars, sometimes called *M. fehi* [FEHI OR FE'I BANANA], is widely cultivated throughout the Pacific.

Some wild bananas are very attractive plants and others have interestingly bizarre features. Thus *M. coccinea* and *M. beccarii* have brilliant scarlet bracts, *M. velutina* has fruits self-peeling at maturity and *M. ingens*, from New Guinea, is the largest herb known to science.
MUSACEAE, about 30 species.

Muscari a genus of herbaceous perennial bulbous plants native to Europe, the Mediterranean and western Asia. A number of species are commonly cultivated in gardens, the best-known being *M. botryoides* [GRAPE HYACINTH], with small globose, sky-blue or white flowers in dense racemes 3cm long on stalks up to 25cm high. *M. armeniacum* is similar to *M. botryoides*; a double form called 'Blue Spike' is often cultivated. Other popular species include *M. comosum*, the strongly scented *M. racemosum* (= *M. moschatum*) and *M. latifolium*, which is unusual in having almost black, fertile flowers surmounted by a head of bright violet-blue sterile ones.
LILIACEAE, about 40 species.

*Seeds of the Mung Bean (*Phaseolus aureus*), which may be eaten whole or germinated to give "Chinese bean sprouts".* (× ¾)

Muscat or **muscatel** a variety of *GRAPE (*Vitis vinifera*) cultivated in wine growing areas throughout the world. The fruit is ellipsoidal in shape with a high sugar content. It is one of the forms of grape which are dried to form *raisins. Most muscatel wine is sweet, including dessert wines such as Beaumes de Venise, except that produced in Alsace and Bulgaria.
VITACEAE.

Mushrooms see Fungi, p.152.

Musk or **musk plant** common names for the herbaceous *Mimulus moschata*. The MUSK MALLOW is *Malva moschata*.

Mustard a condiment derived from the seeds of *Brassica nigra* [BLACK MUSTARD], *B. juncea* [BROWN MUSTARD], both with dark-colored seeds, and *Sinapis alba* [WHITE MUSTARD] with yellow seeds. BROWN and WHITE MUSTARD are the main sources of seed currently used in the manufacture of table mustard. BLACK MUSTARD has now been more or less completely replaced by BROWN MUSTARD, a smaller, more compact plant whose fruits do not shatter so easily, and are therefore much more suited for mechanical harvesting.

Mustard seed has been used as a spice for thousands of years. The pungency of mustard is due to essential oils (mustard oils, isothiocyanates) which are released by the enzyme hydrolysis of mustard oil glucosides when crushed or powdered mustard seed is mixed with water. English mustard is made with cold water, not with vinegar, which allows the full development of its characteristic clean, hot flavor.

French mustards are of two basic types – the pale yellow Dijon and the darker Bordeaux. Dijon mustard is made from BLACK or BROWN MUSTARD seeds from which the coats have been removed; they are then ground with verjuice (from sour green grapes) into a fine paste. Bordeaux mustard is made from the whole seeds, including the coats, and contains vinegar, sugar, tarragon and other herbs and spices. The vinegar inhibits myrosinase enzyme action and gives a milder mustard lacking in mustard oils. German mustard is similar to Bordeaux mustard in composition. The mild mixed mustard used in the USA (and in piccalilli sauce) is made mainly from WHITE MUSTARD seeds.

Mustard seed, especially that of *S. alba*, is used in curries and pickles in the East. Mustard oil is obtained by pressing the seeds of various species of *Brassica* and is widely used in India as a cooking oil. The mustard in mustard and cress, the salad vegetable, is the seed-leaf stage of *Sinapis alba* or more commonly today of *RAPE, *Brassica napus*.
CRUCIFERAE.

Mutisia an exclusively South American genus of perennial herbs or, more rarely, low shrubs, which occur mainly in the Andes from Colombia to Patagonia. The flowers, borne in heads at the apex of short branches,

are frequently very colorful, varying from orange or yellow to red or purple (rarely white), but only five species, *M. decurrens*, *M. ilicifolia*, *M. latifolia*, *M. clematis* and *M. acuminata*, are at all frequently cultivated. COMPOSITAE, about 60 species.

Myosotis [FORGET-ME-NOTS] a genus of annual, biennial or perennial herbs from temperate regions of all continents. The plants are often hairy and the small, blue, pink, yellow or white flowers are solitary or in terminal cymes. There are many cultivars of *M. scorpioides* [FORGET-ME-NOT], *M. alpestris* [ALPINE FORGET-ME-NOT] and *M. sylvatica* [GARDEN or WOOD FORGET-ME-NOT], all of which flower best in partly shaded positions. Many of the New Zealand species in cultivation, including *M. angustata* and *M. albida*, have white flowers, and *M. australis* is yellow or white flowered.
BORAGINACEAE, 40–50 species.

Myrica an almost cosmopolitan genus of small, aromatic, deciduous or evergreen trees or shrubs. The best-known species is *M. gale* [SWEET GALE], from north temperate regions. Its strongly aromatic leaves are used medicinally against dysentery, as a moth-repellent, and as a flavoring for beer. *M. rubra* (= *M. nagi*), from Japan, southern China, Korea and the Philippines, is cultivated in China for its edible seeds. The evergreen North American tree *M. cerifera* [WAX MYRTLE], the shrubs *M. californica* [CALIFORNIA BAYBERRY] and *M. pensylvanica* [BAYBERRY] are three of the species which produce wax on the surface of their fruits, which is used for candles.
MYRICACEAE, about 35 species.

Myriophyllum a cosmopolitan genus of mainly aquatic perennial herbs commonly

The characteristic cinnamon-colored flaking bark of Myrtus luma (= Amomyrtus luma). (×⅛)

*Today, Black Mustard (*Brassica nigra*) is grown mainly for soil cover or grazing and produces an amazing visual display in spring.*

called milfoils. The plants have either a free-floating habit or a submerged one, in which the plant is anchored by rhizomes to the substrate. In all forms the inflorescences are borne aerially. Several species, including *M. hippuroides* and *M. heterophyllum* [WATER MILFOILS] are used as oxygenating plants in aquaria.
HALORAGACEAE, about 45 species.

Myristica a predominantly Southeast Asian genus of evergreen trees of which the most important commercially is *M. fragrans* (= *M. aromatica*) [NUTMEG TREE] which has aromatic leaves, pale yellow flowers and brown seeds with a thin scarlet aril. It is from the seeds and aril that the spices nutmeg and mace are obtained, but the species is also grown for ornament. It is native to the Moluccas but is grown commercially mainly in Indonesia and the West Indies (Grenada) (see also Nutmeg and Mace).

The wood of the Javanese *M. inermis* is used as a fumigant and the oil of *M. simianrum*, native to the Philippines, is used to cure skin ailments.
MYRISTICACEAE, about 80 species.

Myroxylon a small genus of tropical American trees. The trunks of *M. balsamum* yield a *balsam (*Peru balsam from variety *pereirae*, Tolu balsam from variety *balsamum*) used for medicinal purposes and as a fixative in the perfume industry. *M. balsamum* also yields a fragrant, hard, heavy, reddish-brown timber which is exported from Brazil. It is cultivated in the tropics.
LEGUMINOSAE, 2 species.

Myrrh or **sweet cicely** common names for *Myrrhis odorata*, a herbaceous perennial native to Europe. It is sometimes used as a potherb, the pinnate aromatic leaves being eaten as salad, and the roots boiled and used as a vegetable. The white flowers are borne in umbels and followed by dark brown ridged fruits about 2.5cm long. This old-time herb was also used in home remedies and for flavoring brandy.

The myrrh used in medicine, incense etc is an oleo-gum-resin obtained from the trunks of species of *Commiphora, mainly C. molmol, native to Arabia and northeast Africa.

Myrtus [MYRTLES] a genus of fragrant evergreen shrubs with some trees, from tropical and warm-temperate regions, particularly America. The leaves are fragrant when crushed, and the small, usually white flowers are also fragrant.

M. communis [COMMON MYRTLE] the only European native (unless it is naturalized from Asia), has ovate to lanceolate leaves which are dark shining green above; distillation yields the perfume "eau d'ange". The globose fruit is purple-black (white in 'Albocarpa'). Var *tarentina* [TARENTUM MYRTLE] was worn by Athenian judges and victors at the Olympic games. Among ornamental species hardy in warmer temperate areas are: *M. bullata* (= *Lophomyrtus bullata*), a small tree with puckered leaves; *M. luma* (= *Luma apiculata, Amomyrtus luma*), a bush or tree to 20m with flaking rusty bark; *M. ugni* (= *Ugni molinae*) [CHILEAN GUAVA or MURTILLO], a small shrub with fragrant, rounded and concave, rose-tinted petals, and ripe fruits used sometimes in jam making; and, perhaps the hardiest, the prostrate shrub *M. nummularia* (= *Myrteola nummularia*).
MYRTACEAE, about 100 species.

N

Najas a cosmopolitan genus of submerged annual and perennial herbs. Some species such as *N. flexilis* when they grow profusely can be used as fertilizer or, when dried, as packing material. In Hawaii, *N. major* is eaten as a salad vegetable.
NAJADACEAE, about 35–50 species.

Nandina an East Asian genus represented by a single half-hardy evergreen ornamental shrub species, *N. domestica* [SACRED BAMBOO]. It is bamboo-like in habit with alternate, twice or thrice pinnate leaves which turn red in autumn. The small white flowers are borne in large terminal panicles and the fruit is a red berry.
BERBERIDACEAE, 1 species.

Naranjilla the common name for the sub-shrub *Solanum quitoense*, not known in the wild state but grown commercially in Colombia, Ecuador and Costa Rica at 1 000–2 500m, for its small, round, bright orange, tomato-shaped, thick-skinned fruits. The rather acidic pulp is used for making drinks and sherbets.
SOLANACEAE.

> . . . Daffodils,
> *That come before the swallow dares, and take,*
> *The winds of March with beauty . . .*
>
> William Shakespeare

Narcissus a genus of bulbous herbs of Europe and North Africa, centered on Spain and Portugal and occupying a wide range of habitats from lowlands to exposed mountain sites at elevations up to 2 000m. The flowers have six free perianth segments, uniting into a cylindrical tube above the ovary. A distinctive shallow ring or deep cup projecting at the junction of perianth segments and tube is called the corona; it is often more brightly colored than the white or yellow perianth.

There are about 60 wild species and most hybridize freely. Most grow wild in the Pyrenees or around the Mediterranean but some occur further north, *N. pseudonarcissus* [TRUMPET NARCISSUS] extending to England and Wales. *N. tazetta* [POLYANTHUS NARCISSUS] has the widest range, from Spain eastwards to Japan. Most species are spring-flowering but *N. viridiflorus* and *N. serotinus* flower in September and October.

The popular name for *Narcissus* species is DAFFODIL, although, especially in the UK, a distinction may be drawn between those with long, trumpet-shaped coronas [DAFFODILS] and those with small or medium-sized coronas [NARCISSUS]. The genus is divided into two subgenera, *Hermione* and *Narcissus*. The former includes *N. tazetta*. The subgenus *Narcissus* includes most of the commercially important species: *N. hispanicus*, *N. bulbocodium* [PETTICOAT DAFFODIL], *N. jonquilla* [JONQUIL], *N. triandrus* [ANGEL'S TEARS], *N. pseudonarcissus*, *N. poeticus* [POET'S NARCISSUS] and *N. cyclamineus*. The most frequently grown cultivars arose from *N. hispanicus*, *N. pseudonarcissus*, *N. poeticus* and *N. tazetta*.

Narcissus became an important ornamental crop at the end of the 19th century and has increased in popularity ever since. In the UK, the Royal Horticultural Society list contains nearly 10 000 named cultivars classified into 11 divisions based on appearance, trumpet shape and parent species. Relatively few cultivars are widely grown commercially, perhaps three-quarters of the world cultivated area being made up of fewer than 20.

Narcissuses are grown commercially in a number of countries throughout the world with a suitable temperate climate and light soils, but the UK area is larger (5 300 hectares) than all the others together. There is considerable international trade in narcissus bulbs, mainly from the Netherlands, but increasingly from the UK.

The narcissus is deservedly popular as a garden plant and requires little attention to produce an annual show of flowers. The main commercial importance is for sales of bulbs and of outdoor-grown or greenhouse-forced flowers.

Bulb growing in the field is a highly mechanized, efficient operation, with mechanical planting, harvesting and grading. Bulbs are grown in ridges at planting densities of 1.5–2kg/m² and left for two years before lifting in the UK or annually in the Netherlands.
AMARYLLIDACEAE, about 60 species.

The curious-shaped leaves of the Naranjilla (Solanum quitoense). The tomato-like fruits of this plant make a refreshing drink. (×2)

Narcotic plants see Medicinal and narcotic plants, p. 220.

Nardostachys a small genus of perennial herbs from the Himalayas and western China, with fragrant rhizomes. The rose-purple-flowered *N. jatamansi* (= *N. grandiflora*) yields an essential oil, spikenard, of ancient fame and still valued as a perfume. It is also used as a black hair dye and tonic, and to treat nervous disorders.
VALERIANACEAE, 3 species.

Nasturtium a small genus of perennial herbs, native to the Northern Hemisphere, though now widely naturalized. It is sometimes included in the genus *Rorippa*. *N. officinale* and *N. microphyllum* are often cultivated as GREEN or SUMMER *WATERCRESS and the hybrid *N. × sterile* as BROWN or WINTER WATERCRESS. The name Nasturtium was applied by a number of the old herbalists to species of the unrelated genus *Tropaeolum*, and is still one of the common names [GARDEN NASTURTIUM] of *T. majus*, a strong-growing annual climber much cultivated in gardens for its attractive bright orange or yellow flowers.
CRUCIFERAE, 6 species.

Navicula a very common genus of diatoms usually found living on mud or sand surfaces. The cell is motile (free swimming) and this enables it to move out into the light after being buried by any disturbance of the substrate.
BACILLARIOPHYCEAE.

Neckera a genus of mosses from which some species have been transferred to other genera such as *Neckeradelphus*, *Neckeropsis*, *Calyptrothecium*, *Pinnatella* and others. Although *Neckera crispa* is a conspicuous moss (especially on tree trunks) across most

of the north temperate zone and *N. complanata* is distributed across four continents, the genus is primarily tropical. It is among the genera which account for the "pendent" or festooning habit found in some mosses in damp conditions in warm climates. *N. crispa*, on rock substrata, is a notable basic (alkaline) soil indicator.
NECKERACEAE, about 75 species.

Nectarine the reddish yellow fruit produced by the species *Prunus persica* var *nucipersica*, a form of peach. Although nectarines are usually smaller than peaches and have a distinctly different flavor, the only factor which distinguishes them without ambiguity is their smooth skin. It is believed that the nectarine first arose thousands of years ago after the spread of the peach from China to Central Asia.
ROSACEAE.

Nectria a cosmopolitan genus of ascomycete fungi, many species of which cause important cankers. Probably the most thoroughly studied is *N. galligena*, which causes a perennial canker on a number of hardwoods, notably apples and pears. The flask-shaped ascocarps (perithecia) are usually brightly colored and are formed in clusters on outbreaks of the canker or scattered on bark. Infection occurs in wounds of the tree, the most important being leaf scars formed when the leaves are shed in wet weather; pruning wounds, growth cracks and scab lesions are also important. *N. galligena* also causes "eye rot" of apple or pear fruit.
NECTRIACEAE, about 50 species.

Nelumbo a genus of two aquatic species, the pink-petaled *N. nucifera* [INDIAN, HINDU, CHINESE or SACRED LOTUS], from Asia and northeastern Australia, and the yellow-petaled North American species *N. lutea* (= *N. pentapetala*) [AMERICAN LOTUS, WATER CHINQUAPIN]. The Asian species is sacred to Buddhists; both are prized ornamentals.

N. lutea has leaves 30–60cm across, raised

Canker of an apple branch caused by the ascomycete fungus Nectria galligena, *which enters through wounds.* ($\times\frac{1}{4}$)

The Petticoat Daffodil, one of the commonly cultivated dwarf forms, is a naturally occurring variety of Narcissus bulbocodium. ($\times\frac{1}{3}$)

above water level. The flowers are large, up to 25cm wide, yellow and fragrant. The leaves of *N. nucifera* are even larger than those of the AMERICAN LOTUS (up to 90cm) and the fragrant, white-tipped rose or pink flowers may be up to 30cm wide.

There are numerous old cultivars of *N. nucifera*, including ones known in the west such as 'Kermesina,' 'Alba Grandiflora', 'Alba Striata' with striped petals, as well as gold-fringed, gold-freckled, double-flowered and dwarf forms suitable for container growth.

As well as being highly valued ornamentals, both species have rhizomes that are edible, either raw or cooked. They are juicy, crisp and slightly sweet, and if ground up yield a fine, easily digested form of starch. The leaves are also used in steamed meat dishes, and the fruits (achenes) are regarded as a luxury dessert in China, also being eaten in the Americas. The achenes may also be candied or preserved in China. (See also *Lotus*.)

The Lotus of the ancients and of the myth of the Lotus-Eaters is thought by some authorities to have been *Ziziphus lotus*.
NYMPHAEACEAE, 2 species.

Nemesia a genus of annual or perennial herbs and subshrubs mostly native to South Africa, with a few in tropical Africa. A number are used as garden bedding plants. The markedly asymmetrical, two-lipped flowers are usually borne in racemes or sometimes solitary. The annual species *N. versicolor* and *N. strumosa* are commonly grown in a range of cultivars with petals in shades of white, yellow, blue, orange, scarlet, and pink.
SCROPHULARIACEAE, about 50 species.

Nemophila a genus of annual herbs native

to North America, some of which are cultivated as spring and summer bedding plants. The most popular are *N. maculata* [FIVESPOT], which has flowers up to 4.5cm across with white petals each tipped with a purple blotch, and *N. menziesii* (= *N. insignis*) [BABY-BLUE-EYES], with flowers of a similar size, but blue or white and lacking blotches, although cultivars with blue flowers margined with white and with blue veined with purple are also grown.
HYDROPHYLLACEAE, about 11 species.

Neoregelia is a genus of South American evergreen herbaceous plants, mostly native to Brazil, with one species in Colombia and Peru. They produce strap-shaped leaves in a basal rosette in such a way as to form a depression or almost a tube at the center. The compact inflorescence is sunk in the center of this depression. The leaves at the center of the rosette are often bright red and this color persists after the white, blue or violet flowers

Narcissus requienii *growing in the Pyrenees, where it is often found in meadows, along stream banks or on cliff ledges.* ($\times\frac{1}{2}$)

have withered (the flowers last a single night). Several species are grown as ornamental greenhouse and pot plants. *N. carolinae* (= *N. marechalii*) has many cultivated varieties and strains, and var *meyendorffii* is probably the typical form; *N. spectabilis* is also widely grown.
BROMELIACEAE, about 40 species.

Neottia a genus of saprophytic leafless orchids indigenous to temperate Eurasia. *N. nidus-avis* [BIRD'S-NEST ORCHID] is found in deciduous woods on calcareous soils. It has no chlorophyll and is entirely of a brownish color. Species of *Neottia* have short, thick, much-branched roots which are mycorrhizal. The outer zone of each root is permeated with a network of fungal hyphae. The hyphae are able to survive and grow

near the surface of the root but towards the inside they are digested by enzymes produced by the orchid which, obtaining all its nutrients from the fungus, thus compensates for its photosynthetic deficiency. ORCHIDACEAE, 3 species.

Nepenthes a genus of carnivorous pitcher plants which are common throughout the jungles of the Old World tropics (excluding Africa but including Madagascar). Most are herbaceous vine-like plants, often epiphytic, which climb by means of tendrils which are extensions of the midrib. Mature leaves may be from 30cm to 1m long. The tendrils develop to form pitchers which are generally green, often blotched with red or purple, and sometimes variously ribbed. The pitchers are variable in shape and size according to species. Insects are attracted to honey and the bright color of the pitcher; once inside, the slippery surface prevents them from climbing out, and they drown in the water

Part of a Bull Kelp (Nereocystis sp) showing the top of the tubular stipe (stalk) which provides buoyancy. Attached to this are a number of flattened blades. (× $\frac{1}{10}$)

Pitcher Plants (Nepenthes sp), growing in the upper montane rain forests of Malaya. (× $\frac{1}{3}$)

contained in the base of the pitcher. The plant then absorbs the products of decay.

Many beautiful species and hybrids are cultivated, including *N. albomarginata* (a plant suitable for hanging baskets), the robust *N. mirabilis*, *N. hookerana* and their hybrids *N. × lawrenceana* and *N. × morganiana*.
NEPENTHACEAE, about 70 species.

Nepeta a genus of annual or perennial herbs occurring more or less throughout warm-temperate regions of the Northern Hemisphere, as well as in the mountains of tropical Africa. Their flowers have a curved, narrow corolla-tube and a two-lipped limb, the lower lip saucer-shaped and often tesselated. The best-known species is the aromatic *N. cataria* [CATMINT, CATNIP] from Europe and North Africa, a popular garden plant with gray leaves and white purple flowers, much liked by cats. *N. × faassenii* (= *N. mussinii* × *N. nepetella*) is used as an edging plant with silvery leaves and whorls of lavender flowers. *N. nervosa* has light blue flowers and is also cultivated for ornament.
LABIATAE, 200–250 species.

Nephrolepis a genus of terrestrial ferns which are widely dispersed throughout the tropics and subtropics of both hemispheres. They are commonly called LADDER or SWORD FERNS, the attractive crowns of long drooping leaves being a common feature. The genus reproduces rapidly by producing numerous runners and is one of the most important commercial greenhouse ferns. Some, such as *N. biserrata* cultivar 'Furcans', are suitable for growing in baskets.

Other commonly cultivated species include *N. cordifolia* [ERECT SWORD FERN], all cultivars of which, including 'Compacta', 'Plumosa' and 'Variegata', bear erect leaves,

and *N. exaltata*, whose cultivars 'Bostoniensis' [BOSTON FERN] and its crested-leaved derivatives are extremely popular house ferns. Two other cultivars, 'Smithii' and 'Whitmanii' are known as LACE FERNS. POLYPODIACEAE, about 30 species.

Nereocystis [BULL KELP] a genus of brown algae which grows in the subtidal zone along the western coast of North America. The plant consists of a sturdy holdfast from which a long tubular stipe arises. This ends in a bladder (containing carbon monoxide and carbon dioxide), bearing flattened fronds. PHAEOPHYCEAE.

Nerine a genus of bulbous perennials native to southern Africa. The leaves are basal and strap-like. A number of species, eg *N. sarniensis* [GUERNSEY LILY] are cultivated as garden and greenhouse plants for their attractive inflorescences of red to crimson flowers. Other popular cultivated species are *N. bowdenii* (rose-pink flowers), *N. pumila* (bright scarlet flowers), *N. humilis* (rose-purple flowers), and *N. flexuosa* cultivar 'Alba' (pure white flowers).
AMARYLLIDACEAE, about 30 species.

Nerium a small genus of very ornamental evergreen shrubs, native from the Mediterranean eastwards to Japan. *N. oleander* [COMMON OLEANDER, ROSE BAY] and "*N. odorum*", a sweet-scented form of it, are widely grown as longliving street plantings in mild climates, for their attractive white, pink, rose or purple flowers. However, the plants are poisonous throughout and can be fatal if eaten. They contain oleandrin, a cardiac glycoside which is used medicinally and in rat poisons.
APOCYNACEAE, 2–3 species.

Nest ferns a general term applied to certain ferns where the leaves wrap around trunks and branches producing a space in which humus accumulates, eg in the Stag's Horn Fern (*Platycerium bifurcatum*).

Common Oleander (Nerium oleander). (× $\frac{1}{4}$)

*The berries of the Woody Nightshade (*Solanum dulcamara*). The berries have a bitter-sweet taste and are slightly poisonous. (×1)*

Nettle a common name for numerous un-related species of plants which usually have stinging hairs. *Urtica dioica [COMMON STINGING NETTLE] is a common temperate weed which is used as a source of fiber; the leaves can be eaten as a vegetable. Several other *Urtica* species are also commonly called nettles. *U. urens* [DOG NETTLE, SMALL NETTLE] has similar uses to *U. dioica* and the leaves are used medicinally as a diaphoretic. A number of *Lamium* species are called DEAD NETTLES since the leaves look like those of the stinging nettle but do not sting. The name FALSE NETTLE is applied to several species of *Boehmeria*. The WOOD NETTLE is *Laportea canadensis*, a North American perennial herb which yields a fine strong fiber comparable to *ramie.

Neurospora a widespread genus of terrestrial saprophytic fungi especially common in the tropics on soil and vegetation exposed to fire. The fungus is visible as a pink powdery mass made up of chains of wind-dispersed conidia. It can be a serious contaminant in bakeries, where its ability to grow rapidly at fairly high temperature enables it to colonize warm bread; hence its common names of RED BREAD MOLD and BAKERY MOLD. It is also a contaminant of wood-drying kilns. *N. crassa* is much used in genetical studies, as many biochemically deficient mutants have been isolated.
LASIOSPHAERIACEAE, 4 species.

Nicandra [APPLE OF PERU] a genus comprising a single species, *N. physaloides*, a glabrous annual herb from Peru. The flowers are solitary, pale blue and drooping, and the calyx is inflated and bladder-like when in fruit. It is a weed in tropical America and the USA and has been used in the production of fly poison. It is sufficiently hardy to be grown as an ornamental border plant.
SOLANACEAE, 1 species.

Nicotiana a large New World genus of annual and perennial herbs (rarely shrubs), including the tobacco plant and several ornamental species. About half the species originated in South America; the remainder are natives of western North America and Mexico, Australia and certain islands in the South Pacific.

Appreciable amounts of nicotine are present in about 10 species. Of great economic importance as a stimulant, nicotine is also extracted for insecticidal purposes. Only *N. tabacum* [COMMON *TOBACCO] and, less importantly, *N. rustica* (both from South America) are widely cultivated, being grown throughout the world except in arctic and near-arctic zones. Indian tribes of North America used to smoke, sniff or chew for ritual or pleasure the leaves of *N. bigelovii* and *N. attenuata*.

Several species are cultivated for their attractive inflorescences of tubular flowers ("flowering tobaccos"), particularly *N. alata* (= *N. affinis*), a hybrid called N. × *sanderae* (= *N. alata* × *N. forgetiana*), and *N. glauca* [TREE TOBACCO].

Classic studies in genetics, cytology and physiology have been done with species of *Nicotiana*. Recently, pioneering studies on culturing and hybridization of isolated cells, generation of plants from anthers and tumor formation in hybrids have been performed with members of this genus.
SOLANACEAE, 64 species.

Nidularium a small Brazilian genus of epiphytic herbs with rosettes of attractively colored (often reddish or purple) spiny-margined leaves. The flowers are red, blue, purple or white, in a central compound inflorescence, surrounded by bracts which are usually colored. Many of the species are cultivated as greenhouse ornamentals, or outdoors in warmer climates, including *N. innocentii*, with several varieties, *N. fulgens*, with bright scarlet flowers, and *N. procerum* (= *Karatas purpurea*).
BROMELIACEAE, 20–25 species.

Nierembergia a genus of subshrubs and

*The Stag's Horn Fern (*Platycerium bifurcatum*) forms a "nest" around the branch of a tree in which humus collects. The branched fertile leaves give this species its popular name. (×1/12)*

*The flowers and stinging leaves of the Common Stinging Nettle (*Urtica dioica*), a common temperate weed. (×3/4)*

perennial herbs native to Mexico and South America. A number of the species, such as *N. scoparia* (= *N. frutescens*) [TALL CUP FLOWER], *N. repens* (= *N. rivularis*) [WHITE CUP FLOWER] and *N. gracilis*, are grown as border and pot plants for their large conspicuous flowers, usually whitish, blue or violet in color.
SCROPHULARIACEAE, about 35 species.

Nigella a small genus of annual herbs, occurring naturally from Europe across the Mediterranean to Central Asia. *N. damascena* [LOVE-IN-A-MIST, DEVIL-IN-A-BUSH] has large blue flowers surrounded by an involucre of finely dissected upper stem leaves. Some cultivars have white, pink or mauve flowers. *N. hispanica* [FENNEL FLOWER] has deep blue flowers and red stamens, but lacks the showy involucre. *N. sativa* [BLACK CUMIN], a blue-flowered species, with solitary flowers without an involucre, is cultivated for its seeds which are used for seasoning.
RANUNCULACEAE, about 20 species.

Nightshade a common name for various species of *Atropa and *Solanum, some of which are poisonous. The most commonly occurring species are *A. belladonna* [DEADLY NIGHTSHADE], *S. nigrum* [BLACK NIGHTSHADE] *S. dulcamara* [WOODY NIGHTSHADE, BITTERSWEET]. Another species, *Circaea lutetiana*, is known as ENCHANTER'S NIGHTSHADE.

Nitella one of the most common genera of stoneworts, often found forming subaquatic meadows at the bottom of shallow lakes. The plants consists of extremely large translucent internodal cells with branches of small cells arising at the nodes.
CHAROPHYCEAE.

Nitrobacter a genus of soil-inhabiting bacteria, several species of which are involved in the conversion of nitrites to nitrates, the latter being available for uptake by roots of plants.

Nitrosomonas a genus of soil-inhabiting bacteria, several species of which are involved in the conversion of ammonia and ammonium salts to nitrites.

Nolana a genus of low growing, sometimes prostrate and fleshy-leaved annual or perennial herbs or subshrubs native to semidesert waste regions of America from Peru to Patagonia (Chile). A few species, including N. humifusa (= N. prostrata) and N. paradoxa, are cultivated as garden ornamentals for their attractive white, pink or blue flowers.
NOLANACEAE, about 60 species.

Nomocharis a genus of bulbous herbs native to the high meadows of northern Burma, western China and southeast Tibet. They are closely related to *Lilium with beautiful, more or less open flowers resembling *Odontoglossum orchids. Among popular cultivated species are the white-flowered N. mairei (sometimes flushed purple), the yellow-flowered N. euxantha (tinged red at the base) and the yellowish-white-flowered N. saluenensis (flushed rose-pink).
LILIACEAE, about 10 species.

Nopalea a genus of large cacti native to Mexico, Guatemala and Central America. They are mostly tree-like with definite trunks and flat, fleshy branches bearing spines that are very like those of the PRICKLY PEAR (*Opuntia). The most important species is N. cochenillifera [COCHINEAL CACTUS] of Mexico, the principal food plant of the cochineal insect from which a scarlet food-dye is obtained. N. cochenillifera is also widely cultivated for its foliage and its red flowers.

Other cultivated species are N. brittoni and N. dejecta.
CACTACEAE, about 8 species.

Norfolk Island pine the common name for *Araucaria heterophylla (= A. excelsa), a tall impressive conifer native only to Norfolk Island in the South Pacific. It is, however, widely cultivated as an ornamental tree, especially in the Mediterranean region, and as a greenhouse pot plant.
ARAUCARIACEAE.

Nostoc a genus of blue-green algae having simple chains of cells, which may be differentiated as spores or thick-walled heterocysts, embedded in mucilage. Nostocs have a wide distribution in marine, freshwater and terrestrial habitats and are abundant from polar regions to coral atolls. They often appear as gelatinous masses on damp rocks or soil, where they are important as nitrogen fixers. They may form symbiotic associations, with fungi (as lichens), liverworts, cycads and the flowering plant *Gunnera. Some species are eaten as a delicacy in the Far East.
NOSTOCACEAE.

Nothofagus an important genus of deciduous or evergreen trees ranging in South America from about latitude 33° southwards to Cape Horn and from New Zealand northwards through Tasmania and eastern Australia to New Caledonia and New Guinea. Many species are dominants or codominants in the tropical montane and lowland cool temperate forests of the Southern Hemisphere, though some grow in subtropical lowland conditions.

Several species are important providers of hardwood timber for furniture and con-

The Yellow Water Lily (Nuphar lutea) *is common in ponds and slow-moving rivers throughout north temperate regions.* ($\times \frac{1}{12}$)

The Roble Beech (Nothofagus obliqua), *a native of Chile and Argentina but sometimes cultivated in north temperate regions.*

struction. Some deciduous species, principally the South American N. antarctica, N. obliqua [ROBLE BEECH] and N. procera, are grown as garden ornamentals for their form and fine autumn coloring. A number of evergreen species are also favored as ornamentals, notably the New Zealand N. menziesii [SILVER BEECH], N. truncata [HARD BEECH] and N. solandri [BLACK BEECH], and the Australian N. moorei [AUSTRALIAN BEECH] and N. cunninghamii [MYRTLE BEECH].
FAGACEAE, 35–40 species.

Notholaena a group of rock-inhabiting xerophytic ferns [sometimes called GOLD, SILVER or CLOAK FERNS] mainly native to warm-temperate and tropical America. They are sometimes included in the genus *Cheilanthes. A few are cultivated, especially in rock gardens, including N. aurea [GOLDEN CLOAK FERN], N. newberryi [COTTON FERN] and N. sinuata [WAVY COTTON FERN], while N. trichomanoides with its pendulous fronds is popular for hanging baskets.
POLYPODIACEAE, about 60 species.

Notocactus a South American genus of small, globular to short-cylindric cacti, much in demand by specialist cactus growers. The flowers, borne near the center, are mostly large and yellow; N. uebelmannianus has yellow or deep wine-purple flowers, and N. graessneri combines golden spines with lime-green blooms. N. ottonis and N. leninghausii are old favorites, now rivaled by some superb recent discoveries, such as N. brevihamatus, and N. magnificus.
CACTACEAE, about 30 species.

Nuphar a genus of northern temperate aquatic plants often known as YELLOW WATER-LILIES. Many species are cultivated, usually in still water, although *N. advena* and *N. lutea* [YELLOW WATER-LILY, BRANDY BOTTLE] can be grown in slow-flowing streams. Both of these species and others such as *N. japonica* and *N. polysepala* have peltate floating leaves and large, open yellow flowers. The starchy rhizomes of some species such as *N. advena* [YELLOW POND LILY] are eaten as food by North American Indians.
NYMPHAEACEAE, about 25 species.

Nutmeg the common name for *Myristica fragrans* [NUTMEG TREE], from the Moluccas, from which the spices nutmeg and *mace are obtained. Its fruit resembles a small leathery skinned peach with soft yellow flesh, containing one large shiny brown seed enveloped in a bright red network of tissue, the aril. When ripe, the fruit splits into two or four segments and opens out like a star to reveal the spectacular arilate seed.

The fruit is gathered at this stage and separated into pericarp, aril and seed. The former is used to make a jelly preserve. The other parts are slowly dried and cured; the aril then becomes mace and the seed kernels nutmegs. Any broken or inferior seeds are pressed to make "nutmeg butter" for perfumery. Nutmeg and mace are toxic if taken in large quantities due to the presence of the aromatic oil myristicin.

Today, although nutmegs are grown in various countries, the best come from the species' original home in the Spice Islands. MYRISTICACEAE.

Nuts see p. 238.

Nymphaea [WATER-LILIES] a genus found in shallow waters of the Northern Hemisphere and the tropics, a few reaching southern Africa and Australia. The leaves are more or less circular or oval and the

European White Water-Lilies (Nymphaea alba) in a natural pond. (× 1/15)

Nymphaea 'Gladstoniana,' one of the largest hardy-hybrid water-lilies in cultivation. (× 1/3)

flowers are large, showy and, like the leaves, usually float on the water's surface.

Many species have been cultivated for millennia; they show a wide range of colors and are often hybrids, derived from *N. alba* [EUROPEAN WHITE WATER-LILY], *N. lotus* [WHITE LOTUS], *N. rubra*, *N. caerulea* [EGYPTIAN BLUE LOTUS], *N. capensis* [CAPE BLUE WATER-LILY] and *N. tetragona* [PYGMY WATER-LILY], which is small enough for aquarium use.

The seeds of many species are eaten, those of *N. lotus* being made into a kind of bread in India, and in Africa the rhizomes are eaten. The rest of the plant contains the toxic alkaloid nupharin.
NYMPHAEACEAE, about 50 species.

Nymphoides a genus of annual or perennial aquatic herbs, superficially resembling *Nymphaea* [WATER-LILIES]. *Nymphoides* species are found throughout the tropics and temperate regions, some of them having very wide distribution, eg *N. indica* [WATER SNOWFLAKE], which is pantropical and has yellow and white flowers. *N. aquatica*, of eastern North America, produces curious swollen banana-like root tubers. A common Eurasian species is the yellow-flowered *N. peltata*. The fruits (capsules) of all species mature under water and split by an internal swelling of mucilage.

Several species, sometimes commonly called FLOATING HEARTS, are cultivated as ornamentals, including *N. cordata*, of eastern North America, and the three species mentioned above, but their rapid growth makes them a potential weed menace in the tropics.
MENYANTHACEAE, 20 species.

Nypa palm the common name for *Nypa fruticans*, the only member of the genus *Nypa*. It is unique in its abundance and local dominance in mangrove forests from Sri Lanka and the Bay of Bengal eastwards to the Solomons, Micronesia and the Ryu Kyu Islands. This palm has a stem that lies horizontally below the surface of the semiliquid mud and often branches dichotomously, but no erect trunk. Stout roots are

Rootstocks of a water-lily (Nymphaea sp), from a dry-season waterhole in Australia. Such roots were formerly a staple carbohydrate food for the local aboriginal people.

borne on the lower surface. *Nypa* is second only to the *coconut in its importance in the domestic economies of the Malay archipelago. The fronds make a very durable thatch and the "meat" of the young nuts is edible. The young inflorescences, bound and beaten, give a copious yield of sugary liquid, which is boiled to produce a dark treacle. This exudate is then fermented into toddy.
PALMAE, 1 species.

Nyssa a genus of Southeast Asian and North American trees and shrubs. The timber of some species is used commercially, especially *N. aquatica* [COTTON GUM, TUPELO GUM] and *N. biflora* (= *N. sylvatica* var *biflora*) [TUPELO, WATER GUM, TWIN-FLOWERED NYSSA]. *N. sylvatica* [TUPELO, PEPPERIDGE, BLACK or SOUR GUM] is cultivated for ornamental purposes. It is a broadly conical tree, reaching a height of 25m and producing scarlet and gold autumnal coloring. The fruit of *N. ogeche* [OGEECHE LIME, OGEECHE PLUM] is edible.
NYSSACEAE, 6 species.

The Fringed Water-Lily (Nymphoides peltata). (× 3/4)

Nuts

Man has eaten nuts since the earliest times, firstly gathered from wild trees but progressively a number of species have been brought into cultivation. The nutritive value of nuts is high, most being rich in protein, oil and fat and they are often high in vitamins. In general, nuts are eaten raw, but some such as the *ALMOND (*Prunus amygdalus) and *MACADAMIA (Macadamia integrifolia and M. tetraphylla) are also eaten cooked. The PEANUT (*Arachis hypogaea) is normally eaten roasted and/or salted.

In a strict botanical sense, a nut is dry type of indehiscent fruit in which the single seed is enclosed in a hard bony or woody pericarp. In this sense few of the seeds known colloquially or in commerce as nuts are correctly described. Most are the stones of drupes such as ALMONDS (Prunus amygdalus) or *WALNUT (Juglans regia) or are simply seeds from other kinds of fruit container such as PINE KERNELS (*Pinus pinea etc) and *BRAZIL NUTS (Bertholletia excelsa).

The true nuts which are commercially important are found in the families Betulaceae and Fagaceae. Several species from the *HAZEL genus (Corylus) (Betulaceae) provide nuts which are enclosed in a leafy involucre. The principle species are the HAZELNUT or COB (C. avellana), and the FILBERT (C. maxima). Most of today's varieties originated in the 19th century. They are primarily used as dessert nuts but are also used extensively in confectionery.

The SWEET *CHESTNUT (Castanea sativa) (Fagaceae) and its relatives in the same genus (see table) are true nuts that are completely enclosed in a very spiny involucre, which eventually opens to allow nuts to fall out. The SWEET CHESTNUT is native to southwestern Asia; it was introduced to southern Europe by the Greeks and later to Britain by the Romans. It did not reach North America until the 18th century, where it has now virtually replaced the AMERICAN CHESTNUT (C. dentata), which was all but exterminated by chestnut blight. CHESTNUTS have many uses, being ground into flour normally for use in soups, stews etc, or eaten whole, after either roasting or boiling. Preserved in sugar or syrup, they are used in dishes such as the famous French delicacy, marrons glacés. CHESTNUTS have also found use as livestock feed, notably for pigs. Other true nuts are the *WATER CHESTNUTS (Trapa natans and other species) which can be consumed, boiled or their starchy contents ground into flour.

Among the most popular nuts derived from drupes is the ALMOND (Prunus amygdalus). The tree is much like a PEACH in appearance, but the fruit is not fleshy, but dry and leathery. ALMONDS originated in the Near East but spread early to Europe and subsequently to North America. Two forms exist, the BITTER ALMOND (var amara), which is the main source of almond oil, and the SWEET ALMOND (var amygdalus), which is grown for its edible nuts.

The family Juglandaceae contains two genera, Juglans and *Carya, that provide WALNUTS, HICKORY NUTS and PECANS where the fruits are indehiscent drupes which contain stones considered as nuts but in which the hard shell is the inner fruit wall.

The WALNUT (Juglans regia) native from southeastern Europe to China, is widely cultivated throughout north temperate regions. The nuts are mainly used for dessert purposes, but also in confectionery and cooking. Young green fruits are often pickled in vinegar. The North American BLACK WALNUT (J. nigra) produces nuts that have a tough shell, although softer-shelled cultivars have also been developed. The strongly flavored nuts are mainly used in confectionery and ice cream. The nuts of the BUTTERNUT (J. cinerea), another North American tree, have a rich flavor, but also suffer from having a hard shell. Several Carya species yield edible nuts, but the most widely cultivated is the North American PECAN (C. illinoinensis), which has long enjoyed popularity in North America as a dessert and confectionery nut, and is now increasing in popularity in Europe.

The PISTACHIO (*Pistacia vera) has long been cultivated in the Mediterranean area and western Asia and more recently in the southern United States. PISTACHIO NUTS, which are the seeds contained in the bony shell of ovoid drupes, are expensive to produce but are valued for their flavor and ornamental color.

The aforementioned nuts are grown mainly in temperate regions, but a number of tropical species are also important. CASHEW NUTS are produced by *Anacardium occidentale, native to Brazil but now widely grown for their edible kernels in South America, India and East Africa. The nuts, which are mainly used for dessert purposes, have a very high fat and protein content. The CASHEW APPLE is the swollen juicy pedicel – an accessory false fruit. One of the other best-known tropical nuts is the South American BRAZIL NUT. As with the closely related *SAPUCAIA (Lecythis sabucayo) and MONKEY NUT (L. usitata), the nuts of commerce are harvested from wild trees.

BRAZIL NUTS, actually seeds, are produced in quantities of 12–22 inside a large woody, thick-walled capsular fruit that weighs up to 2kg. They have a very high fat content and are mainly used as dessert nuts, particularly around the Christmas period. The tree grows wild in the rain forests of the Amazon Basin in Brazil, Venezuela and Guiana and is never cultivated commercially. The *COCONUT (Cocos nucifera) is commercially important for the fiber (coir) extracted from the fruit wall, the dried endosperm or flesh (copra), the oil extracted from the copra, and as a dessert nut. The edible part is the endospermous lining to the stone or nut which encloses a central cavity containing milk when ripe. It is eaten raw and the shredded, desiccated flesh is used extensively in confectionery. Nuts used mainly as masticatories are the *BETEL NUT (*Areca catechu) and *KOLA NUT, from various *Cola species.

Possibly the most widely cultivated tropical "nut" is the PEANUT (Arachis hypogaea), an annual member of the family Leguminosae, in which fruits (pods) develop beneath the surface of the soil, hence its alternative name of GROUND NUT. The "nuts" of commerce are the seeds, but some appear on the market still enclosed in their fragile fruit cases. PEANUTS are rich in both oil and protein and are chiefly grown as an oil crop; only the best-quality product reaches the dessert market.

The only commercially important nuts contributed by Australian species are MACADAMIA NUTS. MACADAMIAS have long been a source of food for aborigines, but are now mainly consumed as dessert nuts in the United States, with some being used in confectionery. MORETON BAY CHESTNUTS (Castanospermum australe), the product of another Australian tree, are collected from the wild and only have local importance.

OYSTER NUTS, the seeds of a member of the cucumber family, Telfairia pedata, are mainly consumed in East Africa. PINE NUTS or PINE KERNELS are mainly obtained from the STONE or UMBRELLA PINE (Pinus pinea), grown in Italy, where they are eaten raw or roasted and salted, and also in confectionery. Many other pines yield edible seeds which are eaten locally.

PILI NUTS (Canarium luzonicum and C. ovatum), from the Philippines, and JAVA ALMONDS (C. commune), from Java, are the stones of drupes from trees belonging to the family Burseraceae. Along with MACADAMIA NUTS, both PILI NUTS and JAVA ALMONDS have the highest known fat content of any nut (70–72%). The seeds of the PILI NUT yield an oil (pili nut oil) that is used in confectionary, as are the whole roasted nuts. Uncooked nuts are a purgative, but they are eaten locally after cooking. Similarly the JAVA ALMOND is also used in confectionary and the oil is used in cooking and for lighting. These species are not cultivated, the nuts being collected from wild trees and only very small quantities enter international trade.

Common edible nuts: 1 Hazelnut; 2 Walnut; 3 Black Walnut; 4 Giant Filbert; 5 Butternut; 6 Pecan; 7 Almond; 8 Pistachio; 9 Sweet Chestnut; 10 Macadamia Nut; 11 Brazil Nut; 12 Pine Nut; 13 Water Chestnut; 14 Cashew Nut; 15 Coconut; 16 Betel Nut; 17 Kola. 1 to 4, 6 to 8, 10, 12, 14, 17 (×1); 5, 9, 11, 13, 16 (×⅔); 15, 17 (pod) (×½).

EDIBLE NUTS

Common name	Scientific name	Main areas of cultivation
*HAZELNUT, COB, EUROPEAN FILBERT	Corylus avellana	Turkey, Italy, Spain, France, England, Oregon
GIANT FILBERT	C. maxima (C. americana)	(as above)
TURKISH COBNUT	C. colurna	Turkey
SWEET *CHESTNUT	Castanea sativa	S Europe, N America
AMERICAN CHESTNUT	C. dentata	N America
JAPANESE CHESTNUT	C. crenata	Japan, N America
CHINESE CHESTNUT	C. mollissima	China, Korea, N America
*ALMOND	Prunus amygdalus (= P. dulcis)	Mediterranean, SW Asia, N America
SWEET ALMOND	var amygdalus	America
BITTER ALMOND	var amara	
*WALNUT	Juglans regia	Europe, Asia, N America
BLACK WALNUT, EASTERN WALNUT	J. nigra	N America
BUTTERNUT	J. cinerea	N America
JAPANESE WALNUT	J. ailanthifolia	Japan, N America
CHINESE WALNUT	J. cathayensis	China, N America
PECAN	*Carya illinoinensis	N America
SHAGBARK HICKORY	C. ovata	N America
SHELLBARK HICKORY	C. laciniosa	N America
*BRAZIL NUT, PARANUT	Bertholletia excelsa	Amazon region (wild)
*SAPUCAIA, SAPUCAYA	Lecythis sabucayo	S America (wild)
MONKEY NUT	L. usitata	S America (wild)
CASHEW NUT	*Anacardium occidentale	Tropical S America, India, E Africa
*COCONUT	Cocos nucifera	India, Sri Lanka, Malaysia, Indonesia, Philippines
*MACADAMIA NUT, AUSTRALIA, or QUEENSLAND NUT, (smooth shell)	Macadamia integrifolia	Australia, California
MACADAMIA NUT (rough shell)	M. tetraphylla	Australia, California
MORETON BAY CHESTNUT	Castanospermum australe	Australia (wild)
OYSTERNUT	Telfairia pedata	E Africa
PEANUT, GROUND NUT	*Arachis hypogaea	India, tropical Africa, China
PILI NUT	Canarium luzonicum C. ovatum	Philippines
JAVA ALMOND	C. commune	Java
PISTACHIO	*Pistacia vera	E. Mediterranean, India, S USA
*BETEL NUT	Areca catechu	Old World tropics
*KOLA	Cola nitida	W Africa, Caribbean
	C. acuminata	W Africa, Brazil
WATER CHESTNUT	Trapa natans T. bicornis T. maximowiczii	E Asia, Malaysia, India
PINE NUT, PINE KERNEL	*Pinus pinea P. pinaster	Mediterranean Mediterranean
SWISS STONE PINE	P. cembra	Europe
MEXICAN STONE PINE	P. cembroides	Mexico

Oak the common name for the large and economically important genus *Quercus*, which consists of deciduous, semi-evergreen or evergreen trees and some shrubs prized for their noble aspect and autumn colors. Oaks occur predominantly in the Northern Hemisphere, with the majority in North America, but a large number in Europe, the Mediterranean and Asia. In South America, oaks occur only on the Andes of Colombia. The relatively small tropical distribution is virtually confined to high mountains.

The leaves are rarely entire, and the margins are usually cut or lobed in various ways. The fruit is a large, solitary, characteristic nut (acorn) more or less enclosed from the base upward by a cup (cupule) composed of variously shaped scales usually more or less imbricate.

Oak provides one of the finest hardwoods and is in the forefront of temperate timbers because of its great strength and durability. It is not easy to distinguish the various commercial oak timbers, but a distinction is made between the WHITE OAKS and RED OAKS, the former being somewhat harder and more durable. Both kinds are used for the same purposes: furniture, and bridge, ship and many other types of building. The wood will take a high polish and when radially cut it is a favorite for paneling because of the fine "silver grain" formed by the wood rays. Important WHITE OAK species are *Q. alba*, *Q. macrocarpa*, *Q. robur*, *Q. petraea* (= *Q. sessiliflora*); important RED OAKS are *Q. rubra* (= *Q. borealis*), *Q. velutina* and *Q. palustris*. *Q. virginiana* [LIVE OAK] is considered to be the most durable of all oak timbers. It is used for trucks, ships and toolmaking, but unfortunately is now in short supply.

The timber of some oaks is used for inlay work. For example, "brown oak" is oak timber stained by the mycelium of the bracket fungus *Fistulina hepatica*. Similarly, oak timber may be stained a deep emerald green by the mycelium of the cup fungus *Chlorosplenium aeruginascens*. The green coloring substance has been isolated and characterized as a rare type of green pigment named xylindein.

*Cork is obtained from the outer bark of *Q. suber* [CORK OAK, LIVE OAK] from southern Europe. *Q. coccifera* [KERMES OAK, GRAIN TREE] is the host for a scale insect, *Coccus ilicis*, which since medieval times has been extracted to give a fine scarlet dye. A bright yellow dye, quercitron, is obtained from the inner bark of *Q. velutina* [BLACK or YELLOW-BARKED OAK]. *Q. faginea* and the closely related *Q. infectoria* are important sources of tannins from the bark and from galls.

Both *Q. coccifera* and *Q. ilex* [HOLM OAK] are evergreen and are characteristic of the Mediterranean region, the former making impenetrable holly-like bushes, the latter trees

Ripe Common Oats (Avena sativa) ready for harvesting. This species is particularly suited to cool, wet climates. (×1)

Two ornamental oaks – Red Oak (Quercus rubra) in winter dress, with behind it an evergreen Holm Oak (Q. ilex).

that will form forests if left alone. The HOLM OAK is often considered a reliable indicator of the Mediterranean climate.

Acorns are not generally of much economic importance, but they are eaten by small game birds and have been used to fatten pigs and poultry. They are not much used for human consumption but those of *Q. rotundifolia* are quite palatable.

Many oaks reach noble proportions and are long-lived. For this reason a number of species are grown as specimen ornamentals. Added to this, many have rich autumn tints and/or attractively cut leaves. In the former group are *Q. ellipsoidalis* [NORTHERN PIN OAK], with crimson autumn foliage, *Q. phellos* [WILLOW OAK], with yellow and orange autumn foliage, *Q. coccinea* [SCARLET OAK] and *Q. palustris* [PIN OAK], both with scarlet autumn foliage, and *Q. rubra* [RED OAK], whose leaves turn through red, mixed yellow and brown to red-brown before falling. The last three species are native to the eastern USA and make a major contribution to the magnificent autumn display of that region.

Other attractive ornamentals include *Q. alnifolia* [GOLDEN OAK], with leaves yellow beneath, *Q. castaneifolia* [CHESTNUT-LEAVED OAK], with coarsely toothed leaves, grayish beneath, *Q. frainetto* [HUNGARIAN or ITALIAN OAK], with deeply lobed leaves, and *Q. marilandica* [BLACKJACK OAK], which has almost triangular leaves. Finally, mention should be made again of the HOLM OAK, which is widely grown in temperate regions for its rounded form and glossy evergreen leaves. It thrives in most well-drained soils and tolerates clipping, shade and coastal situations.
FAGACEAE.

Oat the common name for several species of the mostly annual grasses of the genus *Avena, predominantly native to temperate regions and to mountains in the tropics. Oats are a major temperate cereal and occur in the cultivated form in three groups: the diploid *strigosa-brevis* [FODDER OATS], the tetraploid *abyssinica* and, most importantly, the hexaploid *sativa-byzantina-nuda* group. The first oats to be cultivated were those of the

strigosa-brevis group and these continued in cultivation until the 20th century at the extremes of soil and climate conditions.

The two main species of oat cultivated are the hexaploid *A. sativa* [COMMON OAT], grown in cooler and wetter regions, and *A. byzantina* [RED or ALGERIAN OAT] which is grown in hotter, drier areas. Most cultivated oats are grown in the USA, southern Canada, the USSR and Europe, particularly in Mediterranean countries. Another cultivated hexaploid species, *A. nuda* [NAKED OAT], is Chinese in origin and may well have been derived from the other two cultivated species of this group. There is evidence that these three cultivated species were derived from *A. sterilis* [ANIMATED OAT]. *A. fatua* [WILD OAT, TARTARIAN OAT, POTATO OAT, FLAVER, DRAKE] is a weedy species evolved from either *A. sterilis*, *A. sativa* or *A. byzantina*. *A. abyssinica*, confined to Ethiopia, is a member of the *barbata* group of tetraploids and may well have evolved by autopolyploidy from the *strigosa* group diploids.

Only wheat, maize and barley are more important as temperate cereal crops than oats. As well as being adapted to cool, moist conditions, oats grow and mature rapidly, so can be grown where the season is short. The straw is a valuable winter feed for livestock and some of the crop is cut for hay or silage. The oat grain contains about 15% of protein and 8% fat, the remainder being mostly carbohydrate. It is a valuable crop for Man and his livestock, and has been esteemed for centuries as a food for horses.
GRAMINEAE.

Ochna a genus of deciduous trees and shrubs native to South Africa and tropical Africa and Asia. Although the yellow or green flowers are attractive, it is the clustered fleshy fruits which are the main reason for growing such species as the South African *O. atropurpurea* and *O. serrulata* as ornamental shrubs in gardens within the tropics or in temperate regions under glass.
OCHNACEAE, about 85 species.

A marble gall (oak apple) on oak (Quercus sp) containing the larva of the cynipid wasp Andricus kollari. This gall contains tannins used in the manufacture of blue-black ink. ($\times 1\frac{1}{2}$)

A temperate mixed woodland of southern England comprising oak (Quercus sp), Silver. Birch (Betula pendula) and Bracken (Pteridium aquilinum).

Ochroma see BALSA.
BOMBACACEAE, 1 species.

Ocimum a genus of herbs and subshrubs mainly from tropical and subtropical regions, and containing several potherbs. The best-known is *O. basilicum* (see BASIL). *O. sanctum* [HOLY BASIL] is a sacred plant in the Hindu religion. *O. viride* [FEVER PLANT] is used as a tea-like infusion in West Africa and is locally used as a treatment for feverish conditions.
LABIATAE, about 150 species.

Ocotea a tropical and subtropical genus of trees mostly from the New World but with some species in southern and East Africa and the Atlantic islands. Several are economically important for their timber. *O. rodiaei* [BIBISI TREE, GREENHEART] yields a very hard wood used for marine work (see GREENHEART). Other New World species include the West Indian *O. leucoxylon* [MAZA LAUREL] with a soft wood used for general carpentry. *O. rubra*, from Guiana and Brazil, and the hard, heavy *O. spathulata*, from the West Indies. *O. bullata* [BLACK STINKWOOD] is locally important in South Africa. Several species, notably *O. indecora* and *O. squarrosa* (both natives of Brazil), also yield medicinal preparations.
LAURACEAE, about 300 species.

Odontoglossum a genus of tropical Central and South American epiphytic, pseudo-bulbous orchids, many species of which are prized as ornamental plants, eg *O. crispum*, with many cultivars, and *O. pulchellum* [LILY-OF-THE-VALLEY ORCHID].

Flower color ranges from purest white to deep, dull maroon-purples and bright shining chestnut browns. Usually large (up to 10cm in diameter) and frequently with striking color combinations, the flowers have made odontoglossums very popular plants for cultivation both as specimen plants and cut-flowers. Many hybrids between *Odontoglossum* species, and with species of related genera such as *Oncidium*, *Miltonia*, and *Brassia*, have produced the full spectral range of flower colors.
ORCHIDACEAE, about 200 species.

Oedogonium a genus of widespread filamentous green algae found in ponds and streams. When young they are usually attached by a holdfast but when mature they form free-floating masses.
CHLOROPHYCEAE.

Oenothera a genus of herbs and shrubs native to temperate and subtropical North and South America and also represented in the West Indies. It includes the EVENING PRIMROSES, which are night-flowering, and the SUNCUPS or SUNDROPS, which are day flowering.

The flowers are either solitary or in long racemes and are normally large, showy, and white, yellow or reddish in color. Although a number of species have become widespread weeds, many species are cultivated either in rock gardens (such as *O. acaulis*, a white to

rose flowered tufted perennial) or in garden borders. Popular species include *O. biennis* [COMMON EVENING PRIMROSE], whose roots and shoots are edible, *O. nocturna, O. missour-ensis* (= *Megapterium missourense*), *O. tetra-gona* (= *O. youngii*), *O. bistorta* [SUNCUP] and *O. fruticosa* [SUNDROPS].
ONAGACEAE, about 80 species.

Oil crops see p. 244.

Okra the common name for **Hibiscus esculentus* (= *Abelmoschus esculentus*), a stout, annual herb which grows up to 2m, native to central Africa but widely cultivated throughout the tropics and subtropics for its soft edible pod. Among other common names are GUMBO, SYRIAN MALLOW and LADIES FIN-GERS. The showy yellow flowers ripen to produce fruits (pods) which when full grown are up to 25cm long and contain numerous brown or black seeds. The wall of the unripe pod is fleshy and extremely mucilaginous.

The fruit is usually used while still unripe and green and while in this state may also be preserved by slicing and drying. After drying it is often powdered. Okra fruits are also preserved by canning, salting, pickling and so on, but are more often used fresh or cooked as a vegetable ingredient of many regional dishes; in this form it is very slimy because of the mucilage it contains and is perhaps an acquired taste. The powdered okra is used as a thickening for soups, stews etc.

The seeds of okra are often used as a substitute for coffee while the leaves and young immature fruits have long been used as emollient poultices. Decoctions of the fruit are widely used for a variety of medicinal purposes such as the treatment of catarrh, eye complaints and diseases of the urogenital system.

The stem of okra also contains long useful

Olearia stellulata has a mass of white daisy-like flowers, from which the common name, Daisy Bush, is derived.

Okra fruits (Hibiscus esculentus); *the pods are picked while still green and are eaten fresh or cooked. Okra is grown throughout the tropics.* ($\times \frac{1}{10}$)

fibers which can be used for rope- and sail-making, and the leaves can also be used as a culinary herb. Many varieties have been produced through its long history in several continents. These varieties differ in the size and form of the fruit, its flavor and the amount of mucilage contained in its flesh. *H. abelmoschus* (= *Abelmoschus moschatus*) [MUSK OKRA, MUSK MALLOW], is a closely related species which comes from the Indian subcontinent and, like okra, is cultivated throughout the tropical regions of the world. It is grown for its showy flowers as well as its seeds, called ambretta seeds, which have an oleo-resinous oil in the seed-coats. This has a musky odor and is often used in perfumery as a substitute for musk.
MALIACEAE.

Olea a small genus of evergreen trees or shrubs, native in warm temperate or tropical regions, from southern Europe and the Mediterranean to Africa (the main center), southern Asia, eastern Australia and New Caledonia. The best-known species is *O. europaea* [COMMON OLIVE], very widely grown since early times in areas of Mediter-ranean climate for its edible fruits and oil (see OLIVE). Other cultivated species include *O. ferruginea* (= *O. cuspidata*) [INDIAN OLIVE] and *O. africana* [WILD OLIVE]. The wood of these and other species such as *O. laurifolia* [BLACK IRONWOOD] is often valued.
OLEACEAE, about 20 species.

Olearia [DAISY BUSHES, DAISY TREES, TREE ASTERS] a genus of evergreen pubescent small trees and shrubs native to Australia, Tasmania and New Zealand, with a few species in New Guinea. The foliage is almost heath-like in *O. solandri*, extremely narrow in *O. lineata*, and in *O. nummulariifolia* the small leaves are thick and rounded and packed closely on the stems. Most species, however, have typical flat leaves usually with entire margins. In *O. macrodonta* and *O. ilicifolia* the leaves are coarsely toothed and holly-like. In *O. argyrophylla* they may be over 15cm in length but in most species they are much

smaller, for example 5cm in *A. allomii*.

The flower heads are of typical daisy form. In *O. colensoi, O. forsteri* and *O. traversii*, outer ray florets are absent. Flower color is usually white or yellowish in *O. gunniana* pink and blue forms are known and in *O. semi-dentata* the typical form is purple. The latter is one of the largest-flowered species, with flower heads as much as 5cm in diameter. Most of the above and many other species are grown in gardens as ornamental shrubs.
COMPOSITAE, about 100 species.

Olive the common name for *Olea europaea, widely cultivated in areas with a Mediter-ranean climate. Trees may survive for 1 500 years or more and are among the oldest trees known in Europe; it is thought that some of the clones in existence today have survived from Roman times.

Olives are hand picked when straw-colored for green table olives and when black and ripe for table use, cooking or more usually for oil. Olives are inedible as gathered from the trees, and they are subjected to various treatments such as steeping in a potash or salt solution which gives rise to a lactic acid fermentation, and they are then preserved in brine. Black olives may be stored direct in a brine solution but the best-quality table ones are preserved in a marinade of olive oil, often with various herbs such as species of thyme, rosemary etc. Green olives are often sold stoned and stuffed with sweet red pepper, anchovy or almonds.

The main producers of olive oil are Spain, Italy and Greece. The finest oil is obtained from the first cold pressing and is of delicate flavor (virgin oil); subsequent pressings are made under heat and are of poorer quality. The press cake formed may be treated with solvents to give a final extract of oil and the residue can be processed for use as cattle cake, fertilizer or in soap manufacture. Apart from a major use in garnishing salads, olive oil is widely used for medicinal purposes, for canning sardines and other preserves, in dressing wool, and in soap manufacture and cosmetics. Olive wood is decorative and can be finely worked.

The name OLIVE is also given to many

The spectacular Oncidium pulvinatum *is one of the many orchid species that grow in the Brazilian rain forest.* ($\times 1$)

Cumin (Cuminum cyminum) sown in between olive trees (Olea europaea) in the Middle East.

unrelated species such as *Canarium album* [CHINESE OLIVE], *Elaeocarpus serratus* [CEYLON OLIVE], *Sterculia foetida* [JAVA OLIVE] and *Dodonaea viscosa* [SAND OLIVE] to name but a few.

Olpidium a widespread genus of plant-parasitic chytrid fungi. *O. brassicae* in the roots of crucifers is thought to carry certain plant viruses such as tobacco necrosis virus. At maturity, the total protoplasm of the thallus is converted into one spherical, ellipsoid or cylindrical zoosporangium, one to several in a host cell. A discharge tube usually penetrates the host cell wall to the exterior. The released zoospores reinfect other plants.
OLPIDIACEAE, 25–30 species.

Omphalodes a genus of annual, biennial or perennial herbs found in damp shady places in mountain crevices or maritime rocks and sands through southern Europe, North Africa, southwestern Asia and Mexico. The flowers, which are solitary and axillary or in terminal cymes, are blue or white in color with a short tube and five prominent scales at the throat. Several species are cultivated, particularly the perennial *O. verna* [BLUE EYED MARY, VENUS NAVELWORT, CREEPING FORGET-ME-NOT], a popular plant of cottage gardens, and *O. cappodocica*.
BORAGINACEAE, about 24 species.

Oncidium a large tropical Central and South American genus of epiphytic pseudo-bulbous orchids all with markedly similar flowers, both in shape and color pattern, but of considerable vegetative variability. The basic flower color is bright canary-yellow for the petals and sepals, while their bases are a unique combination of the basic yellow, overlain, speckled, blotched or dotted with shades of brown. The plants can be upright, pendulous or scrambling, generally with a long strap-shaped leaf arising from each pseudobulb. Densely mottled blackish-green leaves are characteristic of plants such as *O. krameranum* [BUTTERFLY ORCHID], from the damper and warmer regions from Costa Rica to Ecuador. Species such as the West Indian *O. altissimum* and *O. papilio* (also BUTTERFLY ORCHID), from Venezuela to Peru and Brazil, are widely grown and hybridized. Other attractive cultivated species include *O. cheirophorum* [COLOMBIA BUTTERCUP], from Costa Rica to Panama, and the Brazilian *O. flexuosum* [DANCING-DOLL ORCHID].
ORCHIDACEAE, about 400 species.

Onion the common name for species of *Allium*, the best-known of which is *A. cepa* [COMMON ONION], grown throughout the world as an annual crop, although it is ordinarily a biennial. The COMMON ONION probably originated in Central Asia, possibly Afghanistan, and there are references to its cultivation in the Middle East dating back at least 3 000 years.

With the spread of onion culture, numerous cultivars (such as the *shallots) have appeared, differing in size, shape, color, flavor, adaptability in bulb formation in relation to temperature and day-length, keeping qualities, pungency and resistance to pathogenic diseases such as *Botrytis* neck-rot and leafblight, *Sclerotium* white rot, and onion yellow dwarf virus. World production of onions is approximately 12 million tonnes, principal producers being the USA, Spain, Japan, Turkey, Italy and Egypt.

The COMMON FIELD ONION is propagated from seeds, but other races are propagated as sets, ie small onions arrested in their development from seed, ripened off and planted out the following spring as multipliers where the bulb can be divided into separate plantable portions. The onion bulb consists of a mass of swollen leaf-bases. They form by the lateral expansion of leaf-base cells and the subsequent differentiation of swollen and bladeless bulb scales instead of bladed leaves. Compared to other fresh vegetables, onions are high in digestible carbohydrate and intermediate in protein content. When onion tissue is wounded, an enzyme reaction releases sulfur-containing volatile compounds that give onions their characteristic flavor and lachrymatory properties. This reaction takes some time and treatment which rapidly disrupts enzyme proteins (eg boiling) will prevent the development of full flavor.

Another widely cultivated species, also Asian in origin is *A. fistulosum* [WELSH ONION, JAPANESE BUNCHING ONION], which is leafier and has a softer bulb; the long leaf-bases form the edible part. It is the principal garden onion of China and Japan. (See also CHIVES, GARLIC, LEEK and SHALLOT.)
LILIACEAE.

Onions (Allium cepa) growing commercially in Lincolnshire, England.

Oil Crops

Economically important vegetable oils form two main groups: the essential volatile oils (see p.132) and the fixed or fatty oils. It is probable that all seeds contain some fixed oil but only about a dozen contain sufficiently large quantities of oil of a suitable type, and are cultivated easily enough to be major crops (see accompanying table). The fixed oils are divided into: (a) drying oils which are used in the manufacture of paints and varnishes; (b) semidrying oils used for foods and in soap manufacture; and (c) nondrying oils which remain liquid at ordinary temperatures and are used for edible purposes.

In organized industrial situations vegetable oil is extracted from plant material by hydraulic and screw presses or solvent extraction systems, but throughout the tropics there are many local simple methods of extraction. The residue remaining after oil extraction (oilcake) is rich in protein and is therefore employed in the manufacture of animal foods. In certain cases the residues contain toxic factors and then the material can only be used as a

Major oil-yielding plants: 1 Castor Oil; 2 Coconut; 3 Sesame; 4 Olive; 5 Oil Palm; 6 Sunflower. Fruits: 1, 3, 4, 5, 6 ($\times\frac{1}{2}$); 2 ($\times\frac{1}{4}$). Fruiting head: 5 ($\times\frac{1}{6}$). Shoots: 1 ($\times\frac{1}{15}$); 3 ($\times\frac{1}{6}$); 5 ($\times\frac{1}{120}$); 6 ($\times\frac{1}{8}$).

fertilizer. Although oil has been extracted from the seeds of many species, there are only about twelve major oil crops which, in the main, are cultivated in the tropics.

Oils used mainly as cooking and salad oils and in margarine manufacture include those from the *SUNFLOWER (Helianthus annuus). *MAIZE or CORN (Zea mays), GROUND NUT (Arachis hypogaea), *SOYBEAN (*Glycine max), *OLIVE (Olea europaea), and SESAME (*Sesame indicum). *Rape oil (Brassica campestris and B. napa) is also used in margarine manufacture and in industry as a lubricant. *Castor oil (*Ricinus communis), in addition to its medicinal properties, is largely used in the

OIL CROPS

Common name	Scientific name	Family	Common name	Scientific name	Family
I Major oil crops			CASHEW NUT KERNEL OIL	*Anacardium occidentale	Anacardiaceae
*SUNFLOWER OIL	Helianthus annuus	Compositae	*CHAULMOOGRA OIL	Hydnocarpus wightiana H. anthelmintica Taraktogenos kurzii	Flacourtiaceae
*RAPE OIL	*Brassica campestris, (B. rapa) B. napus	Cruciferae			
*CASTOR OIL	Ricinus communis	Euphorbiaceae	*COCOA or CACAO BUTTER OIL	Theobroma cacao	Sterculiaceae
*MAIZE or CORN OIL	Zea mays	Gramineae	*COLZA	Brassica rapa B. napus	Cruciferae
PEANUT, GROUND NUT OIL	*Arachis hypogaea	Leguminosae			
*SOYBEAN, SOYA OIL	Glycine max	Leguminosae	CROTON OIL	*Croton tiglium	Euphorbi- aceae
*LINSEED OIL	Linum usitatissimum	Linaceae	ENG OIL	Dipterocarpus tuberculatus	Dipterocarp- aceae
*COTTON SEED OIL	Gossypium species	Malvaceae			
*OLIVE OIL	Olea europaea	Oleaceae	GORLI OIL	Oncoba echinata	Flacourtiaceae
*COCONUT OIL (COPRA)	Cocos nucifera	Palmae	GURJUM, GURJUN OIL	*Dipterocarpus lamellantus D. turbinatus D. alatus	Dipterocarp- aceae
PALM OIL	*Elaeis guineensis	Palmae			
AMERICAN PALM OIL	E. oleifera	Palmae	HEMPSEED OIL	*Cannabis sativa	Moraceae
SESAME OIL	*Sesamum indicum	Pedaliaceae	ILLIPE NUT OIL	*Shorea macrophylla S. stenoptera	Dipterocarp- aceae
II Minor oil crops			*KAPOK SEED OIL	Ceiba pentandra	Bombacaceae
BABASSU OIL	*Orbignya barbosiana (O. speciosa)	Palmae	*MACASSAR OIL	Schleichera oleosa	Sapindaceae
			MAW SEED OIL	*Papaver somniferum	Papaveraceae
COHUNE OIL	O. cohune	Palmae	*MUSTARD OIL	Sinapis alba Brassica nigra B. juncea	Cruciferae
BEN OIL	*Moringa pterygosperma (M. oleifera)	Moringaceae			
			NIGER SEED OIL	*Guizotia abyssinica	Compositae
*BRAZIL NUT OIL	Bertholletia excelsa	Lecythidaceae	OITICICA OIL	Licania rigida	Chrysobalan- aceae
CANDLENUT	*Aleurites moluccana	Euphorbiaceae			
CHINA WOOD OIL	A. fordii	Euphorbiaceae	*PEACH OIL	Prunus persica	Rosaceae
TUNG OIL	A. montana	Euphorbiaceae	PERILLA OIL	*Perilla frutescens (P. ocymoides)	Labiatae
JAPAN WOOD OIL	A. cordata	Euphorbiaceae			
CARAPA, ANDIROBA	Carapa guianensis, C. procera	Meliaceae	PISTACHIO OIL	*Pistacia vera	Anacardiaceae
			*WALNUT OIL	Juglans regia	Juglandaceae

Two major oil crops in mass cultivation.
Above Olive groves (Olea europaea) *near Toledo,*
Spain. The oil, which is extracted from the fruit wall
and seed kernel, is mainly used for cooking, salad
dressings, soap manufacture and in food preservation.
Right Sunflowers (Helianthus annuus), *the seeds of*
which yield oil that is used in cooking and in the
manufacture of margarine.

preparation of paints, enamels, soap and
lubricants.

*Linseed oil (*Linum usitatissimum*) is the
most important of the drying oils and is
used in the manufacture of paints and
varnishes. *Cotton oil (*Gossypium* species)
is a by-product of commercial cotton
production for fiber and is used in
foodstuffs. The dried kernel or copra of
Cocos nucifera [*COCONUT] is one of the
major sources of vegetable oil in world
trade. Fruit of the West African *Elaeis
guineensis* [OIL PALM] yields two oils: palm
oil from the mesocarp and palm kernel oil
from the kernel. Both are used in food and
soap manufacture. Some other
commercially valuable minor oil crop
species are listed in the accompanying
table.

Onobrychis a genus of herbs or spiny sub-shrubs from Europe, North Africa and western Asia. They have tufts of odd-pinnate leaves and racemes or spikes of pink, purple or white, pea-like flowers, and are an important constituent of grazing pastures and poor vegetation in southeastern Europe. *O. vicii-folia* [SAINFOIN, ESPARCET or HOLY CLOVER] is used as a fodder crop, although its use has declined with the introduction of improved strains of *CLOVER and *LUCERNE. Some species including *O. laconica* and *O. radiata* are grown as garden ornamentals.
LEGUMINOSAE, about 60 species.

Onoclea a genus represented by a single species of fern, *O. sensibilis* [SENSITIVE FERN], native to northern Asia and North America, and cultivated in gardens for its two kinds of foliage: deeply pinnatifid sterile leaves up to 1m long and the smaller erect bipinnate fertile fronds with the pinnules rolled up into bead-like segments which open to discharge the spores.
ASPIDIACEAE, 1 species.

Ononis a genus of annual, biennial and perennial herbs and subshrubs mainly from Europe and the Mediterranean region. Most have pink or purplish pea-like flowers. They will thrive in the poorest stony and sandy soils and species like *O. spinosa* [SPINY REST-HARROW] were at one time a serious weed of arable land. Two attractive garden sub-shrubs are *O. rotundifolia* and *O. fruticosa*, with pink flowers and grayish foliage.
LEGUMINOSAE, about 70 species.

Onopordum a genus of annual and biennial herbs native to Europe, North Africa and western Asia. Several species are cultivated in garden borders, such as *O. acanthium* [COMMON COTTON or SCOTCH THISTLE] and

O. tauricum (= *O. virens*). They are tall, erect plants, up to 3m high, with silvery jagged leaves and reddish purple or blue, thistle-like flower heads.
COMPOSITAE, about 40 species.

Onosma a genus of evergreen annual biennial or perennial herbs or subshrubs with rough bristly hairs, native from the Mediterranean region to the Himalayas and China. They have tubular, pendent flowers and usually gray-green leaves. Of several cultivated species *O. alboroseum*, with white flowers turning deep red then bluish, *O. tauricum*, with yellow flowers, and *O. echioides*, with pale yellow flowers, are among those grown in rock gardens.
BORAGINACEAE, about 150 species.

Oomycetes see Fungi, p.152.

Ophioglossum a cosmopolitan genus of primitive terrestrial ferns, the most common species, including *O. vulgatum* and *O.* TONGUE]. In *Ophioglossum* the stem is short with a sterile frond which is normally ovate although in some species, such as the epiphyte *O. pendulum*, the sterile frond may be elongated into a long strap-like hanging leaf. The fertile frond in this genus is of interest and appears as a double row of large sunken sporangia and usually arises from the junction of a sterile blade and its petiole. Several species, including *O. vulgatum* and *O. pendulum*, are cultivated for their attractive foliage.

Local medicinal uses of members of the genus include a cough remedy (*O. pendulum*) and an aphrodisiac (*O. spicatum*). *O. ovatum* and *O. reticulatum* are used as fuel in Madagascar and parts of the Moluccas respectively.
OPHIOGLOSSACEAE, 30–50 species.

The Restharrow (Ononis repens) *is a widespread weed of pastures and coastal areas, with attractive pea-like flowers.* ($\times \frac{1}{3}$)

Ophiopogon a small genus of perennial stemless herbs with grass-like leaves, native from India to Japan, including the East Indies. A number of species are cultivated for their attractive racemes of white or bluish flowers followed by conspicuous berries. They include *O. intermedius* (= *O. spicatus*), *O. jaburan*, with various cultivars, and *O. japonicus* [LILYTURF, MONDO GRASS]. The tubers of the latter species are consumed as a vegetable in Japan.
LILIACEAE, 10–20 species.

Ophrys a small genus of herbaceous perennial orchids with ovoid tubers, native to Europe, North Africa and East Asia. The flowers are similar to those of *Orchis, but are without a spur. They often bear a striking resemblance to insects, and some species are in fact pollinated by attracting male insects which try to copulate with the flowers, a phenomenon described as pseudocopulation.

O. apifera [BEE ORCHID], is a particularly attractive species about 30–40cm tall with few flowers. The sepals are pink, the petals greenish, and the brown labellum is curiously marked and very hairy. It is very widely distributed but always rare. *O. insecti-fera* [FLY ORCHID] has rather small flowers showing a striking resemblance to a fly. *O. speculum* [MIRROR ORCHID, MIRROR OF VENUS] is a common, attractive Mediterranean orchid with a labellum of reflective brilliant metallic-blue surrounded by gold (hence the common name).
ORCHIDACEAE, about 30 species.

Opium poppy the common name for *Papaver somniferum*, an erect, glaucous annual with coarsely lobed or toothed leaves, sometimes spiny, and large showy flowers with white, red or purple petals which have a dark spot at the base.

The OPIUM POPPY probably originated as a cultivated plant in Turkey and has been

In the orchid genus Ophrys, *the flowers of different species resemble female insects of the particular species involved in pollination. Shown here are:*

Left the Fly Orchid (O. insectifera), *from Europe* ($\times 2$) *and* Right *the* Late Spider Orchid (O. fuciflora), *from central and southern Europe.* ($\times 4$)

cultivated since ancient times in much of Europe and Asia for its medicinal narcotic latex, opium. The seeds (sometimes called maw seeds), which contain no alkaloids, are used as a condiment and in baking, often being sprinkled on bread or confectionery. They are also the source of a drying oil used in the manufacture of paints, varnishes and soaps. The cultivars used for seeds belong to the subspecies *hortense*, which more recently has gained popularity as a garden plant.

Today the main areas of opium production (obtained from subspecies *somniferum*) are the Middle East, Turkey, India, China and the Balkan peninsula. Latex is extracted from the capsules by making incisions with a special knife. The latex exudes from the cuts as a white fluid and rapidly coagulates, turning brown. The following day the coagulated drops of latex are scraped off the capsule and then dried or mixed with water and boiled till thick. They are then kneaded into balls of crude opium which can be kept for years.

Opium contains over 25 alkaloids, and also oils and resins. Morphine, the most important alkaloid, is a powerful and medically important analgesic and narcotic. Opium, however, poses very serious problems of addiction, leading to socially harmful effects and often mental and physical deterioration and breakdown. This has led to restrictions on its cultivation but with only limited success.
PAPAVERACEAE.

Opuntia a large genus of prostrate to tree-like cacti native from temperate North America to the southern tip of South America. Most opuntias have at least some of their stems jointed. The shape of the stems may be obovate, cylindrical, globular or flattened into round pads, and the leaves are cylindrical to conical, usually small and deciduous.

Species like Opuntia basilaris, *seen here growing in Death Valley, California, can survive temperatures as high as 70°C.* (× $\frac{1}{3}$)

Opuntias are stem-succulents – their leaves are reduced to spines and the role of photosynthesis is taken over by the fleshy stems. (× $\frac{1}{5}$)

Most species bear spines and all bear tufts of barbed bristles (glochids) on each areole. The flowers are usually large and spreading, with yellow the predominant color, although orange, white and purple forms occur.

Economically *Opuntia* has a mixed reputation. As a weed, certain species invaded Australia and by 1920 dominated considerable areas of agricultural land. Parts of South Africa became similarly infested. However, in frost-free countries, species of *Opuntia* can be grown as hedges, and African farming communities commonly plant a few as fodder reserves in times of drought (the spines are cut off or burnt). *O. ficus-indica* [INDIAN FIG] is a widespread species now extensively cultivated commercially in the tropics, subtropics and Mediterranean countries for its large, juicy fruits which are eaten raw after peeling, or made into jam, jelly or various confections. The young stems are diced after removal of the spines, and eaten in salads or cooked as a vegetable. The reticulate woody skeletons of the cylindrical-stemmed CHOLLAS are used for making trinkets and rustic ornaments.

The genus has been divided into four subgenera: *Brasiliopuntia*, *Consolea*, *Cylindropuntia* and *Opuntia*, on the basis of vegetative, floral and fruit characters. *Brasiliopuntia* contains species with a tree-like habit with unjointed primary stems. The jointed branches close to the main stems are cylindrical, those further away are flattened. Seeds are solitary or few in number and have a woolly texture. These features are to be seen in such species as the Brazilian *O. brasiliensis* which grows to a height of about 6m and bears yellow flowers 5cm in diameter and subglobose yellow fruits.

Consolea also consists of mostly tree-like species with cylindrical stems bearing oblong or flattened joints. The seeds bear hairs on the sides only. Members of this subgenus are native to Florida and the West Indies and include the cultivated *O. rubescens*, which grows to about 6.5m, the joints bearing close-set areoles but few or no spines. The flowers are about 2cm across, yellow to red in color and give rise to obovoid to globose reddish fruits.

Cylindropuntia contains prostrate to tree-like species with globose to cylindrical jointed or unjointed stems, smooth or tubercled. The seeds are glabrous. This subgenus is further subdivided into five sections, of which *Austrocylindropuntia*, *Cylindraceae* [CHOLLA], and *Grusonia* contain mostly erect species. Species in *Corynopuntia* have a prostrate and spreading habit and those of *Tephrocactus* are clump-forming with stems having globose to short-cylindrical joints.

Austrocylindropuntia includes such cultivated species as *O. cylindrica* [DANE CACTUS], from Ecuador and Peru, *O. subulata* [EVE'S PIN CACTUS], from Argentina, and the Bolivian

Opuntia leucotricha *is a tree-like species with brilliant yellow flowers and white to red edible fruits.* (× $\frac{1}{4}$)

attractive ornamentals, including flat-padded species such as *O. leucotricha*, from Mexico, and *O. erinacea* var *ursina* [GRIZZLY BEAR CACTUS], from the southwest USA, with white curling bristles, and *O. scheeri* with golden-yellow bristles. Especially commendable for miniature bowls and window sill culture are certain dwarf cultivars of the Mexican *O. microdasys* [RABBIT EARS, BUNNY EARS], with tiny pads covered in soft white-yellow bristles that can be handled without discomfort.

CACTACEAE, about 300 species.

Orach or **orache** the common name for *Atriplex hortensis*, an annual herb which grows to a height of 2m, with angled, toothed leaves. It is native to Asia, and is sometimes cultivated in Europe and North America for the leaves and young stems which are used as a vegetable. Cultivars with red or coppery tints are grown as ornamentals.
CHENOPODIACEAE.

Orange or **sweet orange** the fruit of *Citrus sinensis*, by far the most important commercial citrus fruit. It originated in northeastern India and adjacent regions of China and is now grown in subtropical and tropical zones all over the world, principally in the Mediterranean region (Spain, Italy, Israel), Florida and California (USA), Brazil and South Africa. World production of oranges represents over 82% of the world's total production of citrus fruit.

All sweet orange varieties can be eaten fresh or pressed for their juice, but eating oranges of high quality must be easy to peel and juicing oranges must have tough membranes and adherent peel so that debris does not easily pass into the juice. Typical examples of eating and juicing oranges are 'Shamouti' and 'Valencia Late' respectively. Most of the world's crop is eaten fresh, but over 40% of the US crop is pressed.

More than 160 orange cultivars have been described, but only a few are commercially important. Oranges can be classified into four main groups.

O. vestita [OLD MAN OPUNTIA, COTTON POLE CACTUS], whose areoles bear long white hairs almost covering the stem. *O. clavarioides* [CRESTED OPUNTIA, SEA CORAL, FAIRY CASTLES, GNOME'S THRONE, BLACKFINGERS], possibly Chilean in origin, is most often grafted on another *Opuntia* or *Cereus*. *O. salmiana*, from Brazil, Paraguay and Argentina, with its freely branching habit and sprawling cylindrical stems which bear yellow bristles, short spines and numerous white flowers, is an attractive houseplant.

Most of the *Cylindraceae* are North American in origin but a number are cultivated as ornamentals there and elsewhere. Their primary stems are normally unjointed but they bear cylindrical, jointed branches. Cultivated species include *O. acanthocarpa*, a densely spiny shrub with laterally compressed tubercles, straw-colored spines and red to yellow flowers, *O. cholla*, with sheathed brownish spines and large pink flowers, *O. bigelovii* [TEDDY-BEAR CACTUS], with a short erect trunk bearing tubercles hidden by a dense mat of pale yellow spines which later turn black. The flowers are greenish yellow and about 4cm across. Other species in this group include *O. tunicata*, *O. prolifera* [JUMPING CHOLLA] and *O. ramosissima* [PENCIL CACTUS].

Species in the section *Grusonia* have erect stems, more or less divided into cylindrical joints and the tubercles are confluent into ribs. They are predominantly north Mexican in origin.

Corynopuntia species are mainly native to

On the Canary Island of Fuerteventura, the Indian Fig (Opuntia ficus-indica) is commonly used as a hedging or barrier plant.

Mexico and the southwestern USA. The better-known cultivated species include *O. schottii* [DEVIL CACTUS], with yellow flowers, and slender brownish-gray spines tinged pink or red, *O. vilis* [LITTLE TREE OPUNTIA, MEXICAN DWARF TREE CACTUS], with purple flowers and club-shaped joints bearing white to red spines, and *O. stanlyi* with yellow flowers and short club-shaped joints bearing flattened brown or reddish spines.

The clump-forming species of *Tephrocactus* include the Argentinian *O. articulata*, with many cultivars such as 'Inermis' [SPRUCE-CONE], with few spines (sometimes none), and 'Syringacantha' [PAPER CACTUS, PAPER-SPINED PEAR], whose brownish spines are papery in texture. Other cultivated species in this section include *O. sphaerica* [THIMBLE CACTUS], native to the Andes of Peru and Chile, with broad, low tubercles, areoles brown and woolly with spines brown becoming gray, and orange flowers (violet in cultivar 'Violaciflora', and *O. floccosa* [CUSHION CACTUS, WOOLLY SHEEP], native to the Andes of Peru and Bolivia, the oblong joints of which have areoles covered with long white hairs that almost conceal the stems.

The largest number of cultivated species in this genus belong to the subgenus *Opuntia* [PRICKLY PEAR, TUNA]. *O. ficus-indica* belongs to this group as well as a large number of

The fruit of the Sweet Orange (Citrus sinensis) is known botanically as a hesperidium; its flesh is rich in vitamin C.

The Bird's Nest Orchid (Neottia nidus-avis) contains no chlorophyll and relies for nutrition on its symbiotic association with fungi. (× ¼)

I. COMMON ORANGES. A group including 'Bernia' ('Vernia'), an important late Spanish variety of fair quality, also grown in North Africa, 'Biondo' ('Comuna'), the common name for several seedy, local varieties, still widespread in Italy and Spain, 'Ovale', the oval, seedless Italian cultivar of high standard and 'Shamouti' (better known as Jaffa orange) the early to mid-season variety of Israel. 'Valencia Late' (including Jaffa late) is the most important seedless late orange cultivar grown in both hemispheres, accounting for probably more than 15% of the total world orange production.

II. SUGAR or ACIDLESS ORANGES. This group includes local cultivars (eg 'De Nice') insipid in taste and of minor importance.

III. PIGMENTED (BLOOD) ORANGES. This group owes its pigmentation to antho-cyanins dissolved in the cell sap. Its cultivars succeed well in cool areas (Italy and Spain), where the color becomes deep wine red. Important cultivars are 'Moro' and 'Tarocco' (Italy); 'Sanguinilla Negra' (Spain); 'Ruby' (California).

IV. NAVEL ORANGES. The name derives from a small rudimentary fruitlet embedded at the stylar end of the main fruit. It is not an unusual feature in other species, but here it becomes recognized as a "trade mark" of this group which originated as sport of a common cultivar. 'Washington' ('Bahia'), which originated in Brazil, is now one of two main commercial oranges of California. RUTACEAE.

Orbignya a South American genus of slow-growing slender palms with dense crowns of large pinnate leaves. Although widely applied to the genus, the common name BAB-ASSÚ PALM refers in fact only to *O. barbosiana*

(= *O. speciosa*). This species is very produc-tive but the seeds have a very hard endocarp which creates problems in commercial ex-traction. However, when processed they yield a valuable vegetable oil (babassu oil) which is used primarily in margarine pro-duction and also in the local soap industry. *O. cohune* [COHUNE PALM], of Central America, is also a source of kernel oil. PALMAE, 25 species.

Orchid a term applied to members of the family Orchidaceae, many of which are culti-vated as garden, house or greenhouse plants or for cut-flowers. Most orchid greenhouse plants are of tropical origin; a few are terrestrial (ie ground-inhabiting) but most are epiphytes growing naturally on the bark of trees. In greenhouse cultivation the epi-phytic orchids can be grown successfully in pots containing substantial amounts of peat or other natural fibrous materials.

Under natural conditions, development of orchids grown from seed depends on the infection of the root with specific fungi and the establishment of a symbiotic relationship. Commercially orchids are vegetatively propagated by division of the rhizome (which may be pseudobulbous), by cuttings, by apical meristem culture, or by germination from seeds, which are extremely small and

The Lady Orchid (Orchis purpurea); each flower has a large deep-purple veined lip (labellum), two back-swept lateral petals, with one of the sepals forming a hood. (× 1)

The Green-winged Orchid (Orchis morio) is a locally plentiful, though declining, species of old calcareous pastures. (× ⅓)

undifferentiated. Interspecific and inter-generic hybridization is commonly used by growers to obtain a greater richness and beauty of floral form and color. ORCHIDACEAE.

Orchis a genus of temperate terrestrial or-chids native to North America and Eurasia, usually found on open grassy slopes and in woodlands. *O. simia* [MONKEY ORCHID] and *O. militaris* [MILITARY ORCHID] are two species in which the common names are an accurate description of the flowers' appearance.

Morphologically, all species are typical of the average terrestrial temperate orchid with few- to many-flowered racemes of purple, mauve, lilac, crimson or white flowers more or less surrounded by a basal rosette of oblong-ovate to lanceolate and elliptical leaves. In some species the leaves are dark purplish-black spotted and in *O. mascula* [EARLY PURPLE ORCHID] spotted and unspotted leaf forms occur.

The Dwarf Orchid (Orchis ustulata) is also known as the Burnt Orchid since the hood is brown or dark purple. (× 1)

Species grown in damp situations in the garden include the European *O. militaris*, with large fragrant rose or red-violet flowers, the Eurasian *O. morio* [SALEP ORCHID, GANDER-GOOSE, GREEN-WINGED ORCHID] with purplish (rarely white) flowers, the North American *O. rotundifolia* [SMALL ROUND-LEAVED ORCHID, SPOTTED KIRTLE PINK], with white to mauve flowers, and *O. spectabilis* (= *Galeorchis spectabilis*) [SHOWY ORCHID, WOODLAND OR-CHID, PURPLE HOODED ORCHID, KIRTLE PINK], with showy, pink to mauve (rarely white) flowers.

ORCHIDACEAE, about 50 species.

Origanum a genus of perennial herbs and small shrubs native to Europe, the Mediter-ranean region and Central Asia. *O. majorana* (= *Majorana hortensis*) [SWEET MARJORAM, COMMON WILD MARJORAM] and *O. vulgare* [POT MARJORAM] are popular kitchen herbs, also known as OREGANO. Other species used as herbs include *O. compactum* and *O. glandulosum* in North Africa, *O. creticum* [SPANISH HOPS], *O. heracleoticum* and *O. onites* [POT MARJORAM] in southern Europe. *O. dic-tamnus* [DITTANY OF CRETE] is a popular ornamental white-woolly dwarf shrub with fragrant, pink, drooping flowers and reddish shiny bracts.

LABIATAE, 15–20 species.

Ornithogalum a largely temperate genus of bulbous herbs mainly native to the Mediter-ranean region and South Africa. The narrow leaves are produced in a basal rosette with a racemose or corymbose inflorescence of white, greenish-white or yellow to orange-red flowers arising from the center. Most species have white flowers with a green central stripe on each segment. *O. thyrsoides* [CHINCHERINCHEE, CHIN-CHIN], has beautiful white, long-lasting flowers in a dense raceme, and is popular as a cut-flower or pot plant.

Other cultivated species, usually known as STAR OF BETHLEHEM, include *O. arabicum*, *O. nutans*, *O. pyrenaicum* and *O. umbellatum*.

LILIACEAE, about 150 species.

The parasite Orobanche crenata, *one of the largest and most destructive of the Broomrapes, seen here completely killing its host, a Broad Bean* (Vicia faba) *plant.* (× 1/5)

The Military Orchid (Orchis militaris), *so called because of the resemblance of the hanging labellum to a soldier-like figure.* (× 1)

Orobanche [BROOMRAPE] a genus of tem-perate Eurasian and subtropical annual to perennial herbs parasitic on the roots of other flowering plants (usually herbaceous dicot-yledons), for example *O. ramosa* on cultivated plants such as *Cannabis sativa* [*HEMP], *Nicotiana tabacum* [TOBACCO], *Solanum* spe-cies etc. They contain little or no chlorophyll and thus derive all or most of their organic nutrients from their host.

OROBANCHACEAE, about 140 species.

Orris root the common name for the rhi-zome of the perennial herb *Iris* × *germanica* var *florentina* (= *I. florentina*), which is culti-vated on a commercial scale in some Mediter-ranean countries and in Iran and northern India. Orris, the dried and powdered tissue, which has the sweet scent of violets, is used in the perfumery and cosmetics industries.

IRIDACEAE.

Ortanique a hybrid derived from a chance cross in Jamaica between *Citrus sinensis* [SWEET *ORANGE] and *C. reticulata* [*TAN-GERINE, *MANDARIN]. The fruit looks like a thin-skinned orange, is juicy and has a pleasant flavor. It is now grown on a con-siderable commercial scale.

RUTACEAE.

Oryza a genus of annual, sometimes peren-nial, tropical grasses centered in the wet tropics of Asia and Africa, and characterized by their large spreading paniculate inflores-cences with flexuous branches and small, one-flowered spikelets. The important cereal *O. sativa* [*RICE] belongs to this genus.

GRAMINEAE, 15–20 species.

Osage orange the common name for *Mac-lura pomifera* (= *M. aurantiaca*) [also known as BOW WOOD], a fast-growing spiny de-ciduous tree reaching up to 18m, from the southern and southeastern USA. It is the only species of its genus. The fruits are orange-shaped, yellowish green and very

decorative, and it is grown in the USA and Europe as an ornamental tree and hedging plant. The wood is bright orange-colored, strong, hard and durable, and was formerly used by the Osage Indians and other tribes for bows and clubs. The bark yields a yellow dye.

MORACEAE.

Oscillatoria a genus of blue-green algae having simple, unbranched, unsheathed fila-ments of similar cells. The filaments are motile and often show an oscillatory move-ment, to which the name is due. Species are common in soil, freshwaters and in littoral marine habitats.

OSCILLATORIACEAE.

Osmanthus a genus of evergreen shrubs or small to medium trees native mainly to East Asia and the USA, with a few species in the Pacific area (Polynesia). They have white or yellowish flowers in axillary or terminal clusters. The fruit is a drupe, more or less oval, usually blackish-blue or violet. Species such as *O. armatus*, *O. delavayi* and *O. suavis* are useful ornamentals for their evergreen

Fruits and leaves of the Osage Orange (Maclura pomifera). (× 1/6)

habit and holly-like appearance. The flowers of *O. fragrans* [SWEET OLIVE] are used in China to flavor tea. It has long been cultivated.

OLEACEAE, 30–40 species.

Osmunda a genus of temperate and tropi-cal ferns of Europe, Asia and North and South America. They are intermediate be-tween the primitive and the more advanced ferns. The leaves are bipinnate and have a separate distal part of the frond for the fertile area, which is dark brown and reduced. The massive sporangia are borne marginally on short, thick stalks on the reduced fertile part of the frond. The most common species, *O. regalis* [ROYAL FERN], is widely cultivated as an ornamental. The young shoots (cro-ziers) of *O. cinnamomea* [CINNAMON FERN] are eaten locally.

OSMUNDACEAE, 8–10 species.

Osteospermum a genus of herbs and sub-shrubs native to southern Africa, tropical East Africa, with others in southwest Arabia

Harvesting rice (Oryza sativa) by hand in the Philippines, a method still widely employed by subsistence farmers of Southeast Asia.

and Jordan. Several are cultivated as ornamentals, such as *O. ecklonis*, a subshrub or shrub with narrow, shallowly toothed leaves and daisy-like flower heads up to 7cm across; the outer ray florets are bluish to violet below and white above, the inner disk florets blue. COMPOSITAE, about 70 species.

Ostrya [HOP HORNBEAMS] a genus of medium-sized deciduous trees with wide-spreading horizontal branches. They are found throughout the temperate regions of the Northern Hemisphere. The leaves closely resemble those of the true *HORNBEAMS (Car-

The Royal Fern (Osmunda regalis), showing the brown fertile fronds with the larger sterile fronds behind. ($\times\frac{1}{20}$)

pinus) while the fruit clusters have a hop-like appearance.

Species occasionally cultivated are *O. carpinifolia* [EUROPEAN HOP HORNBEAM], native to southern Europe and Asia Minor, and *O. virginiana* [AMERICAN HOP HORNBEAM] from the eastern USA. The wood of the AMERICAN HOP HORNBEAM is extremely hard and tough – hence its popular name of IRONWOOD – and is widely used for tool handles and fence posts.
BETULACEAE, about 10 species.

Ourisia a Southern Hemisphere genus of herbaceous, tufted or rhizomatous plants occurring in temperate forests and alpine areas of South America, New Zealand and Tasmania. The New Zealand and Tasmanian species are characterized by delicate, irregular, white, yellow or purple-centered flowers. The South American species include red, long-tubed, hummingbird-pollinated species. The white-flowered *O. macrocarpa* and *O. macrophylla*, both from New Zealand, and the red-flowered *O. coccinea* are widely cultivated.
SCROPHULARIACEAE, about 25 species.

Oxalis a very large genus of annual to perennial herbs with creeping or erect, sometimes woody stems; they are sometimes stemless, often bulbous, tuberous or rhizomatous, and rarely shrubs or subshrubs. They are widespread in temperate regions, but most abundant and diverse in South America and South Africa. The leaves are trifoliolate to many-foliolate, or sometimes reduced to one or two leaflets. The leaflets can often move in response to changes of temperature or light. The yellow or white to

pink or purple flowers are borne singly or in umbellate cymes on leafless stalks. Some species can reproduce by bulbils and may, like the South African *O. pes-caprae* (= *O. cernua*) [BERMUDA BUTTERCUP], become pernicious weeds of cultivated ground. In the tropics, subtropics and the Mediterranean area, the bulb of this species, like the tuberous stems of the high Andean *O. tuberosa* [OCA], are used as food.

Several species are cultivated as garden ornamentals, including the stemless, lilac-pink-flowered *O. adenophylla*, and the erect, much-branched, rose-flowered *O. rosea* and *O. rubra* (white-flowered in cultivar 'Alba'). *O. corniculata* [CREEPING OXALIS, CREEPING LADY'S SORREL], a profusely branched perennial is a cosmopolitan weed often found on the floor of greenhouses. The trifoliolate leaves of *O. acetosella* [CUCKOO BREAD, WOOD SORREL] are sometimes substituted for *SHAMROCK.
OXALIDACEAE, about 800 species.

Oxydendrum a genus comprising a single species, *O. arboreum* (= *Andromeda arborea*) [SOURWOOD, SORREL TREE], native to the eastern USA. It is a small deciduous tree with deeply fissured bark and bears many pendulous panicles of small, white flowers. It is widely cultivated in warmer temperate climates for its foliage which turns brilliant scarlet in autumn.
ERICACEAE, 1 species.

Oxytropis a genus native to mountainous and cold regions of North America, Asia and Europe. The species are perennial herbs, with purple, violet, white or yellow pea-like flowers, produced in spikes, racemes or borne directly on the stem. A number of dwarf species including *O. campestris*, *O. jacquinii*, and *O. pilosa* (= *Astragalus pilosus*) are cultivated as rock-garden plants.
LEGUMINOSAE, 250–300 species.

Oxalis enneaphylla in flower in Tierra del Fuego, Chile. This species is also grown in cultivation. ($\times\frac{1}{5}$)

P

Pachysandra a small genus of perennial herbs or subshrubs native to East Asia, with one species in North America. Two species are grown for ground cover: *P. procumbens* [ALLEGHANY PACHYSANDRA], with spikes of purplish-white flowers, and *P. terminalis* [JAPANESE PACHYSANDRA], with shorter spikes of greenish-white flowers, tinged purple.
BUXACEAE, about 5 species.

Paddle wood a very hard, compact, close-grained wood, derived from the tropical American tree *Aspidosperma excelsa*. It is yellowish-white, tinged with pink, and has an elasticity which makes it suitable for paddles.
APOCYNACEAE.

Paeonia a genus of perennial herbs and shrubs mainly native to the north temperate region. The genus includes a number of very popular ornamental species such as the European *P. officinalis*, which has large, showy flowers up to 14cm across, varying in color from white through pink to crimson, and single or double depending on cultivar, the European and southwest Asian *P. arietina*, with reddish-pink flowers, the Caucasian *P. mlokosewitschii* with yellow cup-shaped flowers, and the Siberian and Chinese *P. lactiflora* (= *P. albiflora*, *P. chinensis*) [CHINESE PEONY, COMMON GARDEN PEONY] and its hundreds of garden derivatives.

Of a number of attractive shrubby species ("tree peonies"), the most popular are the Asiatic *P. suffruticosa* [MOUTAN, TREE PEONY], with large bluish-purple flowers, and the Chinese *P. lutea* with fragrant yellow flowers.
PAEONIACEAE, 33 species.

Paliurus a genus of deciduous shrubs or small trees native from southern Europe to East Asia, some of which are grown as ornamentals. The stipules are usually modified as spines which in *P. spina-christi* (= *P. aculeatus*) [CHRIST'S THORN, JERUSALEM TREE] are either straight or bent backwards. The flowers are small, greenish-yellow, borne in axillary clusters or terminal panicles.
RHAMNACEAE, 6–8 species.

Palmetto the common name given to palms belonging to the genus *Sabal*, native to warm America and the West Indies. Many are cultivated, eg *S. minor* [DWARF, SCRUB or BUSH PALMETTO], *S. mexicana* (= *S. texana*) [TEXAN PALMETTO] but the most useful is *S. palmetto* [CABBAGE TREE, BLUE, CABBAGE or COMMON PALMETTO], whose leaves are used for thatching houses and supply an edible palm cabbage.
PALMAE, about 25 species.

Panax a small genus of perennial herbs mainly native to East Asia and North America. The best-known species is *P. pseudoginseng* (= *P. ginseng*) [GINSENG], which is cultivated for its rhizomes and fleshy roots, especially in Manchuria and Korea which is the main exporter. The North American *P. quinquefolius* [AMERICAN GINSENG] is also cultivated for its roots and rhizomes.

The dried and powdered rhizomes and roots of these species have long been esteemed, particularly in Chinese medicine, for their restorative powers. Ginseng is gain-

The fragrant flowers of Pancratium maritimum, *a Mediterranean species known as the Sea Lily or Sea Daffodil.* ($\times \frac{1}{3}$)

*Modern Garden Pansies (*Viola × wittrockiana*) are all highly complex hybrids derived principally from* Viola tricolor *(Heartsease) and* V. lutea. ($\times \frac{1}{2}$)

ing increasingly in popularity as a stimulant in the Western world, where it is marketed under various trade names. The active principles are glycosides, one of which is panaquillon.

The original wild ginseng root (known locally as zansami or man root) was first found in the Korean mountains some 5 000 years ago. In Korea, intensive botanical research, coupled with selection of roots of the appropriate age (about six years) and careful quality control, has led to the world-wide marketing of a superior product. Ginseng may be consumed from the powdered root, compressed into a palatable tablet, as a tea-like infusion, or as an elixir in an aqueous alcohol base.
ARALIACEAE, about 6 species.

Pancratium a genus of bulbous perennial herbs mainly from the Mediterranean region to tropical Asia and tropical Africa. Several species are cultivated for their white flowers but are scarcely hardy. *P. canariense* from the Canary Islands is 60cm tall with flowers 10cm across; *P. illyricum* is shorter with very fragrant flowers, and *P. maritimum*, from the Mediterranean also has very fragrant flowers. *P. zeylanicum* is only 30cm tall, with solitary flowers.
AMARYLLIDACEAE, about 15 species.

Pandanus a large genus of trees and shrubs from the Old World tropics. The common name of SCREW PINES given to some species derives from the fact that the branched stems are often twisted and bear cone-like fruits that resemble *pineapples. *Pandanus* species have aerial prop roots and are conspicuous trees growing to a height of 20m or more.

The best-known species is *P. odoratissimus* [PANDANG, BREADFRUIT, THATCHSCREW PINE], which is found on sea coasts from Southeast Asia to the Polynesian islands, and many cultivated forms are known. They make handsome greenhouse plants, as do other species, including *P. vertchii*, *P. baptistii* and *P. candelabrum* [CANDELABRA or CHANDELIER TREE]. In addition to their ornamental value,

certain species such as *P. furcatus* (Bengal to Burma), *P. odoratissimus* and *P. utilis* (Madagascar) are valued for their leaves which are woven into hats, floor coverings, baskets etc. The fruits of some species are soft, sweet and edible.
PANDANACEAE, about 650 species.

Pandorea a genus of woody climbing shrubs native to Australia and Malaysia, two of which are cultivated as ornamentals. *P. pandorana* (= *Bignonia australis*) [WONGA-WONGA VINE] produces panicles of numerous pinkish-white flowers and *P. jasminoides* (= *B. jasminoides*) [BOWER PLANT] has much larger pinkish-white flowers in fewer-flowered panicles.
BIGNONIACEAE, about 10 species.

Panicum a large genus of tropical and warm temperate annual and perennial grasses. Many of these species are termed *millets, and as such are of economic importance, especially in southern Europe and India, as human food and animal fodder. *P. miliaceum* [COMMON OR PROSO MILLET] is the true millet of history and is still grown in eastern and southern Asia, and in the USA. The grain is used in the same way as *rice, or as porridge, and as flour. The leaves and stems are also used for fodder. *P. sumatrense* [LITTLE MILLET] is grown extensively in India as a substitute for rice, for flour and for animal fodder. *P. maximum* [GUINEA GRASS] is a 3m high perennial grown in tropical Africa, India and America as one of the best fodder grasses of the tropics. *P. molle* [MAURITIUS GRASS] and *P. purpurascens* [PARA GRASS] are also important fodder plants. A few species are grown for ornamental purposes, including *P. virgatum* [SWITCH GRASS]. In contrast *P. repens* is one of the most troublesome weeds of tea plantations in Asia.
GRAMINEAE, about 500 species.

Paphiopedilum rothschildianum. Seen here are two of the three striped sepals, one of the dotted lateral petals and the pouch-shaped, red-colored dorsal petal or labellum. (× ⅓)

Bud of a Corn Poppy (Papaver rhoeas), showing the two protective sepals still folded round the developing flower. (× 1½)

Pansy the common name for a group of species, varieties and hybrids of the genus *Viola*, favorites of gardeners for centuries. PANSIES have many common names, such as HEARTSEASE, LOVE-IN-IDLENESS, FLOWER OF LOVE, CUDDLE-ME-TO-YOU, HERB TRINITY, THREE FACES UNDER A HOOD. The GARDEN PANSY is *V. × wittrockiana*, a hybrid of mixed origin, derived from *V. tricolor* [TRICOLOR PANSY] and including *V. lutea* [YELLOW MOUNTAIN PANSY], and possibly *V. altaica*.
VIOLACEAE.

Papaver a genus of mainly annual and a few perennial herbs mainly native to Europe, Asia and other parts of the Old World, but with a few in western North America. The flowers are borne solitary on leafless stalks as in *P. nudicaule* [ICELAND POPPY, ARCTIC POPPY] or on leafy branching stems as in *P. rhoeas* [FLANDERS, COMMON or CORN POPPY], *P. glaucum* [TULIP POPPY] and *P. somniferum* [*OPIUM POPPY]. Although many species, such as *P. rhoeas* and *P. dubium*, have flowers with red petals, the numerous cultivated varieties of other species, such as *P. nudicaule*, *P. alpinum* and *P. orientale* [ORIENTAL POPPY], have petals in a range of colors, including white, yellow, orange, red, crimson and purple. The fruit is a capsule opening by pores immediately below the persistent stigmatic disk. (See also OPIUM POPPY.)

The SHIRLEY POPPIES are a strain of *P. rhoeas* with white and pink, single or double flowers. Pigments from these have been used to color medicines and wine.
PAPAVERACEAE, about 100 species.

Papaw, pawpaw, papaya or **melon tree** common names for *Carica papaya*, a sparsely branched small tree perhaps originating in Mexico or Costa Rica, though now cultivated throughout the tropics for its delicious fruits, which may weigh up to 9kg. The fresh fruit is eaten with lemon or lime juice or in fruit salad; it may be tinned, crystallized or made into jam, ice cream, jellies, pies or pickles; when unripe, it is used like marrow or apple sauce. The young leaves of *C. papaya* are

sometimes eaten as a vegetable and the seeds used as a vermifuge, counter-irritant and abortifacient.

The green fruit is scarified for the latex which contains papain, a proteolytic enzyme much in demand as a tenderizer for meat, and in the manufacture of chewing gum and cosmetics, in tanning, in degumming silk and in giving shrink-resistance to wool. Papain is widely produced in tropical Africa, notably Tanzania and Uganda and in Sri Lanka, although most of the exported latex comes from Hawaii.

Asimina triloba [AMERICAN PAWPAW], which belongs to the family Annonaceae, is a small tree of the eastern USA. It bears purple flowers and small yellow-green pulpy fruits which are a local delicacy.

Paphiopedilum a genus of tropical and subtropical, mostly terrestrial or epiphytic Asiatic orchids without pseudobulbs and closely related to the temperate northern hemisphere genus *Cypripedium* [LADY'S SLIPPER ORCHIDS]. They are tufted plants bearing strap-shaped to ovate leaves. The flowers are borne on upright stems, and colors vary from white, yellows, greens and browns to violet, deep crimson to purple. As well as being streaked and shaded in various colors, the flowers are further distinguished by their hairy margins and deep red or green shining warts.

Paphiopedilums have been extensively hybridized to produce larger and larger, more

A Papaw tree (Carica papaya), showing a cluster of developing fruits which are large berries containing numerous seeds.

heavily colored flowers, and are, perhaps, the most popular of all cultivated orchids. *P. insigne*, from the Himalayas, contains a range of cultivars and is a parent of several hybrids in cultivation.
ORCHIDACEAE, about 50–60 species.

Paprika a bright red powder made from the dried and ground fruits of certain varieties of *Capsicum annuum*. Types produced range from sweet to pungent but typically paprika lacks the pungency of *chillies. Lower quality paprikas include seed and sometimes parts of the flower, which increase the pungency. Paprika is produced mainly in Spain (where it is known as pimentón), Portugal and Hungary. The finely ground powder is used to flavor and garnish food, including ketchup and other sauces. It is the national spice of Hungary.
SOLANACEAE.

Paradisea a small genus of rhizomatous herbs native to the mountains of Europe, and eastern Tibet. *P. liliastrum* [ST. BRUNO'S LILY, PARADISE LILY] is cultivated for its racemes of conspicuous funnel-shaped white flowers about 5cm long.
LILIACEAE, 2 species.

Parmelia is one of the many genera of crust-forming lichens that grow on bare rocks slowly building up an organic substrate. (×1)

Paris a small genus of perennial rhizomatous herbs native to Europe and temperate Asia, with a single stem bearing four or more net-veined leaves in a whorl at the top. The solitary, erect flowers are greenish and inconspicuous. *P. quadrifolia* [HERB PARIS] was formerly used medicinally for such complaints as headaches and rheumatism. This species and *P. polyphylla* are sometimes grown as ornamentals, particularly for their attractive bluish-black or bright scarlet fruits.
LILIACEAE, about 20 species.

Parkia a genus of tropical leguminous trees, some with edible pods or seeds. The seeds (beans) of *P. biglobosa* (= *P. africana*) [AFRICAN LOCUST, NITTA TREE] are rich in **protein** and are embedded in a mealy pulp **which is eaten raw**. The seeds are used to

make a drink, or like those of *P. filicoidea* [FERN-LEAVED NITTA, WEST AFRICAN LOCUST], *P. javanica* of Indonesia and *P. speciosa* of Malaysia, powdered as condiments.
LEGUMINOSAE, about 40 species.

Parkinsonia a small genus of American and African tropical and subtropical trees or shrubs bearing short racemes of large yellow flowers. The only cultivated species is the fragrant-flowered *P. aculeata* [JERUSALEM-THORN, MEXICAN PALO VERDE, RETAMA], from tropical America, which requires greenhouse conditions in cooler temperate regions. However, it is widely cultivated as an ornamental shrub in tropical and subtropical gardens.
LEGUMINOSAE, 2–4 species.

Parmelia a large genus of lichens, either leaf-like or forming leafy crusts on trees and rocks. The genus has a worldwide distribution. *P. omphalodes* is very abundant in western parts of Scotland and was once a common source of dye for tartans.
PARMELIACEAE, about 550 species.

Parnassia a north temperate and arctic genus of perennial herbs with saucer-shaped flowers, characteristic of marshland and upland bogs. Among the cultivated species are *P. palustris* [GRASS OF PARNASSUS, WHITE BUTTERCUP], *P. glauca* (= *P. caroliniana*) and *P. parviflora*. All of these species have white flowers with spreading petals and contrastingly colored green, pink or purplish veins.
SAXIFRAGACEAE, about 15 species.

Paronychia a cosmopolitan genus of small annual or perennial herbs, often tuft- or carpet-forming. Several species are grown in rock gardens, such as *P. argentea* and

Virginia Creeper (Parthenocissus quinquefolia) *is a fast-growing creeper with spectacular autumn colors.*

St. Bruno's Lily (Paradisea liliastrum) *growing in an alpine meadow. This hardy species is commonly cultivated. (×⅓)*

P. kapela (including *P. serpyllifolia*), both with conspicuous silvery bracts concealing the flowers. *P. argentea* and *P. capitata* yield Algerian tea, which has various medicinal uses.
CARYOPHYLLACEAE, about 50 species.

Parrotia a genus comprising a single species, *P. persica* [PERSIAN IRONWOOD or IRON TREE], native to Iran. It is a small deciduous tree, often grown as an ornamental for the attractive autumn tints, gold and crimson red, of its ovate leaves.
HAMAMELIDACEAE, 1 species.

Parsley the common name for *Petroselinum crispum*, a biennial or short-lived perennial with aromatic leaves, rich in vitamin C. It is probably native to southern Europe and western Asia but is now widely cultivated in most temperate countries as a herb, garnish and flavoring. Most modern cultivars have curled and crisped leaves but varieties with

Grass of Parnassus (Parnassia palustris), *a plant of marshes and bogs with beautiful, faintly honey-scented flowers.* (× 1)

flat leaf segments (eg var *neapolitanum*) are still widely grown. Var *tuberosum* [TURNIP-ROOTED PARSLEY], with swollen roots which are eaten boiled, is grown in some European countries.
UMBELLIFERAE.

Parsnip the common name for *Pastinaca sativa*, a hardy biennial usually grown as an annual root vegetable, with long tapering cream-colored roots with a sweet taste and somewhat larger and coarser than the related carrot. The plant is native to the Mediterranean region and western Asia but is now cultivated throughout the temperate regions of the world. It is normally eaten boiled or roasted.
UMBELLIFERAE.

Parthenocissus a small genus of climbing vines native to North America, Mexico and temperate Asia. It is closely related to the genera *Vitis* and *Ampelopsis*. They are deciduous, vigorous plants with tendrils or adhesive pads. Many of them develop rich autumn colors on the leaves and some are widely grown to cover walls and fences, especially *P. quinquefolia* (= *A. quinquefolia*, *V. quinquefolia*) [VIRGINIA CREEPER, WOODBINE, AMERICAN IVY, FIVE-LEAVED IVY]. This is a vigorous species native to the USA and Mexico, with five-lobed leaves, brilliantly colored in the autumn. *P. tricuspidata* (= *A. tricuspidata*) [BOSTON IVY, JAPANESE IVY], from China and Japan, is a widely planted, self-clinging climber with adhesively tipped tendrils, and is also endowed with brilliant autumn colors. *P. henryana* (= *V. henryana*), native to China, is another self-clinging species, bearing three- to five-lobed leaves with silvery-white veins over dark green and bronze.
VITACEAE, about 15 species.

Paspalum a large genus of mainly tropical American, mostly perennial grasses with erect stems and an inflorescence consisting of one or more spike-like racemes. *P. scrobiculatum* [KODO MILLET], grown in India, especially in the barren hill regions, is hardy and drought-resistant. *P. notatum* [BAHIA GRASS], grown as fodder from Mexico to South America, has been introduced, as var *saurae* [PENSACOLA BAHIA GRASS], into the southern USA as a lawn grass.

The South American *P. dilatatum* [DALLIS GRASS], known as a pasture grass in the USA, has also been introduced into India, and *P. mandiocanum* is an important fodder crop in Brazil. The Mexican and South American *P. malacophyllum* [RIBBED PASPALUM] has been introduced into the southern USA for use as a hay crop and for soil conservation.
GRAMINEAE, about 400 species.

Passiflora a large tropical genus of vines which climb by means of tendrils. They are chiefly native to tropical America, with a few species in Asia, Australia and Madagascar. The genus is noted for its fruit and its flowers, thought by early Spanish Christian missionaries to represent the instruments of the Crucifixion. The most striking features of the flower is the corona, consisting of one to six rows of filamentous structures arising from the base of the corolla, each conspicuously banded with contrasting colors.

Among the most common cultivated species, *P. caerulea* [BLUE PASSION FLOWER] is a vigorous climber with fragrant white, blue and purple flowers 10–18cm across. The flowers are followed by attractive orange fruits. Garden hybrids between this species and others including *P. alata* and *P. racemosa* have been produced. *P. coccinea* [RED PASSION FLOWER, RED GRANADILLA] has flowers 10–12cm in diameter with vivid scarlet to orange sepals and petals. The corona is purple, pink or white. *P. manicata*, with its vivid scarlet petals, is also commonly called RED PASSION FLOWER. *P. edulis* [PURPLE GRANADILLA], from southern Brazil, and *P. quadrangularis* [GIANT GRANADILLA] are commonly cultivated in tropical and subtropical countries for their edible fruit. *P. quadrangularis* is also cultivated for its flowers which have an attractive shaggy corona, white with zones of blue and purple. The yellow or purple berries of *P. edulis* are the source of passion fruit juice, and the species is grown commercially for this purpose in Australia, New Zealand, South Africa, Hawaii and Kenya. It is the variety *flaviocarpa* [YELLOW PASSION FRUIT] which is particularly cultivated for this purpose. A juice used in beverages is also extracted from the fruit of *P. maliformis* [SWEET CALABASH, SWEETCUP, CONCH APPLE], which is cultivated for this purpose in the West Indies.
PASSIFLORACEAE, about 400 species.

Paulownia a small genus of vigorous, fast-growing, deciduous trees from China which are often grown as ornamentals. *P. tomentosa* (= *P. imperialis*) [PRINCESS, KARRI, EMPRESS or FOXGLOVE TREE] has large, felty, heart-shaped leaves whose size decreases with the maturity of the tree, and huge clusters of purple-blue, foxglove-like flowers. *P. lilacina* has a similar habit but bears drooping lilac

Flower and fruit of Passiflora vitifolia. (× ½)

foxglove-like flowers with yellow throats. The wood of *P. tomentosa* is used in Japan for furniture making and other purposes. The genus is sometimes placed in the Scrophulariaceae.
BIGNONIACEAE, 6–8 species.

Pavonia a large genus of tropical, subtropical and temperate herbs and shrubs. A number of species, including the shrubby tropical American *P. spinifex* and *P. hastata* are grown as ornamentals, while the bark of *P. bojeri*, from Madagascar, and *P. shimperiana*, from tropical Africa, yields a fiber used to make cloth. The tropical African tree *P. hirsuta* yields a mucilage from its roots which is used in butter-making by the natives because of its capacity to coagulate milk.
MALVACEAE, about 200 species.

Pea a term applied to the plants and the seeds of many members of the family Leguminosae (itself often called the PEA family). The most familiar are from the genus *Pisum*: *P. sativum* var *sativum* (= *P. hortense*) [CULINARY, GARDEN or COMMON PEA] and *P. sativum* var *arvense* [FIELD PEA]. The former has white flowers and green or yellow seeds, the latter has red or purple flowers and dark seeds. Var *macrocarpon* is the SUGAR or SNOW PEA with edible young pods.

The FIELD PEA is an important crop, providing valuable feed for stock and improving the soil by the action of its nitrogen-fixing root nodules. It is grown either alone or mixed with a cereal, and may be cut green for fodder or left until maturity and the seeds harvested for a protein-rich dry feed.

The dried peas of *P. sativum* var *sativum* were once one of the staple foods in Western Europe, but nowadays, with proteins obtained mainly from dairy products, peas are used immature as a green vegetable. The recent increased demand for tinned and

A field of the Common or Cultivated Pea (Pisum sativum var sativum), showing the white flowers which distinguish it from the Field Pea. ($\times \frac{1}{3}$)

frozen vegetables has led to the selection of cultivars which will retain the shape, color, flavor and texture of the peas and ripen at different seasons so that canning factories can receive a steady supply. There are innumerable named cultivars of peas based on particular character combinations such as tall, dwarf, color (green or yellow) texture (smooth or wrinkled) etc.

Other peas include *Psophocarpus tetragonolobus* [*ASPARAGUS PEA], *Cicer arietinum* [*CHICK PEA], *Vigna* [*COWPEAS], *Lathyrus* [*GRASS PEA, SWEET PEA] and species of *Abrus* [ROSARY PEA], *Cajanus* [PIGEON PEA], *Chorizema* [FLAME PEA], and *Clitoria, Centrosema* [BUTTERFLY PEAS].
LEGUMINOSAE.

Peach a drupe fruit similar to the *plum and *cherry, produced by *Prunus persica*. The peach, after the apple, is the world's most widely grown fruit tree and they may also be grown as ornamentals for their blossom. Although the peach is native to China it is now grown in the warmer areas of the temperate zones, in the USA, the Mediterranean countries, the Far East, Central and South America, South Africa and Australia.

Peaches are of two main types – "freestones" (used mostly for fresh consumption) and "clingstones" (used mostly for canning); they may also be classified on the basis of flesh color – yellow or white. Since peaches can only be stored for a few weeks a very high proportion is processed. Most peach varieties are grafted on rootstocks grown from peach seeds – either wild types or selected commercial varieties of other *Prunus* species such as *P. tomentosa*. In France peach × almond (*P. amygdalus*) hybrids have done well as rootstocks for peaches on poor soils subject to lime-induced iron deficiency.

New genetic material from China has been a major contributor to changes in varieties. Thus CHINESE CLING, introduced to America from China via England in 1850, was the parent of such well-known varieties as 'J. H. Hale' and was involved in the ancestry of a high proportion of present-day varieties. ROSACEAE.

Pear the common name for the popular edible fruits produced by the 15–20 species of the genus *Pyrus*. They evolved in Central Asia with secondary centers arising in both China and the Caucasian region. In China and Japan *P. pyrifolia* (= *P. serotina*) [ORIENTAL OR SAND PEAR], a species with hard crisp fruit, formed the source, with *P. nivalis* and *P. communis*, from which cultivated varieties were selected, with perhaps *P. syriaca*, *P. ussuriensis* and *P. longipes* as additional donors. Because cultivars of *P. pyrifolia* are gritty (due to stone cells) and have little flavor, they are mostly used for culinary purposes.

In Asia Minor the important pear type now known as the EUROPEAN or COMMON PEAR, was first selected from *P. communis*. *P. pyrifolia* crossed with *P. communis* (*P.* × *lecontei*) has given rise to important commercial cultivars such as 'Leconte', 'Kieffer' and 'Garber'. These are more resistant to fireblight (caused by the bacterium *Erwinia amylovora*) than the standard European pears but have fruit of poorer quality, which makes them more suitable for canning than for fresh consumption.

Ripe Peach fruits (Prunus persica) ready for harvesting. ($\times \frac{1}{4}$)

In North America, pears have usually been grafted on pear seedlings obtained from commercial varieties such as 'Bartlett' (known in Europe as 'Williams' Bon Chrétien') and 'Beurre Rose'. On such rootstocks large trees are produced which are able to withstand cold winter conditions. In contrast, in Europe *Cydonia oblonga* [QUINCE], including vegetatively propagated selections such as 'Quince A', are used as rootstocks because of their ability to dwarf and to induce trees to fruit early in their life.

Pear decline caused by a mycoplasma has caused the death of millions of trees in the USA. Seed from the Chinese species *P. calleryana* and *P. betulifolia* and from European pear varieties has provided a source of rootstocks that render grafted trees resistant to the mycoplasma.

The present world production of pears is over 7 million tonnes per year. Major producers are Italy, China, West Germany, the USA, France and Japan.
ROSACEAE.

Pearl barley the hard, pearly pellets left after removal of the adhering floral organs or husk from the *barley grain. It is used in soups, stews and casseroles.
GRAMINEAE.

Pecan nut the fruit of the tree *Carya illinoinensis* (= *C. pecan*), a native of the southeastern USA and Mexico. The wild trees produce edible fatty nuts (dry drupes), which were eaten by the Indian tribes in these areas. Now on a commercial scale, the pecan industry in America is of considerable importance and large plantations of trees are

A dwarf Pear tree (Pyrus communis) in full flower and trained along wires in the form of an espalier. Such dwarf forms are high yielding.

cultivated in states of the southern USA.

More than 500 varieties are now grown, mainly selections from wild seedlings, and named cultivars are usually propagated by budding or grafting on seedling stocks. The quality of the nut is dependent on a high kernel percentage. Most of the harvest is shelled mechanically and used for confectionery, sweets, ice cream and the fresh and salted nut trade.

The food value of pecans is high, and the nuts contain a higher fat content than any other vegetable product (over 70%). Oil from rejected and moldy nuts is used for cooking and cosmetic production.
JUGLANDACEAE.

Pediastrum a genus of microscopic colonial green algae with a distinctive plate-like form, found in freshwater ponds. A colony consists of 16 or 32 cells, each with a cup-shaped chloroplast and large pyrenoid. The marginal cells are drawn out into two projections.
CHLOROPHYCEAE.

Pedicularis a large genus of semiparasitic, usually perennial herbs native to the north temperate regions, with one species in Andean Ecuador and Colombia. A few species are cultivated for ornament in the border or rock garden but the seed must be sown with that of a suitable host plant – usually grasses. Most have cylindrical or swollen tubular, hooded, purplish or reddish flowers, although white and yellow are also known.

Well-known cultivated species include *P. delavayi*, with white markings on the rose-purple corolla, *P. palustris* [SWAMP OR MARSH LOUSEWORT], an annual suitable for damp places, *P. sylvatica* [LOUSEWORT], with bright red flowers, and *P. verticillata* with pink or white varieties.

Cultivated North American species include the perennials *P. canadensis* [COMMON LOUSEWORT, WOOD BETONY], with yellow or reddish (rarely white) flowers and *P. densiflora* [INDIAN WARRIOR], with crimson flowers.
SCROPHULARIACEAE, about 400 species.

Pedilanthus a genus of succulent cactus-like shrubs native to warm regions of North, Central and South America and the West Indies. Two species, *P. macrocarpus* and *P. tithymaloides* (= *Euphorbia tithymaloides*) [REDBIRD CACTUS, RIBBON CACTUS, SLIPPER SPURGE or FLOWER, JEW BUSH, DEVIL'S BACKBONE, JAPANESE POINSETTIA], are cultivated as ornamentals, the latter with zigzag stems and densely clustered flowers enclosed by bright red or purple bracts.
EUPHORBIACEAE, about 30 species.

Peganum a small genus of shrubs native from the Mediterranean region to Mongolia and from the southern USA to Mexico. The seeds of *P. harmala* [HARMAL, HARMELA SHRUB] are narcotic and in parts of the East they are burnt and the smoke inhaled. Other local uses are to treat eye diseases and nervous complaints, as an aphrodisiac and to stimulate the appetite. A red dye is extracted from the fruits.
ZYGOPHYLLACEAE, 4–5 species.

Pelargonium a large genus including the "geraniums" of cultivation. Almost pantropical, the genus is particularly abundant in southern Africa, and extends into some warm-temperate areas. Mostly herbaceous and often somewhat succulent, many of the wild species have small flowers; breeders have enlarged them in cultivation.

The use of pelargoniums as ornamentals is varied, and they are grown as subshrubs, bedding plants or in hanging baskets. Almost all found today in gardens are of hybrid origin. *P. fulgidum* hybrids such as *P. × ignescens* are the commonest subshrubs. The 'Regal' or 'Domesticum' pelargoniums listed under *P. × domesticum* [SHOW GERANIUM, FANCY GERANIUM, MARTHA WASHINGTON

Pecan nuts (Carya illinoinensis) are grown almost exclusively in the USA, where they are extremely popular. (× 1)

Pelargonium species are commonly called Geraniums. The red Geraniums, such as shown here, belong to the group of "zonal" hybrids known as Pelargonium × hortorum. (× $\frac{1}{5}$)

GERANIUM, SUMMER AZALEA, REGAL GERANIUM], with beautiful flowers and often colored leaves, are mostly developed from *P. cucullatum*, *P. angulosum* and *P. grandiflorum*. Further hybridization with *P. quercifolium* has produced more cultivars.

P. × hortorum [FISH GERANIUM, ZONAL GERANIUM, HOUSE GERANIUM, BEDDING GERANIUM], of complex hybrid origin, mainly derived from *P. inquinans* and *P. zonale*, constitutes another important section in the genus, often with patterned leaves, and both single and double-flowered. The IVY LEAVED or HANGING GERANIUM group of pelargoniums are based on *P. peltatum* and are the most popular for hanging baskets.

Some species are grown commercially, mainly in Madagascar, Sri Lanka, East Africa and Algeria. The young shoots are distilled to produce an oil, geranium oil, used in perfumery.
GERANIACEAE, about 250 species.

Pellia a small genus of thallose liverworts, all of which occur in the temperate and boreal climates of much of the Northern Hemisphere. They are found, often in great abundance, in moist habitats in forests, along streams and in open marshes and fens. The thallus is flat, bright green or with a purplish tinge, irregularly branched, and with an indistinct broad midrib which is simply a thicker area. The sporophytes are often numerous, with a long seta and a spherical capsule.
PELLIACEAE, about 4 species.

Peltigera a genus of leafy lichens, mostly found growing on the ground. In almost all species the algal symbiont is blue-green. *P. canina* and other species sometimes infest lawns on poorly drained soils.
PELTIGERACEAE.

Pelvetia a genus of brown algae found at the top of the zonation on temperate rocky shores. It is remarkably tough and resistant to desiccation. It grows in dense tufts with bifurcating fronds and is generally similar to *Fucus* except that the frond has no midrib. At least one species, *P. canaliculata* [CHANNELLED WRACK, "COW-TANG"] has been used as animal fodder.
PHAEOPHYCEAE.

Penicillium a genus of imperfect fungi species which are familiar and conspicuous on all sorts of decaying material as "blue" and "green" molds (eg the blue mold of damaged or rotting oranges, caused by *P. italicum*).

The contamination of a bacterial culture by a stray conidium of *P. notatum* and the resultant inhibition of bacterial growth led to Alexander Fleming's discovery of the penicillins, the most important compounds to have been isolated from fungi. In culture, *P. notatum*, and in particular, *P. chrysogenum* commonly produce two or more of four types of penicillin, which are referred to as F, G, X and K.

The penicillins are produced commercially by fungal fermentation, but the particular penicillin synthesized by the fungus can be influenced by the nature of the precursor added to the culture medium. The production of penicillin G, which is more stable and more useful clinically than the other natural penicillins, can be enhanced by adding phenylacetic acid to the medium. Many other compounds can be incorporated into the side-chain, provided that the fungus is presented with the appropriate precursor. In this way, novel penicillins with additional antibiotic properties have been produced biologically. Others can be made semisynthetically.

Other economically important species include *P. griseo-fulvum* and *P. patulum*, which produce the antifungal antibiotic, griseofulvin. This causes marked distortion of fungal hyphae and has proved useful in the systemic treatment of fungal infections of the skin, hair and nails.

P. roqueforti is responsible for the ripening of Roquefort cheese. The curd is inoculated

Fountain Grass (Penisetum setaceum) grows wild in Africa and the Middle East and is cultivated as a garden ornamental. ($\times \frac{1}{25}$)

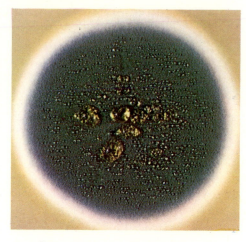

Penicillium chrysogenum, a species used in penicillin production. The colony is blue with a sterile margin. ($\times 1$)

with the fungus, which produces a unique flavor and odor. In Camembert cheese, the role of the fungus *P. camemberti* is to alter the texture of the ripening cheese by producing proteolytic enzymes, which break down some of the proteins.
HYPHOMYCETES, about 140 species.

Pennisetum a genus of generally stout annual or perennial grasses found throughout the tropics and subtropics with dense, spike-like inflorescences. *P. americanum* (= *P. typhoides*, *P. glaucum*) [BULRUSH, SPIKED MILLET, CAT-TAIL MILLET, PEARL MILLET], which probably originated in tropical Africa, is grown as a food crop in the Sahel, Sudan and India. The grain may be cooked like rice or ground to flour and used for porridge, cakes and unleavened bread. In Africa, malted seed is important for beer. The green plant is sometimes grown for fodder.

P. clandestinum [KIKUYU GRASS], from the East African Highlands, has been introduced at higher altitudes throughout the tropics. It requires a rainfall of 1000mm per year, has a high leaf yield rich in protein and low in fiber, and can tolerate heavy grazing. *P. purpureum* [ELEPHANT OR NAPIER GRASS], from the Rift Valley in Uganda, has also been introduced throughout the tropics, but requires higher rainfall and will not withstand prolonged, heavy grazing. It is generally cut for silage. *P. unisetum* [SILKY GRASS, NATAL GRASS], *P. alopecuroides* (= *P. japonicum*) [CHINESE PENNISETUM], *P. clandestinum*, *P. macrostachyum*, *P. villosum* (= *P. longistylum*) [FEATHER TOP] and *P. setaceum* (= *P. ruppelii*) [FOUNTAIN GRASS] are occasionally cultivated for ornament.
GRAMINEAE, 130 species.

Pennywort the common name usually given to the aquatic perennial umbelliferous herb *Hydrocotyle vulgaris* [sometimes also called MARSH PENNYWORT]. The name derives from the roundish leaves and is also used for a number of other species such as *Cymbalaria muralis*, *H. asiatica* [ASIATIC PENNYWORT], *Obolaria virginica* [AMERICAN PENNYWORT],

Umbilicus rupestris (= *U. pendulinus*) [WALL PENNYWORT].

Penstemon [BEARD-TONGUE] a large genus of perennial herbs or shrubs with its main center of distribution in the western USA, but with one species native to northeastern Asia. Many have large, showy flowers, and numerous species are popular garden ornamentals. Many, such as the pink- to scarlet-flowered *P. barbatus* and the lilac-flowered *P. spectabilis*, make good bedding plants.

Several hybrids are cultivated, particularly involving *P. cobaea*, with white, violet-purple or deep purple flowers, and *P. hartwegii*, with scarlet flowers, as parents (so-called "*P. gloxinioides*"). Several dwarf or prostrate species make good rock-garden plants, including *P. davidsonii* and *P. rupicola*, both with ruby-red flowers.
SCROPHULARIACEAE, about 250 species.

Peperomia a very large genus of low annual or perennial herbs with succulent leaves, native chiefly to the humid forests of tropical America. A number of species are cultivated for their attractive foliage as greenhouse or house pot plants. Among the more popular species are *P. arifolia*, with ovate, gray and green variegated leaves, and *P. argyreia* (= *P. sandersii*) [WATERMELON BEGONIA], with ovate white leaves with green veins. *P. rotundifolia* (= *P. nummulariifolia*) [YERBA LINDA] with round brown and light green variegated leaves is a useful species for hanging baskets.

P. caperata [EMERALD-RIPPLE PEPEROMIA, GREEN-RIPPLE PEPEROMIA, FANTASY PEPEROMIA] has a compact, tufted habit with glossy dark green, ovate or suborbicular-ovate

The Barbados Gooseberry (Pereskia aculeata), shown here in flower among the sword-like leaves of another cactus species. ($\times \frac{1}{4}$)

A pepper plantation, with the pepper vines (Piper nigrum) growing on posts and the peppercorns drying in the sun.

Peppercorns in a sack ready for export. India and Indonesia are the chief pepper-producing regions of the world.

leaves with reddish stalks. The leaves of cultivar 'Variegata' have a white margin surrounding a central green zone. *P. griseoargentea* [IVY LEAF PEPEROMIA, PLATINUM PEPEROMIA, SILVER-LEAF PEPEROMIA] has a similar compact habit but with more rounded leaves, silvery gray on the upper surface.

There are several cultivars of *P. obtusifolia* [BABY RUBBER PLANT, AMERICAN RUBBER PLANT, PEPPER FACE], including 'Alba' in which the young leaves are yellowish-white, marked bright scarlet, and 'Variegata' with broad whitish margins to the leaves surrounding a green and grayish blotched central zone.
PIPERACEAE, about 1000 species.

Pepino the common name for *Solanum muricatum* (also known as the MELON PEAR), an erect, spiny herb from the Andes which produces an ovoid yellow/purple fruit with a firm yellow edible flesh. Pepino is also the Spanish common name for *CUCUMBER.

Pepper the name given to the aromatic fruits of several different plant species. CAYENNE, CHILI, NEPAL, BELL, RED, GREEN and SWEET PEPPERS and PAPRIKA all belong to the genus *Capsicum*. JAMAICA PEPPER, also known as *ALLSPICE or PIMENTO, is obtained from *Pimenta dioica* and MELEGUETA PEPPER, "GRAINS OF PARADISE" or GUINEA GRAINS, from *Aframomum melegueta*. NEGRO or AFRICAN PEPPER comes from *Xylopia aethiopica*.

BLACK and WHITE PEPPER, the everyday condiments, come from *Piper nigrum* [PEPPER PLANT], which is a woody perennial vine native to the rain forests of the mountains of southwestern India. Each fruit is a small spherical red drupe with a fleshy outer layer enclosing the seed. Black peppercorns are whole dried fruit; white ones have had the fleshy mesocarp removed before drying. The pungency of the spice persists even after long storage. There is also an aromatic oil whose fragrance is only noticeable in fresh peppercorns. Most pepper, of course is used to flavor food, but some is distilled for pepper oil which is used in perfumery. Pepper is historically a southern Asian crop, being produced mainly in India and Indonesia, but it has recently been introduced into cultivation in Africa, South and Central America and the Pacific islands. The dried fruits of *P. cubeba* [CUBEB PEPPER], from Malaya, have antiseptic properties.

Peppermint the common name for *Mentha × piperita*, a hybrid between *M. aquatica* [WATERMINT] and *M. spicata* [SPEARMINT, *MINT]. There are two main varieties, 'black' and 'white', cultivated for the distillation of oil, the latter being the peppermint oil with the more delicate aroma. Peppermint oil is very widely used in pharmacy and as a flavoring agent. Many toothpastes, chewing gums, flatulence tablets, sweets and cordials owe their pleasant smell and taste to the essential oil distilled from peppermint. It is a widespread plant in Europe and the USA where it is cultivated commercially in numerous clones.

Japanese or Chinese peppermint oil is different in nature, being rich in menthol. It is extracted from *M. arvensis* var *piperascens*. Certain species of *Eucalyptus* are commonly called "peppermint", eg *E. dives* [BROAD-LEAVED PEPPERMINT].

Pereskia a small genus of trees or shrubs native to the West Indies, Mexico and Central and South America. The best-known are *P. aculeata* [BARBADOS GOOSEBERRY, LEAFY CACTUS, LEMON VINE], which is a climber, and *P. grandifolia* [ROSE CACTUS], which is an erect shrub growing to a height of about 5m. Both species produce white or pinkish flowers. The fruit of *P. aculeata* is edible and its leaves are sometimes used as a vegetable in tropical America.
CACTACEAE, about 20 species.

Peridinium an important genus of dinoflagellates found mainly in the marine plankton although there are a few freshwater species. The cell shape has many variations but basically it has a median transverse groove and a longitudinal groove. The two flagella lie mainly within these grooves. The cell is covered by a pattern of plates, which may be variously ornamented.
DINOPHYCEAE.

Perilla a small genus of annual herbs mainly native to the mountains of Southeast Asia. *P. frutescens* (= *P. ocimoides*) and its cultivar 'Crispa' (= *P. nankinensis*) are cultivated in Japan, China, Korea and elsewhere

The Scarlet-bugler (Penstemon centranthifolius) growing wild in the San Gabriel Mountains, California, USA. (×$\frac{1}{10}$)

Cultivated Petunias (Petunia × hybrida) come in many colors, as can be clearly seen from the display of color at this breeding station.

as a source of a drying seed oil (perilla oil) used in the manufacture of varnishes, printing inks and linoleum. Some cultivars, such as 'Crispa' and 'Atropurpurea', are sometimes grown as bedding plants in warmer regions for their purple or rose, variegated leaves.
LABIATAE, about 6 species.

Pernettya a small genus of prostrate to erect shrubs mainly native to the Southern Hemisphere, particularly New Zealand, Tasmania and temperate South America. Several species are cultivated as garden ornamentals, the commonest being the Chilean *P. mucronata* [CHAURA], which has white flowers and cultivars with large decorative berries varying in color from white through pink to purple or crimson.
ERICACEAE, about 20 species.

Peronospora a large and widespread genus of parasitic fungi belonging to the general group called "downy mildews". *P. parasitica* (white mold of brassicas), *P. tabacina* (blue mold of tobacco), *P. farinosa* (downy mildew of beet and spinach) are important pathogenic species.
PERONOSPORACEAE, about 75 species.

Perovskia a small genus of hardy and shrubby plants native to western Asia, through Iran to Baluchistan. A few species, such as *P. atriplicifolia*, are cultivated in gar-

dens as border plants. The leaves emit an aromatic odor not unlike sage when crushed and the smallish flowers appear in late summer in panicles and are blue, violet or purple in color.
LABIATAE, 5–7 species.

Persea a large genus of tropical trees and shrubs mainly native to South America, the most important species being *P. americana* [*AVOCADO, AGUACATE, PALTA, ALLIGATOR PEAR]. However, the genus contains a number of other useful species, notably *P. borbonia* [RED BAY, SWEET BAY, LAUREL TREE, TISSWOOD, FLORIDA MAHOGANY], which is not only an attractive ornamental tree but also provides a hard, heavy, close-grained timber useful for cabinet-making. The bark of the two Chilean species, *P. lingue* and *P. meyeriana*, is used for tanning. These and *P. indica* (= *P. teneriffae*), from the Azores, Madeira and the Canary Islands, are also grown as ornamental trees.
LAURACEAE, about 150 species.

Persimmon the common name for the round, usually orange, edible fruit and the plant of certain species of *Diospyros. D. kaki* [*KAKI] is grown in the subtropics throughout the world, but especially in China and Japan. The wild North American species *D. virginiana* [COMMON or AMERICAN PERSIMMON] is present in hardwood forests from New Jersey to Texas. In the USA *D. virginiana* is often used as a rootstock to produce dwarf trees which are convenient to handle but often last as little as 10 years. The strong, hard wood ("North American ebony") is also used, often in golf club heads.
EBENACEAE.

Petasites a genus of coarse perennial rhizomatous or stoloniferous herbs native to the north temperate zone, particularly in the Old World, but also in North America. The leaves are large and rhubarb-like, and the flowers are borne on fleshy spikes. Their bold foliage is very effective beside water, but once intro-

A cup fungus, known as the Red Elf Cup Fungus (Peziza coccinea), which grows on decaying branches on damp ground in woodlands. (×¼)

Phacelia campanularia, a hardy annual growing in a garden border. The leaves are often red margined and scented if bruised. (×1/10)

duced into a garden they can be difficult to control. *P. fragrans* [WINTER HELIOTROPE], with vanilla-scented, pale lilac flowers, *P. hybridus* (= *P. vulgaris*) [BOG RHUBARB, BUTTERBUR] and *P. albus* [WHITE BUTTERBUR] are perhaps the most popular cultivated species. *P. japonicus* [FUKI] is grown as a vegetable in Japan for its edible petioles.
COMPOSITAE, about 15 species.

Petrea a genus of tropical American, Mexican and West Indian twining shrubs with opposite or whorled leathery leaves. *P. volubilis* [PURPLE WREATH, QUEEN'S WREATH] and *P. arborea* are among several species cultivated for their attractive racemes of violet-heliotrope or white flowers.
VERBENACEAE, about 30 species.

Petunia a small genus of annual or perennial herbs from South America. They are generally straggling plants with funnel- or salver-shaped solitary flowers on terminal or axillary peduncles.

The cultivated petunias, which have veined, striped, star-marked, fringed, giant and double flowers, have been obtained by crosses involving *P. axillaris* [LARGE WHITE PETUNIA], *P. inflata* and *P. violacea* [VIOLET-FLOWERED PETUNIA] and are classified as *P. × hybrida*. They can be divided into four main groups: *P. × hybrida* 'Pendula', free-flowering trailing forms for window boxes and hanging baskets; *P. × hybrida* 'Nana-compacta', free-flowering compact forms for bedding and pot plants; *P. × hybrida* 'Grandiflora' with large flowers; and *P. × hybrida* 'Flore-pleno', with double flowers.
SOLANACEAE, 30–40 species.

Peziza a widespread genus of saprophytic fungi forming sessile or shortly stalked, cup-shaped, fleshy, colored fruiting bodies, 1–10cm across. Most species grow on the ground, on rotting wood or sawdust, bonfire sites, dung, straw and plaster. Forms in which the broken flesh of the apothecium exudes a colored fluid are sometimes classified in *Galactinia*.
PEZIZACEAE, about 50 species.

Phacelia a large genus of New World (mainly western USA and Mexico) hairy, often glandular, annual to perennial herbs noted for their delicate flowers. The flowers are bell-shaped to tubular, sometimes with a short tube, and purple or blue to white or yellow. Several species are grown in gardens as hardy annuals including the violet-flowered (rarely white) *P. minor* [CALIFORNIA BLUEBELL, WHITLAVIA], *P. divaricata*, with blue or violet flowers, and *P. tanacetifolia*, with blue or lavender flowers. The last named species is sometimes cultivated as a nectar plant for honey bees.
HYDROPHYLLACEAE, about 200 species.

Phalaenopsis a small genus of tropical epiphytic orchids found growing wild from India and Malaya to the Philippines, New Guinea and northern Queensland but widely cultivated throughout the rest of the world as greenhouse plants. The very short stems bear two rows of densely arranged large, broad or oval succulent leaves. Many of the species, the so-called "MOTH ORCHIDS", have spectacular flowers of purest white, clear pink, sulfur-yellow or candy-striped petals like mothwings, and slightly smaller sepals and labellum. Many hybrids have been made between different species of *Phalaenopsis* and intergenerically with *Vanda* and *Aerides*.
ORCHIDACEAE, about 55 species.

Phalaris a small genus of annual and perennial grasses, widely dispersed in North America, Europe and North Africa.

A double-flowered variety of the Mock Orange (Philadelphus coronarius) growing in a medieval street in a town in southern France.

A bumblebee visiting the flower of a Scarlet Runner Bean (Phaseolus coccineus). (×4)

P. arundinacea [CANARY REED GRASS] is a tall reed species of fens and slow-moving waterways and is sometimes cultivated as var *picta* [RIBBON GRASS, GARDENER'S GARTERS, LADY'S LACES], which has longitudinally white-striped leaves. *P. canariensis* [CANARY GRASS] is a Mediterranean annual widely grown for bird seed. *P. tuberosa*, also of Mediterranean origin, is an important fodder grass in Australia, and *P. caroliniana* is sometimes cultivated for forage in the southern USA.
GRAMINEAE, about 20 species.

Phaseolus a most important New World genus including a number of species important as food, forage, cover and green-manure crops as well as some ornamentals. Old World species previously referred to this genus are now assigned to *Vigna*. The plants vary in habit from short shrubby forms to extensive climbers or creepers. Both annual and perennial forms occur. Usually the rootstock is quite woody and arises from a tuberous root. The flowers are pea-like, and the fruits are typical legumes containing from two to many variably sized and shaped seeds with a rough or smooth seed coat.

Seeds of *P. vulgaris*, the well-known COMMON, NAVY, SNAP, *HARICOT, KIDNEY or FRENCH BEAN, are used to produce baked beans. MADAGASCAR, SIEVA, CAROLINA, BUTTER or LIMA beans are the seeds of *P. lunatus* (including *P. limensis*). Pods of several species, such as *P. vulgaris*, *P. acutifolius* var *latifolius* [TEPARY BEAN] and *P. coccineus* [SCARLET RUNNER BEAN], are used as a green vegetable. Beans form an important item of food in many subtropical and tropical areas, notably in Latin America, often being the main protein food. *P. caracalla* (= *Vigna caracalla*) [SNAIL FLOWER] and *P. coccineus* are ornamentals.
LEGUMINOSAE, about 200 species.

Phellodendron a small East Asian genus of deciduous trees bearing yellowish flowers, berry-like, pea-sized fruits and most attractive pinnate leaves which turn yellow in autumn. *P. amurense* [CORK TREE] grows to about 16m, and has fissured corky bark when mature. Its seeds yield an insecticide and its bark is used in local medicine in a preparation for treating skin diseases. It also makes a fine ornamental, as does *P. sachalinense*. *P. chinense* has aromatic leaves when crushed, and its bark is also used medicinally.
RUTACEAE, 7–9 species.

Philadelphus a genus of deciduous shrubs mainly native to temperate and subtropical Asia, North America and southern Europe. Many of the species make attractive ornamental shrubs, bearing very fragrant, cup-shaped flowers with four white petals, sometimes tinged with purple at the base.

Dwarf French Beans (Phaseolus vulgaris), ready for picking. (×$\frac{1}{10}$)

The most widely grown garden plants of the genus are *P. coronarius* [MOCK ORANGE, SYRINGA] and its hybrids with such North American or Mexican species as *P. microphyllus* and *P. coulteri* [ROSE SYRINGA], including *P. × lemoinei* and its many cultivars.
SAXIFRAGACEAE, 65–75 species.

Phillyrea [JASMINE BOX, MOCK PRIVET] a genus of three species of evergreen shrubs or trees native to the Mediterranean region and southwest Asia. *P. decora* (= *P. vilmoriniana*), *P. angustifolia* and *P. latifolia* are cultivated for their deep green foliage. They make excellent hedges because of their ability to withstand clipping well. They all bear clusters of small white or greenish-white flowers and fleshy bluish-black drupaceous fruits.
OLEACEAE, 3 species.

Philodendron a large genus of climbing shrubs, small trees, epiphytes and (rarely) stemless herbs indigenous to tropical America. The leaves are alternate, sheathed at the base, and vary from heart-shaped to lobed or finely dissected. The inflorescence is spathaceous, white, yellow or red, often fragrant. Several species of *Philodendron*, particularly ones with an epiphytic habit, are popular foliage plants, grown outdoors in warm climates, otherwise as house or greenhouse plants.

Popular ornamental species include *P. bipennifolium* [HORSEHEAD PHILODENDRON, PANDA PLANT], with glossy dark green reflexed leaf blades, *P. domesticum* (= *P. hastatum*) [SPADE-LEAF PHILODENDRON], with elongate-triangular, sagittate, undulate, bright green, reflexed leaf blades, *P. erubescens* [RED-LEAF PHILODENDRON, BLUSHING PHILODENDRON], with reddish-purple stems, ovate-triangular, sagittate-cordate leaf blades, shiny dark green above, brownish below, *P. giganteum* [GIANT PHILODENDRON], with ovate leaves, with blades up to 1m long on petioles 1m long, and

Some of the many garden varieties of Phlox drummondii, *from Texas, a commonly cultivated half-hardy annual.* ($\times \frac{1}{5}$)

P. scandens [HEART-LEAF PHILODENDRON], with ovate, reflexed leaf blades, green above and green or red-purple beneath in subspecies *scandens* but with variations in other cultivars.

The fruits of *P. bipinnatifidum* are eaten in its native Brazil. Various medicinal preparations for treating rheumatism are made largely from *P. radiatum* (= *P. dubium*) [DUBIA] and *P. pedatum* (= *P. laciniatum*, *P. laciniosum*).
ARACEAE, about 200 species.

Philonotis a genus of mosses which shares with the related genus *Bartramia* an almost spherical capsule form. Representatives are found throughout Europe and much of Asia, North and Central Africa, North America and the extreme southern part of South America.
BARTRAMIACEAE, about 170 species.

Phleum a small genus of annual and perennial grasses from temperate regions of both hemispheres. *P. pratense* [TIMOTHY GRASS, CAT'S TAIL] has been introduced to all temperate grasslands throughout the world, being extensively cultivated for grazing, silage and hay making. Another relatively widespread species, with short hairy spikes, found naturally in mountain districts of Europe and introduced to many other parts of the world, is *P. alpinum* [ALPINE TIMOTHY or CAT'S TAIL].
GRAMINEAE, 10 species.

Phlomis a genus of perennial herbs, sub-shrubs or shrubs native from the Mediterranean region to Central Asia and China. Many are adapted to dry, arid climates by having densely hairy or woolly leaves and stems and compact dense growth.

Among the cultivated ornamentals, *P. fruticosa* [JERUSALEM SAGE] is a densely branched shrub up to 130cm tall with gray-green leaves and whorls of bright yellow flowers. *P. russelliana* (= *P. viscosa*) is a robust, striking, shrubby plant with large heart-shaped hairy leaves and whorls of yellow flowers. *P. tuberosa* is a sparsely to densely hairy perennial herb up to 2m tall with large, coarse leaves and purple or pink flowers.
LABIATAE, about 100 species.

Phlox a genus of annual and perennial herbs and shrubs mainly native to North America and Mexico but with one Siberian species. Two important species in temperate horticulture are *P. subulata* [MOSS or MOUNTAIN PHLOX], suitable for rock gardens, and the easily grown, free-flowering perennial border phlox, *P. paniculata* [SUMMER, FALL or PERENNIAL PHLOX].

Many cultivars have been raised with white, purple, red, orange and lilac flowers, some with variegated foliage. Perhaps the best-known cultivated species is *P. drummondii* of Texas and New Mexico, a commonly grown annual now widely naturalized. It exists in forms with flowers of the same color range as *P. paniculata*, but also in yellow and blue, some with fringed petals. The genus also includes cushion-forming creeping perennials such as *P. douglasii*, another rock garden favorite.
POLEMONIACEAE, about 60 species.

Phoenix a genus of palms, the best-known being *P. dactylifera* [DATE PALM] (see DATES). *Phoenix* ranges from the Canary Islands through the tropics and subtropics of Africa to Vietnam. The trunk is often stout and usually suckers at the base to form clumps. There are numerous spreading leaves in the

Philodendron species make popular houseplants. Shown here are Heart-leaf Philodendron (P. scandens) (outer left and right), P. tuxla (center left) and P. pedatum (center right). The flowers of P. scandens have a red spathe. ($\times \frac{1}{25}$)

Flowers and leaves of New Zealand Flax (Phormium tenax). ($\times \frac{1}{20}$)

crown (up to 200 in *P. canariensis*). The fruit is fleshy but sometimes rather dry and the familiar elongate stone is the seed.

The DATE PALM is restricted to dry places (although it requires ground water if it is to flower and fruit), as indeed is most of the genus. The notable exceptions are *P. sylvestris* [WILD DATE, INDIA DATE], which is widespread in India and a source of palm sugar, and *P. reclinata* [SENEGAL DATE PALM] and *P. paludosa* which are inhabitants of mangrove forests in West Africa and the estuarine shores of Asia respectively. Excellent edible dates are produced by interspecific hybrds of *P. dactylifera* × *P. reclinata*.

All the above species and several others, notably *P. rupicola* [CLIFF DATE, WILD DATE PALM, INDIA DATE PALM], of Himalayan India, Sikkim and Assam, are cultivated as ornamentals.
PALMAE, about 17 species.

Phormium a genus represented by two herbaceous perennials with basal sword-shaped leaves, endemic to New Zealand, including Norfolk Island. *P. colensoi* [MOUNTAIN FLAX] grows up to 2m high and bears coarse leaves 1.5m long, and flower stems up to 2m long, bearing tubular yellowish flowers. *P. tenax* [NEW ZEALAND FLAX, NEW ZEALAND HEMP], which yields a commercially valuable leaf fiber, includes a number of ornamental cultivars including 'Variegatum' with creamy-white striped leaves and 'Atropurpureum' with reddish-purple leaves. This species grows to a height of up to 5m, with leaves up to 3m long, and bears tubular dull-red flowers.
AGAVACEAE, 2 species.

Photinia a genus of deciduous or evergreen shrubs or small trees native to the Himalayas, Southeast Asia and Japan. A few species are cultivated as garden ornamentals, notably *P. villosa*, from Japan, a deciduous shrub with white flowers borne in

axillary inflorescences. *P. beauverdiana* from China is similar but may reach a height of about 10m. The evergreen species include *P. serrulata* [CHINESE HAWTHORN], a shrub with large oblong leaves which are deep red when young, and *P. glabra* [JAPANESE PHOTINIA], a small tree with long (wedge-shaped) leaves and flowers with long soft hairs at the base of the petals.
ROSACEAE, about 40 species.

Phragmidium a genus of rust fungi with all their spore stages on Rosaceae. *P. mucronata* is sometimes troublesome on roses, while *P. violaceum* is very common on brambles.
PUCCINIACEAE, about 60 species.

Phragmites a small cosmopolitan genus of grasses, the most important of which is *P. australis* (= *P. communis*) [COMMON REED, CARRIZO]. The plant bears erect narrow, leaves and tall stems reaching to 10m in height. It is an abundant, dominant species in waterways all over the world and frequently forms huge beds or "fens" in permanently wet alkaline waters. It is an economically important grass, used in thatching, matting and basketwork. Its network of roots and rhizomes on river banks and fens assists greatly in the prevention of erosion and the reclamation of flatland.
GRAMINEAE, 3–4 species.

Phygelius a very small South African genus of small erect evergreen shrubs. *P. capensis* [CAPE FUCHSIA, CAPE FIGWORT] has narrow, tubular, showy red flowers about 5cm long on loose spreading inflorescences, while *P. aequalis* has salmon-colored flowers. They are often grown as border plants.
SCROPHULARIACEAE, 2 species.

Phyllanthus a genus of tropical and subtropical trees, shrubs and perennial herbs. A number of species are cultivated, for example *P. myrtifolius*, from Sri Lanka, a dense ever-

Young leaves of the Hart's Tongue Fern (Phyllitis scolopendrium). ($\times \frac{1}{5}$)

A Date Palm (Phoenix dactylifera), *showing the bunches of fruits.*

green shrub, which with its small myrtle-like leaves makes an excellent hedge plant, and *P. arbuscula* (= *P. speciosus*) [FOLIAGE FLOWER], from Jamaica, with flattened leaf-like branches (cladophylls). At least two species, the Asiatic *P. acidus* [GOOSEBERRY TREE, INDIAN GOOSEBERRY] and *P. emblica* [MYROBALAN, EMBLIC], are grown for their fleshy berries which may be eaten raw or preserved.
EUPHORBIACEAE, about 600 species.

Phyllitis a small genus of tropical, subtropical and warm-temperate ferns. The leaves are entire or slightly lobed, with linear sori in pairs on each vein, covered when young by a membranous indusium. *P. scolopendrium* (= *Scolopendrium vulgare*) [HART'S TONGUE FERN], with its attractive strap-shaped leaves, is cultivated indoors and in the garden. The Brazilian species *P. brasiliensis* (= *Scolopendrium brasiliense*) usually requires to be grown in a warm greenhouse.
POLYPODIACEAE, about 8 species.

Phyllodoce a genus of evergreen, compact, hardy, heath-like shrubs from the alpine–arctic regions of the Northern Hemisphere.

Several species are cultivated, including *P. aleutica* from East Asia, the Aleutian Islands and Alaska, a dwarf, spreading, carpeting shrub with white and pale yellow, bell-shaped flowers, *P. breweri* [RED HEATHER], from South Nevada and California, a semi-procumbent plant with comparatively large, rose-purple, saucer-shaped flowers, *P.* × *intermedia* (= *P. empetriformis* × *P. glanduliflora*), a variable dwarf hybrid of vigorous

growth, with reddish-purple flowers, *P. nipponica*, from Japan, a neat, compact shrub with red to pink bell-shaped flowers, and *P. caerulea*, a dwarf circumpolar species with purple flowers.
ERICACEAE, about 8 species.

Phyllostachys a genus of tall evergreen bamboos native to the Himalayas and eastwards to Japan. The woody stems are flattened on either side alternately. Whangee or Wangee canes are produced from the stripped stems of *P. nigra* [BLACK BAMBOO]. Important timber bamboos in cultivation include *B. bambusoides* [TIMBER BAMBOO, GIANT, HARDY or JAPANESE TIMBER BAMBOO], whose stems reach a height of approximately 22m, and *P. vivax*, a vigorously growing species whose stems attain a height of about 15m.

The young shoots of *P. aurea* [FISHPOLE BAMBOO, GOLDEN BAMBOO], *P. aureosulcata* [YELLOW-GROOVE BAMBOO, STAKE BAMBOO, FORAGE BAMBOO] and *P. dulcis* [SWEETSHOOT BAMBOO] are used as food in China and Japan. However, *P. pubescens* (= *P. edulis*) [MOSO BAMBOO], whose stems have many uses, is the major source of edible bamboo shoots in Japan. The larger ones are used for making stakes and fishing rods.

Some species, including *P. aurea*, *P. nigra* and *P. flexuosa*, are grown for ornament, and *P. meyeri* [MEYER BAMBOO] is sometimes used for hedges.
GRAMINEAE, about 35 species.

Physalis a genus of mostly American annual or perennial herbs known chiefly for *P. peruviana* [CAPE GOOSEBERRY, GROUND CHERRY, GOLDENBERRY] and the decorative *P. alkekengi* [WINTER CHERRY, BLADDER CHERRY, CHINESE LANTERN PLANT]. Although

Pieris formosa, an evergreen shrub whose young leaves are red; those of var forrestii *'Wakehurst', shown here, are exceptionally brilliant.* ($\times\frac{1}{60}$)

native to South America, the CAPE GOOSEBERRY is now grown mainly in the Cape region of South Africa (hence the common name) and in Australia.

Other species grown for their edible fruits include *P. ixocarpa* [MEXICAN HUSK TOMATO, TOMATILLO, JAMBERRY] and *P. pruinosa* [DWARF CAPE GOOSEBERRY, GARDEN HUSK or STRAWBERRY TOMATO]. These are rather acid, yellow or purple fruits which are eaten raw or cooked as a vegetable in stews and sauces. Small quantities of canned fruit are also exported.

P. alkekengi has a calyx which, when mature, becomes brilliant orange-red, a color it retains on drying; for this reason the species is widely used in dried flower arrangements. The scarlet fruit is edible and in the Middle Ages was used against gout and as a diuretic.
SOLANACEAE, 50–100 species.

Physoderma a genus of chytrid fungi widespread in distribution and sometimes parasitic on vascular plants. *P. maydis* produces brown spot of corn/maize and *P. alfalfae* attacks *Medicago*.
PHYSODERMATACEAE, about 50 species.

Physostegia a North American genus of perennial herbs preferring low-lying or swampy habitats. Several species are cultivated, notably *P. virginiana* [FALSE DRAGONHEAD, OBEDIENT PLANT, LION'S HEART] and many cultivars with showy rose-pink, white or carmine flowers. If individual flowers are pushed to one side they remain in that position, hence OBEDIENT PLANT.
LABIATAE, 13–15 species.

Phytelephas a genus of stemless or short-trunked palms found in northwestern South America and Panama. Several species, particularly *P. macrocarpa* [TAGUA, IVORY NUT PALM, NEGRO'S HEAD], produce vegetable

Above Potato Tuber Blight *caused by the fungus* Phytophthora infestans. *The disease initially infects aerial parts of the plant, but in severe cases gets into the tubers, causing rotting and premature sprouting. Left* White Rot of leeks *caused by* P. porri. ($\times\frac{1}{5}$)

ivory, formerly used for buttons, billiard balls, etc. The fruits develop a spiny, woody outer coat and the mature endosperm consists of extremely hard cellulose, which provides the vegetable ivory.
PALMAE, about 15 species.

Phyteuma a genus of perennial herbs native to Europe and temperate Asia. The small flowers are numerous and clustered into round heads or dense spikes. They can be white, yellowish, blue or purplish in color. *Phyteuma* species are often cultivated in the rock garden or in the alpine house, among the best being *P. comosum*, *P. orbiculare* [RAMPION] and *P. spicatum* [SPIKED RAMPION].
CAMPANULACEAE, about 40 species.

Phytolacca [POKEWEED, POKEBERRY] a genus of mainly herbs, shrubs and a few trees mostly native to the tropics and subtropics of America, with a few occurring in Africa, Madagascar and Asia.

Some species, such as *P. americana* (= *P. decandra*) [VIRGINIAN POKEWEED, POKEBERRY, PIGEON BERRY] and *P. acinosa* [INDIAN POKE], cultivated for the edible leaves which taste rather like spinach. They require special preparation in cooking because saponins are present.

In the Mediterranean region and western Asia *P. americana* is also cultivated for its rich purple berries which are used to "improve" the color of red wine and to color confectionery. On account of this they are known in many countries as cochineal berries.
PHYTOLACCACEAE, about 30 species.

Phytophthora a fungal genus widespread and of great economic importance as a parasite of vascular plants. *P. infestans* causes potato late blight and *P. palmivora* causes cacao black pod. Other species cause diseases of rubber, palms, cotton, etc, often in the form of basal stem or root rots. They require moist conditions for growth.
PYTHIACEAE, about 20 species.

Piassava fiber the long, coarse, brown to black, fiber obtained from the leaf sheaths or fibrous stems of various tropical palms. In the producing countries of Africa, the fiber is employed for tying thatched roofs and for the framework of walls of houses and market stalls. Piassava is also used for weaving fish traps, hat, screens, baskets and toys or used as rope for fencing or training yam vines. As exported to Europe, it is used extensively in industrial and domestic products such as brooms, brushes and even mechanical road sweepers.

In West Africa, piassava is obtained by retting the leaf sheaths of *Raphia hookeri*, *R. vinifera* and *R. farinifera*. Two types of Brazilian piassava are also important: Bahia piassava and Para piassava are obtained from *Attalea funifera* and *Leopoldinia piassaba* respectively, the fibers being plucked from the stems. The piassava of Malagasy is obtained from *Vonitra fibrosa* (= *Dictyosperma fibrosum*).
PALMAE.

Picea see SPRUCE.
PINACEAE, 40–60 species.

Picris a genus of annual, biennial or perennial herbs with yellow dandelion-like flower heads. They are commonly called OX-TONGUES (although this name is sometimes only given to *P. echioides*) and are native to Europe, particularly in the Mediterranean countries, and in Asia and Ethiopia, and naturalized in North America.
COMPOSITAE, about 40 species.

Pieris a genus of evergreen shrubs native to North America, the Himalayas and East Asia.

The Pigeon Pea (Cajanus cajan), a shrub up to 3m tall, is grown throughout the drier tropics, especially in India.

Pillwort (Pilularia globulifera), which produces slender green leaves at the base of some of which can be seen the green and brown globular sporocarps. ($\times \frac{1}{2}$)

The species are cultivated for their neat and compact habit and numerous urn-like flowers. Three of the most attractive species are *P. floribunda* (= *Andromeda floribunda*) [FETTERBUSH] native to the southeast USA, *P. japonica* [LILY-OF-THE-VALLEY BUSH], native to Japan, and the very popular *P. formosa*, especially var *forrestii* 'Wakehurst' and 'Forest Flame'.
ERICACEAE, about 10 species.

Pigeon pea one of the common names given to *Cajanus cajan* (= *C. indicus*) an important leguminous shrub cultivated in many tropical countries (mainly in India) for its gray or yellow seeds which are consumed in soups and curries. It is also used for green manure or as a cover crop in the southeastern USA, northwest Australia, Italy and Hawaii.
LEGUMINOSAE.

Pilea an interesting widespread tropical genus of annual and perennial herbs and subshrubs with inconspicuous flowers but including several ornamentals valued for their leaves and grown as pot plants or in hanging baskets. *P. cadierei* [ALUMINUM PLANT, WATERMELON PILEA] is grown as a houseplant for its beautiful silver-marked leaves, and *P. microphylla* (= *P. muscosa*) is the ARTILLERY, GUNPOWDER or PISTOL PLANT, so called because the male flowers discharge their pollen explosively.
URTICACEAE, about 200 species.

Pilobolus a genus of pin molds that grow on dung. The sporangia have an unusual explosive mechanism of spore dispersal.
PILOBOLACEAE, about 7 species.

Pilocarpus a genus of tropical American trees. The leaves of a number of species, such as *P. pennatifolius* and *P. jaborandi*, are the source of the drug pilocarpine (jaborandi) used to treat eye conditions.
RUTACEAE, about 17 species.

Pilularia a genus of ferns found in Europe, North America, Australia and New Zealand in aquatic habitats, with their rhizomes embedded in the mud of marshes and shallow pools. The best-known species, *P. americana* and *P. globulifera* [PILLWORT], are of no ornamental value, although the latter is sometimes grown in aquaria.
MARSILEACEAE, 6 species.

Pimelea a genus of profusely-flowering evergreen shrubs or perennial herbs mainly native to Australia, New Zealand and the Philippines. Many species, including *P. ferruginea*, *P. ligustrina* and *P. spectabilis*,

Yellow flowers of the Pigeon Pea (Cajanus cajan), with the standard purplish-red outside. ($\times 1\frac{1}{2}$)

are grown for ornament, usually bearing white or pink (sometimes yellow) flowers in small terminal heads. Some of the species are commonly called RICEFLOWERS.
THYMELEACEAE, about 80 species.

Pimenta, pimento or **pimiento** common names used for two unrelated plants: *Pimenta dioica* [*ALLSPICE*], a member of the family Myrtaceae, and non-pungent, edible cultivars of *Capsicum annuum*, the *GREEN*, *RED* or *BELL* *PEPPERS*.

Pimpinella see ANISE.
UMBELLIFERAE, 140–150 species.

Pin molds see Fungi, p.152.

Pine a name popularly used for species of the coniferous genus *Pinus*, although not exclusively. Members of the genus *Callitris*, for example, are known as CYPRESS PINES and *Dacrydium* species are also commonly called PINES, eg *D. cupressinum* [RED PINE] and *D. franklinii* [HUON PINE].

Pineapple the common name for *Ananas comosus*, a rosette-forming perennial monocotyledon, native to the New World, but cultivated throughout the world in subtropical and tropical regions. Each "fruit" is actually a multiple fruit, formed from the entire inflorescence of 100–200 flowers and borne terminally on the stem. After initiating the inflorescence, the stem apex produces more leaf primordia, which form the "crown" of the fruit.

Most commercially grown pineapples are canned or made into fruit juice, and Hawaii produces about 70% of the world's canned pineapple. The most widely grown cultivar today is 'Cayenne' which originally came

A Pineapple (Ananas comosus) plantation in the central lowlands of Trinidad, showing the typical rosette form of these plants.

from French Guiana. In the Philippines and Taiwan pineapple leaf fibers are used for cloth and cordage. Forms with variegated leaves are often grown as ornamentals.
BROMELIACEAE.

Pinguicula [BUTTERWORTS] a genus of small carnivorous herbaceous plants distributed in wet habitats in temperate and cool temperate localities, largely in the Northern Hemisphere. The leaves are in basal rosettes, entire, and usually sticky above. Insects are trapped on the leaf-surface by sticky secretions and stimulate the secretion of mucilage, which alters the tension in the leaf, producing an inrolling of the leaf to enclose the insect. Sessile glands secrete a ferment to digest the prey and the product of decay is finally absorbed by the plant, the leaves unrolling to receive another insect.

Several species are cultivated, including the tender *P. bakeriana*, with carmine pink flowers, and *P. caudata*, with violet-purple flowers.
LENTIBULARIACEAE, about 46 species.

Pinus [PINES] a genus of tall evergreen conifers widely distributed over the north temperate zones of the Old and New Worlds. Pines are predominantly trees of pyramidal habit but a few shrubs do occur. Leaves in the adult plants are of two kinds: scale-like and needle-shaped. The scale-like leaves are borne spirally, usually on lower parts of young shoots ("long shoots"); they are commonly deciduous and bear in their axils so-called undeveloped or "short shoots". On those shoots the needle-shaped leaves arise in

clusters of two to five (usually two, three or five but at times up to eight or reduced to one). These needle leaves persist for up to five years or more. The male and female flowers are separate but on the same tree. The male flowers are in short cylindric "catkins"; the females form cones.

The timber of several species, including that of *P. palustris* [LONGLEAF PINE, SOUTHERN PINE], *P. lambertiana* [SUGAR PINE, GIANT PINE], *P. radiata* [MONTEREY PINE], *P. monticola* [WESTERN WHITE PINE] and *P. parviflora* [JAPANESE WHITE PINE], are among the most important of all softwoods. The timber is utilized in every kind of constructional work and carpentry, often to provide flooring and the interior finish of buildings, shingles and cabinetwork. The resin helps to preserve the wood considerably and when coated with creosote it is ideal for outdoor use for telegraph poles, railway sleepers and road blocks. The wood of *P. sylvestris* [SCOTS PINE] is known as yellow deal; its resin, and that of *P. palustris*, *P. nigra* [AUSTRIAN PINE] and, the most important resin species, *P. pinaster* [MARITIME PINE], is obtained by tapping the trees. The resin may be distilled to give turpentine and rosin. Destructive distillation in closed (airless) vessels yields tar and pitch. Pine oils are obtained by distillation of leaves and shoots and are used as disinfectants; tar is used to treat skin diseases. Industrial uses include the preparation of printer's ink.

Pines are also planted extensively for ornament and in plantations, especially on poor soils. *P. pinea* [ITALIAN STONE PINE, UMBRELLA PINE] produces seeds which are large and soon lose their vestigial wing. They are eaten as a delicacy in the Mediterranean region where in France, for example, they are known as "pignons".
PINACEAE, 70–100 species.

Piper a very large and important genus of

Carnivorous pitcher plants of the genus Sarracenia growing in a bog in Sarawak, which supplement their nutrition with trapped insects.

A new shoot of a Scots Pine showing clusters of yellow male cones at the base with two pinkish female cones at the tip. (× 1)

tropical small trees, shrubs and woody climbers. Its best-known member is *P. nigrum* [*PEPPER PLANT]. Other important species include *P. betle* [BETLE PEPPER], used in conjunction with the *betel nut, and *P. methysticum* [KAVA PEPPER], source of the Polynesian drink kava. Several species, including *P. cubeba* (= *Cubeba officinalis*) and *P. excelsum*, are used medicinally, principally for treating stomach ailments. *P. aduncum* is cultivated for soil conservation and *P. ornatum* is used as an ornamental.
PIPERACEAE, 1 500–2 000 species.

Piptanthus a small genus of deciduous or evergreen shrubs and small trees native to Asia, with racemes of pea-like yellow flowers. The best-known of the several ornamental species (which require sheltered sites) is *P. nepalensis* (= *P. laburnifolius*) [NEPAL or EVERGREEN LABURNUM], which nevertheless loses its dark green foliage in severe winters.
LEGUMINOSAE, 8–9 species.

Piqueria a small genus of annual or perennial herbs or shrubs native to Mexico and northern Chile, with one species in Haiti. *P. trinervia*, with fragrant white flowers, is the florists' "STEVIA", grown for its winter blooms and also as a bedding plant.
COMPOSITAE, 20 species.

Pisonia a widespread genus of tropical and subtropical trees and a few climbing shrubs. One of the best-known is the small tree *P. umbellifera* [BIRD-CATCHER TREE, PARAPARA], which occurs from Mauritius to Australia and New Zealand. The leaves of *P. grandis* [BROWN CABBAGE TREE], from Polynesia, Sri Lanka, Indonesia and East Africa,

and *P. alba* [LETTUCE TREE] are used locally as a vegetable. Decoctions of the leaves of *P. aculeata*, from southern Florida to tropical South America, the West Indies and southern Asia, and of the fruits of *P. capitata* (= *Cryptocarpus capitatus*), from Mexico, are used in local medicine, especially to treat rheumatism and fevers.
NYCTAGINACEAE, about 50 species.

Pistacia a small genus of resinous shrubs and trees extending from the Mediterranean regions eastwards and southwards to Afghanistan and Malaysia. *P. vera* [PISTACHIO] is the source of the pistachio nut, which is the seed enclosed in the bony shell (endocarp) of the drupe, which partly splits along the ventral suture.

It is a small tree, native to the Near East and West Asia where it is still cultivated, as it is in several east Mediterranean countries, India and the southern USA. Pistachios are valued for their green cotyledons and are used in cake decoration, or are eaten raw or roasted and salted. *Mastic and varnishes are made from resins produced by other *Pistacia* species, notably *P. mutica* [BOMBAY MASTIC] and *P. lentiscus* [MASTIC]. *P. terebinthus* [TURPENTINE TREE, TEREBINTH] is a source of insect galls used in tanning, of a sweet-scented gum and was formerly a source of turpentine.
ANACARDIACEAE, 10 species.

Pistia a tropical and subtropical aquatic genus represented by a single species, *P. stratiotes* [RIVER LETTUCE, WATER LETTUCE, TROPICAL DUCKWEED]. The floating plant consists of a rosette of sessile leaves, the lower side of each bearing a swelling which consists of a spongy air-containing tissue which serves as a float. It is used to coat the surface of tropical ponds and water tanks, thus keeping the water cool and fresh. However, the plant propagates very rapidly and may choke water courses in the tropics.
ARACEAE, 1 species.

Unopened flower of the Common Butterwort (Pinguicula vulgaris), on a glandular stalk. (× 1½)

The Water Lettuce (Pistia stratiotes).

Pisum a very small but important genus of annual herbs, consisting of *P. sativum* and *P. fulvum*. The former species constitutes a complex group which includes the cultivated CULINARY *PEA and FIELD PEA as well as their wild relatives. *P. fulvum*, which is native to the Mediterranean area, has yellow or orange flowers, whereas *P. sativum* has red or white flowers. *Pisum* is closely related to the much larger genera *Vicia* and *Lathyrus*.
LEGUMINOSAE, 2 species.

Pitcairnia a large genus of stemless herbs native to tropical America and the West Indies, with one species (*P. feliciana*) in tropical Africa. Most are terrestrial; the leaves are borne in rosettes and are narrow, usually spiny. The inflorescence is usually a raceme; the flowers are mostly red or yellow. Several species are cultivated in the open in warm countries or under glass in temperate climates, such as *P. corallina* from the Andes of Columbia and Peru with coral-red petals with a narrow white margin.
BROMELIACEAE, 220–250 species.

Pitcher plant the common name given to the carnivorous plants belonging to three families: Sarraceniaceae, Nepenthaceae, and Cephalotaceae, which contain the genera *Sarracenia, *Darlingtonia, *Heliamphora, *Nepenthes and *Cephalotus. Pitcher plants are characterized by their highly modified leaves, which form a pitcher-like vessel with a lid. Insects and other small animals are trapped inside and dissolved to supply the plant with nutrients.

Pithecellobium a genus of mimosoid trees and shrubs of warm temperate, subtropical and tropical regions, chiefly in America and Asia. The flowers are usually white, in heads or spikes, and the fruit is a leguminous pod with thickened valves containing numerous compressed fleshy seeds.

*The crinkled foliage of the Tawhiwhi (*Pittosporum tenuifolium*) which makes it a popular foliage plant for flower arranging. ($\times \frac{1}{2}$)*

Several species are grown for shade in coffee and cacao plantations. The pods may be used as cattle fodder. Others, such as *P. jiringa*, from Borneo, and *P. minahassae*, from Indonesia, yield a useful timber, and some produce tannin. The pods and seeds of *P. dulce* [MANILA TAMARIND, OPIUMA, HUAMUCHIL] and *P. flexicaule* [TEXAS EBONY] are edible. Several species are cultivated as ornamentals.

The South American *P. saman* is now classified as *Samanea saman*.
LEGUMINOSAE, about 170 species.

Pittosporum a large genus of evergreen shrubs and small trees native to warm temperate, subtropical and tropical regions of the Old World. Several species are cultivated in sheltered warm temperate places, such as the New Zealand species *P. crassifolium* [KARO] and *P. tenuifolium* (= *P. nigricans*) [TAWHIWHI, KOHUHU], both with attractive foliage, the Australian *P. undulatum* [VICTORIAN BOX, MOCK ORANGE] and the Chinese and Japanese *P. tobira* [AUSTRALIAN LAUREL, MOCK ORANGE, HOUSE-BLOOMING MOCK ORANGE], all with fragrant, yellowish flowers.

The timber of *P. crassifolium* is tough and is used in New Zealand for cabinet and inlay work.
PITTOSPORACEAE, 100–120 species.

Plagiothecium a genus of mosses found in all five continents but absent from tropical Africa.
HYPNACEAE, about 80 species.

Plane trees the common name for members of the genus *Platanus*. They constitute the most widely planted of large deciduous ornamental trees. Most species, including the ornamental *P. racemosa*, are native to the southwest of North America. *P. occidentalis* [BUTTONWOOD, BUTTONBALL TREE, AMERICAN SYCAMORE OR PLANE] is one of the two major ornamental species, a fast-growing and beautiful tree to 40m high and more, grown also for timber in the southeastern USA.

P. orientalis [ORIENTAL PLANE], from southeastern Europe to western Asia, is smaller and less often planted in the West, notably in the towns of the Balkan peninsula. But the best-known plane tree throughout the world is *P. × hybrida* (= *P. × acerifolia*) [LONDON PLANE], a common sight in many European town squares and avenues and also a common street tree in North America. It probably resulted from a cross between *P. orientalis* and *P. occidentalis*. Like the BUTTONWOOD, the LONDON PLANE has a tall trunk with attractively scaling bark and a beautiful winter silhouette characterized by arching branch tracery and the hanging fruiting balls. The leaves resemble those of the MAPLE.

P. kerrii is an anomalous species from Indo-China, with entire pinnate-veined leaves.
PLATANACEAE, 7–10 species.

Plankton the name used to describe collectively the free-living and mainly microscopic plant and animal organisms found in the sea, in lakes and larger rivers. The plant organisms (phytoplankton) are mainly algae, the animal organisms are known as the zooplankton.

Plantago a large genus of annual or perennial herbs, rarely subshrubs, distributed throughout the temperate regions of the world and on mountains in the tropics. The genus has a number of characteristic species which are endemic to oceanic islands and some of these have developed a shrubby habit. *P. robusta*, from St. Helena, and *P. princeps*, from Hawaii, for example, have pole-like stems crowned by a rosette of long, narrow, leathery leaves.

Some species, such as *P. maritima* [SEA PLANTAIN], have very wide natural distributions which, in the case of *P. media* [HOARY

*The Ribwort Plantain (*Plantago lanceolata). ($\times \frac{1}{4}$)

*The unmistakable crown of the London Plane (*Platanus × hybrida), with its characteristic scaling bark.*

PLANTAIN], *P. major* [COMMON PLANTAIN] and *P. lanceolata* [NARROW-LEAVED PLANTAIN, RIBGRASS, RIBWORT PLANTAIN], have extended to most areas of the world due to introduction by Man. The last three species can be troublesome weeds, especially in lawns and other improved grassland.

Species occasionally cultivated in gardens include the tufted *P. alpina*, the low-growing perennial herb *P. nivalis* and the evergreen shrub *P. cynops*. The young leaves of some species, such as *P. coronopus* [CROWFOOD PLANTAIN, BUCK'S HORN PLANTAIN], are used locally in salads. The seeds of *P. ovata* and other species have medicinal properties.
PLANTAGINACEAE, about 250 species.

Plasmodiophora a small but widespread genus of obligate plant parasites, usually causing an enlargement of existing cells (hypertrophy). The most important species is *P. brassicae*, which causes club root of *Brassica* and other crucifers.
PLASMODIOPHORACEAE, 4–6 species.

Platanus see PLANE TREES.
PLATANACEAE, about 10 species.

Platycerium [STAG'S HORN FERNS, ELKHORN FERNS] a genus of epiphytic ferns native to the tropics and subtropics of Africa, Malaysia, Australia and South America. They bear two kinds of leaves, the sterile ones standing more or less erect and shield-like, and the fertile ones pendulous and normally branched. Most species are cultivated as greenhouse ornamentals, one of the most popular being *P. grande*. Also very popular is *P. bifurcatum*, which has many variants and is useful in hanging basket arrangements.
POLYPODIACEAE, about 17 species.

Platycodon a genus represented by a single species, *P. grandiflorus* (= *P. glaucus*) [BALLOON FLOWER, CHINESE BELLFLOWER], of northeast Asian origin. It is a showy perennial herb cultivated in several varieties for its large, attractive bell-shaped blue or white flowers.
CAMPANULACEAE, 1 species.

Plectranthus a genus of annual to perennial herbs and shrubs from tropical Africa to Japan, Malaysia, Australia and the Pacific islands. They are closely allied to *Coleus.* Several southern African species, such as *P. esculentus*, are cultivated locally for their edible root tubers (Kaffir or Hausa potato). A few are cultivated as greenhouse ornamentals, including *P. oertendahlii* [BRAZILIAN COLEUS, CANDLE PLANT], which was described from cultivation, but is probably African, a prostrate foliage plant with whitish-purple flowers, cultivated for its hanging habit, and the tropical African *P. myrianthus*, a somewhat hairy shrub with large racemes of tubular blue flowers.
LABIATAE, about 250 species.

Pleione a genus of terrestrial pseudobulbous orchids native from the Himalayas eastwards to Taiwan and Thailand. They bear pink, white or yellow flowers, and are imported as bulbs (actually corm-like pseudobulbs) into many countries, where they are a favorite greenhouse or houseplant ["INDIAN CROCUSES"].
ORCHIDACEAE, 10 species.

Plum the name commonly used to describe cultivars of a number of *Prunus* species and hybrids, many of which are of uncertain geographic origin. They are drupe fruits, marketed commercially on a large scale, and now grown in the temperate regions of both hemispheres. The world crop is estimated to be 4 million tonnes.

Plum fruits are globular or oblong, very juicy, yellow, red or purple, with smooth or occasionally slightly rough skins. The stone of these drupe fruits is compressed and longer than it is broad. Cultivated plums evolved in several centers, including North America for *P. americana* [AMERICAN PLUM], Europe for *P. domestica* [EUROPEAN PLUM], western Asia for *P. insititia* [DAMSON PLUM], western and Central Asia for *P. cerasifera* [CHERRY PLUM] and China for *P. salicina* [JAPANESE or ORIENTAL PLUM]. These, as well as many wild species, have been successfully hybridized in various breeding programs to improve such features as fruit quality and environmental adaptibility.

The EUROPEAN PLUM is the most widely grown species. Considered by many authorities to have evolved by hybridization with *P. insititia*, *P. cerasifera* and *P. spinosa* [SLOE], the often large, high quality fruits are very perishable and are therefore usually eaten fresh. However, where the climate permits, many cultivars are dried and marketed as *prunes*. In cooler areas the cultivars tend to bear smaller fruits, more suitable for canning or making into jam. Yugoslavia produces slivovitz (plum brandy) from distilled juices of such plums.

The JAPANESE PLUM is grown principally in those areas enjoying freedom from spring frosts, such as parts of the USA, Italy and South Africa. Most cultivars of this species produce very large, highly colored fruit of inferior quality to the best European types. Their lower perishability, however, permits export of fresh fruit over long distances.

The *DAMSON PLUM is a native of many parts of Europe and Asia. It bears flowers in clusters and produces small, round [BULLACE or MIRABELLE] or oval [DAMSON] fruits. *P. insititia* var *italica* is the popular GREENGAGE.

P. americanus and other North American species, including *P. angustifolia*, *P. hortulana*, *P. maritima*, *P. nigra* and *P. subcordata*, yield rather inferior fruit used principally for jams and jellies. However, plant breeders have produced several successful cultivars from hybrids of American and Japanese species,

Tussock grassland (Poa spp) in New South Wales, Australia, with Eucalyptus woodland beyond.

Victoria Plums (Prunus domestica), one of the most popular of cultivated varieties. (× $\frac{3}{4}$)

combining the hardiness of some of the types with the higher quality and larger fruit size of the Japanese.
ROSACEAE.

Plumbago a genus of shrubs or perennial (rarely annual) herbs occurring in the warmer regions of the world. The flowers are borne in terminal spikes and have a white to blue or reddish salver-shaped corolla. Several species are grown as garden ornamentals, although only *P. europaea*, a perennial herb with violet flowers, and *P. micrantha*, a white-flowered annual, are hardy in cool temperate conditions. *P. auriculata* (= *P. capensis*) [CAPE LEADWORT], a climbing shrub popularly grown for its beautiful pale blue or white flowers, needs protection from frost in temperate conditions.
PLUMBAGINACEAE, about 10 species.

Plumeria a tropical American genus of small trees or shrubs. Some of these are exceedingly elegant and are used as ornamental trees in warm climates. Of these, *P. rubra* (= *P. acutifolia*) [FRANGIPANI, TEMPLE TREE, WEST INDIAN JASMINE, NOSEGAY], native from Mexico to Panama, is by far the most important. It is a small tree growing to about 5m and has fleshy branches with light gray bark. The waxy white flowers, which are often flushed with pink or carmine, are beautifully fragrant. The short tubular corolla opens out into five oval lobes which form a disk almost 5cm in diameter, often with a yellow "eye". *P. alba*, from Puerto Rico and the Lesser Antilles, is a larger tree, bearing white flowers with a yellow "eye".
APOCYNACEAE, about 7 species.

Poa a large cosmopolitan genus of annual and perennial grasses. Economically, meadow grasses are widely cultivated for pastures and lawns. Seeds of *P. trivialis* [ROUGH MEADOW GRASS, ROUGH BLUEGRASS] and *P. pratensis* [SMOOTH MEADOW GRASS, KENTUCKY BLUEGRASS, JUNE GRASS], for example, are mixed with seeds of various *fescues and *rye grasses for sowing artificial permanent

pastures. *P. pratensis*, *P. arachnifera* [TEXAS BLUEGRASS] and *P. compressa* [CANADA BLUE-GRASS, WIREGRASS] are among several important species for cattle on ranges.

Common lawn grasses include the hard-wearing *P. annua* [ANNUAL BLUEGRASS, ANNUAL MEADOW GRASS, DWARF MEADOW GRASS], *P. trivialis*, *P. pratensis* and *P. arachnifera*. Some species, notably *P. annua*, can, under certain conditions, become troublesome weeds.
GRAMINEAE, about 250 species.

Podocarpus a genus of coniferous trees and shrubs commonly called podocarps. Some species were formerly placed in the genus *Taxus* [*YEW] on account of their similar fleshy, edible, cup-like aril. *P. andinus* of southern Chile, in fact, is known as the PLUM-FRUITED YEW. Podocarps are natives of the Old and New Worlds, mainly the temperate Southern Hemisphere, and mountains and highlands of the tropics, northwards to Japan and the West Indies. The genus is unique in having the only known coniferous parasitic species, *P. ustus*, which selects *Dacrydium taxoides* and other members of the same family as its host.

Many species, including *P. totara* [TOTARA PINE, MAHOGANY PINE], from New Zealand, *P. dacrydioides* [KAHIKA, WHITE PINE], from New Zealand, *P. elongatus* [AFRICAN YELLOW-WOOD, FERN PODOCARPUS, WEEPING PODO-CARPUS], from southern and tropical Africa, and *P. gracilior* [AFRICAN FERN PINE], from East Africa, produce valuable timber used in building and for furniture.

A number of these species are also grown as ornamentals, as are several varieties of the Asian *P. macrophyllus* [SOUTHERN YEW, JAPANESE YEW, BUDDHIST PINE].
PODOCARPACEAE, about 100 species.

Podophyllum a genus of perennial herbs from moist, wet or shaded habitats of North America and Asia. They produce mottled foliage and single, white or purple, rose-like flowers on terminal drooping stems. The succulent berries are edible in *P. peltatum* [AMERICAN MANDRAKE, WILD LEMON, RACCOON

Poison Ivy (Rhus radicans) *is one of the most toxic species of* Rhus *and causes severe dermatitis to any who touch it.* ($\times \frac{1}{8}$)

Jacob's Ladder (Polemonium caeruleum), *a hardy and attractive garden perennial about 60cm tall, flowering in early summer. Many varieties and subspecies are in cultivation.* ($\times \frac{2}{3}$)

BERRY, MAY APPLE] and *P. hexandrum* (= *P. emodi*) [HIMALAYAN MAY APPLE]. The latter, with white flowers and scarlet fruit, is commonly grown in gardens. A resin obtained from the dried rhizome and roots of *P. peltatum* is used to treat warts; it also has a violent purgative action.
BERBERIDACEAE, 8–10 species.

Pogostemon a genus of herbs and shrubs mainly native to Indomalaysia and China. The best-known is the commercially grown *P. patchouli* (= *P. heyneanus*, *P. cablin*) [PATCHOULI], a herbaceous soft-wooded shrub native to the Philippines and the East Indies, but cultivated in India and Malaya. The shoots and young leaves when distilled yield an oil (patchouli oil) which is used in the manufacture of soaps and perfumes, although the scent is not universally appreciated.
LABIATAE, about 40 species.

Poinsettia the common name for *Euphorbia pulcherrima*, a shrub native to Central America and Mexico. In cooler climates it is a favorite indoor pot plant, with bright scarlet flowers [CHRISTMAS STAR, CHRISTMAS FLOWER, PAINTED LEAF, LOBSTER PLANT, MEXICAN FLAMELEAF], and it is cultivated throughout the tropics and subtropics as an ornamental garden shrub. The large, highly-colored inflorescence bracts enclose small green or yellowish flowers. There are numerous cultivated varieties with white, pink, or red bracts which may be in double whorls. Some also have variegated foliage.
EUPHORBIACEAE.

Poison ivy the common name for two species of the genus *Rhus*, native from Canada to Guatemala, *R. radicans* [POISON IVY] and *R. toxicodendron* [OAKLEAF POISON IVY].

R. radicans (also called POISON OAK, MER-

CURY, COW ITCH) can grow as an upright shrubby plant to 3m without support or as a trailing or climbing vine developing roots whenever it comes into contact with a potential support.

R. toxicodendron (also called HIEDRA) is nearly always found as a low-growing shrub (to 2m), with upright stems and slender, often downy branches. The leaflets are lobed in a similar fashion to those of some species of *OAK (hence the common name).

The leaves, stems, flowers, fruits and roots of these and related species (eg *R. diversiloba*) contain large quantities of a toxic, volatile, phenolic substance called urushiol. This may be released by bruising or crushing any part of the plant. Even smoke from burning plants can cause cases of poisoning. The most common symptom is a severe skin rash or dermatitis.
ANACARDIACEAE.

Polemonium a genus of usually perennial herbs, mostly restricted to western North America, but with a few Eurasian and one South American species. They mostly grow in meadows and moist ground. A number of species are cultivated as ornamental plants for their bowl- or saucer-shaped, mainly bluish or purple flowers.

P. caeruleum [JACOB'S LADDER, GREEK VAL-ERIAN, CHARITY], and several of its cultivars, are widely cultivated in temperate regions as border plants. Other species are less commonly grown; the smaller ones, such as

A Poinsettia (Euphorbia pulcherrima) *growing in the gardens of the Central University, overlooking Caracas, Venezuela.*

P. reptans, are sometimes grown in rock gardens. *P. caeruleum* and *P. reptans* have local medicinal uses, notably to induce sweating.

POLEMONIACEAE, about 23 species.

Polianthes a small genus of perennial herbs native to Mexico. *P. tuberosa* [TUBEROSE], which is unknown in the wild, is common in cultivation for its spikes of very fragrant, white, funnel-shaped, tubular flowers borne in pairs and used as a cut-flower. There is a popular double cultivar called 'Pearl'. It is grown in southern France for an essential oil derived from the flowers, which is used in high-quality perfumes.

AGAVACEAE, about 12 species.

Polyanthus a very popular race of spring garden flowers (*Primula* × *polyantha*) (= *P.* × *variabilis*), which appear to be a complex group of hybrids of three species: *P. vulgaris* (= *P. acaulis*) [COMMON PRIMROSE], *P. veris* (= *P. officinalis*) [COMMON COWSLIP] and *P. elatior* (= *P. veris* var *elatior*) [OXLIP]. Further selective breeding of *P. elatior* with *P. vulgaris* or *P. veris*, or both, probably produced the POLYANTHUS.

They are hardy herbaceous perennials. Several funnel-shaped flowers are borne in a cluster at the head of a single leafless stem. The flowers are similar to those of other primroses and may be yellow, red, orange, bronze, maroon and white in color. Each flower may have at the center of its corolla a lighter color than the outer part, giving a target arrangement. There are also double-flowered varieties.

PRIMULACEAE.

Polygala an almost cosmopolitan genus of herbs, shrubs and trees. The flower has an overall resemblance to the pea flower. Some species are cultivated for ornament. *P. calcarea* [MILKWORT], from Europe, is a creeping rock plant with purple, pink and

Amphibious Bistort (Polygonum amphibium) growing in water, showing its floating leaves and upright flower-spikes. (× 1/15)

blue flowers; *P. chamaebuxus* [BASTARD BOX], a dwarf evergreen creeping shrub from Europe with purple and yellow flowers, and *P. vayredae* from the Pyrenees, with red and yellow flowers, are also grown in rock gardens. The North American *P. paucifolia* [FRINGED POLYGALA, FLOWERING WINTERGREEN, BIRD-ON-THE-WING] is a more robust dwarf rock shrub with deep red flowers.

Some other species, such as *P. myrtifolia*, from South Africa, and *P.* × *dalmaisiana* (= *P. oppositifolia* var *cordata* × *P. myrtifolia* var *grandiflora*), are more suitable for greenhouses. Some species have local uses, many to treat snakebite, perhaps the best-known being *P. senega* [SENECA SNAKEROOT] from eastern North America.

POLYGALACEAE, about 500 species.

Polygonatum [SOLOMON'S SEAL] a genus of perennial herbs from the north temperate zone. The simple stems arise from a fleshy, horizontally jointed sympodial rhizome. The pendulous, white or greenish flowers are solitary or in few-flowered racemes. The most commonly cultivated varieties are hybrids between the Eurasian species, *P. multiflorum* [COMMON SOLOMON'S SEAL, DAVID'S HARP] and *P. odoratum* (= *P. officinale*).

LILIACEAE, about 30 species.

Polygonum [KNOTWEED, SMARTWEED, FLEECE FLOWER] a very diverse genus of mostly annual and perennial herbs with a very wide distribution, particularly in temperate regions. The often pink or white flowers are clustered in spikes or panicles and the prolific fruits are three-sided. Among the common and persistent farm and garden weeds are *P. persicaria* [RED SHANK], a sprawling branched annual, *P. aviculare* [KNOT GRASS], a dwarf, prostrate, fast-growing annual, and

P. convolvuloides, a climbing plant with small white flowers. The North American *P. coccineum* [WATER SMARTWEED] is an aquatic or semi-aquatic perennial with spikes of small rose-colored flowers.

A number of handsome, vigorous, perennial herbaceous species with pink and crimson flowers are cultivated in gardens. These include *P. cuspidatum* [MEXICAN BAMBOO, JAPANESE POLYGONUM], *P. chinense*, *P. baldschuanicum* [RUSSIAN VINE, BUKHARA FLEECE FLOWER], a rampant, deciduous, woody climbing species which flowers in foaming white masses, and *P. aubertii* [CHINA FLEECE VINE, SILVER LACE VINE], a twining perennial vine with long erect panicled racemes of white flowers. *P. affine* and *P. vacciniifolium* provide mats of excellent ground cover with their deep pink upright spikes of flowers. *P. campanulatum* and *P. bistorta* [BISTORT, SNAKEWEED], are of moderate size and attractive growing habit for borders and intershrub planting.

P. sachalinense [GIANT KNOTWEED, SACALINE], a coarse perennial herb from Sakhalin, is sometimes grown as forage or as a screen plant in parts of North America. The young shrubs are used as food in its native habitat.

POLYGONACEAE, about 150 species.

Polypodium a very complex, cosmopolitan genus of ferns which some authorities have now subdivided into at least 20 separate genera. Under such a system, *Polypodium* now consists of only a relatively small number of species which differ from each other in frond shape, number of indurated (thick walled) sporangial cells and other microscopic characters. Most species have creeping rather than erect rhizomes and bear simple or compound leaves with one or more rows of circular, non-indusiate sori on each side of the midrib. Both epiphytic and terrestrial species have evergreen, rather leathery fronds.

One of the most popular species for culti-

The Common Milkwort (Polygala vulgaris) growing on a sand dune. It is also found on heathland and grassland. (× 1/2)

vation out of doors is *P. vulgare* [EUROPEAN POLYPODY, WALL POLYPODY, WALL FERN, ADDER'S FERN], with deeply pinnatifid leaves, variously crested, dissected and plumed in the numerous cultivars. The North American species *P. virginianum* [ROCK POLYPODY, AMERICAN WALL FERN] and *P. glycyrrhiza* (= *P. vulgare* var *occidentale*) [LICORICE FERN] are somewhat similar in habit to *P. vulgare*.

The most commonly cultivated American species is *P. aureum* (= *Phlebodium aureum*) [RABBIT'S-FOOT FERN, HARE'S FOOT FERN, GOLDEN POLYPODY], with surface-creeping rhizomes and pinnatifid leaves up to 1.3m long with sori often bright yellow in color. There are numerous cultivars, including 'Mandaianum' [BLUE FERN] and the crested-leaved 'Cristatum'.
POLYPODIACEAE, about 35 species.

Polyporus a genus which formerly included most of the so-called bracket fungi but is now greatly reduced. Most occur on wood and are worldwide in distribution. A number of species have a central stipe. An important parasitic species is *P. squamosus* [DRYAD'S SADDLE], a handsome fungus, but one which causes a serious heart rot of elms. The young fruiting bodies of many species, including *P. frondus* and *P. squamosus*, are edible. Extracts from *P. officinalis* and some other species have medicinal properties.
POLYPORACEAE, about 50 species.

Pomegranate trees (Punica granatum) *in an irrigated orchard in Tunisia.*

Polytrichum commune, *a common moss of wet acid woodland, showing mature capsules.* (× 3)

Polysiphonia a genus of filamentous red algae found on the seashore, often growing as epiphytes upon other algae. The plant is constructed of a single column of axial cells around which are arranged a variable number of pericentral cells (siphons).
RHODOPHYCEAE.

Polystachya a pantropical genus of epiphytic and rock-inhabiting orchids, reaching their greatest variety and abundance in tropical Africa. They have thickened or pseudobulbous stems, two-ranked leaves and racemose inflorescences. The majority are green- or white-flowered and are rarely cultivated, with the exception of the pantropical *P. flavescens* (= *P. luteola*, *P. minuta*), with bright yellowish-green flowers, the yellow-flowered *P. bella* from Kenya, and *P. foliosa* (= *P. stenophylla*) from the West Indies and tropical America.
ORCHIDACEAE, about 200 species.

Polystichum a genus of terrestrial woodland ferns, cosmopolitan in distribution. They have erect rhizomes and bear elongate pinnatifid to bipinnate leaves with serrated margins. Sori, borne on the underside of the leaves, are protected by orbicular indusia. A number of wild species are cultivated as garden or greenhouse plants.

P. aculeatum [HARDSHIELD FERN, PRICKLY SHIELD FERN, HEDGE FERN] and *P. setiferum* [SOFT SHIELD FERN, ENGLISH HEDGE FERN] are widespread. *P. acrostichoides* [CHRISTMAS FERN, DAGGER FERN, CANKER BRAKE] is popular in the USA for Christmas decoration. Other popular cultivated species include the North American *P. scopulinium* [WESTERN HOLLY FERN, EATON'S SHIELD FERN] and *P. andersonii* [ANDERSON'S HOLLY FERN].
POLYPODIACEAE, about 120 species.

Polytrichum an important genus of mosses, worldwide in distribution. *P. juniperinum* is cosmopolitan and several others occur in every continent. *P. commune* and *P. strictum* are important tussock-formers in moorland and several species are prominent in the Arctic and Antarctic. A kind of rope fashioned from the long stems of *P. commune* was once used in prehistoric boat construction.
POLYTRICHACEAE, 70–220 species.

Pomaderris a genus of evergreen trees or shrubs from Australia and New Zealand, with usually downy rust-colored leaves, and inflorescences of small greenish flowers. Some species, including *P. apetala* and *P. rugosa*, are grown as ornamentals or street trees in warm regions of North America.
RHAMNACEAE, 40–50 species.

Pomegranate the common name for the small tree *Punica granatum* (also known as APPLE OF CARTHAGE) and its edible fruit. It is native to Iran and has been cultivated since ancient times. It is now widely grown in the Mediterranean, where it is also naturalized, and has been introduced into California and to most of the tropics and subtropics, especially in semi-arid conditions.

The POMEGRANATE has spiny-tipped branches and produces clusters of up to five flowers on axillary shoots which are yellow, orange or red, bell-shaped and about 2.5cm in diameter. The round fruits are about 9cm across, with tough, leathery, orange to red skins containing numerous seeds each of which is surrounded by a juicy pink to red-purple pulp.

The fruit is often eaten fresh or alternatively the juice may be used as a refreshing drink known as grenadine. The peel of the fruit is rich in tannins which can be used to produce excellent leather, while the bark is said to make a useful vermifuge. There are many cultivars, such as 'Nana' [DWARF POMEGRANATE]. The variety *legrellei* is an ornamental form with large, double, coral-red flowers and interesting foliage.
PUNICACEAE.

Pomelo the common name for *Citrus maxima* (= *C. grandis*) (also known as POMPELMOUS, PUMELO and SHADDOCK), a small tree probably native to tropical Asia but cultivated throughout the tropics, especially in south and Southeast Asia, for its large, round or ovoid fruits (sometimes weighing up to 10kg), which are like outsize *grapefruits* with very thick rind. It is regarded as one of the parents of the grapefruit and the seedlings are often used as stocks for grafting on orange scions.
RUTACEAE.

Pompom the name used for certain varieties of border *dahlias* with small globular inflorescences. The same term is used for the varieties of *Chrysanthemum* with small globular flower heads consisting of tightly packed florets.

*The Alpine Poppy (*Papaver alpinum*); cultivated varieties have a range of colors from white to yellow, orange and pink. (× ⅓)*

Poncirus a genus consisting of a single species, *P. trifoliata* [TRIFOLIATE ORANGE, GOLDEN APPLE, HARDY ORANGE], from China. A citrus-like deciduous spiny shrub bearing white flowers and yellow, globose fruits, it is sometimes grown as an ornamental or hedging plant in many countries. It provides an important cold-resistant rootstock for *Citrus* breeding.
RUTACEAE, 1 species.

Pontederia a genus of emergent aquatics, distributed throughout the New World. The genus is of little economic significance although the widely spread *P. cordata* [PICKEREL WEED] occurs as a weed of rice in South America and may be cultivated in ponds and on damp ground.
PONTEDERIACEAE, 5 species.

*The Californian Poppy (*Eschscholzia californica*), a hardy annual much grown in gardens, and widely naturalized in warm, dry areas. (× ¼)*

Poppy the common name given not only to the herbaceous annual and perennial members of the genus *Papaver* (true poppies), but also to species of other genera in the same family, such as *Argemone*, *Eschscholzia*, *Glaucium*, *Hunnemannia*, *Macleaya*, *Meconopsis* and *Romneya*, many of which are cultivated as garden ornamentals.

In addition to their value as ornamentals some poppies are economically important. *Papaver somniferum* is the source of opium (see OPIUM POPPY). The seeds of *Glaucium flavum* are the source of an oil used in food and in soap making. *Argemone glauca* [HAWAIAN POPPY] has edible seeds and the seeds of *A. mexicana* [MEXICAN POPPY] are the source of argemone oil used in soap manufacture.
PAPAVERACEAE.

Populus [ASPENS, POPLARS] a genus of deciduous trees with soft white wood, found from Alaska to Mexico, from North Africa through Europe, Asia Minor and the Himalayas to China and Japan. They are nearly all fast-growing trees which can grow to a considerable height.

Many of the species are cultivated as ornamentals for their attractive growth habit and foliage. They include *P. tremula* [EUROPEAN ASPEN], *P. tremuloides* [AMERICAN or QUAKING ASPEN], *P. alba* [SILVER-LEAVED or WHITE POPLAR], *P. nigra* [BLACK POPLAR], *P. nigra* 'Italica' [LOMBARDY POPLAR], the hybrid *P.* × *canescens* [GRAY POPLAR], *P. lasiocarpa* [CHINESE NECKLACE POPLAR], *P. deltoides* [COTTONWOOD], *P.* × *canadensis* (= *P. deltoides* × *P. nigra*) [CAROLINA POPLAR].

The balsam poplars are native to North America, Siberia and East Asia. They include *P. balsamifera* [BALSAM POPLAR, HACKMATACK, TACAMAHAC], and *P.* × *gileadensis* [BALM-OF-GILEAD], a presumed hybrid of *P. balsamifera* and *P. deltoides*. However, the most handsome species is *P. trichocarpa* [WESTERN BALSAM POPLAR, BLACK COTTONWOOD] which can reach a height of 60m in its native range on the Pacific coast of North America and can grow 2.5m in a year in Britain where it is commonly planted. The decorative 'Aurora' is a balsam poplar with dark green leaves, variegated white, cream or pink over much of their surface.

The chief economic use of poplars is for their light timbers, much used for wood pulp, matches, matchboxes and light packing cases, and in planting as windbreaks or as ornamentals, frequently lining avenues.
SALICACEAE, 35–40 species.

Porphyra an important cosmopolitan genus of parenchymatous red algae found attached to rocks in the intertidal zone of the seashore. The plants of this genus are often eaten as laver bread in Europe or as "nori" in Japan, where it is especially cultivated.
RHODOPHYCEAE.

Portulaca a widespread genus of annual or perennial leaf-succulents. *P. grandiflora* [SUN PLANT, ROSE MOSS, ELEVEN O'CLOCK], a native of

Portulaca grandiflora, popularly called Rose Moss, Sun Plant or Eleven o'clock, is often grown in dry, sunny gardens. (× ½)

Brazil, Uruguay and Argentina, is a bedding annual grown in dry sunny places. It has large single or double flowers in brilliant red, yellow or white and intermediate shades. *P. oleracea* [COMMON PURSLANE, PURSLEY], with its thick soft prostrate stems, is a cosmopolitan weed, possibly originating in India, that has long been used in salads and as a potherb as var *sativa* [KITCHEN GARDEN PORTULACA]. The cultivar 'Giganthes' is sometimes grown as an ornamental.
PORTULACACEAE, about 100 species.

Posidonia a very small genus of Mediterranean and Australian submerged, aquatic herbs. *P. australis* is the source of posidonia fiber, cellonia or lanmar. Mixed with wool or on its own, it is used for making coarse fabrics and sacking.
POSIDONIACEAE, 3 species.

Potamogeton [PONDWEEDS] the largest exclusively aquatic genus of flowering plants. It is found throughout the world; some species are amphibious with leathery floating leaves (eg *P. natans*) and others have thin transparent leaves that cannot tolerate emergence (eg *P. gayii*), while others still are heterophyllous and have floating and submerged leaves (eg *P. gramineus*).

Potamogeton species are important for animal fodder, although they can be a nuisance in irrigation canals and in ditches. The winter bud (turions) of several species and the rootstock of one species (*P. natans*) also provide emergency food for humans. They are sometimes grown in ponds and aquaria.
POTAMOGETONACEAE, about 100 species.

Potato The common name for *Solanum tuberosum*, one of the world's major staple crops. It is a herb which perennates naturally by underground stem tubers and in cultivation is propagated vegetatively by "seed"

Above Top *The Colorado Beetle (*Leptinotarsa decemlineata*) is a serious pest of Potato plants in America; both the larva (left) and the adult beetle (right) eat the leaves. It has been kept out of most of Europe by strict quarantine regulations. (×1½)*
Above *Young Potato sprout showing the embryo leaves and young roots. (×10)*

tubers. Wild potato tubers were first eaten in South and Central America, where they are still collected today. The cultivated potato has arisen as the result of selection of alkaloid-free diploids which are less toxic and bitter, and safer to eat. Possibly, original weed species such as *S. brevicaule*, *S. leptophyes* and *S. canasense* were superseded under domestication by cultivated tetraploids such as *S. andigena* (= *S. tuberosum* subspecies *andigena*), as well as a few triploid potatoes. Continued potato breeding in the 19th century in Europe and North America resulted in the production of numerous new varieties ascribed to *S. tuberosum* subspecies *tuberosum*.

Potatoes are grown as a summer annual in cool countries and as a winter crop in the subtropics. The main areas of cultivation are in the north temperate zone and the world crop is estimated as about 300 million tonnes, of which 90% is grown in Europe and the USSR.

Most of the potatoes grown in Europe are of maincrop maturity, being harvested in late summer or early autumn. "Early" varieties

mature more quickly and may even be lifted as early as June in favored, frost-free places in which very early planting is possible.

Nutritionally the potato is outstanding. It is basically a wet starchy food but the protein is relatively abundant and of excellent quality. It is a major source of vitamin C in north temperate diets. In prosperous countries, the crop no longer retains the immense social and nutritional importance it had 100 years ago but its (probably great) potential for feeding people in warm countries is as yet virtually unexploited.
SOLANACEAE.

Potentilla a large genus of herbs, shrubs and subshrubs widespread over the north temperate and arctic region, with some representatives in the Southern Hemisphere. Several are cultivated as garden ornamentals.

The best-known shrubby species is *P. fruticosa* [SHRUBBY CINQUEFOIL, GOLDEN HARDHACK, WIDDY], with a distribution range extending over the whole of the Northern Hemisphere. The typical form is a twiggy shrub up to 1.5m high with bright yellow flowers, but there is a large number of cultivated forms varying in habit, flower color etc. Other common species are *P. anserina* [SILVERWEED, GOOSE-GRASS, GOOSE TANSY], *P. argentea* [HOARY or SILVERY CINQUEFOIL] and *P. davurica* (= *P. arbuscula*), which has a low spreading habit and large yellow flowers. *P. mandshurica* and cultivar 'Veitchii' of *P. davurica* are white-flowered forms. *P. reptans* [CINQUEFOIL] is a common wild species found over much of Europe and temperate Asia. The white-flowered European species *P. sterilis* is often called the BARREN STRAWBERRY because of its resemblance to the COMMON *STRAWBERRY (*Fragaria*).
ROSACEAE, about 350 species.

Powdery mildews see Fungi, p. 152.

Potentilla eriocarpa, *a native of the western and central Himalayas. (×⅓)*

Primula hirsuta *is a European species found in the Alps and the Pyrenees, usually growing on wet rocks in partial shade. (×1/10)*

Primula [PRIMROSES] a large genus of perennial herbs mostly native to the Northern Hemisphere, with the chief centers of distribution probably being Assam and mountainous areas of Southeast Asia. A few species are endemic in areas of temperate South America.

The flowers are often borne in umbels on leafless stalks, as in *P. veris* [COWSLIP], *P. elatior* [OXLIP] and *P. auricula* [AURICULA], or singly, as in *P. vulgaris* (= *P. acaule*) [PRIMROSE]. There is a wide range of color with white, yellow, pink and purple predominant.

Among the better-known European species of *Primula* are the yellow-flowered *P. vulgaris*, *P. farinosa* [BIRD'S EYE PRIMROSE], with rose-lilac flowers, *P. veris* with deep yellow flowers, and *P. elatior* with pale yellow flowers.

A number of other cultivated species are native to the Alps, including *P. allionii* from the Maritime Alps, *P. marginata* and *P. auricula*, some races of which may be of hybrid origin. *P. sinensis*, *P. obconica* and *P. malacoides* [FAIRY or BABY PRIMROSE] are a few of the better-known species of Chinese origin. *P. japonica* [JAPANESE PRIMROSE] and *P. sieboldii* are native to Japan, while the Himalayas are the source of many species including *P. rosea* and *P. denticulata* [DRUMSTICK PRIMROSE].

Many of these species are cultivated in a range of varieties, as are hybrids between closely related species, eg *P.* × *kewensis* (= *P. verticillata* × *P. floribunda*). A number of species can only be grown successfully in a heated greenhouse, including *P. malacoides*, with many-flowered umbels of star-like flowers in white, rose, lavender and brick-red cultivars, some of which are double-flowered, *P. obconica* [GERMAN PRIMROSE,

POISON PRIMROSE], in predominantly pink-, lilac- or lavender-colored cultivars, and *P. sinensis* [CHINESE PRIMROSE], with cultivars in a wide range of flower colors and form, including 'Fimbriata', with fringed or crested flowers.

Outdoor species are hardy perennials, some of which (alpine species) are grown in rock gardens and alpine houses, while others (border species) require moist soil and generally have more showy flowers. Among popular border primulas with the flowers arranged in whorls up the stem are *P. aurantiaca*, from China, *P. denticulata*, from the Himalayas, *P. japonica* and *P. sikkimensis* [HIMALAYAN COWSLIP], from the Himalayas. (See also POLYANTHUS.)
PRIMULACEAE, about 500 species.

Prionium a shrubby aloe-like perennial genus represented by a single species, *P. serratum* (= *P. palmitum*) [PALMIET], endemic to South Africa, growing beside water. The stout stem, covered with black fibrous remains of old leaves, bears a dense rosette of leaves at its apex. The inflorescence is a large dense panicle of small greenish-gold flowers. The plant is sometimes grown for ornament while the black fibers of the leaf sheath are used for making brushes.
JUNCACEAE, 1 species.

Privet the common name for *Ligustrum*, a genus of shrubs to tall trees, distributed from Europe to northern Iran and from East Asia and Indomalaysia to New Guinea and Queensland, Australia. There are a number of attractive ornamental evergreen species, including *L. vulgare* [COMMON PRIVET] grown normally as a hedging plant. *L. ovalifolium* [GOLDEN PRIVET] with glossy ovate mid-green leaves each with an irregular yellow border, and with panicles of fragrant cream flowers,

is cultivated both as a flowering shrub and as an effective hedging plant. *L. lucidum* [CHINESE, NEPAL or GLOSSY PRIVET, WHITE WAX TREE] is widely cultivated as a street tree for its attractive ovate dark green glossy leaves although most of its cultivars require protection from cold winds.

Other uses include roasting the seeds as a coffee substitute, especially those of *L. japonicum* [WAX-LEAF PRIVET, JAPANESE PRIVET]. Insect damage to the stem induces the formation of a wax in *L. ibota* [IBOTA PRIVET] and *L. lucidum* which is used industrially in China. *L. vulgare* has very hard wood used for tools. It also makes good charcoal.
OLEACEAE.

Prosopis a genus of mainly spiny, mimosoid trees with bipinnate leaves, mostly of western America, from Peru to Colorado, but with a few species from western and tropical Asia and tropical Africa. The pods and seeds and young shoots of many species are edible. The pods contain a sweet mucilage similar to that of the *CAROB TREE.

Edible species include *P. alba* [ALGARROBO BLANCO], *P. juliflora* [ALGARROBO BEAN, MESQUITE] and *P. dulcis*, [CASHAU or PACCAI]. *P. glandulosa* (= *P. chilensis*) [MESQUITE] grows over large tracts of its native southern USA and north Mexico, the shoots and pods of such varieties as *glandulosa* [HONEY MESQUITE] and *torreyana* [WESTERN HONEY MESQUITE] providing a valuable cattle food. The pods of *P. alba* and *P. juliflora* are also ground whole to give a nutritious meal.

In Asia, *P. dulcis* is cultivated extensively for the sweet pulp surrounding the inedible

Flowers of the English Primrose (Primula vulgaris), *showing on the left the thrum-eyed form and on the right the pin-eyed form.* (× 3)

Cowslips (Primulae veris) *bear clusters of flowers on leafless stalks.* (× 1)

black seeds in the long pods. The pulp of the pods of *P. glandulosa* may be fermented to produce an alcoholic beverage, while the twisted pods of *P. pubescens* [TORNILLO, SCREW BEAN] contain so much sugar that boiling produces molasses. The valuable timber of *P. nigra* and *P. alba* and the tannin obtained from *P. pallida* and *P. dulcis* are also called algarrobo.
LEGUMINOSAE, about 30 species.

Prostanthera [MINTBUSHES] an Australian genus of aromatic evergreen subshrubs, shrubs or small trees. A number of species are in cultivation, a few being sufficiently hardy to be grown out of doors in cooler temperate regions. Cultivated ornamental species include the tall shrub *P. lasianthos* [VICTORIAN CHRISTMAS BUSH, DOGWOOD], with fragrant, white, pink or pale blue tinged or spotted flowers, the evergreen shrub *P. rotundifolia* [ROUND-LEAF MINTBUSH], with lilac flowers, the bushy shrub *P. sieberi*, with rose-violet flowers, and the shrub *P. nivea* [SNOWY MINTBUSH], with white, sometimes blue-tinged flowers.
LABIATAE, about 50 species.

Protea a genus of evergreen shrubs and some small trees with colorful flower heads, found mainly in the Cape region of South Africa, but extending into tropical Africa. A number of species are grown for ornament and as quality florists' flowers.

P. cynaroides [GIANT PROTEA, KING PROTEA] has the largest heads, sometimes exceeding 30cm across, with pale red outer bracts having a velvet sheen, and a boss of numerous woolly lilac flowers, yellow at their base. There are many other beautiful proteas such as *P. mellifera* [SUGARBUSH, HONEY FLOWER], *P. grandiceps* [PEACH PROTEA] and *P. pulchella*. In all of these species the colored bracts are an important decorative feature.
PROTEACEAE, about 100 species.

Prune a dried *plum derived from varieties of *Prunus domestica* [EUROPEAN PLUM] which have a high sugar content. Prunes are marketed mainly by the USA (principally in the Great Lakes and Pacific coast areas), France and Yugoslavia. Fresh prunes keep longer than other plums principally because they have a firmer flesh and a higher sugar content. Although prunes may be eaten fresh or as canned fruit they are normally dried after falling from the tree. Fermentation is prevented by dipping in lye and the fruits are either dried directly under the sun or in kilns. ROSACEAE.

Prunella a genus of perennial herbaceous plants native to Europe, Asia, North Africa and North America. A number of species, such as *P. grandiflora* and *P. webbiana*, are grown in rock gardens for their attractive rose-purple tubular flowers. The purplish-blue-flowered *P. vulgaris* [*ALL HEAL, SELF-HEAL], among other species, is reputed to have medicinal properties.
LABIATAE, 7–9 species.

Prunus a large Central Asian genus subdivided into five main subgenera, namely *Prunophora* [*PLUMS], *Amygdalus* [*PEACHES and *ALMONDS], *Cerasus* [*CHERRIES], *Padus* [BIRD CHERRIES] and *Laurocerasus* [PORTUGAL and CHERRY LAURELS]. Species that have made major contributions to commercial fruit production include *P. domestica* [EUROPEAN PLUM], *P. insititia* [*DAMSON PLUM], *P. cerasifera* [CHERRY PLUM], *P. salicina* [JAPANESE PLUM], *P. americana* [AMERICAN PLUM], *P. armeniaca* [APRICOT], *P. persica* [PEACH], *P. amygdalus* [ALMOND], *P. avium* [SWEET CHERRY] and *P. cerasus* [SOUR CHERRY].

The so-called FLOWERING CHERRIES are the most outstanding and widely cultivated of all ornamentals in the genus *Prunus*. The number of cultivars and selections in cultivation run into hundreds, from *P. avium* 'Plena', the finest of the geans, to the outstanding Japanese flowering cherries, such as *P. serrulata* 'Kwanzan', with its massive clusters of double purple-pink flowers. These Japanese cherries are mostly derived from the white-flowered OSHIMA CHERRY (*P. speciosa* × *P. lannesiana* forma *albida*) or from the very closely related *P. serrulata*.

The BIRD CHERRIES and CHERRY LAURELS belong to the subgenera *Padus* and *Laurocerasus* respectively. Both are characterized by having flowers in racemes of more than 10 flowers. The BIRD CHERRIES are typically deciduous and the CHERRY LAURELS always evergreen. The species commonly known as the CHERRY LAUREL (*P. laurocerasus*) is a vigorous shrub, widely planted for hedging and screening, and tolerant of shady positions. The leaves yield hydrocyanic acid on injury, and when torn into small pieces are used as the active agent in killing-bottles for small insects.
ROSACEAE, about 200 species.

Pseudolarix a genus comprising a single species, *P. kaempferi* (= *P. amabilis*) [GOLDEN

Prunella grandiflora *growing in a meadow. This species is native to Europe and is cultivated in rock gardens.* ($\times\frac{1}{5}$)

LARCH], which is native to eastern China (Chekiang and Kiangsi provinces). A second species, *P. pourteli*, from central China, has been proposed but this is known only from vegetative material.

P. kaempferi is a deciduous tree reaching 40m. Its leaves are needle-like and are produced either singly and spirally arranged on long shoots which are rough with persistent bases of shed leaves, or in almost umbrella-like clusters on short shoots which are characteristically club-shaped and curved with persistent scales and distinct, close-set annual rings separated by constrictions. The cones are not unlike those of *Larix, the males clustered in catkins about 2.5cm across, the females on the same tree, about 5 × 1cm, comprising thick, acuminate, woody scales which break up at maturity, releasing the seeds.

The GOLDEN LARCH makes a splendid ornamental tree, the leaves turning a rich golden-yellow towards the end of the season. It is quite hardy in the warmer parts of temperate regions but slow growing. Propagation is best from seed, which is often abundantly set in some areas, for example Italy.
PINACEAE, 1 species.

Pseudomonas a genus of gram-negative, rod-shaped bacteria which possess polar flagella. The species of this genus are among the commonest and most widely distributed bacteria and are found in aquatic and marine environments, in soil and sewage. Some species, eg *P. phaseolica*, are important as causal agents in plant disease.
PSEUDOMONACEAE.

Pseudotsuga see DOUGLAS FIR.
PINACEAE, about 5 species.

Psidium a large genus of shrubs and trees native to tropical and subtropical America and the West Indies. A number are cultivated (notably in California, Florida, Cuba and India) for their edible fruits, especially *P. guajava* [GUAVA, YELLOW GUAVA, APPLE GUAVA], *P. littorale* var *longipes* [PURPLE GUAVA, STRAWBERRY GUAVA] and var *littorale* [YELLOW STRAWBERRY, YELLOW CATTLEY]. The fruits are round or pear-shaped, sweet and aromatic; they are eaten raw or made into jelly or pies and the juice is used in drinks and ices. The fruit of *P. guajava* is also made into a jam known as "guava cheese".
MYRTACEAE, about 130 species.

Psilocybe a genus of gill fungi with a cosmopolitan distribution. A number of species, including *P. mexicana*, turn blue on bruising, cutting or handling and have long been consumed by Mexican Indians for their hallucinogenic properties. At least two such

The Japanese Flowering Cherry (Prunus serrulata) *comes in many varieties, all of which produce an outstanding display of blossom.* ($\times\frac{1}{4}$)

Winged fruits and trifoliolate leaves of the Hop Tree (Ptelea trifoliata), a native of North America and Mexico. (×1)

active principles have been isolated and their formulae determined. They are psilocybin and the closely related psilocin. Some species are poisonous and deadly.
STROPHARIACEAE, about 75 species.

Psilotum see Club mosses and their allies, p. 92.

Psophocarpus a widespread tropical genus of climbing beans. The best-known species, cultivated for its immature pods, is *P. tetragonolobus* (see ASPARAGUS PEA). The young pods of the tropical African *P. palmettorum* are also eaten.
LEGUMINOSAE, 10 species.

Psoralea a widespread genus of leguminous, scented herbs and shrubs from tropical, subtropical and warm temperate zones. The plants are usually warty or covered with glandular dots, and have pea-like flowers, solitary or in racemes, spikes or clusters.
Among several cultivated species are *P. foliosa* [OXFORD AND CAMBRIDGE BUSH], from Kenya, with both light and dark blue flowers, *P. pinnata*, from South Africa, with solitary blue-striped flowers with white wings, and *P. esculenta* [INDIAN TURNIP, BREADROOT, PRAIRIE POTATO], from North America, with yellowish to blue flowers. The taproots of some species, such as *P. esculenta*, are eaten locally.
LEGUMINOSAE, 115–130 species.

Ptaeroxylon a genus consisting of a single tree species, *P. utile* [CAPE MAHOGANY], native to South Africa. Its aromatic timber has local commercial uses for furniture and construction work.
MELIACEAE, 1 species.

Ptelea a small genus of small shrubs and trees native to North America and Mexico.

Only one species, *P. trifoliata* [HOP TREE, WATER or STINKING ASH], is cultivated to any extent although in some regions such as central Europe it has become naturalized. The HOP TREE reaches 8m and bears fragrant, greenish-white flowers. The fruit is a distinctive round and winged samara which becomes greenish-yellow when ripe. It is an extremely variable species in nature and in addition a number of cultivars exist, such as 'Aurea' with bright yellow leaves.
RUTACEAE, 3–6 species.

Pteridium a genus of ferns comprising a single cosmopolitan species, *Pteridium aquilinum* [BRACKEN]. It is extremely variable and is divided into two subspecies and several varieties (see also BRACKEN).

Pteris a large cosmopolitan genus of mostly tropical, terrestrial ferns with leaves that are once to four times pinnate, and sori in linear rows along the margins of the pinnules. The sporangia are protected when young by the overlapping leaf margin. Many species are popular greenhouse plants, such as *P. cretica* [CRETAN BRAKE, RIBBON FERN], *P. dentata* [TOOTHED BRAKE], *P. ensiformis* [SWORD BRAKE], *P. quadriaurita*, especially cultivar 'Argyraea', and *P. umbrosa*.
PTERIDACEAE, about 250 species.

Pterocarpus a genus of trees and climbers, distributed throughout the tropics but chiefly in the Old World. It includes many useful plants which provide valuable timber, used for cabinetwork and decorative paneling as well as for general carpentry and construction work, resin or dyes. Many are known by the common names PADOUK or PADAUK. The most important timber-producing species are *P. erinaceus* [SENEGAL ROSEWOOD, WEST AFRI-

CAN KINO], *P. soyauxii* [AFRICAN PADOUK, BAYWOOD] and *P. angolensis* [MUNINGA, BLOODWOOD], from Africa, the Indian *P. santalinus* [SANDALWOOD PADAUK, RED SANDERSWOOD, RED SANDALWOOD], the east Indian *P. marsupium* [VENGAI PADOUK], *P. indicus* [MALAY PADAUK, AMBOYNA, BENGAL ROSEWOOD, ANDAMAN REDWOOD], from the Malay Peninsula, the Burmese *P. macrocarpus* [RED or BROWN PADOUK] and *P. dalbergioides* [ANDAMAN PADAUK or BASTARD TEAK], endemic to the Andaman Islands but now grown in India as a teak substitute. *P. erinaceus* and *P. marsupium* have a resinous astringent sap called kino. *P. santalinus* and *P. draco* [DRAGON GUM TREE] are two of several species yielding a red dye.
LEGUMINOSAE, 30–36 species.

Pterocarya [WING NUTS] a small genus of large trees mostly Asiatic in distribution, especially China. The WING NUTS grow into large deciduous trees with many-branched, rounded crowns. The fruit is a small one-seeded nut with two leafy wings, hence the common name. The species most often grown as an ornamental is *P. fraxinifolia* [CAUCASIAN WING NUT], from the Caucasus and Iran.
JUGLANDACEAE, about 8 species.

Puccinia the largest genus of the rust fungi, containing numerous species parasitic on many families of angiosperms. The best-known and most harmful to Man is *P. graminis* (black stem rust of cereals).
P. graminis tritici, usually found on wheat, also infects barley and there are other forms on oats, rye, and various grasses. Each form contains numerous physiological races, each capable of infecting a different combination of host cultivars. Nearly 2000 such races are known in *P. graminis tritici* alone.
On barley, *P. hordei*, known in Europe as brown rust and in North America as leaf rust, causes considerable damage. There is also a

Brown Rust (Puccinia recondita) *on the leaf blade of a wheat plant, showing pustules of spores erupting through the leaf surface. (×8)*

brown or leaf rust of wheat and rye, *P. recondita*. On maize, *P. sorghi*, maize rust (alternate host, **Oxalis*), and *P. polyspora*, tropical maize rust (alternate host unknown), can cause considerable damage to the crop.
PUCCINIACEAE, 3000–4000 species.

Puffballs or **stomach fungi** the class Gasteromycetes of the basidiomycetes. It includes the genera **Lycoperdon* and *Langermannia* [PUFFBALLS], *Geastrum* [*EARTH STARS*], *Scleroderma* [*EARTH BALLS*], *Cyathus* and *Crucibulum* [*BIRD'S NEST FUNGI*], *Phallus*, *Clathrus* and *Dictyophora* [*STINKHORNS*].

Pulmonaria a small genus of low-growing perennial herbs mainly from Europe and northwest Asia. The spotted leaves of *P. officinalis* [LUNGWORT, SPOTTED DOG, JERUSALEM SAGE, JERUSALEM COWSLIP] supposedly resemble diseased lungs and were used, in homeopathic medicine, as a cure for lung diseases. A number of other species have flowers which are first pink or red and then turn blue or purple. Apart from *P. officinalis*, other species cultivated as garden plants for their tubular flowers include *P. angustifolia* [BLUE COWSLIP] and *P. saccharata* [BETHLEHEM SAGE], with whitish or reddish-violet flowers.
BORAGINACEAE, about 15 species.

Pulsatilla a small genus of tuft-forming perennial herbs from Europe and Asia, with a stout rootstock and twice to four times pinnately or palmately divided leaves which are often silky when young. The flowers are solitary and showy, ranging from purple or violet to reddish, pink, yellow or white, and the styles elongate and become feathery in fruit. *P. vulgaris* (= **Anemone pulsatilla*) is the PASQUE FLOWER. It and *P. patens* (= *A. patens*) are grown in gardens; both have been used medicinally.
RANUNCULACEAE, about 25 species.

Pumpkin a name given to edible fruits of certain species of the genus **Cucurbita* which are used, when ripe, as a livestock feed or for human consumption in pies. Pumpkins are produced by cultivars of *C. pepo*, *C. mixta*, *C. moschata*, and *C. maxima*. The fruits of *C. mixta* are known as CUSHAW PUMPKINS; those of *C. maxima* include the largest known fruits in the vegetable kingdom, popularly known as HUNDREDWEIGHT GOURDS, which have been recorded as reaching over 2m in circumference and up to 90kg in weight.
CUCURBITACEAE.

Punica a genus of two species of shrubs and

The fruiting body of a Bird's Nest Fungus (Cyathus olla). The peridioles ("eggs") inside the cup are dislodged by raindrops and catch in vegetation, where they are eaten by animals. (×3)

small trees, one native to Socotra (*P. protopunica*) and the other, *P. granatum* [POMEGRANATE], native to Iran and widely cultivated in the Mediterranean and many tropical and subtropical areas for its edible fruits (see POMEGRANATE).
PUNICACEAE, 2 species.

Puschkinia a genus of two species of perennial bulbous herbs from Asia Minor and the Caucasus, with very pale blue flowers borne

Flower heads of Tanacetum cinerariifolium *being harvested in Kenya; they yield pyrethrum, a substance used in insecticides.*

in clusters on short stems. *P. scilloides* [STRIPED SQUILL] is grown in rock gardens and borders; a white-flowered form is also known.
LILIACEAE, 2 species.

Puya a large genus of terrestrial herbs native in drier parts of South America, mostly in the Andes. Several species are giant plants, *P. gigas* from Colombia growing up to 9m. The leaves, which have spiny margins, are borne in dense rosettes. The flowers are borne in solitary or paniculate spikes or racemes.

Several species are cultivated in the open in mild climates or under glass. These include *P. berteroniana*, from Chile, with tapering leaves up to 1m long and metallic-bluish flowers, *P. ferruginea*, from northern South America, with leaves up to 2m long and greenish-white flowers, and *P. hortensis*, of unknown origin, which is stemless with leaves less than 1m long and purplish-white flowers.
BROMELIACEAE, about 140 species.

Pyracantha a small genus of hardy, thorny evergreen shrubs. They are mainly native to Asia and Europe and are popular shrubs,

The anemone-like Pulsatilla rubra *is a tuft-forming perennial found locally in the mountains of central and southern France.* ($\times \frac{1}{2}$)

producing an abundance of globose, yellow, orange or scarlet berries in early autumn.

P. coccinea [FIRETHORN] is a very useful and handsome vigorous shrub, which can be trained on shaded walls, grown as a border shrub or trained as a low hedge. The orange-red berries are prolific and are often persistent in winter. There are many cultivars. *P. rogersiana* (= *P. crenulata* var *rogersiana*) is another attractive species with masses of small flattened orange-red berries (yellow in cultivar 'Knaphill Butterfly'). Other commonly cultivated species with numerous cultivars include *P. crenulata*, with rusty pubescent stems and oblong to oblanceolate leaves, and *P. koidzumii* (= *P. formosana*), a

The bright orange berries of the Firethorn (Pyracantha coccinea), *a vigorous shrub, suitable for many situations in cultivation.* ($\times \frac{1}{2}$)

much-branched shrub with reddish-purplish twigs.
ROSACEAE, 6–7 species.

Pyrethrum a genus now usually regarded as a section of the genus *Tanacetum* (or *Chrysanthemum*). It consists of perennial herbs with the flowers grouped in daisy-like flower heads, either solitary or in clusters.

The garden pyrethrums are probably derived from a Caucasian variant of *T. coccineum* (= *P. roseum*). These are attractive plants with large and colorful flower heads with long ligules (rays) which may be white, pink, lilac or various shades of red. *T. parthenium* (= *P. parthenium*) [FEVERFEW] is a strongly scented plant with small flower heads, short white ligules and yellow disk flowers. In addition to being grown as an ornamental plant in gardens, it is a medicinal herb. The dried flowers are used in various home remedies.

The natural insecticide pyrethrum or Dalmatian pyrethrum is obtained from the dried flower heads of several species of *Tanacetum* section *Pyrethrum*, mainly from *T. cinerariifolium*, which is native to western Jugoslavia and Albania but widely cultivated elsewhere. It is grown as a field crop in East Africa, Japan, South America and New Guinea. Pyrethrum has the advantage of being non-toxic to warm-blooded animals. The active ingredients are pyrethrins and cinerins.
COMPOSITAE, 40–50 species.

Pyrola a genus of north temperate evergreen perennial herbs, rarely dwarf shrubs, with creeping underground rhizomes, commonly called WINTERGREENS or SHINLEAFS. *P. grandiflora* is the ARCTIC WINTER GREEN and *P. rotundifolia* is known as the WILD LILY OF THE VALLEY.
PYROLACEAE, about 20 species.

Pyrostegia a small genus of climbing vines from South America. Only one species, *P. venusta* (= *P. ignea*) [FLAME VINE, FLAMING TRUMPET], from Brazil and Paraguay, is culti-

vated to any extent. It is a large, handsome plant climbing by means of tendrils which terminate the paired leaflets of its leaves. Its flowers are fiery red to orange and are produced in many-flowered clusters. It is commonly grown in the tropics as an ornamental and in conservatories in the southern USA.
BIGNONIACEAE, about 6 species.

Pyrus a small but important genus of mostly deciduous, frequently thorny trees. They are cultivated mainly for their fruits (see PEARS) but a few are grown as ornamentals for their attractive white blossom, such as *P. salicifolia* [WILLOW-LEAF PEAR] from southeast Europe and the Caucasus, which has narrow silver-green leaves, and *P. ussuriensis* [CHINESE PEAR, SAND PEAR] from northeast Asia, with broad ovate to orbicular leaves. The distribution of the genus extends from Europe and North Africa through Asia to Japan. *Pyrus* species are however widely cultivated in temperate regions of both hemispheres.
ROSACEAE, 15–20 species.

Pythium a genus of fungi related to *Phytophthora* and of a similar habit. They mostly live as saprophytes in water or soil. Many are associated with plant diseases, particularly root-rots and damping-off, and a few animal diseases. They are often only secondary invaders. *P. ultimum*, *P. irregulare* and *P. intermedium* are common causes of damping-off and root-rot in plants from temperate areas; *P. aphanidermatum*, *P. graminicola* and *P. arrhenomanes* cause similar effects in warmer regions.
PYTHIACEAE, about 90 species.

The Flame Vine, Flame Flower, Flaming-trumpet or Golden Shower (Pyrostegia venusta), *from Brazil and Paraguay, is an evergreen climber that produces a mass of orange flowers up to 8cm long.* ($\times \frac{1}{6}$)

Q

Quassia a mainly tropical American genus of trees and shrubs, with a characteristically very bitter bark and wood. The most important economic species is *Q. amara* [QUASSIA, BITTER WOOD], growing from the Lower Amazon to Central America and the West Indies, which is the source of quassia wood. It has attractive pinnate leaves with bright pink or red main veins and winged petioles. The scarlet flowers are borne on crimson flower stalks. The yellowish-white wood is the source of quassia chips, from which a bitter extract is made. It is used as a tonic, a treatment for worms and an insecticide. It is planted for ornament in the tropics.
SIMAROUBACEAE, about 40 species.

Quebracho a local name given to *Schinopsis balansae* [WILLOW LEAF, RED QUEBRACHO] and *S. lorentzii* [BREAK-AXE TREE], tall, very hard-wooded trees native to South America. The heartwood when boiled yields a substance used in tanning, and was once of considerable commercial importance.
ANACARDIACEAE.

Quercus see OAK.
FAGACEAE, about 500 species.

Quillaja a South American genus of evergreen shrubs or small trees. The Chilean *Q. saponaria* is commonly known as the SOAP-BARK TREE because the powdered bark lathers in water. The bark contains saponin, which is used commercially in cleaning textiles.

Tannin is also extracted from the bark.
ROSACEAE, 3–4 species.

Quinces the common name for species of two genera: the COMMON QUINCE (see *Cydonia*) and the FLOWERING QUINCES (see *Chaenomeles*).
ROSACEAE.

Quinine a white crystalline alkaloid of the group known as quinolines. The drug occurs, together with several other related alkaloids, in the bark of trees of the South American genus *Cinchona*, such as *C. officinalis* [CASCARILLA VERDE, ICHO CASCARILLA], *C. calisaya* [YELLOW BARK, CALISAYA BARK], *C. ledgeriana* and *C. pubescens* (= *C. succirubra*) [REDBARK, CINCHONA].

Quinine is toxic to all cells and is still used for the treatment of malaria; a single sufficient dose suppresses the malaria plasmodia in the blood at the trophozoite stage of development. It reduces the body temperature by dilating skin blood vessels and in addition has an analgesic action.

Cinchona trees are now cultivated in India and Indonesia from strains of *C. ledgeriana* [LEDGERBARK CINCHONA], probably a variety of *C. calisaya*, grafted on to stocks of the more hardy *C. pubescens*. The former contains a much higher alkaloid content than other species.
RUBIACEAE, about 15 species.

Quisqualis a genus of woody climbers native in tropical and southern Africa and Indomalaysia. Species cultivated in tropical gardens include *Q. falcata*, with its brilliant red bracts, and the well-known RANGOON CREEPER (*Q. indica*), from Burma, the Malay peninsula, the Philippines and New Guinea, with sweetly-scented flowers which open at night and change from white to pink and blood-red by dawn. The roots and half-ripe fruits of *Quisqualis* are used as a vermifuge in Malaysia.
COMBRETACEAE, about 4 species.

*Flowers of the Flowering Quince (*Chaenomeles speciosa*). (× ¼)*

R

Radish the common name for the widely cultivated **Raphanus sativus*, a crop now grown throughout the world, primarily for the swollen roots, although some are cultivated for their leaves. The related species *R. landra* [ITALIAN RADISH] is grown especially for its leaves and the immature fruit pods of *R. sativus* cultivar 'Caudatus' are used as a vegetable in tropical countries. *R. sativus* and the other species probably originated in Asia or the Mediterranean region.

There are four distinct varieties of *R. sativus*: (1) *radicula*, the small short-season radish, often grown early under glass; (2) *niger*, a larger rooted radish, whose fleshy roots and leaves are popular in Japan, China and India; (3) *mougri*, which has no fleshy roots but is grown for its leaves and immature seeds pods in Southeast Asia; (4) *oleifera*, grown mainly in Western Europe for the leaves which provide a useful fodder.

The different cultivars of *radicula* have round, tapering or cylindrical roots, with white or red flesh, and white, red or black skin. Some have red skin on the upper part of the swollen root, and a white skin on the lower part. One of the *niger* cultivars, 'Longipinnatus' [CHINESE RADISH] has hard, durable roots which make it suitable as a winter radish.

Radish roots have a pleasantly pungent taste, due to the presence of volatile mustard oils, which are released by the action of an enzyme when the cells are damaged. These are more concentrated in the skin, so 'hot' radishes can be peeled and the flesh eaten. Radishes have also been grown as a seed crop, for the extraction of oil.
CRUCIFERAE.

Raffia a soft, whitish fiber widely used as a weaving material for handbags, baskets, mats, hats etc. It is obtained from the young leaflets of various palms from the genus **Raphia*, mainly from parts of tropical Africa.
PALMAE.

Rafflesia a genus of plants which parasitize lianas in the forests of Malaysia. The flowers of *R. arnoldii* from Sumatra are the largest known, some being 50cm or more across. There are no leaves, stems or roots; all the normal vegetative plant tissues are represented by microscopically small filaments permeating the host plant.

Rafflesia flowers show very striking "car-

Ramonda nathaliae is a hardy rock-garden plant endemic to north Greece and south Yugoslavia. ($\times \frac{1}{2}$)

rion flower" features. The perianth surface is brick-red, dark brown or purplish-brown and coarsely warted and mottled, while the inner surface of the diaphragm has vivid white spots. The flowers may emit an odor of decaying meat, as in *R. arnoldii*, and are visited by swarms of carrion flies.
RAFFLESIACEAE, about 12 species.

Raisins the sun-dried grapes of *Vitis vinifera* [GRAPEVINE] with a high sugar content and firm flesh. There are various types according to the variety of grape used. Malaga or Muscatel raisins are high-quality large table fruits obtained from the Alexandrian Muscat grape. The main areas of production are Spain, Turkey, Greece, the USA (California) and Australia.
VITACEAE.

Rambutan the fruit of *Nephelium lappaceum*, a common village tree of Malaysia. The fruits are egg-shaped, covered with long, soft spines and borne in large panicles. The large seed is enclosed in a translucent, white, sweet, juicy aril which is the edible part. A fat extracted from the seed is used for making candles.
SAPINDACEAE.

Ramonda a very small genus, native to mountainous regions of southern Europe, of dwarf stemless hairy herbs with leaves in a rosette and showy flowers borne on leafless stalks. The flowers have four to six broad, overlapping lilac to violet petals, according to species. All three species are grown in rock gardens in temperate countries, the commonest being *R. myconi* from the Pyrenees and adjoining mountains.
GESNERIACEAE, 3 species.

Rampion the common name for *Campanula rapunculus*, a biennial herb with a thick taproot, which was formerly cultivated as a vegetable. The first-year roots and leaves were used in winter salads. The name is also applied to species of *Phyteuma*.
CAMPANULACEAE.

Randia a large genus of tropical shrubs and small trees which are often thorny. A number are cultivated as ornamental flowering shrubs, including the spineless, tropical South American *R. formosa*, which bears attractive, solitary, terminal, tubular white flowers.
RUBIACEAE, 200–300 species.

Ranunculus a large genus of annual or perennial, sometimes aquatic herbs, commonly known as BUTTERCUPS or CROWFOOTS. They are cosmopolitan in distribution but most numerous in the temperate and cold regions of the Northern Hemisphere; the tropical species are confined to mountains. The genus is subdivided into subgenera and sections, according to habit and habitat. Examples include subgenera *Ranunculus* (usually terrestrial plants, leaves rarely divided into capillary segments) and *Batrachium* (aquatic or marsh plants, leaves all with a broad lamina and/or leaves divided into capillary segments).

Some species, such as *R. fascicularis* [EARLY BUTTERCUP, EARLY CELANDINE] and *R. lyalii*, have a tufted habit. *R. adoneus* or *R. eschscholtzii* are scapose perennials. Other species, including *R. creticus* and *R. ficaria*, have swollen fleshy roots. Leaves may be small or compound, sometimes lobed and notched, as in *R. aconitifolius*, or deeply divided into thread-like segments, as in the submerged leaves of the amphibious aquatic *R. flabellaris* [YELLOW WATER BUTTERCUP or CROWFOOT]. The flowers are usually yellow, white or reddish in color, with numerous stamens and fruits consisting of a cluster of numerous achenes.

Buttercups are poisonous and are left strictly alone by most grazing animals. When dried and made into hay, however, the acrid properties are lost. *R. bulbosus* [BULBOUS BUTTERCUP or CROWFOOT], *R. acris* [TALL or COMMON BUTTERCUP] and *R. repens* [BUTTER

The Water Buttercup or Water Crowfoot (Ranunculus aquatilis) has dissected submerged leaves and larger aerial ones. ($\times \frac{1}{4}$)

Radishes (Raphanus sativus) come in many varieties. The small variety illustrated here is used raw in salads and as a garnish. ($\times 1$)

DAISY, YELLOW GOWAN, CREEPING BUTTERCUP] are common to meadows and grasslands.

Some species or their varieties, but more especially *R. calandrinioides* and *R. aconitifolius* [WHITE BATCHELOR'S BUTTONS], are cultivated for their showy, cup-shaped white or yellow flowers. A popular florists' species is *R. asiaticus* [GARDEN RANUNCULUS, PERSIAN BUTTERCUP or CROWFOOT], from southeastern Europe and Southeast Asia. It has spectacular double flowers up to 10cm in diameter, which may be yellow, pink, orange, red-purple or white, and is propagated by the tuberous roots. *R. repens*, which is native to Eurasia and naturalized in North America, exists in a double-flowered form ('Pleniflorus') and a dwarf form ('Nanus'). Other cultivated species include *R. alpestris* (from the Alps), *R. bulbous* (from Europe and North Africa), *R. californicus* (from Oregon and California), *R. crenatus* (from Hungary and Macedonia),

A. creticus (from the Greek Islands), *A. fascicularis* (from North America), *A. giganteus* (from Peru), *A. lyallii* (from New Zealand) and *A. pyrenaeus* (from the Pyrenees and Alps).
RANUNCULACEAE, about 300 species.

Raoulia a small genus of shrubby or mat-forming perennial herbs found chiefly in New Zealand. They have minute leaves, often covered in woolly down. The tiny flower heads are borne just above the stems and are yellow or white in color. *R. eximia* [VEGETABLE SHEEP] can produce a considerable mound of gray woolly stems looking, at a distance, like a reclining sheep. A few species are grown for their foliage, such as *R. australis* (= *R. lutescens*) and *R. glabra*.
COMPOSITAE, about 25 species.

Rape the name given to two closely related species of the genus *Brassica*, *B. campestris* (= *B. rapa*) [CONTINENTAL RAPE] and *B. napus* [BRITISH RAPE]. Rape has been cultivated as an oilseed crop in Europe since the Middle Ages, and by the 19th century the leaves and stems of the biennial form of *B. napus* were being used, as they are today, as an autumn forage crop and for early spring pasturing. There is a range of variation in forage rape of this species, including dwarf and giant forms.

Oilseed forms of *B. campestris*, both annual and biennial, are economically important, particularly the annual form widely cultivated in Canada, India and Pakistan. The subspecies *oleifera* [TURNIP RAPE] is an important forage crop and subspecies *dichotoma* [INDIAN RAPE] is the annual oilseed form. Although *B. napus* crosses readily with *B. campestris*, the product *B. napocampestris* has not so far proved agriculturally useful.
CRUCIFERAE.

Raphanus a small genus of herbaceous annuals, biennials or perennials native from Europe to East Asia, although the cultivated

Raoulia hookeri, from New Zealand, is an unusual species forming a dense silvery mat covered with bright yellow flowers. Some raoulias are cultivated in sunny rock gardens mainly for their attractive foliage. ($\times \frac{1}{3}$)

and wild radishes now have a worldwide distribution (see RADISH). *R. raphanistrum*, [WILD RADISH, WHITE CHARLOCK or RUNCH] is a fast growing annual, a common weed of spring cereals which has spread to most arable areas of the World.
CRUCIFERAE, 8–10 species.

Raphia a small genus of palms which are the source of *raffia fiber and palm wine. The genus is indigenous to Africa, south of the Sahara; only one species, *R. australis*, is subtropical, occurring in Mozambique. Another species, *R. taedigera*, occurs both in Africa and South America.

Raphia leaves, especially those of *R. farinifera* (= *R. ruffia*) [RAFFIA PALM], which may be as much as 20m long, are the largest leaves in the plant kingdom, and are the sources of thatch, poles, *piassava fiber and raffia fiber. The inflorescence of *R. hookeri* [IVORY COAST RAFFIA PALM, WINE PALM] and several other species are tapped for sap which is fermented to yield palm wine.

The mesocarp of the fruit between the scales and the seed contains extractable edible oil in some species such as *R. vinifera* [BAMBOO or WINE PALM]. The lower surfaces of unopened leaflets of several species (such as *R. farinifera* and *R. hookeri*) contain wax, similar to the *carnauba wax of commerce.
PALMAE, about 20 species.

Rasamala the trade name for *Altingia excelsa*, a tall tropical Asian tree which yields a heavy timber used in house construction as well as a yellow fragrant resin (rasamala resin or wood oil) used in the manufacture of perfumes.
HAMAMELIDACEAE.

Raspberry the fruit produced by various species of *Rubus*, notably *R. idaeus* (= *R. idaeus* subspecies *vulgatus*) [EUROPEAN RED RASPBERRY, FRAMBOISE] a widely distributed woodland plant in north temperate regions. In North America the majority of cultivated varieties are derived from the above, *R. idaeus* subspecies *strigosus* (= *R. strigosus*) [AMERICAN RED RASPBERRY], *R. occidentalis* [BLACK RASPBERRY, THIMBLEBERRY] and their hybrid *R. × neglectus* [PURPLE RASPBERRY].

Raspberry bushes consist of biennial woody stems or canes. Stems are prickly (at least when young) and are usually vegetative in their first year, producing inflorescences of white flowers on lateral shoots in their second year. Raspberry fruits are composed of aggregates of small one-seeded drupes which come away from the receptacles when ripe. They are usually red but some yellow-fruited varieties are also known. Raspberries are grown commercially, principally for jam-making, canning, freezing and for flavorings. Major centers of RED RASPBERRY production are eastern Scotland, Eastern Europe and western North America. BLACK RASPBERRIES are grown commercially only in the USA.

Many of the varieties that were popular in the early part of this century have now been lost due to widespread virus diseases. Cultivars resistant to the aphid vectors of these diseases have now been produced.

In the UK the most popular raspberry to have been bred in recent years is undoubt-

Raphia species have long-stalked, pinnate leaves, as with this species growing in Ghana, that are used for thatching and yield fiber.

edly 'Malling Jewel', an early to mid-season variety. Cultivars suitable for commercial growing in the southern USA have been produced by hyridization with species such as *R. parvifolius*, *R. biflorus* and *R. kuntzeanus* to incorporate resistance to diseases and tolerance of heat and arid conditions.

The fruits of two species of so-called ARCTIC RASPBERRIES, *R. arcticus* and *R. stellatus*, are used in Scandinavia, especially for making liqueurs.
ROSACEAE.

Rattan the Malay name for the climbing palms that are so important a feature of the tropical rain forests of the Far East. They constitute a group of 10 genera, including *Calamus* and *Daemonorops*, and are important for their long, flexible stems. Commercially, rattans are known as "cane" in sizes larger than 20mm in diameter and as "rattan" in smaller sizes. Canes are used for walking sticks, ski sticks, broom handles etc. The skin of the rattan is used to make the finest and most durable of woven baskets and the core is used as "split cane" for basketwork and wickerwork. Singapore is the principal trade center.
PALMAE.

Rauvolfia or **Rauwolfia** a genus of trees and shrubs with an almost worldwide tropical distribution. The species range from 30m-high trees, as in the Peruvian *R. praecox* and the Indonesian *R. javanica*, to 15cm-high shrubs, as in *R. nana* found in Zimbabwe. The African *R. vomitoria* (= *R. pleiosiadica*) [SWIZZLESTICK] and the tropical American *R. tetraphylla* (= *R. canescens*) [TRINIDAD DEVIL] are some of the species which have been examined as possible commercial sources of important alkaloid drugs.

The original source of these alkaloids, notably reserpine, was the Indian and Southeast Asian *R. serpentina* [INDIAN SNAKEROOT, TRINIDAD DEVIL PEPPER], a small climbing shrub which has been used in Indian medicine for centuries for the treatment of reptile poisoning, fevers, and insanity and for the induction of childbirth. The alkaloids are now used routinely in Western medicine for the treatment of high blood pressure. Now *R. tetraphylla* and *R. vomitoria* are also used commercially as sources of drugs.
APOCYNACEAE, about 70 species.

Ravenala a genus consisting of a single tree-like plant native to Madagascar, *R. madagascariensis* [TRAVELLER'S TREE or PALM]. It gets its common name because water accumulates in the leaf-bases and can be used for emergency drinking. The leaves, which are up to 3m long, are borne in spectacular fan-like arrangements on a sub-aerial stem. Flour is produced from the seed, and sap from the stem. It is widely cultivated in the tropics for ornament.
STRELITZIACEAE, 1 species.

Red algae the common name for the algae belonging to the division Rhodophyta. Some

Raspberries (Rubus idaeus) *are among the most popular temperate soft fruits; the fruit is an aggregate of small, fleshy drupes.* (×1)

3 000 marine species are known and these are found growing on intertidal or subtidal rocks or as epiphytes on other algae. Only about 150 freshwater species have been identified and these are mainly attached to stony substrates in fast-flowing streams. They are normally easily recognized by their reddish coloration, which comes from a pigment called phycoerythrin.

Economically, the red algae are of some considerable importance. A number, such as *Porphyra*, *Rhodymenia* and *Dilsea* have traditionally been eaten, either raw or cooked. The production of the jelly, agar, has become worldwide and is now based on a number of different red algae, such as *Gelidium* in California and Japan, *Gracilaria* in Australia and *Pterocladia* in New Zealand. A similar mucilaginous material, carrageenin, is extracted commercially from *Chondrus* and *Eucheuma*. As yet only *Porphyra* is grown commercially and that for human consumption, in Japan.

Red pepper one of the SWEET PEPPERS, a non-pungent variety of *Capsicum annuum* which may be eaten cooked or raw. It is commonly called *PIMENTA*.
SOLANACEAE.

Red snow a phenomenon caused by certain unicellular algae of the Chlorophyceae, eg *Chlamydomonas*, which contain a reddish pigment (haematochrome). When such algae grow on the surface of snow they give it a red appearance.

Red tide a phenomenon characterized by the red coloration of inshore seas, caused by massive growth of species of certain algae, particularly dinoflagellates. In temperate and subtropical seas the genera usually responsible for red tides are *Gonyaulax* or *Gymnodinium*.

In tropical waters the blue-green alga *Trichodesmium* can produce a similar effect. It is generally assumed that the flow of nutrients off the land or from upwelling water, plus midsummer high temperatures, triggers off the massive growth of algae.

Red tides can be a serious hazard since a water-soluble specific nervous toxin is formed by the cells of the algae, which is then transferred to fish and especially mollusks. If eaten, the affected seafood can be lethal.

Redwood a common name now used almost exclusively for the only species of the genus *Sequoia*: *S. sempervirens*. SIERRA and GIANT REDWOOD, however, are names commonly given to the related tree *Sequoiadendron giganteum* (= *Sequoia wellingtonia*), also known as BIG TREE or GIANT SEQUOIA. DAWN REDWOOD is the popular name given to *Metasequoia glyptostroboides* from China.

In the lumber trade, redwood may also be used as a general term for any wood of a red color, especially tropical trees. These include *Pterocarpus indicus* [ANDAMAN REDWOOD] and *Guarea trichilioides* [WEST INDIAN REDWOOD].

Reed a general term referring to many different species of water-loving grasses. The name is most commonly applied to genera of the family Gramineae, particularly *Arundo* and *Phragmites* which are dominant genera

Raspberries (Rubus idaeus) *under commercial cultivation. Raspberry canes are biennial and fruit forms on the previous year's growth.*

of marshes and along stream and river banks in all parts of the world.

There are also a number of lesser-known reeds, including species of *Calamagrostis*, such as the north temperate *C. epigejos* [SMALL WOOD REED], *C. canescens* (= *C. lanceolata*) [PURPLE SMALL-REED] and *C. stricta* (= *C. neglecta*) [NARROW SMALL-REED], *Glyceria maxima* [REED SWEET GRASS], of European wetlands, and *Phalaris arundinacea* [CANARY REED GRASS].
GRAMINEAE.

Reseda see MIGNONETTE.
RESEDACEAE, about 60 species.

The Rose of Jericho or Resurrection Plant (*Anastatica hierochuntica*). After flowering, the plant rolls up into a detached ball which blows about the desert. When it rains, the seeds inside the apparently dead bundle germinate. (× ⅓)

Resins an exceedingly complex group of substances usually originating as exudates from the wounds of certain plants. They mainly form as oxidation and polymerization products when a mixture of essential oils is exposed to light and air. The resins of commerce and of vernacular speech are in fact solutions of resins in the essential oils from which most of them originate.

Resins are produced by an immense variety of plants but usually in small amounts. Only a few families produce resins in quantitites which make them commercially significant. The Pinaceae produce pine and kauri resins. The Leguminosae produce *copal resins and the Dipterocarpaceae produce *dammar resins. Other commercial resins are asafoetida (see *Ferula*), dragon's blood, *elemi, *guaiacum, *lacquer, ladanum (see *Cistus*), *sandarac and *storax. Amber is a fossil resin from the extinct pine tree, *Pinus succinifera*. There are many other resins which are soft, such as mastic, obtained by tapping the bark of several *Pistacia* species such as *P. lentiscus*.

Despite the production of synthetic resins, large quantitites of natural resins are still used in varnishes, paper sizing, soap making, textile stiffening, perfumery and in medicine.

Resurrection plant a common name applied to several species which appear dry and dead during hot dry weather but which expand and continue to grow when moistened. There are two forms of "resurrection": in *Selaginella lepidophylla* and *Leucobryum glaucum*, for example, the whole plant turns green and may grow but in *Anastatica hierochuntica* [RESURRECTION PLANT, ROSE OF JERICHO], it is the seeds released on wetting which make sudden growth.

Rhacomitrium a genus of mosses, one species of which, the cosmopolitan *R. lanuginosum*, is ecologically important in the Arctic and on mountains in the north temperate zone. Here on mountain top detritus it may be the dominant species, forming *Rhacomitrium* heath.
GRIMMIACEAE, about 75 species.

Rhamnus see BUCKTHORN.
RHAMNACEAE, about 150 species.

Rhaphiolepis a small genus of evergreen shrubs from East Asia. They are grown as garden ornamentals, especially *R. indica* [INDIAN HAWTHORN], from China, and *R. umbellata* (= *R. japonica*) [YEDDA HAWTHORN], from Japan. The former has whitish flowers tinged with pink, the latter fragrant pure white flowers. Their hybrid, *R. × delacourii*, is an attractive rose-pink-flowered shrub. A brown dye is made from the bark of *R. umbellata*.
ROSACEAE, about 15 species.

Rheum an Asian genus of perennial herbs, normally with a thick woody rootstock and large toothed or palmately lobed leaves in basal clumps. *R. palmatum*, *R. nobile* [SIKKIM RHUBARB] and *R. undulatum*, are cultivated as ornamentals. The genus also includes *R. rhabarbarum* [GARDEN *RHUBARB], with its edible petioles, and *R. officinale* [MEDICINAL RHUBARB].
POLYGONACEAE, about 50 species.

Rhinanthus a genus of semiparasitic annual herbs from temperate Eurasia and North America. Characteristic of damp pastures, they have mainly yellow, two-lipped flowers in spike-like racemes. *R. minor* [YELLOW RATTLE] can be a serious weed in cereal crops. *R. crista-galli* [RATTLE BOX] is sometimes cultivated in rock gardens.
SCROPHULARIACEAE, about 50 species.

Rhipsalidopsis a Brazilian genus of two species of epiphytic cacti, *R. gaertneri* (= *Schlumbergera gaertneri*) [EASTER CACTUS] and *R. rosea*. Both species and their hybrid *R. × graeseri*, are cultivated as greenhouse ornamentals, for their showy trumpet-shaped, often pendulous flowers borne at the ends of the jointed stems.
CACTACEAE, 2 species.

Rhipsalis a genus of many epiphytic cacti from tropical and subtropical America, but also occurring in Africa, Madagascar and Sri Lanka where they are presumed to have been introduced by Man. They are almost or quite spineless with cylindrical, ribbed or flat stems with long and short joints, often in whorls. The small flowers are followed by round, white or colored berries. Several species, including *R. baccifera* [MISTLETOE CACTUS] and *R. houlletiana* [SNOWDROP CACTUS] are cultivated as basket or pot plants.
CACTACEAE, about 60 species.

Rhizobium an important genus of bacteria which are found in the root-nodules of legumes and some other plants. They form a symbiotic relationship with the host plant, the bacteria converting atmospheric nitrogen into salts that are then made available to the legume.

Rhizophora a genus of tropical coastal trees commonly called MANGROVES, all of which develop large arching aerial roots from the branches, forming props and trunks. *R. mucronata* and *R. mangle* [AMERICAN MANGROVE, RED MANGROVE] yield extracts from the bark and wood used for tanning and dyeing. Red mangrove wood is hard, durable and close-grained and is used in building construction and for cabinetwork. (See also Mangrove bark.)
RHIZOPHORACEAE, about 6 species.

Rhizopus a well-known and ubiquitous genus of pin molds, characterized by the presence of root-like hyphae (rhizoids) at the base of unbranched sporangiophores and with multispored sporangia. Several species cause molds of fruits or bulbs in storage.
MUCORACEAE, about 5 species.

Rhododendron a very large genus of evergreen or deciduous shrubs and some trees, with entire leaves. Most species are very closely related to each other and may merge botanically and distributionally so that the precise number is subject to dispute. The genus includes azaleas, which are mainly deciduous and have funnel-shaped flowers.

Rhododendron covers huge areas of the temperate Northern Hemisphere and is abundant in the Himalayas and much of Southeast Asia right down to the northern tip of Australia. These plants nearly all occur on acid soil, rich in slowly decaying organic matter, and species may occur at all altitudes from just above sea level to above 5 800m in the Himalayas.

They may form 30m-high forest giants, as in *R. arboreum* [TREE RHODODENDRON] or tiny creeping 2.5cm-high shrublets, as in *R. forrestii*. The leaves range in size from about 2cm long in *R. pubescens* to about 30cm long in *R. rex*. The shape of leaves varies considerably from rhombic to broadly ovate, through elliptic to obovate to linear-lanceolate. The underside of leaves may be glandular, as in *R. rosmarinifolium*, or covered with an indumentum, for example, rusty-brown in *R. russatum*, and grayish-white in *R. sinogrande*.

The size of the flowers varies from the

enormous scented trumpets of *R. nuttallii*, about 12cm long, and the huge multi-flowered trusses of *R. sinograde*, both from Southeast Asia, to the minute flowers of some of the high mountain species. The flower shape varies from the narrow tubes of *R. keysii* to the saucer-shaped *R. aberconwayi* and *R. souliei*. The inflorescences are usually terminal. The flowers are commonly five-lobed and are often heavily spotted and/or blotched in the throat. The flower color ranges from pure white to pink, orange, red, crimson to almost black, many shades of yellow, mauve, purple to almost blue and sometimes different intensities and combinations of these colors.

Rhododendrons are classified into groups called subgenera, then into sections, according to whose classification is used. Thus under one system the genus can be divided into eight natural subgenera: *Rhododendron, Pseudazalea, Pentanthera, Tsutsutsi, Azaleastrum, Pseudorhodorastrum, Rhodorastrum, Therorhodion*, each containing one or more sections.

Many cultivated rhododendrons and azaleas are cultivars of hybrids. Examples of hybrid groups of evergreen rhododendrons are the CATAWBA hybrids, derived mainly from *R. catawbiense* crossed with *R. maximum*, *R. caucasicum*, *R. ponticum* or *R. arboreum*, and the early flowering CAUCASICUM hybrids derived mainly from crosses of various hardy or semi-hardy species with *R. caucasicum*. Other hybrid groups include FORTUNEI, GRIFFITHIANUM, JAVANICUM and THOMSONII hybrids.

The common *R. ponticum*, from Spain, Portugal to the eastern Mediterranean and Caucasus, has naturalized itself all too successfully in many parts of Britain and has become a serious arboricultural weed in places.

Rhododendrons are largely grown for their beauty although some drugs have been extracted from them (especially from *R. chrysanthum* from East and North Asia), and the wood is occasionally used.

Most azaleas come from North America and Japan and its adjacent areas. The better-known hybrid groups of azaleas include GABLE hybrids, derived mainly from crosses between *R. yedoense* var *poukhanense* and *R. kaempferi*, GHENT hybrids, which are derived from *R. luteum* (= *R. flavum*) [PONTIC AZALEA], a common yellow species from around the south Black Sea coast, INDIAN ["INDICA"] hybrids, derived mainly from *R. indicum*, *R. mucronatum* and *R. simsii*, and KURUME hybrids, derived from *R. kaempferi* and *R. kussianum*.

The evergreen, but less hardy "Indian" azaleas are extensively grown as pot plants. Colors here vary from several shades and combinations of red, pink and mauve to white, and many are double. The deciduous azaleas are mostly hardier and the flowers range from red through orange and yellow and pink to white; they are often scented. ERICACEAE, 750–1200 species.

Rhodotypos a genus consisting of a single species, *R. scandens* (= *R. kerrioides*) from temperate East Asia. It is a deciduous shrub growing to 2m high and is cultivated for its white flowers and shiny black autumn berries.
ROSACEAE, 1 species.

Rhoeo a Central American and West Indian genus represented by a single species, *R. spathacea* (= *R. discolor*) [BOAT LILY, OYSTER PLANT, MOSES-ON-A-RAFT, MOSES-IN-THE-BULRUSHES, MAN-IN-A-BOAT, TWO-MEN-IN-A-BOAT], a perennial herb with fleshy leaves attached in a dense spiral to a short stem. Inconspicuous clusters of white flowers are enclosed in boat-shaped spathes, hence the common name for this greenhouse ornamental, widely grown for its foliage.
COMMELINACEAE, 1 species.

Rhoicissus a small genus of evergreen climbing plants native to tropical and southern Africa. *R. capensis* (= *Cissus capensis*) [CAPE GRAPE, EVERGREEN GRAPE] is cultivated as a greenhouse plant or houseplant for its large attractive leaves and its clusters of edible reddish-purple berries.
VITACEAE, 10 species.

Rhubarb the common name for a number of species of the genus *Rheum*, notably *R. rhabarbarum* [PIE PLANT, WINE PLANT, GARDEN RHUBARB], a hardy perennial cultivated in all temperate countries for its edible red leaf-stalks, used as a dessert when stewed. The leaves are poisonous because of their high concentration of calcium oxalate. Among numerous varieties in cultivation are the early thin-stemmed 'Hawkes Champagne' and the large main crop 'Sutton'. The rhizomes of *R. officinale* [MEDICINAL RHUBARB], *R. ribes* [CURRANT RHUBARB] and *R. palmatum* [CHINA RHUBARB] all yield medicinal extracts.
POLYGONACEAE.

Rhus a genus of small deciduous trees, shrubs or vines, native to temperate and subtropical Eurasia and North America, and

Flowers of Rhododendron pulchrum *(= Azalea pulchrum), one of the thousands of species, varieties and cultivars of this genus in cultivation. (× 3)*

The Fuschia-flowered Gooseberry (Ribes speciosum), *a flowering currant from California which is popular as an ornamental shrub.* (× ¼)

commonly known as SUMACS or SUMACHS. Some authorities also include *POISON IVY (Toxicodendron)* and POISON OAK species in the genus.

Several North American species, especially *R. typhina* (= *R. hirta*) [STAGHORN SUMAC], *R. glabra* [SMOOTH SUMAC, SCARLET SUMAC, VINEGAR TREE] and *R. copallina* [DWARF SUMAC, SHINING SUMAC, MOUNTAIN SUMAC, WING-RIB SUMAC], are grown as ornamentals, valued for their shape, vivid autumn coloring and attractive dense fruiting inflorescences or infructescences like dark red pyramids at the branch ends.

The southern European *R. coriaria* [SICILIAN SUMAC, ELM-LEAVED SUMAC, TANNER'S SUMAC] was once used as a spice, and as a medicine. Nowadays it is commercially important as a source of tannin, as are the North American *R. copallina* and *R. typhina,* and the Eurasian *R. chinensis* (= *R. javanica*) [NUTGALL TREE]. In the case of *R. chinensis,* the tanning substances are extracted from the galls which the plant grows in response to attack by a tree louse.

A number of species sometimes ascribed to *Toxicodendron* are highly poisonous. They include *R. radicans* [POISON IVY], *R. diversiloba* [POISON OAK], *R. toxicodendron* [OAKLEAF POISON IVY] and *R. vernix* [POISON SUMAC]. ANACARDIACEAE, about 150 species.

Rhytisma a genus comprising parasitic fungi which cause black lesions (tar spots) on leaves of SYCAMORE or GREAT MAPLE (*R. acerinum*) or WILLOW (*R. salicinum*). Infected leaves on which the fungus has over-wintered shoot off needle-shaped ascospores which infect next year's foliage. *R. acerinum* is sensitive to air-borne pollution and does not occur near to industrial centers. PHACIDIACEAE, about 20 species.

Ribes a genus of deciduous shrubs, widely distributed in the temperate regions of the Northern Hemisphere and extending through the mountains of Mexico and the Andes to temperate South America. The flowers are usually small and inconspicuous, arranged in short, drooping, axillary racemes.

Several species are cultivated for ornament, but the chief importance of the genus lies in the fact that it includes the currants and gooseberries. The cultivated varieties of these scarcely differ from the wild plants except in their larger and better-flavored fruits and disease-resistance.

R. uva-crispa (= *R. grossularia, Grossularia reclinata*) [GOOSEBERRY] has small greenish flowers in very short racemes, and a hairy, green, red or yellow berry (see also GOOSEBERRY).

R. nigrum [BLACKCURRANT] bears drooping racemes of greenish-white flowers, followed by black fruits. *R. dikuscha* (Siberia, Manchuria), *R. ussuriense* (Korea to Manchuria) and *R. bracteosum* (Alaska to California) are the other BLACKCURRANT species.

Red and white currants, which differ only in the color of the berry and in the greater acidity of the red currant, are derived mainly from *R. rubrum* [RED CURRANT, NORTHERN RED CURRANT] which is endemic to Asia and Western Europe, but some varieties probably derive from hybrids between this species and *R. spicatum,* native to temperate Asia and northern Europe. *R. sativum, R. petraeum* and *R. multiflorum* are also commonly called RED CURRANTS.

Of the ornamental species, the most important is *R. sanguineum* [FLOWERING CURRANT], from California and Oregon, which has hanging racemes of pink flowers (deep crimson in some garden varieties) in early spring. *R. speciosum* [FUCHSIA FLOWERED GOOSEBERRY] and *R. aureum* [GOLDEN, BUFFALO

Red Currants (Ribes rubrum); *the juicy, slightly acid berries are used in pies, jams and jellies.* (× ½)

Tar spot lesions on a Sycamore (Acer pseudoplatanus) *leaf caused by the microscopic fungus* Rhytisma acerinum. (× 2)

or MISSOURI CURRANT] are also from the Pacific coast of North America. The former is very spiny, with long, narrow, hanging crimson flowers very suggestive of those of a fuchsia; *R. aureum* is spineless, with glabrous leaves and relatively large golden-yellow flowers. *R. americanum* [AMERICAN BLACKCURRANT], very like *R. nigrum,* is sometimes planted for its autumn color.
SAXIFRAGACEAE, about 150 species.

Riccia a large genus of thallose liverworts. The plants are small and usually inconspicuous (except in wet weather) and characteristically occur in rosettes on fine-grained soil in the open; a very few species are aquatic. *Riccia* is widely distributed throughout the warmer and drier parts of the world and is particularly common and abundant in Mediterranean-type climates.
RICCIACEAE, 100–120 species.

Rice or **paddy** a cereal crop grown in the warmer parts of the world. It is the principal food for about 60% of mankind, world production in 1971/72 being approximately 310 million tonnes. China and India are the major producers but the USA and Thailand are the main exporting countries. Over 90% of the World's rice is grown and eaten in Asia, from Afghanistan to the East Indies and northwards to Japan. In these regions it is the staple food. It is also a staple food in parts of Africa and South America.

Its origins are somewhat obscure. There is evidence of rice cultivation in India in 2 500 BC but it is possible that the plant was first cultivated some 1 100 years earlier in China or in Southeast Asia. The cultivated rices are generally regarded as *Oryza sativa,* a sort of all-embracing name for many different races and perhaps also species. However, AFRICAN RICE, whose distribution is limited to tropical West Africa, is a morphologically distinct species, *O. glaberrima.* At a recent conference of rice specialists it was concluded that there are between 13 and 24 species of *Oryza* and that any of them may have contributed to cultivated rice. Today there are several thousand types in cultivation, each of which has been selected for special properties of yield and special methods of cultivation. *O. sativa* was introduced into Europe and Africa in relatively recent times, and it was later taken to South and Central America from Europe.

Basically, there are two sorts of grains: the glutenous kinds which contain a sugary material, which are soft and much used in the Orient, and the starchy kinds. Among the starchy kinds, long-, medium- and short-grained types are recognized. The grain is rather soft and somewhat sticky when cooked. The long-grain rices have shorter growing periods and are best cultivated in warm regions; the grain is hard and does not turn sticky when cooked. Rice hulls and straw are widely used as animal fodder and the straw alone finds many uses as a fabric.

Two major groups within *O. sativa* are recognized. The *japonica* group from the temperate zone and the *indica* group from the tropics. In suitable climatic regions (tropical lowlands with abundant water supplies), it is possible with breeds of *indica* to obtain two, three, or exceptionally four crops a year.

From the point of view of cultivation, PADDY is divided, somewhat artificially, into two groups: the UPLAND or HILL PADDY which is grown without flooding and the so-called LOWLAND PADDY which is grown in land flooded during the growing season. The hill PADDY is grown as an annual in humid tropical regions during the monsoon rains. It has rather lower yields than lowland PADDY and is frequently grown in uncleared forest. Hill PADDY is used mostly for local consumption.

Although the physiology and anatomy of rice indicates that it is a semi-aquatic plant and much is grown in standing water, there is increasing interest in upland culture with the development of "dryland" cultivars and races, often with heavy grains and/or non-shattering panicles and thick roots.

During recent years the acute need for increased world food supplies has stimulated an interest in improving rice production. This interest has been heightened by the development of new rice varieties by the International Rice Research Institute in the Philippines.
GRAMINEAE.

Ricinus a genus comprising a single species, *R. communis* [CASTOR OIL PLANT, CASTOR BEAN, PALMA CHRISTI, WONDER TREE] a shrub or tree which may reach a height of 10–12m in the tropics. The source of origin is either Africa or India and it can be subdivided into four subspecies. It is widely cultivated in temperate, subtropical and tropical countries, the most important producers being Brazil, India, China and the USSR.

R. communis is grown chiefly for its valuable seed oil but also as a fast-growing ornamental which can reach a height of 3m in one season. There are numerous cultivars with brightly colored, green, orange, red and scarlet stems and leaves. The seeds are poisonous if eaten. (See also Castor oil.)
EUPHORBIACEAE, 1 species.

Rivina a genus represented by a single species from tropical America and southern USA, *R. humilis* [BLOODBERRY, ROUGE PLANT, BABY PEPPER]. It may be grown in temperate zones either in the greenhouse or as an annual border plant. It bears racemes of small pinkish flowers and small attractive red berries. The latter are the source of a red dye.
PHYTOLACCACEAE, 1 species.

Robinia a genus of deciduous trees and shrubs native to North America. A number of

A modern-style Rice (Oryza sativa) paddy with a uniform variety in cultivation.

Traditional cultivation of Rice (Oryza sativa) in segmented paddies. This system has allowed different varieties of rice to mingle and so, over the centuries, great genetic diversity has evolved. Such variety is of vital importance for the future development of the crop.

species are planted for their ornamental value, being particularly attractive in flower; the pea-like flowers of some species are fragrant. One of the most attractive ornamentals is *R. pseudoacacia* [BLACK LOCUST, YELLOW LOCUST, FALSE ACACIA], which encompasses a wide range of varieties and cultivars. The timber of this species, which is hard, durable and close-grained, is used in ship and house construction, as well as for turnery.
LEGUMINOSAE, about 20 species.

Rochea a small genus of South African leaf-succulents. *R. coccinea* (= *Cotyledon coccinea*, *Crassula coccinea*) is a popular florists' flower, grown for its broad heads of scarlet blooms. Other cultivated species include *R. jasminea* (= *Cotyledon jasminea*, *Crassula jasminea*) with white flowers, turning pink, and the white- or pink-flowered hybrid *R.* × *versicolor* (= *R. coccinea* × *R. subulata*).
CRASSULACEAE, 3–4 species.

Rocket the common name for a number of cruciferous species. *Hesperis matronalis* [ROCKET, DAME'S ROCKET, SWEET ROCKET] is a perennial herb cultivated for its flowers. *Eruca vesicaria* subspecies *sativa* [GARDEN ROCKET] is sometimes cultivated for its leaves, eaten as salad. ROCKET and YELLOW ROCKET are two of the common names of *Barbarea vulgaris* [WINTERCRESS]. The name is also given to a wide range of unrelated species including *Erysimum asperum* [PRAIRIE ROCKET] and *Reseda lutea* [DYER'S ROCKET].

Rodgersia a small East Asian genus of perennial rhizomatous herbs with large, peltate, pinnately or palmately compound toothed leaves, and large, showy, terminal panicles of white, pink or red flowers borne on erect stems. Some species, especially *R. pinnata*, are cultivated for ornament.
SAXIFRAGACEAE, about 6 species.

Romneya a genus represented by a single species, *R. coulteri* [CALIFORNIA TREE POPPY], from California, an erect, glaucous perennial herb up to 2–5m tall, woody at the base, and with leaves irregularly pinnately divided and lobed. They are cultivated in borders for their large, white, fragrant flowers, up to 15cm in diameter.
PAPAVERACEAE, 1 species.

Romulea a genus of small perennial herbs native to Europe, the Mediterranean and South Africa, very closely related to *Crocus*. They differ in having the stem appearing above ground at flowering time and a much shorter perianth tube. The flowers vary from deep violet to lilac, yellow and white in color. Cultivated species include *R. bulbocodium* and *R. rosea* (= *Ixia rosea*).
IRIDACEAE, 60–80 species.

Rondeletia a genus of evergreen shrubs or trees found in the tropics of America and the West Indies. The flowers have a corolla tube ending in four or five large lobes, giving the flower a superficial primrose-like appearance. Several species are cultivated, including *R. amoena* (= *R. versicolor*), which grows to 2.5m tall and bears pink flowers, yellow-bearded at the throat, produced in groups in the axils of its branches. *R. cordata* is similar but with deeper pink or red flowers. *R. odorata* (= *R. speciosa*) has brilliant orange-red flowers with yellow throats.
RUBIACEAE, about 100 species.

Root crops see p. 290.

Rorippa a genus of annual or perennial cruciferous herbs, often inhabiting damp places and woods in the Northern Hemi-

The orange-red hips of the Turkestan or Japanese Rose (Rosa rugosa) 'Scabrosa' are as attractive as the flowers. (×1)

Many thousands of Rose varieties and cultivars are in cultivation. Shown here is the yellow-flowered 'Grandpa Dickson'. (×$\frac{1}{12}$)

sphere. They are sometimes known as YELLOW CRESS and are closely related to *Nasturtium* [*WATER CRESS].

Rosa [ROSES] a genus of deciduous or rarely subevergreen shrubs or climbers with prickly stems, native to the temperate and subtropical regions of the Northern Hemisphere but, unlike its close relative *Rubus*, not extending as far south as the equator, except in cultivation. Many of the species, especially those of North America, sucker freely at the base and form dense thickets.

The stems are more or less covered in surface prickles, which may be straight or hooked, conical or flattened (as in *R. sericea*) or fine and needle-like. The color range of the saucer-shaped flowers is from white through yellow to pink and purplish red. The geranium-red color of some modern cultivars has its origin in the pigment pelargonidin, which is not found in any wild species. The vase-shaped receptacle and carpels ripen into a fleshy hep or hip.

The "Queen of Flowers" is the most important and long esteemed of flowering shrubs grown for ornament and at no time in horticultural history have roses been out of fashion. Garden roses are complex hybrids descended in the main from nine wild species. About 5 000 cultivars are estimated to be in cultivation although the constant demand for novelty means that the life of a cultivar is short on average. Some wild species, such as *R. canina* [BRIER ROSE, DOGROSE] and *R. eglanteria* (= *R. rubiginosa*) [EGLANTINE, SWEET BRIER], are cultivated. Others, such as *R. odorata* [TEA ROSE] are used in hybridization; crossed with *R. chinensis* [CHINA ROSE], this has resulted in HYBRID TEA ROSES. The hybrid *R. × noisettiana* [NOISETTE ROSE, CHAMPNEY ROSE], derived from *R. chinensis* and *R. moschata* [MUSK ROSE] has given rise to the double, scented, pink-to-white-flowered group called "NOISETTES". Hybrids between *R. multiflora* [POLYANTHA ROSE] and HYBRID TEA ROSES have given rise to FLORIBUNDAS such as 'Queen Elizabeth'.

Garden roses can be divided into 25 groups

in 3 general categories: 1 CLIMBERS: Noisettes, Ayrshires, Climbing Polyanthas, Hybrid Wichuraianas, Boursaults, Teas, Climbing Hybrid Teas, Kordesiis; 2 OLD SHRUB: Albas, Sweetbriars, Scotch Roses, Gallicas, Centifolias, Mosses, Damasks; 3 MODERN SHRUB: Chinas, Portlands, Bourbons, Hybrid Perpetuals, Hybrid Teas, Hybrid Polyanthas (Floribundas), Miniature (Fairy) Roses, Dwarf Polyanthas (Poly-poms), Hybrid Rugosas, Hybrid Musks.

The hips, long valued for food, were shown in 1934 to be the richest natural source of vitamin C and are now pressed commercially to give rose hip syrup. The flowers of *R. damascena* [DAMASK ROSE], and allied old cultivars, are used for the extraction of *attar of roses in the perfume industry. In the USA, *R. multiflora* and related ramblers have been planted for miles along freeways as an ideal safe, self-regenerating crash barrier.
ROSACEAE, 100–120 species.

Roscoea a small genus of perennial herbs with attractive and distinctive condensed terminal spikes of flowers. They occur naturally from the Himalayas to western China. Several species, such as *R. cautleoides* and *R. humeana*, both from China, are occasionally cultivated outdoors in mild climates for their blue, purple or rarely yellow or white flowers.
ZINGIBERACEAE, about 15 species.

Rosewood wood from a number of species of the genus *Dalbergia*. The useful rose-scented timber of the Australian tree *Dysoxylum fraseranum* is also called rosewood, as is that from *Tipuana tipu*, from southern Brazil to Bolivia, and other species whose wood resembles *D. nigra* to some extent.
LEGUMINOSAE.

Rosin or **colophony** the solid, resinous constituent of the crude oleoresin exuded by many conifer trees, principally pines such as *Pinus palustris* (= *P. australis*) [LONGLEAF PINE, SOUTHERN YELLOW PINE, PITCH PINE], *P. sylvestris* [SCOTS PINE] and *P. pinaster*

The Dog Rose (Rosa canina) is a wild species that is sometimes cultivated; it is native to Europe, western Asia and North Africa, and naturalized in North America. (×$\frac{1}{6}$)

Irregular flower of Roscoea purpurea, *a plant of the Himalayas and southwestern China.* (×1)

[MARITIME PINE]. If this is steam-distilled the volatile constituents are driven off for use as turpentine, leaving the rosin behind at the bottom of the still. After cooling, the rosin becomes a glassy brittle solid.

Rosin is mainly used in the soapmaking trade as it will combine with alkalis to form a soap. It is also used in the manufacture of soldering fluxes, varnishes, printing inks, sealing wax, ointments and plasters. In the form of a cake, it is also used for rosining the bows of stringed instruments.

Rosmarinus a small genus of aromatic evergreen Mediterranean shrubs. The best known is *R. officinalis* [ROSEMARY], which is a conspicuous element of Mediterranean shrub communities and is also widely cultivated for ornament and for its aromatic oil, oil of rosemary, which is used in perfumery, cosmetics and soaps. The fresh or more usually dried leaves are widely used as a flavoring in cooking and in salads. It is also a constituent of potpourris. The flowers are a popular source of honey.
LABIATAE, 3–4 species.

Roystonea a genus of fast-growing columnar palms native to Florida, the West Indies and Central and tropical South America. The central bud of *R. oleracea* [CABBAGE PALM, CARIBBEE PALM, SOUTH AMERICAN ROYAL PALM] is edible and a form of sago is obtained from the trunk. The timber is used for building and the leaves for thatching. This species and *R. regia* [CUBAN ROYAL PALM] are widely planted as ornamentals for their striking appearance.
PALMAE, about 14 species.

Rubber plants the source of latex, which yields, among other constituents, the long-chain hydrocarbon, rubber. Rubber has been used by American Indians for centuries, but was initially regarded by Europeans as a curiosity whose first discovered use, in 1770,

was to erase marks made by pencils.

In 1839 it was discovered that rubber heated with sulfur (vulcanized) became stable, unaffected by temperature, and could be molded. This coincided with the developing use of rubber in bicycle and, later, automobile tyres, which was responsible for the rubber boom of the late 19th and early 20th centuries.

About 1800 species of flowering plants contain rubber, but only a dozen or so are of commercial importance. Among these are:

(i) *Hevea brasiliensis* [PARA RUBBER], the source of about 99% of the world's natural rubber. *H. brasiliensis* produces the best quality and highest yields of rubber and is grown commercially in many tropical countries.

(ii) *Manihot glaziovii* [CEARA RUBBER] and *M. dichotoma* [JEQUIE RUBBER], small trees native to dry parts of northeastern Brazil. Attempts have been made to establish them as a plantation crop in areas too dry for *Hevea*, notably East Africa, but yields have proved low although rubber quality is good.

(iii) *Castilla elastica* (= *C. ule*) [PANAMA RUBBER], the most important source of rubber until the 1850s. Plantations of the rapidly growing trees were established in tropical America and Indonesia. The latex flows more rapidly and is more stable than untreated *Hevea* latex, but the trees cannot be tapped early or often, so overall yields are too low for *Castilla* to compete successfully with *Hevea*.

(iv) *Ficus elastica* [INDIA RUBBER], the main source of Asian wild rubber. It was planted commercially in the 19th century but was soon supplanted by *Hevea*.

(v) *Funtumia elastica* [LAGOS RUBBER], a forest tree, exploited and later planted in West Africa, but yields are so low that tapping is justified only when prices are high.

(vi) *Landolphia kirkii* and *L. gentilii* [LANDOLPHIA RUBBER], exploited in tropical Africa,

*Creeping legumes planted as a ground cover crop to suppress weeds in a plantation of young Para Rubber trees (*Hevea brasiliensis*) in Java.*

*A typical traditional "tapping panel" on a Para Rubber tree (*Hevea brasiliensis*), making use of a coconut shell to collect the latex.*

particularly the Congo basin, but the climbing lianas are unsuitable for cultivation.

(vii) *Cryptostegia grandiflora* and *C. madagascariensis* [MADAGASCAR RUBBER], native to Madagascar and introduced into India. Latex is collected by bleeding the stems. The rubber is of good quality but prohibitively expensive to collect. *Landolphia* and *Marsdenia* species are also called Madagascar rubber.

(viii) *Parthenium argentatum* [GUAYULE], a shrub, native to the USA, which contains rubber as inclusions in individual cells and is extracted by milling the whole plant. Interest in guayule as a crop is currently reviving.

(ix) *Taraxacum kok-saghyz* and *T. megalo-*
(continued on p. 292)

Rootcrops

The term rootcrops is applied to those plants which store food material, such as starch or sugars, in their underground organs and are cultivated on an agricultural scale. Similar plants are grown on a smaller scale, horticulturally, such as *BEETROOT and *PARSNIP, are referred to as root vegetables (see Vegetables, p. 338). Some other agricultural fodder crops such as *KALE and *RAPE play a similar role on temperate zone farms and are often incorrectly, referred to as 'roots'.

The introduction of rootcrops to European agriculture began in the 17th century. In Britain C. T. Townshend (1674–1738), or "Turnip Townshend" as he became known, began to experiment with *TURNIPS to replace the unprofitable bare fallows in the crop rotation with a crop which was not

Temperate rootcrops and vegetables: 1 Carrot (Daucus carota ssp sativus); 2 Celeriac (Apium graveolens var rapaceum); 3 Beetroot; 4 Parsnip (Pastinaca sativa); 5 Salsify (Tragopogon porrifolius); 6 Potato; 7 Jerusalem Artichoke (Helianthus tuberosus). (× ½)

only valuable in itself, but allowed the land to be cleared of weeds as well. The TURNIPS were fed to sheep in the winter months, and the sheep were kept in mobile folds, thus at the same time manuring the land and so improving its fertility. Since Townshend's day several other rootcrops have been developed such as *MANGEL-WURZELS, *SUGAR BEET and *POTATOES. They may be used as winter feed or sold off the farm as cash crops.

Rootcrops have a high water content, ranging from about 77% in SUGAR BEET to over 90% in TURNIP. The dry matter is mainly carbohydrate – largely starch in TURNIP and *SWEDE, but sugar (sucrose) in SUGAR BEET.

SUGAR BEET together with BEETROOT, MANGOLDS or MANGEL-WURZELS and fodder beet are all cultivars of *Beta vulgaris* subspecies *vulgaris*. As a sugar crop plant, beet dates from the mid-18th century when the presence of sugar in the sap of fodder beet was noted. This led to the selection of improved strains and the beginnings of the sugar beet extraction industry in Silesia, SUGAR BEET

is produced mainly in temperate climates, especially Europe, the USSR and North America, unlike cane sugar which is a crop of tropical and subtropical climates.

ROOTCROPS

Common name	Scientific name	Family
I Temperate		
*TURNIP	*Brassica campestris* ssp *rapifera*	Cruciferae
*SWEDE, RUTABAGA	*Brassica napus* var *napobrassica*	Cruciferae
MANGEL, MANGEL-WURZEL, MANGOLD	*Beta vulgaris* ssp *vulgaris*	Chenopodiaceae
BEET, *SUGARBEET, BEETROOT	*Beta vulgaris* ssp *vulgaris*	Chenopodiaceae
*POTATO	*Solanum tuberosum*	Solanaceae
II Tropical		
SWEET POTATO	*Ipomoea batatas*	Convulvulaceae
TOPEE-TAMBU	*Calathea allouia*	Marantaceae
*CASSAVA, MANIHOT	*Manihot esculenta*	Euphorbiaceae
TARO, TANIER, COCOYAMS, ARROWROOTS	See Sugar and Starch plants (p. 318)	
*YAM, WHITE GUINEA	*Dioscorea rotundata*	Dioscoreaceae
YAM, YELLOW GUINEA	*D. cayenensis*	Dioscoreaceae
YAM, GREATER	*D. alata*	Dioscoreaceae
YAM, BITTER	*D. dumetorum*	Dioscoreaceae
YAM, ASIATIC	*D. esculenta*	Dioscoreaceae
YAM, AMERICAN	*D. trifida*	Dioscoreaceae

(See also Vegetables for other root vegetables such as RADISH, SALSIFY, CARROT, PARSNIP, CELERIAC, CHERVIL, ARTICHOKES, ULLUCO, ANU, YAM BEAN, ONION, SHALLOT, GARLIC, LEEK.)

291

Tropical rootcrops and vegetables: 1 Arrowroot (Maranta arundinacea); *2 Sweet Potato; 3 Yam* (Dioscorea sp); *4 Ulluco* (Ullucus tuberosus); *5 Cassava; 6 Oca* (Oxalis tuberosa); *7 Yam Bean* (Pachyrrhizus erosus); *8 Taro* (Colocasia esculenta); *9 Anu* (Tropaeolum tuberosum); *10 Tanier* (Xanthosoma sagittifolium). *Tubers: 1, 2, 3, 7, 10* ($\times\frac{1}{4}$); *4, 6* ($\times\frac{1}{2}$); *5, 8* ($\times\frac{1}{6}$); *9* ($\times\frac{2}{3}$). *Plants* ($\times\frac{1}{10}$).

MANGOLDS or MANGEL-WURZELS are beets in which the storage carbohydrate is mainly starch not sugar.

The TURNIP and SWEDE are both brassica crops. TURNIP is *Brassica campestris* subspecies *rapifera*, and is a biennial grown for its swollen hypocotyl, the storage organ. The

flesh of the slow-maturing winter cultivars is yellow, while that of the rapidly maturing garden cultivars is white. In addition to its use as a fodder crop other early-maturing cultivars are grown as garden vegetables. The SWEDE is *Brassica napus* variety *napobrassica* and probably originated from hybridization between *B. campestris* and *B. oleracea*. It is similar to the turnip and has a yellow flesh although not so tough and fibrous as that of the turnip. (See also Brassicas p. 62.)

The POTATO was grown as livestock feed in Europe by the 16th century, but only became accepted as food for humans in the 18th century. The storage organs are morphologically the swollen underground stem-

tips. It is one of the world's staple crops and a major foodstuff. It is grown in cool climates throughout the world.

In the tropics the distinction between rootcrops and vegetables is not so clear as in temperate climates. Rootcrops are an important source of starchy foods, especially in the wetter parts of the tropics. Many of them are grown and consumed locally and do not appear on world markets.

The most important tropical rootcrops are *SWEET POTATOES, CASSAVA (MANIOC) and *YAMS of various sorts. Strictly speaking they are tubers not roots.

Other tropical rootcrops include the edible aroids, and several others given in the table. These are mainly consumed locally.

rhizon [RUSSIAN DANDELION], plants in which rubber is stored in the roots and extracted by milling.

(x) *Palaquium gutta* [*GUTTA-PERCHA], the source of an isomer of rubber with different physical properties, notably less elasticity and an ability to absorb oxygen from the air, becxoming brittle and acrid-smelling.

(xi) *Manilkara bidentata* [BALATA], the source of balata, another nonelastic rubber similar to gutta-percha, obtained from wild trees in northern South America and used to make machine belting.

Rubia a genus of mainly perennial herbs found in the Mediterranean region, temperate regions of Asia, southern and tropical Africa and America. The best-known species is *R. tinctorum* (see MADDER).
RUBIACEAE, about 40 species.

Rubus a large genus of mostly prickly shrubs, some nearly herbaceous, almost worldwide in distribution, from the Arctic south to Australia and the Falkland Islands, covering both temperate and tropical regions. The genus is important economically since many species produce edible fruits, which are collected either from wild or cultivated plants. It includes *R. idaeus* (= *R. idaeus* subspecies *vulgatus*) [EUROPEAN RED *RASPBERRY], *R. strigosus* (= *R. idaeus* subspecies *strigosus*) [AMERICAN RED RASPBERRY], *R. occidentalis* [BLACK RASPBERRY, THIMBLEBERRY] and various *BLACKBERRIES including *R. laciniatus* [EVERGREEN BLACKBERRY, CUT-LEAF BLACKBERRY], *R. procerus* [HIMALAYA BERRY] and *R. fruticosus* [EUROPEAN BLACKBERRY].

A Ditch Grass (Ruppia maritima) *growing in brackish water. The inflorescence, with unopened stamens and large flat green stigmas, is floating on the water surface.* (× 3)

Broad-leaved Dock (Rumex obtusifolius). (× ¼)

The fruit is generally compound, formed of a number of small one-seeded drupes. In blackberries or dewberries the drupes remain attached to the cone-like receptacle, in raspberries the cone of coherent drupes becomes separated from the receptacle at maturity.

Among the herbaceous species, *R. chamaemorus* [CLOUDBERRY, MALKA, SALMONBERRY, YELLOW BERRY, BAKED-APPLE BERRY] produces orange fruits which are popular in Scandinavia and elsewhere, particularly for making jam. Of the numerous shrub species, *R. idaeus* is probably the most important commercially. *R. phoenicolasius* [WINEBERRY] from China and Japan has red fruits similar to the raspberry though less sweet.

Fruits of at least 40 species are recorded as being eaten in various parts of the world, either fresh or, in the case of commercial crops, after jamming, canning or freezing. A number of hybrid types, sometimes regarded as cultivars of *R. ursinus* [PACIFIC DEWBERRY, PACIFIC BLACKBERRY], are also known and cultivated, eg *LOGANBERRY, YOUNGBERRY and BOYSENBERRY.

A number of species, for example *R. odoratus* [FLOWERING RASPBERRY, PURPLE-FLOWERING RASPBERRY, THIMBLEBERRY], are grown as ornamentals. This species is a shrub to 2m high, native to the USA, with shredding bark and loose clusters of whitish to rose-purple flowers.
ROSACEAE, about 250 species.

Rudbeckia a small genus of annual, biennial or perennial herbs native to North America. The flower heads are often very large and showy with long yellow, orange-yellow, mahogany or bicolored ray florets, and dark brown, purple or black centers. Rudbeckias make good herbaceous border plants and are frequently cultivated in gardens. *R. hirta* [GLORIOSA DAISY, BLACK EYED SUSAN] can have either single or double

flowers. *R. fulgida* [CONE FLOWER] has orange-yellow ray florets with brown-purple centers.
COMPOSITAE, about 25 species.

Rue the common name for *Ruta graveolens*, a strongly smelling small shrub. An essential oil in the divided leaves gives the characteristic odor; the flowers are a dirty yellow color. A native of southern Europe, the Mediterranean region, and the Crimea, it is cultivated to flavor sauces, meats, drinks, vinegar etc, and the distilled oil is used in perfumery; it has also been used medicinally. Other plants superficially resembling rue are often given the name, for example *Thalictrum flavum* is called YELLOW MEADOW RUE.
RUTACEAE.

Ruellia a large genus of evergreen perennials and small shrubs native to tropical America, Africa and Asia, with a few in the temperate USA. Some species are cultivated for their attractive foliage and usually large showy flowers, in the open in warm countries or in heated greenhouses in cold temperate regions. Examples include *R. portellae*, a dwarf herb from Brazil with deep green leaves with grayish veins (reddish purple beneath), and *R. macrantha*, a shrub, also from Brazil, with ovate-lanceolate dark green leaves. Both species bear rose-purple tubular flowers.

The South American *R. graecizans* (= *R. amoena*, *R. longifolia*), a subshrub with ovate to oblong-lanceolate leaves and red flowers, is grown in temperate areas as a

The Gloriosa Daisy (Rudbeckia hirta), *a popular garden plant also known as Black-eyed Susan; cultivar 'Gloriosa Daisy' is shown here.* (× 1/10)

Rye (Secale cereale) is still an important grain crop, especially in colder climates that have poor soils. (×⅛)

houseplant. Another attractive species with large red-purple flowers is the Brazilian *R. makoyana* [MONKEY PLANT, TRAILING VELVET PLANT].

The genus is treated here in a broad sense but some workers have divided it into a number of smaller genera of which *Ruellia* in the narrow sense consists of only five South American species.
ACANTHACEAE, about 250 species.

Rumex a large herbaceous genus native chiefly to temperate regions, some members of which are commonly called DOCKS or SORRELS. The flowers are small and greenish or occasionally reddish or yellowish, produced in large, usually branched inflorescences. Many species, such as the North American and Eurasian *R. acetosella* [SHEEP SORREL], are serious weeds. The leaves and petioles of some species, such as *R. acetosa* [GARDEN SORREL] and *R. scutatus* [FRENCH SORREL] are used locally in salads.

The large leaves of species such as *R. obtusifolius* [BROAD LEAVED DOCK] and *R. alpinus* [MONK'S RHUBARB, ALPINE DOCK] produce a cooling sensation and are traditionally used to soothe nettle stings and wrap butter. *R. venosus* [WILD BEGONIA, WILD HYDRANGEA] is occasionally cultivated for its attractive red fruits. The roots of *R. hymenosepalus* yield tannin.
POLYGONACEAE, about 150 species.

Ruppia a very small but widespread monocotyledonous genus of submerged aquatic herbs with linear bristle-like leaves, commonly known as DITCH GRASSES. They usually grow in brackish water in coastal areas but are also found inland in fresh water in South America and New Zealand. Plants have been collected at 4 000m in the Andes.
RUPPIACEAE, 2–7 species.

Ruscus a small genus of evergreen shrubs native from Madeira, Europe and the Mediterranean region to Iran. The leaves are minute and scale-like, their function being carried out by leaf-like branches called cladophylls. The flowers are small and unattractive, borne on the underside of the cladophylls, but are followed by scarlet berries. *R. aculeatus* [BOX HOLLY, JEW'S MYRTLE, BUTCHER'S BROOM] is occasionally cultivated and is often used in florists' arrangements; it was at one time made into besoms in Italy.
LILIACEAE, about 3 species.

Rush a general term applying to members of the monocotyledonous families Juncaceae, eg *Juncus* [RUSHES], *Luzula* [WOODRUSHES], some members of the Cyperaceae, eg *Scirpus americanus* [CHAIRMAKER'S RUSH], the Typhaceae, eg *Typha* [BULRUSHES], and the Butomaceae, eg *Butomus umbellatus* [FLOWERING RUSH]. Some *Equisetum* species are known as SCOURING RUSHES. The term is frequently applied to all those plants with stiff, often hollow stems or leaves used in the manufacture of baskets, plaited mats and chair bottoms.

Rusts see Fungi, p. 152.

Ruta a genus of aromatic perennial herbs and evergreen subshrubs mainly native to southern Europe and the Mediterranean region but also represented in the Canary Islands and western Asia (see RUE).
RUTACEAE, about 40 species.

Newly cut Perennial Rye Grass (Lolium perenne). The swathes of grass are turned as they dry and then baled for storage as hay.

Rutabaga a common name given to *Brassica napus (napobrassica* group) also known as *SWEDE, SWEDISH TURNIP.
CRUCIFERAE.

Rye a cereal of some importance in northern, central and Eastern Europe. The only cultivated species, *Secale cereale*, is an annual with very tall straw, often 2m high. The genus *Secale* is quite closely related to wheat (*Triticum*) with which it can be crossed.

Rye is primarily cultivated for its grain although it is used in some areas as a valuable spring forage. The dry grain is normally milled and the flour used for a variety of purposes. The traditional use has always been the production of a rather dense dark brown bread, the "black bread" of Eastern Europe. Other uses of rye are animal feed, production of starch, crispbread and, in the USA and Canada, the production of rye whisky.

World production in 1974 was 32 million tonnes which is evidence of the fact that it remains an important grain crop, although somewhat in decline.
GRAMINEAE.

Ryegrass the common name for *Lolium*, a genus of perennial and annual grass species native to Eurasia.

L. perenne [PERENNIAL RYEGRASS, LYMEGRASS, TERRELL GRASS, STRAND WHEAT] and *L. multiflorum* (= *L. italicum*) [AUSTRALIAN or ITALIAN RYEGRASS] are among the most valuable fodder grasses of temperate grasslands. *L. temulentum* [DARNEL, BEARDED RYEGRASS] is poisonous and was once a serious weed of cornfields.
GRAMINEAE, 8–10 species.

S

Saccharomyces [YEASTS] an economically important genus including several species used in the brewing and baking industries. These latter budding yeasts have no fruiting body (ascocarp) surrounding the asci. Forms which do not reproduce sexually but are otherwise similar to *Saccharomyces* are called asporogenous yeasts and are classified as imperfect fungi.
SACCHAROMYCETACEAE, about 30 species.

Saccharum a small genus of tall perennial grasses probably native to Southeast Asia, one of which, *S. officinarum*, is widely cultivated throughout the tropics (see SUGAR CANE).
GRAMINEAE, about 10 species.

Sage the common name for some species of *Salvia*. The best-known is *S. officinalis* [COMMON or GARDEN SAGE], a plant from the Mediterranean region, cultivated as a flavoring and as a spice. *S. fruticosa* (= *S. triloba*) [GREEN SAGE] is used as a substitute for or as an adulterant of GARDEN SAGE. Other useful sages include *S. sclarea* [CLARY SAGE], cultivated for the production of an oil used in perfumery and cosmetics as well as in certain wines and liqueurs, and *S. mellifera* [BLACK SAGE], a shrub whose flowers are excellent producers of honey in the western USA.

A number are cultivated as garden or greenhouse ornamentals, bearing whorls of showy, two-lipped flowers. Popular species include the perennials *S. azurea* [BLUE SAGE] and *S. patens*, which have deep flowers, and the shrubby, scarlet-flowered species *S. elegans* [PINEAPPLE-SCENTED SAGE] and *S. splendens* [SCARLET SAGE], which are cultivated as half-hardy annual bedding plants. *S. guaranitica* (= *S. ambigens*), with dark blue to violet-blue flowers, is also often cultivated as an annual. *S. argentea* [SILVER SAGE] is grown for its woolly gray foliage in winter.

A number of sages including the New World *S. tiliifolia* [LINDEN LEAF SAGE], *S. lyrata* [LYRE-LEAFED SAGE, CANCER WEED] and the European *S. officinalis* and *S. pratensis* [MEADOW SAGE], yield extracts used in herbal medicines.

The name sage is also given to members of other genera such as *Phlomis fruticosa* [JERUSALEM SAGE] and *Lantana camara* [YELLOW SAGE].

Sageretia a small genus of deciduous or evergreen, usually spiny shrubs, mainly native to North America and East and southern Asia. The edible fruits are blackish, about the size of a small grape, round and fleshy. They are gathered commercially in some countries, such as Turkey and Pakistan. One species, *S. thea*, furnishes an inexpensive substitute for black tea in part of Southeast Asia; its Chinese name means "beggar's tea".
RHAMNACEAE, about 35 species.

Sagittaria a mainly temperate and tropical American genus of hardy, usually bog or aquatic perennial stoloniferous herbs rooting in mud. *S. sagittifolia* [SWAN POTATO, COMMON EUROPEAN ARROWHEAD, WATER ARCHER] has

Young Sugar Cane plants (Saccharum officinarum). Sugar Cane is cultivated throughout the tropics as a source of sugar.

The Glasswort or Marsh Samphire (Salicornia fruticosa). (× ¾)

white flowers tinged with purple. Cultivated varieties of this species are often grown as ornamental pool-side plants and the cultivar 'Flore-pleno' has double flowers. Other ornamental species are *S. lancifolia* and *S. latifolia* [DUCK POTATO, WAPATU, ARROW LEAF], both from North America, with white flowers. *S. montevidensis* [GIANT ARROWHEAD], from warm temperate North and South America, grows to about 1.2m tall and has white or purple flowers. *S. subulata* [AWL-LEAF] from North America is sometimes used as an aquarium plant. Some species, such as *S. sagittifolia* and *S. latifolia*, produce edible tubers and are grown for food in the Orient.
ALISMATACEAE, about 20 species.

Sago palm the common name for some members of the genus *Metroxylon*, especially *M. rumphii* [PRICKLY SAGO PALM] and *M. sagus* [SPINELESS SAGO PALM] which are native to the Indonesian archipelago, New Guinea, New Britain and possibly the Moluccas, where they grow in low-lying marshy areas. The pith of the trunks of these palms yield sago, which is almost pure starch and is a staple food in the Southwest Pacific. It is used elsewhere in the world for puddings and sauces, and in industry as a textile stiffener. Other sources of sago include *Arenga pinnata* [GOMUTI PALM], *Caryota urens* [KITUL PALM] and cycads such as *Cycas revoluta*.

Saintpaulia a small East African genus of hairy mostly stemless perennial herbs with attractive shortly tubular violet flowers. It includes the extremely popular house plant *S. ionantha* [AFRICAN VIOLET, USAMBAR VIOLET] with numerous cultivars.
GESNERIACEAE, about 20 species.

Salicornia a genus of temperate and subtropical succulent leafless annuals characteristic of saline habitats such as coastal salt marshes and inland salt pans. *S. europaea* [SAMPHIRE, CHICKEN-CLAWS, PIGEON-FOOT] is sometimes cultivated as an ornamental. *S. fruticosa* [GLASSWORT, MARSH SAMPHIRE] is only rarely cultivated though at one time the soda ash derived from the plant was used in

glass and soap making. The perennial species formerly included in *Salicornia* are now included in *Arthrocnemum*.
CHENOPODIACEAE, about 8 species.

Salix [WILLOWS, SALLOWS, OSIERS] a large genus of trees and shrubs mainly occuring in cool-temperate or colder situations in the Northern Hemisphere; they are rare in most parts of the tropics and the Southern Hemisphere, although they are totally absent only from Australasia. Most species occur in rather open places, the larger species usually in swampy areas or along streams and rivers, the smaller species more often in boggy places on heaths and moors or in damp, stony ground on mountains and in the Arctic.

The genus is divided into three subgenera. The main subgenus, *Salix*, consists of trees or

The Common Arrowhead (Sagittaria sagittifolia) in the shallow water of a stream.

tall shrubs with narrow and pointed leaves [TRUE WILLOWS], subgenus *Caprisalix* of tall or short shrubs with narrow and pointed or broad and rounded or blunt leaves (OSIERS and SALLOWS respectively), and subgenus *Chamaetia* of dwarf, creeping, mountain or Arctic shrubs with small, broad and rounded or blunt leaves [DWARF WILLOWS].

Hybridization between different species of *Salix* is a very widespread phenomenon and many artificial hybrids have been made, to add to the considerable number of natural ones. Since most of these hybrids are fully fertile, and can cross with their parental species or with other species or hybrids, the boundaries between many species have become somewhat blurred so that identification can be very difficult.

Salix timber, although not particularly strong or durable, has a number of special uses. Cricket bats are made from *S. alba* var *calva* [CRICKET BAT WILLOW]. Baskets and other wickerwork are made from several species commonly known as osiers, especially *S. viminalis* [COMMON OSIER, BASKET WIL-

LOW] and *S. purpurea*. These species can be encouraged to put out long, straight, flexible suckers by severe pruning (pollarding) of old trunks, from which the suckers can be harvested annually. *S. nigra* [BLACK WILLOW] is used as a source of paper pulp and the wood of *S. jessoensis* [YELLOW WILLOW] and *S. tetrasperma* [INDIAN WILLOW TREE] is used for a variety of purposes such as tool handles, boxes and furniture.

Many species are prized for the ornamental appearance. Some of the best-known are *S. alba* [WHITE WILLOW], *S. babylonica* [WEEPING WILLOW], *S. fragilis* [CRACK WILLOW, BRITTLE WILLOW], *S. pentandra* [BAY-LEAVED WILLOW], *S. caprea* [SALLOW, GOAT WILLOW, PUSSY WILLOW], *S. lanata* [WOOLLY WILLOW],

S. herbacea [DWARF WILLOW] and *S. reticulata* [RETICULATE WILLOW].
SALICACEAE, about 300 species.

Salpiglossis a genus of annual, biennial and perennial herbs which include some showy ornamentals with large, trumpet-shaped flowers. Native to Chile, they can be successfully cultivated in temperate gardens, particularly *S. sinuata* [PAINTED-TONGUE], which is a valuable half-hardy annual with flower colors varying from purple to reddish or yellowish and often striped.
SOLANACEAE, about 8 species.

Salsify the common name for *Tragopogon porrifolius* [OYSTER PLANT, VEGETABLE OYSTER]. A hardy biennial plant of the Mediterranean region, with stems up to 1m high, bearing purple flowers, it is cultivated in many temperate parts for the roots which are eaten as a vegetable. BLACK SALSIFY and SPANISH SALSIFY are names given to *Scorzonera hispanica* with black-skinned roots.
COMPOSITAE.

The African Violet (Saintpaulia ionantha) is a highly popular house plant. (× ½)

Salsola a genus of perennial herbs usually growing in saline conditions. The leaves of *S. kali* [RUSSIAN THISTLE, PRICKLY SALTWORT] end in spines. The young leaves of this species and some others, such as *S. asparagoides* and *S. soda*, may be boiled and eaten as vegetables.
CHENOPODIACEAE, about 150 species.

Saltbush one of the common names for some members of the genus *Atriplex*, particularly those that grow in saline soils. Species grown as ornamentals include *A. canescens* [FOUR-WING SALTBUSH] and *A. confertifolia* [SPINY SALTBUSH]. Several species, such as *A. halimus* [MEDITERRANEAN SALTBUSH] and *A. campanulata*, are used as stock feed in arid or saline areas, being particularly valuable as an emergency feed during drought.
CHENOPODIACEAE.

Salvia a large genus of herbs and subshrubs distributed throughout tropical and temperate regions, particularly in the New World (see SAGE).
LABIATAE, about 700 species.

The Scarlet Sage (Salvia splendens). (× ⅐)

Kariba Weed (S. auriculata). (×⅕)

Salvinia a genus of free-floating, rootless water ferns, principally tropical and subtropical in distribution and often forming wide mats over still or slowly moving water. At each node there are three leaves – one finely divided submerged leaf, functioning as and instead of roots, and two floating leaves. Some species are serious weeds in tropical lakes, eg *S. auriculata* [KARIBA WEED]. SALVINIACEAE, about 10 species.

Samanea a genus of tropical American and African trees and shrubs, bearing dense, globose heads of yellowish flowers and somewhat curved pod-like fruits. The only species in extensive cultivation is *S. saman* (= *Pithecellobium saman*) [RAIN TREE, SAMAN, COW TAMARIND, MONKEYPOD, ZAMANG], from tropical America. It is a popular and fast-growing ornamental tree grown throughout the tropics and is an ideal "host" for ornamental epiphytes. LEGUMINOSAE, up to 20 species.

Sambucus [ELDERS, ELDERBERRY] a genus of temperate and subtropical shrubs and small trees, and rarely, perennial herbs. The hardy woody species are mainly from Europe and North America. All are typified by flat clustrees, and, rarely, perennial herbs. The hardy berries. *S. nigra* [COMMON or EUROPEAN ELDER] is a common European shrub or small tree. Purgative properties have been attributed to infusions of its roots. The flowers and berries of *S. nigra* and, in North America, *S. canadensis* [SWEET ELDER, AMERICAN ELDER] have long been used for making wines. The berries of many other species are poisonous.

S. ebulus [DANE'S ELDER, DANEWORT, DWARF ELDER, WALLWORT], a herbaceous species of roadsides and waste places, was formerly an important medicinal plant, being recommended for the healing of all manner of ailments from jaundice to gout.

Among the species cultivated as ornamentals are *S. canadensis* and its cultivars 'Maxima', with huge flower heads, 'Aurea', with greenish-yellow foliage, *S. nigra* and its cultivars, *S. pubens* [AMERICAN RED ELDER], and *S. racemosa* [RED-BERRIED ELDER, EUROPEAN RED ELDER]. CAPRIFOLIACEAE, about 20 species.

Sandalwood tree strictly, the common name for *Santalum album*, but several other species yield fragrant wood known by the same name. These include *Adenanthera pavonia* and *Pterocarpus santalinus*.

Sandarac a resin obtained from the trunk of the coniferous tree *Tetraclinis articulata*, which is native to northwestern Africa, southern Spain and Malta. The Australian *Callitris endlicheri* (= *C. calcarata*) [BLACK CYPRESS PINE] is the source of the very similar Australian sandarac. The resin is used in varnishes for paintings and in lacquers for photographic work. It is also used in medicine mainly as a pill coating. CUPRESSACEAE.

Sanguinaria a North American genus represented by a single species, *S. canadensis* [BLOODROOT, RED PUCCOON], a rhizomatous herb producing annually one palmately lobed, scalloped-edged leaf and one leafless flower stem. A double form (cultivar 'Multiplex') of the usual single, white, cup-flower is known in cultivation. PAPAVERACEAE, 1 species.

Sanguisorba a genus of perennial herbs from temperate Eurasia and North America. Some authorities include in this genus species of *Poterium*. They have flowers in compact heads and a basal rosette of pinnate leaves. The best-known species are *S. officinalis* (= *Poterium officinale*) [GREAT BURNET, BLOODWORT], with dull red flower heads, and *S. minor* (= *P. sanguisorba*) [SALAD BURNET], whose leaves can be used in salads. ROSACEAE, 2–3 species.

*Male (left), with yellow stamens, and female (right), with red stigmas, inflorescences of the Salad Burnet (*Sanguisorba minor*), a common herb of chalk and limestone grassland. (×6)*

The floating fern Salvinia cucullata *is an attractive species native to India, Malaya, Indonesia and Western Australia. (×1)*

Sanicula a genus of perennial herbs widely occurring naturally, except in Australasia. The minute, inconspicuous flowers are borne in irregular umbels and are followed by bristly, animal-dispersed fruits. *S. marilandica* [SANICLE, BLACK SNAKEROOT] and *S. europaea* [WOOD SANICLE] have local medicinal uses as astringents and as antispasmodics. The latter and *S. arctopoides* [FOOTSTEPS-OF-SPRING] and *S. bipinnatifida* [PURPLE SANICLE] are grown for ornament. UMBELLIFERAE, 35–40 species.

Sansevieria or **Sanseverinia** a genus of African and Asiatic xerophytes with stout rhizomes and more or less succulent, leathery, elongated, fibrous, flat cylindrical

leaves. The leaves of several species (often commonly called BOWSTRING *HEMP), including those of *S. roxburghiana* [INDIAN BOWSTRING HEMP], used in India, and the South African *S. aethiopica* and *S. hyacinthoides* (= *S. guineensis*) [AFRICAN BOWSTRING HEMP], yield strong white fibers used for cordage and mats. The last-named species is grown commercially in Jamaica and Central America.

S. zeylanica [CEYLON BOWSTRING HEMP] is cultivated in its native Sri Lanka. Most of the material grown under this name is probably *S. aethiopica* or *S. hyacinthoides*. *S. trifasciata* [MOTHER-IN-LAWS TONGUE, SNAKE PLANT], in its variegated forms, is a popular foliage house plant.
AGAVACEAE, about 60 species.

Santalum a genus of small to medium-sized trees, all root-parasites, native from India to Australia, Southeast Asia and the Pacific. *S. album* [WHITE SANDALWOOD] is much valued for the fragrant white sapwood which is used for carvings and fine woodwork, and as a form of incense, while an aromatic oil, used in perfumery and in religious ceremonies, is extracted from the yellow heartwood. It is widely cultivated in the tropics, especially in southern India.
SANTALACEAE, about 10 species.

Santolina [COTTON LAVENDERS] a genus of aromatic evergreen shrubs, subshrubs and herbs from southern Europe and the Mediterranean region. They are fast-growing and hardy, with dissected pinnate foliage, and small yellow or white terminal heads of flowers. Species such as *S. chamaecyparissus* (= *S. incana*) [LAVENDER COTTON] are cultivated mostly for their white, gray or silver, feathery foliage.
COMPOSITAE, about 8 species.

Sanvitalia a genus of annual herbs native to the southwestern USA, Mexico and Central America, with one species native to Bolivia and Argentina. A few species, notably *S. procumbens*, from Mexico and Guatemala, are grown in gardens for their attractive yellow and purple flower heads.
COMPOSITAE, 7–8 species.

Sapele the common name for *Entandrophragma cylindricum*, a tropical West and central African tree which yields a valuable, scented mahogany-like timber widely used for furniture making. The name SAPELE MAHOGANY or HEAVY SAPELE is also given to the related species, *E. candollei*, locally called OMU.
MELIACEAE.

Saponaria [SOAPWORTS] a genus of herbaceous annuals, biennials and perennials, native to Europe and Asia but mainly centered in the Mediterranean region. The leaves of the perennial *S. officinalis* [SOAPWORT, SOAP ROOT, BOUNCING BET] and the annual *S. calabrica* contain saponins which have the property of foaming in water to give a soap-like action. Because of their gentle

action they have been used to clean woollen garments, tapestries etc. Extracts of *S. officinalis* have also been used medicinally.

S. calabrica and the perennial *S. ocymoides* [ROCK SOAPWORT] are grown in rock gardens and borders.
CARYOPHYLLACEAE, about 30 species.

Sapote a name applied to several edible tropical fruits belonging to different families. *Pouteria sapota* (= *Calocarpum sapota*, *C. mammosum*) [SAPOTE, MAMMEE, MAMEY, COLORADO SAPOTE, MARMALADE PLUM] is a large tree, 20m or more in height, cultivated in tropical America, the Caribbean and the Philippines. The edible fruit, up to 2kg in weight, is consumed either fresh or as a preserve. Other fruits named SAPOTE are those of *P. campechiana* (= *Lucuma nervosa*) (see EGG FRUIT), *Casimiroa edulis* [WHITE SAPOTE, ZAPOTE BLANCO] and *Diospyros digyna* [BLACK SAPOTE].

Saprolegnia a genus of aquatic oomycete fungi which parasitize fish. *S. ferax*, *S. parasitica*, *S. delica* and *S. diclina* are some of the species which may cause serious epidemics among fish either in aquaria or in open waters.
SAPROLEGNIACEAE, about 20 species.

Sapucaia the edible seeds of the tropical American genus *Lecythis*, especially *L. zabucayo* and *L. usitata*. The oblong, wrinkled seeds are about 5cm long and enclosed in a large woody fruit with a lid, called "monkey pods". They are considered to be superior to the *Brazil nut which they resemble.
LECYTHIDACEAE.

Sarcococca a small genus of low evergreen shrubs native to India, the Himalayas and

Yellow flowers and silvery foliage of Lavender Cotton (Santolina chamaecyparissus). ($\times \frac{1}{2}$)

Flower heads of the Soapwort (Saponaria officinalis). This species was formerly grown for its leaves, which contain saponins – substances that lather in water like soap. ($\times 3$)

China to western Malaysia. They are similar to *BOX (*Buxus*) but have alternate leaves. Some species, commonly known as SWEET BOX, have very fragrant flowers which are followed by ornamental fleshy black or red berries, and are grown as autumn or winter ornamentals. The most commonly cultivated species are *S. hookerana*, from Tibet and the Himalayas, and its varieties *digyna* and *humilis*, and *S. ruscifolia* [FRAGRANT or SWEET BOX], from southern China.
BUXACEAE, about 16 species.

Sargassum a genus of brown algae which may be found growing attached to rocks in the sublittoral fringe or floating as detached fronds. It is very abundant in the area of the western Atlantic Ocean known as the Sargasso Sea. Several species, including the Pacific *S. echinocarpum*, *S. vulgare* and *S. enerve*, are consumed as food. Many species, including *S. fusiforme* are also used as manure and sources of algin and iodine.
PHAEOPHYCEAE.

Sarracenia a small genus of rhizomatous perennial carnivorous flowering plants occurring in sunny marshy places of eastern North America. All of the species are distinguished by the possession of rosettes of tubular or trumpet-shaped pitcher leaves, especially adapted as traps for the capture of insects. Wild and cultivated species and their hybrids are given a wide variety of common names including PITCHER PLANT, INDIAN CUP, TRUMPET LEAF, SIDE-SADDLE FLOWER. Most cultivated forms including *S. flava*, *S. purpurea* and *S. minor* are grown in moist situations in the open or in the greenhouse for their unusual leaves and bowl-shaped pendulous flowers.
SARRACENIACEAE, about 8 species.

Sarsaparilla the name given to several species of the genus *Smilax* and to the drug extracted from them (see *Smilax*). The common name WILD SARSAPARILLA is also given to the quite unrelated *Aralia nudicaulis* and *Schisandra coccinea*, while *Hardenbergia violacea* is sometimes known as FALSE SARSAPARILLA.

Sasa a genus of dwarf rhizomatous bamboos (rarely exceeding 1.2m in height) native to Japan and temperate East Asia. The woody stems are sparsely branched with the leaves crowded at the ends of the branches. *S. veitchi* [KUMA BAMBOO GRASS] is cultivated as an ornamental for its purple waxy stems and is very fast-growing. *S. tessellata* has large, beautifully tessellated leaves and *S. chrysantha* often has yellow variegated leaves.
GRAMINEAE, about 130 species.

Sassafras a very small genus of aromatic deciduous trees with one species in North America and two species in East Asia. The wood and bark of the American *S. albidum* [SASSAFRAS] is important as the source of the antiseptic, oil of sassafras, which is used in perfumery, medicine and as a flavoring. *S. tzumu* is the CHINESE SASSAFRAS.
LAURACEAE, 3 species.

The Meadow Saxifrage (Saxifraga granulata) is a European species that grows best in base-rich, undisturbed grassland. It is widely grown as an ornamental. (×1)

Scapania undulata; most of the shoots are female, the sex organs enclosed by a distinct flattened tubular "perianth". (×5)

Satinwood the name given to the timber of a number of unrelated tropical trees. One of the most commercially valuable is *Zanthoxylum americanum* (= *Z. flavum*) [WEST INDIAN SATINWOOD], whose golden yellow wood has traditionally been the best for cabinetwork. *Chloroxylon swietenia* [EAST INDIAN SATINWOOD] which is grown in India and Sri Lanka, is another source of commercial satinwood.
   Euxylophora paraensis [BRAZILIAN SATINWOOD] produces a clear yellowish hard-grained timber used in furniture making. The popular name of the tall tree *Ceratopetalum apetalum*, of New South Wales, is SCENTED SATINWOOD because of its characteristic odor; its straight-grained, brownish timber is used for cabinetwork, furniture and plywood veneers. The wood of *Murraya paniculata* [ORANGE JASMINE], an Asiatic tree widely cultivated in the tropics, is also known as satinwood.

Satsuma a variety of *MANDARIN ORANGE* (*Citrus reticulata*). It is grown principally in Japan, and is gaining popularity as a dessert fruit because it is seedless and easy to peel.
RUTACEAE.

Satureja a genus of aromatic annual or perennial herbs or subshrubs, widespread in temperate and warm regions. Two species, *S. hortensis* [SUMMER SAVORY] and *S. montana* [WINTER SAVORY], are cultivated, particularly in the Mediterranean region, for their leaves and stems used for flavoring in cookery and for their essential oil. The extract from *S. hortensis*, known as oil of savory, is used in flavoring sauces and various processed foods.
COMPOSITAE, about 30 species.

Satyrium a genus of predominantly tropi-cal and southern African terrestrial orchids found mainly in open woodland and upland grassland. There are also several more species found in the East Indies and the Mascarene Islands. *Satyrium* species are characterized by their spikes of bright pink, scarlet, crimson, vermilion or occasionally white flowers. *S. nepalense* is one of the few species which has been successfully cultivated.
ORCHIDACEAE, about 120 species.

Sauerkraut a traditional German dish made from *CABBAGE* (*Brassica oleracea*). The leaves are shredded and fermented in brine and spices. The bacterium *Lactobacillus cucumeris* produces lactic acid, which preserves the cabbage and gives it its characteristic flavor.

Sauromatum a small genus of tropical African and Asian tuberous-rooted, stemless perennial herbs. A number of varieties, including *pedatum* and *venosum* of *S. guttatum* [VOODOO LILY, RED CALLA, MONARCH-OF-THE-EAST], are cultivated as greenhouse plants. Their ornamental value lies in the elongated conical inflorescence (spadix) which is enclosed by an inflated, yellowish but variously spotted spathe. On the detrimental side, the sterile terminal appendage to the spadix emits a foul odor. The tuber of the tropical African *S. nubicum* is used locally as a food.
ARACEAE, 4 species.

Saxegothaea a genus represented by a single Chilean and Patagonian species, *S. conspicua* [PRINCE ALBERT YEW]. As the common name suggests it is a yew-like evergreen bushy tree, more or less conical in shape, bearing rather fleshy cone-like fruits.
PODOCARPACEAE, 1 species.

Saxifrage the common name for members of the large genus *Saxifraga*, distributed principally throughout the Arctic and north and south temperate regions. Most saxifrages are mountain plants, growing on screes, cliff-ledges or vertical rock-faces; some of these also occur at low levels on the arctic tundra. A few, however, grow in woods or damp meadows or beside streams. Except for half-a-dozen annuals or biennials they are all perennial. Many species have a neat, cushion-like habit and conspicuous and elegant flowers, and are popular as plants for the rock garden or alpine house.

The saxifrages most popular among gardeners belong to the section *Porphyrion* (often, but less correctly, called *Kabschia*). These have tight cushions of very small leaves and flowers large in proportion to the size of the plant. Most garden plants of this section are hybrids; among the species which have contributed to them are *S. burserana* and *S. marginata* (= *S. coriophylla*), both with white flowers, *S. juniperifolia* (= *S. juniperina*) and *S. ferdinandicoburgi*, with yellow flowers, and *S. lilacina*, with lilac flowers. The encrusted saxifrages (section *Aizoonia*) are also popular, mainly because of the beauty of their silver-gray leaf rosettes which are heavily lime-crusted. Some of them, however, also produce many-flowered panicles of white, yellow, pinkish or veined flowers; these include *S. longifolia*, *S. callosa* (= *S. lantoscana*, *S. lingulata*), *S. cotyledon* and *S. paniculata*. Other widely grown saxifrages include the purple-flowered *S. oppositifolia* [PURPLE MOUNTAIN SAXIFRAGE], an arctic–alpine species, *S. umbrosa* and its hybrids, especially *S. × urbium* (= *S. spathularis × S. umbrosa*) (the well-known *LONDON PRIDE* belongs here), and various "MOSSY SAXIFRAGES". These last are hybrids of complex and obscure parentage derived from various species in the section. Species belonging to the section *Dactyloides*, such as *S. rosacea*, are vigorous, soft-leaved plants forming large mats and bearing pink or white flowers.

S. stolonifera (= *S. sarmentosa*) [STRAW-BERRY GERANIUM, CREEPING-SAILOR, MOTHER-OF-THOUSANDS] is a white-flowered stoloniferous species belonging to the section *Diptera*. It is popular as an indoor pot or basket plant. The cultivar 'Tricolor' [MAGIC-CARPET SAXIFRAGE] has variegated, greenish-white leaves flushed pink or rose. Another widely cultivated species is *S. granulata* [MEADOW SAXIFRAGE, FAIR MAIDS OF FRANCE], a member of the section *Saxifraga*. It is a bulbous perennial with white flowers up to 2.5cm across (double in cultivar 'Flore Pleno'). SAXIFRAGACEAE, about 300 species.

Scabiosa [SCABIOUS] a genus of annual to perennial herbs, rarely subshrubs, many of which make fine ornamentals. They occur naturally in temperate Eurasia, the Mediterranean and the mountains of East Africa. The flowers are borne in heads usually on long stalks with an involucre of numerous bracts, and may be blue, purple, pink, white or creamy in color. *S. caucasica* has large pale blue flower heads (white in cv 'Alba', lilac in 'Goldingensis') 7–10cm in diameter, and is a fine plant for the herbaceous border and for cutting. *S. atropurpurea* [MOURNFUL WIDOW, PINCUSHION FLOWER, EGYPTIAN ROSE, SWEET SCABIOUS, RED INDIAN SCABIOUS] is also useful for providing cut-flowers in winter and under glass, while *S. graminifolia* and *S. silenifolia* are useful rock-garden plants. DIPSACACEAE, about 80 species.

Scapania a genus of leafy liverworts, characterized by having the upper (dorsal) lobe of the folded leaf clearly smaller than the lower (ventral) one. *Scapania* is largely a genus of the Northern Hemisphere, usually in colder climates or at high altitudes; the very few species known from the tropics occur in the cooler climates of high mountain systems. SCAPANIACEAE, about 50 species.

The Pepper Tree (Schinus molle) is often planted as a shade tree in the tropics and subtropics.

Flower head (capitulum) of the Devil's Bit Scabious (Scabiosa succisa = Succisa pratensis), comprising a mass of purple flowers with long-stalked stamens. It is native to North Africa and Europe and naturalized in the USA. (× 2)

Scarborough lily the common name for *Vallota speciosa* (= *V. purpurea*, *Amaryllis purpurea*), a bulbous ornamental lily with scarlet to crimson, funnel-shaped flowers and long (0.6m) strap-shaped leaves, native to South Africa but cultivated in tropical and subtropical gardens in a range of varieties. AMARYLLIDACEAE.

Schefflera a large genus of evergreen trees and shrubs widely distributed throughout the tropics and subtropics. Commonly known as UMBRELLA TREES, RUBBER TREES and STARLEAFS, a number of species, including *S. delavayi*, *S. digitata* and *S. octophylla*, are grown as ornamental foliage plants (indoors in cooler temperate areas). The leaves of *S. aromatica* [LINGKERSAP] are used in Indonesia as a vegetable. ARALIACEAE, about 150 species.

Schinus a mainly South American genus of evergreen trees with panicles of inconspicuous white or yellow flowers. *S. molle* [PEPPER TREE, AUSTRALIAN PEPPER TREE, CALIFORNIA PEPPER TREE, PERUVIAN MASTIC TREE, AMERICAN MASTIC TREE], native to the Peruvian Andes, is widely grown in the tropics and subtropics as a street tree. It has a graceful habit, its pendulous branches producing valuable shade, and it bears attractive reddish fruits (drupes), which in its native Peru are made into a mildly alcoholic drink. The ground seeds may be used as a condiment and/or an adulterant for pepper. The tree also produces a gum resin of the mastic type. Other cultivated species include the Brazilian *S. terebinthifolius* [BRAZILIAN PEPPER TREE, CHRISTMAS-BERRY TREE], used extensively for Christmas wreaths. ANACARDIACEAE, 20–30 species.

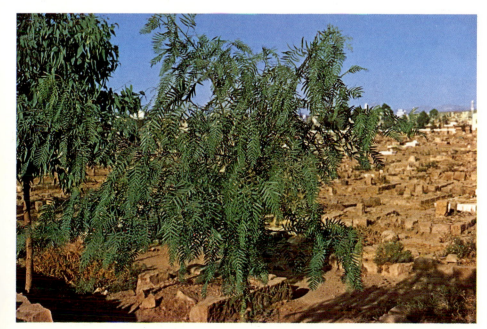

Schisandra a genus of deciduous and ever-green dioecious or monoecious twining shrubs native to East Asia and North America. The fragrant, usually pink or red flowers are followed by conspicuous, pendulous spikes of attractive berries which are edible in some species. The hardy cultivated species mostly come from East Asia. *S. chinensis* is a vigorous climbing plant, with white and pale pink flowers in April and May. The dried wood has a pleasant fragrance. *S. coccinea* [BAY STAR VINE, WILD SARSAPARILLA], from North America, bears crimson flowers and clusters of red berries.
SCHISANDRACEAE, about 25 species.

Schizaea [COMB, RUSH or CURLY GRASS FERNS], a genus of mainly tropical (but some temperate) small ferns in which the large sporangia are borne in double rows on the lower surface of special parts of the pinnae. A number of species, including *S. bifida*, *S. latifolia* and *S. pusilla*, are cultivated as greenhouse ornamentals.
SCHIZAEACEAE, about 30 species.

Schizanthus a small genus of Chilean annual or biennial herbs widely grown under glass or outdoors for ornament. They are usually glandular with pinnatisect leaves. An abundance of showy "butterfly" flowers are borne in terminal cymes. *S. retusus* [BUTTERFLY-FLOWER, POOR MAN'S ORCHID] cultivar 'Grahami', with lilac and orange flowers, and *S. pinnatus* with purple, violet, lilac, rose-pink, white and yellow flowers are thought to be the parents of many beautiful garden hybrid forms, eg *S. × wisetonensis*, with flowers varying from white to red-brown with yellow in the middle of the upper lip of the corolla tube.
SOLANACEAE, 10–15 species.

Schizophragma a very small East Asian genus of vigorous deciduous ornamental climbing shrubs. The genus is closely related to *Hydrangea and members have loose terminal clusters of white flowers which are two sorts: the outer are sterile and showy consist-

The crocus-like flowers of the Kaffir Lily (Schizostylis coccinea), from South Africa, are arranged in racemes. (×1)

In some parts of South America the Totora Reed (Scirpus totora), is used for fodder, fuel and fertilizer and as a boatbuilding material. These two Aymara Indians are collecting the reeds from Lake Titicaca in Bolivia.

ing of one large petaloid calyx lobe, the inner ones fertile, smaller and almost sessile. *S. hydrangeoides* [JAPANESE HYDRANGEA VINE] and *S. integrifolium* are excellent for part-shaded walls or for decorating old trees.
SAXIFRAGACEAE, 3–4 species.

Schizostylis a very small genus of rhizomatous herbs native to South Africa. The leaves are linear and two-ranked. The bell-shaped flowers are borne in terminal racemes each flower subtended by a green spathe. *S. coccinea* [CRIMSON FLAG, KAFFIR LILY] has red flowers on stems 60cm high and is widely cultivated, as is the pink-flowered variety 'Viscountess Byng'.
IRIDACEAE, 2 species.

Schlumbergera a small epiphytic genus native to Brazil, which includes a number of attractive showy-flowered ornamentals, notably the widely cultivated series of hybrids with large cerise or magenta flowers, known as *S. × buckleyi* (= *S. bridgesii × S. truncata*), which are commonly called CHRISTMAS CACTUS. *S. truncata* [CRAB CACTUS, CLAW CACTUS, YOKE CACTUS, THANKSGIVING CACTUS] is a popular houseplant with whitish fuchsia-like flowers in winter (salmon-pink in 'Salmonea' and violet in 'Violacea').
CACTACEAE, 3–5 species.

Sciadopitys a conifer genus comprising a single species, *S. verticillata* [JAPANESE UMBRELLA PINE, PARASOL PINE] an evergreen pyramidal tree native to Japan. The leaves are of two types, the main ones paired, united along their whole length and arranged in definite whorls of 10–30; this arrangement resembles that of the ribs of an umbrella,

hence the popular name. Along the inter-nodes between the whorls are triangular, somewhat overlapping, scale-like leaves which are green at first, becoming brown in the second year. It produces a durable water-resistant wood which makes it a useful boat-building material.
TAXODIACEAE, 1 species.

Scilla a genus of herbaceous bulbous per-ennials native to temperate European and Asian regions, but also plentiful in tropical and southern Africa. At one time the genus included the ENGLISH BLUE BELL and related species (*S. hispanica* and *S. non-scripta*) but these have now been placed in another genus, *Hyacinthoides* (= *Endymion*). Many of the Southern Hemisphere species are sometimes referred to *Ledebouria*. Other spe-cies are sometimes referred to *Brimeura*, eg *S. amethystina* (= *B. amethystina*).

Some species, eg *S. hyacinthoides* [HYA-CINTH] and *S. peruviana* (= *S. hughii*) [CUBAN LILY, PERUVIAN JACINTH, HYACINTH-OF-PERU], have numerous (50–100) flowers in the inflorescence, which is conical or hemi-spherical in shape. Species which have fewer flowers (20 or less) in each inflorescence include *S. verna* [SPRING SQUILL, SEA ONION], *S. bifolia* and the autumn-flowering *S. autumnalis* [AUTUMN SQUILL, STARRY HYACINTH].

The species which are cultivated in gar-dens are hardy dwarf forms, 15–30cm high, suitable for rock gardens or lawns. They include *S. bifolia*, with five to eight lilac, pink or white flowers per stem, *S. monophylla* [DWARF-SQUILL], with six to 20 bright blue, bell-shaped flowers per stem, and *S. peruviana* with about 100 star-shaped blue flowers per stem. Two very early-flowering garden spe-cies are *S. amoena* (= *S. siberica*) [STAR HYA-CINTH], from Asia Minor and southern Europe, and *S. mischtschkenkoana* (= *S. tubergeniana*) from Iran and the Caucasus. The last-named species is particularly widely grown especially as 'Spring Beauty', which has larger flowers than other varieties.
LILIACEAE, about 80 species.

Scirpus a large cosmopolitan genus of rhizomatous or tuberous perennial herbs which characteristically inhabit wet moors, bogs and marshes. Their stems are usually robust, erect and angular, bearing three ranks of very reduced sheath-like leaves. Their inflorescences occur as branched ter-minal tufts at the top of the stem and consist of inconspicuous many-flowered spikelets.

Common north temperate species include *S. lacustris* [COMMON BULRUSH, CLUBRUSH], a tall and often dominant species in rivers and ponds, *S. caespitosus* [DEER GRASS] and *S. fluitans* [FLOATING MUD RUSH]. A number of attractive species including *S. cernuus* are cultivated for ornament. The stems of *S. lacustris*, *S. americanus* [CHAIRMAKER'S RUSH], and other species are used for matting and basketwork.
CYPERACEAE, about 200 species.

Sclerotinia a widespread genus of cup fungi with stalked saucer- or cup-shaped fruiting-bodies (apothecia) arising from either a true sclerotium (eg *S. trifoliorum*) or a pseudo-sclerotium of plant tissue hardened and "mummified" by mycelium (eg *S. fructicola*). *Sclerotinia* is unusual among cup fungi in that it includes some important plant pathogens such as *S. fructigena*, the causal agent of brown rot of apples and pears.
SCLEROTINIACEAE, about 50 species.

Scorzonera a large genus of mainly peren-nial composite herbs with milky sap, native to the Mediterranean region and Eastern Europe to Central Asia. Most species have somewhat grass-like leaves and long flower-stalks subtending large yellow or purple or pink flower heads in which the flowers are all strap-shaped.

The best-known species is *S. hispanica* [BLACK OYSTER PLANT, COMMON VIPER'S GRASS, BLACK OR SPANISH SALSIFY], whose black-skinned roots are cultivated as a vegetable in

*The Sea Lettuce is an edible seaweed (*Ulva lactuca*), here photographed on a sandy beach exposed at low tide. (× ½)*

the Mediterranean region and elsewhere in Europe. The roots are also used as a coffee substitute and the leaves eaten in salads. It is grown as an annual or biennial.
COMPOSITAE, about 150 species.

Scrophularia a large but economically un-important genus of annual, biennial or per-ennial herbs or subshrubs from temperate Eurasia, with a few species in North and tropical America. Many species have brownish-yellow or greenish-purple flowers and some are fetid. *S. nodosa* [FIGWORT], from Europe, and *S. marilandica* [MARYLAND FIG, CARPENTER'S SQUARE] from North America, have local medicinal uses. *S. chrysantha* is a biennial or perennial from the Caucasus that has golden-yellow flowers and is cultivated for ornament. The variegated-leaved forms of *S. aquatica* [COMMON EUROPEAN WATER FIG-WORT] are also cultivated.
SCROPHULARIACEAE, about 250 species.

Scutellaria a large genus of perennial ever-green shrubby plants distributed worldwide except in South Africa. The slender tubular flowers are two-lipped and hooded, with a small lateral lobe. *S. galericulata* (= *S. epilobii-folia*) [COMMON SKULLCAP], cosmopolitan in the north temperate zone, grows up to 30cm high with bright blue-violet flowers, and is found in lowland, damp places, frequently in wet grassland in most of Europe.

Numerous species and varieties are culti-vated as greenhouse or hardy perennials, commonly under the name of HELMET FLOWERS or SKULLCAP FLOWERS, including *S. albida* (small white flowers), from south-east Europe to Central Asia, *S. longifolia* (long scarlet flowers), from Mexico to Gua-temala, *S. alpina* (purple flowers), from the mountains of southern Europe to Siberia, and *S. orientalis* (yellow flowers, rarely pink), mainly from southeast Europe to Siberia.
LABIATAE, about 300 species.

Sea lettuce the common name for species of the genus *Ulva*, especially *U. lactuca*. It is a green alga with a leaf-like plant body grow-ing up to 30cm in length. Sea lettuce can be eaten fresh or cooked and is sometimes used as animal fodder.
CHLOROPHYCEAE, about 30 species.

Seaweed a popular term used to describe any marine alga growing along the coast or in floating masses in shallow inlets, lagoons, etc.

Secale a small genus of annual or perennial grasses indigenous to southeast Europe and southwest Asia, including the cultivated *S. cereale* (see RYE). Species of annual weeds including *S. ancestrale*, *S. dighoricum*, *S. segetale* and *S. afghanicum* so closely re-semble the annual *S. cereale* that they may more correctly be regarded as subspecies of the latter.

At least two other species, the annual *S. sylvestre* and the perennial *S. montanum*, have probably been implicated with weedy

forms of *S. cereale* in the evolution of cultivated RYE.
GRAMINEAE, about 5 species.

Sedge the common name given to species
of grass-like perennial herbs belonging to the
genus *Carex*. However the term is also given
to some species belonging to other genera of
the same family, eg *Cladium mariscus* [FEN
SEDGE], *Cyperus* species and *Rhynchospora*
species.
CYPERACEAE.

Sedum a large and widespread genus
mostly of herbaceous perennial leaf-
succulents, and some subshrubs and a few
epiphytes. Mainly native to north temperate
countries, there are outliers extending to
Africa, Madagascar, South America and the
Philippines. The leaves are mostly small,
overlapping, succulent and stalkless. The
inflorescence is usually terminal and cymose,
and the flowers are usually yellow or white in
color.
    Ornamental species show considerable
variation in habit. For example *S. frutescens*

Flowers and flower buds of Sedum spectabile. *The
flowers are particularly attractice to butterflies
and bees.* (×2)

The popular name for Sedum rosea *is Roseroot, so
named because it has a rose-scented rootstock.* (×½)

and *S. oxypetalum*, the giant Mexican STONE-
CROPS, are thick-stemmed shrubs with smooth
peeling bark. The much-branched fleshy
stems reach a height of 1.5m. *S. dendroideum*,
also from Mexico, may reach 2.2m in height
or, in the trailing form, a length of 6m.
    By contrast, there is a group of hardy or
tender herbs with low decumbent stems,
which die back after flowering. This north
temperate group contains the hardy STONE-
CROPS such as the blue-flowered *S. caeruleum*,
from southern Europe and North Africa, and
the pink-flowered *S. pilosum*, from Asia Minor
and the Caucasus. Some species in this
group, such as the North African, European
and west Asian *S. acre* [STONECROP, WALL-
PEPPER, BITING STONECROP, GOLDMOSS], with
bright yellow flowers, are suited to arid stony
parts of the rock garden where little else will
survive. There are several cultivars of this
species whose small triangular-ovate overlapping leaves contribute to its moss-like
appearance. The white-flowered *S. album*
(= *S. balticum*) [WHITE STONECROP, WALL-
PEPPER], from the Pyrenees and North Africa,
has a similar habit.
    Some of the taller species cultivated as
border ornamentals, including those hardy
perennial herbs with a root crown with
broad-based leaves, such as the circumboreal
S. rosea [ROSEROOT] (sometimes placed in the
genus *Rhodiola*), bear annual erect flowering
stems. *S. spectabile* from Korea and China,
which is widely cultivated in numerous
cultivars, similarly produces simple erect
annual flowering stems. *S. telephium*
(= *S. carpaticum*) [ORPINE, LIVE-FOREVER,
LIVELONG], from Europe, has a similar habit.
    Many of the more tender species, such as
the Mexican *S. dendroideum* and the popular
basket plant *S. morganianum* [BURRO'S TAIL,
BEAVER'S TAIL, DONKEY'S TAIL, HORSE'S TAIL,
LAMB'S TAIL], are grown in cool greenhouses.
The small star-shaped flowers are arranged
in large showy corymbs and range from
white to yellow, pink, red and purple.
    S. acre, boiled in beer, used to be considered
a folk remedy for fevers.
CRASSULACEAE, about 600 species.

Selenicereus a small genus of slender
climbing cacti (often included in the genus
Cereus), native from Texas to South America and the West Indies. *Selenicereus* produces the largest flowers in the Cactaceae (up
to 40cm long and 30cm across in the Peruvian *S. megalanthus* and the Central American *S. macdonaldiae*); they are showy,
trumpet-like white flowers which open up at
night. The most celebrated species is
S. grandiflorus (= *C. grandiflorus*) [QUEEN OF
THE NIGHT].
CACTACEAE, about 20 species.

Sempervivum see HOUSELEEK.
CRASSULACEAE, about 40 species.

Senecio one of the largest genera of flowering plants, including annual and perennial
herbs, shrubs, trees, climbers and succulents,
and even epiphytic and aquatic species,
found throughout the world. The limits of the
genus cannot be satisfactorily defined and
many "satellite" genera, such as *Cacalia*,
Cineraria, *Crassocephalum*, *Kleinia* and
Ligularia, are included in it. The florets of the
compact daisy-like flower heads possess
outer showy ray florets while the bracts
which make up the involucre are mostly in a
single row.
    Many of the herbaceous species have been
long known to be toxic to stock, and in
southern Africa cause diseases including
Molteno cattle sickness, straining disease in
cattle, and stomach staggers in horses. The
many alkaloids of *Senecio* species cause cir-
rhosis of the liver, and can inhibit cell
division, which may be an explanation for
their ancient use in cancer treatment.
    S. jacobaea [COMMON RAGWORT, TANSY] is
usually avoided by cattle in the fields in

*The Cobweb Houseleek (*Sempervivum
arachnoideum*) is a hardy dwarf succulent found in
the mountainous regions of Europe and northern
Asia.* (×½)

Although a native of central and southern Europe, Senecio squalidus was named the Oxford Ragwort because it was first recorded in Oxford, England, in the late 17th century. ($\times \frac{1}{5}$)

Europe. The common annual, *S. vulgaris* [GROUNDSEL], which may become a troublesome weed of cultivation, has diuretic properties, and has long been used in poultices to reduce abscesses. This plant is also used as a rabbit and cagebird food.

Many species are cultivated for ornament, including the South African annual *S. elegans* (= *Jacobaea elegans*) [PURPLE RAGWORT], with purple flower heads, and the Mediterranean *S. cineraria* (= *Cineraria maritima*), the perennial Chinese *S. tanguticus*, with yellow, plume-like panicles, and the evergreen New Zealand shrubs *S. greyi* and *S. monroi*, both with yellow flower heads. *S. rowleyanus* [STRING OF BEADS] and *S. radicans* [CREEPING BERRIES] are succulent perennials, native to South or Southwest Africa, grown in hanging baskets; they have slender prostrate stems and white rayless flower heads.

The highly variable "cinerarias" of greenhouses are hybrids (*S. × hybridus*) derived from *S. cruentus* (= *Cineraria cruenta*) and *S. heritieri* from the Canary Islands. Another cultivated subshrub is the Argentinian *S. vira-vira* (= *S. leucostachys*, *C. candidissima*) [DUSTY-MILLER], which has deeply pinately dissected leaves and white flower heads. *S. cineraria* is also known as DUSTY MILLER.

Cultivated climbers include the South African *S. macroglossus* [CAPE IVY, CLIMBING GROUNDSEL, WAXVINE], which has a remarkable ivy-like habit, and *S. mikanioides* [GERMAN IVY, PARLOR IVY], the latter now naturalized in England and the USA (California).

The woody species often have a regular branching system which gives them a candelabrum-like aspect. It is seen in the "TREE SENECIOS", three species of slowgrowing trees of the East African mountains, related to the coarse herbs of Eurasia, but sometimes treated as a separate genus, *Dendrosenecio*. Some of the woody species are, or were, conspicuous on island floras (though by no means restricted to them), such as the "CABBAGE TREES" of St. Helena, *S. leucadendron* [HE-CABBAGE] and *S. redivivus*

[SHE CABBAGE], both of which have been depleted by wood-cutting and grazing by goats.
COMPOSITAE, over 1 500 species.

Sensitive plants plants that possess organs showing very rapid movements in response to stimuli of a mechanical nature. The movements are generated by sudden changes in cell turgidity either in special organs (pulvini) or groups of motor cells. The best-known example is *Mimosa pudica*, often called THE SENSITIVE PLANT, which responds to touch by the collapse of its bipinnate leaves. *Cassia nictitans* [WILD SENSITIVE PLANT], from North America and the West Indies, has sensitive leaflets.

Some plants have sensitive stamens, for example some species of *Centaurea*. When touched the filaments contract suddenly pulling the anther tube downwards, thus causing the stigma to extrude the pollen shed inside it onto the body of the visiting insect. Other rapid movements are seen in the trap mechanisms of such insectivorous plants as *Dionaea* and *Utricularia*.

Septoria a large genus of the imperfect fungi and one of a number of genera causing leaf spot diseases, such as celery leaf spot (*S. apiicola*) and wheat leaf spot (*S. tritici*). The fungus may be seed-borne and, under high humidities, infected seed produces seedlings whose seed leaves bear flask-shaped pycnidia from which spores (conidia) are extruded in tendrils and are spread by rain-splash onto the growing crop.
SPHAEROPSIDACEAE, about 1 000 species.

Sequoia a genus represented by a single species, *S. sempervirens* [REDWOOD, COAST REDWOOD]. Native specimens of *S. sempervirens* are immense and stately evergreen conifers found wild only on a narrow coastal belt – the "fog belt" – on the Pacific coast of North America from southwestern Oregon through northern and central California to south of Monterey. Trees of this species are probably the tallest in the world, the maximum height being about 120m; the diameter of the trunk at ground level may reach 10m. The COAST REDWOOD may live for nearly 1 000 years, the average range being 400–800 years.

The wood is in great demand, being soft, fine-grained and easy to work. It is used in building construction and carpentry generally, also for paneling, railway sleepers, telegraph poles, road blocks, fence poles etc. Excessive demand has denuded many of the original forests but some have been preserved by conservation efforts. The COAST REDWOOD is an outstanding tree for single specimen planting in parks and large gardens.
TAXODIACEAE, 1 species.

Sequoiadendron a genus represented by a single species of gigantic conifer native only to the western slopes of the Sierra Nevada in California, although it is often grown elsewhere as an ornamental. *S. giganteum* (= *Sequoia gigantea*, *S. wellingtonia*) [BIG TREE,

Ragwort (Senecio jacobaea) is a common weed of neglected pastures where it is avoided by grazing cattle.

GIANT SEQUOIA, SIERRA REDWOOD] is an imposing tree that can grow to 100m or more and attain a girth of up to 30m. This species was originally placed in the genus *Sequoia* along with *Sequoia sempervirens* [REDWOOD], with which it shares the characteristic of a thick spongy inner bark and spirally arranged leaves, but differs in its ovate to lanceolate leaves of one kind only, buds that are not scaly, and longer cones.

The individual specimen GENERAL SHERMAN TREE, which stands in the Sequoia National Park, is alleged to be more than 3 000 years old and is claimed, at 2 000 tonnes, to be the most massive (though not the tallest) tree in the world.

Supplies of timber are limited but when used it is mainly for farm buildings, posts and stakes, as the wood is durable, especially in contact with soil. Although straight-grained, light and soft, the wood is not particularly easy to work.
TAXODIACEAE, 1 species.

Senecio heritieri, a species endemic to the south coast of Tenerife, Canary Islands. ($\times 1$)

Serpula a cosmopolitan genus of basidiomycete fungi known chiefly through the species *S. lacrymans* [DRY ROT], a serious agent of decay of timber inside homes. The fungus establishes itself on roof or structural timbers only if they become damp. Once established, *S. lacrymans* can spread rapidly and generates enough moisture for growth from the water liberated during the hydrolysis of the cellulose in the timber. It causes a light brown rot which splits into cubical fractions and eventually crumbles into powder, hence the name DRY ROT. It appears to be entirely confined to house timbers although other members of the genus are found in woodland habitats.
POLYPORACEAE, about 14 species.

Sesamum a small genus of tropical southern African and Asian herbs, one of which, *S. indicum* (= *S. orientale*) [GINGELLY, GINGILIE, OIL PLANT, SESAME], is grown extensively throughout the tropics and subtropics for its seeds (sesame seeds). The main producers are India, China, Sudan, Mexico, Venezuela and Burma. The seeds have a pleasant nutty flavor and are used in making confectionery and as a garnish on loaves of bread. They are also very rich in oil (sesame or gingili oil) which is commercially important. It is used as a salad and cooking oil, and in the manufacture of soap and margarine. The residue is used for cattle-feed.
PEDALIACEAE, about 37 species.

Sesbania a genus of tropical and subtropical leguminous herbs, shrubs and trees. Some species, such as *S. sesban* (= *S. aegyptiaca*) [SESBAN], a shrub or small tree, and *S. tripetii* [SCARLET WISTERIA], are cultivated as ornamentals for their attractive much-branched habit and pea-like flowers, or, as in

*Newly harvested Sesame (*Sesamum indicum*) in Turkey. On drying, the capsules split to release the oil-rich seeds.*

the case of *S. exaltata* [COLORADO RIVER HEMP], for their fibers. The petals and tender leaves of the small tree *S. grandiflora* [GALLITO, SCARLET WISTARIA TREE, VEGETABLE HUMMING BIRD] are eaten in curries and soups in the West Indies, and the bark and leaves are used medicinally.
LEGUMINOSAE, about 50 species.

Setaria is a large genus of annual and perennial grasses distributed throughout the warm temperate, subtropical and tropical regions. Being closely related to the other large *MILLET genera, *Panicum* and *Pennisetum*, species of *Setaria* are best recognized by their large bristly, brush-like spicate inflorescences.
S. italica (= *Panicum germanicum*) [ITALIAN or FOXTAIL MILLET, HUNGARIAN or BENGAL GRASS] is an important annual widely cultivated in southern Europe, North Africa and Japan. The grain is used almost exclusively for animals and particularly for cagebirds in Western Europe. The genus includes a number of ornamental species, including the perennial *S. palmifolia* [PALM GRASS], which is grown in warm greenhouses for its attractive, long, fan-like leaves which may be variegated. At the other extreme, the widespread *S. viridis*, *S. glauca* and *S. verticillata* [GREEN, YELLOW and ROUGH BRISTLE GRASSES respectively] are dwarf annuals up to a maximum of 10cm in height, with dense inflorescences and narrow leaves.
GRAMINEAE, about 140 species.

Seville orange the common name for the extremely acid citrus fruit *Citrus aurantium,*

also known as BITTER ORANGE, SOUR ORANGE, BIGARADE. It is native to Vietnam but is now widely cultivated or naturalized in the tropics, subtropics and Mediterranean region. Too bitter to be eaten fresh, the oranges are made into marmalade, and neroli oil, an essential oil used for flavoring and in perfumes, is extracted from the flowers. *C. aurantium* subspecies *bergamia* is the BERGAMOT, which produces an oil, bergamot oil, from the rind.
RUTACEAE.

Shaddock one of the popular names for the largest of the citrus fruits, *Citrus maxima* (= *C. grandis*) (see POMELO).
RUTACEAE.

Shallot a form of the common *ONION, *Allium cepa* var *aggregatum* (= *A. ascalonicum*), which has been selected to branch freely and produce many small lateral bulbs rather than large bulbs.
LILIACEAE.

Shamrock a common name for the trifoliolate leaves of *Medicago lupulina* [BLACK MEDICK, HOP CLOVER, YELLOW TREFOIL, NONESUCH] and sometimes for those of *Trifolium dubium* [YELLOW *CLOVER], *T. repens* [WHITE CLOVER], and *Oxalis acetosella* [WOOD SORREL]. The SHAMROCK is the national emblem of Ireland.

Shepherdia a very small genus of North American deciduous shrubs or small trees, two species of which are grown as ornamentals. *S. argentea* (= *Elaeagnus utilis*) [SILVER BUFFALO or RABBIT BERRY], is a thorny shrub growing to about 6m high with silver, densely hairy leaves, and greenish-yellow spring flowers which result in edible, globular, red or yellow berries in the autumn. *S. canadensis* [BUFFALO BERRY, SOAPBERRY] is thornless and has insipid fruits.
ELAEAGNACEAE, 3 species.

Sesamum indicum is a drought-resistant species and so is usually a successful crop in regions where rainfall is erratic. (×$\frac{1}{15}$)

*A long-tongued Hoverfly (*Rhingia campestris*) visiting a Red Campion (*Silene dioica*). (× 7)*

Shorea a large genus of trees extending from Sri Lanka to the Malay archipelago and south China. It is a major source of timber, which is produced in several grades. Most important is meranti in a range of categories such as red meranti from *S. martiana*, dark red meranti from *S. pauliflora* and yellow meranti from *S. bracteola*. Another is balau or selangan batu, a heavy, primary hardwood suitable for outdoor construction. *S. robusta* [SAL TREE] grows in vast, almost pure forests along the base of the Himalayas. Like other species, including *S. laevis* and *S. koordersii*, it is a source of the aromatic gum, *dammar, as well as timber. The fruits of *S. macrophylla* [ILLIPE NUT], and related Malayan and Bornean species, yield edible fat used as a substitute for cocoa butter.
DIPTEROCARPACEAE, about 180 species.

Shortia a small genus of attractive evergreen perennial stemless herbs, native to the mountains of eastern North America and Asia. The long-stalked leaves are heart-shaped or almost round. Several species are cultivated in rock gardens or alpine houses for their large, white, blue or pink, solitary, nodding flowers, for example *S. galacifolia* [OCONEE-BELLS] and *S. soldanelloides* (= *Shizocodon soldanelloides*) [FRINGED GALAX, FRINGE-BELL].
DIAPENSIACEAE, about 8 species.

Sida a large genus of annual and perennial herbs and small shrubs, mainly native to tropical America. Several species have become naturalized in subtropical and warm temperate regions. Some species, such as *S. hermaphrodita* (= *S. napaea*) [VIRGINIA MALLOW], with loose axillary cymes of white flowers, are cultivated as ornamentals. A few species are used in India and China as a *hemp-fiber substitute, for example *S. rhombifolia* [QUEENSLAND HEMP] and

S. cordifolia. The leaves of *S. stipulata* are used locally in the treatment of insect bites and dysentery.
MALVACEAE, about 200 species.

Sidalcea a genus of perennial herbs native to western North America, with rosettes of basal, rounded, usually palmately lobed and divided leaves, and spikes of large, white to pink, red or purple flowers. They make popular garden plants. Cultivated species include *S. candida* with white to yellowish flowers, and *S. malviflora* [CHECKER BLOOM] with pink or lilac flowers. The latter has given rise to a number of good garden cultivars, such as 'Listeri' with satiny pink flowers.
MALVACEAE, 18–22 species.

Silene [CAMPIONS, CATCHFLIES] a large genus (including *Melandrium*) of annual, biennial or perennial herbs or subshrubs, widely distributed in north temperate regions but particularly abundant in eastern Mediterranean countries. Most annual species are weeds of cultivated land and most perennials grow in open sandy or stony ground.

Several species are grown as garden plants, for example *S. armeria* [SWEET WILLIAM, CATCHFLY, NONE-SO-PRETTY], *S. coelirosa* [ROSE OF HEAVEN] which are pink-flowered annuals or biennials. *S. californica* [CALIFORNIA INDIAN PINK] and *S. virginica* [FIREPINK] are perennial crimson-flowered species. Other garden species include the perennial pink-flowered *S. acaulis* [CUSHION FLOWER, MOSS CAMPION] and *S. caroliniana* [WILD PINK], and the white-flowered *S. quadrifida* (= *S. alpestris*) [ALPINE CATCHFLY] and *S. stellata* [STARRY CAMPION, WIDOW'S FRILL].

S. vulgaris (= *S. cucubalus*) [BLADDER CAMPION, MAIDEN'S TEARS] is a widespread perennial species on grasslands and sea cliffs throughout Europe. It owes its name to its inflated and finely veined calyx. Two widespread species, the European *S. nutans* [NOD-

*The Bladder Campion (*S. vulgaris *subspecies* vulgaris*), which grows in fields and on roadsides. (× ½)*

Shorea curtisii, a giant emergent tree of lowland rain forest in Malaya.

DING CATCHFLY] and *S. noctiflora* [NIGHT FLOWERING CATCHFLY, STICKY COCKLE], of Europe, southwest Asia (and naturalized in North America), illustrate admirably the adaptations for moth pollination. Their flowers are droopy and closed by day, but by night are fully opened out, exuding a strong odor similar to that of hyacinths. Other well-known Old World species include *S. alba* (= *Lychnis alba, Melandrium album*) [WHITE CAMPION, EVENING CAMPION, WHITE COCKLE] and *S. dioica* (= *L. dioica, M. dioicum*) [RED CAMPION, MORNING CAMPION].
CARYOPHYLLACEAE, about 500 species.

Silphium a small genus of tall perennial herbs from south and eastern North America, which contain resinous juices in their stems. *S. laciniatum* [COMPASS PLANT, POLAR PLANT, ROSINWEED] turns its leaf edges to north and south to avoid midday radiation. This and other species, such as *S. integrifolium*, are cultivated for their heads of yellow flowers. Both *S. laciniatum* and *S. perfoliatum* [CUP PLANT, INDIAN CUP PLANT, RAGGED CUP] have had uses as tonics and expectorants.
COMPOSITAE, about 23 species.

Simarouba a genus of mainly tropical American evergreen shrubs and trees, of which *S. glauca* [PARADISE TREE, BITTERWOOD, ACEITUNO] is cultivated as an ornamental and sometimes for its seeds which contain an edible oil.
SIMAROUBACEAE, about 10 species.

Sinapis a small European genus of usually annual herbs, with bright yellow flowers. It is closely related to *Brassica* and is sometimes included in this genus. Most species occur as weeds, especially *S. arvensis* [*CHARLOCK]. *S. alba* [WHITE MUSTARD] occurs as a weed in Mediterranean countries, although cultivated forms have been developed. These are sometimes used as fodder crops, but are usually grown for the seed which is ground and used as a component of *mustard. CRUCIFERAE, about 10 species.

Sinningia a small genus of mainly tropical American herbaceous perennials and shrubs usually with tuberous roots. The most commonly grown is the Brazilian *S. speciosa*, popularly known as *GLOXINIA, BRAZILIAN GLOXINIA, VIOLET SLIPPER (not to be confused with the botanical genus *Gloxinia*), and containing numerous cultivars. The 'Fyfiana' group of cultivars are the GLOXINIAS of florists. This and other *Sinningia* species and hybrids, such as *S. regina* [VIOLET SLIPPER GLOXINIA, CINDERELLA-SLIPPERS], make popular pot plants with their long, showy, tubular or bell-shaped flowers. GESNERIACEAE, about 75 species.

Siphonodon a small genus of trees and climbing shrubs native to Southeast Asia, Malaysia and northeastern Australia. The creamy-white wood of *S. australe* [IVORY WOOD] is used for rulers, inlay work and carving. CELASTRACEAE, about 5–6 species.

Alexanders (Smyrnium olusatrum) was once used as a salad and as a potherb but was superseded by celery. ($\times \frac{1}{8}$)

Sisal (Algave sisalana) growing in a plantation in Madagascar. The leaves yield a durable fiber used for ropes, twine, carpets and sacking.

Sisal or **sisal hemp** a fiber from the leaves of plants of certain species of the genus *Agave*, notably *A. sisalana* [SISAL AGAVE], *A. fourcroydes* [HENEQUIN], *A. cantala* [MAGUEY, CANTALA], and *A. letonae* [SALVADOR HENEQUIN]. Sisal hemp consists of long, creamy-white fibers and ranks next to *abaca fiber (Manila hemp) in strength and durability. The fiber is suitable for ropes and twine as well as carpets and sacking. *A. sisalana*, with tough, linear, spine-tipped leaves up to 4m long, is the species most widely grown.

The main producers are Brazil, Tanzania, Mozambique, Angola, Kenya, Madagascar and Haiti. Production has increased in recent years despite competition from synthetic fibers. *A. fourcroydes* is grown only in Mexico, Central America and the Caribbean, with Mexico and Cuba as the main producers. *A. cantala* is grown mainly in the Philippines. AGAVACEAE.

Sisyrinchium a genus of mostly low-growing annual or perennial herbs found in North and South America and the West Indies. Most species have grass-like linear or cylindrical leaves. The flowers are usually blue, yellow or white, but occasionally purplish red, as in the North American *S. douglasii* (= *S. grandiflorum*) [GRASS-WIDOW, PURPLE-EYED GRASS, SPRING BELL].

Many species are cultivated, the commonest being *S. angustifolium* (= *S. anceps*, *S. gramineum*) [BLUE-EYED GRASS], from eastern North America, and *S. bermudiana* (= *S. iridioides*), from Bermuda. These are small species, 30–60cm tall, with blue flowers arising from grass-like leaves. *S. striatum*, from Argentina and Chile, is a larger plant with broad leaves like a flag iris and pale yellow, star-shaped flowers on stems 75cm high. IRIDACEAE, about 80 species.

Skimmia a small genus of low evergreen shrubs native to China, Japan and the Himalayas. The leaves are more or less oval-oblong and give a strongish scent when crushed; the fragrant flowers are white, in compact terminal panicles. The best-known cultivated species, *S. japonica* (= *S. fragrans*), from Japan, has long-lasting brilliant red subglobose fruits. *S. reevesiana* (= *S. fortunei*), from China, Taiwan and Luzon (Philippines), is a smaller plant with dull crimson fruits. *S. × rogersii* and *S. × foremanii* are the presumed hybrids between these two species. The leaves of *S. laurifolia* yield an essential oil used as a perfume in soap manufacture. RUTACEAE, 7–9 species.

Slime fungi see Fungi, p. 152.

Sloe the common name for *Prunus spinosa* (also known as BLACKTHORN), distributed

The bright orange colony of the slime fungus Fuligo septica growing on a rotton log in a temperate woodland. ($\times 2$)

Solanum surratense *is a plant of semi-arid places. The whole plant, apart from the petals, is covered in sharp straw-colored spines.* (× ½)

*The orange-red berries of the False Jerusalem Cherry (*Solanum capsicastrum*), a popular houseplant. The berries are inedible.* (× 1)

throughout Europe and parts of western Asia. It has white flowers followed by small, blue-black, bitter-tasting fruits which are used for flavoring liqueurs. The wood is used for carpentry and turnery.
ROSACEAE.

Smilacina a small North American and Asian genus of herbs with creeping rootstocks and simple stems. A few, such as *S. racemosa* [SOLOMON'S ZIGZAG, TREACLE-BERRY, FALSE SPIKENARD] and *S. stellata* [STAR-FLOWER, STAR-FLOWERED LILY OF THE VALLEY], are cultivated in gardens for their racemose inflorescences of white or pink flowers and their colorful berries.
LILIACEAE, about 25 species.

Smilax a large genus of woody or herbaceous climbing plants from the temperate and tropical zones of both hemispheres. The shoots bear paired stipular tendrils which assist climbing. The flowers are small, greenish, and borne in axillary umbels. The fruit is a black or red berry.

The drug sarsaparilla, used medicinally and as a flavoring, is obtained from the rootstock of several species, including *S. utilis* [JAMAICAN SARSAPARILLA], and *S. aristolochiifolia* [MEXICAN OR TAMPICO SARSAPARILLA]. The Malaysian species *S. myosotiflora* and the North American *S. glauca* [SAWBRIER, WILD SARSAPARILLA] both yield locally used medicinal products.

The young shoots of *S. laurifolia* [LAUREL-LEAVED GREENBRIER, BLASPHEME or BAMBOO VINE] are eaten as a vegetable in parts of the southeastern USA. *S. aspera*, from southern Europe, is a most decorative species with spiny zigzag shoots and narrowly ovate-lanceolate leaves. The "SMILAX" of florists is *Asparagus asparagoides*.
SMILACACEAE, about 200 species.

Smithiantha a small genus of perennial hairy herbs, probably all native to Mexico, some of which are cultivated in greenhouses in cooler temperate zones for their showy tubular flowers and often velvety heart-

shaped leaves. Popular species include the red-flowered *S. cinnabarina* (= *Naegelia cinnabarina*), the red-and-yellow-flowered *S. zebrina* (= *N. zebrina*), and the yellowish-white-flowered *S. multiflora* [NAEGELIA], as well as their hybrids with flowers in various shades of yellow, orange, pink and red.
GESNERIACEAE, about 4 species.

Smoke tree the common name for *Cotinus coggygria* (= *Rhus cotinus*), also known as VENETIAN SUMAC, a deciduous bushy shrub native to central and southern Europe and Asia. The common name aptly describes the appearance of the shrub in late summer, when the masses of silky hair-like inflorescences cover the whole plant in a pinkish-fawn, smoke-like envelope. *C. obovatus* (= *C. americanus*), from North America, is the AMERICAN SMOKE TREE or CHITTAMWOOD.
ANACARDIACEAE.

Smut fungi see Fungi, p. 152.

Smyrnium a small genus of erect herbaceous biennial or perennial umbelliferous herbs native to Europe, western Asia and North Africa, with twice- or thrice-ternately compound leaves and small greenish-yellow flowers in many-rayed umbels. *S. perfoliatum* is occasionally cultivated in herbaceous borders. *S. olusatrum* [BLACK POT HERB, ALEXANDERS, HORSE PARSLEY, BLACK LOVAGE] was formerly cultivated and used as a salad and as a potherb, before *celery was introduced.
UMBELLIFERAE, about 8 species.

Snowdrop the common name for members of *Galanthus*, a genus of horticulturally important bulbous plants, ranging from Western Europe to the Caucasus, Asia Minor and Iran. The attractive flowers are solitary and drooping; the three outer perianth segments are white, the three inner ones shorter with green or yellow markings.

Snowdrops are popular garden plants, particularly valued for their winter flowering season, which can last from September to April. Some species, such as *G. nivalis* subspecies *reginaeolgae*, *G. elwesii* [GIANT SNOWDROP] and *G. caucasicus*, are better grown in full sun, whereas others, such as *G. nivalis* subspecies *nivalis* [COMMON SNOWDROP], prefer half-shade. Many cultivars exist, many with double flowers.
AMARYLLIDACEAE, 15–20 species.

Solanum one of the largest genera of flowering plants, containing mainly annual and perennial herbs. Some are climbers (such as *S. wendlandii*) or shrubs or small trees (such as *S. erianthum*, *S. aviculare* [KANGAROO APPLE] and *S. cernuum*). The genus includes plants of agricultural, ornamental and medicinal use. Although distributed throughout tropical and temperate areas of the world, the majority of species are native to South America.

One of the commonest species in Eurasia and Africa is the herbaceous annual weed by which the genus is typified, *S. nigrum* [POISON

BERRY, BLACK, COMMON or GARDEN NIGHT-SHADE]. In many species, the petals curve backwards, and are white, blue or purple in color; occasionally they are yellow as in *S. cornutum*. The fruits vary in color through green, yellow, orange, red, purple, brown and black and are globular, as in *S. sodomeum* [APPLE OF SODOM, YELLOW POPOLO] and *S. pseudocapsicum* [JERUSALEM CHERRY], ovoid, as in *S. dulcamara* [BITTERSWEET, FELONWOOD, WOODY NIGHTSHADE], obovoid, as in the many varieties of *S. melongena* [*AUBERGINE, BRINJAL, EGGPLANT, JEW'S APPLE, MAD APPLE], or globose to ovoid, as in the cultivated tropical climber *S. wendlandii* [POTATO VINE, GIANT POTATO CREEPER, PARADISE FLOWER]. Typically succulent, glabrous and smooth, these fruits (berries) may be dry and leathery in a few species, as in *S. aculeatissimum* (= *S. ciliatum*) [LOVE APPLE, COCKROACH BERRY, SODA-APPLE NIGHTSHADE] and *S. elaeagnifolium* [SILVER-

The best Snowdrop for gardens is Galanthus nivalis, *which has a robust constitution and quickly forms large clumps.* (× ½)

LEAF NIGHTSHADE], or enclosed by enlarged sepals, which can be spiny (eg S. cornutum).

Among the ornamental specimens, the tropical American S. mammosum [NIPPLE-FRUIT], S. capicastrum [FALSE JERUSALEM CHERRY] and S. giganteum [AFRICAN HOLLY], native to India and Sri Lanka, are cultivated for their attractive but inedible fruits, S. muricatum [MELON SHRUB, MELON PEAR, PEPINO], for their flowers, and the climbers S. jasminoides [POTATO VINE] and S. crispum [CHILEAN POTATO TREE] for their showy inflorescences.

S. tuberosum [IRISH or WHITE POTATO] is a major world vegetable, cultivated for its underground starchy stem tubers (see POTATO). However, solanums are usually cultivated for their edible fruits, one of the most familiar being S. melongena [*AUBER-GINE, EGGPLANT, BRINJAL]. Popular tropical American fruits include *PEPINO (S. muricatum) and COCONA (S. topiro) whilst those of the JERUSALEM CHERRY and particularly those of the North Andean subshrub S. quitoense [*NARANJILLO, LULO] are used in the New World tropics to make a delicious beverage. The berries of several species, including S. ferox in Malaysia and S. diversifolium in the West Indies, are only of local importance. The leaves of some tropical species such as S. aethiopicum and S. macrocarpon, are sometimes used as potherbs.

Many solanums have long been used as medicines to treat swellings, inflammations, fevers, infections of the throat, chest and heart, and as a diuretic. S. nigrum and S. dulcamara are among the solanums extensively employed by herbalists. Many species, such as the North American perennial S. carolinense [CAROLINA HORSE NETTLE] are still widely used in native medicine, especially in Africa and the Americas. Steroid alkaloids are characteristic of many solanums, and some, such as solasodine, are now used as a source of steroid hormones. These alkaloids can also be toxic, however, and solanine poisoning has been recorded since earliest times. Though it can be fatal, major nervous symptoms and/or severe gastro-intestinal irritations are more commonly reported.
SOLANACEAE, about 1500 species.

Soldanella a small genus of European dwarf perennial herbs with simple, leathery, evergreen leaves in basal rosettes. The nodding bell-shaped or funnel-shaped flowers are solitary or borne in umbels, on leafless stalks, and are blue, violet, or rarely white in color. Among the species grown in rock gardens are the pale-blue-flowered S. alpina (white in cultivar 'Alba'), the bluish-purple-flowered S. minima and the blue-violet-flowered S. pusilla.
PRIMULACEAE, about 8 species.

Soleirolia a genus represented by a single species, S. soleirolii [BABY'S-TEARS, POLLYANNA VINE, ANGEL'S TEARS, IRISH MOSS, CORSICAN CURSE, MIND-YOUR-OWN-BUSINESS], which is a creeping herb native to the western Mediter-ranean islands and Italy. It is a useful ground-cover plant in mild climates and as a houseplant, but is prone to be over-vigorous when it can become a serious weed.
URTICACEAE, 1 species.

Solidago [GOLDEN RODS] a genus of herbaceous perennials, most of which are native to the American continent; a few occur in Europe, such as S. virgaurea (= S. brachy-stachys), and Asia. All species possess clusters or racemes of small yellow flowers heads. S. virgaurea (from 15–100cm high according to variety) and S. canadensis [CANADA GOLDEN ROD] (up to 2m high), especially the cultivar 'Golden Wing', are cultivated in gardens, as well as a wide range of hybrids mainly derived from crosses between the species.

A number of species, including the above are used in herbal medicine. An essential oil can be extracted from CANADA GOLDEN ROD and S. odora [SWEET GOLDEN ROD], whose leaves have been used as tea. Some species, notably S. leavenworthii [LEAVENWORTH GOLDEN ROD], are a source of rubber.
COMPOSITAE, about 120 species.

Sonchus a cosmopolitan genus of mainly herbaceous annuals, biennials or perennials. Seventeen species are rosette shrubs belonging to the subgenus Dendrosonchus, which occurs only in the Canary Islands, Cape Verde Islands and Madeira, with one species also growing in Morocco. Most of the cultivated ornamental species such as S. arboreus, S. congestus and S. pinnatus, all with branched small or medium-sized yellow flower heads, belong to this group.
COMPOSITAE, about 50 species.

Flowers and seeds of the Sowthistle (Sonchus asper); the latter are dispersed by the wind ensuring rapid colonization of fields and wasteground which makes this species a widespread weed. (× ⅓)

Sooty molds see Fungi, p.152.

Sophora a genus of evergreen or deciduous trees or shrubs and sometimes spiny, or rarely perennial herbs, occurring in warm and temperate regions of both hemispheres. The flowers are white, yellow or rarely bluish-violet and are borne in terminal panicles or leafy racemes up to 50cm long.

Several species are grown for ornament, especially the deciduous S. japonica [JAPANESE PAGODA TREE, CHINESE SCHOLAR TREE], from China and Korea, which reaches 25m in height and provides leaf and fruit extracts used to adulterate opium in China. The wood, bark and fruits give a yellow dye. S. tetraptera (= S. grandiflora) [KOWHAI, FOUR-WING SOPHORA, OR NEW ZEALAND LABURNUM], from New Zealand and Chile, is a widely grown decorative evergreen or semi-deciduous tree or shrub with golden-yellow flowers, which also provides extremely durable timber used in cabinetwork. S. secundiflora [MESCAL BEAN, FRIJOLITO], an evergreen tree from southwestern North America and Mexico, has attractive and fragrant bluish-violet flowers and also provides fruits (mescal beans) used as intoxicants by various Indian tribes and as necklace beads.
LEGUMINOSAE, about 80 species.

× **Sophrolaeliocattleya** a group of hybrid pseudobulbous orchids resulting from crosses between various species of the tropical American genera *Cattleya, *Laelia and *Sophronitis.
ORCHIDACEAE.

Sophronitis a small genus of small tropical South American epiphytic orchids. The flower stems arise from the base of the one-leaved pseudobulbs, each stem bearing a solitary flower somewhat resembling that of a small *Cattleya. Their major use is in orchid intergeneric hybridization programs, where their brilliant red flower coloration is a genetically dominant feature.
ORCHIDACEAE, about 6 species.

Sorbaria a small genus of deciduous shrubs mostly native to Asia. These shrubs, at one time included in the genus *Spiraea, have attractive pinnate leaves and dense, large racemes or spikes of small white flowers. Several are cultivated in gardens, such as S. aitchisonii (= Spiraea aitchisonii), from western Asia, and S. tomentosa, from the Himalayas.
ROSACEAE, about 10 species.

Sorbus a genus of deciduous trees and shrubs of the Northern Hemisphere. They are found as far south as Mexico and the Himalayas. The genus includes the ROWANS or MOUNTAIN ASHES, the WHITEBEAMS and the SERVICE TREES. Many of the species are grown as ornamentals and can tolerate shade and atmospheric pollution. The flowers are white (or occasionally pink) and are borne in terminal, compound corymbs. The fruits, usually small pomes, but commonly called "berries", are white, yellow, orange, pink or red.

Two of the most attractive trees are *S. americana* [AMERICAN MOUNTAIN ASH, DOGBERRY, MISSEY-MOOSEY, ROUNDWOOD], of eastern North America, and the very similar *S. aucuparia* [QUICKBEAM, ROWAN, EUROPEAN MOUNTAIN ASH] which grows throughout most of Europe. There are several cultivars of the latter species, all bearing subglobose scarlet fruits which in cultivar 'Edulis' are edible and used for making preserves.

S. sargentiana, a native of western China, is commonly grown in parks and gardens, most often as a graft on the stem of *S. aucuparia*. Among other Asian rowans, less commonly grown, are *S. cashmiriana* [KASHMIR ROWAN], *S. commixta* [JAPANESE ROWAN] and *S. vilmorinii* [VILMORIN'S ROWAN]. *S. aria* [COMMON WHITEBEAM, CHESS APPLE], with several cultivars, and *S. intermedia* [SWEDISH WHITEBEAM] bear large white flowers and scarlet fruits. The crimson, brown-speckled fruits of *S. torminalis* [WILD SERVICE TREE], of Europe, North Africa and Southwest Asia, although rather acid, are edible and at one time were sold in Kent, England, as "chequers berries". *S. domestica* [TRUE SERVICE TREE] is found in southern Europe, and is widely planted all over central Europe for ornament, with large pear-shaped greenish-brownish-red fruits which are edible and in some areas are fermented with grain to produce an alcoholic beverage.

Species of *Sorbus* hybridize readily and a

A Mountain Ash (Sorbus aucuparia) growing on a hillside in the Highlands of Scotland. Several cultivars are in cultivation.

number are grown as ornamentals, such as *S. × thuringiaca* [BASTARD SERVICE TREE], which is the result of a cross between *S. aucuparia* and *S. aria*.
ROSACEAE, about 90 species.

Sordaria a common fungus on animal dung, with fruiting bodies (ascocarps) in the form of single, black, flask-shaped perithecia which contain numerous asci. Crosses between white-spored mutant strains and black-spored strains result in hybrid asci containing usually four black and four white ascospores. By studying the arrangement of spores of different colors within the asci, it is possible to interpret the genetics of inheritance of spore color. The best-known and most studied species is *S. fimicola*.
LASIOSPHAERIACEAE, about 4 species.

Sorghum a genus of tall, annual and perennial, tropical and subtropical grasses easily recognized by their dense terminal paniculate inflorescences and broad flat leaves.

The cultivated sorghums, which were initially developed in Africa, are important commercial crops, especially in the tropics, because they are easily grown and produce high yields of grains or fodder with little attention. The many cultivated sorghums are usually considered to be varieties of

S. bicolor (= *S. vulgare*) but some races are so different that they have been regarded by some authorities as distinct species. The five basic races are BICOLOR, CAUDATUM, DURRA, GUINEA and KAFIR.

The most important crops are the seed-bearing sorghums, and perhaps the most commonly cultivated forms are GUINEA CORN or COMMON SORGHUM, GHALLY or INDIAN SORGHUM (widely cultivated in India, Africa and North America) and DURRA or DURHA, the most commonly cultivated race in North Africa and Egypt. All are important sources of human food, being used predominantly for porridge and bread, very often being blended with wheat flour for the latter purpose.

Several forms of KAFIR CORN or AMABELE are cultivated in South Africa for their colored grains used in the brewing of kafir beer or tshwala. The variety *feterita* is highly prized as a grain crop by the Sudanese while the poorer quality *cerevisiae* is utilized extensively by the East Africans for the manufacture of beer.

A number of thick-stemmed strains are also cultivated for their sweet cane, commonly known as SORGO, used for chewing, forage production or for making syrup and sugar. There are also a number of distinctive perennial species grown as useful fodder crops. *S. versicolor* is an important perennial of East and southern Africa. *S. sudanense* [SUDAN GRASS] is another important forage grass cultivated in the drier parts of the world as it is one of the best drought-resistant grasses known in agriculture and all classes of livestock eat it readily, although under drought stress it may produce hydrocyanic acid with consequent poisoning of livestock. *S. halepense* [JOHNSON GRASS, MILLET GRASS] has been in cultivation for a much longer period and is also reckoned to be a good forage grass and is sometimes grown for ornament in garden borders.
GRAMINEAE, about 60 species.

Southernwood the common name for *Artemisia abrotanum* (= *A. procera*), a highly fragrant shrub with aromatic leaves, native to southern Europe. The leaves were once used as a substitute for tea and used locally as a medicine.
COMPOSITAE.

Soybean the common name for *Glycine max*, a small bushy, usually erect annual plant, originally native to northeast China and one of the oldest crops grown by Man. Many different varieties of soybean are now grown commercially throughout the Far East, Africa, India, Central and South America, the USA and many southern European countries.

The pea-like flowers are borne in short axillary racemes, each ultimately producing a narrow, flat, hairy pod, constricted between the seeds. The pods, up to 7.5cm in length, contain two to four small, globose, white, brown, red or black seeds.

Because of the extensive use of the plant, there are now thousands of varieties under cultivation, these often being locally adapted to particular regional soil and climatic conditions.

Soybean is the world's most important grain legume and production is continuing to increase. The USA, China and Brazil together contributed over 90% of the world production of 58 million tonnes in 1973. It is the most important cash crop in the USA, which accounts for over two-thirds of world production. The dramatic increase in importance of this crop is emphasized by the corresponding increase from 1924 to 1973 of 700 000 to 23 000 000 hectares of land used in the USA to grow soybean. Much of the crop in the USA is processed for oil which provides some 35% of the nation's total oil and fats.

The oil at one time was extracted from the seed by hydraulic presses, but is now less wastefully extracted by hydrocarbon solvents. After refining the oil it is put to many uses including its incorporation into salad oil, margarine, and industrially into paints, varnishes, insecticides, disinfectants, cleaning compounds, linoleum and other products.

The protein cake left after oil extraction is either processed into soybean flour or incorporated into animal feed. Lecithin, another soybean extract, with emulsifying and wetting properties, is also widely used in the food and pharmaceutical industries as well as in rubber, textile, and petroleum products.

The protein in the seeds of all varieties is highly nutritious, and for this reason, and because of their high iron, calcium and vitamin content, soybeans have for centuries been an essential component of the diet of many people in East Asia. The soybean is probably the richest of all natural vegetable foods. The plant also provides animal forage. LEGUMINOSAE.

Left Soybeans (Glycine max) are the most important legume grain and are an increasingly important source of vegetable protein. Above Herbicide being applied to a field of soybeans.

Spaghetti, vegetable a variety of VEGETABLE MARROW, which is becoming increasingly popular. The very easily grown, trailing plant produces several marrow-like fruits 20–25cm long. When boiled, the white to orange edible flesh separates into spaghetti-like strips. Spaghetti proper is made from hard wheat (*Triticum durum*).

Sparaxis a small South African genus of herbaceous perennials with stately tubular or funnel-shaped flowers (sometimes called WANDFLOWERS). *S. grandiflora*, with purple-white flowers (yellow-throated in var *lineata*) and *S. tricolor* [HARLEQUIN FLOWER], with multicolored red, orange, yellow, purple or white flowers, are grown from corms as garden plants in warm countries.
IRIDACEAE, 4–6 species.

Sparmannia a small genus of evergreen shrubs from Africa and Madagascar, which may be grown as ornamentals in greenhouses or outside in warm regions, reaching a height of 3–6m. When *S. africana* [AFRICAN HEMP] is grown as a pot plant it reaches a height of about 1m, bearing large hairy, heart-shaped leaves and clusters of white, open flowers (double in cultivar 'Flore Pleno'). In its native habitat it is also a source of a fiber.
TILIACEAE, about 4 species.

Spartina [CORD GRASSES] a small genus of robust, deep-rooting coarse perennial grasses native to coastal regions of North and South America, Europe and North Africa.

S. alterniflora is an American species introduced into Europe in the early 19th century, where it has been hybridized with a southern European species, *S. maritima*, to create the male-sterile *S. × townsendii* [TOWNSEND'S CORD GRASS]. *S. anglica* [COMMON CORD GRASS] is a fertile amphidiploid hybrid derived from the sterile primary hybrid *S. × townsendii*.

Both these species now cover vast areas of British and French coastline. By means of their long fleshy rhizomes, they are able to stabilize mud, while the filtering action of their stems and leaves collects debris and moving silt to raise the level of the land. They have also been planted to protect foreshores from erosion.

The North American *S. pectinata* [PRAIRIE OR FRESH WATER CORD GRASS] is sometimes cultivated as an ornamental at the edges of lakes and ponds.
GRAMINEAE, about 16 species.

Spartium a genus consisting of a single species of shrub, *S. junceum* (= *Genista juncea*) [SPANISH BROOM, WEAVERS' BROOM], native to the Mediterranean. The fragrant, showy, yellow pea-like flowers are large and were once a source of dye. It is widely grown as an attractive garden shrub.
LEGUMINOSAE, 1 species.

Spathiphyllum a genus of tropical evergreen rhizomatous perennial herbs, mainly native to Central and South America, but also represented in Southeast Asia. Many species, including *S. blandum*, *S. cannifolium*, *S. 'Clevelandii'* (= *S. clevelandii*) [WHITE ANTHURIUM] and *S. floribundum* [SNOWFLOWER], are cultivated for the ornamental appearance of the expanded, often whitish or greenish-yellow spathe which encloses the often fragrant inflorescence (spadix).
ARACEAE, about 35 species.

Spathodea a small tropical African genus of evergreen trees. *S. campanulata* [SCARLET BELL, FOUNTAIN TREE, FLAME OF THE FOREST, UGANDA or NILE TREE, TULIP TREE, NANDI-FLAME] is widely planted throughout the tropics as a street or specimen tree and is now naturalized in Malaysia. It bears terminal bunches of orange-scarlet flowers which make it one of the most striking of all tropical trees. The yellow-flowered specimens extensively cultivated in Kenya are derived by root cuttings from a single tree. BIGNONIACEAE, 2–3 species.

Spergula a small genus of temperate, annual or rarely perennial herbs of which one, *S. arvensis* [TOADFLAX, CORN SPURREY], is a general weed common in cornfields. It is sometimes cultivated on poor sandy soils as a green manure or forage crop. CARYOPHYLLACEAE, about 5 species.

Sphaerotheca a small but important cosmopolitan genus of powdery mildews, several species of which are of economic importance as parasites of higher plants, in particular *S. macularis* (= *S. humuli*) on hops, *S. morsuvae*, which causes American gooseberry mildew, and *S. pannosa*, on roses. ERYSIPHACEAE, about 6 species.

Sphagnum a large genus of peat or bog mosses, widely distributed in all five continents. Although great tracts of land dominated by *Sphagnum* (peat bogs) are found most extensively in north temperate to subarctic latitudes, the genus is also present in the tropics as exemplified by the Old World tropical species *S. luzonense*.

Some species, such as *S. cuspidatum* and certain varieties of *S. subsecundum*, habitually grow submerged in pools; others, such as *S. papillosum*, are among the principal peat-formers, while the characteristically wine-red clumps of *S. rubellum* and pale yellowish cushions of *S. compactum* are associated with the drier parts of the bog.

The stems bear branches in tufts

Sphagnum tenellum is a characteristic moss of damp heathland, showing here its ripe capsules which are produced profusely in early summer. (×3)

(fascicles), a proportion of them hanging wick-like down the stem. The leaves are veinless and composed of a single layer of cells of two types – narrow, green cells and much larger, colorless cells which hold water. Between them the two types of cells form an unmistakable network or pattern. The capsules open and discharge the spores explosively by an "air gun" mechanism unique among bryophytes and the spores themselves are distinctive.

Economically, *Sphagnum* is perhaps best-known for its use as a constituent of field dressings in World War I – the direct result of its good water-holding and antiseptic properties. It also has a wide range of uses in horticulture, especially as a packing material and in seed beds. In the form of peat, dead *Sphagnum* has long furnished an important fuel. In recent years living *Sphagnum* has been finding a new use as a sensitive indicator of the precise level of certain types of pollution, as for instance of radioactive fallout. SPHAGNACEAE, about 150 species.

Spices see Flavoring plants, p. 142.

Spinach, spinach beet and **New Zealand spinach** the common names for plants of different genera cultivated for their edible, green, succulent, ovate to triangular leaves. SPINACH (*Spinacia oleracea*) is an erect, smooth, annual herb of southwestern Asian origin, with large, alternate, petiolate, triangular-ovate or arrow-shaped leaves. Two types are cultivated, distinguished according to whether they produce smooth or prickly seeds.

SPINACH BEET or SWISS *CHARD (*Beta vulgaris*) is a hardy biennial resembling *S. oleracea*, but with broader leaves. NEW

Spanish Broom (Spartium junceum), growing wild along the northern shore of the Peloponnese, Greece, with the Gulf of Corinth beyond. It is also a popular garden ornamental.

ZEALAND SPINACH (*Tetragonia tetragonioides* (= *T. expansa*) of Australasia, Japan and South America is similar to *S. oleracea* but has smaller leaves; it is a branching, spreading plant grown in Europe as summer greens. CHENOPODIACEAE.

Spiraea a genus of deciduous shrubs of the Northern Hemisphere formerly including species now placed in such genera as *Aruncus, *Filipendula, *Holodiscus and *Sorbaria. Many species and hybrids are grown as ornamentals and can be divided into spring-flowering or summer-flowering groups. Flower color is predominantly white, pink or reddish and although the flowers are small

Spinach (Spinacia oleracea) leaves of which are eaten as a vegetable. (×⅕)

they are aggregated into showy umbels, racemes, panicles or corymbs.

Some garden species, such as the pink, summer-flowering *S. salicifolia* [BRIDEWORT], have been known since the 17th Century, while others are comparatively new to cultivation. Spring-flowering spiraeas include *S. prunifolia* [BRIDAL WREATH], with umbels of white double flowers, *S. cantoniensis* [REEVES SPIRAEA], *S. chamaedryfolia* and *S. nipponica* [TOSA SPIRAEA], all with more or less hemispherical corymbs of white flowers, and white-flowered hybrids such as *S. × vanhouttei* (*S. cantoniensis* × *S. trilobata*) [BRIDAL-WREATH] and *S. × arguta* (*S. × multiflora* × *S. thunbergii*) [FOAM OF MAY, BRIDAL-WREATH], the latter with a number of cultivars.

Although many of the summer-flowering group are pink to red flowered, some species, including *S. albiflora* (= *S. japonica* var *alba*), which grows to a height of 60cm, and the taller (up to 4m) *S. veitchii* produce white flowers in mid to late summer. Other attractive and popular summer-flowering spiraeas include the erect *S. douglasii*, with elongated panicles of rose-colored flowers, *S. japonica* (= *S. callosa*) [JAPANESE SPIRAEA], with large terminal corymbs of pink flowers (dark pink in cultivar 'Ruberrima', dark red in cultivar 'Atrosanguinea' and white in cultivar 'Ovalifolia') and *S. tomentosa* (= *S. rosea*) [HARDHACK, STEEPLEBUSH], with short spike-like racemes of purple-rose flowers.

Summer-flowering hybrids include *S. × bumalda* (*S. albiflora* × *S. japonica*), with corymbs of white to deep pink flowers (crimson in cultivar 'Froebelii'), and *S. × billiardii* (*S. douglasii* × *S. salicifolia*), with narrow panicles of bright pink flowers.
ROSACEAE, about 100 species.

The Hedge Woundwort (Stachys sylvatica). *This and other* Stachys *species are well known to herbalists since the leaves contain an oil with antiseptic properties. This species is not generally cultivated in gardens.* (× 1½)

Splachnum luteum *is a delicate-looking moss with characteristic umbrella-shaped capsule bases (apophyses).*

Spiranthes a genus of terrestrial orchids with tuberous roots distributed throughout the world, but absent from tropical and southern Africa. The flowers are usually rather small in spirally twisted spikes or racemes. Cultivated species include *S. cernua* [COMMON LADY'S TRESSES, SCREW-AUGUR] and *S. gracilis* [SOUTHERN or SLENDER LADY'S TRESSES]. *S. spiralis* [AUTUMN LADY'S TRESSES] occurs naturally in Europe on pastures and downs.
ORCHIDACEAE, about 200 species.

Spirogyra a genus of filamentous green algae common in freshwater ponds and lakes. Each cell contains a very characteristic spiral chloroplast and a single central nucleus.
CHLOROPHYCEAE.

Splachnum a genus of mosses which grow mainly on fairly fresh animal faeces and are chiefly north temperate and subarctic in distribution. They are notable for the large, inflated, umbrella-like apophysis (sterile tissue at the base of the capsule) which is sometimes colored red (*S. rubrum*) or yellow (*S. luteum*). In certain cases the apophysis seems to be capable of attracting dung flies and other insects able to assist in spore dispersal.
SPLACHNACEAE, 8–10 species.

Spondias a small genus of tropical trees from Southeast Asia and tropical America. *S. tuberosa* produces curious soft tubercular organs called cunca which develop on the main roots just below the soil surface; under pressure they release water which is used by travellers. Several species are widely cultivated in the tropics for their fruits, which have a sweet or acid pulp surrounding a large single seed. They include the tropical American *S. mombin* (= *S. axillaris*, *S. lutea*) [HOG

PLUM, YELLOW MOMBIN, JOBO], *S. cytherea* (= *S. dulcis*) [OTAHEITE, GOLDEN APPLE, AMBARELLA], from the Society Islands, and the tropical American *S. purpurea* [RED, SPANISH or JAMAICA PLUM, JOCOTE]. The fruits are processed into sherbets, preserves and jellies.
ANACARDIACEAE, about 12 species.

Spongospora a small genus of slime fungi, an important member of which, *S. subterranea*, attacks the potato plant underground causing powdery scab on the tubers and small galls on other parts. The scabs are shallow oval or circular lesions edged with ragged periderm and filled with a brown powder (the spore balls). Forma *nasturtii* of this species causes a similar disease, crook root, in watercress.
PLASMODIOPHORACEAE, 1–2 species.

Sprekelia a genus represented by a single species, *S. formosissima* (= *Amaryllis formosissima*) [JACOBEAN LILY, ST. JAMES LILY, AZTEC LILY], a bulbous plant with linear leaves, from Mexico. It is cultivated for its very large showy bright crimson flowers which are solitary, borne on a leafless stalk and subtended by a tubular spathe.
AMARYLLIDACEAE, 1 species.

Spruce the name used generally to refer to members of the coniferous genus *Picea* which is widely distributed over the cooler areas of the Northern Hemisphere of both the Old and New Worlds, from the Arctic Circle to the high mountains of the more southerly warm temperate latitudes of the Tropic of Cancer. Spruces are evergreen trees of more or less conical outline with a reddish-brown furrowed bark and irregularly branched horizontal to pendulous branches. When firmly rooted they can serve as useful windbreaks.

Many species are important timber trees, including the European *P. abies* (= *P. excelsa*) [NORWAY SPRUCE], *P. asperata*, native to West China, and *P. jezoensis* [YEDDO SPRUCE], of North Asia and Japan. Four North American species are of great economic importance –

P. rubens (= *P. rubra*) [RED SPRUCE], *P. mariana* (= *P. nigra*) [BLACK SPRUCE], *P. sitchensis* [SITKA SPRUCE], and *P. glauca* (= *P. alba*) [WHITE SPRUCE] – being widely used for paper pulp. Spruce wood is soft and without odor, is easy to work and takes a good finish. It is used in general carpentry, for propping poles, packing cases and the sounding boards of stringed instruments.

The resin of *P. abies* is purified to yield Burgundy pitch and the leaves and shoots are distilled to give Swiss turpentine. Extracts of shoots and leaves, mixed with various sugary substances, can be fermented to make spruce beer. *P. rubens* is the main source of spruce gum.

Although various conifers may be sold as Christmas trees, NORWAY SPRUCE is the one most commonly chosen. In addition to the numerous cultivars of *P. abies* and *P. glauca*, other spruces extensively planted as ornamentals include cultivars of *P. pungens* [COLORADO SPRUCE] and *P. engelmannii* [ENGELMANN SPRUCE, ROCKY MOUNTAIN SPRUCE].

The common name *HEMLOCK SPRUCE is given to species of *Tsuga*.
PINACEAE.

Spurge an ancient and common name for some herbaceous or subshrub members of the large genus *Euphorbia*. The common names SPURGE LAUREL and SPURGE NETTLE are given to *Daphne laureola* and *Cnidoscolus urens* (= *Jatropha urens*) respectively.

Squash a name given to edible fruits of certain species of the genus *Cucurbita* which are used as table vegetables or in pies. Squashes are produced by cultivars of *C. pepo*, *C. mixta*, *C. moschata* and *C. maxima*. Summer squashes are fruits of *C. pepo* (also called PUMPKIN, MARROW) used when immature as a table vegetable. Winter squashes are fruits of all four species used ripe as livestock feed, as

table vegetables, or in pies. The distinction, however, is not absolute. (See also p. 216.)
CUCURBITACEAE.

Squill the common English name for plants of the genus *Scilla* such as *S. autumnalis* [AUTUMN SQUILL] and *S. natalensis* [BLUE SQUILL] and also the name used in pharmacy for bulbs of a related plant *Urginea maritima* (= *Scilla maritima*), which contains glycosides which have been used as heart stimulants, and also a toxic principle used in rat poisons.
LILIACEAE.

Stachys a large genus of annual or perennial herbs of temperate and subtropical regions and tropical mountains. They have opposite leaves and stems square in section. The flowers are borne in clusters in spike-like inflorescences.

S. officinalis (= *S. betonica*) [WOOD BETONY, BISHOP'S WORT] and other species such as *S. palustris* [MARSH WOUNDWORT], from Europe and North America, have had various local medicinal uses, including dressings for wounds. *S. affinis* (= *S. sieboldii*) [CHINESE or JAPANESE ARTICHOKE, CHOROGI, KNOTROOT] has tuberous roots which are cultivated in Japan, Belgium and France as a vegetable.

Several species are grown for ornament for their attractive two-lipped flowers, some as rock plants. Among these are the mat-forming *S. corsica*, from Corsica and Sardinia, with pale pink flowers, the widely grown *S. byzantina* (= *S. lanata*) [LAMB'S EAR, LAMB'S TONGUE], from southwest Asia, with woolly gray leaves and pink or purple flowers, and

Taiga landscape in Labrador, Canada. The slopes up to the tree line have a scattered cover of White Spruce (Picea glauca) with, lower down, Black Spruce (Picea mariana).

In the Carrion Flower (Stapelia variegata) the appalling smell, brownish-purple colors and fleshy petals combine to attract carrion flies. (×1)

S. officinalis, with red-purple, pink or white flowers.
LABIATAE, about 300 species.

Stachyurus a genus of small deciduous spring-flowering trees and shrubs native from the Himalayas to Taiwan and Japan. *S. chinensis* (China) and *S. praecox* (Japan) are grown for their pendulous racemes of pale yellow flowers and globose greenish-yellow fruits.
THEACEAE, about 6 species.

Stangeria a genus of cycads comprising a single species, *S. eriopus* [HOTTENTOT'S HEAD], native to southern Africa. The stem is subterranean and the leaves have a remarkably fern-like appearance. The female cones are ovoid-elliptical and up to 18cm long.
STANGERIACEAE, 1 species.

Stanhopea a genus of tropical South and Central American and West Indian epiphytic orchids. All species have hard, ridged, egg-shaped pseudobulbs bearing a single tough leaf. The inflorescences are pendulous and when the plants are grown in a hanging basket in a greenhouse the flowers usually poke through the base of the container. The two to 10 flowers are large, of a shiny, tough, waxen appearance and usually powerfully scented. The flower structure appears very complex because the petals are strongly reflexed.
ORCHIDACEAE, about 200 species.

Stapelia a genus of almost leafless stem succulents from tropical and southern Africa. They have soft, fleshy, more or less square stems and bear five-lobed, cup or bell-shaped flowers, extremely diverse in size and markings, from small and inconspicuous to up to 30cm from tip to tip as in *S. gigantea* [ZULU-GIANT, GIANT TOAD PLANT]. The generic common name of CARRION FLOWER comes from the leathery, usually dull-colored corolla which, combined with a fetid odor,

attracts blowflies, the agents of cross-pollination. In *S. hirsuta* [HAIRY or SHAGGY TOAD PLANT], fringes of brownish-purple hairs add to the illusion of an animal carcass. *S. variegata* [TOAD CACTUS, STARFISH PLANT] is the commonest, easiest to grow and longest-known in cultivation.
ASCLEPIADACEAE, about 90 species.

Star apple the common name for the tropical American and West Indian tree *Chrysophyllum cainito*, growing up to 15m, whose purplish, round, edible fruit when ripe yields a white, jelly-like sweet pulp. When the fruit is cut transversely it shows the cells radiating from the center in a star-like form.
SAPOTACEAE.

Stellaria an economically unimportant genus of annual to perennial plants. A few are weeds of almost cosmopolitan distribution, the best-known being *S. media* [CHICKWEED]. The large-flowered species are commonly called STITCHWORTS, such as *S. holostea* [GREATER STITCHWORT, ADDER'S MEAT, MOON FLOWER]; the smaller-flowered species are known as CHICKWEEDS.
CARYOPHYLLACEAE, about 120 species.

Stephanandra a small genus of temperate East Asian ornamental deciduous shrubs. The genus includes *S. incisa* (= *S. flexuosa*) [LACE SHRUB], from Japan and Korea, and

The tufted perennial grass Stipa calamagrostis *colonizing eroded Alpine shale; the feathery inflorescences are just visible.*

*Chickweed (*Stellaria media*) is a creeping annual that can produce several generations in a single growing season. (× $\frac{1}{8}$)*

S. tanakae, from Japan, which grows to a height of about 2m, producing dense clusters of small, greenish or white, star-shaped flowers. The leaves are ovate or triangular and deeply toothed, turning bright yellow or orange in the autumn.
ROSACEAE, 4 species.

Stephanotis a small genus of climbing shrubs native to Madagascar, Peru, Malaya and China. *S. floribunda* [MADAGASCAR JASMINE, MADAGASCAR CHAPLET FLOWER, CLUSTERED WAX FLOWER, FLORADORA] is frequently cultivated in temperate greenhouses for its ovate, leathery green leaves and scented, white, waxy, tubular flowers.
ASCLEPIADACEAE, about 15 species.

Stereum a large cosmopolitan genus of basidiomycete fungi characterized by having thin, somewhat leathery fruiting bodies with a smooth spore-bearing surface. Several species are parasites of trees and cause decay. *S. purpureum* often causes spectacular "silver-leaf" symptoms in plums and related trees after air-borne spores have infected pruning wounds or natural breakages. Progressive infection causes die-back of branches and eventual death of the tree.

S. sanguinolentum occurs naturally on lopped branches and stumps of conifers and in Canada it causes a severe stem rot in BALSAM FIR, while in the USA it infects large pruning wounds on EASTERN WHITE PINE. In Europe it is becoming an increasingly important cause of rot, especially in spruces. *S. gausapatum* is the main agent of "pipe-rot" in oaks, entering the trunks through the branches that have died because of shading or through wounds exposing the heartwood.
STEREACEAE, about 100 species.

Sternbergia a genus of bulbous plants native to southeastern Europe and Southeast Asia. A number of species are cultivated for their attractive flowers. These are large, yellow, without a corona, and appear either in autumn or spring and either before or with the leaves. *S. lutea* is claimed by some to be the biblical LILY OF THE FIELD (also called WINTER DAFFODIL).
AMARYLLIDACEAE, 4–8 species.

Stewartia or **Stuartia** a small genus of deciduous trees and shrubs, native to East Asia and eastern North America. They are characterized by their smooth, peeling, reddish bark, leaves which change color to attractive yellow-reddish tints in the autumn, conspicuous cup- or saucer-shaped white flowers and glossy red capsular fruits. Several species are cultivated as ornamentals, including the shrubby *S. ovata* (= *S. pentagyna*, *Malachodendron pentagynum*) [MOUNTAIN CAMELLIA] and *S. malacodendron* (= *S. virginica*) [SILKY CAMELLIA].
THEACEAE, about 6 species.

Stinkhorns a group of gasteromycete fungi related to the true *puffballs (*Lycoperdon*) but adapted for spore dispersal by flies. Most species are tropical, but *Phallus impudicus* [COMMON STINKHORN] is widespread in temperate countries.

Clathrus ruber is quite frequent in southern Europe. It has a red spherical network of spongy tissue, on the inside of which the spores are formed. Flies enter through the interstices of the network and consume the spores. *Aseroe rubra* [AUSTRALIAN STINKHORN] has a flower-like fruiting body. The "petals" of the "flower" have their ends drawn out into long dangling processes, just like those of the true petals of many fly-pollinated flowers. The spore mass is situated at the center of the "flower" and is eaten and dispersed by flies.

Stinkwood the name given, especially in warm countries, to certain species which emit an objectionable odor (often when the wood is worked), including *Terminalia melanocarpum* [STINK TREE], of Queensland, *Celtis cinnamonea* [STINKWOOD], of Sri Lanka, *Foetidia mauritiana* [STINKWOOD], of Mauritius, and several South African species: *Celtis africana* [WHITE or CAMDEBOO STINKWOOD], *Cryptocarya latifolia* [BASTARD STINKWOOD] and *Prunus africana* [RED STINKWOOD].

Stipa a large cosmopolitan genus of tropical and temperate grasses native to xerophytic grasslands, commonly known as SPEAR GRASSES, PORCUPINE GRASSES, FEATHER GRASSES, or NEEDLE GRASSES. Most species are tufted perennials (or rarely annuals) recognized by their inrolled leaves and the slender feathery appearance of their compound inflorescences.

The genus is of great importance in the steppe regions of the USSR where *S. capillata*, *S. joannis* and *S. tirsa* (= *S. stenophylla*) are dominant component species of the flora. It is also important in the drier steppes of Hungary, Rumania, Jugoslavia and the Iberian peninsula. The west Mediterranean *S. tenacissima* is the famous ESPARTO or ALFA GRASS, which is made into fine paper.

Flowers and unripe fruits of the Wild Strawberry (Fragaria vesca). (× 1)

Many of the North American prairies and the grasslands of South America are dominated by species of *Stipa*, eg *S. leucotricha* [TEXAS NEEDLE GRASS] and *S. viridula* [GREEN NEEDLE GRASS, FEATHER BUNCHGRASS]. One of the better-known South American species, *S. ichu* [ICHU GRASS], and *S. pennata* [FEATHER GRASS], from Europe and Asia, and the Australian *S. elegantissima*, are cultivated for their handsome feathery inflorescences.
GRAMINEAE, about 150 species.

Stocks a small group of herbaceous, mostly heavy-scented, garden flowers in the closely related genera *Malcolmia* and **Matthiola*. *Malcolmia maritima* (= **Cheiranthus maritimus*) [VIRGINIA STOCK], is a small annual with white, lilac or red flowers, which is either sown in the spring for summer flowering or the autumn for spring flowering.
CRUCIFERAE.

Stokesia [STOKES' ASTER] a genus compris-

ing a single species, *S. laevis* (= *S. cyanea*), a perennial herb native to the USA. The flower heads, which resemble those of CHINA ASTERS, have distinctly tubular disk florets and deeply lobed outer florets. A number of varieties with white, blue or purple flowers are grown as garden-border plants.
COMPOSITAE, 1 species.

Stoneworts the common name for the Charophyta, a group of green algae which have a unique structure, consisting of large internodal cells and small nodal cells bearing whorls of branches. They are normally found growing on the beds of lakes, where they may form subaquatic meadows. Some species have calcified walls giving the plants a brittle stony texture.
CHAROPHYTA.

Storax a fragrant gum resin extracted from the stem of **Styrax officinalis*, a small Mediterranean tree. It is used as an antiseptic, as an inhalant and expectorant, and sometimes in incense, soaps and perfumes. A similar product is obtained from species of **Liquidambar*, including *L. orientalis* [LEVANT STORAX, ORIENTAL SWEET GUM] and *L. styraciflua* [AMERICAN SWEET GUM, RED GUM, BILSTED].

Stranvaesia a small genus of evergreen shrubs or small trees from East and Southeast Asia. They bear a close resemblance to **Cotoneaster* and have a fairly dense bushy habit. The leaves are leathery and lanceolate and the flowers white. The brilliant crimson berries make such species as *S. davidiana* attractive garden ornamentals.
ROSACEAE, about 5 species.

Stratiotes a European genus represented by a single species, *S. aloides* [WATER ALOE, WATER SOLDIER, SOLDIER'S YARROW], a lime-

The Water Aloe or Water Soldier (Stratiotes aloides) remains submerged during the winter and rises to the surface only at flowering time during the late spring and summer. (× $\frac{1}{10}$)

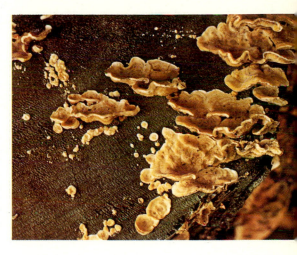

The basidiomycete fungus Stereum hirsutum is a parasite of deciduous trees and fallen logs, branches and twigs. (× $\frac{1}{2}$)

loving aquatic herb. The plant spends the winter submerged, surfacing in spring to produce stolons bearing buds and flowers. It is sometimes grown in ponds and aquaria.
HYDROCHARITACEAE, 1 species.

Strawberry the common name for the commercially very important edible soft fruit of plants of the rosaceous genus *Fragaria*, whose species are distributed throughout Eurasia and America extending southwards to Brazil and Chile. Strawberries, once mainly cultivated in temperate climates such as northern parts of Europe, America and Japan, are now increasingly grown in warmer climates.

The modern strawberry (*F. × ananassa*) is probably a hybrid of *F. virginiana* and *F. chiloensis* [CHILOE STRAWBERRY]. Intensive selection and breeding have given many improved modern varieties. Previously, only small wild species such as *F. vesca* [WILD STRAWBERRY] and *F. moschata* [MUSK or HAUTBOIS STRAWBERRY] were available for European gardens.

Strawberries are perennial herbaceous rosette plants that propagate vegetatively by elongated branches (called stolons or runners) which terminate in young runner plants. These runner plants root and then produce further runners, forming clones of like individuals. Strawberry plants sold for cultivation are vegetatively propagated by runners. The berry is a false fruit, an enlarged fleshy receptacle bearing the tiny true fruits.

Italy, Poland, France and the UK are Europe's largest producers of strawberries, together producing around 300 000 tonnes annually compared with about 200 000 tonnes in the USA. The fruit is eaten fresh or processed for preserves, canning and freezing.

The related *Waldsteinia sterilis* and **Potentilla sterilis* are called BARREN STRAWBERRY and *Duchesnea indica*, which is unrelated, is the MOCK or INDIAN STRAWBERRY.

Strelitzia a small genus of perennial banana-like herbs native to South Africa.

The best-known is *S. reginae* [BIRD OF PARA-DISE FLOWER, BIRD'S TONGUE FLOWER, CRANE FLOWER, CRANE LILY], often cultivated, in a range of varieties, in tropical gardens and temperate greenhouses for its showy, exotic, orange-yellow and blue flowers, and large banana-like leaves. *S. nicolai* [WILD BANANA, BIRD OF PARADISE TREE] is tree-like, up to 9m tall, and is sometimes cultivated.
STRELITZIACEAE, about 5 species.

Streptomyces a genus of actinomycete fungi common in soil, but also found in fresh waters and seas. *Streptomyces* species are characterized by a well-developed branching mycelium and long chains of spores on the aerial hyphae.

They are important producers of medically useful antibiotics. Strains of *S. griseus* and *S. olivaceus* have been used for the fermentative production of vitamin B12, and *S. olivochromogenes* is used in the conversion of starch into glucose–fructose syrups. Common scab of potatoes is caused by *S. scabies*.
STREPTOMYCETACEAE, 70–100 species.

Streptosolen a genus comprising a single species, *S. jamesonii* [FIRE BUSH, ORANGE BROWALLIA, MARMALADE BUSH, YELLOW HELIOTROPE], an evergreen straggling shrub native to Colombia and Ecuador and cultivated outside in mild climates or otherwise in greenhouses. It produces large clusters of attractive orange-red tubular-funnelform flowers.
SOLANACEAE, 1 species.

Striga an Asian, Australian and tropical and southern African genus of semiparasites mainly of grass crops, although some legumes, eg BROAD BEAN and COWPEA are also parasitized. After seed germination, the host roots are invaded by haustoria and sub-

The attractive clusters of white, fragrant flowers make Styrax hemsleyana *a popular garden ornamental.* (× 1)

*The Bird of Paradise Flower (*Strelitzia reginae*), from South Africa, has a type of inflorescence known as a cincinnus.* (× ½)

stantial development occurs before the aerial shoots of the parasite are formed. These flower rapidly and produce numerous minute seeds about 0.4mm in diameter. In Africa and Asia *Sorghum* crops are the most severely affected.
SCROPHULARIACEAE, about 40 species.

Strobilanthes a large genus of herbs and subshrubs from tropical Asia and Madagascar, including some fine ornamentals. The flowers are blue, violet, white or occasionally yellowish, and are borne in spikes, panicles or singly in the leaf axils. Among many attractive species are the shrubs *S. isophyllus* (= *Goldfussia isophylla*) [BEDDING CONEHEAD], with long willow-like leaves, and *S. dyeranus* (= *Perilepta dyerana*) [PERSIAN-SHIELD], whose young leaves are crimson and purple in color.
ACANTHACEAE, 250–300 species.

Strophanthus a genus of evergreen shrubs and small trees, often climbing, native to tropical South Africa and Asia. The seeds of *S. sarmentosus*, *S. gratus* and other species from Africa yield strophanthin, a heart drug. It is also the active principle in native arrow-poisons.

Species cultivated as ornamentals are all climbers and include *S. gratus*, with white to purplish flowers over 5cm long, without petal-tails but with corona-like appendages, and the attractive *S. preussii* with terminal cymes of small flowers, cream when first open, with purple markings and petal-tails up to 30cm long.
APOCYNACEAE, 40–50 species.

Strychnos a large tropical and subtropical genus with a wide variety of habit including erect, sometimes thorny trees and climbing shrubs. The flowers are small, borne in cymes, and the fruit is a berry with seeds which contain the deadly poisonous alkaloid strychnine in their testa.

The most valuable commercial source of strychnine is the seed of *S. nux-vomica* [STRYCHNINE, NUX-VOMICA TREE], a tree native to India, Burma, and Sri Lanka. The large fruits contain three to five, flat, circular seeds (nux vomica) from which strychnine and another alkaloid, brucine, can be extracted. Nux vomica is used as a tonic and as a stimulant of the nervous system. *S. toxifera* [CURARE POISON NUT] is one of the sources of *curare, which is extracted from the bark and roots, and is used by South American Indians as an arrow poison. Although curare

A laboratory culture of Streptomyces *species. The search continues for antibiotic-producing actinomycete strains.* (× ½)

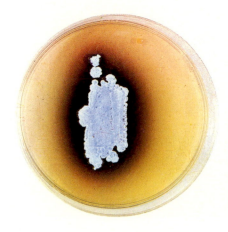

is lethal in quite small doses, causing almost instant paralysis of the motor nerves, one of its constituent alkaloids (tubocurarine) is used medicinally as a muscle relaxant.

S. potatorum [CLEARING NUT TREE, WATER-FILTER NUX-VOMICA], of Burma and India, has many uses. Not only is its timber hard and durable enough to be used in house construction, and for agricultural implements, but its seeds when ground, can be added to water to clarify it before drinking. Extracts from the seeds are also used locally to treat certain eye diseases, while the fleshy part of the fruit is edible.

Edible fruits are obtained from a number of tropical and South African species, the most important of which is *S. madagascariensis* [MKWAKWA], which grows to 6m; its edible flesh provides one of the most useful fruits in southern Africa. Another popular fruit is the berry of the spiny South African shrub *S. spinosa* [MSALA, NATAL ORANGE], although it is only the flesh that is edible; the seeds are poisonous. The genus also includes a number of useful timber trees, including the South African *S. henningsii* [NATAL TEAK] and *S. atherstonei* [CAPE TEAK].
LOGANIACEAE, about 200 species.

Styrax a genus of deciduous and evergreen trees and shrubs from warm temperate and tropical regions, mainly of the Northern Hemisphere, some of which are very beautiful, with characteristic pure white, pendulous bell-shaped flowers. *S. benzoin* yields the gum resin benzoin, used medicinally as an antiseptic and as a remedy for respiratory complaints. Other species yielding resin include the South American *S. ferrugineus* and the Malaysian *S. taninensis*. *S. officinalis* is a small Mediterranean tree which yields the gum resin *storax. It is also cultivated for its large fragrant flowers.

Other ornamentals are *S. americanus* [MOCK ORANGE], an attractive shrub with small racemes of white flowers, and *S. obassia* [FRAGRANT SNOWBELL], a large-leaved shrub with fragrant many-flowered racemes.
STYRACACEAE, about 100 species.

Sugar and starch crops see p. 318.

*An aerial view of a large Sugar Cane (*Saccharum officinarum*) plantation in central Trinidad. Sugar Cane is grown throughout the tropics and is still the world's most important source of sugar.*

*Sugar Beet (*Beta vulgaris*) is an important crop in most European countries, the USA and China, providing nearly half of the world's sugar.*

Sugar beet one of the cultivars of *Beta vulgaris*, an important crop plant which serves as a source of sugar in almost every country in Europe, and in the USSR, North America and many other countries in both the Northern and Southern Hemispheres. In the subtropics, it is grown under irrigation. It normally completes its vegetative cycle in two years. A large, succulent taproot is developed during the first year of its growth. In this is stored reserve food supplies and during the second year the plant produces flowers and fruits.

The root comprises the bulk of the tissues of a sugar beet plant. It is cone-shaped and narrows downwards into a slender taproot. Most cultivated beets have an ivory-white flesh but in poor strains (ie poor from the standpoint of total weight and/or sucrose content) the tissues are watery and greenish or yellowish. Highly developed strains contain 15% or more of sucrose. Nearly half of the world's production of sugar now comes from sugar beet.
CHENOPODIACEAE.

Sugar cane the common name for a number of species and hybrids of *Saccharum*, notably *S. officinarum*, the cultivated 'Noble' canes, *S. spontaneum*, a variable wild species, extending from Africa through India to Southeast Asia, *S. sinense*, the old cultivated canes (natural hybrids of *S. officinarum* and *S. spontaneum*), and *S. robustum*, a wild species indigenous from Borneo to the New Hebrides. In addition, both wild interspecific hybrids (*S. spontaneum* × *S. robustum*) and intergeneric hybrids (*Saccharum* × *Miscanthus*) have been recognized.

It is probable that *S. robustum* is the ancestor of sugar cane, the likely source of early selection and cultivation being New Guinea, which is still the greatest area of diversity of

Sugar and Starch Crops

are various palms, notably the SAGO and SUGAR PALMS already mentioned, the AMERICAN CABBAGE PALM (*Oreodoxa oleracea*) and several cycads, including *Cycas circinalis* and *C. revoluta*, both known as SAGO PALM and species of *Encephalartos* and *Zamia*. *MAIZE (*Zea mays*) is the source of corn starch, and *RICE (*Oryza sativa*) is used to produce rice starch (cornflour).

Although sugar (in the form of sucrose) is manufactured by all green plants, it is only extracted on a commercial or substantial scale from a relatively small number of species, the most important of which are *SUGAR CANE (*Saccharum officinarum*) and *SUGAR BEET (*Beta vulgaris*). Starch is likewise produced by the vast majority of green plants, and all cereals as well as some vegetables provide us with starch. Pure starch is extracted commercially from a number of species for nutritional purposes, the most important being *ARROWROOT (*Maranta arundinacea*), *CASSAVA (*Manihot esculenta*) and SAGO PALMS (*Metroxylon rumphii*, *M. sagu* and other genera).

In several tropical countries, commercial and local supplies of sugar are obtained from various species of palms, including the DATE PALM (*Phoenix dactylifera*), the GOMUTE or SUGAR PALM (*Arenga pinnata*) and the

HONEY or SYRUP PALM (*Jubaea chilensis*). A number of thick-stemmed strains of *Sorghum* are cultivated for their sweet cane known as sorgo and may be used for chewing or for making syrup or sugar. The North American *SUGAR MAPLE (*Acer saccharum*) and BLACK MAPLE (*A. nigrum*) are sources of maple syrup and sugar.

Starch is obtained commercially from the *POTATO (*Solanum tuberosum*) and used in manufacturing, from the root tubers of CASSAVA (*Manihot esculenta*), which is also a staple food, and from the rhizomes of ARROW-ROOT (*Maranta arundinacea*). In addition, a number of Aroids are important as food crops in the tropics because of their edible rhizomes which contain large amounts of starch, such as the *TARO (*Colocasia esculenta*), DASHEEN (*C. esculenta* var *globifera*) and Tanier (*Xanthosoma* species). They rarely enter commerce except through local markets although their potential importance is great since their starch is easily digested and suitable for children. Several *yams (*Dioscorea* species) are a source of starch as

Common sugar and starch crops: 1 Sugar Cane ($\times \frac{1}{30}$, $\frac{1}{5}$); 2 Sugar Maple; 3 Wild Date Palm; 4 Sugar Palm; 5 Sugar Beet ($\times \frac{1}{5}$); 6 Sago Palm.

SUGAR AND STARCH CROPS

Common name	Scientific name	Family
I Sugar plants		
*SUGAR CANE	*Saccharum officinarum*	Gramineae
*SUGAR BEET	*Beta vulgaris*	Chenopo-diaceae
SUGAR MAPLE	*Acer saccharum*	Aceraceae
BLACK MAPLE	*A. nigrum*	Aceraceae
*BARLEY (germinating)	*Hordeum vulgare*	Gramineae
SWEET *SORGHUM, SORGO	*Sorghum bicolor*	Gramineae
WILD *DATE PALM	*Phoenix sylvestris*	Palmae
PALMYRA PALM	*Borassus flabellifer*	Palmae
TODDY PALM, SAGO PALM, JAGGERY PALM	*Caryota urens*	Palmae
*COCONUT PALM	*Cocos nucifera*	Palmae
GOMUTI PALM, SUGAR PALM	*Arenga pinnata*	Palmae
HONEY PALM, SYRUP PALM	*Jubaea chilensis*	Oleaceae
*NYPA PALM	*Nypa fruticans*	..
MANNA ASH	*Fraxinus ornus*	..
II Starch plants		
*POTATO	*Solanum tuberosum*	Solanaceae
*CASSAVA, MANIOC	*Manihot esculenta*	Euphor-biaceae
*ARROWROOT	*Maranta arundinacea*	Marantaceae
QUEENSLAND ARROWROOT	*Canna edulis*	Cannaceae
TARO	*Colocasia esculenta*	Araceae
GIANT TARO	*Alocasia macrorrhiza*	Araceae
DASHEEN	*Colocasia esculenta* var *globifera*	Araceae
GIANT SWAMP *TARO	*Cyrtosperma chamissonis* (*C. edule*)	Araceae
TANIER, COCOYAM	*Xanthosma atrovirens* X. sagittifolium X. violaceum	Araceae
EAST INDIAN ARROWROOT	*Curcuma angustifolia*	Zingiber-aceae
FIJIAN ARROWROOT, TAHITIAN ARROWROOT	*Tacca leonto-petaloides* (*T. pinnatifida*)	Taccaceae
GREATER ASIATIC *YAM	*Dioscorea alata*	Dioscore-aceae
WHITE GUINEA YAM	*D. rotundata*	..
YELLOW GUINEA YAM	*D. cayenensis*	..
AIR POTATO	*D. bulbifera*	..
CUSH-CUSH, YAMPEE	*D. trifida*	..
*SAGO PALM	*Metroxylon rumphii* M. sagu	Palmae
SAGO PALM, GOMUTI PALM	*Arenga pinnata*	Palmae
SAGO PALM	*Caryota urens*	Palmae
AMERICAN CABBAGE PALM, CARIBEE PALM	*Oreodoxa oleracea* (*Roystonea oleracea*)	Palmae
KAFFIR BREAD	*Encephalartos caffer*	Zamiaceae
BREAD TREE	*E. altensteinii*	Zamiaceae
SAGO PALM, QUEEN SAGO	*Cycas circinalis*	Cycadaceae
JAPANESE SAGO PALM	*C. revoluta*	Cycadaceae
*MAIZE	*Zea mays*	Gramineae
WHEAT	*Triticum* species	Gramineae
*RICE	*Oryza sativa*	Gramineae

the numerous 'Noble' clones derived from *S. robustum*. After the establishment of *S. officinarum* canes in continental Asia, it is probable that they then hybridized with *S. spontaneum*.

These large perennial grasses are cultivated in the tropics, and also in a few areas of the temperate zone. The largest producers are Cuba, India, and Brazil, but other warm countries, including the West Indies, Egypt, Natal, Java, Australia, the Philippines, several Central and South American countries and the southern USA also grow significant amounts. It is the world's most important source of sugar; world production was around 44 million tonnes in 1972.

The sugar cane plant is a giant, clumpforming perennial grass characterized by stems which can attain 3m or more in height. The culm, stem or cane is the important part of the plant in the sugar cane industry, since it stores sucrose in its mature state. The toughness of the rind is a particularly important feature. The famous 'Bourbon' cane for example, grown exclusively in West Africa, is considered highly desirable for its soft chewing rind. Uba rinds, by contrast, are extremely hard and favored by sugar producers for their excellent milling qualities and resistance to disease and attacks by animals.

Extensive breeding and selection programs for the improvement of such features as sugar yield and disease resistance involve essentially the widening of the genetic base from which new populations of 'Noble' clones can be raised. This has meant the introduction of new derivatives of *S. spontaneum* into hybridization programs with existing 'Noble' × *S. spontaneum* lines. In addition to being the source of sugar, sugar cane is also the basis of numerous other products including cane syrup and *molasses which are used in cookery. Molasses is a source of industrial alcohol and of

Sunflowers (Helianthus annuus) are cultivated for their fruits, which contain an oil used in cooking and in margarine. Shown here is a group of fruits in the center of the fruiting head. (× 3)

The massive heads of yellow flowers make the Sunflower (Helianthus annuus) a popular ornamental for gardens. (× $\frac{1}{3}$)

the many types of rum produced in tropical America and the West Indies. Some of the best-known are the light-bodied rums of Cuba, Trinidad and Venezuela and the heavier, more pungent Jamaica rum. Demerara rums are produced from the molasses from cane grown in the Demerara River area of Guiana. The difference in flavors of the various types of rum is partly due to differences in the quality of the sugar cane and variations in the soil and climate. (See also Bagasse.)
GRAMINEAE.

Sugar maple the common name for *Acer saccharum*, a tree native to northeastern North America, having a sweet sap from which sugar and syrup are obtained. The sap, which flows for about a month between March and April, is extracted through incisions in the bark. The sap is evaporated into a syrup, from which the sugar crystallizes.
ACERACEAE.

Sultanas the sun-dried fruits of *Vitis vinifera* var *sultanina*. They are medium-sized and seedless and widely grown in Australia and California.
VITACEAE.

Sunflower the common name for members of the North American genus of annual and perennial herbs, *Helianthus*. The best-known species is the annual *H. annuus* [COMMON SUNFLOWER, MIRASOL], which is the world's second most important source of vegetable oil. The crop is mainly grown on a large scale in the more temperate countries such as Chile, Uruguay, Argentina, Yugoslavia, Turkey and particularly the USSR (with twothirds of the world production), but there is some cultivation in the tropics. Resistance to some diseases has been achieved by hybridizing *H. annuus* with *H. tuberosus* [JERUSALEM ARTICHOKE].

Sunflower seeds contain 25–32% oil. In the USSR, plant breeding has practically doubled the oil yield. Sunflower oil is light yellow and is useful as a salad or cooking oil and, when hydrogenated, can be employed in the manufacture of margarine. The oil contains a high percentage of polyunsaturated fatty acids, such as linoleic acid. The seed residue, left after oil extraction, is used as a constituent of animal food.

Sunflower seeds and their products have been utilized in a variety of other ways. The seed is a common constituent of bird food and, particularly in the USSR, is chewed by people. The outer covering (pericarp) or hull of the seed may be removed by decortication. In Canada the hulls have been made into logs to be used as fuel while in the USSR, alcohol has been produced from them. The hulls may also be used as a bedding for livestock.

There are numerous varieties in cultivation, many of them popular as garden ornamentals. The well-known flower heads, with bright yellow ray florets and purplish or brown disk florets may attain a diameter of 40cm.

Amongst other attractive ornamental species cultivated in a range of varieties are *H. debilis* [CUCUMBER-LEAVED SUNFLOWER], *H. angustifolius* [SWAMP SUNFLOWER], *H. atrorubens* [DARKEYE SUNFLOWER] and *H. laetiflorus* [SHOWY SUNFLOWER].
COMPOSITAE.

Swainsona a small genus of pea-like herbs and subshrubs from Australia and New Zealand. The most important of these, cultivated as ornamentals, are *S. galegifolia* [SWAN FLOWER, WINTER SWEET PEA], growing to 1m or more and bearing deep red pea-like flowers, and *S. greyana* (= *S. grandiflora*) [DARLING RIVER PEA, POISON PEA], with large pink flowers.
LEGUMINOSAE, about 50 species.

Swede or **Swedish turnip** the swollen root, hypocotyl and stem of *Brassica napus* (napobrassica group). It is used as a vegetable, usually peeled, sliced and then boiled and mashed, but it can be eaten raw. It is frequently used as winter feed for animals. The annual form [SWEDE RAPE] is used as animal feed and varieties have been developed to produce a high yield of seed from which an edible oil is extracted.
CRUCIFERAE.

Sweet peppers the fruits of *Capsicum annuum* (grossum group). The many cultivated

The Sweet Potato (Ipomoea batatas) forms an important part of the staple diet for many people in the tropics, being rich in carbohydrates, protein and vitamins. Unlike potatoes, the tubers of the Sweet Potato do not keep well and are usually eaten within a few days of harvesting. Most of the crop is grown for domestic consumption.

varieties vary markedly in the color, size and shape of their fruits.
SOLANACEAE.

Sweet potato the highly nutritious edible tuber of the American *Ipomoea batatas* [SWEET POTATO VINE], an essential food source in many warm temperate regions, the subtropics and tropics. The plant is a vine-like, perennial herb but in the subtropics it is cultivated as an annual. In the tropics the crop is virtually perennial, in that stem cuttings are taken from the tips of growing vines of standing crops in a continuous planting regime. About 10 tubers develop on each plant, cultivars being of two types: tubers with dry, mealy yellow flesh, and a second type with softer, watery, gelatinous flesh, richer in sugar.

In the tropics, sweet potatoes are grown on a domestic scale rather than commercially. After harvest they do not keep so well as some root crops, and are normally used within a few days, although some storage of mature tubers is practiced in subtropical countries. They are usually eaten as a table vegetable, either boiled or baked, but are also preserved by canning or dehydration, and used as a source of flour, starch, glucose syrup and alcohol. The green parts are used as a potherb, and for feeding animals, especially pigs.
CONVOLVULACEAE.

Swertia a genus of annual and perennial herbs mainly native to East Asia, with some species in North America, Africa and Eurasia. *S. chirata* of India is used as the basis of the bitter tonic chiretta (or chirata). Some other species are cultivated for their blue, purplish, greenish, or yellow to white gentian-like flowers, including *S. perennis* and *S. perfoliata*.
GENTIANACEAE, about 50 species.

Swietenia see MAHOGANY.
MELIACEAE, 6–7 species.

Sycamore a common name given to members of a number of families of deciduous trees. In North America the name SYCAMORE is applied to members of the genus *Platanus*, such as the large spreading tree *P. occidentalis* [EASTERN SYCAMORE] from eastern and central North America, although other common names are used, including BUTTONWOOD, BUTTONBALL and AMERICAN PLANE (SEE PLANE TREES).

In England the name SYCAMORE is given to the Eurasian native *Acer pseudoplatanus*, a member of the maple family, often planted as an ornamental.

The SYCAMORE, EGYPTIAN SYCAMORE, or SYCAMORE FIG, of the Bible, is *Ficus sycomorus*, a shade tree with edible fruit, native to South Africa, Asia Minor and Egypt.

Symphoricarpos a genus of deciduous, ascending or prostrate shrubs, often with shredding bark, native to North America and China. Several of the North American spe-

The bell-shaped, purple flowers of the Russian Comphrey (Symphytum × uplandicum), a hybrid species (S. asperum × S. officinale) of Comfrey native to the Caucasus but now extremely common in Britain and other parts of Europe. (×1)

cies, eg *S. albus* [SNOWBERRY, WAXBERRY], *S. occidentalis* [WOLFBERRY], both with white fruits, and *S. orbiculatus* (= *S. vulgaris*) [INDIAN-CURRANT, CORAL BERRY], with coral-red fruits, are widely cultivated. The flowers are white or pink and borne in clusters or spikes.
CAPRIFOLIACEAE, about 16 species.

Symphytum a small genus of perennial herbs from Europe and western Asia. The pendulous bell-shaped flowers are well known in *S. officinale* [HEALING HERB, BONESET, COMFREY], a common weed in Europe. Often the leaves are distinctly prickly, as in *S. asperum* [ROUGH RUSSIAN OR PRICKLY COMFREY], which is grown for animal fodder. Some species are cultivated as garden ornamentals, eg *S. grandiflorum* and *S. orientale* [TURKISH COMFREY].
BORAGINACEAE, about 25 species.

Symplocos a large genus of small trees native to tropical and subtropical Eurasia, Australia and America. The bark of many species, including *S. tinctoria* [COMMON SWEETLEAF, HORSE SUGAR], *S. racemosa* and *S. fasciculata*, yields red, brown or yellow dyes. The fruits of *S. racemosa* are edible and its flowers are a source of honey. The soft close-grained wood of *S. tinctoria* is used for turnery. *S. paniculata* (= *S. crataegoides*) [SAPPHIRE BERRY, ASIATIC SWEETLEAF] is cultivated for its attractive small white flowers and bright blue berries.
SYMPLOCACEAE, about 300 species.

Syringa see LILAC.
OLEACEAE, about 30 species.

T

Tabebuia a genus of trees or shrubs native to tropical America, with showy yellow, white, pink, red or purple flowers in racemes, panicles, umbels or heads. Several species yield valuable, durable and heavy timber, such as the yellow-flowered *T. guayacan*, from Central America, and *T. serratifolia* [YELLOW POUI], from the Caribbean and South America, and the purple-flowered *T. impetiginosa*, from Brazil.

The genus contains some of the most spectacular flowering trees known and several are cultivated in parks, gardens and as street trees. These flowers all appear at the same time, before the leaves, just before the rains, and cover the whole tree in a mass of color. *T. chrysantha* [ARAGUANEY], with yellow flowers, is the national tree of Venezuela. *T. serratifolia*, also with yellow flowers is widely grown, as is *T. rosea* [PINK POUI, APAMATE], which has purple-pink flowers. *T. donnell-smithii*, along with five or six other species, is now placed in a separate genus *Cybistax*. It is an important timber tree in Mexico and Central America.
BIGNONIACEAE, about 80–100 species.

Tabernaemontana a genus of tropical shrubs and trees often with conspicuous flowers and rubbery latex. It includes *T. malaccensis*, whose milky latex is used as an arrow-poison in Malaya, and *T. elegans* [TREE TOAD], an African species, whose fruits have an outer covering like the skin of a toad

Below *The starch-rich tubers of the Polynesian Arrowroot* (Tacca leontopetaloides). ($\times \frac{1}{4}$) Right *The flower of* Tacca palmata, *growing in lowland rain forest, Malaya. The long hanging projections are supposed to act as attractants and guides to pollinating flies.*

and a slimy pulp, which is fried by the natives.

A number of shrubs are cultivated for their salver-shaped yellow or white flowers, including *T. divaricata* (= *T. coronaria*) [ADAM'S APPLE, CRAPE JASMINE, EAST INDIAN or BROAD-LEAVED ROSEBAY, NERO'S CROWN], and *T. holstii* from tropical Africa. A decoction of the bark of *T. corymbosa* is used in Malaya to treat syphilis.
APOCYNACEAE, about 150 species.

Tacca a small pantropical genus of perennial rhizomatous or tuberous herbs which bear large leaves on long stalks. Economically, the family is important for *T. leontopetaloides* (= *T. pinnatifida*) [EAST INDIAN, FIJIAN, TAHITIAN or POLYNESIAN ARROWROOT], whose rhizomatous tubers yield a starch known as East Indian *arrowroot. It is used in the Pacific islands and Africa for bread-making and as a starch in laundry work. Several species, including *S. chantrieri* [DEVIL FLOWER, BAT FLOWER, MAGICIAN'S FLOWER, DEVIL'S TONGUE, CAT'S-WHISKERS, JEW'S BEARD], are cultivated in greenhouses for their umbel-like inflorescences subtended by attractively colored bracts.
TACCACEAE, about 10 species.

Tagetes a genus of erect, branched or diffuse, strongly-scented annual or perennial herbs, mainly native to subtropical areas of North, Central and South America. The gland-dotted leaves are usually opposite and often pinnately dissected. The yellow, orange or reddish-brown flower heads have tubular disk florets and strap-shaped ray florets.

Several species and numerous single and double cultivars are grown as garden bedding plants, or for cut-flowers. One of the popular taller species is the annual *T. erecta* [AFRICAN MARIGOLD, BIG MARIGOLD, AZTEC MARIGOLD], from Mexico and Central America. Its stout stems reach a height of 1m and bear flower heads up to 12cm across in various shades from light yellow to orange, depending on cultivar. Another widely cultivated annual species with numerous culti-

French Marigolds (Tagetes patula). ($\times \frac{1}{8}$)

vars is *T. patula* [FRENCH MARIGOLD], from Mexico and Guatemala. The plants are bushy in habit and between 20–50cm high, with flower heads up to 5cm across, in various shades and combinations of yellow, orange and reddish-brown. African and French marigolds readily hybridize, giving rise to many cultivars intermediate in form and color, including dwarf and double-flowered types.

The dwarf cultivars are used as edging plants, as are various dwarf compact cultivars grouped under cultivar 'Pumila' of the Mexican and Central American *T. tenuifolia* (= *T. signata*) [SIGNET MARIGOLD, STRIPED MEXICAN MARIGOLD], which has yellow flower heads. *T. filifolia* [IRISH LACE] is a much-branched dwarf annual with white flowers. Native to Mexico and Costa Rica, it is cultivated there and elsewhere for its attractive linear, thread-like finely cut leaves, particularly as an edging plant. *T. minuta* [MUSTER-JOHN-HENRY] is a tall (up to 1m) annual, native to South America but naturalized in parts of the eastern USA and South Africa. It yields an essential oil with insect-repellent properties, and is sometimes grown for seasoning or as a medicinal plant; the leaves are used to make an infusion with diuretic and diaphoretic properties.
COMPOSITAE, about 30 species.

Tallow, vegetable the thick outer layer of chalky white, oily fat on the seeds of *Sapium sebiferum* [CHINESE TALLOW TREE], a native of China and Japan which is also cultivated elsewhere and naturalized in the southern USA. In China, candles and soap are made solely from the tallow, but in India the seeds are crushed to extract the kernel oil. In Asia, a black dye is obtained from the leaves and the wood is used for musical instruments, eg drums, and for wooden sandals. It is also cultivated as an ornamental and shade tree.

The SIERRA LEONE TALLOW TREE is *Pentadesma butyracea*, a tropical West African tree whose seeds yield an edible fat (sometimes called lamy or Sierra Leone butter). The fat is also used in the manufacture of soaps, candles and margarine.

Tamarind the fruit of *Tamarindus indica*, a tropical tree possibly originating in India and

*The "clock" of the Common Dandelion (*Taraxacum officinale*) comprises many achenes, each surmounted by a long stalk and a pappus. (×4)*

frequently grown and naturalized there and in other tropical areas as a shade tree, an ornamental and for the fruits. The pinnate leaves and highly perfumed flowers with reduced, bristle-like petals are extremely attractive.

The typical pod-like fruits produce tamarind, the acid pulp surrounding the seeds, which is widely used in Indian cooking and is employed as a souring agent in curries and chutneys. It is also used as a basis for fruit drinks (particularly in the New World) and medicinally as a laxative.
LEGUMINOSAE.

Tamarix a genus of maritime shrubs and small trees with small scale-like leaves, native to Europe, Africa and Asia. They are grown widely for seaside planting as hedges or windbreaks, being exceptionally tolerant of saline soils and atmospheres. Cultivated species include *T. chinensis* (= *T. amurensis*), *T. capsica*, *T. elegans*, *T. japonica*) [CHINESE TAMARISK] and *T. gallica* (= *T. anglica*) [FRENCH TAMARISK, MANNA PLANT], all with white or pinkish small flowers in spikes or racemes.
TAMARICACEAE, about 54 species.

Tamus a genus of herbaceous perennial climbing plants, native to southern Europe, western Asia, North Africa, the Canary Islands and Madeira. *T. communis* [BLACK BRYONY] is a common plant of scrub and hedgerows in Europe, with shining heart-shaped leaves, and small yellowish flowers which are followed by attractive bright red, very poisonous berries.
DIOSCOREACEAE, about 5 species.

Tanacetum a genus of north temperate species of often aromatic annual to perennial herbs, sometimes woody at the base. The flower heads are solitary, or in corymbs, and may be ligulate, with yellow to orange ligules, or without ligules as in *T. vulgare* (= *Chrysanthemum vulgare*) [TANSY, GOLDEN-BUTTONS].

Several species are cultivated in gardens, such as TANSY, which is used for flavoring and garnishes and also in herbal medicine, tansy tea being a traditional tonic and stimulant, and *T. densum*, a silvery-leaved, mat-forming, low-growing subshrub with orange-colored ligules. As currently interpreted, *Tanacetum* includes the species previously referred to *Pyrethrum*.
COMPOSITAE, 70 species.

Tamarix nilotica growing near a well in the Yemen. Tamarix species are drought- and salt-tolerant evergreen shrubs.

Tangelo a citrus fruit *Citrus × tangelo* (= *C. × paradisi* × *C. reticulata*), orange-yellow in color and smooth-skinned. It is a cross between the *mandarin orange and the *grapefruit or *pomelo and is grown commercially in a range of varieties in the tropics and subtropics.
RUTACEAE.

Tangerine the common name for one of the many varieties of *mandarin orange (*Citrus reticulata*).
RUTACEAE.

Taphrina a genus of parasitic fungi causing leaf curls, twig proliferation (witches' brooms) and other diseases of plants. The best-known are *T. deformans* (peach leaf curl), *T. populina* (golden leaf spot of poplar) and *T. betulina* (witches' broom of birch).
TAPHRINIACEAE, about 100 species.

Taraxacum [DANDELIONS, BLOWBALLS] a genus of perennial rosette-forming herbs with solitary heads of usually yellow (occasionally white or pink) strap-shaped florets borne on hollow, milky, leafless stems. A feature of many members of this genus, for example the *Taraxacum officinale* group, is apomixis, that is the development of seed without fertilization by a male gamete. By this method, various genetically distinct and self-perpetuating stocks of *Taraxacum* are formed, each showing variation amongst its members and tending to obscure the distinction between species. Many of these stocks have been described as new species.

Dandelions are found native in six continents, but most species are temperate, montane or arctic species from Eurasia, the greatest concentration being in northwestern Europe. *T. officinale* [COMMON DANDELION], a cosmopolitan weed, is used in

salads, and to make a variety of beverages, including tea (dried leaves), coffee (ground roots – they are related to chicory), wine (flowers) and schnappes (flowers). In any form it is diuretic, thus giving rise to many common names. The milk (latex) of *T. kok-saghyz* [RUSSIAN DANDELION] is a potential source of rubber.
COMPOSITAE, 50–60 species.

Taro one of the common names of *Colocasia esculenta* (= *C. antiquorum* var *esculenta*), a perennial cultivated throughout the tropics, producing starchy tubers which are edible when boiled or fried. The GIANT TARO is *Alocasia macrorrhiza*.
ARACEAE.

Tarragon or **estragon** the common name for *Artemisia dracunculus* (= *A. redowskii*), a perennial herb cultivated in the USA, Asia and Europe for its narrow aromatic leaves which are used as a herb for flavoring meat dishes, salads and soups. They can also be pickled to form the culinary tarragon vinegar.
COMPOSITAE.

Tau-saghyz a herbaceous plant (*Scorzonera tau-saghyz*), native to Russia, which yields a latex containing up to 30% rubber.
COMPOSITAE.

Taxodium a small genus of deciduous or more or less evergreen coniferous trees, native to eastern North America and Mexico. The most commonly cultivated form is *T. distichum* var *distichum* [BALD CYPRESS, DE-CIDUOUS CYPRESS], a tall handsome tree whose

pale green foliage gives it a surprisingly delicate feathery appearance. *T. distichum* var *nutans* (= *T. ascendens*) [POND CYPRESS] and *T. mucronatum* (= *T. mexicanum*) [MONTEZUMA or MEXICAN CYPRESS] are both less hardy than *T. distichum* and so are not so widely cultivated.

The timber most commonly used is that of *T. distichum*; the wood is soft and does not shrink, is resistant to insect attack and is little affected by damp conditions. These qualities make it a suitable packaging material and it is also used for piping, ventilators, fencing and garden furniture.
TAXODIACEAE, 2 or 3 species.

Taxus see YEW.
TAXACEAE, 8 species.

Tea a stimulant beverage made from the leaves of the evergreen shrub *Camellia sinensis* (= *Thea sinensis*) native to Southeast Asia. Many varieties of *C. sinensis* have been described, but a useful means of subdividing the species is by geographical races, which fall into two main groups – China and Assam teas. The two types readily hybridize and as tea is generally raised from seed it is a very heterogeneous crop. Assam teas (*C. sinensis* var *assamica*), allowed to grow naturally, quickly reach a height of about 15m, but in cultivation are pruned to a low bush 0.5–2m tall. China teas (*C. sinensis* var *sinensis*) are dwarf trees which grow more slowly and have much narrower leaves. It has been proposed that they constitute two distinct species. In both groups the leaves are evergreen, glossy on the upper surface and slightly hairy on the lower.

Green tea leaves being plucked by hand in Sri Lanka. Tea (Camellia sinensis) is by far the most important of the beverages made from plant leaves. After plucking the leaves are either steamed (green tea) or fermented (black tea).

Other types are *C. assamica* subspecies *lasiocalyx*, a southern form which is not widely cultivated but is used in tea breeding, and *C. irrawadiensis*, from Burma, which has been involved in the origin of Darjeeling tea.

The terminal bud and top two leaves yield the best tea; the quality is lower if more leaves are plucked. After plucking, most of the crop is processed by one of two methods, to give green tea or black tea. Other preparations such as pickled tea are used only locally. Black tea provides most of the world's supply and is mainly processed in factories on the estates. The processing allows enzymic oxidation of the polyphenols which, with caffeine and other alkaloids, and essential oils, give tea its characteristic flavor. The shoots are first withered, by spreading them out in open lofts or in a current of warm air, so that they become limp and permeable. They are then passed between rollers so that the enzymes are liberated, the leaves are sieved and fermentation is allowed to proceed for about two hours. After the initial processing, the leaves are dried, graded according to color and fineness, and packed into tea chests for export. In the production of green tea, the leaves are not fermented but steamed; this denatures the enzymes, preventing any appreciable oxidation and blackening of the leaves.

C. sinensis is the main commercial source of *caffeine, and enough can be extracted out

of the waste from tea manufacture to satisfy world demand. Caffeine is used medicinally as a stimulant. Tea-seed oil is a minor product, and less oil occurs in the seeds of *C. sinensis* than in the related species *C. sasanqua*. The residual cake contains saponins and therefore has no potential as fodder, although it can be used as a manure that also acts as a worm-killer; it is used in Indochina to stupefy fish.
THEACEAE.

Tecoma a small genus of shrubs and trees native from the southeastern USA to Argentina. Many species formerly included in *Tecoma* are now placed in *Tabebuia*, *Campsis*, *Tecomaria*, *Podranea* and *Pandorea*.

Several species are cultivated, including *T. stans* (= *Bignonia stans*) [YELLOW ELDER, YELLOW BELLS, YELLOW BIGNONIA], a shrub up to 6m high. It has bright yellow, bell-shaped flowers, 5cm long, and a decorative seed pod up to 20cm long; var *velutina* has more finely divided leaflets which are hairy beneath. *T. leucoxylon* provides a hard wood used in construction work and is the source of the yellow-green timber known as Green Ebony, *Greenheart Ebony or Surinam Greenheart.
BIGNONIACEAE, about 16 species.

Tectona [TEAK], a small genus of trees from the Philippines, Southeast Asia and Indo-malaysia. The large deciduous *T. grandis* [TEAK TREE] is the most important species, being the source of the valuable timber, teak. Very variable in size and form according to locality, soil and climate, it can attain 40–45m in height and 2.5m in diameter, particularly in parts of Burma.

The value of teak lies in its durability, strength, weight, ease of working, and appearance. It is a yellow-brown wood turning a rich dark brown on exposure to air, and is sticky or greasy to touch, due to the high resin content. It is used where resistance to water and durability are of prime importance, as in shipbuilding, decking and plank-

A stand of Terminalia phellocarpa *in central Malaya. The trees lose their leaves in the dry season.*

ing, and for a decorative veneer for plywood. Teak is also resistant to fire and termite attack. The bark of the roots yields a yellow dye, and oil extracted from the wood is used as a substitute for *linseed oil in varnishes and paints.
VERBENACEAE, about 4 species.

Tellima a genus represented by a single species, *T. grandiflora* (= *T. odorata*) [FRINGE CUPS], a rhizomatous perennial herb occurring naturally in North America from Alaska to California. It is often cultivated as a rock-garden plant for its racemes of rather tubular flowers, greenish white in color, but turning red on ageing.
SAXIFRAGACEAE, 1 species.

Telopea a small, very ornamental genus of evergreen trees or shrubs native to Australia (New South Wales) and Tasmania. The flowers are numerous and densely crowded in terminal, usually globose heads, surrounded by an involucre of colored bracts, deep coral-red in *T. speciosissima* [WARATAH] and crimson in *T. truncata* [TASMANIAN WARATAH].
PROTEACEAE, 3–4 species.

Teosinte the common name for *Euchlaena mexicana* (= *Zea mexicana*), a tall, annual, maize-like grass of Mexico which bears large angular grains. It is occasionally cultivated as a cereal or fodder plant and under good conditions will tiller well and produce large crops.
GRAMINEAE.

Tephrosia [HOARY PEA] a genus of herbs and shrubs widely distributed but centered in the tropics and subtropics. A number of species are grown as ornamentals, including the red-flowered *T. grandiflora*, from South Africa, the yellowish-flowered *T. pubescens*,

from northern Australia, and the yellowish- and pink-purple-flowered *T. virginiana* [GOAT'S RUE, CATGUT, RABBIT'S PEA], from the southern USA. Some species, such as *T. candida* from India and *T. vogelii* from tropical Africa, are grown for green manure in the tropics, the latter also yielding a fish poison.
LEGUMINOSAE, 300–400 species.

Terebinth the common name given to *Pistacia terebinthus* (also known as CYPRUS TURPENTINE), a small tree native to the Mediterranean region, used as a source of *turpentine.
ANACARDIACEAE.

Terminalia a large genus of tropical trees, a number of which produce strong, close-grained timber, such as *T. chebula* [TROPICAL ALMOND, BLACK MYROBALAN] and *T. bellirica* [BELERIC MYROBALAN]. *T. ivorensis* [IDIGBO] and *T. superba* [AFARA] are two West African species which provide valuable commercial timber. The fruits of such species as *T. catappa* [MYROBALAN, INDIAN ALMOND, TROPICAL ALMOND, OLIVE BARK TREE] are edible. The bark and unripe fruit of some species, notably *T. chebula*, *T. bellirica* and *T. catappa*, are important sources of commercial tannin. Various species yield astringent extracts from the bark, leaves or fruits, used in local medicine.
COMBRETACEAE, about 200 species.

Tetraphis pellucida; *the vegetative shoots often terminate in an apical leafy cup in which numerous gemmae are produced. (×2)*

Tetraphis a small but interesting genus of mosses found in the Northern Hemisphere, distinguished by the possession of a capsule with a peristome consisting of only four teeth of solid construction. At present it contains the two species, *T. pellucida* and *T. geniculata*. *Tetraphis* is also notable for the abundant diskoid gemmae which it bears in terminal "cups" on the leafy shoot system.
TETRAPHIDACEAE, 2 species.

Teucrium a large genus of herbs or sub-shrubs of wide distribution but with the greatest representation in the Mediterranean area. The foliage is often very hairy, with a gray appearance. Many species, including *T. polium* (= *T. aureum*), *T. chamaedrys* [WALL GERMANDER] and *T. marum* [CAT THYME], have been used for medicinal purposes; others are popular garden ornamentals, such as the large shrub *T. fruticans* [TREE GERMANDER], which has racemes of pale blue flowers, and the dwarf rock garden cultivar 'Prostratum' of the WALL GERMANDER, with spikes of pinkish flowers.
LABIATAE, about 30 species.

Thalictrum [MEADOW RUE] a genus of herbaceous perennials with ternately compound leaves, mostly from temperate regions of the Northern Hemisphere. A few species are cultivated as garden ornamentals. *T. aquilegifolium*, from central Europe, has fragrant flowers with stamens with broadened whitish, yellowish or purplish filaments. *T. alpinum* [ALPINE MEADOW RUE] has small scentless flowers, which bear very slender filaments, allowing the anthers to be fluttered by the breeze. *T. delavayi* is a Chinese species with lilac or reddish sepals and yellow anthers.
    T. flavum [YELLOW MEADOW RUE] is an attractive border plant with its deeply divided blue-gray leaves and crowded sprays of small yellow flowers. Extracts of stems, leaves or roots of the eastern European species *T. simplex* (= *T. angustifolium*) are used locally as a diuretic and for other medicinal purposes.
RANUNCULACEAE, about 100 species.

Thaumatococcus a genus consisting of a single species, *T. daniellii* [MIRACULOUS FRUIT], from tropical West Africa. It is a rhizomatous perennial with a very short main stem and one (sometimes two) foliage leaves situated below the terminal inflorescence which is borne almost at ground level. Inside the fruits are two or three seeds, each containing a protein which is the world's sweetest natural substance. It is currently being investigated commercially.
MARANTACEAE, 1 species.

Theobroma a genus of shrubs and small trees native to Central and South America. Characteristically, flowers and then the fruit (a berry) are produced on the old wood. A number of species, including *T. angustifolium* *T. leiocarpa*, and *T. subincanum* produce seeds which are sources of cocoa. Commercially, however, only one species *T. cacao*, is of major economic importance (see Cocoa).
STERCULIACEAE, about 22 species.

Thermopsis a genus of perennial rhizomatous herbs from North America and northeast Asia. They have digitately trifoliolate leaves and resemble LUPINS (*Lupinus*). Several species, such as *T. montana* and *T. lupinoides* (= *T. lanceolata*), are cultivated for their large showy racemes of pea-like, yellow (rarely purple) flowers.
LEGUMINOSAE, about 20 species.

Thistle a name given to various spiny species of the family Compositae. The majority belong to the genera *Cirsium* and *Carduus*, such as *Cirsium vulgare* [BULL THISTLE] and *C. arvense* [CANADIAN THISTLE], and *Carduus crispus* [WELTED THISTLE] and *C. nutans* [MUSK or SCOTCH THISTLE]. Members of the genus *Sonchus* are commonly known

A forest of Western Red Cedar or Giant Arbor-vitae (Thuja plicata) near Prince George, British Columbia, Canada.

Foliage of the Western Red Cedar (Thuja plicata) cultivar 'Zebrina', which is banded creamy-yellow. Several other cultivars exist.

as SOW THISTLES. The name SCOTCH THISTLE is also used for *Onopordum acanthium*, although this species is in fact not native to the British Isles.
    Silybum marianum (= *Carduus marianus*) [HOLY, BLESSED or MILK THISTLE] is given the latter name because of the whitish veins on the glossy leaves, giving a marbled effect. *Cnicus benedictus* (= *Carduus benedictus*) [BLESSED THISTLE] was formerly used as a cure for gout.
COMPOSITAE.

Thrinax a genus of West Indian fan palms, also native from southern Florida and Mexico to British Honduras. They are handsome, slender, unarmed, solitary palms, some of which are cultivated, including *T. morrisii* (= *T. ekmanii*) [KEY PALM] and *T. parviflora* [PALMETTO THATCH, ROYAL PALMETTO PALM, REEF THATCHES]. In addition to their ornamental value, the leaves of some species are a source of fiber and thatch.
PALMAE, 4–10 species.

Thuidium a genus of mosses notable for the tripinnate branching of some of the species, which imparts a beautiful frond-like character to the shoot. The most widely distributed species, *T. tamariscinum*, occurs in every continent except Australasia.
THUIDIACEAE, about 170 species.

Thuja [ARBOR-VITAE] a small genus of evergreen coniferous trees and shrubs from East Asia and North America. They are usually of pyramidal habit and numerous cultivars of *T. occidentalis* [AMERICAN ARBOR-VITAE, NORTHERN WHITE CEDAR], *T. plicata* [GIANT or WESTERN ARBOR-VITAE, WESTERN RED CEDAR]

and *T. standishi* (= *T. japonica*) [JAPANESE ARBOR-VITAE] are among those species commonly planted as ornamentals. *T. orientalis* (= *Biota orientalis*) [ORIENTAL ARBOR-VITAE] from China and Korea is sometimes placed in a seperate monotypic genus, *Platycladus* (as *P. orientalis*). It is cultivated in a wide range of cultivars.

The wood, which is light, easy to work and without resin canals, is used for general building, furniture, telegraph poles etc. The outer bark makes a useful roofing material. There is an inner, more fibrous bark which serves as a stuffing for upholstery.
CUPRESSACEAE, 5–6 species.

Thujopsis a genus separated from *Thuja* to accommodate the single Japanese species of evergreen tree, *T. dolabrata* [FALSE ARBOR-VITAE, HIBA ARBOR-VITAE]. It is a handsome pyramidal tree growing to 15–30m but is often shrubby in cultivation. Its wood, which is soft, durable and elastic, is used in Japan for house, bridge and boat construction as well as for cabinetwork.
CUPRESSACEAE, 1 species.

Thunbergia a large genus of erect or climbing herbs or shrubs, that are mainly native to Asia, Madagascar and tropical and southern Africa. The tubular showy flowers can be white, yellow, red, purple or blue in color. *T. alata* [BLACK EYED SUSAN] is frequently

Flowers of Black Eyed Susan (Thunbergia alata). (× ½)

grown as a pot plant, being an easily raised annual. It climbs to 60cm and is soon covered in bright yellow or white flowers with a violet-black center. The perennial *T. laurifolia* and the rather similar *T. grandiflora* [BLUE TRUMPET VINE, CLOCK VINE, SKY VINE] are excellent greenhouse climbers with their beautiful racemes of blue flowers.
ACANTHACEAE, about 150 species.

Thyme the common name for *Thymus vulgaris* [COMMON or GARDEN THYME], a hardy aromatic evergreen dwarf shrub from the Mediterranean and the very similar *T. × citriodorus* [LEMON THYME]. The dried leaves are used to flavor meat dishes. An essential oil (thyme oil) is distilled from the leaves and flowers of *T. vulgaris* and *T. capitatus* [CONE

HEAD THYME] and is used for stomach upsets and as a stimulant; it is also a commercial source of the disinfectant, thymol. Many cultivars of *T. praecox* [MOTHER OF THYME, WILD CREEPING THYME] and other species, such as *T. caespititius*, *T. hirsutus* and *T. serpyllum*, are cultivated in rock gardens. *T. herba-barona* has an aroma like caraway seed while *T. mastichina* is widely used in Spain as a flavoring and in marinating olives.
LABIATAE.

Tiarella a small temperate genus of slender erect perennial herbs from North America, with one species in Asia.

T. cordifolia [FALSE MITERWORT, FOAM-FLOWER] is one of the species cultivated as a ground-cover or rock-garden plant. It bears racemes of starry white flowers (rose in cultivar 'Purpurea') and spreads rapidly by means of stolons.
SAXIFRAGACEAE, 6 species.

Tibouchina a large genus of shrubs and subshrubs (rarely perennial herbs) native to tropical America. The best-known of a number of species grown as ornamentals is *T. urvilleana* (= *T. grandiflora*, *T. semidecandra*) [GLORY BUSH, LASIANDRA, PLEROMA, PRINCESS FLOWER], which grows to a height of about 3m and produces large showy rose-purple flowers and attractive, ovate, prominently nerved leaves.
MELASTOMATACEAE, about 350 species.

Tigridia a small genus of bulbous perennial herbs found mainly in Mexico and Guatemala, with a few species from the Andes of Peru and Chile. The flowers have a concave center, three broad outer lobes and three narrow inner ones; color is often brilliant with striking spotted centers. They are borne on few-flowered stems and last only a day. *T. pavonia* [TIGER FLOWER, PEACOCK or MEXICAN TIGER FLOWER] is by far the most commonly cultivated species and includes many cultivars. It has broad leaves and its flowers are 15cm across and of exceptional brilliance. The starchy bulbs are used as a vegetable in parts of Central America.
IRIDACEAE, about 27 species.

Tilia [BASSWOOD, LIME, LINDEN, WHITEWOOD] a genus of ornamental deciduous trees bearing small cymes of white or yellow flowers, that produce abundant nectar. The genus is widespread in the north temperate region. *T. cordata* [SMALL-LEAVED, or EUROPEAN LIME or LINDEN] is a tree to 35m tall with small cordate leaves, slightly silvered beneath, and numerous small flowers spreading or erect from their bracts. It is abundant as a tree of city streets, particularly in eastern North America. *T. platyphyllos* [LARGE-LEAVED LIME or LINDEN] is used as understock for grafts of some rarer species. *T. × europaea* [EUROPEAN COMMON LIME or LINDEN] is a natural hybrid between the two foregoing limes; it is abundant in British streets and is the tallest tree in Britain, at 45m.

The Caucasian *T. × euchlora* [CRIMEAN

Wild Thyme (Thymus serpyllum), *a fragrant herb related to the Garden Thyme* (T. vulgaris). (× 1)

LIME or LINDEN] is now often planted instead of the COMMON LIME; it has handsome glossy leaves but its habit is less attractive. *T. americana* [BASSWOOD, AMERICAN LIME or LINDEN] makes a splendid, conical tree with bright green, very large leaves. However, it grows poorly in Europe and is better replaced by *T. petiolaris* [SILVER PENDENT LIME, WEEPING WHITE LINDEN]. *T. tomentosa* (= *T. argentea*) [SILVER, WHITE LIME or LINDEN] is a more sturdy, domed tree with hardy leaves, which thrives in cities.
TILIACEAE, about 30 species.

Tillandsia a large genus of evergreen perennials, mostly epiphytes, but varying greatly in habit, widespread in tropical and subtropical America. *T. usneoides* [SPANISH MOSS, FLORIDA MOSS, WOOD CRAPE, GRAY BEARD] has no roots, does not require soil, and absorbs water from the atmosphere through specialized leaf hairs. It is commonly cultivated for its attractive, long, gray, lichen-like festoons of branches, and is sometimes used as a packing and stuffing material.

Other cultivated species include *T. fasciculata* [WILD PINEAPPLE] as well as various hybrids. These have basal leaves and showy inflorescences with colored (usually pink, rose or red) bracts and variously-colored flowers.
BROMELIACEAE, about 400 species.

Timber see p. 328.

Tithonia a small genus of tall, annual or perennial herbs or shrubs native to Mexico and Central America, some of which are grown for their showy capitulate inflorescences, eg *T. rotundifolia* (= *T. speciosa*), with dahlia-like red or orange-yellow flower heads.
COMPOSITAE, about 10 species.

The fruits, which are rich in vitamin C, may be eaten raw or cooked. Relatively small quantities are marketed fresh, the great bulk being processed as canned whole fruit or juice, paste, ketchup or powder. World commercial production is almost entirely as a field crop in countries below latitude 45°. In higher latitudes, production is possible only under protection and for the fresh market. Europe and North America account for 75% of world production, European production being concentrated mainly in Italy, Spain, Portugal, Bulgaria, Romania and Greece. Glasshouse production is only 0.05% of the total and is most important in the Netherlands, the UK, Belgium and northern France. The name CURRANT TOMATO is given to *L. pimpinellifolium*, which is largely grown for ornament. *Cyphomandra betacea* is known as the TREE TOMATO.
SOLANACEAE.

Tmesipteris see Club mosses and their allies, p. 92.

Toadflax the common name applied to some members of *Linaria* such as *L. vulgaris* [COMMON TOADFLAX]. The name is also given to species of other genera, notably *Cymbalaria muralis* (= *Linaria cymbalaria*) [IVY-LEAVED TOADFLAX] and the root parasite *Thesium humifusum* [BASTARD TOADFLAX].

Tobacco a name applied to a number of species of the genus *Nicotiana* which are taken as stimulants in many forms: snuff, chew, liquid extract, pipe tobacco, cigars and cigarettes. Tobacco leaves were first used by the Indians of both Americas and by Australian aborigines.

Two species, *N. tabacum* and *N. rustica*, have achieved wide distribution, but only the

Flowers and young fruits of the cultivated Tomato (Lycopersicon esculentum). The bulk of the cultivated tomato crop is processed and relatively little is marketed fresh. (×2)

Tobacco (Nicotiana tabacum) is one of the few members of the family Solanaceae used for its leaves. Through its extensive use for smoking, chewing and snuffs it is one of the most popular and yet most harmful plants in the world. It remains a major crop in international commerce. (×1/20)

former is now a major crop in international commerce. The best production areas are subtropical and maritime; the Atlantic seaboard of the USA, the coasts of the eastern Mediterranean and the shores of the Black and Caspian Seas are major centers.

As with other crops, a great many forms have been developed which differ in their agronomic characteristics, eg flue-cured varieties (named after the flues used in curing-barns), and air-cured oriental tobaccos (both of which are used in cigarettes), air-cured burley (a mainstay of pipe tobaccos), and cigar-filler and cigar-binder types.

The tobaccos owe their attraction to the alkaloid nicotine, which is synthesized in the roots and translocated to the leaves. The leaves of *N. rustica* are used as a commercial source of nicotine for insecticides and citric acid.
SOLANACEAE.

Tolmiea a genus represented by a single species, *T. menziesii* [PICKABACK OR PIG-A-BACK PLANT, YOUTH-ON-AGE, THOUSAND MOTHERS], a hardy evergreen rhizomatous perennial herb, native to the western USA, with long-stalked cordate leaves and reddish flowers. It is cultivated as a pot plant and grown in gardens as ground cover.
SAXIFRAGACEAE, 1 species.

Tomato the common name for *Lycopersicon esculentum* (= *L. lycopersicum*) and its fruit, second only to the potato as a vegetable in world food production. The tomato is a weak-stemmed herbaceous perennial, although in commercial production it is treated as an annual and cannot withstand any frost.

Above *The large fruits of the European tomato cultivar 'Marmande', which are popular in continental Europe for salads.* Below *Tomatoes in commercial cultivation, under glass. The weak shoots require support from the roof.*

Timber

Wood or timber is one of the cheapest, most readily available and versatile of raw materials available to Man. It may be used directly or converted into a wide array of products with diverse properties and uses. Commercially, timbers are divided into hardwoods and softwoods. "Softwood" is the term applied to any coniferous wood such as *PINE, *LARCH, *SPRUCE and *YEW. The term "hardwood" is used for timbers which are the wood of broad-leaved, ie dicotyledonous tree species. The terms apply irrespective of whether the woods are hard or soft in the physical sense, some softwoods being harder than many hardwoods.

Most of the commercial timber used is softwood: it is easy to handle and is readily exploited, growing naturally over wide areas of the Northern Hemisphere in pure stands or mixed with a few other species, as in Europe where it constitutes 75% of the standing forest, North America and the USSR where it is 80% of the forest cover. Hardwood makes up most of the African and tropical American forests and 60% of the Asiatic forests. Hardwoods are much more diverse than softwoods and indeed there are many more commercially important hardwood species, not to mention those that have not yet been exploited. They are more specific in their properties and therefore not so interchangeable in commerce as are softwoods. Particularly in the tropics where the vast majority of hardwoods are found, the forests are made up of an array of different species with only a relatively small number of individuals of any one species per hectare. Only a few hundred of the thousands of known hardwood species enter world trade – there are, for example, 105 commercial wood species in Central and West Africa, 634 in Southeast Asia and 210 in tropical South America. Some of these are listed in the accompanying table together with a summary of their main uses.

Both softwood and hardwood timbers are marketed as *roundwood* for ties, posts, mining timber and fencing, *sawnwood* for building, lumber and packaging, *pulpwood* for paper and board manufacture, and *veneerwood* for veneers, plywoods and panels. It can also be converted into its chemical components, such as cellulose, and used as a raw material for synthetic materials, plastics and even sugar. Tropical hardwoods supply many of the luxury timbers used for furniture and veneers such as *MAHOGANY, TEAK, *ROSEWOOD and *AFRORMOSIA. The hardwood forests of the temperate zones are unable to meet the increasing consumer demand for quality timber and have either been substantially depleted or are under greater environmental protection in response to conservation pressures. This is leading to substantially increased demands on tropical hardwood forests, especially by developed countries. Consumption has increased rapidly during the last three decades. Imports of tropical hardwood by the developed countries have expanded 15-fold since 1950, while consumption of hardwood timber by the tropical regions themselves has little more than doubled in the same period. The largest consumers of tropical hardwoods are Japan and the United States, most of the supply coming from Southeast Asia, directly or indirectly. Western Europe accounts for about one-third of the world's imports of tropical hardwoods, but for historical reasons most of this derives from West and central Africa although in recent years the supply from Southeast Asia has increased. World consumption is expected to continue to show a substantial increase in the future and attention will increasingly be directed to the forests of South America which today produce only about 10% of the world's hardwood timber although it has stocks at least three times as great as Southeast Asia. Similar pressure will be faced by the forests of central Africa.

Another source of pressure on the forests of tropical America stems from the great increase in demand for paper products which are traditionally made from softwoods. Recent technological advances now permit wood chips from certain hardwoods to be pulped for paper and paperboard production, so that the present low contribution of tropical hardwoods to the industry can be expected to grow substantially.

The cultivation of timber trees for wood production has long been practised on a small scale, but large-scale plantations or afforestation programs date from the 16th and 17th centuries in Japan and Europe, much later in North America, Australia, South America and Africa. Natural management of forests is sometimes practised, but is expensive and not always reliable. Reafforestation often involves softwoods such as various fast-growing species of *Pinus*, *Picea* and *Pseudotsuga* and *Eucalyptus*. In the tropics, forest plantations are gradually being established, but they are often expensive to establish and difficult to maintain. Moreover, relatively few broad-leaved tropical hardwoods grow well in monoculture, a notable example being TEAK (*Tectona grandis*), a Southeast Asian species which has been successfully planted within its natural area and elsewhere in Asia and in Africa.

The rapidly increasing exploitation of tropical hardwoods, without adequate replacement through natural regeneration or reafforestation, will inevitably lead to the disappearance of certain luxury timbers. A good example is the West African tree *Pericopsis elata* (= *Afrormosia elata*), whose timber was introduced into international trade about 30 years ago as AFRORMOSIA. It is a highly valued hardwood as it is the most valuable, native wood of Ghana and more expensive than AFRICAN *MAHOGANY, yet because of the large-scale exploitation of natural stands and the trees' poor powers of regeneration, it is already in serious danger of becoming extinct. Plantations have only been made on a small scale and little is known of the biology of the tree. Unfortunately, this is not an isolated example of exploitation greatly exceeding replacement.

It is not always realized that nearly 50% of the wood cut worldwide is used for fuel, four-fifths of this in the developing world. This firewood is usually obtained, not from the moist tropical forests, but from dry or savanna woodlands, scrub and brushwood areas and local plots.

The timber of different species differs widely in its properties, and consequently its uses. Important factors are density, durability, grain and figure, hardness, luster, color, polish, moisture content, porosity, shrinkage, texture, toughness and resistance to disease. The special characteristics of some woods, especially hardwoods, are such that they have become virtually irreplaceable. MAHOGANY, for example, was not only used for furniture and carving, but was an irreplaceable material for the construction of parts of machinery before light metal alloys or plastics became available. Even today, some hardwoods cannot adequately be replaced by other materials in industry and design.

If a timber is to realize its full potential and to be utilized in an optimum fashion, it needs to find the highest value use for which there is a market and to be manufactured into the finished article with a minimum of waste and in such a way as to render the finished articles as durable as possible in order to make replacement necessary as infrequently as possible. Appearance becomes more important as the value of the end product increases and, for such things as carvings, veneers, paneling and high-grade furniture, it is paramount in importance. Appearance is not a single property but a combination of many; the difference between timber and its substitutes is primarily that no two pieces of wood are identical. Color is determined partly by the quantity of cell wall in relation to cell lumen but also by the colors of various substances contained in the cells, normally of the heartwood. Changes in color are provided by growth rings and the patterns produced can be varied by cutting the timber in different ways. OAK, for instance, is normally more attractive quarter-sawn and ELM flat-sawn. Strong differences in color patterns are often found in softwoods with a large contrast between early wood and late wood.

About a quarter of all industrial timber is used structurally in building where strength is an important property. Here the timber must be used in its natural form to make use of its peculiar property of being strong in resisting forces at right angles to the grain direction. It is unlikely that reconstituted wood will ever improve on solid timber, used structurally, because its alignment in the tree is as near ideal as it is possible to get. Shrinkage and movement are particularly important where timber components need to fit closely together as in such uses as furniture, joinery and flooring. Wood is also a good insulator, electrically, thermally and against sound.

Because of its wide array of properties and versatility, timber is almost certain to retain its leading position in world trade and on the domestic front for the foreseeable future.

IMPORTANT COMMERCIAL TIMBERS

Popular name	Scientific name	Uses
AFARA, LIMBA	*Terminalia superba	Plywood, furniture and interior joinery and general utility
AFRORMOSIA; ASAMELA; AFRICAN TEAK	Pericopsis elata (= Afrormosia elata)	Wide range of exterior and interior work, boats, furniture, joinery, floors and decorative veneer
*ASH, EUROPEAN	Fraxinus excelsior	Tool handles, sports goods, furniture, vehicles and for bent parts of boats
ASH, AMERICAN or WHITE	F. americana	
ASH, RED	F. pennsylvanica	
ASH, JAPANESE or MANCHURIAN	F. mandshurica	
AVODIRÉ	Turraeanthus africanus	High class furniture and joinery, decorative veneer
*BALSA	Ochroma pyramidale	Insulation, buoyancy, packaging, models
*BEECH, EUROPEAN	Fagus sylvatica	Furniture, tool handles, turnery, small woodware, floors. Often steamed and bent
BEECH, AMERICAN	F. grandifolia	
BEECH, JAPANESE	F. crenata	
BEECH, ORIENTAL	F. orientalls	
*BIRCH, EUROPEAN or SILVER	Betula pendula (= B. verrucosa)	Plywood, veneer, pulp, furniture, turnery, small woodware
BIRCH, WHITE	B. pubescens	
BIRCH, CHERRY or BLACK	B. lenta	
BIRCH, YELLOW or GRAY	B. alleghaniensis (= B. lutea)	
CEDAR-OF-LEBANON	*Cedrus libani	Furniture, joinery
CEDAR, ATLAS	C. atlantica	
DEODAR	C. deodara	
CEDAR, WEST INDIAN; CIGAR BOX CEDAR	*Cedrela odorata	High quality joinery, boat building and cigar boxes
CEDAR, CHINESE	C. sinensis	
CEDAR, RED, PENCIL or VIRGINIAN	Juniperus virginiana (see Juniper)	Pencils, linings for chests
CEDAR, E. AFRICAN PENCIL	J. procera	
CEDAR, WESTERN RED	*Thuja plicata	Roof shingles, weatherboards, greenhouses, general exterior work
CEDAR, WHITE; GIANT or AMERICAN ARBOR-VITAE	T. occidentalis	
*DOUGLAS FIR	Pseudotsuga menziesii	General joinery and building purposes
*EBONY, E. INDIAN	Diospyros ebenum	Decorative inlay, musical instruments, handles
EBONY, CEYLON	D. reticulata	
MARBLE WOOD	D. marmorata, D. kurzii, D. oocarpa D. mannii	
EBONY, BLACK CARVING		
EBONY, GABOON	D. dendo, D. evila	
*ELM, ENGLISH	Ulmus procera	Furniture, boats, docks and bridges, coffins
ELM, SLIPPERY	U. rubra	
ELM, ROCK or HICKORY	U. thomasii	
ELM, DUTCH or HOLLAND	U. × hollandica	
FIR, SILVER	*Abies alba	Structural work, general joinery, musical instruments
FIR, CASCADE	A. amabilis	
FIR, BALSAM	A. balsamea	
FIR, GIANT	A. grandis	
FIR, RED or CALIFORNIA RED	A. magnifica	
FIR, NOBLE	A. procera	
GABOON; OKOUME	Aucoumea klaineana	Plywood, blockboard, for doors, paneling
GREENHEART	*Ocotea rodiaei	Lock and dock gates, harbors, piers, floors
GUM, BLUE; BASTARD MAHOGANY	*Eucalyptus botryoides	General purposes, boxes, posts; widely planted worldwide
GUM, BLUE; SALIGNA	E. saligna	
*HEMLOCK, SPRUCE	Tsuga canadensis	General building and joinery
HEMLOCK, WESTERN	T. heterophylla	
HICKORY	*Carya ovata C. glabra, C. tomentosa	Handles, sports goods, and any purposes requiring strength and toughness
IDIGBO; FRAMIRÉ	*Terminalia ivorensis	Furniture, joinery, domestic floors
*IROKO; MVULE; AFRICAN TEAK	Chlorophora excelsa	Furniture, interior and exterior joinery, boat building, marine piling
*IRONBARK	Eucalyptus spp	Railway sleepers, bridges, heavy construction work
JARRAH	*Eucalyptus marginata	Marine piling and decking, floors
JELUTONG	Dyera costulata	Pattern making, carving, handicrafts, latex-gum
KAPUR	Dryobalanops aromatica D. lanceolata	Construction work, decking, exterior joinery
KARRI GUM	*Eucalyptus diversicolor	Heavy construction, beams, rafters
*KAURI	Agathis robusta A. australis	Joinery, plywood and formerly boats
KERUING; GURJUN	*Dipterocarpus turbinatus D. alatus	General construction, housing frames, floors, boats
*LARCH, EUROPEAN	Larix decidua	Exterior building work, poles, posts, boats
LARCH, EASTERN	L. laricina	
LARCH, WESTERN	L. occidentalis	
LAUREL, INDIAN	*Terminalia alata	Furniture and joinery, boats, house and boat building, posts
LIGNUM VITAE	Guaiacum officinale G. sanctum	Bushes and bearings, mallet heads, bowls, textile-gin components
LIME, EUROPEAN	Tilia × vulgaris (= T. × europaea)	Turnery, carving, small woodware items
LIME, AMERICAN BASSWOOD	T. americana	Turnery, carving, small woodware items
LIME, JAPANESE	T. japonica	
MAHOGANY, AFRICAN	*Khaya senegalensis	High quality furniture, joinery, veneer, boats
MAHOGANY, RED	K. ivorensis	
MAHOGANY, NYASALAND	K. nyasica K. grandiflora	
*MAHOGANY, TRUE, CUBAN or WEST INDIES	Swietenia mahagoni	Superior furniture, cabinet work, boatwork, valued for stability and finishing qualities
MAHOGANY, HONDURAS or MEXICAN	S. macrophylla	
MAHOGANY, VENEZUELAN	S. candollea	
MAPLE (various); SYCAMORE	*Acer spp A. rubrum, A. saccharinum, A. saccharum, A. pseudoplatanus, A. platanoides	Furniture, joinery, turnery, veneer; rock maple excellent for floors and rollers
MAPLE, QUEENSLAND	Flindersia brayleyana F. australis	Cabinet work, interior joinery, veneer
MERANTI; SERAYA; SAL; LAUAN; ALAN	*Shorea spp S. martiana S. pauliflora S. bracteola	Wide range of uses in construction, joinery, veneer, plywood, boats, floors, according to properties
MORA; MORABUKEA	Mora excelsa	Heavy construction work, marine piling, boats, sleepers
MUHIMBI	Cynometra alexandri	Heavy duty flooring, construction, mining
MUNINGA	*Pterocarpus angolensis	Furniture, high class joinery, veneer, floors
PADAUK, ANDAMAN	P. dalbergioides	
PADAUK, BURMA	P. macrocarpum	
PADAUK, W. AFRICAN	P. soyauxii	
AMBONYA	P. indicus	
*OAK, ENGLISH or PEDUNCULATE	Quercus robur	Furniture, high class joinery, veneers, gates, fences, construction, floors
OAK, DURMAST or SESSILE	Q. petraea	
OAK, RED	Q. rubra	
OAK, SPANISH RED	Q. falcata	
OAK, BAR	Q. macrocarpa	
OAK, MONGOLIAN	Q. mongolica	
OAK, SPANISH	Q. palustris	
OAK, BASKET	Q. prinus	
OBECHE; WAWA	Triplochiton scleroxylon	Interior joinery, plywood
PINE, WESTERN YELLOW	*Pinus ponderosa	General construction, interior joinery, plywood, pulp
PINE, WESTERN WHITE	P. monticola	
PINE, LOBLOLLY	P. taeda	
PINE, SCOTS	P. sylvestris	
PINE, WHITE	P. strobus	
PINE, LONGLEAF	P. palustris	
PINE, PARANA	*Araucaria angustifolia	General construction and joinery, plywood
PINE, MORETON BAY	A. cunninghamii	
MONKEY PUZZLE	A. araucana	
POPLAR, BALSAM	*Populus balsamifera	Matches, baskets, turnery, toys
POPLAR, BLACK	P. nigra	
POPLAR, WHITE	P. alba	
COTTONWOOD	P. deltoides	
RAMIN	Gonystylus bancanus	Furniture, interior joinery, toys, general purposes
RAULI; COIGUE; TASMANIAN MYRTLE	Nothofagus spp	Furniture, doors, interior joinery
ROSEWOOD, BRAZILIAN; JACARANDA	*Dalbergia nigra	High class furniture, veneer, musical instruments, handles
ROSEWOOD, E. INDIAN or MALABAR	D. latifolia	
ROSEWOOD, NICARAGUA	D. retusa	
ROSEWOOD, THAILAND	D. cochinchinensis	
ROSEWOOD, MALABAR	D. sissoides	
ROSEWOOD, HONDURAS	D. stevensonii	
*SAPELE; UTILE	Entandrophragma utile	Furniture, veneer, plywood, joinery
SAPELE, HEAVY	E. candollei	
GEDU NOHOR	E. angolense	
SATINWOOD, CEYLON or EAST INDIAN	Chloroxylon swietenia	Decorative veneer, cabinet work, furniture, interior joinery
SATINWOOD, WEST INDIAN	Zanthoxylum flavum	High class furniture, cabinet work, or joinery
SEQUOIA; REDWOOD	*Sequoia sempervirens	General purpose building and joinery
*SPRUCE, NORWAY	Picea abies	General purpose building and joinery, plywood, pulp
SPRUCE, WHITE	P. glauca	
SPRUCE, RED	P. rubens	
SPRUCE, SITKA	P. sitchensis	
SPRUCE, ENGELMANN	P. engelmannii	
TEAK	*Tectona grandis	Wide variety of interior and exterior furniture and joinery, ship building
WALLABA	Eperua falcata E. grandiflora	Wharf timber, decking, poles, posts, sleepers, general utility
*WALNUT, BLACK	Juglans nigra	High class furniture and veneer, gun stocks
WALNUT, JAPANESE	J. ailanthifolia	
WHITEWOOD, AMERICAN	*Liriodendron tulipifera	Lightweight utility, interior joinery, plywood

Tonka bean or **tongue bean** the seed of *Dipteryx odorata* (= *Coumarouna odorata*), a large tree of the South American tropics. It has been commercially important for several centuries as a pleasant-smelling spice which was used to flavor tobacco and cosmetics, and as a vanilla substitute, but recent concern over the nutritional safety of coumarin, its principal ingredient, has led to the demise of the trade.
LEGUMINOSAE.

Toon the common name for *Cedrela toona* (also known as RED CEDAR or CEDRELA TREE), a tall tree native to India, Malaysia and Australia which yields a valuable, red, light and durable timber used for house construction, furniture, tea chests and cigar boxes. The flowers also yield a yellow and red dye.
MELIACEAE.

Topee-tambu a common name for *Calathea allouia* (also known as SWEET CORN ROOT) a New World tropical herbaceous perennial which grows to a height of 60–120cm and produces root tubers with a yellow periderm and inner starchy flesh which is edible when cooked. It rarely flowers or seeds and is propagated vegetatively.
MARANTACEAE.

Torenia a genus of annual and perennial herbs native to tropical and subtropical Asia and Africa. Some are grown as annual bedding plants, notably *T. fournieri* [BLUE-WINGS, WISHBONE FLOWER]. This species has a much-branched stem and bears tubular, irregularly lobed flowers with a pale, violet-blue corolla (yellow on the back).
SCROPHULARIACEAE, about 40 species.

Torreya a small genus of evergreen trees from East Asia and the USA. Closely related to *Taxus* [*YEW], they are sometimes called STINKING YEWS because of the strong odor emitting from the stems and leaves when they are bruised.

Foliage of the Tree fern Dicksonia antarctica. ($\times \frac{1}{2}$)

The best-known ornamental members of the genus are *T. californica* [CALIFORNIA NUTMEG], *T. grandis* [CHINESE TORREYA], and *T. nucifera* [KAYA, JAPANESE TORREYA]. The roasted seeds of the latter have an aromatic flavor and are used in confectionery and as a source of oil; the wood is valued for general construction work and fencing.
TAXACEAE, 6–8 species.

Tortula a genus of mosses in which most of the species are distinguished by leaves with hair-points and erect cylindrical capsules. *T. muralis* is cosmopolitan and a most abundant and conspicuous moss on walls. *T. ruraliformis* is a notable sand-dune moss.
POTTIACEAE.

Townsendia a small genus of low, often thickly clustering herbs native to western North America and Mexico. The plants have small, alternate leaves and daisy-like flower heads with white or pink to blue or violet (rarely yellowish) rays. A few species, including *T. exscapa* (= *T. sericea*, *T. wilcoxiana*) [EASTER DAISY] and *T. grandiflora*, are occasionally cultivated in rock gardens.
COMPOSITAE, about 21 species.

Trachelospermum a genus of evergreen shrubby climbers mainly native to East Asia, but with one species in North America. In the cultivated species *T. asiaticum* (= *T. divaricatum*), *T. jasminoides* [STAR JASMINE] and *T. lucidum* the flowers are fragrant and white in color, salver-form or tubular in shape and arranged in loose clusters.
APOCYNACEAE, about 10 species.

Trachycarpus a genus of hardy fan palms native to the Himalayas and East Asia. The most widely planted species is *T. fortunei* [CHINESE WINDMILL PALM]. Growing to a height of 3–12m, it has a stout trunk sheathed by old black, fibrous leaf-bases and bearing a fan of leaves at the top of the trunk. *T. martianus* (= *T. khasianus*) reaches a height of about 15m but is of much slower growth, and the leaf sheaths adhere only to the top of the trunk. The fibers from the trunks of some species are used for cordage.
PALMAE, about 6 species.

Tradescantia a genus of North and South American perennial herbs with erect to creeping or pendulous fleshy stems. It includes several popular garden, greenhouse and house ornamentals.
T. virginiana [COMMON SPIDERWORT] and its close relatives, and hybrids between them, provide the common hardy garden tradescantias with long glossy lance-shaped leaves. Several species are popular houseplants; these include the white-flowered *T. fluminensis* and *T. albiflora* (both called WANDERING JEW), whose leaves are green or green-and-white striped, and the more robust, hairy *T. blossfeldiana* [FLOWERING INCH-PLANT], which has purple-tinged stems and leaves and pinkish flowers. *T. navicularis* [CHAIN PLANT] is a succulent, creeping spe-

The spiny, succulent cactus Trichocereus candicans, *which is cultivated for its large, highly-scented, white flowers.* ($\times \frac{1}{6}$)

cies with short, much-thickened leaves and large rose-pink flowers.
COMMELINACEAE, about 40 species.

Tragopogon a genus of biennial or perennial herbs from temperate Eurasia and North Africa. They have grass-like leaves and exceptionally large seed "parachutes". In most species the flowers close at midday, hence the common names of, for example, *T. pratensis* [JACK or JOHN-GO-TO-BED-AT-NOON]. The best-known species is *T. porrifolius* [*SALSIFY, VEGETABLE OYSTER], which is grown as a root vegetable. Other species are cultivated for ornament, some having rose-red florets (eg *T. ruber*), others yellow (eg *T. dubius*).
COMPOSITAE, about 50 species.

Trapa see WATER CHESTNUT.
TRAPACEAE, 3–4 species.

Tree ferns tropical, subtropical and temperate ferns in which the stem is analogous to the trunk of a tree. This condition occurs in several families and quite a few genera of ferns, especially in the families Cyatheaceae (eg *Alsophila*, *Cyathea*) and Dicksoniaceae (eg *Dicksonia*, *Cibotium*, *Thyrsopteris*). In many parts of the world, such as Southeast Asia, tropical America, Africa and Australia, tree ferns are an important constituent of many plant communities especially in mountainous regions. They are also especially common as outliers of montane grasslands. Because of the extremely fibrous coating of the trunk, tree ferns are capable of surviving fires that tend to destroy almost everything else. The growing point is protected by scales and growth can recommence after quite severe fire damage. The stems of some species, including *Cyathea caniculata*, of Madagascar, and *C. medullaris* [BLACK STEMMED TREE FERN] and *Alsophila australis*, both of Australia, contain an edible starch consumed

TRILLIUM 331

by the natives. The hairs, leaves and other parts of tree ferns, including the Malaysian *Dicksonia blumei* and the East Asian and Pacific *Cibotium barometz*, are used in native medicine, particularly to prevent bleeding. The scales from the stems of *C. chamissoi* (= *C. menziesii*) are used to stuff pillows in Hawaii, while in Brazil and Mexico, and other parts of tropical and subtropical America, the stems of various types of tree fern are cut down to provide fiber for planters' pots.

Trefoil the name given to several different species, mostly perennial herbs, with leaves that have three leaflets, including *Lotus corniculatus* [BIRD'S-FOOT TREFOIL], *Medicago lupulina* [YELLOW TREFOIL], *Trigonella caerulea* [SWEET TREFOIL] and *Trifolium campestre* [HOP TREFOIL]. Species of the unrelated genus *Ptelea* (Rutaceae) are commonly known as SHRUBBY TREFOIL.

Trichocereus a genus of cacti native to subtropical and temperate South America. The plants have columnar, ribbed stems and bear large, fragrant, usually white, nocturnal flowers. Cultivated species include the stout-stemmed *T. pasacana* (= *Cereus pasacana*), which also bears edible fruits, and *T. candicans* (= *Cereus candicans*), with particularly large and highly scented flowers. CACTACEAE, about 30 species.

Trichoderma a cosmopolitan genus of the imperfect fungi comprising soil- or bark-inhabiting fungi with a septate, colorless mycelium and branched conidiophores. Some *Trichoderma* species are the conidial states of the ascomycete genus *Hypocrea*. Some, such as *T. viride* produce volatile and non-volatile antifungal antibiotics. HYPHOMYCETACEAE, 2–4 species.

Trichomanes a genus of filmy ferns, largely tropical and subtropical in distribution and growing in extremely moist conditions (in almost continual rain). Many are epiphytes on damp jungle vegetation. The fronds are

The erect, purple-spotted flowers of the Toad Lily, Tricyrtis formosana. (× ½)

Red Clover (Trifolium pratense) growing in a meadow, amongst Yellow Rattle (Rhinanthus minor) and other meadow flowers.

simple and delicately membranaceous in texture, with the sori always at the end of a vein protected by tubular indusia. A number of tropical species, such as the West Indian *T. alatum*, *T. scandens* and *T. capillaceum*, are cultivated in warm, moist greenhouses in temperate countries. HYMENOPHYLLACEAE, about 300 species.

Trichoscyphella a genus of cup fungi commonly found on conifers. Most species are harmless saprophytes but *T. willkommii* causes a serious perennial canker of *Larix decidua* [LARCH]. Infection is through wounds and may be associated with frost damage. When it occurs on small branches, it leads to die-back. Infections on the main stem develop into cankers and eventually kill the tree. HYALOSCYPHACEAE, about 12 species.

Trichosporum see *Aeschynanthus*.

Tricyrtis [TOAD LILY] a small genus of herbaceous perennials with short creeping rhizomes, from the Himalayas and East Asia. The flowers are few, but rather large, often dull and purple-spotted, giving rise to their common name. Several species, including *T. hirta* [JAPANESE TOAD LILY], are cultivated in warm sheltered spots in temperate gardens. LILIACEAE 10–15 species.

Trientalis a genus of small perennial herbs growing in the northern temperate regions of Europe, Asia and North America. *T. europaea* [CHICKWEED WINTERGREEN] produces upright stems 10–25cm high, with four to seven lanceolate leaves from almost the same point, and a terminal white flower about 2cm across with five to nine widespread petals. *T. borealis* (= *T. americana*) [STAR-FLOWER] is very similar. PRIMULACEAE, 2–4 species.

Trifolium see CLOVER. LEGUMINOSAE, about 300 species.

Triglochin a genus of rhizomatous herbs growing in marshes and damp grassland of temperate regions of both hemispheres. There are two species with edible parts: the narrow elongated leaves of *T. maritima* [SHORE PODGRASS], of the north temperate zone, provide an emergency food and the rhizome of *T. procerum* is eaten by Australian aborigines. JUNCAGINACEAE, about 15 species.

Trigonella a genus of annual herbs widespread throughout the Old World. A characteristic feature of these plants is their strong but pleasant scent of coumarin. *T. caerulea* [SWEET TREFOIL] is cultivated in Europe for hay, and in some parts of Switzerland it is used to flavor cheese, soups and potatoes. A decoction of the leaves provides a substitute for tea.

The best-known member of the genus is *T. foenum-graecum* [FENUGREEK], whose powdered seeds are an important curry spice. It is grown in the Mediterranean region and southern Asia as a fodder crop and its seeds, besides being used for flavoring, are employed in veterinary medicine and as a dye. LEGUMINOSAE, about 80 species.

Trillium a small genus of rhizomatous, spring-flowering, perennial herbs found in North America and from the western Himalayas into Japan and the Kamchatka Peninsula. The most commonly cultivated species are all North American: *T. grandiflorum* [WHITE WAKE-ROBIN, SNOW TRILLIUM], up to 45cm tall, with pink or white flowers, *T. sessile* [TOADSHADE, WAKE-ROBIN], with stalkless purple flowers and variegated leaves; and *T. undulatum* [PAINTED TOADSHADE] with white, pink-striped flowers, and undulate leaf margins. The rhizome and

The White Wake-Robin (Trillium grandiflorum) is a widely cultivated ornamental growing to about 45cm in height. (× ½)

roots of *T. grandiflorum* and *T. erectum* (= *T. flavum*) [BIRTHROOT, LAMB'S QUARTERS, PURPLE TRILLIUM] are used in native medicine.
LILIACEAE, about 30 species.

Triteleia a small genus of cormous perennial herbs from western North America with one or two narrow linear leaves. The tubular or funnel-shaped flowers are borne in umbels on leafless stems and resemble those of the closely related genera, *Brodiaea* and *Ipheion*. Some of the species are cultivated, including *T. hyacintha* (= *Brodiaea hyacinthina*) [WILD HYACINTH], *T. ixioides* (= *B. ixioides*) [PRETTY-FACE, GOLDEN BRODIAEA, GOLDEN STAR] and *T. laxa* (= *B. laxa*) [GRASS NUT, TRIPLET LILY].
LILIACEAE, about 14 species.

Triticale a new small-grain annual cereal (*Triticosecale*) obtained by crossing *Triticum* [WHEAT] and *Secale* [*RYE]. The synthesis of TRITICALE represents the first success by plant breeders in their efforts to create a new grain crop by combining species from two distinct genera. It combines, in part at least, the high yield potential of wheat with the ruggedness of rye, and the higher protein content of wheat with the better quality protein of rye. It also appears that some of the Triticales have inherited the winter-hardiness and disease resistance of rye. The flour is not as good for making bread as that of wheat but some improvement has been made by selection. It is used for grain or forage.
GRAMINEAE.

Triticum [WHEATS] a small genus of European, Mediterranean and southwest Asian annual grasses. The species, and particularly the cultivars, fall into three major groups.

Wheat (Triticum aestivum var aestivum) is self-pollinating and fertilization often occurs before the flowers have opened. (× ½)

A Wheat field in northern Tunisia showing clear signs of soil erosion which will result in drastic loss of yield.

The first group includes EINKORN WHEAT, *T. monococcum* var *monococcum*, a diminutive species with single-grained spikelets apparently derived from the rare, wild, southwest Asian *T. monococcum* var *boeoticum*.

The second group includes the EMMER WHEATS. These include the southwest Asian *T. dicoccoides* (= *T. turgidum* var *dicoccoides*) [WILD EMMER] and the cultivated species *T. dicoccum* (= *T. turgidum* var *dicoccum*), the wheat of ancient Egyptians and early Christians, *T. orientalis* [PERSIAN KHORASAN WHEAT] and *T. polonicum* [POLISH WHEAT]. The most important member of this group is *T. durum* (= *T. turgidum* var *durum*) [DURUM WHEAT], which is widely grown in drier regions and produces large, hard grains with a low gluten content.

The third group includes the SPELT or BREAD WHEATS, one of which is represented in the wild. All the thousands of cultivars of this group have almost certainly been derived by hybridization of an EMMER WHEAT with the closely related *Aegilops squarrosa*. *T. aestivum* var *aestivum* [COMMON BREAD WHEAT] is the main source of flour for the world's bread. Close relatives include *T. compactum* [CLUB WHEAT] and the ancient Germanic *T. spelta* [SPELT or DINKEL WHEAT].

The world's major wheat-growing areas are the USA, Canada, southern USSR, the Mediterranean, north and central China, India, Argentina and southwest Australia. Wheat provides the staple food of about one-third of the world's population, being rich in starch but also containing between 8–15% of its weight in the form of protein, and up to 2% of its weight in fats. Its high nutritive value is increased by its content of essential amino acids, mineral salts and vitamins. World production in 1972 was around 350 million tonnes.

Wheat grain is mainly ground into flour used for bread-making, its high gluten content making it especially suitable for this purpose. Wheat is also used to make many familiar foods such as biscuits, pies, cakes, spaghetti, pasta and noodles. Unfortunately, the white bread demanded by Western society removes both the embryo, rich in proteins, oils and vitamins, and the seed coat, rich in proteins and minerals, leaving only the starchy endosperm.

Classification of bread wheats is based on the mode of cultivation and on the milling and baking quality. WINTER WHEATS are sown in the autumn for harvesting the following late summer in areas where the winter climate is not severe. Most of these need vernalization before they flower. SPRING WHEATS are sown in early spring for harvesting in the autumn. Both groups are further subdivided into either HARD WHEATS, which are flinty in texture yielding a high protein (gluten) flour ideal for making bread, or SOFT WHEATS which have a low-gluten, high-starch grain more suitable for biscuits, pastry and breakfast cereals than for making bread. The whole grain is also used for the manufacture of breakfast cereals, alcoholic beverages and animal feed. The straw has its uses in animal bedding, packaging of fragile material and to some extent in animal feed. Traditionally it is used for thatching.
GRAMINEAE, about 20 species.

Tritonia a genus of cormous perennial herbs from tropical and South Africa, mostly with linear leaves, closely related to *Ixia* and *Freesia*. *T. crocata* (= *T. aurantiaca*, Mon-

bretia crocata) grows to about 30cm high and has flowers in simple two-ranked spikes with the flowers turned upwards, as in the freesias. The color is bright orange-scarlet but several variants are known. They are often grown in European gardens and in the north are given greenhouse protection.
IRIDACEAE, about 55 species.

Trollius a small genus of north temperate and arctic perennial herbs with large, showy globose flowers. The showy perianth parts (yellow in most species) are petaloid and completely surround the 5–10 linear leaves or "petals". Several species are cultivated, including *T. europaeus* [GLOBE FLOWER] and *T. asiaticus*, the latter bearing orange flowers and bronze-green leaves.
RANUNCULACEAE, about 25 species.

Tropaeolum a genus more commonly known as NASTURTIUMS by gardeners. It consists of annual or perennial herbs native to Central and South America. The quite different genus *Nasturtium, often included in *Rorippa, is native to north temperate regions. Some species of *Tropaeolum* are climbers, supporting themselves by twining petioles. The irregular flowers can be yellow, orange, red, blue or purple and they are often extremely showy. The petals are often dissected or fringed, the upper two smaller.

Among the main species cultivated are *T. majus* [GARDEN NASTURTIUM, TALL NASTURTIUM, INDIAN CRESS], a tender annual climbing species, with several cultivars. Variety *nanum* is dwarf, but has the same large brilliant yellow or orange flowers. The leaves can be used in salads and the pickled fruits are a substitute for *capers. *T. minus* [DWARF NASTURTIUM] is a much smaller tender annual, non-climbing species. *T. peltophorum* (= *T. lobbianum*) is another tender annual climber with medium-sized red flowers.

Ears of Wheat (Triticum aestivum *var* aestivum) *in flower.* (× 3)

Tropaeolum tricolorum is a perennial climber from Bolivia and Chile that is becoming popular as a greenhouse plant. (× 2)

T. peregrinum (= *T. canariense*) [CANARY BIRD FLOWER, CANARY CREEPER] is a tender annual climbing species with long-stalked yellow flowers. The perennial *T. speciosum* [FLAME NASTURTIUM or CREEPER] will scramble up through shrubs to give a blaze of color with its showy vermillion-red flowers. *T. tuberosum* [ISANU, ANU, ANYU] is grown in the Andean regions of Bolivia and Colombia for its edible tubers.
TROPAEOLACEAE, about 60 species.

Truffles cup fungi with subterranean, mostly roundish, closed fruiting bodies (perithecia). The fruiting bodies are scented and are eaten by rodents which, by breaking the ascocarp, disperse the spores. Truffles are associated with the roots of certain trees, such as oaks, with which they may form mycorrhizal associations. Truffles are a much-prized delicacy, particularly in France and Italy. The truffles of commerce are *Tuber aestivum* [SUMMER TRUFFLE] and *T. magnatum* [WHITE PIEDMONT TRUFFLE]. False truffles are species of *Elaphomyces.*
TUBERACEAE.

Tsuga see HEMLOCK SPRUCE.
PINACEAE, about 10 species.

Tulip the common name for members of *Tulipa*, a genus of bulbous plants which have a center of distribution in western Asia, particularly the southern USSR and neighboring Turkey, Iran and Afghanistan. The distribution, however, stretches eastwards as far as Kashmir and westwards into southern Europe.

The bulb is generally ovoid with a pointed apex and is covered by one or several layers of brownish, papery or leathery tunic. The bulbs themselves may propagate by simple division, or in some instances they produce long underground stolons which in turn develop a new bulb at the tip – whole colonies

may be formed in this way. Young bulbs produce a single large leaf but mature plants have two or more leaves which decrease in size up the stem, often markedly so.

The leaves are generally fleshy and oblong or elliptical to linear, flattish or channelled. The flowers are erect, often large and showy, and may be solitary or several together, though seldom more than five to a stem. The perianth consists of six separate petaloid parts (tepals), the outer three often narrower and more pointed than the inner, but all six opening widely in strong sunshine. The predominant colors are red or yellow, although pink, purple and white also occur. The inside of each tepal often has a colored zone, sometimes in the form of a dark blotch, at the base.

Tulipa is divided by botanists into two main sections. The *Eriostemones* have flowers that are distinctly waisted below the middle, and filaments with a hairy boss at the base; most of the small-flowered species belong here. The larger section, the *Leiostemones*, on the other hand, have non-waisted flowers and hairless filaments, and contain most of the spectacular large red-flowered species.

Horticulturalists recognize up to 11 groups of cultivated tulips as follows: 1 Single Early, 2 Double Early, 3 Mendel, 4 Triumph, 5 Darwin Hybrid, 6 Darwin, 7 Lily-flowered, 8 Cottage, 9 Rembrandt, 10 Parrot, 11 Double Late. Groups 1–2 are early-flowering, 3–5 midseason and 6–11 late-flowering. The wild species and their cultivars are grouped as Kaufmanniana, Fosterana, Greigii and other species.

The tulip has been extensively cultivated, especially in Western Europe, where it was introduced about the early 16th century. The genus exhibits the taxonomic difficulties, as in *Crocus and *Narcissus, associated with long-established cultivation, hybridization and selection. A number of species, particularly *T. sylvestris* and *T. gesnerana*, are widely naturalized.

T. gesnerana is the parent species from which many of the modern garden tulips have arisen, and the staggering range of colors and forms currently available bears witness to the genetic complexities of this taxon. The introduction of this species (or

Tropaeolum majus is popularly known as the Garden Nasturtium. (×$\frac{1}{6}$)

rather various forms of it) into Holland in the early 17th century led to tulipomania; vast sums of money were paid for single bulbs by dealers searching for new colors and forms, and the Dutch government eventually had to intervene to stop speculation in tulips.

Tulips are characteristic of rocky ground, often in the mountains, or of short grassy areas or occasionally stabilized screes. They are typical of habitats where there are winter or early spring rains or snow, but a long hot dry summer. Certain species, such as *T. boeotica* and *T. micheliana*, are frequently associated with cultivated land, particularly wheat fields, the bulbs growing sufficiently deep in the soil to survive normal ploughing. LILIACEAE, 50–150 species.

Tulip tree the common name for the genus *Liriodendron* which consists of the deciduous trees *L. tulipifera* [TULIP TREE, WHITEWOOD], from North America, and *L. chinense* [CHINESE TULIP TREE], from central China. The TULIP TREE is a tall, handsome tree up to 60m tall, whose fragrant white or greenish-yellow flowers are somewhat tulip-like. It is prized as an ornamental, particularly for parkland.

It also has important timber value, producing a soft, fine-grained, light yellow wood (Canary whitewood) much used for carpentry, furniture and boat-building. It does not split easily and is readily worked. The inner bark is said to have medicinal properties. It is also known as yellow poplar, tulip wood and whitewood.

The name TULIP TREE is sometimes given to *Magnolia × soulangiana*.
MAGNOLIACEAE, 2 species.

Fruiting spears of the Cattail Bulrush (Typha latifolia), with the remains of the male flowers at the tip of each spike. ($\times \frac{1}{5}$)

Part of a flower of a Tulip (Tulipa sp) opened to show the reproductive structures; the white gynoecium is crowned by a lobed stigma (female) and this is surrounded by the stamens with white filaments and chocolate-colored anthers (male) covered in pollen. ($\times 5$)

Tulip wood a name given to several unrelated species of trees, eg *Liriodendron tulipifera* [*TULIP TREE] (Magnoliaceae), *Harpullia arborea* (Sapindaceae) and *Dalbergia cearensis, D. decipularis* and *D. frutescens* (Leguminosae). Most of these yield valuable timber used for construction, cabinetwork and general carpentry.

Tung oil a valuable drying oil (also known as China wood oil) derived from the seeds of several *Aleurites* species, notably *A. montana* [MU-OIL, MU TREE, TUNG], *A. fordii* [CHINA WOOD-OIL TREE] and *A. cordata* [JAPAN WOOD-OIL TREE] and *A. moluccana*. *A. fordii* is grown in the southeast USA as well as in Asia. The oil has many industrial uses such as varnish, paint and linoleum manufacture as well as waterproofing fabrics.
EUPHORBIACEAE.

Turmeric a constituent of curry powder derived from the peel and finely grated underground stems (rhizomes) of the species *Curcuma domestica*. Turmeric not only flavors curry powder and other foods, but also helps color it, being a dye and cosmetic as well as spice plant.
ZINGIBERACEAE.

Turnera a genus of herbs and low shrubs from tropical America and Africa. The Mexican species *T. diffusa* (= *T. aphrodisiaca*) [DAMIANA] is exported in the form of dried leaves to the USA for use as a laxative and is used locally as a substitute for tea, to flavor wines and as an aphrodisiac. *T. ulmifolia* [WEST INDIAN HOLLY, YELLOW ALDER] from

tropical America is grown as an ornamental shrub with showy yellow flowers.
TURNERACEAE, about 60 species.

Turnip the swollen root, hypocotyl and stem of the biennial form of *Brassica campestris* (= *B. rapa*) subspecies *rapifera*. It is similar to the *swede, but the leaves arise as a rosette from the top of the root, and not on a short stem as in the swede.

The swollen "root" is used as a vegetable, boiled and mashed. The leaves can be used for salads and pickling or can be boiled and eaten as a green vegetable similar to *spinach. The plants are also used for stock feed in Europe and New Zealand.

There are three types of turnip. The white-fleshed succulent turnips are fast-growing and produce roots with a dry matter content of about 8%. The soft yellow turnips are intermediate between the whites and the hard yellows, which have the highest dry matter content and are the best keepers.

Stubble turnips are a leafy type of white turnip developed for sowing after cereals are harvested. They produce animal feed for use in autumn and early winter.
CRUCIFERAE.

Turpentine a colorless, inflammable oil (also known as oil of turpentine or turpentine oleoresin). Crude turpentine is obtained from coniferous trees such as *Larix decidua* [*LARCH], *Pinus pinaster* [MARITIME PINE] and *Abies balsamea* [BALSAM FIR]. Turpentine can also be produced from tar, the dark, viscous product of destructive distillation of wood.

Turpentine is used as a solvent in the manufacture of paints and varnishes although it is less used nowadays since development of acrylic and other plastic paints. Spirit of turpentine is a refined form of turpentine used in the manufacture of wax polishes as well as in the paint and chemical industries. Oil of terebinth is a very pure form of turpentine used in medicines and veterinary medicines as a constituent of liniments for sprains, fibrositis and for the external

Tulip Breaking Virus, a virus disease welcomed by gardeners since it causes segregation of the color pigments in tulip petals.

U

Tulipa tarda, an attractive species from eastern Turkestan. ($\times \frac{1}{4}$)

treatment of pleurisy and bronchitis. Veterinarians administer it as an internal vermifuge.

Tussilago a genus consisting of a single species, *T. farfara* [COLTSFOOT], an Old World perennial herb. The flowering stems are a few centimetres high and are covered in woolly bracts. They bear a single, yellow, erect flower head at the apex. The leaves are used as a substitute for *tobacco and as such have anti-asthmatic properties. The flowers as well as the leaves are used in a variety of herbal medicines, particularly as expectorants.
COMPOSITAE, 1 species.

Typha a widespread genus of perennial grass-like marsh herbs, the most common being *T. latifolia* [COMMON CATTAIL, GREAT REED-MACE, CATTAIL BULRUSH], and *T. angustifolia* [SMALL REED-MACE, NARROW LEAF CATTAIL]. The leaves of several species are used for stuffing mattresses and pillows and for basketwork and woven chair seats; those of *T. elephantina* [ELEPHANT GRASS] are made into ropes and baskets in India.
TYPHACEAE, about 15 species.

Tulipa micheliana is the commonest tulip in northeast Iran, often growing in Wheat fields. ($\times \frac{1}{4}$)

Ucuhuba the common name for the tropical South American tree, *Virola surinamensis*. The seeds are the source of a solid fat (ucuhuba butter), resembling cacao butter, used in the manufacture of soap and candles. A similar product is obtained from the seeds of *V. guatemalensis* in Guatemala and other species.
MYRISTICACEAE.

These Pennyworts or Navelworts (Umbilicus rupestris) *have colonized the crevices of a dry stone wall, their typical habitat.* ($\times \frac{1}{5}$)

Ugli a fruit similar to the *tangelo (*Citrus × tangelo) in origin in that it is the result of a cross between *TANGERINE and *GRAPEFRUIT. It differs from the tangelo in being larger and coarser skinned.
RUTACEAE.

Ulex a small genus of very spiny, dense-growing evergreen shrubs from Western Europe and North Africa. *U. europaeus* [COMMON GORSE, FURZE or WHIN] forms impenetrable thickets to 2m or more high in sheltered sites and bears masses of bright-yellow pea-like flowers. It is widely cultivated for ornament and there are several cultivars.
The compact, spreading *U. gallii*, which produces golden-yellow flowers in autumn, and *U. minor* (= *U. nanus*) [DWARF FURZE], a more prostrate weaker-growing species with small, bright yellow autumn flowers, are also grown. GORSE has been used as a fodder crop and as a source of lectins.
LEGUMINOSAE, about 26 species.

Ulloco or **ullucu** the common name for

Ullucus tuberosus, a perennial herb native to the high Andes. It is cultivated there for its small underground tubers which are cooked and eaten by the natives as a substitute for *potatoes.
BASELLACEAE.

Ulmus see ELMS.
ULMACEAE, about 18 species.

Ulothrix a genus of unbranched filamentous green algae found in fresh water and on damp soil.
CHLOROPHYCEAE.

Ulva see SEA LETTUCE.
CHLOROPHYCEAE, about 30 species.

Umbilicus a small genus of succulent perennial herbs native to southern Europe, North and Central Africa and western Asia. They are plants of little ornamental value, with very soft and fleshy deciduous leaves, and tall slender leafy spikes of inconspicuous, bell-shaped, greenish-white flowers. The best-known species is *U. rupestris* (= *U. pendulinus*, *Cotyledon umbilicus*) [NAVELWORT, PENNYWORT].
CRASSULACEAE, about 16 species.

Umzimbeet the common name for *Millettia grandis*, a South African tree which yields a hard durable timber used for implements. The fruits are used in native medicine. *M. sutherlandii* is the GIANT UMZIMBEET.
LEGUMINOSAE.

Upas tree the common name for *Antiaris toxicaria*, a tall tree native to tropical Africa

Common Gorse (Ulex europaeus) *growing among sand dunes.*

In damp, unpolluted areas lichens may cover whole branches, as shown by this Usnea *in a wood in northeastern England.* (× 1)

and Asia. It yields a toxic latex, containing cardiac glycosides, which is used as an arrow poison, and a strong fiber from the inner bark.
MORACEAE.

Urceolina a small genus of bulbous herbs native to the Andes of South America. The flowers are numerous, in an umbel on a leafless stem, and are tubular-cylindrical at the base and abruptly widened at the middle, with short spreading lobes. *U. peruviana* (= *U. miniata*), with bright scarlet flowers, and *U. latifolia*, with yellow-red flowers, are two of the species sometimes cultivated as ornamentals.
AMARYLLIDACEAE, 4–5 species.

Urginea a genus of bulbous herbs with basal leaves, native to Europe, the Mediterranean region, Africa and India. *U. maritima* (= *Scilla maritima*) [SEA ONION, RED SQUILL, or SQUILL] has bulbs which yield a diuretic, emetic, expectorant and heart stimulant. A rat poison is also produced from the bulbs of this species. This and a few other species, including *U. physodes* and *U. undulata*, are grown as garden ornamentals for their racemes of small, whitish, yellowish or pinkish flowers.
LILIACEAE, about 40 species.

Ursinia a South African genus of annual or perennial herbs which have dandelion-like leaves and daisy-like flower heads. Several species are grown as half-hardy annual bedding plants, including *U. versicolor*, which has orange ray florets and dark disk florets, the whole head being up to 5cm in diameter, and *U. anthemoides* (= *U. pulchra*), with yellow flower heads up to 6cm across, with purplish ray florets. *U. anethoides* is a more shrubby plant with bright orange-yellow flower heads up to 2.5cm in diameter.
COMPOSITAE, about 40 species.

Urtica a genus of annual or perennial herbs found mostly in north temperate regions, with a few representatives in tropical and south temperate areas. The whole plant is usually covered with stinging hairs, as in the case of the Asian and European *U. dioica* [COMMON STINGING NETTLE] and the more virulent *U. urentissima* [DEVIL'S LEAF] of Timor.

Although not cultivated and, in the case of *U. dioica*, a widespread and widely naturalized weed, several species have their uses. The commonest use is for fishing nets and cordage, as the stems contain strong fibers; *U. breweri* from North America, *U. thunbergiana* from Japan, and *U. dioica* are all used for these purposes. The latter species has many other uses: medicinal extracts are made from leaves and roots and a hair-wash from the seeds. The plant is also a commercial source of chlorophyll. Young shoots of this and several other species can be cooked as a vegetable. The annual *U. pilulifera* [ROMAN NETTLE], native to southern Europe, is sometimes grown as an ornamental under glass.
URTICACEAE, about 50 species.

Usnea a large genus of filamentous lichens found growing on trees, rocks and occasionally bare earth. Collectively known as "OLD MANS BEARD" or "BEARD LICHENS", they are very sensitive to air pollution, so their abundant presence is an indicator of pure air. Specimens of *U. longissima* on trees may reach 12m.
USNEACEAE.

Ustilago the principal genus of the smut fungi affecting a range of flowering plants, mainly grasses. Important pathogenic species include *U. maydis* on maize and *U. nuda* on wheat and barley.
USTILAGINACEAE, about 300 species.

Utricularia [BLADDERWORTS] a genus of carnivorous plants which produce bladders or utricles to trap and digest tiny animals.

These ears of Barley infected with an Ustilago *species look as if dusted with soot, hence their name "smut fungi".* (× ½)

Red Squill, Squill or Sea Onion (Urginea maritima) *growing in northwestern Tunisia. This species inhabits stony areas that are grazed by sheep and goats, surviving after other plants have been consumed.* (× 1/10)

Utricularia has an almost worldwide distribution, with most of the species found in the tropics. Many species are submerged aquatics; others grow in damp mud or are epiphytes on mossy rain-forest trees. Some even live in the water which collects in the leaf axils of *Tillandsia* and other members of the Bromeliaceae.

Although the traps vary in shape from one species to another they are basically similar in construction. Each is a hollow bag borne at the end of a stalk, with a small entrance near to or opposite the stalk. Around the entrance are usually some projecting bristles, so arranged that an insect or crustacean passing the bladder will tend to be guided towards its mouth. After the creature enters, the trapdoor closes, and it is then digested.

A number of species, including *U. minor* and *U. vulgaris*, are grown in aquaria, and others, such as *U. endressii*, in hanging baskets.
LENTIBULARIACEAE, about 120 species.

Uvularia a genus of small herbaceous rhizomatous perennials native to the woodlands of Atlantic North America. The flowers are terminal, bell-shaped and pendent, usually pale yellow. Three commonly grown ornamental species are *U. grandiflora*, *U. sessilifolia* [WILD OATS] and *U. perfoliata* [STRAWBELL]. They are normally grown in wild gardens and prefer a shady location that has rich soil.
LILIACEAE, 4–5 species.

V

Vaccinium a large genus of deciduous or evergreen shrubs of arctic and north temperate regions and of high altitudes in tropical regions. They are plants of woodland, mountains, acid moorlands and bogs. The leaves are simple, and the flowers white, greenish, red or purple, either solitary or in racemes. *Vaccinium* includes several species with edible fruits (see BILBERRY, BLUEBERRY, CRANBERRY and HUCKLEBERRY).
ERICACEAE, about 150 species.

Valeriana a large genus of perennial herbs, subshrubs and shrubs from Eurasia, South Africa, North America and the Andes.

Lamb's Lettuce (Valerianella locusta), a widespread weed that was once grown as a salad plant for its young leaves. (×¼)

Several species, including *V. capensis* and *V. celtica*, are used locally to treat nervous complaints. The dried roots of *V. officinalis* [COMMON or CAT'S VALERIAN, GARDEN HELIOTROPE] contain bornyl valerate and other chemicals, which act as antispasmodics. Some species, including the dwarf, matforming *V. arizonica* with heads of pink, funnel-shaped flowers, are cultivated in rock gardens. Taller species, such as *V. sitchensis* with fragrant, white funnel-shaped flowers, are grown as border plants.
VALERIANACEAE, about 200 species.

Valerianella a genus of small annual herbs most of which are native to North America, Europe, North Africa, Asia and especially to the Mediterranean region. They produce clusters of white to pale-blue or pink flowers. *V. locusta* (= *V. olitoria*) [LAMB'S LETTUCE, CORN SALAD] is cultivated for its leaves which are used in salads. The southern European species *V. eriocarpa* [ITALIAN CORN SALAD] is very similar.
VALERIANACEAE, about 50 species.

Vallisneria a small but intriguing genus of submerged waterplants, being found throughout the warmer waters of the world. Rooted in the mud at the bottom of the water, the plants have a tuft of ribbon-shaped leaves, as well as stolons which grow out to form new plants. *V. americana* [WILD CELERY, WATER CELERY], from New Brunswick to North Dakota and south to Florida, *V. gigantea* from south and East Asia, and from the Philippines to Australia and Tasmania, and the southern European and west Asian *V. spiralis* [EEL GRASS, TAPE GRASS] are grown as oxygenating plants in aquaria or fishponds.
HYDROCHARITACEAE, 8–10 species.

Vallota see SCARBOROUGH LILY.
AMARYLLIDACEAE, 1 species.

Valonia a small genus of coenocytic green algae found in tropical and subtropical seas. The thallus consists of a bladder, anchored to the substrate by rhizoidal outgrowths. *Valonia* is often used for experimental work on salt uptake and water balance in cells.
CHLOROPHYCEAE.

Vanda a genus of evergreen epiphytic orchids found throughout tropical Asia and Malaysia. The strap-shaped or cylindrical leaves are usually arranged in two opposite rows. The erect or pendulous flower stems, alternating with the aerial roots, arise from the leaf axils and bear large fleshy flowers. They are widely grown, especially in the

The climbing stems and cylindrical green leaves of the epiphytic orchid Vanda cooperi. (×1/10)

The European Bilberry or Whortleberry (Vaccinium myrtillus) is common on mountains, moorlands and in shady woods. (×1)

tropics. Vandas have been hybridized among themselves and with species of other genera to produce an enormous range of flower types.
ORCHIDACEAE, about 60 species.

Vanilla a genus of tropical American but widely cultivated orchids with lily-like, green, white or pink flowers. Many species are leafless but the majority have very thick fleshy spear-shaped or oblong leaves. The spice vanillin is obtained from the dried, fermented and processed seed pods of *V. planifolia* (= *V. fragrans*) [VANILLA], from tropical America, which is grown in several tropical countries on a commercial scale, particularly in Mexico, Indonesia, Seychelles, the French Pacific Islands and Madagascar. The latter is the main exporting country, producing 60–80% of the world's vanilla. Hand pollination is required to obtain fruits when the plants are grown outside their native habitat.
  V. pompona (= *V. grandiflora*) [WEST INDIAN VANILLA, POMPONA VANILLA], also native to tropical America, is another commercial source of vanilla. Some species, including *V. barbellata* (= *V. articulata*) [LINK VINE, WORM VINE, WORMWOOD], from the West Indies and Florida, are grown for ornament.
ORCHIDACEAE, about 90 species.

Vaucheria a genus of acellular (siphonous) branched filamentous algae. The typical habitats are moist soil and salt marshes but *Vaucheria* is also found in freshwater streams.
XANTHOPHYCEAE.

Vegetables see p. 338.

× Venidio-arctotis a genus of perennial herbs produced in cultivation as a result of (continued on p. 341)

Vegetables

units, compared with 30 units in *PARSNIP and only traces in POTATO. The *SOYBEAN is the richest vegetable in vitamin B₁ (thiamine), while the fresh fruits of the *GREEN or BELL PEPPER (*Capsicum species) are a very rich source of vitamin C, with 253mg per 100g. Vegetables rich in iron are spinach leaves, turnip tops and parsley leaves. Algae have high iodine content.

The group of plants known as "vegetables" is virtually impossible to define in a satisfactory way. It refers to a very diverse collection of species which provide roots, stems, leaves, flowers, flower buds, fruits, or seeds which are eaten raw or cooked in some way. Most vegetables belong to the flowering plants, but some are found in the algae and fungi. Examples of algae used as vegetables are LAVER and NORI derived from species of *Porphyra, *SEA LETTUCE (Ulva lactuca), and species of OARWEED (*Laminaria), known as KOMBU or KOBU, used in Japan. Numerous macrofungi are

used as vegetables such as mushrooms, ceps, chanterelles, *morels, *truffles etc.

Nutritionally, vegetables vary widely. Some, such as legumes and pulses, provide a rich source of protein, some, such as *POTATOES and *SWEET POTATOES, contain large quantities of carbohydrate, while the *AVOCADO is rich in fat. Apart from these examples, vegetables are normally eaten as a source of vitamins and minerals and roughage. The vitamin content varies widely; raw *CARROT contains, for example, 11 000 international units of vitamin A per 100g, closely followed by *KALE with 10 000

Generally speaking, the degree of sweetness determines whether a fruit should be considered as such or as a fruit vegetable, for example AVOCADO, *TOMATO, *CUCUMBER. Sweetness is not, however, a sufficient criterion for fruits since some vegetables are quite high in sugar content, such as SWEET POTATOES and GARDEN *PEAS. In the accompanying table we have divided

Some common temperate vegetables: 1 Chives; 2 Shallot; 3 Onion; 4 Garlic; 5 Leek; 6 Tomato; 7 Globe Artichoke; 8 Spinach; 9 Lettuce; 10 Rhubarb; 11 Asparagus; 12 Florence Fennel; 13 Chicory; 14 Celery. 1, 2, 3, 5, 6, 11, 12, 13, 14 ($\times\frac{1}{2}$); 4 ($\times\frac{2}{3}$); 7, 8, 9, 10 ($\times\frac{1}{3}$).

VEGETABLES

Common name	Scientific name	Family
I Brassicas		
CABBAGE, SPRING		Cruciferae
CABBAGE, SAVOY		Cruciferae
CAULIFLOWER		Cruciferae
BROCCOLI, CALABRESE	Brassica oleracea	Cruciferae
KALE		Cruciferae
BRUSSELS SPROUTS		Cruciferae
TURNIP, SWEDE	Brassica campestris	Cruciferae
PAK-CHOI	Brassica campestris subspecies chinensis	Cruciferae
PE-TSAI	Brassica campestris subspecies pekinensis	Cruciferae
(See Brassicas p. 62 for further details.)		
II Leaf and stem vegetables		
ASPARAGUS	*Asparagus officinalis	Liliaceae
WILD ASPARAGUS	A. acutifolius	Liliaceae
*CHIVES	Allium schoenoprasum	Liliaceae
*CELERY	Apium graveolens	Umbelliferae
*FENNEL	Foeniculum vulgare var vulgare	Umbelliferae
FLORENCE FENNEL, FINOCCHIO	F. vulgare var azoricum	Umbelliferae
*CHICORY, ASPARAGUS CHICORY, WITLOOF, BELGIAN ENDIVE	Cichorium intybus	Compositae
RADICCHIO, RED VERONA CHICORY, TREVISO CHICORY, CASTELFRANCO CHICORY	Cichorium intybus	Compositae
GRUMOLO, BROAD-LEAVED CHICORY	Cichorium intybus	Compositae
*ENDIVE, ESCAROLLE, BATAVIAN ENDIVE	Cichorium endivia	Compositae
*LETTUCE, CABBAGE LETTUCE, COS LETTUCE	Lactuca sativa	Compositae
WILD LETTUCE	Lactuca taraxaciflora	Compositae
*SPINACH, SUMMER OR ROUND-SEEDED, WINTER OR PRICKLY-SEEDED	Spinacia oleracea	Chenopodiaceae
*SPINACH BEET	Beta vulgaris var cicla	Chenopodiaceae
SEAKALE BEET, SWISS CHARD	*Beta vulgaris var cicla	Chenopodiaceae
*ORACHE	Atriplex hortensis	Chenopodiaceae
NEW ZEALAND *SPINACH	Tetragonia expansa	Aizoaceae
AMARANTH SPINACH	*Amaranthus caudatus, A. hybridus, A. tricolor	Amaranthaceae
SEA KALE	*Crambe maritima	Cruciferae
*BAMBOO SHOOTS	*Bambusa arundinacea, B. beecheyana, B. vulgaris, *Phyllostachys dulcis, P. pubescens, etc.	Gramineae
GLOBE *ARTICHOKE	Cynara scolymus	Compositae
*CARDOON	Cynara cardunculus	Compositae
*OKRA, GUMBO, LADY'S FINGERS	Hibiscus esculentus (= Abelmoschus esculentus)	Malvaceae
JEW'S MALLOW	*Corchorus olitorius	Tiliaceae
*JUTE	C. capsularis	Tiliaceae
WATER SPINACH	*Ipomoea aquatica	Convolvulaceae
*RHUBARB, GARDEN	Rheum rhabarbarum	Polygonaceae

Common name	Scientific name	Family
III Root vegetables		
*RADISH	Raphanus sativus	Cruciferae
WINTER RADISH	R. sativus cv 'Longipinnatus'	Cruciferae
BLACK SALSIFY	Scorzonera hispanica	Compositae
*SALSIFY, OYSTER PLANT	Tragopogon porrifolius	Compositae
*CARROT	Daucus carota subspecies sativus	Umbelliferae
*PARSNIP	Pastinaca sativa	Umbelliferae
CELERIAC, TURNIP-ROOTED *CELERY	Apium graveolens var rapaceum	Umbelliferae
ARRACACHA	*Arracacia xanthorrhiza	Umbelliferae
TURNIP-ROOTED *PARSLEY, HAMBURG PARSLEY	Petroselinum crispum var tuberosum	Umbelliferae
CHERVIL, TURNIP-ROOTED	*Chaerophyllum bulbosum	Umbelliferae
JERUSALEM *ARTICHOKE	Helianthus tuberosus	Compositae
CHINESE *ARTICHOKE	*Stachys tuberifera	Labiatae
OCA	*Oxalis tuberosa	Oxalidaceae
*ULLUCO, ULLUCU	Ullucus tuberosus	Basellaceae
ANU, ANYU	*Tropaeolum tuberosum	Tropaeolaceae
*YAM BEAN	Pachyrhizus erosus	Leguminosae
YAM BEAN, POTATO BEAN	P. tuberosus	Leguminosae
SACRED OR EAST INDIAN LOTUS	*Nelumbo nucifera	Nymphaeaceae
KAFFIR POTATO, HAUSA POTATO	*Plectranthus (Coleus) esculentus	Labiatae
*ONION	Allium cepa	Liliaceae
*SHALLOT	A. cepa var aggregatum (= A. ascalonicum)	Liliaceae
WELSH *ONION, JAPANESE ONION	A. fistulosum	Liliaceae
*GARLIC	A. sativum	Liliaceae
*LEEK	A. porrum	Liliaceae
(See Root Crops p. 290 for TURNIP, SWEDE, POTATO, SUGAR BEET and Sugar and Starch Crops p. 318 for CASSAVA, YAMS, ARROWROOTS.)		
IV Fruit vegetables		
*TOMATO	Lycopersicum esculentum	Solanaceae
*AUBERGINE	Solanum melongena	Solanaceae
*CUCUMBER	Cucumis sativa	Cucurbitaceae
*GHERKIN	C. anguria	Cucurbitaceae
BITTER GOURD, BITTER CUCUMBER	*Momordica charantia	Cucurbitaceae
BOTTLE GOURD, *CALABASH GOURD, WHITE GOURD	Lagenaria siceraria	Cucurbitaceae
SNAKE *GOURD	Trichosanthes cucumerina	Cucurbitaceae
WAX, ASH *GOURD	Benincasa hispida	Cucurbitaceae
*CHAYOTE, CHRISTOPHINE	Sechium edule	Cucurbitaceae
PUMPKINS, MARROWS, SQUASHES	(see Marrows and Squashes p. 216)	
BREADFRUIT	*Artocarpus altilis	Moraceae
JACKFRUIT	*A. heterophyllus	Moraceae
*PEPPER, *SWEET PEPPER	*Capsicum annuum	Solanaceae
*AVOCADO, ALLIGATOR PEAR	Persea americana	Lauraceae
V Legumes and pulses		
(See Legumes and Pulses p. 194.)		

*Some common subtropical and tropical vegetables:
1 Aubergine (×½); 2 Okra (×⅓, ×⅙); 3 Breadfruit
(×⅓); 4 Avocado (×⅓, ×1/20); 5 Bamboo Shoots
(Bambusa sp) (×⅓); 6 Endive (temperate) (×¼);
7 Jackfruit (×1/12).*

the vegetables into a number of classes, some of which are treated elsewhere as rootcrops, sugar or starch crops, oil plants, legumes and pulses, etc. The first group is the brassicas which provide such a range of vegetables from leaves, stems and inflorescences that they are dealt with separately. They provide the most important source of leafy vegetables in temperate countries. The second group is the leaf and stem vegetables. These include several of the more select and expensive products such as *ASPARAGUS (*Asparagus officinalis*), GLOBE, *ARTICHOKE (*Cynara scolymus*) and the various forms of *CHICORY and *ENDIVE (*Cichorium intybus* and *C. endivia*). Also in this group are the various forms of *SPINACH and BEET and the salad vegetables, the most important of which is the *LETTUCE (*Lactuca sativa*) along with CHICORY and *CELERY.

The root vegetables constitute another group. It includes those species which store their reserve food in underground storage organs such as root, hypocotyl, stem, bulb, tuber and corm. Salad "root" vegetables include *BEETROOT (mainly swollen hypocotyl), widely used boiled and dressed with oil and vinegar and as a component of the Russian soup borscht, and the *RADISH

and WINTER RADISH. Umbelliferous rootcrops include the CARROT, the only umbelliferous crop that is grown on a major scale and used for human consumption and for feeding livestock, the PARSNIP and the less common CELERIAC (*Apium graveolens* var *rapaceum*), TURNIP-ROOTED OR HAMBURG PARSLEY (*Petroselinum crispum* var *tuberosum*), TURNIP-ROOTED CHERVIL (*Chaerophyllum bulbosum*), and ARRACACHA or PERUVIAN PARSNIP (*Arracacia xanthorrhiza*), a tropical American rhizomatous vegetable cultivated in the Andes, Central America and the West Indies. *ONIONS and related species are another important group of "root" vegetables known since very early times. They are bulbs composed of swollen, fleshy leaf-bases which vary from globose to ovoid as in the ONION and SHALLOT, to elongated-cylindrical as in the *LEEK.

The most important root vegetable is the POTATO (*Solanum tuberosum*), one of the world's staple crops which along with the *TURNIP (*Brassica campestris* subspecies *rapifera*), *SWEDE (*B. napus* var *napobrassica*) and *SUGAR BEET (*Beta vulgaris*), are considered further in the separate article on Rootcrops (p. 291). A number of tropical vegetables which also store food in underground tubers or modified stems, for example rhizomes or corms, such as SWEET POTATO (*Ipomoea batatas*), *CASSAVA (*Manihot esculenta*) and *YAMS (*Dioscorea* species), are also considered under Rootcrops (p. 291) or Sugar and starch crops (p. 318). Other tropical species such as the *YAM BEAN (*Pachyrhizus erosus* and *P. tuberosus*) and OCA (*Oxalis tuberosa*) are grown as minor vegetable crops.

The term "fruit vegetable" is applied to a group of plants whose fruits are eaten raw or cooked like other vegetables. This group includes the TOMATO and CUCUMBER, which are extensively used as salad vegetables. Other examples are the *SWEET OR BELL PEPPER (*Capsicum* species), which is used fresh as a salad vegetable or cooked, the *AUBERGINE (*Solanum melongena*), widely grown in the tropics and subtropics and increasingly in temperate countries, and the AVOCADO PEAR (*Persea americana*). Tropical fruit vegetables also include the BREADFRUIT (*Artocarpus altilis*) and the JACKFRUIT (*A. heterophyllus*).

Finally there is the very important group of vegetables, eaten fresh or dried, the legumes or pulses, which include some of the most important of the world's crops such as *PEAS, *BEANS and *LENTILS. These are also treated separately (p. 194).

Many of the vegetables grown, especially in temperate zones, are highly selected and are offered under numerous cultivars. Many hundreds of the older vegetable cultivars are in danger of disappearing in the face of modern legislation, and attempts are being made to preserve samples of some of them in seed banks so that they may still be available for future breeding programs.

Leaves and inflorescence of the White Hellebore (Veratrum album), a plant found in mountain pastures throughout central and southern Europe. (×⅙)

crossing *Arctotis stoechadifolia* var *grandis* [BLUE-EYED AFRICAN DAISY], *A. breviscapa* and *Venidium fastuosum* [CAPE DAISY]. There are a number of varieties in cultivation as half-hardy border plants with showy daisy-like flower heads, including the wine-purple 'Bacchus' and the orange-yellow 'Tangerine'.
COMPOSITAE.

Venidium a small genus of woolly or cobwebby annual or perennial herbs endemic to South Africa. The solitary flower heads, which may be large or small, are purple, brown or yellow in color. *V. decurrens* (= *V. calendulaceum*), *V. fastuosum* [CAPE DAISY] and *V. hirsutum* [CAPE DAISY] are among the cultivated species. There are also a number of hybrids with the closely related genus *Arctotis*, in which *Venidium* is sometimes placed.
COMPOSITAE, 20 species.

Venturia a cosmopolitan genus of plant parasitic fungi. The best-known are *V. inaequalis* which causes scab of apples, *V. pyrina* which causes scab of pears and *V. cerasi* the causal agent of cherry scab.
VENTURIACEAE, about 60 species.

Veratrum a small genus of perennial herbs from Europe, Asia and North America. They are handsome, showy plants with broad, plantain-like leaves and stout panicles of feathery flowers. The roots in particular are very poisonous. The dried and powdered roots of *V. album* [WHITE HELLEBORE, LANGWORT] produces hellebore powder, a powerful insecticide. This and some other species yield alkaloids used to treat high blood pressure.
Several species are cultivated as ornamentals, including *V. album* with whitish flowers,

V. nigrum with black-purple flowers and *V. viride* [AMERICAN WHITE HELLEBORE, ITCHWEED, INDIAN POKE] with green-yellow flowers, and *V. californicum* [CORN LILY, SKUNK CABBAGE] with greenish-white flowers.
LILIACEAE, about 20 species.

Verbascum see MULLEIN.
SCROPHULARIACEAE, about 250 species.

Verbena a large genus of annual and perennial herbs, including small shrubs which are native mainly to tropical and subtropical North and South America. The fragrant, white, pink or mauve flowers are borne in dense clusters. Some of the wild verbenas (commercially called VERVAINS) are imported in herbal remedies and are reputed to be beneficial in many medical disorders, especially those of the eye.
Several species are cultivated for their ornamental value, notably *V. canadensis* (= *V. aubletia, Glandularia canadensis*) [ROSE VERBENA, CLUMP VERBENA, CREEPING VERVAIN, ROSE VERVAIN], with a wide range of cultivars used as edging and bedding plants, *V.* × *hybrida* (= *V.* × *hortensis*) [GARDEN VERBENA], a variable species with pink, red, white, blue, purple or variegated flowers, in various cultivars, and *V. tenera* (= *V. pulchella*), with rose-violet flowers (white in cultivar 'Albiflora').
LEMON or SWEET SCENTED VERBENA is *Aloysia triphylla* (= *Lippia atriodora*), a deciduous shrub or small tree belonging to the same family, native to Chile but cultivated elsewhere, particularly in the tropics and subtropics as the source of verbena oil, used in perfumery.
VERBENACEAE, 75–100 species.

Vernonia a very large genus of rather scruffy herbs, shrubs and trees or even succu-

Verbascum dumulosum, a small subshrub species of mullein from Asia Minor but also often grown in cultivation. (× ½)

Germander Speedwell (Veronica chamaedrys), a weak-stemmed perennial herb common in grassland, hedges and woods. (× ½)

lents or lianas, common in grassy places in the tropics and subtropics, particularly in North and South America, where they are known as IRON WEEDS, but also in Southeast Asia, Africa and Australia. They have reddish-purple, bluish-white or even green flowers in terminal heads.
Most species are generally considered weeds, but some are cultivated for ornament, notably *V. baldwinii* [WESTERN IRONWEED], a perennial herb, native to the USA, and *V. conferta*, a tree of the Ugandan swamp forests, with leaves up to 70cm long. The wood of another tree species, *V. arborea* of Southeast Asia, has been used for matches in Java, and that of *V. amygdalina* [BITTER LEAF], of East Africa, has been widely used for stakes in lining out plantations, as it is resistant to termite attack. *V. cinerea* is eaten as a potherb in Java and also in southern Africa where, as in India and Malaysia, its bitter root is used as an effective vermifuge.
COMPOSITAE, about 1000 species.

Veronica [SPEEDWELLS, BROOKLIMES] a large genus of annual or perennial herbs found mostly in north temperate regions, particularly in montane areas, with a few from south temperate parts and tropical mountains.
A few species provide folk remedies for a mixture of ailments. *V. officinalis* [COMMON SPEEDWELL, GYPSYWEED] has been used to treat coughs and skin diseases, and as a substitute for tea. Two other species have culinary uses: *V. beccabunga* [EUROPEAN BROOKLIME], from Eurasia, provides young shoots and leaves used as a vegetable locally and *V. anagallis*, also a temperate species, is cultivated in Japan for salads.
Many species are cultivated for ornament, including *V. incana* (= *V. candida*) which has gray leaves and terminal racemes of dark blue flowers, and *V. longifolia* (= *V. exaltata, V. flexuosa*), with dense racemes of

preparation of perfumes, soap and cosmetics. The tufted rhizomatous grasses are used for contour planting, eg in terraced fields, as an anti-erosion measure.
GRAMINEAE, 2 species.

Viburnum an American, Asian and European genus of shrubs and small trees, many of which are cultivated for their attractive ornamental features, including habit and showy clusters of tubular flowers followed by red or black berries.

Among the most popular cultivated species and hybrids are the deciduous *V. opulus* [GUELDER ROSE, ROSE ELDER, CRANBERRY BUSH, WHITTEN TREE], *V. davidii* and the evergreen *V. tinus* [LAURUSTINUS], all white flowered species; the fragrant creamy white-flowered *V. lentago* [SHEEP BERRY, NANNY BERRY, BLACK HAW, COWBERRY, SWEETBERRY], *V.* × *carlcephalum* (= *V. carlesii* × *V. macrocephalum*), *V. farreri* (= *V. fragrans*), with pink-tinged flowers, and *V. grandiflorum* (= *V. nervosum*).

The fruits of some species, including those of *V. opulus*, the closely related *V. trilobum* [CRANBERRY BUSH, CRANBERRY TREE, HIGHBUSH CRANBERRY, CRAMPBARK, GROUSEBERRY, SQUAWBUSH, SUMMERBERRY, PIMBINA], *V. cassinoides* [SWEET VIBURNUM, SWAMP HAW, TEABERRY, WILD RAISIN, APPALACHIAN TEA] and *V. pauciflorum* [MOOSEBERRY] are edible.
CAPRIFOLIACEAE, about 225 species.

Vicia [VETCHES, TARE] a genus of annual and perennial herbs with pinnate leaves and pea-like flowers, found throughout temperate Eurasia and in temperate America, but particularly in the Mediterranean region.

The most important species is *V. faba* [BROAD, FIELD, HORSE or TICK BEAN], which today is widely cultivated throughout cool regions of the Old World, with China the largest producer, and some production in high-altitude areas of subtropical America. Little is grown in North America or Australia. Its value lies in the high protein content of the seeds which are either consumed directly by Man or by his animals. It also has value as a forage or green manure crop and provides a valuable rotation crop with cereals. Perhaps its most important characteristic is its upright habit and easily threshed pods which makes harvesting easier. Most other vetches are scandent plants. Another unusual feature is its lack of tendrils.

Vicia faba is unknown in the wild, but is thought to have originated in southwestern Asia, as did many of our crop plants. From there it radiated and was certainly in cultivation in Western Europe by the Iron Age.

Other fodder plants are *V. villosa* [HAIRY WINTER or RUSSIAN VETCH], a rampant perennial widespread in the north temperate region on margins and in hedgerows, *V. sativa* [COMMON VETCH, SPRING VETCH, TARE], *V. benghalensis* (= *V. atropurpurea*) [PURPLE VETCH], *V. ervilia* [BITTER VETCH] and *V. angustifolia*.
LEGUMINOSAE, about 150 species.

lilac, blue or white flowers according to cultivar. There are several low-growing species suitable for rock gardens, eg *V. pectinata*, which forms a gray mat with deep blue flowers with a white eye (also white and rose-flowered cultivars), *V. fruticans* (= *V. saxatilis*) [ROCK SPEEDWELL], with deep blue flowers with a red eye, *V. prostrata* (= *V. rupestris*), with many-flowered terminal racemes of deep blue flowers (although many cultivars in a range of other colors are grown). *V. spicata* (= *V. australis*) is an erect or ascending species with taller stems up to 0.5m high, bearing long dense racemes of blue, pink or white flowers depending on cultivar.

A number of cultivated species can, under certain conditions, become pernicious weeds as in the case of the prostrate *V. filiformis*, which is widespread in Europe and parts of North America as a weed of gardens and lawns. Other weedy species include *V. agrestis* [PROCUMBENT or FIELD SPEEDWELL], *V. arvensis* [WALL SPEEDWELL] and *V. polita* [GRAY SPEEDWELL], all of which are common in arable cultivated land in Europe. *V. persica* is the most abundant weed in the whole genus and is naturalized throughout Europe

Viburnum plicatum *var* tomentosum, *a popular ornamental shrub.* (× ¼)

and elsewhere.
SCROPHULARIACEAE, about 250 species.

Vetch the common name for many species of the pea family native to temperate Eurasia and temperate America, particularly members of *Vicia*. Some VETCHES (or TARES), provide valuable forage, hay and green manure crops (see *Vicia*).

Other vetches are MILK VETCH (*Astragalus* species), KIDNEY VETCH (*Anthyllis vulneraria*) and HORSE-SHOE VETCH (*Hippocrepis comosa*). Several species of *Lathyrus* are known as VETCHLINGS.
LEGUMINOSAE.

Vetiveria a genus of two species of coarse perennial grasses native to tropical Africa, Asia and Australia, the more important of which is *V. zizanioides* [KHAS-KHAS, *KHUS KHUS, VETIVER], a native of river banks in India and Sri Lanka. Vetiver is cultivated for its aromatic roots, which are woven into mats, screens and baskets, and which yield an essential oil (oil of vetiver) used in the

Above Left *The dark-blotched flowers of the Broad Bean* (Vicia faba). (× ⅕)

Above Right *Broad Bean fruits ready for harvesting.* (× ¼)

Victoria a very small genus of tropical South American giant water lilies. There are two main species, *V. amazonica* (= *V. regia*) [AMAZON OR ROYAL WATER LILY, WATER MAIZE, AMAZON WATER-PLATTER], from the Amazon and Guiana, and *V. cruziana* [SANTA CRUZ WATER LILY, SANTA CRUZ WATER-PLATTER], from Paraguay, Bolivia and northern Argentina. Hybrids occur and they are frequently confused in cultivation. Victorias are commonly grown in botanic gardens throughout the world for their enormous circular floating leaves and attractive fragrant flowers (white but later pink or red in color).
NYMPHAEACEAE, 2–3 species.

Vigna see COWPEA.
LEGUMINOSAE, about 170 species.

Vinca a small genus of erect or trailing perennial herbs native to Europe, North Africa and western Asia. *V. major* [GREATER PERIWINKLE, BLUE-BUTTONS, BAND PLANT] and *V. minor* [COMMON or LESSER PERIWINKLE, MYRTLE] are cultivated for their attractive funnel-shaped flowers ranging from white through blue to purple in color. There are forms with double flowers and variegated leaves. They are useful ground cover plants tolerating the shade of trees and shrubs.
APOCYNACEAE, about 12 species.

The Teesdale Violet (Viola rupestris) occurs throughout Europe but in Britain it is confined to 11 localities on limestone turf or bare ground in the northeast. (×¾)

Vine a climbing plant whose stems are too weak or flexible to support themselves and therefore attach themselves to some support by stems, tendrils, petioles, aerial roots etc. The name derives from *Vitis vinifera* [GRAPE VINE] to which it is often applied in the strict sense.

Viola a large genus (which includes VIOLAS, VIOLETS and *PANSIES) of perennial herbs (rarely small shrubs) with a wide distribution

Garden Violas (Viola × williamsii) are popular garden subjects. (×⅓)

in northern and southern temperate zones. The flowers are showy, nodding, the lower petal prolonged into a spur. *V. odorata* [SWEET VIOLET, GARDEN VIOLET, FLORIST'S VIOLET, ENGLISH VIOLET], with deep blue or white flowers, is cultivated in southern France for the essential oil in the flowers which is used in perfumes, toiletries and crème de violette liqueur.

A number of varieties of garden violets are cultivated. Very large-flowered cultivars appeared in the mid-19th century, the most famous being 'Czar'; some of these were derived from *V. odorata* but the famous 'Parma' or 'Neopolitan' violets of this time, with heavily scented, large double flowers, are probably derived from the closely related *V. alba*.

Other cultivated species include *V. labradorica* [LABRADOR VIOLET], with purple bronze leaves, *V. canina* [DOG VIOLET], *V. riviniana* [DOG or WOOD VIOLET], *V. septentrionalis* [NORTHERN BLUE VIOLET], a good foliage plant for heavy soils in shade, and, of course, the showy-flowered *V. × wittrockiana* [PANSY, GARDEN PANSY, SHOW PANSY, LADIES DELIGHT, STEPMOTHER'S FLOWER], one of the most popular of all garden plants. This hybrid is of mixed origin, derived largely from *V. tricolor* [EUROPEAN WILD PANSY, MINIATURE PANSY, FIELD PANSY, JOHNNY-JUMP-UP, HEARTSEASE], native to Europe and naturalized in North America, with variously colored petals, yellow to reddish-purple, violet-blue or white or various combinations. Some characters are derived from *V. lutea*, from Europe, and possibly from *V. altaica*, from Asia Minor. First developed at the beginning of the 19th century numerous varieties of the "show" and "fancy" pansies, with very large "faces", have remained undiminished in popularity. Winter-flowering as well as summer-flowering cultivars are now available, the former being very hardy and flowering in sheltered positions outdoors during winter and early spring.

V. × williamsii, which includes a range of plants commonly known as VIOLAS, VIOLETTAS and TUFTED PANSIES, were derived by crossing the tufted Pyrenean species *V. cor-*

nuta [HORNED VIOLA], with varieties of *V. × wittrockiana*. The various cultivars have a characteristically tufted and more compact growth than PANSIES.

Another group of hybrids, *V. × visseriana*, derived from crosses between *V. cornuta* and *V. gracilis*, are also sometimes given the common name VIOLETTA. These are miniature plants suitable for rock gardens or for edging borders. The flowers vary in color according to cultivar from white through yellow to various shades of blue and purple.
VIOLACEAE, about 400 species.

Viscum see MISTLETOE.
LORANTHACEAE, about 20 species.

Vitex a genus of tropical and subtropical deciduous trees and shrubs, a few of which are used ornamentally. The flowers are usually borne in many-flowered cymose inflorescences and are usually tubular, spreading out into five lobes or teeth. The best known is *V. agnus-castus* [CHASTE, HEMP or SAGE TREE, INDIAN SPICE, MONK'S PEPPER TREE], from the Mediterranean region, which has pale blue flowers (white in cultivar 'Alba') and reaches about 6m tall. The tree is aromatic due to the presence of essential oils and the drupaceous fruits can be used as a substitute for pepper. *V. negundo* [CHINESE CHASTE TREE], is very similar, with lilac or lavender blue flowers.
VERBENACEAE, about 270 species.

Viola cheiranthifolia only occurs in the subalpine zone of the Pico de Teide in Tenerife, Canary Islands. (×½)

Black Grapes (Vitis vinifera). (× ½)

Vitis [VINES] a genus of woody clinging plants of the Northern Hemisphere which includes *V. vinifera* [GRAPE VINE]. The fruit is a globose to ovoid berry with two or four seeds in a soft pulp. This species from the Caucasus is estimated to have given rise to about 10 000 cultivars. The wild subspecies *sylvestris* is quite palatable and produces an acceptable wine. The cultivated vine (subspecies *vinifera*) contains three principal groups of cultivars: *orientalis*, the large-berried table grapes, *occidentalis*, the small-berried wine grapes and *pontica*, intermediate forms from southwestern Asia and Eastern Europe.

On the west coast of the USA vineyards largely consist of cultivars of the native species *V. riparia*, *V. vulpina* [FROST GRAPE, RIVER BANK GRAPE], *V. rupestris* [BUSH GRAPE, SAND GRAPE, SUGAR GRAPE, MOUNTAIN GRAPE] the celebrated *V. labrusca* [FOX GRAPE, SKUNK GRAPE], and *V. rotundifolia* [MUSCADINE GRAPE, SCUPPERNONG, BULLACE GRAPE, FOX GRAPE], with a strongly musk-flavored pulp.

In addition to the wine and fruit grapes, several species of *Vitis* are grown for their ornamental foliage, such as *V. coignetiae* [JAPANESE CRIMSON GLORY VINE], which has brightly colored leaves in the autumn, and *V. thunbergii* from East Asia.
VITACEAE, about 60 species.

Volvox a cosmopolitan genus of green algae. The plants consist of colonies made up of thousands of small biflagellate cells resembling those of the genus *Chlamydomonas*. Some cells reproduce the colony vegetatively, others sexually.
CHLOROPHYCEAE.

Vriesea a tropical American genus of evergreen terrestrial but mainly epiphytic perennials with glossy leaves in dense rosettes, sometimes variously spotted or striped. The attractive foliage and the showy, long-lasting inflorescences of tubular yellow or white flowers have led to the cultivation of a number of species, such as *V. fenestralis* and *V. splendens* (= *V. speciosa*) [FLAMING-SWORD] and their hybrids, as ornamentals.
BROMELIACEAE, about 240 species.

Wahlenbergia a genus of annual or perennial herbs or subshrubs, with slender, graceful growth and bell-shaped flowers, occasionally red but predominantly white or blue to lilac. They are mostly from the southern temperate zone but a number are native to Europe and other areas of the Northern Hemisphere. The genus is closely allied to *Campanula* and includes a number of attractive ornamental species such as *W. trichogyna*, an erect much-branched blue-flowered perennial, *W. albo-marginata* (= *W. saxicola*) and *W. hederacea*. Some botanists include the genus *Edraianthus* in *Wahlenbergia*.
CAMPANULACEAE, about 100 species.

Wallflower the common name for members of the genus *Cheiranthus*, which consists of perennial herbs or subshrubs native to the Mediterranean region. The most commonly cultivated and best-known species is *C. cheiri* (= *C. fruticulosus*) [COMMON WALLFLOWER] which is a favorite garden plant.

C. cheiri was probably originally cultivated for its medicinal properties, as it contains a specific glucoside, cheirotoxin. The flowers of

Water Chestnuts (Trapa natans) *being cultivated on a pond at Thausi, India. The plant floats on the water surface but roots into the bottom mud. The plots are demarcated by* Ipomoea reptans.

the wild form are single and bright yellow, but a wide range of variously colored single and double cultivars now exist, many of which seem to have arisen as hybrids between *C. cheiri* and other closely related *Cheiranthus* or *Erysimum* species, and their ancestry is difficult to trace. Modern cultivated color forms now range from yellow and orange to red, purple and brown or various combinations of these.

C. scoparius [CANARY ISLANDS WALLFLOWER] is an attractive species from high mountain regions of the Canary Islands. It is a branched woody perennial with magenta flowers turning white on maturity. The SIBERIAN WALLFLOWER (*C. × allionii*) is a bright orange, free flowering biennial or perennial first raised in 1846.
CRUCIFERAE.

Walnut the name commonly used for the timber and fruit of species of *Juglans*. Walnut is a highly valued hardwood timber and several species are used, including *J. nigra* [BLACK WALNUT] and *J. ailanthifolia* (= *J. sieboldiana*) [JAPANESE WALNUT].

The species most widely exploited for its edible nuts is *J. regia* [COMMON, PERSIAN or ENGLISH WALNUT, MADEIRA NUT] which is cultivated commercially in the USA (especially California), France, Italy, China and India. *J. regia* also produces a valuable timber. Other species of local economic importance are *J. cinerea* [NORTH AMERICAN BUTTERNUT] and *J. cathayensis* [CHINESE BUTTERNUT, CHINESE WALNUT].

The name AFRICAN WALNUT is applied to two different species, namely *Lovoa klaineana* and *Coula edulis*, both of which produce a useful timber often used as a substitute for mahogany.

Water chestnut the common name for the tropical and Asian aquatic floating plants *Trapa bicornis* and *T. maximowiczii*, and the Eurasian and African *T. natans*, cultivated for their starchy seeds which can be consumed

A Water Chestnut (Trapa natans) with fruits. The starchy fruits are eaten fresh, or dried and ground for use in cooking. ($\times \frac{1}{6}$)

after boiling, or ground into a flour. The name WATER CHESTNUT is also given to the perennial East Asian herb *Eleocharis dulcis* [CHINESE WATER CHESTNUT], whose tubers are eaten by the Chinese.

Water-lilies the common name for several genera of perennial aquatic herbs of the Nymphaeaceae, grown in ponds and pools for their leaves and showy flowers. Species of *Nelumbo* [LOTUS], *Nuphar* [YELLOW WATER-LILIES], *Nymphaea* and *Victoria*, are all considered to be WATER-LILIES.
NYMPHAEACEAE.

Water melon the common name for *Citrullus lanatus* (= *C. vulgaris*, *Colocynthis citrullus*) and its fruit, an important food crop in tropical and subtropical areas. It is of southern African origin, and fruits of wild plants, known as "tsamma", provide food and water for the Bushmen.

It is a vigorous, usually softly hairy annual vine bearing fruits which vary greatly in size and color. Commonly they are subglobose, dark green with darker longitudinal mottlings, smooth, and 5–25kg in weight. The flesh is whitish, yellow or pink, extremely juicy and refreshing to the taste. The seeds are white, brown, red or black, edible, and rich in oil and protein (up to 40%); the yellowish-green oil is used in cooking and in oil lamps.
CUCURBITACEAE.

Watercress the common name for *Nasturtium officinale* (= *Rorippa nasturtium-aquaticum*), a hardy perennial herb native to most of Europe, where it grows wild in ditches and small streams. It is widely cultivated in the UK, France and Germany for its sharp, clean taste in salads. It has also been introduced into the USA where it is grown as a salad plant, and is naturalized.

Formerly, the sterile triploid hybrid between *N. officinale* and *N. microphyllum* (= *Rorippa microphylla*) [BROWN CRESS] was grown but it is susceptible to fungal and viral diseases and *N. officinale* has largely replaced it. Watercress is grown from seed or from cuttings which are used to inoculate beds

which remain productive for three years. Crops can be cut up to 10 times a year.
CRUCIFERAE.

Watsonia [BUGLE LILY] a genus of corm-bearing perennials from South Africa and one species from Madagascar. It includes species such as *W. pyramidata*, which make their growth during winter and flower in spring or early summer. Among other cultivated species are *W. bulbillifera*, which bears orange to red flowers on branched stems in spring and early summer. *W. densiflora* bears dense red-flowered spikes (white in cultivar 'Alba') late on into the summer. Another late-summer-flowering species is *W. beatricis*, usually with an unbranched inflorescence of closely packed orange-red, funnel-shaped flowers 5cm across.
AMARYLLIDACEAE, about 60 species.

Wattle a term which means a wicker rod or bundle. It is also a name widely applied to the Australian species of *Acacia*.

Wattle and daub construction, in which mud or plaster is applied to a framework of wicker, is one of the earliest forms of building used by primitive man and is still used today as a cheap form of house-making.

Wax a term used to denote a class of substances having certain physical characteristics in common rather than defining a particular group of chemical compounds. The total amount and form in which wax develops on the plant surface are both very dependent on the environment. A plant's waxy surface has many physiological roles to play, some probably not yet appreciated, but its dominant role is the control of the plant's

A wattle and daub house by a roadside in northern Venezuela. The wickerwork frame can be seen where the fill-in of dried mud has crumbled away.

intake and transpiration of water.

Plant waxes were Man's first plastic; they have been used since the earliest times for candles, varnishes and polishes and for the "lost wax" process in metal casting. Their more recent uses include sealing wax, electrical insulation, dental impressions, waxed paper and cartons, pharmaceuticals and cosmetics. The finest of all waxes is *carnauba wax obtained from *Copernicia prunifera* [SOUTH AMERICAN WAX PALM], but many plants have been used as commercial wax sources. The richest is probably *Ceroxylon andicola* [ANDEAN WAX PALM], in which the wax on the bole may be 0.5cm thick.

Weigela a small Asian genus of hardy deciduous shrubs with showy pink, purplish, reddish or white, foxglove-like flowers. They are exceptionally hardy and adaptable shrubs widely cultivated in gardens and parks. *W. florida* is probably the most common species in cultivation and has many cultivars and hybrids derived from it.
CAPRIFOLIACEAE, about 12 species.

Welwitschia a genus comprising a single species, *W. mirabilis*, one of the most unusual and isolated of all land plants, found only in a coastal strip of desert in Namibia and Angola. It has a very long taproot and a very thick trunk up to 1m in diameter but rising only to about 0.3m above ground. The stem terminates in a two-lobed disk-like structure which bears either male or female cones around its inner margin. The leaves (only two or three) are thick, leathery, strap-like and split at the ends, and grow continually throughout the life of the plant.

The plant is of no economic or ornamental value but its unusual character has excited the imagination of botanists interested in its place in the evolution of the more conventional seed plants. It is difficult to grow

A male plant of Welwitschia mirabilis, *a species of great botanical interest that is restricted to the narrow coastal strip of desert of Namibia and Angola. The genus belongs to the order Gnetales, a group of gymnosperms that is significantly different from the conifers and cyads.* (× ⅓)

and is rarely found in cultivation except in botanic gardens.
WELWITSCHIACEAE, 1 species.

Whisk ferns see Club mosses and their allies, p. 92.

Wild oats species of *Avena* that are serious weeds throughout the world. *Avena fatua* (also known as TARTARIAN *OAT, POTATO OAT, FLAVER) is the most widespread species in cooler climates but *A. sterilis* (also known as ANIMATED OAT) and *A. ludoviciana* (the latter usually regarded as a subspecies of *A. sterilis*) are also very serious. They flourish wherever a high proportion of the land is planted to

Wild Oats (Avena fatua) *growing in a field of* Barley (Hordeum vulgare). *Wild Oats are a serious and persistent weed of Wheat, Barley and Cultivated Oats.*

wheat, barley or cultivated oats. Intensification of cropping worsens the problem.

Wild oats are weeds because they compete with the crop plants; with cereal crops, yield losses of up to 50% have been recorded. Wild oats are of no agricultural value themselves except as forage if they are cut before the seeds are mature. At maturity the seeds are not retained on the panicle, as are those of cultivated cereals, but shed to the ground. This process of maturation and shedding takes place over a period of about three to four weeks, and by the time the cereal crops are harvested, most wild oat seeds will be shed. However, some seed does find its way into the harvested grain where it is regarded as a serious contaminant. The wild oat seed has a very hard husk and this reduces its value for livestock or human food.
GRAMINEAE.

Willow bark the by-product of the use of the wood of species of willow. Willow wood has many purposes, such as the making of cricket-bats, baskets, garden-furniture, clog-soles. All of these trades yield willow bark as a by-product. Most is used in the tanning industry, especially that of *Salix viminalis* [OSIER, BASKET WILLOW] and *S. caprea* [GOAT or PUSSY WILLOW, FLORIST'S WILLOW, SALLOW]. These barks yield good-quality tannins which are particularly useful for making glove leathers. Willow bark also contains salicin which is sometimes used medicinally for the treatment of rheumatism.
SALICACEAE.

Wineberry or **wine raspberry** the common name for *Rubus phoenicolasius*, a close relative of the *RASPBERRY, *BLACKBERRY, *LOGANBERRY and others of the genus *Rubus*. It is a native of China and Japan but has been introduced to North America and parts of Europe. It produces arching branches which root at their tips in the same way as blackberry branches. The fruit is red and like a

small raspberry and although edible is rather insipid by comparison.
ROSACEAE.

Their masses of drooping racemes of fragrant flowers make wisterias among the most beautiful and popular climbers for walls and trellises. Pictured here is a well-established specimen of the Chinese Wisteria (Wisteria sinensis).

Wisteria a small genus of attractive robust woody climbers with twining stems, native to eastern North America and eastern Asia. Wisterias have large leaves and long drooping racemes of blue, mauve or white fragrant, pea-like flowers. Several species have been in cultivation since the early 19th century. The largest is *W. sinensis* [CHINESE WISTERIA] which can climb to 30m. *W. floribunda* [JAPANESE WISTERIA], with numerous cultivars, and the white-flowered, downy-leaved *W. venusta* are also commonly grown. All these species, particularly *W. floribunda*, have numerous cultivars.
LEGUMINOSAE, about 9 species.

Woad a name applied to the cruciferous biennial herb *Isatis tinctoria*, native to Europe and western Asia. The plant has waxy pale green leaves, yellow flowers and pendent purple seed pods. It was cultivated in Britain for centuries for a blue dye first used by the Ancient Britons. It was also cultivated elsewhere, including Egypt, in the early Christian era as a source of imperial blue dye.

In the production of woad, the leaves were kneaded and rolled into balls before being dried and ground to form a powder. The next stage, fermentation for up to nine weeks, was an unpleasantly smelly process leading to the production of a clay-like substance, the dye.

Woad was highly esteemed for dying fabrics since the intense blue color was fast under extreme conditions, for example of sun, rain and sea-water. For this reason woad was used in sailor's garments, policemen's uniforms and the gowns of students at Christ's Hospital School, London (the "Blue-coat Boys") until replaced by *indigo.

Although woad was eventually superseded by indigo, a superior blue dye obtained from the leguminous *Indigofera tinctoria* and other species, it was used to assist the fermentation process of indigo.
CRUCIFERAE.

Wolffia a small genus of widely distributed tropical, subtropical and warm temperate floating aquatic herbs. Considered to be the smallest known flowering plants they consist of a globular, fleshy rootless thalloid plant body. Flowers are produced infrequently and consist of either a single stamen or a single carpel. *W. columbiana* [COMMON WOLFFIA] is native to North and South America; *W. arrhiza* is more widespread and is found in Europe, Asia, Africa and Australia.
LEMNACEAE, about 8 species.

Woodsia a genus of small tufted ferns with pinnately divided fronds, found in alpine and arctic areas in temperate and cool temperate regions of both the Northern and the Southern Hemisphere. Several species are cultivated in rock gardens, such as *W. alpina* [NORTHERN WOODSIA], *W. ilvensis* [RUSTY or FRAGRANT WOODSIA] and *W. scopulina* [ROCKY MOUNTAIN WOODSIA].
POLYPODIACEAE, about 40 species.

Woodwardia a small genus of large ferns with pinnately divided or lobed fronds, native to Europe, Asia and North America. A few are cultivated in woodland gardens or under glass, such as the common European species *W. radicans* [EUROPEAN CHAIN FERN], the Asian *W. unigemmata* and the North American *W. virginica* [VIRGINIA CHAINFERN].
BLECHNACEAE, about 12 species.

Wulfenia a small genus of perennial rhizomatous European, Asian and Himalayan herbs with basal leaves and blue or purple flowers in spikes. A few are cultivated as rock-garden plants.
SCROPHULARIACEAE, 3–5 species.

Wyethia a small genus of perennial herbs native to western North America. They have entire, sometimes woolly leaves and large yellow solitary heads of flowers, with ray florets over 4cm long. *W. angustifolia* and *W. amplexicaulis* are two of the species sometimes cultivated in garden borders.
COMPOSITAE, about 14 species.

Xanthium [COCKLEBUR] a small genus of New World herbs, introduced into Europe and elsewhere. The hooked prickles on the "fruit" of the COCKLEBUR are remarkably efficient in assisting distribution of the seeds by becoming entangled in the wool and fur of animals. The removal of the fruits from the wool of sheep is very difficult and even the presence of hooks in the wool seriously downgrades its quality.

The annual *X. strumarium* (probably originally from South America) is capable of rapid growth up to 1m in height and is found on waste ground. Its leaves are covered with short hairs and are poisonous to livestock. *X. strumarium* subspecies *italicum* (= *X. echinatum*) is rough-stemmed and somewhat aromatic with yellowish glands, while *X. spinosum* is largely hairless. Although probably native to America, both these species have now achieved worldwide distribution.
COMPOSITAE, 2 species.

Xanthoceras a genus of two species of

Part of the succulent, curved inflorescence (spike) of Xanthorrhoea hastilis, *which bears numerous small flowers. (×1)*

deciduous shrubs or small trees with dark green pinnate leaves, native to northern China. One species, *X. sorbifolium*, is grown chiefly for its attractive white flowers with a yellow or red spot at the base of each petal.
SAPINDACEAE, 2 species.

Xanthorrhoea an Australian genus of slow-growing long-lived plants, similar in habit to *Aloe. The linear leaves are borne in tufts at the top of a thick woody stem. The leaf bases are persistent, and all species bear long, bulrush-like spikes of white flowers. The leaf-bases of *X. australis* [AUSTRALIAN GRASS TREE] and *X. hastilis* [SPEARLEAF GRASS TREE] yield resin used in varnishes.
XANTHORRHOEACEAE, about 12 species.

Xanthosoma a genus of large rain forest herbs native to tropical America and the West Indies. Some species with their arrow- or spear-shaped leaves are ornamental, and variegated forms, such as *X. lindenii* [INDIAN KALE, SPOON FLOWER], are in cultivation. *X. violaceum* [BLUE TARO, BLUE APE, TANIER, YAUTIA, COCOYAM] is grown for its edible tubers but the most important tuberous species is *X. sagittifolium* [TANIER, YAUTIA, COCOYAM]. This has been cultivated in the New World since pre-Columbian times and was introduced to West Africa in the last century, where the tubers are eaten roasted or boiled, and spread throughout the Pacific and Asia. The burnt stems have been used as poultices for strained muscles of working elephants. *X. brasiliense* is grown for its edible leaves.
ARACEAE, about 40 species.

Xeranthemum a genus of annual herbs native from the Mediterranean region to Iran. They are densely hairy with erect branching stems and compact heads of flowers surrounded by papery and persistent

Yam tubers (Dioscorea *sp*).

bracts. *X. annuum* [IMMORTELLE] has several different varieties with white pink or violet heads which when dried are used for decoration as everlastings; it is widely cultivated.
COMPOSITAE, about 6 species.

Xylaria a cosmopolitan genus of wood-rotting fungi which form finger-like leathery, black stromata in which are embedded numerous flask-shaped fruiting bodies (perithecia). The best-known species are *X. hypoxylon* [CANDLE-SNUFF FUNGUS], in which the young stromata are branched and covered at the tips by white, powdery conidia, and *X. polymorpha* [DEAD-MENS' FINGERS], with black swollen stromata.
XYLARIACEAE, about 100 species.

The Candle-snuff Fungus (Xylaria hypoxylon)
showing spore dispersal when the fruiting bodies are
knocked by a finger.

Yam the common name for *Dioscorea* species cultivated for their starchy tubers which may be boiled, roasted or fried. They are tropical herbaceous twining climbers. The most important species is *D. alata* [GREATER or ASIATIC YAM] which originated in Southeast Asia and spread in early times to the Pacific. It was carried by the Portuguese and Spaniards to the New World and is now the most popular yam in the eastern Caribbean. The two species most commonly cultivated in West Africa are *D. rotundata* [WHITE GUINEA YAM], of the savanna regions, and *D. cayenensis* [YELLOW GUINEA YAM] of the forest zone. All the above species usually produce single tubers and have cordate leaves.

D. trifida [CUSH-CUSH YAM] produces a number of small tubers and originated in the New World. *D. esculenta* [LESSER YAM] is grown in Asia and the Pacific. The large tubers of the South African *D. elephantipes* [HOTTENTOT BREAD] protrude from the ground. *D. bulbifera* [AERIAL YAM] occurs in Asia and Africa and has edible aerial tubers borne in the axil of the leaves. Two wild yams, *D. dumetorum* in Africa and *D. hispida* in Asia, are used in times of famine; they contain a poisonous alkaloid, which must be removed by soaking and boiling.

Some species, including *D. deltoides, D. elephantipes* and *D. composita* are the so-

called "drug yams" yielding diosgenin, used to produce oral contraceptive drugs.
DIOSCORACEAE.

Yam bean the Mexican and tropical American climbing bean *Pachyrhizus erosus*, which is cultivated in Central America and extensively in the Old World tropics, and is naturalized in southern China and Thailand. It produces pods eaten when immature as a vegetable and swollen, starchy, edible roots. The leaves, stems, roots, ripe pods and seeds have insecticidal properties.

P. tuberosus [YAM, POTATO or MANIOC BEAN] is grown as a root crop but is more restricted and grown principally in Ecuador, the West Indies and China. The name YAM BEAN is also given to the tropical African climbing bean, *Sphenostylis stenocarpa* which yields edible seeds and tubers.
LEGUMINOSAE.

Common Yew (Taxus baccata) *is often used for topiary. These specimens are the product of several hundred years of clipping. The Common Yew is the most widely cultivated species of the genus* Taxus.

Yew the popular name for *Taxus*, a genus of evergreen trees, shrubs and subshrubs. Yews are widely distributed throughout the north temperate zone of the Old and New World, with one species *T. celebica* (= *T. chinensis*) [CHINESE YEW], virtually on the equator on the Indonesian island of Celebes.

All parts of the plant, except the scarlet aril (often called the yew berry), are highly poisonous. The enclosed seed or "stone" of the aril also contains poison. The poison is a mixture of alkaloids collectively referred to as taxine. Yew poisoning, resulting in heart and respiratory failure, is extremely serious and fatal results in both humans and animals are well documented.

The wood of the yew is close-grained. durable and hard but elastic. In Britain it was the traditional material for making bows and

Young Yam plants (Dioscorea sp) beginning to twine round their supports in a cultivated plot on Dauan Island, Torres Straight, Queensland, Australia.

is still used today for archery sports, being combined with HICKORY (*Carya), the latter for the side facing the "string", the former on the side away from it. In spite of its high quality the wood is less popular now than formerly; it is used mainly for floor blocks, panels, fence posts, mallet heads etc, and as a veneer in cabinet-making. About five species are known in cultivation, including the hybrid *T. × media, T. cuspidata* (= *T. sieboldii*) [JAPANESE YEW], *T. celebica* and *T. canadensis* [CANADA or AMERICAN YEW], but the commonest is *T. baccata* [ENGLISH or COMMON YEW]. This and its many cultivars, including 'Hibernica' [IRISH YEW], make excellent subjects for topiary.
TAXACEAE.

Ylang-ylang or **ilang ilang** the common name for *Cananga odorata* (= *Canangium odoratum*), a large quick-growing, soft-wooded tree, native to the Pacific islands and tropical Asia. It is cultivated in many tropical countries for its large, greenish-yellow, strongly scented flowers which yield by distillation an essential oil which is used in high-grade perfumes, soaps and other cosmetics. A mature tree produces about 9kg of fresh

Seeds of the Common Yew (Taxus baccata) surrounded by a scarlet aril. ($\times \frac{1}{2}$)

flowers in a season which would give about 30g of the oil.
Artabotrys hexapetalus (= *A. odoratissimus*) [CLIMBING YLANG-YLANG] is a woody vine with large greenish or yellowish flowers. It is native to southern India and Sri Lanka where an infusion from the fragrant flowers is drunk as tea, and it is widely cultivated in the tropics.
ANNONACEAE.

Yucca a genus of hardy and tender North and Central American xerophytic shrubs and trees of distinctive appearance; the best-known is *Y. brevifolia* [*JOSHUA TREE]. The trunks and stems are thick and sparsely branched and grossly enlarged in some species, eg *Y. elephantipes*. The leaves are sword-like and usually stiff, entire or toothed, with a sharp point and commonly with curling fibers splitting from the margins. They are assembled in large terminal rosettes. The flowers are borne in stout panicles.
Yuccas are symbiotically linked with a moth, *Pronuba*, that lays its eggs inside the ovary and deposits a ball of pollen in the tubular stigma. Fertilization results in a mass of seeds, some of which the moth larvae feed upon.
The hardy yuccas are suitable for hot, dry, stony but sheltered sites, where they add an exotic touch to the garden or park. Their spectacular flower display commands attention. The most common species cultivated are *Y. baccata* [SPANISH BAYONET, BLUE YUCCA DATIL], *Y. filamentosa* [ADAM'S NEEDLE], *Y. flaccida, Y. glauca* [SOAPWEED], *Y. whipplei* [OUR LORD'S CANDLE], *Y. recurvifolia* and *Y. gloriosa* [SPANISH-DAGGER, PALM LILY, ROMAN CANDLE], the last two being among the most popular. The berry-like fruits of *Y. baccata* are collected and eaten dried by some of the American Indians.
AGAVACEAE, about 40 species.

Zamia the largest genus of living cycads, widely distributed in subtropical and tropical America. They have a palm-like habit with short or subterranean trunks. Some species, including *Z. integrifolia* [SAGO CYCAS, COONTIE, SEMINOLE-BREAD], and *Z. pumila* (= *Z. furfuracea, Z. silvicola* [FLORIDA ARROWROOT, SAGO CYCAS, COMPTIE, COONTIE, SEMINOLE-BREAD], yield a starchy, edible sago.
ZAMIACEAE, about 40 species.

Zannichellia a cosmopolitan genus of completely submerged aquatic perennial herbs with linear, thread-like leaves. Usually, only one species is recognized: the polymorphic *Z. palustris* [HORNED PONDWEED]. It usually shows a preference for water rich in nutrients and can withstand considerable pollution. *Zannichellia* can be considered a beneficial plant as it stabilizes mud and by absorbing ions contributes to the purification of the water which it inhabits.
ZANNICHELLIACEAE, 1 species.

Zantedeschia a small genus of South African herbaceous perennials from damp, wetland marshy sites. The leaves are succulent, lustrous green, and arrow-shaped on long fleshy stalks. The inflorescence is in the form of an erect spadix subtended by a longer showy expanded spathe.
Z. aethiopica (= *Z. africana, Richardia africana, R. aethiopica*) [ARUM LILY, FLORIST'S CALA, CALLA, TRUMPET LILY] is cultivated as a

The Arum Lily (Zantedeschia aethiopica) has inconspicuous flowers massed on a yellow spadix and surrounded by a white spathe. ($\times \frac{1}{5}$)

*Inflorescence of the Golden Calla (*Zantedeschia elliottiana*), an outstanding species native to South Africa. (× ½)*

cool greenhouse plant or as a marginal aquatic in sheltered ponds and lakes outdoors. It possesses an attractive, yellow, fragrant spadix and a flaring, recurved, white spathe. *Z. elliottiana* (= *R. elliottiana*) [GOLDEN

Canopy of Zanthoxylum ailanthoides, *a native of Japan and China and occasionally cultivated for its large attractive pinnate leaves. (× 1/30)*

or YELLOW CALLA] is a greenhouse species with yellow spadix and a flaring, recurved, yellow spathe (greenish yellow outside). ARACEAE, 6–8 species.

Zanthoxylum a temperate and subtropical genus of trees and shrubs with aromatic bark and glandular-dotted leaves, from East Asia, east Malaysia, the Philippines, Africa and North and South America. The fruits of several Asian species, including *Z. piperitum* [JAPAN PEPPER], which is cultivated in Japan and elsewhere, yield a condiment used as a substitute for pepper.

Z. americanum [NORTHERN PRICKLY ASH, TOOTHACHE TREE], of North America, is an aromatic shrub, the dried bark of which, like that of another North American species, *Z. clavaherculis* [SOUTHERN PRICKLY ASH, HERCULE'S CLUB], has medicinal properties. The light, yellow, fine-grained wood of *Z. flavum* [WEST INDIAN SATINWOOD] polishes well and is used in furniture manufacture. RUTACEAE, about 200 species.

Zauschneria a small genus of low-growing glandular perennial herbs from dry coastal hills in southern North America. The most common species is *Z. californica* [CALIFORNIAN FUCHSIA] with orange and red fuchsia-like flowers, globose at the base and narrowed into a long tube, with eight petaloid lobes, four erect and four deflexed. ONAGRACEAE, 4 species.

Zea see MAIZE.
GRAMINEAE, 2–3 species.

Zebrina a small genus of trailing perennial

herbs native to Mexico and Guatemala. *Z. pendula* (= *Tradescantia zebrina*) [WANDERING JEW, INCH PLANT] is a popular houseplant bearing small rose-purplish flowers and attractive ovate-oblong leaves varying according to cultivar but typically purplish beneath and green and silver-white striped above; they are dark red or reddish green, but not striped, in cultivar 'Purpussii' (= *Z. purpusii*) and metallic-green, striped white, red and green in cultivar 'Quadricolor' (= *T. multicolor*). COMMELINACEAE, 2–4 species.

Zelkova a genus of deciduous trees and shrubs native to East and western Asia. *Z. serrata* (= *Z. keakii*) [JAPANESE ZELKOVA, SAW-LEAF ZELKOVA], is a large tree native to China and Japan, where it forms lowland forest with maple, beech and oak. It produces a hard, fine-grained golden wood like that of *elm and is highly valued in Japan for special building purposes such as temples; it is also a popular bonsai subject. It is grown in Europe and the USA for ornamental purposes.

Z. carpinifolia is being used in Europe as a replacement tree for those lost to Dutch elm disease. Wood from all species of *Zelkova* is used in cabinet-making and inlay work. ULMACEAE, about 5 species.

Zephyranthes a genus of half-hardy bulbous plants from the warm parts of the Western Hemisphere. The leaves are basal, narrow and strap-shaped. The yellow, white or pink flowers are funnel-shaped, recurved, solitary on a hollow leafless stalk.

The majority of the species are not hardy in cooler temperate regions and if grown outdoors the bulbs should be stored over vents. Otherwise, they are cultivated as cool greenhouse or conservatory ornamental plants, flowering in spring and summer. Cultivated species include the red-flowered Bolivian *Z. pseudocolchicum* and *Z. atamasco* (= *Amaryllis atamasco*) [ATAMASCO LILY], from the southern USA, which produces white, purple-tinged flowers.

One of the cultivated South American species (from the La Plata region) is *Z. candida*, with thick, stiff, rush-like leaves and white flowers, sometimes tinged pink. The widely cultivated southern Mexican and Central American *Z. grandiflora* bears large pink or rose flowers which are up to 10cm across. AMARYLLIDACEAE, about 40 species.

Zigadenus a small genus of perennial herbs native to Asia and North America. Some of the bulbous species, including *Z. elegans* (= *Anticlea elegans*) [WHITE CAMAS, ALKALI GRASS], *Z. glaucus* [WHITE CAMAS] and *Z. venenosus* (= *Toxicoscordion venenosum*), are cultivated as garden plants, producing aerial stems 30–90cm high with racemes of whitish-yellow or greenish flowers. *Z. nuttallii* (= *Toxicoscordium nuttallii*) [DEATH or POISON CAMAS], like most other species, is poisonous to livestock. LILIACEAE, about 15 species.

Zingiber a tropical Asian genus of perennial aromatic herbs with branched fleshy rhizomes. The flowers are borne usually in short, compact terminal inflorescences. The rhizomes of *Z. officinale* [GINGER PLANT] and related species are an important source of commercial ginger and an essential oil. Many cultivars of *Z. officinale* are widely cultivated in tropical and warm temperate countries. ZINGIBERACEAE, 80–90 species.

Zinnia a small predominantly Mexican, Central and South American genus of annual or perennial herbs or low shrubs, some of which are popular as garden-border plants or as cut-flowers.

The radiate flower heads are showy, with the ray florets occurring in an almost unlimited range of color combinations according to species and cultivar. Cultivated species include the low subshrub *Z. acerosa* (= *Z. pumila*), with yellow disk florets and white ray florets, and a number of annuals such as *Z. angustifolia* (= *Z. linearis*), with black-purple disk florets and bright orange ray florets with a yellow stripe down the center, *Z. elegans* [YOUTH-AND-OLD-AGE], with large flower heads up to 15cm in diameter in nearly all the numerous cultivars, which show a wide range of variation in the color of the ray florets, and *Z. haageana* [MEXICAN ZINNIA], with yellow disk florets and ray florets orange, red and yellow, or red and orange, according to cultivar. COMPOSITAE, about 17 species.

Zizania a small genus of annual or perennial aquatic grasses from Central and East Asia, and North America. *Z. aquatica* [AN-

Below Left The attractive scaling bark of Zelkova sinica. *(×$\frac{1}{12}$)*
Below Right The rhizomes of the Ginger Plant (Zingiber officinale), *which are dried and ground to make the spice ginger. (×$\frac{1}{2}$)*

*Eel Grass (*Zostera angustifolia*) growing on mud in the intertidal zone of a river estuary. It is pollinated under water. (×1)*

NUAL WILD RICE, INDIAN RICE] was once the staple diet of North American Indians, and is still cultivated on a small scale for its grains. The leaves of *Z. latifolia* are used for mat-making in Japan.
GRAMINEAE, 2–3 species.

Ziziphus a genus of tropical and subtropical trees and shrubs from both hemispheres. The horticulturally most important species include the large fruited INDIAN and COMMON or CHINESE JUJUBES, *Z. mauritiana* and *Z. jujuba* (= *Z. vulgaris*) [CHINESE JUJUBE, CHINESE DATE] respectively. The former is a shrub or small tree, commonly evergreen, native to India but widely cultivated elsewhere for its globose, fleshy, acid fruits up to 2.5cm in diameter.

Z. jujuba, a deciduous tree native from southeastern Europe to China, is more widely cultivated in many cultivars, particularly in East Asia. Its fleshy fruits are subglobose or oblong to obovoid and up to 3cm long. Jujube fruits are used in confectionery, stewed, dried or pickled. The smaller and less sweet fruits of

Z. lotus are thought to have been the lotus fruits of antiquity, from which was produced a wine that gave rise to blissful forgetfulness. RHAMNACEAE, about 40 species.

Zostera a genus of perennial grass-like herbs found growing beneath the surface of shallow salt water on gently sloping shores. Most species are found in temperate seas of both hemispheres. Members of the genus are commonly known as SEA-GRASSES, EEL GRASSES or GRASS-WRACKS.. Although *Zostera* itself is not eaten by molluscs, crustaceans or fish, the luxuriant flora and fauna that grow on it are the main source of food for many of these animals.
ZOSTERACEAE, about 12 species.

Zoysia a genus of slender creeping rhizomatous grasses native to coastal regions in Southeast Asia and New Zealand. They are much used for lawns in tropical and subtropical countries, notably the southern USA, being particularly successful in sandy ground. Widely grown, sometimes in mixtures with grasses of other genera, are strains of *Z. japonica* [JAPANESE or KOREAN GRASS], *Z. tenuifolia* [MASCARENE GRASS, KOREAN VELVET GRASS] and *Z. matrella* (= *Z. pungens*) [MANILA GRASS, JAPANESE CARPET GRASS, ZOYSIA GRASS] which forms a very close turf. GRAMINEAE, 5–6 species.

Zygopetalum a small genus of epiphytic pseudobulbous orchids from tropical America. They are all fairly robust plants with thick lanceolate leaves and erect flower stems arising from the base of the pseudobulbs. The flowers are usually a yellowish-green, variously ornamented with brown or purple markings. Species such as *Z. mackayi* are widely grown and they have been used to produce bigeneric hybrids with related genera such as *Zygosepalum* and *Chondrorhyncha*.
ORCHIDACEAE, about 20 species.

Glossary

ABAXIAL   On the side facing away from the stem or axis.

ACCESSORY FRUITS   Fruits derived from both tissues outside the ovary and ovarian tissues.

ACHENE   A small, dry, single-seeded fruit that does not split open.

ACUMINATE   Narrowing gradually to a point.

ADAXIAL   On the side facing the stem or axis.

ADVENTITIOUS   Arising from an unusual position, eg roots from a stem or leaf.

ALGINATES   Colloidal substances derived from members of the brown algae.

ALKALOID   A class of natural substances containing amine nitrogen, found mainly in the flowering plants, but also in a few other plant groups and in some animals. They are important because they are poisonous but they also have medicinal uses.

AMPHIBIOUS   Able to live both on land and in water.

AMPHIDIPLOID   (=Allotetraploid) A hybrid, the nuclei of the cells of which contain two diploid sets of chromosomes, one set derived from each parent.

ANGIOCARP   A fungal fruiting body which is closed at least until the spores are ready for dispersal.

ANNUAL   A plant that completes its life cycle from germination to death within one year.

ANODYNE   A pain-killing or soothing medicine.

ANTHELMINTIC   A drug capable of destroying or expelling intestinal worms.

ANTHERIDIUM   The male gamete-forming organ (sex organ) of many lower plants, such as mosses, ferns and fungi. Normally it produces motile sperms.

ANTHOCYANIN   The pigment usually responsible for pink, red, purple, violet and blue colors in flowering plants.

ANTISCORBUTIC   An agent that prevents scurvy.

APERIENT   A mild laxative.

APICAL MERISTEM   A group of actively dividing cells which occurs at the tip of an organ, such as a shoot, from which new growth is initiated.

APOMIXIS   (adj. apomictic)   Reproduction by seed formed without sexual fusion.

AQUATIC   Living in water.

ARCHESPORIUM   The layer of cells on the inside of a fern sporangium which normally give rises to the spore mother cells.

AREOLE   A group of hairs or spines, as found on the stems of cacti.

ARIL   A fleshy or sometimes hairy outgrowth from the hilum or funicle of a seed.

ASCOMYCETES   A class of fungi which reproduce by means of an ascus.

ASCUS   A cylindrical, round, club-shaped or pear-shaped organ of certain fungi (ascomycetes), within which the spores (ascospores) are produced.

AUTOPOLYPLOID   An organism which has more than the normal two sets (diploid) of chromosomes, all derived from the same parent species (cf Amphidiploid).

BASIDIOMYCETE   A class of fungi which reproduce by means of basidiospores. It includes the mushrooms, toadstools and puffballs as well as micro-fungi such as rusts and smuts.

BASIDIOSPORE   The unit of dispersal of basidiomycete fungi.

BASIDIUM   A specialized reproductive cell of basidiomycete fungi, often club-shaped or cylindrical. Each cell produces usually four spores (basidiospores) on peg-like appendages (sterigmata).

BAST FIBER   Fibrous elements within the sugar conducting tissues (phloem) of thickened stems and roots.

BERRY   A fleshy fruit without a stony layer and containing one or more seeds.

BIENNIAL   A plant that completes its life cycle in more than one year, but less than two, and which usually flowers in the second year.

BIPINNATE   (of leaves)   A pinnate leaf with the primary leaflets themselves divided in a pinnate manner.

BIPINNATIFID   (of leaves)   A pinnatifid leaf the segments of which are also pinnatifid.

BISERIATE   In two rows.

BOREAL   Relating to or growing in the cold northern latitudes dominated by coniferous forests.

BRACT   A leaf, often modified or reduced, which subtends a flower or inflorescence in its axil.

BULB   An underground organ comprising a much-shortened stem enclosed by fleshy scale-like leaves. It acts as a perennating organ and is a means of vegetative reproduction.

BULBIL   A small bulb or bulb-like organ often produced on above-ground organs.

CALCAREOUS   Containing or rich in lime.

CAPITULUM   (adj. capitulate)   A head of closely-packed, stalkless flowers.

CAPSULE   1 A dry fruit which normally splits open to release its seeds. 2 The sporangium of bryophytes.

CARCINOGEN   An agent that causes or creates a predisposition to cancer.

CARDIAC GLYCOSIDE   A class of drug that increases the force of contraction of the heart without increasing its oxygen consumption.

CARMINATIVE   An agent that relieves flatulence.

CARPEL   One of the flower's female reproductive organs, comprising an ovary and a stigma, and containing one or more ovules.

CATKIN   A pendulous inflorescence of simple, usually unisexual, flowers.

CHYTRID   A member of an order of fungi (Chytridales) which are microscopic, water- or soil-inhabiting species often parasitic on algae or higher plants.

CILIUM   A small hair-like structure on the surface of certain cells which, by waving movements, brings about locomotion.

CIRCUMBOREAL   Growing throughout the boreal regions of all Northern Hemisphere continents.

CLONE   A group of plants that have arisen by vegetative reproduction from a single parent and which are therefore all genetically identical.

CODOMINANT   One of two or more species that together dominate a vegetation type.

COENOCYTIC   Not having cross walls, so that many individual nuclei occupy the same area of cytoplasm; non-cellular.

CONIDIOPHORE   The stalk (hypha) on which one or more asexual spores (conidia) are borne.

CORDATE   (of leaves)   Heart-shaped.

CORM   (adj. cormous)   A bulbous, swollen, underground, stem-base, bearing scale-leaves and adventitious roots; it acts as a perennating or storage organ and is a means of vegetative propagation.

COROLLA   All the petals of a flower; it is normally colored.

CORONA   A series of petal-like structures in a flower, either outgrowths from the petals, or modified from the stamens, eg a daffodil 'trumpet'.

CORYMB   (adj. corymbose)   A rounded or flat-topped inflorescence like a raceme but the flower stalks are longer on the outside so that all the flowers are at about the same level.

CRENATE   (of leaf margins)   Shallowly round-toothed with indentations no further than one-eighth of the distance to the midrib.

CULM   The stem of a grass or sedge.

CULTIVAR   (abbreviation cv)   Cultivated variety. A taxonomic rank used to denote a variety that is known only in horticultural cultivation. Cultivar names are nonlatinized and in living languages; typographically they are distinguished by a non-italic typeface, with a capital initial letter and enclosed in single quotation marks, for example Betula pendula 'Fastigiata'.

CV   Abbreviation for cultivar.

CYTOLOGY   The study of cells and their internal structure.

DECIDUOUS   (of plants)   Shedding their leaves seasonally.

DECOCTION   An extract obtained by boiling.

DECUMBENT   Prostrate; lying flat, but with the growing tip extended upwards.

DEHISCE   To split open to discharge its contents.

DEMULCENT   A mucilaginous or oily substance capable of soothing damaged mucous membranes.

DIAPHORETIC   An agent that increases perspiration.

DICHOTOMOUS   Branching into two equal forks.

DICOTYLEDON   A member of one of two subclasses of angiosperms; a plant whose embryo has two cotyledons (cf monocotyledon).

DINOFLAGELLATE   A member of the algal class Dinophyceae – single-celled aquatic algae which move by the action of two flagella. They are notable components of the phytoplankton of the oceans.

DIPLOID   Having two sets of chromosomes in the nucleus of its cells, one from each parent.

DIURETIC   An agent that tends to increase the flow of urine.

DRUPE   A fleshy fruit containing one or more seeds, each of which is surrounded by a stony layer.

EMETIC   An agent that induces vomiting.

EPIPHYTE   (adj. epiphytic)   A plant that grows on the surface of another, without deriving food from its host.

ESSENTIAL OIL   A vegetable oil made up of complex mixtures of volatile organic compounds that impart the characteristic fragrance or taste to plants. They are often extracted for use in perfumes and flavorings.

EUKARYOTE   (adj. eukaryotic)   Any organism that has a membrane-bound nucleus; this group includes all algae (except blue-green algae), fungi and green plants (cf prokaryote).

EUSPORANGIUM   (adj. eusporangiate) One of the two main types of sporangia found in ferns. They are stalked and have walls at least two cells thick (cf leptosporangium).

EXPECTORANT   An agent that tends to promote discharge of mucus from the respiratory tract.

FASCIATION   The coalescing of stems, branches etc into bundles.

FEBRIFUGE   An agent that reduces the temperature of the body.

FLOWER   The structure concerned with sexual reproduction in the flowering plants (angiosperms). Essentially it consists of the male organs (androecium) comprising the stamens and the female organs (gynoecium) comprising the ovary, style(s) and stigma(s), usually surrounded by a whorl of petals (the corolla) and a whorl of sepals (the calyx). Male and female organs may be in the same flower (bisexual) or in separate flowers (unisexual).

FORMA   or form   A taxonomic division ranking below variety and used to distinguish plants with trivial differences.

FRUIT Strictly the ripened ovary or group of ovaries of a seed plant and its contents. Loosely, the whole structure containing ripe seeds, which may include more than the ovary.

GAMETE The haploid reproductive cell (male or female) that is produced during sexual reproduction. At fertilization the nucleus of the gamete fuses with that of another gamete of the opposite sex to form a diploid zygote that develops into a new individual.

GEMMA A cell or group of vegetative cells produced by some algae, fungi and bryophytes which are dispersed and may grow into new individuals – a form of asexual reproduction.

GLABROUS Without hairs.

GLAUCOUS With a waxy, grayish-blue bloom.

GLOBOSE Spherical, rounded.

GLYCOSIDE A class of compound found in many plants, yielding a sugar on hydrolysis along with other substances.

HAUSTORIUM A peg-like fleshy outgrowth from a parasitic plant, usually embedded in the host plant and drawing nourishment from it.

HERB (adj. herbaceous) A plant that does not develop persistent woody tissue above ground and either dies at the end of the growing season or overwinters by means of underground organs, eg bulb, corm, rhizome.

HETEROCYST (of blue-green algae) A thick-walled, transparent cell in which fixation of nitrogen takes place.

HETEROPHYLLY (adj. heterophyllous) Having leaves of more than one type on the same plant.

HETEROTROPHIC Unable to synthesize food from simple inorganic substances and thus requiring a source of organic food, as in the fungi.

HEXAPLOID An individual whose cells contain six sets of chromosomes, not the normal two.

HILUM The scar left on a seed marking the point of attachment to the stalk of the ovule.

HYPHA The tubular filamentous units of construction of fungi which together make up the mycelium.

HYPOCOTYL The part of an embryo or seedling that lies between the seed leaves (cotyledons) and the primary root (radicle).

IMBRICATE (of sepals and petals) Overlapping, as in a tiled roof.

INDUMENTUM A covering, usually of hairs.

INDURATE Hardened.

INDUSIUM (adj. indusiate) A flap of tissue that covers the sporangium cluster (sorus) of some ferns.

INFLORESCENCE Any arrangement of more than one flower, eg capitulum, corymb, cyme, panicle, raceme, spadix, spike and umbel.

INFRUCTESCENCE A cluster of fruits, derived from an inflorescence.

INTERGENERIC HYBRID A hybrid with parents belonging to different genera.

INTERSPECIFIC HYBRID A hybrid with parents belonging to different species.

INVOLUCRE A whorl of bracts beneath or surrounding an inflorescence.

ISOMER A chemical compound containing the same elements in the same numbers as another compound but differing in structural arrangement.

KARYOTYPE The characteristic size, shape and number of the set of chromosomes, of a somatic (non-reproductive) cell.

KERNEL A general term applied to the abundant nutritive tissue of a large seed, as with the seed inside the stone of a peach.

LABELLUM Either of the two parts, usually upper and lower lip, into which the corolla of orchids is divided.

LANCEOLATE Narrow with tapering ends, as a lance.

LEPTOSPORANGIUM (adj. leptosporangiate) One of the two main types of sporangia found in ferns. They have little or no stalk and walls that are only one cell thick (cf eusporangium).

LIANA A woody climbing vine.

LIGULATE Strap-shaped or tongue-shaped.

LIGULE (of leaves) A scale-like membrane on the surface of a leaf; (of flowers) the strap-shaped corolla in some Compositae.

LITHOPHYTE A plant which grows on stones and not in the soil.

LITTORAL Living on the seashore.

MEIOSIS Two successive nuclear divisions of a diploid cell in the course of which the chromosome number is reduced from diploid to haploid; the haploid cells often subsequently develop into gametes (cf mitosis).

MERISTEM A group of cells capable of dividing indefinitely.

MICROPHYLL A small leaf.

MITOSIS A form of nuclear division in which the chromosome complement is duplicated exactly and the daughter nuclei remain haploid, diploid or polyploid. The form of division that occurs during normal vegetative growth (cf meiosis).

MONOCARPIC Fruiting only once and then dying.

MONOCOTYLEDON One of two subclasses of angiosperms: a plant whose embryo has one cotyledon (cf dicotyledon).

MONOTYPIC Having only one member (species, genus etc).

MOTILE Able to move.

MYCELIUM The mass of hyphae which constitutes the body of a fungus.

MYCORRHIZA The association of the roots of some vascular plants with fungal hyphae.

NAKED (of flowers) Lacking a perianth.

OBOVATE (of leaves) Having the outline of an egg, with the broadest part above the middle and attached at the narrow end.

OBOVOID (of solid objects) Inversely egg-shaped, with the widest point above the middle.

OOGONIUM The unicellular female sex organ of certain fungi and algae.

OVATE (of leaves) Having the outline of an egg with the narrow end above the middle.

OVOID (of solid objects) Egg-shaped.

PALMATE (of leaves) With more than three segments or leaflets arising from a single point, as in the fingers of a hand.

PANICLE (of inflorescences) Strictly a branched raceme, with each branch bearing a further raceme of flowers. More loosely, it applies to any complex, branched inflorescence.

PAPILLA (adj. papillate) A small blunt hair or projection.

PARASITE A plant that obtains its food from another living plant to which it is attached.

PARENCHYMA (adj. parenchymatous) Tissue made up of thin-walled living photosynthetic or storage cells which are capable of division even when mature. The commonest type of plant tissue.

PELTATE (of leaves) More or less circular and flat with the stalk inserted in the middle.

PENDULOUS Hanging down.

PERENNATING Persisting from year to year.

PERENNIAL A plant that persists for more than two years and normally flowers annually.

PERICARP The wall of a fruit that encloses the seeds and which develops from the ovary wall.

PERISTOME The fringe of teeth that occur at the tip of the capsule of mosses.

PERITHECIUM A closed, spherical or flask-shaped fruiting body (ascocarp) of certain ascomycete fungi; each has a small pore at the top to allow release of the ascospores.

PHOTOSYNTHESIS The process by which green plants manufacture sugars from water and carbon dioxide by converting the energy from light into chemical energy with the aid of the green pigment chlorophyll.

PINNA Each segment of a fern leaf.

PINNATE (of leaves) Compound, with leaflets in pairs on opposite sides of the midrib.

PINNATIFID (of leaves) Pinnately divided at least as far as the midrib.

PINNATISECT (of leaves) Pinnately divided, but not as far as the midrib.

PINNULE Each segment of a pinnate leaf.

PISTIL The female reproductive organ consisting of one or more carpels, comprising ovary, style and stigma; the gynoecium as a whole.

PLASMODIA Naked multinucleate masses of moving protoplasm which feed in amoeboid fashion, as in some stages of the life history of certain slime fungi.

PLOIDY Referring to the number of full chromosome sets in a nucleus, ie haploid = one set, diploid = two sets, polyploid = more than two sets.

POLYPLOID An organism that has more than the normal two sets of chromosomes in its somatic cells.

POME A fleshy false fruit, the main flesh comprising the swollen receptacle and floral parts surrounding the ovary, as in the apple and other members of the family Rosaceae.

PROKARYOTIC Organisms lacking a membrane around their nuclei, plastids and other cell organelles, as in the bacteria and blue-green algae (cf eukaryote).

PROPHASE An early stage in nuclear division when the chromosomes become visible and distinct from each other.

PROTEOLYTIC ENZYME A substance that aids the breakdown of proteins.

PROTHALLUS The more or less independent sexual phase (gametophyte) of certain primitive plants such as ferns and club mosses.

PROXIMAL Near to.

PSEUDOBULB (adj. pseudobulbous) A swollen, bulb-like part of the stem of orchids.

PSYCHOACTIVE DRUG A drug capable of affecting mental activity.

PSYCHOTOMIMETIC DRUG A drug capable of inducing psychotic symptoms (mental disturbance involving distortion of contact with reality).

PTERIDOPHYTE A unit of classification of plants which encompasses the lower vascular plants, ie ferns, club mosses and their allies, and horsetails.

PURGATIVE An agent that causes evacuation of the bowels.

PYCNIDIUM A minute hollow globose or flask-shaped asexual fruiting body of some fungi containing conidiospores.

PYRENOID A starch-rich body found in some algae and liverworts.

RACEMATE An optically inactive form of a chemical compound that also exists in optically active forms.

RACEME (adj. racemose) An inflorescence consisting of a main axis, bearing single flowers alternately or spirally on stalks (pedicels) of approximately equal length. The apical growing point continues to be active so there is usually no terminal flower and the youngest branches or flowers are nearest the apex. This mode of growth is known as monopodial.

RECURVED Curved or bent backwards.

REFUGIA   Habitats that have escaped drastic changes in climate, enabling species and populations to survive, often in isolation.

RESTITUTION NUCLEUS   A single nucleus instead of two resulting from a failure in meiosis.

RETTING   The breaking down of fibrous material into separate fibers by soaking in water and allowing microbial activity to take place.

RHIZOID   Root-like or roothair-like structures found in most lower plants, which help to anchor the plants and absorb water and nutrients. Not true roots.

RHIZOME   (adj. rhizomatous)   A horizontally creeping underground stem which persists from season to season (perennates) and which bears roots and leafy shoots.

RHIZOMORPH   A thick strand of fungal hyphae.

ROSETTE   A group of leaves arising closely together from a short stem, forming a radiating cluster on or near the ground.

ROTATE   Wheel-shaped.

SAGITTATE (of leaves)   Shaped like an arrowhead with two backward-directed barbs.

SALVERFORM   Trumpet-shaped.

SAMARA   A dry fruit in which the wall is extended to form a flattened membrane or wing.

SAPONINS   A toxic soap-like group of compounds which are present in many plants.

SAPROPHYTE   A plant that cannot live on its own, but which needs decaying (non-living) organic material as a source of nutrition.

SCANDENT   Climbing.

SCAPOSE   With a solitary flower on a leafless stalk (scape).

SCLEROTIUM   A hardened resting body produced by some fungi as a means of surviving harsh environmental conditions.

SCRAMBLER   A plant with a spreading, creeping habit usually anchoring by hooks, thorns or tendrils.

SEPTUM (adj. septate)   A cross wall dividing cells or compartments.

SESSILE   Without a stalk, eg leaves without petioles or stigmas without a style.

SETA   The stalk of the capsule in mosses and liverworts.

SHRUB   A perennial woody plant, normally with well developed side-branches that appear near the base so that there is no trunk.

SIMPLE (of leaves)   Not divided or lobed in any way.

SOLITARY (of flowers)   Occurring singly in each axil.

SORUS   A cluster of sporangia, as in ferns.

SPADIX   A spike of flowers on a swollen fleshy axis.

SPATHE (adj. spathaceous)   A large bract subtending and often ensheathing an inflorescence. Applied only in the monocotyledons.

SPIKE   An inflorescence of simple racemose type in which the flowers are stalkless.

SPORANGIUM   A hollow structure in which spores are formed.

SPOROPHYLL   A fertile leaf or leaf-like organ.

STAND   A uniform group of plants growing in a continuous area.

STERILE   Unable to reproduce sexually.

STIPE   A stalk as in the stalk of large fungi or the stalk of a fern leaf.

STIPULAR   Relating to stipules.

STIPULE   A leafy appendage, often paired, and usually at the base of the leaf-stalk.

STROBILUS   A cone; a shortened section of a stem bearing modified leaves or scales (appendages) on which sporangia are produced.

STROMA (pl. stromata)   A dense mat of hyphae from which fungal fruiting bodies are produced.

STYLAR   Referring to the style.

SUBGLOBOSE   Almost round or spherical.

SUBLITTORAL   Growing near the sea but not within the tidal limits.

SUBSHRUB   A shrub which has tender new growth that dies back seasonally.

SUBSPECIES   A taxonomic division ranking between species and variety. It is often used to denote a geographic variation of a species.

SUCCULENT   With fleshy or juicy organs containing reserves of water.

SYMBIONT   An organism that lives in a symbiotic relationship with another.

SYMBIOSIS   The non-parasitic relationship between living organisms to their mutual benefit.

SYMPODIAL (of stems or rhizomes)   With the apparent main stem consisting of a series of usually short axillary branches (cf cyme).

TAXON   A taxonomic group of any rank, eg species, genus, family.

TAXONOMY   The science of classification of living organisms.

TERATOGEN   A substance that can cause malformation of the fetus.

TERMINAL   Situated at the tip.

TERNATE (of leaves)   Compound, divided into three parts more or less equally. Each part may itself be further subdivided.

TERRESTRIAL   Only living on the land.

TESTA   The outer protective covering of a seed.

TETRAPLOID   Having twice the normal (diploid) number of chromosomes in the nuclei of its cells.

THALLOSE   Flattened; not distinctly divided into root, stem and leaves.

TOPIARY   The practice or art of trimming trees and shrubs into unusual shapes.

TREE   A large perennial plant with a single branched and woody trunk and with few or no branches arising from the base (cf shrub).

TRIFOLIOLATE (of leaves)   Having three leaflets.

TRIPLOID   Having three times the haploid number of chromosomes in the nuclei of its cells.

TUBER   An underground stem or root that lives over from season to season and which is swollen with food reserves.

TUBERCLE   A rounded swelling or protuberance.

UMBEL   An umbrella-shaped inflorescence with all the stalks (pedicels) arising from the top of the main stem. Umbels are sometimes compound, with all the stalks (peduncles) arising from the same point and giving rise to several terminal flower stalks (pedicels).

UNDERSHRUB   A perennial plant with lower woody parts, but herbaceous upper parts that die back seasonally.

VAR   Abbreviation for variety.

VARIETY   A taxonomic division ranking between subspecies and forma, although in the past often used as the major subdivision of a species. Such taxa are named by adding the italicized variety name, for example *Pinus ponderosa* var *arizonica*. It was once used to designate variants of horticultural origin or importance, but the rank of cultivar should now be used for this category, although many names of horticultural origin still reflect the historical use of the variety rank.

VASOACTIVE DRUG   A drug that acts on blood vessels.

VERNALIZATION   The process of exposing seedlings to low temperatures which is necessary if subsequent flowering and fruiting is to be effective.

VERTICILLASTER (of inflorescences)   With groups of flowers arranged in whorls at the nodes of an elongated stalk.

XEROPHYTE   A plant which is adapted to withstand extremely dry conditions.

Index of Scientific Names

The following is an index to genera that do *not* have entries of their own.

Abelmoschus 170, 242
Achyranthes 180
Aerides 261
Aframomum 73, 259
Agastache 178
Alcea 172
Aleuria 152
Alhagi 213
Alkanna 17, 122
Alliaria 168
Aloysia 168, 202, 341
Alpinia 150
Alsophila 108, 330
Altingia 282
Amomum 73
Amomyrtus 231
Amorpha 179
Amyris 125
Anabasis 26
Anadenanthera 226
Ananas 122
Anastatica 284
Anthyllis 145, 342
Antiaris 335
Anticlea 350
Arceuthobium 225
Archangelica 143
Argyranthemum 87
Armoracia 143, 173
Arrhenatherum 145
Artabotrys 349
Arthrocnemum 295
Ascroe 314
Asimina 253
Asphodeline 45
Aspidium 121
Aspidosperma 252
Attalea 49, 141, 265
Aucoumea 329
Auricularia 182
Aurinia 25
Austrocedrus 197
Austromyrtus 181
Averrhoa 147
Avicennia 212

Backhousia 181
Bahia 129
Baileya 213
Balanocarpus 112
Balsamita 87, 102, 168
Banisteriopsis 220
Barbarea 104, 287
Bartramia 262
Beilschmiedia 43
Bellevalia 177
Belliolum 121
Beloperone 185
Benincasa 159, 216, 338
Bertholletia 61, 239
Betonica 168
Bixa 122
Blighia 16, 147
Bordetella 51
Botrychium 51
Bowenia 108
Brassavola 77
Brassolaeliocattleya 190
Brimeura 177, 301
Brugmansia 220
Bryophyllum 185
Bubbia 121
Buchloe 145
Bursera 125

Cacalia 126, 302
Calamagrostis 284
Calea 220
Calocarpum 147, 297
Calocedrus 197
Calycanthus 19, 74

Calyptrothecium 231
Canangium 349
Canarium 24, 125, 239, 243
Caoutchoua 72
Carapa 244
Casearia 60
Casimiroa 297
Castanospermum 239
Catamoches 77
Catamodes 77
Catha 56, 67, 187
Centrosema 145, 256
Cephaelis 180, 221
Ceratocystis 126
Ceratonia 17, 74
Ceratopetalum 298
Ceratozamia 108
Ceriops 212
Chamaemelum 36, 83
Chamaespartium 156
Chelidonium 26
Chimonobambusa 43
Chlorella 22
Chlorophora 122, 181
Chondrodendron 107, 221
Chondrorhyncha 351
Choricarpia 181
Chrysophyllum 112, 314
Chukrasia 78
Cibotium 330
Cladium 302
Clathrus 278, 314
Claviceps 128, 221
Cliftonia 65
Closterium 23
Clostridium 51
Cnidoscolus 313
Coccinia 216
Coffea 95, 221
Colocasia 318, 323
Colocynthis 89, 218
Copaifera 49, 100, 181
Copernicia 345
Coriandrum 100, 143, 168
Corynanthe 26
Corynebacterium 51
Coula 344
Coumarouna 330
Crassocephalum 302
Crocanthemum 165
Crucibulum 58, 278
Cryptocarpus 267
Cryptocarya 54
Cryptostegia 208, 289
Cubeba 267
Cuminum 107, 143
Curcuma 122
Curtisia 45
Cusparia 143
Cyamopsis 195
Cyathus 58, 278
Cylicodiscus 160
Cynoches 77
Cynometra 181, 329
Cyrilla 181
Cyrtosperma 318

Dacryodes 125
Dendranthema 87
Dendrosenecio 303
Dentaria 73
Derris 115
Dichanthium 145
Dichelostemma 64
Dictyophora 278
Dictyosperma 265
Didymoglossum 137
Digitaria 145, 224

Dilsea 283
Dimocarpus 197
Dinochloa 49
Dipladenia 211
Diplazium 136
Diplococcus 51
Diplotaxis 168
Dipteryx 330
Dodonaea 243
Doxantha 55
Drepanocladus 228
Dryobalanops 71, 329
Duchnesnea 315
Durio 123, 147
Dyera 329
Dysoxylum 288

Echinochloa 224
Echinospartium 156
Echioides 41
Elaphomyces 100
Empusa 127
Enallagma 104
Entandrophragma 209, 297, 329
Eperua 329
Epicampes 64
Epicattleya 127
Eremophila 296
Eruca 168, 287
Eucheuma 283
Euchlaena 324
Eudorina 22, 23
Eusideroxylon 181
Euxylophora 298
Exospermum 121

Festuca 135
Flindersia 329
Fortunella 89, 147
Funkia 173
Funtumia 289
Furcraea 140

Galactinia 260
Galeobdolon 190
Galeorchis 250
Garcinia 122
Geastrum 123, 278
Githago 16
Glandularia 341
Gloxinia 306
Goldfussia 316
Gonioma 60
Gonium 22
Gonystylus 329
Gracilaria 283
Grossularia 286
Guibourtia 97
Gymnocarpium 136
Gymnotheca 176
Gyromitra 226

Haematoxylon 122
Halleria 172
Hedyotis 176
Heliocereus 128
Himanthalia 20
Hippocrepis 342
Holosteum 85
Hopea 118, 181
Hunnemannia 273
Hyacinthella 177
Hydnocarpus 83
Hydrastis 56
Hydrodictyon 22
Hyparrhenia 145
Hyssopus 168, 178

Indigofera 145, 179, 347
Inga 69
Intsia 181
Ionotus 152
Isatis 122

Ismene 177
Isoetes 92, 93
Isoplexis 144

Jacobaea 303
Jacobinia 185
Jatropha 313
Juniperus 184, 329

Kaempfera 185
Kandelia 212
Karatas 235
Kerstingiella 195
Kitchinia 185

Lactobacillus 298
Lagerta 189
Lambertia 172
Langermannia 152, 278
Laportea 235
Lawsonia 122
Lecanora 213
Lecythis 239, 297
Ledebouria 301
Lens 193, 195
Leopoldinia 265
Lepidozamia 108, 208
Leucaena 145
Levisticum 168, 205
Licania 244
Licuala 210
Lippia 341
Lobivia 124
Lobularia 25
Lochneria 77
Lophomyrtus 231
Lovoa 344
Lucuma 124, 297
Luma 231
Lycopodium 92
Lysiloma 69

Macadamia 208
Macroptilium 145
Macrotomia 41
Majorana 213, 250
Malcolmia 315
Mammea 147, 211
Mandragora 212, 221
Marsdenia 289
Matricaria 83
Megapterium 242
Melachodendron 314
Melandrium 305
Melanorrhoea 189
Melocanna 49, 140
Merremia 227
Methysticodendron 220
Microcycas 108
Microcystis 50
Millettia 335
× Miltassia 224
× Miltonidium 224
Momordica 159
Montbretia 332
Monostroma 21
Mora 329
Morchella 226
Mormodes 77
Muhlenbergia 64
Mulgedium 189
Murraya 298
Mycobacterium 51
Myosoton 85
Myrrhis 168, 231
Mystacidium 31
Myrteola 231

Naegelia 30
Narthecium 45
Neckeradelphus 232
Neckeropsis 232
Nectandra 160
Neisseria 50
Neoglaziovia 140
Nephelium 147, 202, 281
Nitrobacter 51
Nitrosomonas 51
Nopalxochia 128
Nypa 237

Obolaria 258
Ochlandra 49, 140
Olneya 181
Oreodoxa 318

Pachyrrhizus 198, 338, 348
Palaquium 161, 292
Panaeolus 220

Pandorea 324
Pandorina 22
Parahebe 164
Parthenium 289
Pastinaca 255, 338
Paullinia 56, 67
Pausinystalia 26
Payena 161
Peganum 220
Peltophorum 114
Pentadesma 321
Pentaglottis 17
Pericopsis 14, 329
Perilepta 316
Petroselinum 143, 168, 255, 338
Phalaenopsis 201
Phallus 278, 314
Phanerophlebia 110
Phlebodium 272
Phoradendron 225
Phylloglossum 92
Phyllostylon 60
Phymatotrichum 26
Pimenta 19, 266
Platanthera 111
Podranea 324
Poinciana 114
Poinsettia 134
Poterium 9, 296
Pouteria 124, 147, 297
Protium 125
Pseudowintera 121
Psilotum 92, 93
Psychotria 26, 180
Psylliostachys 201
Pteridium 122
Pterocladia 283
Pueraria 145

Ramelina 199
Reseda 122
Reynosia 181
Rhamnidium 181
Rhaphanus 338
Rhizobium 57
Rhizocarpon 199
Rhodanthe 166
Rhodiola 302
Rhodymenia 21, 283
Rhynchospora 302
Richardia 349
Rickettsia 51
Rivularia 51
Roccella 122, 199
Rubia 122

Sabal 141
Salmonella 51
Sanseverinia 296
Sapium 122, 321
Sarothamnus 110
Saxifraga 204
Scenedesmus 22
Schaefferia 60
Schinopsis 280
Schleichera 189, 208, 244
Schotia 173
Scleroderma 123
Scolopendrium 263
Scopolia 26
Scorzonera 338
Scytonema 51
Sebastiana 183
Sechium 216, 338
Selaginella 92, 284
Shigella 51
Shizocodon 305
Sicana 216
Sideroxylon 181
Silybum 325
Sphaeropteris 108
Sphenostylis 195, 348
Spinacia 311, 338
Sterculia 243
Streptococcus 51
Struthiopteris 214
Stylites 92, 93
Stylosanthes 145
Synsepalum 225
Syzygium 91, 143, 221

Tabellaria 51
Tamarindus 143, 147, 195, 321
Taraktogenos 83
Tasmannia 121
Tecomaria 172, 324
Telekia 66
Telesonix 60

Telfairia 239
Teline 156
Tetraclinis 296
Tetragonia 311, 338
Thea 323
Themeda 145
Thermopolyspora 44
Thesium 327
Thymus 326
Thyrsopteris 330
Tipuana 288
Tmesipteris 92, 93
Toddalia 181
Tofieldia 45
Torresea 27
Tortella 228
Toxicodendron 286
Toxicophloea 11
Toxicoscordion 350
Trachylobium 177
Trachyspermum 75
Trebouxia 198
Trentepohlia 20
Treponema 50
Trichodesmium 283
Trichosanthes 159, 216, 338
Triplochiton 329
Tripsacum 145
Triticosecale 332
Tritonia 105
Tuber 333
Tulipa 333
Turbina 220
Turraeanthus 329

Ugni 231
Ullucus 335, 338
Ulothrix 22
Uncaria 108, 122, 150
Urena 140

Vatica 118
Virola 335
Voandzeia 195
Vonitra 265

Waldsteinia 315

Xeranthemum 179
Xylopia 259

Zygogynum 121
Zygosepalum 351

Index of Common Names

The following is an index to the common names of plants and plant products cited in the text. A **bold** number indicates *either* a main entry heading in the book *or* a popular name that is generally applied to all or many members of a genus, and appears as a subheading to the genus title. An *italic* number indicates an illustration. Page numbers are not given where the genus name is also applied to the popular name; for example, the popular name for the genus *Dahlia* is also DAHLIA; information on this group will be found under the genus entry *Dahlia*.

Aaron's Beard 178
Aaron's Rod 227
Abaca fiber 8, *8*, 141, 167, 230, 306
Abir 165
Abrin 8
Absinthe 18, 31, 42, 56
Acacia *10*
  Bull-horn 9
  False 287
  Sweet 9, 76
Acajou 10, 161
Acajou gum 28
Aceituno 305
Aconite 11, 221
  Indian 11
  Winter 128
Aconitine 26, 221
Adam's Apple 321
Adam's Flannel 227
Adam's Needle 349
Adder's Meat 314
Adder's Tongue 246
  Yellow 131
Adonis
  Autumn 12
  Spring 12, *13*
  Summer 12
Afara 324, 329
Afriemengras 131
Afrormosia 329
Agapanthus, Blue *15*
Agar 20, 156
Agaric
  Fly *154*, 220
Agrimony 15
  Common 15, *16*
  Scented 15
Aguacate 47, 147, 260
Aguardiente 31
Air Plant 186
Ajowan oil 75
Akavit 18
Akee 147
Akoloa 59
Al 226
Alan 329
Albizia
  Fragrant 17
  Tall 16
Alcohol 18
Alder 17, 122
  Black 17, 171
  Common 17
  European 17
  Gray 17
  Red 17
  White 17
  Yellow 334
Alder bark 17
Alder Buckthorn 17, 144
Ale 18
Alecost 168
Alehoof 158
Alexanders 168, *306*, 307
Alfa 131
  False 131
Alfalfa 145, **205**, *205*, 215
  Common 205
  Yellow-flowered 205
Alfilaria 130
Alga
  Frog-spawn 53
Algae **20**
Algaroba 17, 74

Algarrobo 17
Algarrobo Bean 275
Algarrobo Blanca 275
Alginates 20, 190, 208
Alizarin 208
Alkalois **26**
Alkanet 122
  Dyer's 17
All-good 84
All-heal 18, 276
Alligator Pear 47, 147, 338
Allspice 19, *142*, 143
  Carolina 19, **74**
  Japanese 19
Almond 19, *19*, 239, *239*, 276
  Bitter 24, 239
  Country 24
  Ground 109
  Indian 24, 324
  Java 24, 239
  Oil 24
  Sweet 24, 239
  Tropical 24, 324
Aloe 221
  American 15
  Barbados 24
  Cape 24
  Curacao 24
  Green 150
  Partridge Breast 24
  Water 315, *315*
Alum Root 157, 167
Aluminium Plant 265
Alyssum
  Golden 25
  Sweet 25
Amaranth, Grain 81
Amarillo 124
Amaryllis 171
Ambarella 312
Amber 284
Amberbell 131
Amboyna 277, 329
Amchur 212
Anabasine 26
Andiroba 244
Andromeda
  Marsh 29
Anemone **29**, *30*
  Crown 30
  Japanese 30
  Poppy 30
  Scarlet 30
  Wood 30
Angelica 39, 143, 168
  Japanese 39
  Lovage **205**
  Purple 30
  Wild 30, *30*, 168
Angelica Tree
  Chinese 38
Angel's Fishing Rod 116
Angel's Tears 232, 308
Angel's Trumpet 113, 220
Angostura 30, 108, 143
Anis 30
Anise 31, *179*, *142*, 143, 168
  Purple 179
  Star 31, 179
    Japanese 179
    Yellow 179
Anise Tree 179
Aniseed 132, 221
Anisette 31

Annatto 31
Annona Blanca 31
Anthraquinone 221
Anthrax 49
Anthurium, White 310
Anu *291*, 333, 338
Anyu 333, 338
Aonori 21
Apamate 321
Ape, Blue 347
Apio 41
Apple 37, *38*, 56, 147, *148*
  Adam's 321
  Balsam 226
  Beef 213
  Bitter 89, 106
  Cashew 147, *238*
  Chess 309
  Conch 255
  Crab **104**
    American 104
    Chinese 104
    Plum-leaved 104
    Prairie 104
    Showy 104
    Siberian 104
    Southern Wild 104
    Wild 104
    Wild Sweet 104
  Custard 147
  Golden 273, 312
  Indian 113
  Jew's 46, 307
  Kangaroo 307
  Love 307
  Mad 46, 307
  Mammee 147
  Mammey 211
  May 270
    Himalayan 270
  Melon 216
  Pond 31
  Rose
    Yellow 133
  Sugar 31
  Thorn 220, 221
    Downy 220
Apple of Peru **235**
Apple of Sodom 307
Appleberry 57
Applejack 18
Apricot 38, 147, *148*, 276
  St. Domingo 211
Araguaney 321
Aralia
  Chinese 38
  False **118**
Aramina 39, 140
Arbor-vitae **325**
  American 325, 329
  False 326
  Giant 325, *325*, 329
  Hiba 326
  Japanese 326
  Oriental 326
  Western 325
Archangel, Yellow 190
Archer Plant 97
Arecoline 26, 221
Armagnac 18
Arnicine 41
Arrack 18, **41**
Arrow Leaf 294
Arrowhead
  Common European

Arrowhead (cont)
  294, 295
  Giant 294
Arrowroot 41, 213, 290, *291*, 318
  African 41
  Bermuda 41
  Bombay 41
  Brazilian 75
  Brazilian Para 41
  East Indian 41, 107, 318, 321
  Fijian 318, 321
  Florida 349
  Guinea 68
  Indian 107
  Polynesian 321, *321*
  Portland 41, 42
  Purple 71
  Queensland 41, 71, *71*, 318
  Rio 41
  St Vincent 41
  Tacca 41
  Tahitian 318, 321
  Tous-les-mois 41
  Tulema 41
  West Indian 41
Artichoke 42, 338
  Chinese 42, 313, 338
  French 42
  Globe 42, *42*, 83, 109, 338, *339*
  Japanese 42, 313
  Jerusalem 42, 290, 319, 338
Artillery Gunpowder 265
Arum
  Bog 68, *68*
  Dragon 120
  Water 68
  Whitespot Giant 27
Arum Lily 42, *42*, 349
Asafoetida *135*, 221
Ash 4, 122
  Alpine 43, 131
  American 329
  Black 43
  Blue 43
  Canary 43
  European 43, 329
  Flowering 43, *43*
  Japanese 329
  Manchurian 329
  Manna 43, *43*, 213, 318
  Mountain 43, 131, **227**, 308, *309*
    American 309
    Blue 43
    European 309
  Prickly 39, 43
    Northern 351
    Southern 351
  Quaking 43
  Red 43, 329
  Stinking 277
  White 43, 329
Asparagus 43, 44, 338, *339*
  Common 43
  Fern 43, 44
  Garden 43
  Wild 338
Asparagus Bean 44
Asparagus Pea 44
Aspen 44, **273**
  American 44, 273
  Chinese 44
  European 44, 273
  Japanese 44
  Large-toothed 44
  Quaking 44, 273
  Trembling 44
Asphodel **44**
  Bog 45
  Giant 45
  Scottish 45
  White 45
  Yellow 45
Assegai 45
Assegai wood **45**
Aster **45**
  Beach 129
  China 45, *45*, **69**, *69*, 315
  Garden 45
  New England 223
  New York 223
  Stoke's **315**
  Tree **242**
Atropine 26, 221
Attar of Roses 46, 132
Aubergine 46, 307, 338, *340*

Aubrietia 47, *47*
Auricula 47, 274
Auricula Tree 70, *70*
Australia Nut **208**
Avens
  Mountain 121
  Water 157, *157*
  Wood 157
Avocado 47, *48*, 147, 149, 260, 338, *340*
Avodiré 329
Awl-leaf 294
Azalea 48, *48*
  Pontic 285
  Summer 257

Babassu oil 244, 249
Baby-blue-eyes 233
Baby's Breath 161
Baby's-tears 308
Bacteria **50**
Bacteriophage 51
Bael 13
Bagasse 49, 225
Bahia piassaba **49**, 141
Balata 49, 213, 224, 292
Balau 305
Balisier 165
Balloon Flower 209
Balloon Vine 73
Balm 49, 168, 218
  Bastard 218
  Bee **55**, 108, 226
  Field 158
  Fragrant 226
  Lemon 143, 168, 169, 218, *218*
  Molucca 226
  Sweet *218*
Balm of Gilead 8, 97, 273
Balsa 49, 140, 329
  Garden 179
Balsam 49, 231
  Garden 179
  Himalayan 179, *179*
  Indian 179
  Peru 231
  Tolu 231
Balsam Apple 226
Balsam of Peru 221
Balsam Pear 226
Balsam-bog 48
Balau 305
Bamboo 49, 140
  Black 264
  Calcutta 114
  Fishpole 264
  Forage 264
  Giant 114, 264
  Golden 264
  Hardy 264
  Hedge 52
  Himalayan 43
  Meyer 264
  Moso 264
  Sacred 232
  Simon 43
  Spiny 52
  Stake 264
  Sweetshoot 264
  Timber 264
    Japanese 264
  Tonkin 43
  Yellow-groove 264
Bamboo shoots 338
Banana 52, *52*, 147, 149, 230,
  Abyssinian 127, *127*, 141
  Fehi 147, 230
  Fe'i 230
  Flowering 52
  Japanese 52
  Wild 316
Baneberry 12
Banksia
  Bull 52
  Coast 52
  Swamp 52
Banyan 32, **52**
Banyan Tree 138
  Indian 138
Baobab 12
Barberry 26, **52**
  Alleghany 52
  European 52
  Indian 52
  Magellan 52
  Salmon 53
Barley 53, *53*, 56, 81, 318, *336*, 346
  Pearl **256**
  Six-rowed 81, *81*
  Two-rowed 81

Barrel
  Golden 123
  Mexican Giant 123
Barren Wort 127, *128*
Basil 53, 143
  Holy 241
  Sweet 53, 168, *169*
    Oil of 53
  Sweet Bay 168
Basket Plants **13**
Basketflower 79, 177
Basswood **326**, 329
Bastard Balm 218
Bastard Cypress **82**
Bastard Jasmine **82**
Bastard Teak 66
Bat Flower 321
Batchelor's Button 79, 281
  White 281
Batu, Senegal 305
Bay
  Bull 209
  Red 260
  Rose 234
  Sweet 192, 209, 260
Bay fat 192
Bay rum **53**
Bayberry 231
  California 231
Bayonet Plant 11
Baywood 277
Bdellium 97
Beachgrass
  American 27
  European 27
Bead Tree 218
Bean **53**
  Adzuki 195
  Algarrobo 275
  Asparagus 44, 103
  Baked 163, 261
  Bengal 145, 227
  Bonavist 189
  Broad 145, *194*, 195, 342, 342
  Butter 194, *195*, 201, 261
  Butterfly 90
  Carob 195
  Carolina 195, 261
  Castor 287
  Cherokee 130
  Civet 195
  Cluster 195
  Common 261
  Dry Shell 163
  Dwarf 163
  Field 342
  French 163, *194*, 195, 261, *261*
  Goa 44, 195
  Green 163, 195
  Green Shell 163
  Haricot 194, 195, 261
  Horse 342
  Hyacinth 119, 189, 195
  Indian 77
  Jack 71, *195*, 195
  Jumping **183**
  Kidney 119, 163, *194*, 195, 261
  Lima *194*, 195, **201**, 261
  Locust 195
    African 195
  Madagascar 195, 261
  Manioc 348
  Mescal 308
  Moth 195
  Mung 119, 195, **230**, *230*
  Navy 261
  Potato 338, 348
  Rice 195
  Runner 163
    Scarlet *194*, 195, 261, *261*
  Salad 195
  Sieva 195, 261
    Dwarf 201
  Snap 163, 261
  String 195, 163
  Sweet 158
  Sword 71, 195
  Tepary 195, 261
  Tick 342
  Tongue **330**
  Tonka **330**
  Velvet 145, 227
  Wax 163, 195
  Yam 195, *291*, 338, 348
  Yardlong 103, 195

Bean sprouts 230, *230*
Bean-tree 189
Bearberry 39, *39*, 75
Beard Lichen 336
Bear's Breeches **10**
Bear's Ear *47*
Bear's Tail, Cretan 79
Bearwind 70
Bearwood 75
Beauty Leaf, Ceylon 70
Beauty Berry 69
Beauty-of-the-night 225
Beaver's Tail 302
Bedding Conehead 316
Bedstraw
  Dye 150
  Ladies 122
  Our Lady's 150
  Yellow-flowered 150
Bee Balm **55**
Beech **54**
  American 54, 329
  Australian 236
  Black 236
  Blue 54, 173
  Brown 54, 106
  Copper 54
  European 54, 329
  Hard 236
  Japanese 329
  Myrtle 236
  Oriental 54, 329
  Roble 236, *236*
  Silver 236
Beefwood, Red 76
Beer 18, 56
Beerseem 91
Beet 290
  Field **212**
  Garden **54**
  Sea, Wild 55, *55*
  Seakale 54, 83, 338
  Spinach 54, **311**, 338
  Sugar 290, *291*, **317**, *317*
Beetroot **54**, *54*, 290, *290*, *291*
Beggar's tea 294
Beggarweed 115, 145
Begonia 54
  Angel-wing 54
  Grape-leaf 54
  Holly Hock 54
  Holly-leaf 54
  Maple-leaf 54
  Painted-leaf 54
  Watermelon 258
  Wild 293
  Winter 55
Bel Fruit 13
Bell, Scarlet 311
Belladonna 26, 46, 220, 221
Bellflower **54**
  Adriatic 54
  Chilean 191
  Chimney 54
  Chinese 269
  Clustered 54
  Creeping 54
  Giant 54
  Marsh 54
  Scottish 55
  Tall 54
  Tussock 54
  Willow 54
Bells of Ireland 226
Bells, Yellow 324
Ben oil 226, 244
Bent
  Black 16
  Brown 16
  Colonial 16
  Common 16
  Creeping 16
  Rhode Island 16
  Velvet 16
  White 16
Benteak 190
Bent-grass **16**, *16*, 145
Benzoin 221, 317
Berberidine 26
Berberine 26
Bergamot **55**, 304
  Lemon 168
  Oil of 55
  Red 168
  Sweet **55**, 226
Bergamot oil 132, 304
Berry
  Baked-apple 292
  Buffalo 304
  Cockroach 307
  Coral 320
  Himalaya 292
  Honey 218

Berry (cont)
  Male 207
  Nanny 342
  Rabbit 304
  Raccoon 270
  Sapphire 320
  Sheep 342
  Yellow 292
Berry Tree, China 218
Betaxanthin 26
Betel 108
Betel Nut **55**, *239*, *239*
Betel Palm 55
Betel Pepper 55, 73
Betony 168
  Wood 313
Beverage plants 56
Bezetta rubra 86
Bibisi Tree 241
Big Tree 283, 303
Bigarade 304
Bignonia, Yellow 324
Bilberry **57**, *57*, 147, *148*
  European *337*
Bilimbi 147
Bilsted 315
Bilva 13
Bindweed **57**
  Black 57
  Field 57, *97*
  Greater *70*
  Hedge 57, 70
  Japanese 70
  Lesser *97*, *97*
  Pink 97
  Riviera 97
Birch **57**, 122, 141
  Black 57
  Canoe 57
  Cherry 57
  Common 57
  Downy 57
  European 329
  Himalayan 58
  Paper 57
  Indian 58
  Silver 57, *57*, *241*, 329
  White 57, 329
Bird of Paradise Flower 315, *316*
Bird of Paradise Tree 316
Bird-catcher Tree 267
Bird-on-the-wing 271
Bird's Foot Trefoil 205, *205*
  Slender 205
  Southern 205
Bird's Head 40
Bird's Nest Fungi **58**, *278*, *278*
Bird's Tongue Flower 316
Bird's-nest, Yellow 32, 226, *226*
Birthroot 332
Birthwort, European 40
Bishop's Caps **225**
Bishop's Head 127
Bishop's Weed 27
Bishop's Wort 313
Bishop's-weed 13
Bistort 271, *271*
Bitter Cress **73**
Bitter Leaf 341
Bitter Root 197
Bitter Wood 280
Bitternut 75
Bittersweet 235, 307
  Climbing 78
Bitterweed 165
Bitterwood 305
Bixin 31
Black drink 171
Black Eyed Susan 292, 326, *326*
Black Pot Herb 307
Blackberry **58**, *58*, 122, 147, 292
  American 204
  Cut-leaf 292
  European 292
  Evergreen 147, *292*
  Pacific 292
Blackbutt 131
Blackfellows Bread 152
Blackfingers 248
Blackgram 195
Blackthorn **58**, *58*, 122, 306
Blackwood
  African 112
  Australian 9
Bladder Fucus 146

Bladder Senna **96**
  Common 96, *97*
Bladderworts **336**
Bladderwrack 146
Blaeberry 57
Blanket Flower 150, *150*
Blanket Leaf *227*
Blanket-weed 89
Blaspheme 307
Blazing Star **197**, 219
Bleeding Heart 116
Blimbing 147
Blister Cress **130**
Blonde Lilian 131
Blood Flower *43*
Bloodberry 58, 287
Bloodleaf **180**
Bloodroot 122, 296
Bloodwood 277
Bloodwort 296
Blowball **322**
Blue Blossom 77
Blue Devil 124
Blue Weed 124
Bluebell 35, 54
  California 261
  English *176*, *177*, 301
  Scottish 55
  Spanish 177, *177*
Blueberry **58**, 147
  European 57
  Highbush 58, 147, *148*
  Late Sweet 58
  Low Sweet 58
  Lowbush 58, 147
  Rabbit-eye 58
  Swamp 58
  Velvet-leaf 58
Bluebottle 79
Blue-buttons 188, 343
Blue-eyed Mary 96, 243
Bluegrass
  Annual 270
  Canada 145, 270
  Kentucky 145, 269
  Rough 269
  Texas 145, 270
Blue-green algae 50
Bluelips 96
Bluestem 145
  Angleton 145
  Brahman 145
  Diaz 145
Bluet 176
  Creeping 176
  Mountain 79
Bluewings 330
Blusher 27
Bog Rosemary 29, *29*
Bog Rush **183**
Bogbean 219, *219*
Bois de Rose 97
Bolbonal 206
Boneset 320
Bootlace Fungus 40, *41*
Borage 59, *60*, 108
Borecole 62
Boronia **59**
  Brown 59
  Scented 59
Boronia oil, Scented 132
Botany Bay Tea Tree 101
Bottle Tree 12
Bottlebrush **69**, *69*
Botulism 51
Bouncing Bet 297
Bouquet Garni **60**
Boura 18
Bow Wood 250
Bower Actinidia 12
Bower Plant 253
Bowman's Root 157
Box **60**
  Bastard *271*
  Cape 60
  Common 60
  Flooded 133
  Fragrant 297
  Jasmine **262**
  Rattle 284
  Sweet 297
  Victorian 268
Box-elder 10
Boxwood 60
  Florida 60
  Knysna 60
  San Domingan 60
  Venezuelan 60
  West Indian 60
Boysenberry 147, 292
Bracken **60**, *61*, 122, 137, *241*, 277

Brake
  Cretan 277
  Toothed 277
Bramble 58, **61**, 147
Brandy 18, 56
Brandy Bottle 237
Brass Buttons 103
Brazil Nut **61**, 239, *239*
Brazil nut oil 244
Brazil Wood 67, 122
Brazilian Plume 185
Bread Tree 126, 318
Breadfruit 42, 147, 252, 338, *340*
  Mexican 226
Breadnut 42
Breadnut Tree 64
Breadroot **277**
Bread-tongues **258**
Break-axe Tree 280
Breastwort 167
Briar root **61**
Bridal Wreath **144**, 312
Bridewort 312
Brinjal 46, 307
Broccoli 62, **64**, *64*
  Heading 62
  Italian 62
  Sprouting 62, 64
    Green *63*
    Italian 67
    Purple *63*
    Summer 67
  Turnip 62
Brodiaea
  Harvest 64
Brome
  Awnless 64
  Field 64
  Hungarian 64
  Prairie 145
  Smooth 64, 145
  Soft 64
Brooklime **341**
  European 341
Broom **64**, 110, 122, 221
  Butcher's 293
  Common 110, *110*
  Dalmation 156
  Dyer's 122
  Genoa 156
  Hedgehog 129
  Madeira 156
  Spanish 310, *311*
  Warminster 110
  Weaver's 310
  White 110
Broomrape **250**, *250*
Broomroot **64**
Browallia, Orange 316
Brown algae 64, *64*
Brucine 26, 316
Brussels Sprouts 62, *63*, **65**
Bryony 65
  Black 322
  Redberry 65
  White 65, *65*
Buckbean 219
Buckeye **14**
  Bottle-brush 14
  California 14
  Dwarf 14
  Ohio 14
  Red 14
  Sweet 14
  Yellow 14
Buckrams *151*
Buckthorn **65**, 221
  Alder 221
  Common 65
  European 65
  Purging 65
  Redberried 65
  Sea **65**
Buckwheat **65**, 81, **135**
  Bush 65
  Common 135, *135*
  Kangra 65
  Siberian 65, 135
  Tartary 65, 135
  Wild 65, 129
Buddleia 65
Bugbane 12, 87
Bugle 16, *17*
Bugloss 17, 29
  Viper's 124
Bukhara 271
Bullace 112, 147, 269
Bullock's Heart 31, 147
Bulrush **66**, 258, 293
  Cattail **334**, 335
  Common 301
  Reedmace 66
Bunchgrass

Bunchgrass (cont)
  Feather 315
Bunny Ears 248
Bunya Bunya 39
Bur Marigolds **55**
Burgundy pitch 313
Burlap 185
Burnet 296
  Great 296
  Salad 168, 296, *296*
Burning Bush 66
Burr
  New Zealand **9**
Burro's Tail 302
Bursting-heart 133
Bushmans-poison 11
Busy Lizzy 179
Butcher's Broom 293
Butterbur 260
  White 260
Buttercup **66**
  Bermuda 251
  Bulbous 66, 281
  Colombia 243
  Common 281
  Creeping 281
  Early 281
  Meadow 66
  Persian 281
  Tall 281
  Water 66, *281*
  White 66, 254
  Yellow 281
Buttercup Winter Hazel 102
Butterfly Flower 53, 300
Butterfly Tree 53
Butterfly Weed 43
Butternut 75, 122, 239, *239*
  Chinese 344
  North American 344
Butterwort **266**, *267*
Buttonball 320
Buttonball Tree 268
Buttonwood 268, 320

Caapi 220
Cabbage 62, **67**, *67*, 298
  Chinese 85, *86*
  Flowering *63*
  Portuguese 62
  Red *63*
  Round *63*
  Savoy 62, *63*
  Skunk 207, 341
  Wild *63*
Cabbage Tree 100, 252, 303
  Brown 267
Cacao 26, 94
Cacao black pod 264
Cacao butter oil 244
Cacao Tree 56
Cachi 48
Cactus
  Ball, Golden 123
  Barrel 67, 123, *124*, 135
  Candelabra 134
  Christmas 128, 300
  Claret-cup 124
  Claw 300
  Cochineal 236
  Cotton Pole 248
  Crab 300
  Cushion 248
  Dane 247
  Devil 248
  Dumpling **205**
  Dutchman's Pipe 128
  Easter 128, 284
  Easter Lily 124
  Eve's Pin 247
  Feather 211
  Fishbone 128
  Giant 74
  Grizzly Bear 248
  Hedgehog 135
  Leafy 259
  Mexican Dwarf Tree 248
  Mistletoe 284
  Old Lady 211
  Organ-pipe 67
  Paper 248
  Pencil 248
  Rainbow 124
  Redbird 257
  Ribbon 257
  Rose 259
  Snowdrop 284
  Strawberry 124
  Teddy-bear 248
  Thanksgiving 300

Cactus (cont)
  Thimble 248
  Toad 314
  Torch 124
  Yoke 300
Cade 132
Caffeine 26, 56, **67**, *221*, 323
Caffeol 95
Cajan 195
Cajang 103
Cajuput 221
Cajuput oil 215
Cajuput Tree 215
Calabash Gourd **67**
Calabash, Sweet 147, 255
Calabash Tree 67
Calabazilla 216
Calabozo 67
Calabrese 62, *63*, **67**
Calamint 168
  Lesser 68
Calf's Snouts **36**
Calico Bush 186
California Rose 70
Calisaya Bark 221, 280
Calla 350
  Florist's 350
  Golden *350*
  Red 298
  Yellow 350
Calliopsis 100
Calvados 18
Camas
  Death 350
  Poison 350
  White 350
Camass 70
Camellia 70
  Mountain 314
  Silky 314
Camel's Foot Tree 53
Camembert cheese 258
Camomile 108
Camphor **71**, 132, 221
  Ngai 58
Camphor oil 132
Camphor Tree 87
Campion **305**
  Bladder 305, *305*
  Evening 305
  Morning 305
  Moss 305
  Red 305, *305*
  Rose 207
  Starry 305
  White 305
Canada balsam 8
Canary Bird Bush 106
Canary Bird Flower 333
Canary Creeper 333
Cancer Weed 294
Candelabra Tree 39, 252
Candelilla 134
Candelilla wax 134
Candle Plant 187, 269
Candleberry Tree 17
Candlenut 17, 244
Candle-snuff Fungus *348*
Candytuft 178
Cane **71**
  Dumb 116
  Malacca 141, **210**
  Small 43
  Rattan 141
Canella bark 71
Canistel 124
Canker Brake 272
Cannabinoids 71
Cannonball Tree 103
Cantala 306
Cantaloupe 106, 216
Canterbury Bell 54
Caoutchouc **72**, 138
Cape Daisy 341, *341*
Cape Lance Tree 45
Cape Leadwort 269
Cape Marigold **117**
Caper Bush 72
Capers **72**, *142*, 143
Cappan 122
Caprifig 138
Caramba 147
Carambola 147, *149*
Carapa 244
Caraway 107, 132, *142*, *143*, 168
Caraway oil 132
Cardamom 56, **73**, *142*, *143*
Cardinal, Red 130
Cardol 28
Cardoon 42, **73**, 109, 338

Caricachi 48
Carnation 74, 115
    Border 74, 74
    Perpetual 74, 74, 115
    Spray, American 74
Carnauba wax 74
Caroa 140
Carob 74
Carob Tree 17, 203, 275
Carolina Allspice 74
Carpenter's-square 301
Carpet Grass 48, 145
Carrageenan 20, 157, 283
Carragheen 75
Carrion Flower 313, 313
Carrizo 263
Carrot 75, 75, 290, 290, 338
Carvone 221
Casabanana 217
Cascara 75, 221
Cascarilla Verde 280
Cashau 275
Cashew Apple 147, 238
Cashew Nut 238, 239, 239
Cashew nut kernel oil 244
Cashew Nut Tree 10, 28, 238
Cassava 75, 76, 81, 213, 290, 291, 291, 318
Cassia 76, 87
    Padang 88
    Purging 76
    Tanner's 76
Cassia-flower Tree 88
Cassie 9, 76
Cassina 56, 172
Cassine 171
Cast Iron Plant 45
Castor cake 76
Castor oil 76, 77, 244, 244
Castor Oil Plant 26, 77, 287
Catalpa
    Common 77
    Hybrid 77
    Western 77
Catawba Tree 77
Catchfly 305
    Alpine 305
    Night Flowering 305
    Nodding 305
Catchweed 150
Catechu 9, 122, 150
    Black 150
    Pale 150
Catgut 324
Cathedral Bell 91
Catmint 168, 234
Catnip 234
Cat's Ear 178
    European 178
    Spotted 178
Cat's Foot 31
Cat's Tail 262
    Alpine 262
Cat's Whiskers 321
Catta 186
Cattail 66
    Common 335
    Narrow Leaf 66, 335
    Red-hot 10
Cattail Bulrush 334
Cattleya, Yellow 276
Caucho 72, 76
Cauliflower 62, 63, 77, 77
Cayenne 72
Cayenne pepper 77
Ceanothus 77
    Catalina 77
    Feltleaf 77
Cedar 78
    Alaska 82
    Algerian 78
    Atlas 78, 329
    Blue 78
    Burma 329
    Chinese 329
    Cigar Box 329
    Incense
        California 197, 197
        Japanese
            Red 106, 106
    Lebanon 98
    Pencil 123, 184, 329
        East African
    Red 78, 184, 330, 329
        Colorado 184

Cedar-Red (cont)
    Japanese 106, 106
    Southern 185
    Western 325, 325, 329
    West Indian 78, 329
    White 82, 123, 329
    Northern 325
Cedar wood oil 78, 132
Cedar-of-Lebanon 78, 329
Cedrela Tree 330
Celandine
    Creeping 281
    Early 281
    Tree 58, 208
Celeriac 78, 78, 290, 338
Celery 78, 78, 168, 338, 339
    Turnip-rooted 78, 338
    Water 337
    Wild 337
Celery oil 132
Cellonia 273
Century Plant 15
Cepe, Edible 59
Cephaeline 180
Ceriman 226
Chain Plant 330
Chamomile 36, 83, 168
    Lawn 83
    Ox-eye 36
    Roman 213
    Stinking 83
    Sweet 83
    Wild 83
    Yellow 36, 122
Champaca oil 223
Champedak 42
Chandelier Plant 186
Chandelier Tree 252
Chard 83
    Swiss 54, 83, 338
Charity 270
Charlock 83, 83, 306
    White 282
Chaste Tree 343
    Chinese 343
Chat 186
Chatterbox 127
Chaulmoogra oil 83, 244
Chaura 260
Chayote 83, 83, 217, 338
Checker Bloom 305
Checkerberry 151
Cheirotoxin 344
Chenille 10
Chenille Honey-myrtle 215
Cherimoya 26, 147, 149
Cherry 84, 85
    Barbados 211
    Bird 276
    Bladder 264
    Brazil 133
    Cooking 147
    Cornelian 119
        Japanese 101
    European Dwarf 84
    Flowering
        Japanese 276
    Fluted 133
    Ground 264
    Jerusalem
        False 307, 307
    Maraschino 84, 213
    Mazard 84
    Oshima 276
    Sour 84, 147, 276
    Spanish 224
    Surinam, Red 133
    Sweet 84, 147, 148, 276
    West Indian 211
    Winter 264
Cherry Pie 166
Cherimoya 31
Chervil 82, 122, 168, 169
    Bulbous 82
    Garden 36
    Salad 36, 82
    Turnip-rooted 82, 338
    Wild 36
Chestnut 84, 122, 141
    American 84, 126, 239
    Chinese 84, 239
    Dwarf 84
    Horse 14, 86

Chestnut-Horse (cont)
    Chinese 14
    Common 14
    Japanese 84, 239
    Moreton Bay 239
    Spanish 84
    Sweet 85, 85, 126, 239
    Water 239, 239, 344, 344, 345
    Chinese 345
Chewing gum 85, 213
Chickasaw Lima 71
Chicken-claws 294
Chickweed 85, 314, 314, 331
    Common 85
    Mouse Ear 80
    Field 85
    Umbellate 85
    Water 85
Chicle 49, 85, 213
Chicory 56, 85, 85, 87, 338, 339
    Asparagus 338
    Broad-leaved 338
    Castelfranco 338
    Treviso 338
    Verona, Red 338
    Wild 126
Chiku 213
Chile Ancho 85
Chile Mulato 85
Chile Pequin 85
Chilean Fire Tree 126, 126
Chili Pepper 72
Chilies 85
Chilitos 211
Chilli 142, 143
China Grass 58
China Jute 85
China wood oil 244, 334
Chincherinchee 250
Chin-chin 250
Chinese Fir 107
Chinese Silk Plant 58
Chinquapin 84
    Water 233
Chipire 134
Chirata 320
Chiretta 320
Chittamwood 307
Chives 19, 85, 143, 168, 338, 339
    Chinese 19
Choco 83
Chocolate 221
Cholla 67, 247
    Jumping 248
Chorogi 313
Christmas Bush, Victorian 275
Christmas Flower 134, 270
Christmas Rose 86, 166
Christmas Star 270
Christmas Tree 86
    New Zealand 222, 222
Christmas-berry Tree 299
Christophine 83, 217, 338
Christ's Thorn 252
Chrysanthemum 87
Chufa 109
Cicely 36
    Sweet 168, 231
Cider 18, 37, 86, 87
Cigar Box Tree 78
Cigar Flower 107
Cinchona 280
    Ledgerbark 280
Cinchonine 26
Cinderella-slippers 306
Cineole 221
Cineraria 87, 303
Cinerins 279
Cinnamein 221
Cinnamon 87, 142, 143
    Batavia 88
    Ceylon 87
    Chinese 87
    Saigon 88
    True 87
    White 71
    Wild 71
Cinquefoil 274
    Hoary 274
    Shrubby 274
    Silvery 274
Citron 89, 147, 216
Citronella oil 132
Citrus fruits 89

Claret-cup Cactus 124
Clary 89
Clary Wort 89
Clearing Nut Tree 317
Cleavers 150, 151
Clementine 90, 147, 211
Clitocybin 90
Cloudberry 147, 292
Clove oil 91, 132
Clove Tree 132, 133
Clover 90
    Alsike 91, 145
    Aztec 91
    Bokhara 218
    Buckhara 218
    Bush 196
    Calvary 215
    Crimson 91, 145
    Dutch 91
    Egyptian 91
    Holy 246
    Hop 304
    Italian 91
    Japanese 196
    Korean 196
    Musk 130
    Pin 130
    Red 90, 91, 118, 145, 331
    Sour 218
    Strawberry 91, 145
    Subterranean 91, 145
    Suckling Yellow 91
    Swedish 91
    Sweet 168, 218
        White 145, 218
        Yellow 145
    Tick 145
    Water 214, 215
    White 91, 145, 304
    Yellow 304
Cloves 91, 142, 143, 221
Club Moss 92, 92, 93, 207
    Common 207
    Sand
        Small 92, 93
        True 92
Club root 268
Clubrush 66, 301
    Turkish 239
Cobra Plant 113
Coca 91
    Truxillo 91
Coca Tree 26
Cocaine 26, 91, 221
Cochineal 236
Cochineal berries 264
Cochineal Cactus 236
Cockle
    Sticky 305
    White 305
Cocklebur 347
Cockroach Berry 307
Cockscomb 78, 79
Cock's-foot 111, 145
Coco Bolo 112
Coco de Mer 119
Coco Plum 87
Cocoa 56, 67, 94, 325
Cocoa butter oil 244
Coconut 94, 94, 122, 141, 147, 239, 239, 244, 244
    Double 119, 119
Coconut Apple 95
Coconut oil 244
Coconut water 56
Cocoyam 318, 347
Codeine 26, 221
Coffee 26, 56, 67, 95, 95, 221
    Fig 56
    Liberica 95
    Mogdad 76
    Negro 76
Coffee Plant 56
Coffee Tree, Kentucky 161
Coffeeberry 75
Cognac 18
Cohosh
    Black 12, 87
Coigue 329
Coir 94, 95, 141, 141
Cola 26, 56, 96, 188
Cola Tree 56
Colanin 188
Colchicine 26, 96
Cole 62
Coleus, Brazilian 269
Colewort 62
Collards 62, 96

Colocynth 86, 216
Colophony 288
Coltsfoot 335
Columbine 38
    Canadian 38
    Sitka 38
Colza 62, 97
Colza oil 244
Comb Hedgehog 124
Comfrey 320
    Prickly 320
    Russian 320
    Rough 320
    Turkish 320
Compass Plant 305
Comptie 349
Conchita 145
Cone Flower 292
    Purple 123
Congo jute 39
Conifers 98
Coniine 26, 221
Contra Hierba 119
Contrajerva 119
Convallatoxin 200
Conyza 129
Coolabah 133
Coolwort 225
Coontie 349
Copals 97, 186
    East African 177
Copihue 191
Copper-leaf 10
Copra 244
Coral Bells 167
Coral Flower 167
Coral fungi 89
Coral Gem 205
Coral Plant 182
Coral Tree 130, 131
    Hawaiian 130
Coralberry 40
Coreopsis 122, 168, 169
    Green 101
Coriander oil 101
Cork 101
Cork Tree 261
Cork-bark Tree 162
Corkwood 140
Corn 81, 210
    Dent 210
    Flint 210
    Guinea 309
    Indian 210
    Kafir 309
    Squirrel 309
    Sweet 210
Corn Cockle 16
Corn on the Cob 210
Corn Root, Sweet 330
Corn Salad 337
    Italian 337
Cornel 101
    Dwarf 101
    Japanese 101
Cornelian Cherry 119
Cornflour 101
Cornflower
    American 79
    Corsican Curse 308
Costmary 102, 168, 169
Cotton 103, 103, 140, 141, 244
    Devil's 141
    Egyptian 103
    Lavender 297, 297
    Sea Island 103
    Tree 159
    Wild 170
Cotton Grass 129
Cotton Rose 170
Cotton seed oil 244
Cottonweed 159
Cottonwood 273, 329
    Black 273
Couch
    Sand 15
Couch-grass 15, 16
Coumarin 36, 221
Courbaril 177
Courgette 103, 216, 217
Cowberry 147
Cow Parsnip, Common 167
Cow Tree 64
Cowage 227
Cowberry 104, 342
Cow-itch 227, 270
Cowpea 103, 119, 195, 195, 256, 316
    Common 103
Cowslip 274, 275

Cowslip (cont)
    Blue 278
    Cape 189
    Common 271
    Himalayan 275
    Jerusalem 278
    Virginian 222
Cow-tang 258
Crab Apple 104
Crab's Eyes 8
Crakeberry 126
Crampbark 342
Cranberry 104, 147, 148
    American 104
    European 104
    Highbush 342
    Mountain 104
    Rock 104
    Small 104
Cranberry Bush 342
Cranberry Tree 342
Crane Flower 316
Cranesbill 156
    Bloody 157
    Dove's Foot 157
    Meadow 157
    Wild 157
Cream of Tartar Tree 12
Creambush 172
Creeper
    New Guinea 227
    Rangoon 280
Creeping Berries 303
Creeping Jenny 207
Creeping Sailor 299
Crème de violette 343
Creosote Bush 191
Cress 62, 63, 104
    American 104
    Bitter 73
    Blister 130
    Brown 345
    Common 196
    Garden 196
    Indian 333
    Land 104
    Lebanon 14
    Rock 38
        Hairy 39
    True 104
    Violet 180
    Water 104
    Yellow 288
Cretan Bear's Tail 79
Crocus 105
    Autumn 26, 96
    Indian 269
Crook root 312
Crookneck, Summer 216
Cross-vine 55
Crotal 122
Crotalism 105
Croton 106
    Purging 106
Croton oil 244
Croton Oil Plant 106
Crotons 95
Crottle 122
Crowberry 126
Crowfoot 125
    Bulbous 281
    Persian 281
    Water 281, 281
Crown Imperial 146, 146
Crown of Thorns 134
Cubbra Borrachero 226
Cuckoo Bread 251
Cuckoo Flower 73, 206, 207
Cuckoo Pint 42
Cucumber 106, 157, 216, 217, 338
    Bitter 338
    Serpent 217
    Squirting 123, 217
    Wild 106
Cucumber Tree 209
    Large-leaved 209
    Yellow 209
Cuddle-me-to-you 253
Cudweed 159
Cumin 142, 143, 243
    Black 235
Cumin seed 107
Cup and Saucer Plant 91
Cup Flower
    Tall 235
    White 235
Cup Plant 305
    Indian 305
Cupid's Dart 77
Curare 107, 221, 316

Curare Poison Nut 26, 107
Curarine 26
Curcuma **107**
Currant **107**
  Black 56, *107*, 108, *147*, *148*, 286
    American 286
  Buffalo 286
  Common 108
  Flowering 108, 286
  Garden 108
  Golden 286
  Missouri 286
  Northern 286
  Red 108, 147, *148*, 286, *286*
Curubá 147, 217
Cushaw 107
Cush-cush 318
Cushion Bush 69
Cushion Flower 305
Custard Apple 31, 147
  Common 31
  Wild 31
Cutch 9, **108**, 122
  Black 9
Cycads **108**
Cyclamen
  Florists' 108, *109*

Cypress **109**
  Arizona 107
    Rough-barked 109
    Smooth 110
  Bald 323
  Bastard **82**
  Bhutan 110
  False **82**
  Funeral 109
  Himalayan 110
  Hinoki 82
  Hybrid **107**
  Italian 109
  Lawson 82
  Mediterranean 109
  Mexican 107, 323
  Monterey 107, 109
  Montezuma 323
  Nootka 82, 107
  Patagonian 139
  Pond 323
  Portuguese 109
  Sawarra 83
  Summer 188
  Swamp 82
  White 83
  Yellow 82, 107
Cypress Pine **69**
  Black 69
  White 69
Cytisine 221

Daffodil **111**, 232, *232*
  Peruvian 177
  Petticoat 232, *233*
  Winter 314
Dahlia
  Bedding 112
  Candelabra 112
  Garden 111, *111*
  Tree
    Bell 112
    Flat 112
Daisy 54, **112**
  African 40
    Blue-eyed 341
  Alpine 112
  Barberton 157
  Blue 135
  Bush 112
  Butter 281
  Cape 341, *341*
  Common 112
  Double Orange 129
  Easter 330
  Globe **158**
  Gloriosa 292, *292*
  Kingfisher 135
  Michaelmas 45, 112, **223**
  Moon 196
  Mount Atlas 28
  Ox-eye 87, 112, 196
  Paris 87
  Rock Tree 112
  Seaside 129
  Shasta 87, 112, 197
  Swan River 60, *61*, 112
  Transvaal 157
Daisy Bush **242**
Daisy Tree **242**
Dal **193**
Damask oil 167
Damiana 334

Dammar **112**, 305
  Almond 112
Damson **112**, *147*, *148*, 269
  Butter 112
  West African 112
Damson Plum 160
Dandelion **112**, **322**
  Common 112, *112*, 322, *322*
  Russian 292, 323
Dandelion coffee 56
Danewort 296
Dasheen 318
Date **113**, 113, 147
  Chinese 351
  Cliff 263
  Jerusalem 53
  Wild 113
Date Palm 113, *113*, 262
  India 263
  Senegal 263
  Wild 263
Date Plum 117
Datura, Sacred 113
David's Harp 271
Dawn Flower, Blue 180
Day Flower 97
Daylily
  Fulvous 166
  Lemon 166
  Orange 166
  Tawny 166
  Yellow 166
    Dwarf 166
Dead-men's Fingers 348
Dead-nettle **190**
  Giant 190
  Red 190
  Spotted 190
  White 190, *190*
Deal, yellow 266
Death Cap 25
Delphinium **114**, *114*
Deodar 78, 329
Derris Root **115**
Desert Candle **128**
Destroying Angel 25
Devil Flower 321
Devil-in-a-bush 235
Devil's Backbone 257
Devil's Cotton 8
Devil's Fig 40
Devil's Herb 46
Devil's Leaf 336
Devil's Paintbrush 171
Devil's Tongue 27, 135
Devil's Walking Stick 39
Dewberry, Pacific 147, 292
Dhak 66
Dhal 193
Diatoms **116**
Digitalin 144
Digoxin 144
Dill **117**, *117*, 157, 168, *169*
  Indian 117
  Japanese 117
Diosgenin 348
Diptheria 51
Dittany 66
Dittany of Crete 250
Divi-divi 67, **118**
Divine Flower 115
Dock
  Alpine 293
  Broad-leaved 292, 293
Dodder 32, **118**, *118*
  Common 118
  Greater 118
Dogbane
  Hemp 37, 140
  Spreading 37
Dogberry 119, 309
Dog-hobble 197
Dogrose 288, *288*
Dog's Tail, Crested 109
Dogwood **118**, 275
  Blood-twig 119
  Flowering 69, *101*, 119
    Western 119
  Giant 119
  Mountain 119
  Red Barked 119
  Red Osier 119
  Tartarian 119
Donkey's Tail 302
Doub 109
Douglas Fir **119**, *120*, **329**
Dove Tree 114

Dove's Foot 157
Downy mildew 260
Dragon Arum *120*
Dragon Head **120**
Dragon Mouth **172**
Dragon Root 40
Dragon Tree 120
Dragonhead, False 264
Dragon's Blood 106, **120**
  Malayan 111
Dragon's Blood Tree 120
Dropwort 138
Dry rot 303
Dryad's Saddle 272
Dubia 262
Duckweed **121**, *121*
  Common 121
  Ivyleafed 121
  Lesser 121
  Star 121
  Tropical 267
Dulse 21
Dumb Plant 116
Durian **123**, *147*, *149*
Dusty Miller 79, 303
Dutch elm disease 126
Dutch Man's Pipe 40
Dutchman's Breeches 116
Dyer's Greenweed 156
Dyer's Woad 181
Dyes **123**

Earth Smoke 150
Earth balls **123**, 278
Earth stars 106, **123**, 278
Earthnut 38, 97
Eau d'ange 231
Eau de créole 211
Ebony 117, **123**
  Calabar 123
  Carving, Black 329
  Ceylon 329
  Coromandel **123**, 123
  East Indian 117, 329
  Gaboon 323, 329
  Green 324
  Greenheart 324
  Macassar 117, 123
  North American 260
  Senegal 112
  Texas 268
Ebony wood 53
Edelweiss **124**, 196
Eel Grass 351, *351*
Egg Fruit 46, **124**
Eggplant 307
  Chinese 46
  Snake 46
Eglantine 288
Elaterium 123
Elder 47, **296**
  American 296
  Common 296
  Dane's 296
  Dwarf 296
  European 296
  Ground 13
  Prickly 39
  Red, American 296
    European 296
  Red-berried 296
  Rose 342
  Sweet 296
  Yellow 324
Elderberry **296**
Elecampane 179
Elemi 66, **125**
  African 125
  American 125
  Manila 125
  Mexican 125
  West Indian 125
Eleven O'Clock 273
Elm **125**
  American 125
  Chichester 125
  Common *126*
  Cornish 125
  Dutch 125, 329
  English 125, 329
  European Field 125
  Hickory 329
  Holland 329
  Huntingdon 125
  Red 125
  Rock 126, 329
  Scotch 125
  September 125
  Slippery 125, 329
  Spanish 100
  White 125
  Witch 125

Elm (cont)
  Wych 125
Emblic 263
Emetine 26, 180, 221
Empress Tree 255
Endive 85, 87, **126**, 338, *339*
  Batavian 338
  Belgian 338
Eng oil 118, 244
Eng Tree, Indian 118
Enzian 156
Ephedra 127
Ephedrin 127, 221
Epiphyllum *128*
Ergot 26, **128**, 220, 221
Ergotamine 26, 221
Ergotism 26
Escarolle 87, 338
Esparcet 145, 246
Esparto 138, 140, *141*
Essence of roses 46
Estragon 323
Eye rot 233
Eyebright **134**
Eucalyptus 122, 221
Eucalyptus oil 132
Eugenol 221
Eupatorin 134
Everlasting
  Mountain 31

Fair Maids of France 299
Fairy Castles 248
Fairy Clubs **89**
False Aralia 118
False Cypress **82**
False Nettle
  Hawaiian 59
Farmer's lung 44
Fat Hen 84, *84*
Feather Top 258
Felonwood 307
Felt Bush 186, *186*
Fennel 31, **135**, 168, *169*
  Florence 135, 338, *339*
  Giant 135, *135*
  Sweet 135
Fennel Flower 135, 235
Fenugreek 26, *142*, *143*, 331
  Asparagus 43, 44
Fern **136**
  African 270
  Ball 113
  Beech 47
  Bird's Nest 45, *45*, 137
  Bladder 110
    Berry 110
    Brittle 110
  Blue 272
  Boston 234
  Buckler 136
    Crested 121
    Narrow 136
  Chain
    European 347
    Virginia 347
  Christmas 272
  Cinnamon 250
  Cliff, Fragrant 121
  Climbing 137
  Cloak, Golden 236
  Comb 300
  Cotton, Wavy 236
  Curly Grass **300**
  Dagger 272
  Deer 58, 59
  Elkhorn **268**
  Fragrant 121
  Gold 236
  Grape 236
  Hard 58, 59
  Hare's Foot 113, 272
  Hart's Tongue 262, *263*
  Hedge 272
    English 272
  Holly
    Anderson's 272
    Japanese 121
    Western 272
  King 213
  Ladder 234
  Lady 46, *47*, 136
  Leather 12
  Licorice 272
  Maidenhair **12**
    American 12
    Black 12
    Common 12, *12*, 137

Fern–Maidenhair (cont)
    True 12
  Male 121, 136, 137
  Marsh 12
  Mosquito **48**
  Necklace 45
  Oak *47*
    European 136
  Ostrich 214
  Parsley 106
  Rabbit's-foot 114, 272
  Ribbon *137*, 277
  Royal 250, *251*
  Rush **300**
  Savannah 158, *158*
  Sensitive **246**
  Shield
    Eaton's 272
    Hard 272
    Marginal 121
    Prickly 272
    Soft 272
  Silver 236
  Squirrel's-foot 113
  Stags Horn *136*, *137*, *235*, **268**
  Sun **158**
  Sword 234
    Erect 234
  Tree *116*, **330**, *330*
    Black-stemmed 331
  Wall, American 272
  Whisk 92
  Wood
    Coastal 121
    Crested 121
    Leather 121
Fescue **135**
  Blue 135
  Creeping 135
  Meadow 145
  Reed 145
  Red 135
  Sheep's 135, 145
  Tall 145
Fetterbush 265
Fever Plant 241
Feverfew 279
Fiber
  Istle **181**
  Istli 140, **181**
  Kitul 75, 141
  Ozone 43, 140
Field Balm 158
Fig **138**, *147*, *148*
  Adriatic 138
  Climbing 138
  Cluster 138
  Common 138
  Devils 40
  East Indian 138
  Edible 138
  India Rubber 138
  Indian 247, *248*
  Maryland 301
  Smyrna 138
  Sycamore 320
  Weeping 138
  Wild 138
Figwort
  Cape 263
  Water 301
  European
    Common 301
Filaree
  Red-stemmed 130
  White-stemmed 130
Filbert 164
  European 239
  Giant 239, *239*
Finocchio 135, 338
Fiorin 16
Fir **8**
  Balsam 8, 314, 329, 334
  Cascade 329
  Caucasian 8
  China **107**
  Chinese **107**
  Colorado 8
  Douglas 8, **119**, *120*
  Giant 8, 329
  Grand 8
  Gray 120
  Himalayan 8
  Korean 8
  Noble 8, 329
  Oregon 120
  Red 8, *8*, 329
    Californian 329
  Scotch 8
  Silver 8, 329
    Common 8
    White 8
Fire Bush 316
Fire Tree, Chilean 126, *126*

Firecracker, California 64
Firepink 305
Firethorn 279, *279*
Fireweed 83, 127
Fishtail Palms **75**
Fivespot 233
Flag 181
  Blue 181
    Western 181
  Crimson 300
  Poison 181
  Spiral 102
  Sweet 11, *12*, 220
  Yellow 181
  Water 181
Flag Root 11, 220
Flageolets 163
Flamboyant 114
  Yellow 114
Flame Creeper 333
Flame Flower 279
Flame of the Forest 311
Flame Vine 279
Flamegold 188
Flameleaf, Mexican 270
Flaming Trumpet 279, *279*
Flaming-sword 344
Flamingo Flower *37*
Flamingo Plant 185
Flaver 346
Flaver Drake 241
Flavoring plants **142**
Flax **139**, *139*, 140, *141*, 201, 202, *202*
  Blue 139
  Fairy 139
  False 139
  Mountain 263
  New Zealand 139, 140, 263, *263*
  Pink 139
  Spurge 139
Fleabane **129**
  Daisy 129
  Running 129
Fleece Flower **271**
Fleece Vine, China 271
Fleur-de-lis 181
Floating Hearts 237
Floppers 186
Floradora 314
Flour **139**
Flower of Love 253
Flowering Inchplant 330
Flowering plants **98**
Flowering Rush 66, *66*
Flower-of-an-hour 170
Fly Agaric 25
Flycatcher
  Australian 79, *80*
Foam of May 312
Foamflower 326
Fodder crops **145**
Foliage Flower 263
Footsteps-of-spring 296
Forget-me-not **231**
  Alpine 231
  Chinese 109
  Creeping 243
  Garden 231
  Wood 231
Forsythia 144
Foulbrood 49
Fountain Tree 311
Four O'clock Plant 225
Foxglove 144, 221
  Common 144
  Yellow
    Large 144
    Small 144
  Spanish Rusty 144
Foxglove Tree 255
Foxtail **24**
  Alpine 24
  Marsh 24
  Meadow 24
  Slender 25
Framboise 282
Framiré 329
Frangipani 269
Frankincense **144**
  Bible 60
Fraxinella 66
Fret-lip 162
Frijolito 308
Fringe Cup 324
Fringebell 305
Fritillary
  Snake's Head 146
Frog-bit 177
Frog-spawn Alga **53**
Frostweed 165
Fuchsia

Fuchsia (cont)
Californian 351
Cape 263
Tree 173
Fuki 260
Fumitory
Common 150
Yellow 102
Fundi 224
Fungi 152
Bootlace 41
Honey 41, 152
Saddle 166
Slime 306
Smut 336
Furze 335
Dwarf 335
Needle 156
Fustic 122

Gaboon 329
Gage 147
Galanga 185
Galangal 150
Greater 150
Lesser 150
Galax, Fringed 305
Galbanum 135
Galingale, English 109
Gallant Soldier 150
Gallito 304
Gamalote 48
Gambier 122, 150
Bengal 108, 150
Gamboge 122, 151
Gamboge Tree 151
Siamese 151
Gandergoose 250
Garbancillo 46
Garbanzo 85
Garden Egg 46
Gardener's Garters 261
Gardenia oil 132
Gari 76
Garland Flower 112,
164
Garlic 19, 151, 143,
338, 339
Bear's 151
Hog's 151
Wild 151
Wood 151
Garlic Mustard 168
Gay Feather 197
Gedu nohor 329
Geiger Tree 100
Genepi 42
Genip 218
Gentian
Alpine 156
Bottle 156
Catesby's 156
Closed 156
Fringed 156
Great Yellow 156
Pine-barren 156
Stemless 156, 156
Gentian bitter 156
Geranium
Alpine 130
Bedding 257
Fancy 257
Fish 257
Hanging 257
House 257
Ivy-leaved 257
Martha Washington
257
Regal 257
Rock 167
Rose 132
Show 257
Silver-leaved 157
Strawberry 299
Zonal 257, 257
Geranium oil 132, 257
Germander
Tree 325
Wall 325
Getony 257
Gherkin 106, 157, 216,
217, 338
Bur 216
West Indian 106, 216
Ghost Tree 114
Giant Cactus 74
Gill-over-the-road 158
Gillyflower 215
Gin 18, 56
Gingelly 304
Ginger 56, 142, 143,
157, 221, 351, 351
Butterfly 164
Kahili 164
White 164
Ginger beer 18

Ginger Plant 351
Gingerol 157
Gingili oil 304
Gingilie 304
Ginseng 221, 252
American 252
Gladiolus 157, 157
Garden 158
Glasswort 294, 294
Globe Amaranth 159
Globe Flower 333
Glory Bower 90
Glory Bush 326
Glory Flower
Chilean 123
Glory Vine
Japanese Crimson
344
Glory-of-the-snow 85
Gloxinia 158, 159, 306
Brazilian 306
Tree 188
Violet Slipper 306
Glycyrrhizin 197
Gnome's Throne 248
Goat's Rue 150, 324
Goatsbeard, False 45,
46
Godetia 159, 159
Gold and Silver Flower
204
Gold Dust 25, 25
Gold Dust Tree 47
Gold Flower 178
Golden Brodiaea 332
Golden Buttons 322
Golden Eardrops 116
Golden leaf spot 322
Golden Rain Tree 188,
189
Golden Rod 122, 308
Canada 308
Leavenworth 308
Sweet 308
Golden Seal 26
Golden Star 332
Goldenberry 264
Golden-chain 189
Golden-shower 76, 279
Goldfields 191
Goldmoss 302
Goldwire 178
Gonorrhea 50
Goober 38
Good King Henry 84
Good Luck Plant 100
Gooseberry 147, 148,
159, 286
Barbados 258, 259
Cape 147, 149, 264
Dwarf 264
Chinese 12, 147
Fuchsia Flowered
286, 286
Indian 263
Gooseberry Tree 263
Goosefat 84, 84
Goosegrass 150, 151,
274
Gooseneck 207
Gorli oil 244
Gorse 118
Common 335, 335
Spanish 156
Gourd 159
Ash 217, 338
Bitter 89, 159, 226,
338
Bottle 67, 159, 216,
338
Calabash 67, 216,
338
Club 217
Dipper 216
Dishcloth 206, 217
Fig-leaf 107, 216
Goareberry 216
Gooseberry 159, 216
Hedgehog 216
Hercules'-club 216
Hundredweight 278
Ivy 217
Knob-kerrie 216
Malabar 216
Missouri 216
Ornamental 107
Scallop 216
Serpent 159, 217
Siamese 216
Silver-seed 216
Snake 159, 217, 338
Sour 12
Sponge 159, 206, 217
Sugar-through 216
Teasel 216
Towel 206, 217

Gourd (cont)
Trumpet 216
Viper's 217
Wax 159, 217, 338
West Indian 159
White 159, 217, 338
White-flowered 159,
216
Wild 159
Yellow-flowered 216,
216
Goutweed 13
Gouty Stem 12
Gowan, Yellow 281
Grain amaranth 81,
159
Grain Tree 240
Grains of Paradise 259
Gram 195
Black 195
Green 195
Horse 119, 195
Madras 195
Red 195
Grama
Black 60
Blue 60, 145
Side-oats 60
Granadilla 56, 147
Giant 147, 255
Purple 255
Red 255
Sweet 147
Yellow 147
Granjeno 79
Granny's Bonnet 38
Grape 147, 159, 230
Black 344
Bullace 147, 344
Bush 147, 344
Cape 285
Evergreen 285
Fox 147, 344
Frost 147, 344
Mountain 344
Muscadine 344
Oregon 209
River Bank 344
Sand 147, 344
Sea 94
Seaside 94
Skunk 147, 344
Sugar 344
Grapefruit 89, 147, 159,
160, 211, 335
Grapevine 159, 281
Grass
Alfa 314
Alkali 350
Bahia 145, 255
Pensacola 255
Bamboo
Kuma 298
Bengal 304
Bermuda 109, 145
Bitter 220
Blue-eyed 306
Bristle 304
Green 304
Rough 304
Yellow 304
Buffalo 145
Buffel 145
Canary 261
Reed 145
Carpet 48, 56, 145
Japanese 351
China 58, 141
Citronella 109
Coco 109
Cord 310
Common 310
Freshwater 310
Prairie 310
Townsend's 310
Cotton 129
Narrow-leaved 130
Couch 109
Crab, Hairy 145
Dallis 145
Deer 301
Ditch 292
Eel 337, 351
Elephant 145, 258,
335
Feather 314, 315
Fountain 258, 258
Goose 125, 274
Grama 60
Guatemala 145
Guinea 145, 253
Hungarian 304
Ichu 315
Japanese 351
Johnson 309
June 269

Grass (cont)
Kikuyu 258
Knot 271
Korean 351
Lemon 132, 168
Lyme, Sea 126
Manila 351
Manna 158
Marram 27, 68, 214
Mascarene 351
Meadow
Annual 270
Dwarf 270
Rough 269
Smooth 269
Millet 309
Mondo 246
Moor 226
Napier 145, 258
Natal 258
Needle 314
Green 315
Texas 315
Nut 109
Oat 165
Oil 108
Orchard 145, 111
Pampas 101
Pangola 145
Para 145, 253
Pearl 64
Pin 130
Plume 128
Porcupine 314
Purple-eyed 306
Quaking
Common 61
Small 64
Ravenna 128
Reed, Canary 261,
284
Rescue 64, 145
Rhodes 86, 145
Ribbon 261
Rye, Alkali 126
Scurvy 94
Early 94
Common 94
Silky 258
Soft, Creeping 171
Spanish 131
Spear 314
Spring 36
Star 178
Sudan 309
Sugar 158
Sweet
Floating 158
Reed 158, 284
Switch 253
Tape 337
Terrell 293
Timothy 145, 262
Tufted Hair 115
Tussock 115
Umbrella 109
Vasey 145
Velvet 171
Korean 351
Vernal, Sweet 36
Viper's, Common 301
Whitlow 120
Zoysia 351
Grass Nut 64, 332
Grass of Parnassus 254,
255
Grass oils 132
Grass Pea 160
Grass Tree
Australian 347
Spearleaf 347
Grass-widow 306
Grass-wrack 351
Grassy-bell 116
Grassy-bells 124
Gray Beard 326
Green Dragon 40
Greenbrier, Laurel-leaved
307
Greengage 147, 160,
269
Greenheart 160, 241,
329
African 160
Surinam 160, 324
Greenweed, Dyer's 122
Grenadille Wood, African
112
Grevillea 160
Grindol 160
Gromwell, Blue 203
Ground Elder 13
Ground Ivy 158
Ground Nut 38, 195,
195, 239, 244
Banbara 195

Ground Nut (cont)
Kersting's 195
Ground nut oil 244
Groundnut Pea Vine
192
Ground-pine 16
Groundsel 203, 303,
303
Bush 49
Climbing 303
Tree 49, 205
Grouseberry 342
Grumulo 338
Guaiacum 200
Guaiacum resin 161
Guanabana 147
Guar 195
Guarana 26, 56
Guava 122, 147, 149,
276
Apple 276
Chilean 231
Pineapple 135
Purple 276
Strawberry 276
Yellow 276
Guava cheese 276
Guayule 289
Guelder Rose 342
Guinea Grains 259
Gum
Black 237
Blue 131, 329
Tasmanian 131
Cotton 237
Karri 131
Lemon-scented 131
Mountain, Silver-
Leaved 133
Murray 133
Red 315
Forest 133
River 133
Silver-dollar 133
Snow, Alpine 133
Sour 237
Swamp 131
Tupelo 237
Water 237
Gum ammoniacum 119
Gum Arabic Tree 9
Gumbo 170, 242, 338
Gumbo-limbo 66
Gumweed 160
Gurjum oil 244
Gurjun oil 118, 244
Gurjun 329
Gutta sandek 161
Gutta-percha 49, 161,
292
Gymnosperms 98
Gypsyweed 341

Hackberry
Chinese 79
Common 79
European 79
Mississippi 79
Hackmatack 273
Haematoxylin 204
Hag Taper 227
Hagaed Wood 112
Halfa 167
Hanged Man 11
Hanover Salad 62
Hardhack, Golden 274
Harebell 54, 55, 177
Hare's Ears 66
Hare's Tail 190
Haricot Bean 163
Harlequin Flower 310
Hardhack 312
Haricot Bean 163
Harmal 257
Harmela Shrub 257
Hashish 72
Hausa potato 269
Haw
Black 221, 342
Summer 163
Swamp 342
Sweet 221
Hawksbeard 104
Beaked 104
Smooth 104
Hawkweed 171
Mouse Ear 171
Shaggy 171
Haworthia, Zebra 163
Hawthorn 163
Chinese 164, 263
Common 163
Indian 284
May 163, 163
Water 37, 164
Yedda 284

Haymaids 158
Hazel 164
Beaked 164
Buttercup Winter 102
Chinese 164
Common 164, 164
Corkscrew 164
Tibetan 164
Turkish 164
Witch 162
Chinese 163
Hazelnut 239, 239
Healing Herb 320
Heartsease 252, 252,
343
Heartseed 73
Heath 129, 164
Cape 129
Cornish 129
Cross Leaved 129
Fringed 129
Hardy 129
Irish 111
Prickly 10
St. Daboec's 111
Spring 129
Tree 129
Heather 69, 118, 129,
164
Bell 129, 164, 164
Bog 129, 164
Christmas 164
Cornish 129, 164
Cross-leaved 164
Everblooming French
164
Mediterranean 164
Purple 129, 164
Red 263
Scottish 69, 164
Snow 129, 164
White 76
Winter, White 164
He-cabbage 303
Heliotrine 221
Heliotrope 166, 221
Common 166
Garden 337
Summer 166
Winter 260
Yellow 316
Hellebore
Green 166
Stinking 166
White 341, 341
American 341
Helleborein 166
Helleborine 79, 166
Giant 127
Marsh 127
Red 79
Sword-leaved 79
White 79
Helmet Flower 301
Hemlock 26, 166, 166,
167, 221
Canada 167
Carolina 167
Chinese 167
Common 167
Ground 167
Japanese 167
Mountain 167
Water 87
Western 167, 329
Hemlock Spruce 122,
167, 329
Indian 167
Hemp 140, 141, 141,
167
African 310
Bowstring 140, 167,
297
African 140, 297
Ceylon 140, 297
Indian 140, 297
Colorado River 304
Deccan 167, 170
Dog bane 140
Indian 71, 140, 170
Manila 8, 141, 141,
167
Mauritius 140, 150,
167
New Zealand 263
Queensland 140, 305
Sann 105
Sisal 167
Sun 105, 140, 167
Water 134
Hemp Nettle 150
Common 150
Downy 150
Hemp Tree 343
Hempseed oil 244
Hempweed, Climbing 223

Hen-and-chickens **123**, 176
Henbane 26, **177**, *177*, 220
  Black 177
Henequin *35*, *140*, 221, 306
  Salvador 306
Hengal 112
Henna 122, **167**
Herb Christopher 12
Herb Paris 254
Herb Robert 157
Herb Trinity 253
Herba Militaris 11
Herbal tea 56
Hercules Club 39
Herniary 167
Heron's-bill **129**
Hessian 185
Hibiscus **170**
  Chinese 170
  Hawaiian 170
Hickory 329, 349
  Big-bud 75, *75*
  Mockernut 75
  Pignut 75
  Shagbark 75, 239
  Shellbark 75, 239
  Smoothbark 75
Hills of Snow 177
Hogweed 167
  Giant 167
Holcus, Woody 171
Holly 123, **171**
  African 308
  American 171
  Box 293
  Canary Island 171
  English 171
  European 171
  Horned 171
  Japanese 171
  Madeira 171
  Oregon 171
  Sea 130
  Singapore 211
  Tarajo 171
  Tea
    Carolina 172
  West Indian 334
Hollyfern 110
  Japanese 121
Hollyhock **172**
  Figleaf 172
  Sea 170
Honesty 206
  Perennial 206
Honey Berry 218
Honey Flower 275
Honey Fungus 40, *41*
Honey Locust 158
Honeysuckle **172**, *172*
  African 172
  Australian 172
  Bush 204
  Cape 172
  Common 204, *204*
  Coral 204
  French 165
  Giant 204
  Goat-leaf 204
  Japanese 204
  Native 52
  Tatarian 204
  Trumpet 204
Hop 56, 118, 143, **172**, *172*, 176
  False 185
  Field 11
  Japanese 176
  Spanish 250
Hop Tree 277, *277*
Horchata de Chufa 56, 109
Horehound 214
  Black 49, *49*
  Fetid 49
Horn-a-plenty 113, 220
  Hop 181, **251**
    American 251
    European 251
  Japanese 173
  Oriental 173
Hornbeam 123, **173**, *173*, 181
  American 54, 173
  Chinese Henry's 173
  Common 173
  Cordate 173
  Witch 162
Hornwort 36, 228
Horse Chestnut **14**, 86
  Chinese 14
  Common 14, *14*
Horse Nettle, Carolina 308

Horse Sugar 320
Horse' al 179
Horsemint 55
Horseradish 63, 143, **173**
Horseradish Tree 226
Horse's Tail 302
Horsetail **174**
  Field *174*, *175*
  Water *174*, *175*
Horsetail Tree 76
Hortensia 177
Hottentot Bean **173**
Hottentot Bread 348
Hottentot's Head 313
Hound's Tongue
  Common 109
Houseleek **176**
  Cobweb 176, *302*
  Common 176
  Roof 176
  Spiderweb 176
Huamuchil 268
Huanaco 221
Huckleberry **176**
  Bear 176
  Black 176
  Blue 176
  Box 176
  Dwarf 176
  Hairy 176
  Sugar 176
Huisache 9
Humble Plant 224
Hupulone 143
Hyacinth **177**, 301
  Common 177, *177*
  Dutch 177
  Grape 177, 230, *230*
  Peacock 125, *125*
  Roman 177
  Star 301
  Starry 301
  Summer **150**
  Water 125, *125*, 184
  Wild 70, *176*, 177, 332
Hyacinth of Peru 301
Hybrid Cypress **107**
Hydnocarpus oil **83**
Hydrangea
  Climbing 114
  French 177
  Wild 293
Hygrine 26
Hyoscine 26, 113, 221
Hyoscyamine 26, 113, 177, 221
Hyssop 168, **178**, *178*
  Anise 178
  Giant 178
    Fragrant 178
  Nettle-leaved 178
  Water 49

Icacillo 87
Icaco 87
Ice Plant **178**, **222**
Iceland Moss 82
Icho Cascarilla 280
Idigbo 324, 329
Iigiri Tree 178
Ilama 31
Ilang Ilang 349
Illipe nut oil 244
Immortelle 130, **179**, 347
  Mountain 130
  Swamp 130
Inca Wheat 27
Incense Tree, Chilean 197
Inch Plant 351
Inchworm 187
India Rubber 72
Indian Bean 77
Indian Cup 298
Indian Pipe 226
Indian Spice 164
Indian Warrior 257
Indian-currant 320
Indigo 122, 145, **179**, 347
  False 52
  Yellow 52
  Yoruba 204
Ingera 128
Ink Cap **100**
  Glistening *153*
  Magpie *100*
Ipecac 180, 221
  American 157
  False 180
Ipecacuanha **180**, 221
Iris *181*
  Butterfly 181

Iris (cont)
  Crested 181
  Dwarf 181
  English 181
  Lamance 181
  Peacock 226
  Persian 181
  Roof 181
  Spanish 181
  Wall 181
  Wild 181
  Yellow 181
Irish Lace 321
Irish Moss 308
Iroko **181**, 329
Iron, Red 205
Iron Tree 254
Ironbark **181**, 329
  Bloodwood 181
  Broadleafed 181
  Narrowleafed 181
  Pink 181
  Red 133, 181
  White 131
Ironweed 341
  Western 341
Ironwood **181**, 222, 251
  Black 242
  Borneo 181
  Dwarf 181, 205
  Giant 181
  Persian 254
  Scrub 181
  Sonora 181
  South-sea 181
  White 181
Isanu 333
Istle **181**
  Jamauve 181
  Tula 181
Istli **181**
Itchweed 341
Ivory Wood 306
Ivy **181**
  American 255
  Boston 255
  Canary Island 182
  Cape 303
  Common 182
  Five-leaved 255
  Geranium 303
  Ground 158
  Irish 182
  Japanese 255
  Kenilworth 108
  Madeira 182
  Parlor 303
  Persian 182
  Poison 270, *270*, 286
    Oakleaf *270*, 286
Ixtle 181
Ixtli 181

Jaborandi 265
Jacaranda 112, 329
Jacinth
  Peruvian 301
  Spanish 177
Jack in the Pulpit 40
Jack Wood 106
Jackfruit 42, 122, 147, 338, *340*
Jack-go-to-bed-at-noon 330
Jacob's Ladder 270, *270*
Jacob's Staff 227
Jade Plant 104
Jaggery 40, 59, 75
Jalap *180*, 225
Jamberry 147, 264
Jambolan 133
Jamestown Weed 113
Japan Wood Tree 67
Japonica **82**
Jarrah 131, 329
Jasmine 56, 132, 143, **182**
  Arabian 182
  Bastard **82**
  Cape 151
  Catalonian 182
  Chilean 211
  Cinnamon 164
  Common White 182
  Crape 321
  Italian 182
  Madagascar 314
  Orange 298
  Primrose 182
  Rock 29
  Star 330
  West Indian 269
  Willow-leaved 82
  Winter 182, *182*

Jelutong 329
Jerusalem Artichoke 42, 319
Jerusalem Cross 207
Jerusalem Thorn 254
Jerusalem Tree 252
Jessamine, Night-blooming 82
Jew Bush 257
Jew's Apple 46
Jew's Beard 321
Jew's Ear Fungus **182**, *183*
Jimsonweed 113
Jobo 312
Job's Tears 95, *95*
Jocote 312
Joey Hooker 150
John-go-to-bed-at-noon 330
Johnny-jump-up 343
Jonquil 232
Joshua Tree **182**, 349
Jove's Fruit 201
Judas Tree 80
Jujube
  Chinese 351
  Common 351
  Indian 351
Jumping beans **183**
Juneberry **27**
Juniper *54*, 122, *142*, 143, **184**
  Californian 185
  Chinese 184
  Coffin 184
  Common 184
  Creeping 184
  Grand 185
  Greek 185
  Needle 185
  Plum 185
  Prickly 184
  Sargent 184
  Sierra 185
  Syrian 185
Juniper tar 184
Jupiter's Beard 79
Jute 141, *141*, *184*, **185**, 338
  Bastard 170
  China *85*, 140
  Congo 39, 140
  Tossa 185
  Upland 185
  White 185

Kaffir beer 309
Kaffir Boom 130
Kaffir Bread 126, 318
Kaffir Lily **90**, *300*
Kaffir Potato 269
Kahika 270
Kajoo Lahi 79
Kakee 185
Kaki **185**, 260
Kalanchoe
  Flame 186
  Pen-wiper 186
Kale 62, *63*, 145, **186**
  Borecole 186
  Cabbage 62
  Chinese 62
  Cottager's 186
  Crimped 186
  Curled 62, *63*
  Curly 62, 186
  Flowering 62
  Hungry Gap 62, 186
  Indian 347
  Italian 62
  Marrow Stem 62, 186
  Portuguese 62
  Ragged Jack 186
  Ruvo 62
  Scotch 62
  Sea 63, **104**, *104*, 338
  Siberian 62, 186
  Thousand Headed 186
Kalmia, Bog 186
Kangaroo Paw 31
Kangaroo Vine 88, *88*
Kapa 141
Kapok 59, 140, *141*, **186**, *187*
  White 140
Kapok seed oil 244
Kapok Tree 78, 186, *187*
  White 59
Kapur 329
Kariba Weed 186, 296, *296*
Karo 268

Karri 329
Karri Tree 187, 255
Kat **186**
Katsura Tree 80
Kauri **186**, 329
  Queensland 186
Kauri gum 186
Kauri Pine **186**
Kava 18, 122, 267
Kawaka 197
Kaya 330
Keck 36, 167
Kelp 21, **186**, *191*
  Bull 20, 186, 234, *234*
  Giant 20, 65, 186, 208
Kenaf *140*, 170
Kentucky Coffee Tree 161
Keruing 118, 329
Khair 9
Khas-khas **187**, 342
Khat 56, **186**
Khesari 160
Khus Khus 132, **187**, 342
King Nut 75
Kingcup 80
King's Crown 185
Kingwood, Brazilian 112
Kino **187**, 277
  Bengal 66
  Gum 66
  Malabar 187
  West African 277
Kiri wood **187**
Kirsch 18
Kitchingia 186
Kitul Fiber 75
Kiwi Fruit 12
Kiwiberry 147
Knapweed 79
  Greater 79, *79*
Knotroot 313
Knotweed **271**
  Dyer's 122
  Giant 271
Kohl rabi 62, *63*, **188**, *188*
Kohuhu 268
Kola 56, **188**, *188*, 221, 239, *239*
  Bitter 188
Komatsu-na 62
Kombu 21
Korakan 125
Kowhai 308
Kudzu 145
Kümmel 75
Kumquat 89, 147, **188**
  Oval 188
  Round 188
Kurrat 193
Kutai 39
Kvass 18

Labdanum 88
Lablab 119, **189**, 195, 195
Laburnine 189
Laburnum 221
  Common 189
  Evergreen 267
  Indian 76
  Nepal 267
  New Zealand 308
  Scotch 189
Lac 138, **189**
Lac Tree 189, 208
Lace Bark 171, 189
Lace Leaf 37
Lace Shrub 314
Lace Vine 271
Lacquer **189**
  Burmese 189
Lacquer Tree 189
Ladanum 88
Ladies Delight 343
Ladies Fingers 242
Lad's Love 47
Lady of the Night 65, 82
Lady's Finger 170, 338
Lady's Laces 201
Lady's Mantle **17**, *17*
  Alpine 17
Lady's Smock 73, *73*
Lady's Tresses
  Autumn 312
  Common 312
  Slender 312
  Southern 312
Lager 18
Lamb's Ear 313
Lamb's Quarters 332

Lamb's Tail 302
Lamb's Tongue 313
Lamy 321
Lanamar 273
Lancewood
  Cuban 121
  Guianan 121
  Jamaican 121
Lantern Plant, Chinese 264
Larch *190*, **191**, 331, 334
  Dahurian *191*
  Eastern 191, 329
  European 191, 329
  Golden 276
  Japanese 191
  Western 191, 329
Larkspur 114
  Alpine 114
  Forking 114
  Rocket 114
Lasiandra 326
Lasiocarpine 221
Latex *191*, 289
Lathyrism 160
Lattice Leaf 37
Lauan 329
Laudanoscine 26
Laurel **192**
  Alexandrian 70, 192
  Australian 268
  Bay *142*, 143, 192, *192*
  Californian 192
  Canary Island 192
  Cherry 276, 192
  Indian 329
  Japanese 192
  Madrona 39
  Maza 241
  Mountain 186, *186*
  Portugal 276
  Sheep 186, 192
  Spotted 47, 192
  Spurge 112, 192
  Swamp 192
  Western 186
Laurel Tree 260
Laurelberry fat 192
Laurustinus 342
Lavatera, Tree 192
Lavender **193**
  Common 192, 193
  Cotton **297**
  French 192
  Sea 193, **201**, *201*
  Spanish 192
  Spike 192
  Wild 193
Lavender Cotton **297**, *297*
Lavender oil 132, 192
Laver bread 21, **193**
Lawyer's Wig 100, *100*
Leadwort, Cape 269
Lectin 335
Leechee 202
Leek 19, **193**, *193*, 264, 338, *339*
Lemon 56, 89, 143, 147, **193**, *193*
  Garden 216
  Wild 270
Lemon Balm 218, *218*
Lemon oil 132, 193
Lenten Lily *167*
Lenten Rose 166, *167*
Lentil **193**, *194*, 195
Leopard Bane *119*
Lettuce 196, *196*, 338, *339*
  Bitter 196
  Cabbage 338
  Common 189, *189*
  Cos 338
  Lamb's 337, *337*
  Opium 196
  Poison 196
  Prickly 189, 196
  River 267
  Sea 21, **301**, *301*
  Water 267, *267*
  Wild 338
  Willow-leaved 196
Lettuce Tree 267
Lichen **198**
  Pine 198
  Snow 82
  Stone, Red 122
Licorice *142*, 143, **197**, 221
  Indian 197
  Wild 46, 197
Licorice Root 165, 197
Life Plant 186

Lignum Vitae 161, 161, 200, 329
Lilac 200
  Common 200
  Hungarian 200
  Persian 200, 218
Lily 200
  African 14, 15
  Amazon 133
  Arum 42, 42, 349, 350, 350
  Atamasco 351
  Avalanche 131
  Aztec 312
  Barbados 171
  Belladonna 27
  Blood 162, 162
  Boat 285
  Bugle 345
  Butterfly 164
  Canada 200
  Corn 341
  Crane 316
  Cuban 301
  Day 167
  Easter 200
  Fire 110
  Flamingo 36
  Formosa 200
  Fox-tail 129, 129
  Ginger 102, 164
  Glory 158, 158
  Golden-rayed 200, 201
  Guernsey 234
  Ifafa 110
  Jacobean 312
  Kaffir 90, 300
  Kamchatka 146
  Lenten 167
  Madonna 200
  Martagon 200
  May 210
  Nankeen 200
  Orange 200
  Palm 349
  Panther 200
  Paradise 254
  Peruvian 25, 25
  Plantain 173
    Blue 173
    Fragrant 173
    Narrow-leafed 173
  Pond, Yellow 237
  Prairie 219
  Pyjama 105, 105
  Resurrection 185
  St. Bernard's 36, 36
  St. Bruno's 254, 254
  St. James 312
  Scarborough 299
  Spear 119
  Spire 150
  Tiger 200
  Toad 331, 331
    Japanese 331
  Torch 188
  Triplet 332
  Trout 131
  Trumpet 350
  Turk's Cap 200
    Scarlet 200
  Voodoo 298
  Water (see Water-lily)
Lily of the Field 314
Lily of the Valley 200, 221
  False 210
  Star-flowered 307
  Wild 279
Lily-of-the-Nile 15
Lily-of-the-palace 171
Lily-of-the-valley Bush 265
Lily-of-the-valley Tree 90
Lilyturf 246
Lima Bean 201
Limba 329
Lime 56, 89, 147, 201, 326
  American 326, 329
  Crimean 326
  European 326, 329
  European Common 326
  Japanese 329
  Key 201
  Large-leaved 326
  Mandarin 147
  Ogeeche 237
  Persian 201
  Rangpur 147
  Silver 326
  Silver Pendent 326
  Small-leaved 326

Lime (cont)
  Spanish 218
  Tahiti 201
  West Indian 201
  White 326
Lime oil 132
Limonene 143, 221
Limu Eleele 127
Linaloe oil 66
Linden 326
  American 326
  Crimean 326
  European 326
  European Common 326
  Large-leaved 326
  Silver 326
  Small-leaved 326
  White 326
  Weeping 326
Linen 140
Ling 69, 164
Lingkersap 299
Linseed 139, 201, 244
Linseed Cake 139
Linseed oil 201, 202, 244
Lint 103
Lion's Heart 264
Lion's Mouth 36
Lipstick Plant 13
Liquorice 197
Litchi 147, 149, 202
Litmus 199
Live Forever 302
Livelong 302
Liverleaf 167
Liverwort 228
Living Stone 202, 203
Lobelia 203
  Edging 203
Lobeline 203, 221
Lobster Plant 270
Locust 203
  African 254
  Black 203, 287
  Caspian 158
  Honey 158, 203
  Water 158
  West African 254
  Yellow 287
Locust Bean 203
  West African 203
Locust Tree 177, 203
Loganberry 147, 204, 292
Logwood 122, 162, 204
London Pride 204, 299
Longan 147
Loofa 206 206, 217
  Angled 217
Loosestrife
  Garden 207
  Purple 207, 207
  Spiked 207
  Yellow 207, 207
Loquat 129, 147, 149
Lords and Ladies 42
Lotus 205, 345
  American 233
  Chinese 233
  East Indian 338
  Egyptian Blue 237
  Hindu 233
  Indian 233
  Sacred 233, 338
  White 237
Lousewort 257
  Common 257
  Marsh 257
  Swamp 257
Lovage 168, 169, 205
  Black 307
  Scotch 200, 205
Love grass 128, 145
  Weeping 128
Love-in-a-mist 135, 235
Love-in-a-puff 73
Love-in-idleness 253
Love Lies Bleeding 25
LSD 26
Lucerne 118, 145, 205, 205, 215
  Common 205
  Wild 145
Luffa 217
Lulo 308
Lungwort 278, 341
Lupin 195, 206, 221
  Blue 206
  Pearl 206
  Russell 206
  Sundial 206
  Tree 206
  White 206
  Yellow 26, 206

Lupinine 26, 221
Lupulin 172
Lupulone 143
Lychee 147, 202
Lymegrass 293
Lysergic acid 26

Macadamia 208, 239, 239
  Rough Shell 208, 239
  Smooth Shell 208, 239
Macassar oil 208, 244
Mace 142, 143, 208, 220, 237
Macroptilium 145
Mad Apple 46
Madagascar Chaplet Flower 314
Madagascar rubber 208
Madder 122, 208, 292
  Field 208
  Indian 208
  Levant 208
  Wild 208
Mader 70
Madia oil 208
Madrona 39
Madrono 39
Madwort 25
Maerl 21
Magician's Flower 321
Magnolia 208
  Chinese 209
  Great-leaved 209
  Saucer 209
  Southern 209
  Star 209
  Umbrella 209
Maguey 15, 140, 306
Mahala Mat 77
Mahoe, Blue 170
Mahogany 209, 260
  African 14, 187, 209, 329
  American 329
  Australian 123
  Bastard 131, 329
  Big Leaved 209
  Borneo 70
  Cape 209, 277
  Cuban 209, 329
  Espave 28
  Honduras 209, 329
  Ivory Coast 209
  Mexican 209, 329
  Nyasaland 187, 209, 329
  Pod 14
  Red 131, 187, 209, 329
  Sapele 297
  Senegal 209
  Swamp 131
  True 209, 329
  Venezuelan 209, 329
  West Indies 209, 329
Maidenhair Fern 12, 12
Maidenhair Tree 98, 99, 210
Maiden's Breath 161
Maiden's Tears 305
Maiden's Wreath 144
Maiwein 150
Maize 56, 81, 81, 101, 145, 210, 210, 211, 244, 318
  Water 343
Malacca Cane 210
Malka 292
Mallee 131
  Brown 131
  Congoo 131
Mallow 210, 211
  Common 211
    European 210, 210
  Curled 210
  Indian 85, 140, 210
  Jew's 186, 338
  Marsh 25, 210
  Musk 211
  Rose
    Common 170
    Swamp 170
  Sea 210
  Syrian 242
  Tree 192
  Virginia 305
Mamey 211, 297
Mammee 147, 211, 297
Mammey Apple 147
Mamoncillo 218
Man root 252
Mandarin 89, 147, 211, 250
  King 211

Mandrake 212, 220, 221
  American 270
Mangabeira rubber 163
Mangel 290
Mangel-wurzel 212, 290, 291
Mango 147, 149, 212
Mangold 290, 291
Mangosteen 147, 149, 151, 212
Mangrove 212
  American 284
  Red 212, 284
Mangrove bark 122, 212
Manihot 290
Manila Hemp 8, 230, 306
Manila Tamarind 268
Man-in-a-boat 285
Manioc 75, 81, 213, 318
Manna 213
Manna Plant 322
Manuka 196
Manzanilla 213
Manzanilla Real 213
Manzanita
  Eastwood 39
  Great Berried 39
  Pine-mat 39
Maple 10
  Amur 10
  Ash-leaved 10
  Black 318
  Cretan 10
  Downy Japanese 10
  Field 10
  Great 10, 286
  Hers's 10
  Hornbeam 10
  Italian 10
  Japanese 10
  Norway 10
  Oregon 10
  Queensland 329
  Red 10
  Rose 106
  Sugar 10, 318, 318, 319
  Trident 10
Maraschino Cherry 213
Marble wood 329
Marc grappa 18
Mare's Tail 171, 171
Marguerite 196
  Blue 135
  Golden 36
Marigold 213
  African 213, 321
  Big 321
  Bur 55, 213
  Cape 117, 213
  Common 68, 213
  Corn 213
  Desert 213
  French 213, 321, 321
  Marsh 70, 70, 214
  Pot 68, 168, 213
  Signet 321
  Striped Mexican 321
  Sweet Scented 220
Marijuana 71, 220
Marjoram 143, 214
  Pot 168, 214, 214, 250
  Sweet 168, 169, 214, 250
  Wild, Common 250
Marmalade 304
Marmalade Bush 316
Maroochy Nut 208
Marrow 103, 216
  Custard 216, 217
  Orange 216
  Vegetable 107, 107, 216, 217, 310
Marsh Mallow 25, 170, 210
Marshmallow 25
Marvel of Peru 225
Maskflower 24
Massoia Bark 88
Masterwort 46
  Greater 46, 46
  Lesser 46
Mastic 214, 267, 284
  American 214
  Bombay 214, 267
  Chios 214
Mastic Tree
  American 299
  Peruvian 299
Matai 125
Maté 56, 172, 214

Matrimony Vine 207
Maw seed 247
May 163, 163
  Pink 163
May Wine 150
Meadowsweet 138, 138
Mealies 215
Medick
  Black 145, 215, 304
Medlar 147, 148, 215, 224
  Japanese 129, 147
Melilot 168, 218
  Common 218
  Small Flowered 218
  White 218
Melon 106, 218, 219
  Cantaloupe 218
  Casaba 216, 218
  Dudaim 216, 219
  Honeydew 172, 216, 217, 219
  Horned
    African 216
  Malabar 216
  Mango 216, 219
  Musk 106, 216, 217, 218
  Netted 216, 217, 218
  Nutmeg 216, 217, 218
  Ogen 218
  Orange 216
  Oriental Pickling 216, 219
  Persian 216
  Pomegranate 216
  Preserving 216
  Queen Anne's Pocket 216
  Rock 216
  Serpent 216, 219
  Snake 216, 219
  Stink 216
  Sweet 218
  Water 216, 217, 345
    Chinese 216
  Winter 172, 216, 218
Melon Shrub 308
Melon Tree 73, 253
Melongene 46
Meni oil 205
Meranti 305, 329
Mercury 270
  Annual 219
  Dog's 219, 219
Mermaid's Wine Glass 11
Mescal beans 308
Mescal Button 205
Mescaline 26, 204, 205
Mesembryanthemum 221
Mesquite 275
  Honey 275
  Western 275
Messmate 131
Methoxsalen 27
Mexican Firecracker 123
Mezcal 18, 56, 56, 220
Mezereon 112, 112, 221
Mezereum 270
Michaelmas Daisy 45, 112, 223
Miedra 270
Miel de palma 183
Mignonette 223
  Common 223
  Upright, White 223
  Wild 223
Mignonette Tree 167
Milfoil 11
  Water 231
Milk Tree 64
Milkweed 140
  Common 43
  Swamp 43, 140
Milkwort 274, 271
  Sea 158
Millet 56, 81, 223, 304
  African 81, 125, 224
  Bajra 81
  Barnyard, Japanese 81, 224
  Broomcorn 145
  Bulrush 81, 224
  Cat-tail 224, 258
  Common 81, 81, 223, 253
  Finger 81, 81, 125, 224
  Fonio 81
  Foxtail 81, 81, 224, 304
  Fundi 81
  German 81
  Golden 223

Millet (cont)
  Italian 81, 224, 304
  Koda 81, 224
  Kodo 81, 224, 255
  Little 253
  Pearl 81, 145, 224, 258
  Proso 81, 224, 253
  Spiked 224, 258
  Teff 81
Mimosa 9, 132
  Bailey's 9
  Golden 9
Mind-your-own-business 308
Minnie Bush 219
Mint 168, 169, 223, 224
  Apple 168, 224
    American 224
  Bergamot 168, 224
  Common 224
  Corn 218
  Eau de Cologne 168
  Field 218, 218, 224
  Ginger 168
  Horse 168
  Japanese 224
  Lemon 168
  Orange 224
  Round-leafed 224
  Scotch 168
  Water 168
Mintbush 275
  Round-leaf 275
  Snowy 275
Mirabelle 112, 147, 269
Miracle Leaf 186
Miracularin 143
Miraculin 143
Miraculous Berry 143, 225
Miraculous Fruit 225, 325
Mirasol 319
Mirror of Venus 246
Missey-moosey 309
Mist Flower 134
Mistletoe 19, 225, 225
  Dwarf 225
  Red 225
Miterwort 225
  False 326
Mkwakwa 317
Mochi Tree 172
Mockernut 75
Mohintli 185
Molasses 225, 319
Mold
  Bakery 235
  Blue 258
  Bread, Red 235
  Green 258
Mombin, Yellow 312
Monarch-of-the-east 298
Money Tree 133
Moneywort 207
Monkey Musk 224
Monkey Nut 38, 239
Monkey Plant 293
Monkey Puzzle Tree 39, 39
Monkey-bread Tree 12
Monkey-flower 224, 224
  Allegheny 224
Monkeypod 296
Monkshood 11, 12, 26
Moon Flower 314
Moonflower 180
Moonwort 60, 60, 136
Mooseberry 342
Mora 329
Morabukea 329
Morel 226, 227
  False 226
Morello 147
Morning Glory 180, 180, 220, 227
  Beach 180
  Common 180
  Dwarf 97
  Imperial Japanese 180
Morphine 26, 221, 247
Moses-in-the-bulrushes 285
Moses-on-a-raft 285
Moss 228
  Bog 229
  Florida 326
  Iceland 82, 198, 199
  Irish 75, 308
  Reindeer 89, 122, 198
  Rose 273, 273
  Spanish 140, 141, 326

Moss (cont)
Stag's-horn 207
Mother of Thyme 326
Mother-in-law 186
Mother-in-law Plant 116
Mother-in-laws Armchair 124
Mother-in-laws Tongue 297
Mother-of-thousands 299
Moulded Wax 123
Mountain Ash 43, 131, 227, 308, 309
American 309
European 309
Mountain Bluet 79
Mountain Everlasting 31
Mournful Widow 299
Mouse-ear 80
Alpine 80
Moutan 252
Msala 317
Mu Tree 17, 334
Mudar 70
Mugwort 42
Western 42
White 42
Muhimbi 329
Mulberry 147, 227
American 227
Black 147, 148, 227, 227
Downing 227
English 227, 227
French 69
Indian 226
Paper 64, 227
Persian 227
Red 147, 227
Russian 227
White 227
Mullein 227, 341
Black 230
Cretan 79
Dark 227
Great 227
Purple 230
White 227
Mundu 122
Mung Bean 230, 230
Muninga 277, 329
Mu-oil 334
Murtillo 231
Muscadine 147
Muscarine 90
Muscat 230
Muscatel 230
Muscle Tree 173
Mushroom 15
Field 15
Fool's 25
Horse 15
Sacred 220
Musk 224, 230
Lavender Water 224
Monkey 224
Wild 130
Musk Clover 130
Musk Plant 230
Musquash Root 87
Mustard 62, 143, 221, 230
Black 62, 63, 230, 231
Broad-leaved 62
Brown 62, 230
Celery 62
Chinese 62
Garlic 168
Indian 62
Japanese 62
Leaf 62
Potherb 62
Treacle 130
White 62, 63, 230, 306
Mustard Greens 62
Muster-John-Henry 321
Mvule 329
Mycetoma 44
Myristicin 237
Myrobalan 122, 263, 324
Beleric 324
Black 324
Myrrh 168, 231
Abyssinian 97
African 97
Myrtle 231, 343
Common 231
Crape 190
Queen 190
Jew's 293

Myrtle (cont)
Sand 193
Tarentum 231
Tasmanian 329
Wax 231

Naegelip 307
Nanan Wood 190
Nandiflame 311
Nanny Berry 342
Naranjilla 232, 232, 308
Narcissus 132, 232, 233
Poet's 232
Polyanthus 232
Trumpet 232
Narcotine 26
Nardoo 137
Naseberry 213
Nasturtium
Dwarf 333
Flame 333
Garden 232, 333, 333
Tall 333
Natal Tak 317
Navelwort 335, 335
Venus 243
Nectarine 147, 233
Spanish 87
Needle Bush 162
Needle Furze 156
Negro's Head 264
Nemu Tree 17
Neroli oil 304
Nero's Crown 321
Nettle 235
Dead 235
Dog 235
False 235
Hawaiian 58
Hemp 150
Roman 336
Small 235
Stinging 235
Common 235, 235, 336
Wood 235
Nettle Tree 79
New Jersey Tea 78
New Zealand Burr 9
Niam fat 205
Nicotine 26, 221, 327
Niger seed 161
Night-blooming
Jessamine 82
Nightshade 235, 307
Black 235, 307
Common 307
Deadly 26, 46, 47, 235
Enchanter's 88, 235
Alpine 88
Garden 307
Silverleaf 307
Woody 235, 235, 307
Nile Tree 311
Ningro 136
Nipplefruit 308
Nipplewort 191
Nispero 213
Nitronga Nut 106
Nitta Tree 254
Fernleaved 254
None-so-pretty 305
Nonesuch 304
Noog 161
Norfolk Island Pine 236
Nori 21
Nosegay 269
Nupharin 237
Nut
Barbados 182
Betel 239, 239
Butter 75
Cashew 238, 239, 239
Curare Poison 26, 107
Earth 38
Grass 64, 332
Ground 38
Illipe 305
King 75
Macadamia 208, 239, 239
Madeira 244
Maroochy 208
Monkey 38, 239
Pecan 256, 257
Physic, French 182
Pili 239
Pine 239, 239
Pistachio 267
Purging 182, 183
Rush 109

Nut (cont)
Souari 75
Tiger 109
Nutgall Tree 286
Nutmeg 142, 143, 208, 220, 237
Brazilian 106
California 233
Peruvian 192
Nutmeg butter 237
Nutmeg oil 132
Nutmeg Tree 132, 231, 237
Nux vomica 316
Nux-vomica Tree 316
Water-filter 317
Nypa Palm 237
Nyssa
Twin-flowered 237

Oak 122, 240, 240, 241
African 205
Australian 131
Black 240
Blackjack 240
Bull 76
Bur 329
Chestnut-leaved 240
Cork 101, 240
Durmast 329
English 229, 329
Golden 240
Holm 240, 240
Hungarian 240
Italian 240
Kermes 240
Live 240
Mongolian 329
Pedunculate 329
Pin 240
Northern 240
Poison 270, 286
Red 240, 240, 329
Spanish 329
River 76
Scarlet 240
Sessile 329
She 76
Silk 160
Swamp 76
White 240
Yellow-barked 122, 240
Oak apple 241
Oarweed 20, 186, 190
Oat 81, 81, 240, 240
Algerian 241
Animated 47, 241, 346
Cultivated 47, 48
Diploid 81
Fodder 240
Hexaploid 81
Naked 241
Potato 241, 346
Red 241
Wild 47
Tartarian 241, 346
Tetraploid 81
Wild 241, 346, 346
Common 47
Oatgrass 145
Obeche 329
Obedient Plant 264
Oca 251, 290, 291, 338
Oconee-bells 305
Ocotillo 67, 144, 144
Oil 244
Ajowan 75
Argemone 273
Babassu 244
Balsam 118
Ben 244
Brazil nut 244
Cacao 244
Cashew nut kernel 244
Castor 244
Champaca 223
Chaulmoogra 244
Chinawood 244
Citronella 109
Cocoa 244
Coconut 244
Cohune 244
Colza 244
Corn 244
Cotton seed 244
Croton 244
Eng 118, 244
Gingergrass 109, 244
Groundnut 244
Gurjum 244
Gurjun 244, 329
Hempseed 244

Oil (cont)
Illipe nut 244
Kapok seed 244
Linseed 244
Lemongrass 109
Macassar 244
Maize 244
Maw seed 244
Mustard 244
Niger seed 244
Nitronga 106
Oiticica 244
Palma Rosa 109
Peach 244
Perilla 244, 260
Pistachio 244
Rasamolo 282
Walnut 244
Wood 282
Oil Flax 139
Oil Grass 108
Oil of anise 179
Oil of cade 184
Oil of chamomile 83
Oil of citronella 132
Oil of marjoram 214
Oil of meadowsweet 138
Oil of sassafras 298
Oil of savin 184
Oil of savory 298
Oil of terebinth 334
Oil of vetiver 132, 342
Oil Palm
American 125
Oil Plant 304
Okoume 329
Okra 170, 242, 242, 338, 340
Olas Paper 59
Old Man 42
Old Mans Beard 336
Oleander 221
Common 234, 234
Oleandrin 234
Oleaster 162
Olibanum Tree, Indian 60
Olive 147, 242, 243, 244, 244, 245
Ceylon 243
Chinese 243
Common 242
Indian 242
Java 243
Sand 243
Sweet 256
Wild 242
Olive Bark Tree 324
Olive oil 242
Olololiuqui 220
Omu 297
Onion 19, 19, 122, 243, 243, 304, 339, 388
Bunching
Japanese 243
Common 243
Field, Common 243
Gypsy 151
Japanese 388
Sea 301, 336, 336
Spring 19
Welsh 19, 243, 388
Opium 26
Opium Poppy 26, 221, 246
Opiuma 268
Opopanax 9
Oppossum-wood 162
Opuntia 247
Crested 248
Little-tree 248
Old Man 248
Orach 248
Orache 46, 248, 338
Stalked 162
Orange 56, 248
Acidless 249
Bitter 147, 304
Blood 249
Common 249
Hardy 273
King 147
Mandarin 211, 298, 322
Mexican 86
Mock 261, 262, 268, 317
House-blooming 268
Natal 317
Navel 249
Osage 122, 250, 250
Pigmented 249
Seville 55, 89, 147, 304

Orange (cont)
Sour 147, 304
Sugar 249
Sweet 89, 147, 211, 248, 248, 250
Trifoliate 273
Vegetable 216
Orange oil 132
Orange-fringe 162
Orange-plum 162
Orchid 249
Bee 246
Bird's Nest 233, 249
Butterfly 243
Clam-shell 127
Club-spur, Little 162
Cockle-shell 127, 127
Dancing-doll 243
Dwarf 249
Early Purple 249
Fly 246, 246
Frog 95
Giant 127
Green Rein 162
Green Woodland 162
Green-winged 249, 250
Jewel 31, 159
Lady 249
Lady's-slipper 110, 110, 253
Late Spider 246
Lesser Butterfly 162
Lily-of-the-valley 241
Lizard 171
Man 10, 11
Military 249, 250
Mirror 246
Monkey 249
Moth 261
Musk 242
Nun 207
Pansy 224
Poor Man's 300
Purple Fringeless 162
Purple Hooded 250
Purple-spire 162
Pyramidal 28, 28
Round-leaved 250
Small 250
Salep 250
Showy 250
Spotted
Common 111
Woodland 250
Yellow Fringed 162
Orchid Tree 53, 53
Orchis
Green Man 11
Oregano 250
Wild 168
Orpine 302
Orris Root 181, 250
Ortanique 211, 250
Orseille 122
Osier 295, 346
Common 295
Oswego tea 55, 168, 226
Otaheite 312
Otto of Roses 46, 132
Ouabain 11
Our Lady's Lace 36
Our Lord's Candle 349
Ouzo 18, 31
Oxalis, Creeping 251
Oxeye, Yellow 66
Oxford and Cambridge
Bush 277
Oxlip 271, 274
Ox-tongues 265
Oyster Plant 222, 285, 295, 338
Black 301
Oysternut 239
Ozone fiber 43

Paccai 275
Pachysandra
Alleghany 252
Japanese 252
Padauk 277, 329
Andaman 277, 329
Burma 329
Sandalwood 277
West African 329
Paddle wood 252
Paddy 286
Hill 287
Lowland 287
Upland 287
Padouk
African 277
Brown 277
Red 277

Padouk (cont)
Vengai 277
Pagoda Tree, Japanese 308
Painted Lady 123
Painted Leaf 270
Painted-tongue 295
Pak-choi 62, 63, 85, 86
Paku 136
Palas 66
Palissander, Javanese 112
Palm
Assai 134
Babassu 249
Bamboo 282
Betel-nut 26, 221
Butterfly 86
Cabbage 134, 289
American 318
Caribbee 289, 318
Chilean Wine 183
Coconut 318
Cohune 249
Coquitos 183
Curly 176
Date 113, 113, 262, 263
India 263
Wild 263, 318, 318
Fan
Blue 61
Dwarf 141, 141
Fishtail 75
Flat 176
Gebang 102
Gomuti 294, 318
Honey 318
Ivory Nut 222, 264
Jaggery 318
Japanese 318
Prickly 294
Spineless 294
Key 325
Kitul 75, 294
Leopard 27
Nypa 141, 318
Oil 244
American 125, 244
Palmetto
Royal 325
Palmyra 59, 141, 318
Panama Hat 140
Parlor 83
Princess 116
Raffia 282
Ivory Coast 282
Rattan 68, 210
Royal
Cuban 289
South American 289
Sago 75, 108, 108, 141, 222, 294, 318, 318
Sugar 40, 40, 141, 318, 318
Syrup 318
Talipot 102
Thatch-leaf 176
Toddy 75, 318
Traveller's 283
Wax 82
Andean 345
South American 345
Windmill
Chinese 141, 330
Wine 75, 282
Palm Grass 304
Palm Hearts 134
Palm Lily 100
Palm Oil 125
Palm wine 18
Palma Christi 287
Palma de Macetas 117
Palmetto 252
Blue 252
Bush 252
Cabbage 252
Common 252
Dwarf 252
Scrub 252
Texan 252
Palmetto Thatch 325
Palmiet 275
Palmita 73
Palmyra Palm 59
Palo Campechio 162
Palo Muerto 28
Palo Negro 112
Palo Verde
Mexican 254
Palta 260
Pampas Grass 101
Panama Hat Plant 73

Panaquillon 252
Panda Plant 186, 262
Pandang 252
Pansy **253**, 343
  European Wild 253,
    253, 343
  Field 253, 343
  Garden 253, 343,
  Miniature 343
  Mountain
    Yellow 253
  Shirley 253
  Show 343
  Tricolor 253
  Tufted 343
Panther Cap 25
Papain 253
Papaverine 26, 221
Papaw 73, 73, 147,
  149, **253**, 253
  American 253
Papaya 73, **253**
Paper 140
Paper Plant 135
Paper Reed 109
Paperback Tree 215
Paprika 72, **254**
Papyrus 109, 109, 140
Para Rubber Tree 72
Paradise Flower 67, 307
Paradise Plant 185
Paradise Tree 305
Paraguay Tea 172
Paranut 239
Para-para 267
Parrot's Beak 205
Parrot's Bill 90
Parrot's Claws 165
Parrot's Flower 165
Parsley 143, 168, 169,
  254
  Cow 36, 37
  Hamburg 338
  Horse 307
  Turnip-rooted 255,
    338
Parsnip 255, 290, 338
  Cow
    Common 167
  Peruvian 41
Partridge Berry 151
Paspalum, Ribbed 255
Pasque Flower 278
Passion Flower
  Blue 255
  Red 255
Passion Fruit 50, 147,
  149
  Yellow 255
Pasta althaea 25
Pastis 31
Patchouli 132, 270
Patchouli oil 270
Pawpaw 147, **253**
  Mountain 73
Pea **255**
  Asparagus **44**, 103,
    194, 195, 256
  Australian 119
  Black-eyed 195
  Butterfly 90, 145, 256
  Chick **85**, 195, 195,
    256
  Chickling 160, 193
  Common 255, 296
  Congo 195
  Culinary 255, 267
  Cultivated 256
  Darlington River 320
  Earthnut 192
  Everlasting 193, 193
  Field 195, 255, 267
  Flame 86, 256
  Garden 194, 195,
    255
  Glory 90
  Grain **85**
  Grass **160**, 193, 256
  Heart 73
  Hindu 103
  Hoary **324**
  Jerusalem 103
  Kaffir 195
  Kordofan 90
  Perennial 192
  Pigeon 195, 195,
    256, **265**, 265
  Poison 320
  Rabbit's 324
  Rosary 8, 256
  Sea 192
  Snow 255
  Sturt's Desert 90, 91
  Sugar 255
  Sweet 193, 256
  Winter 320

Pea (cont)
  Tangier 193
  Winged 195, 205
Pea Shrub **72**
  Russian 73
  Siberian 72
Peach 19, 147, 148,
  256, 256, 276
  Cling 256
    Chinese 256
  Freestone 256
  Vine 216
Peach leaf curl 322
Peach Wood 67
Peacock Flower 67, 114
Peacock Plant 185
Peanut 38, 239, 244
Pear 147, 148, **256**,
  257
  Alligator 47, 147,
    260, 338
  Balsam 226
  Chinese 279
  Common 256
  European 256
  Melon 259, 308
  Paper-spined 248
  Prickly 236, 248
  Sand 279
  Willow-leaf 279
Pearl Barley **256**
Pearl Bush 134
Pearl Plant 163
Pearly Everlasting Flower
  29
Peavine, Grass 160
Pebble Plant **202**
Pecan 239, 239
Pecan nut **256**, 257
Peepal Tree 138
Peepul 138
Pelican Flower 40, 40
Pellitory 28
Penicillin 258
Pennisetum
  Chinese 258
Penny Flower 206
Pennyroyal 168
Pennywort 258, 335,
  335
  American 258
  Asiatic 258
  Marsh 258
  Wall 258
Peony
  Chinese 252
  Garden
    Common 252
  Tree 252
Peperomia
  Emerald-ripple 258
  Fantasy 258
  Green-ripple 258
  Ivy Leaf 259
  Platinum 259
  Silver Leaf 259
Pepino 259, 308
Pepper 26, 142, 143,
  221, **259**, 259
  African 259
  Baby 287
  Bell 259
  Betel 55, 267
  Black 259
  Cayenne **77**, 259
  Chilli 72, 72, 259
  Cubeb 259
  Green **160**, 259
  Jamaican 19, 259
  Japan 31, 350, 351
  Kava 267
  Melagueta 259
  Negro 259
  Nepal 259
  Paprika 259
  Red 259, **283**
  Sweet 259, **320**, 338
    Green 72
    Red 72
  Tabasco 72
  Trinidad Devil 283
  White 259
Pepper Face 259
Pepper oil 259
Pepper Plant 259, 267
Pepper Tree 214, 221,
  299, 299
  Australian 299
  Brazilian 299
  California 299
  Monk's 343
Pepper Vine 28
Pepperbush
  Sweet 90
Peppercorn 19, 259,
  259

Pepperidge 237
Peppermint 132, 143,
  168, 169, 221, 224,
  259
  Broadleaved 131, 259
  Chinese 259
  Japanese 259
Peppermint oil 132, 259
Peregrina 182
Periwinkle
  Common 343
  Greater 343
  Lesser 343
  Madagascar 77, 221
  Rose 77
Persimmon 117, 122,
  147, 149, **260**
  American 117, 260
  Common 117, 260
  Kaki 185
Pe-tsai 63, 85
Petty Whin 156
Petunia 260
  Large White 260
  Violet-flowered 260
Peyote 204, **205**
Peyote Button 220
Peziza
  Orange-peel 152
Pheasant's-eye 12, 221
Philodendron
  Blushing 262
  Giant 262
  Heart-leaf 262, 262
  Horsehead 262
  Red-leaf 262
  Spade-leaf 262
Phlox 262
  Fall 262
  Moss 262
  Mountain 262
  Perennial 262
  Summer 262
Photinia
  Japanese 263
Physic, Indian 157
Phytoplankton 22, 268
Piassava fiber **265**
Pickaback Plant 327
Pickerel Weed 273
Pie Plant 285
Pig-a-back Plant 327
Pigeon Berry 264
Pigeon Pea 256, **265**,
  265
Pigeon-foot 294
Pignons 266
Pignut 97
Pig's Ear 103
Pilea, Watermelon 265
Pili Nut 239
Pillwort 265, 265
Pilocarpine 26, 265
Pimbina 342
Pimenta 19, 72, **266**
Pimento 19, 72, **266**
Pimienta 19
Pimiento 72, **266**
Pimpernel **28**
  Bog 28
  Scarlet 28, 29
  Wood 207
  Yellow 207
Pincushion Flower 162,
  299
Pine 141, **266**
  Austrian 266
  Buddhist 270
  Chile 39, 329
  Cypress **69**, 266
    Black 296
  Giant 266
  Golden 160
  Ground 16
  Hoop 39
  Huon 111, 266
  Japanese Umbrella
    300
  Lace Bark 189
  Longleaf 266, 288,
    329
  Mahogany 270
  Maritime 266, 334
  Monterey 98, 266
  Moreton Bay 39
  Mountain 111
  Norfolk Island 39,
    236
  Parana 39, 329
  Parasol 300
  Pitch 288
  Red 111, 266
  Rottnest Island 69
  Scots 8, 39, 266, 267,
    288, 329
  Screw 252

Pine-Screw (cont)
  Thatch 252
  Southern 266
  Stone 239
    Mexican 239
    Swiss 239
  Sugar 266
  Totara 270
  White 270, 329
    Eastern 314
    Japanese 266
    Western 266, 329
  Yellow
    Southern 288
    Western 329
Pine Kernel 239
Pine resin 56
Pineapple 56, 122, 140,
  141, 147, 149, **266**,
  266
  Wild 326
Pineapple Flower 133
Pineapple Guava 135
Pinene 71
Pink
  California Indian 305
  Cheddar 115
  Clove 115
  Cluster-head 115
  Cottage 115
  Grass 115
  Kirtle
    Spotted 250
  Maiden 115, 115
  Rainbow 115
  Wild 305
Piperidine 26, 221
Piperine 26
Pipe-rot 314
Pisco 18
Pistachio 239, 239, 267
Pistol Plant 265
Pita 140
Pitanga 133
Pitcher Plant 234, 266,
  267, 298
  Californian 113, 113
Pith Plant 13
Plane **268**
  American 268, 320
  London 268, 268
  Oriental 268
Plankton **268**
Plantain
  Buck's Horn 268
  Common 268
  Crowfoot 268
  Edible 147, 230
  Grass 268
  Hoary 268
  Narrow-leaved 268
  Rattlesnake 159
  Ribwort 268, 268
  Sea 268
  Water 17
    Common 17
    Great 17
    Ribbon-leaved 17
  Wild 165
Pleroma 326
Pleurisy Root 43
Plum 56, 147, 148, 269
  American 147, 269,
    276
  Cherry 147, 269, 276
  Coco 87
  Damson 112, 160,
    269, 276
  Date
    Chinese 185
    Japanese 185
  European 160, 269
  Hog 312
  Jamaica 312
  Japanese 147, 269,
    276
  Java 133
  Marmalade 213, 297
  Ogeeche 237
  Oriental 269
  Pigeon 94
  Red 312
  Spanish 312
  Victoria 269
Plum brandy 269
Plum Yew 79
  Chinese 79
  Japanese 79
Plume Flower 185
Plush Plant 186
Poached Egg Flower
  201
Pocket Handkerchief
  Tree 114, 114
Podgrass, Shore 331
Podocarpus

Podocarpus (cont)
  Fern 270
  Weeping 270
Pohutukawa 222
Poinciana
  Dwarf 67
  Royal 114
Poinsettia 134, 134,
  270, 270
  Japanese 257
Poison Berry 307
Poison Bulb, Asiatic 105
Poison Ivy 270, 270, 286
  Oakleaf 270, 286
Poison Nut 316
Poke, Indian 204, 220,
  341
Pokeberry **264**
Pokeweed **264**
  Virginia 264
Polar Plant 305
Polyanthus **271**
Polygala, Fringed 271
Polygonum
  Japanese 271
Polypody 137
  European 272
  Golden 272
  Rock 272
  Wall 272
Pomace 76
Pombe 18
Pomegranate 147, 149,
  272, 272, 278
  Dwarf 272
Pomelo **272**, 322
Pompelmous 272
Pompom **272**
Pondweed **273**
  Cape 37
  Horned 349
Poor Man's Weather
  Glass 28
Popinac 9
Poplar 141, **273**
  Balsam 273, 329
  Black 273, 329
  Carolina 273
  Gray 273
  Lombardy 273
  Necklace,
    Chinese 273
  Silver-leaved 273
  Western 273
  White 273, 329
  Yellow 334
Popolo, Yellow 307
Poppy 143, **273**
  Alpine 273
  Arctic 253
  Blue 215
  Bush 115
  Californian 131, 273
  Chinese, Yellow 215
  Common 253
  Corn 253
  Crested 40
  Flanders 253
  Hawaiian 273
  Himalayan 215, 215
  Horned
    Red 158
    Yellow 158
  Iceland 253
  Mexican 273
    Prickly 40
  Opium 26, 221, **246**,
    253
  Oriental 253
  Plume 208
  Sea 158
  Tibetan 215
  Tree
    California 288
  Tulip 253
  Welsh 215
Portulaca
  Kitchen Garden 273
Posidonia fiber 273
Possum Apple 117
Possum Wood 117
Potato **273**, 274, 290,
  290, 308, 318
  Air 318
  Chosui 195
  Duck 294
  Hausa 96, 269, 338
  Irish 308
  Kaffir 269, 338
  Prairie 277
  Sudan 96
  Swan 294
  Sweet 180, 180, 290,
    291, 291, 320, 320

Potato (cont)
  Tennessee 216
  Telingo 318
  White 308
Potato Creeper, Giant
  307
Potato late blight 264
Potato Tree
  Chilean 308
Potato tuber blight 264
Poui
  Pink 321
  Yellow 321
Powderpuff, Red 69
Prairie Flower 150
Prairie Rocket 130
Prayer Plant 213
Prickly Ash 39
Prickly Elder 39
Prickly Pear 236, 248
Pretty-face 332
Pride of Barbados 67
Pride of India 188, 218
Pride-of-the-peak 162
Primrose **274**
  Baby 274
  Common 271
  Drumstick 274
  English 275
  Evening 241
    Common 242
  Fairy 274
  German 274
  Japanese 274
  Poison 274
Prince Albert Yew 298
Princess Flower 326
Princess Tree 187, 255
Privet 144, 200, **275**
  Chinese 275
  Common 275
  Glossy 275
  Golden 275
  Ibota 275
  Japanese 275
  Mock **262**
  Nepal 275
  Swamp 144
  Wax-leaf 275
Prophet Flower 41
Proso 224
Protea
  Giant 275
  King 275
  Peach 275
Prune **276**
Psilocin 277
Psilocybin 277
Psilophytes 92
Puccoon 203
  Red 296
Puffball 153, 207, **278**
  Giant 152
Pulque 15, 18
Pummelo 147, 272
Pumpkin 107, 216,
  217, **278**
  America 216
  Autumn 216
  Brazilean 216
  Canada 216
  Cushaw 216, 278
  Summer 216
  Wild
    Fetid 216
  Winter 216
Purslane
  Rock 68
  Sea 46, 162
Pussy-ears 186
Putty 201
Pyrethrin 279
Pyrethrum 278, 279
  Dalmatian 279
  Garden 87

Qat 56, **186**
Quack-grass **15**, 16
Quaking Grass
  Common 61
  Small 64
Quamash 70
Quass 18
Quassia 280
Quassia Root 280
Quebracho 122, **280**
  Red 280
Queen Anne's Lace 36
Queen of the Meadow
  138
Queen of the Night 221,
  302
Queen of the Prairie 138
Queensland Nut 239
Quickbeam 309
Quickthorn 163

Quihuicha 27, **159**
Quillwort 92
  Sand 92, 93
Quince 147, *148*, 256, **280**
  Bengal 13
  Common 82, 108, 280
  Flowering **82**, 280, *280*
  Japanese **82**, *82*
  Maules 82
Quinidine 221
Quinine 26, 221, **280**
Quinoa 84
Quinones 56

Rabbit Ears 248
Radicchio 338
Radish 62, *63*, **280**, *281*, 338
  Chinese 280
  Italian 280
  Wild 282
  Winter 338
Raffia 141, *141*, 280, 282
Ragged cup 305
Ragged Robin *206*, **207**
Ragi 125
Ragweed **27**
Ragwort 26, 303, *303*
  Common 302
  Oxford *303*
  Purple 303
Railroad Road 180
Rain Tree 296
Raisin 230, **281**
  Malaga 281
  Muscatel 281
  Wild 342
Raki 18, 31
Rakkyo 19
Rambutan 147, *149*, **281**
Ramie 58, 141, *141*,
Ramin 329
Rampion 54, 264, **281**
  Spiked 264
Ram's Horn 214
Ramsons *151*
Rand Plant 343
Rangoon Creeper 280
Ranunculus, Garden 281
Rape 62, *145*, 230, 244 **282**
  British 282
  Continental 282
  Indian 62, 282
  Oilseed 62, 63
  Swede 320
  Turnip 62, 282
Rasamala **282**
Rasamala resin 282
Raspberry 147, *148*, **282**, *283*
  Arctic 283
  Black 147, 282, 292
  European 204
  Flowering 292
    Purple 292
  Purple 282
  Red
    American 147, 282, 292
    European 282, 292
  Wine **346**
Rati 8
Rattan 111, *111*, **283**
Rattan Palm **68**
Rattle Bell **105**
Rattle Box **105**
Rattle Weed 52
Rattle, Yellow 284, *331*
Rattlesnake-master 130
Rauli 329
Red algae **283**
Red Bud 80
Red Elf Cup Fungus *260*
Red Hot Poker **188**
Red Pepper **283**
Red Robin 157
Red Root 203
Red tide 21, **283**
Redbark 280
Redgram 195
Red-top **16**
  Western 16
Redwood **283**, 303, 329
  Andaman 277, 283
  Brazil 64
  Coast 303
  Dawn 222, 283
  Giant 283

Redwood (cont)
  Sierra 283, 303
  West Indian 283
Reed **283**
  Common 263
  Giant 43
  Paper 109
  Small
    Narrow 284
    Purple 284
  Small Wood 284
  Spanish 43
  Totora *300*
Reed Grasses 43
Reed-mace
  Great 335
  Lesser 66
  Small 335
Reef Thatches 325
Reindeer Moss 89
Rescue Grass 64
Reserpine 221
Resins **284**
Restharrow 246
  Spiny 246
Resurrection Plant **284**, *284*
Retama 254
Retsina 56
Rheumatic fever 51
Rhododendron *285*
  Catawba 285
  Caucasicum 285
  Fortunei 285
  Ghent 285
  Griffithianum 285
  Indica 285
  Kurame 285
  Thomsonii 285
  Tree 284
Rhubarb **285**, 221, *222*, 339
  Bog 260
  Chile 161
  China 28**5**
  Currant 285
  Garden 284, 285, 338
  Medicinal 284, 285
  Sitkim 284
Rice 56, 81, *81*, 101, 250, *251*, **286**, *287*, 318
  African 81, 286
  Indian 351
  Wild
    Annual 351
Rice spirit 41
Rice Tree, Formosa 135
Riceflowers 266
Ricinine 26
River Tea Tree 21**5**
Robustic acid 160
Rock Cress **38**
  Hairy 39
Rock Jasmine 29
Rock Purslane 68
Rock Rose **88**
Rockberry 126
Rocket 168, **287**
  Dame's 287
  Dyer's 223, 287
  Garden 287
  Prairie 130, 287
  Sea 67, 68
    American 67
  Sweet 167, 287
  Wall 168
  Yellow 287
Rock-rose 89, 165, *165*
  White 165
Rod, Silver 45
Roman Candle 349
Roquefort cheese 258
Rose 288, *288*
  Brier 288
  Cabbage 46, 132
  California 70
  Champney 288
  China 288
  Christmas **86**, 221
  Damask 46, 132, 288
  Dog 288, *288*
  Egyptian 299
  Floribunda 288
  French 46
  Guelder 342
  Hybrid Tea 288
  Japanese *288*
  Lenten 166, *167*
  Musk 46, 132, 288
  Noisette 288
  Sweet 288
  Tea 288
  Turkestan *288*
Rose Bay 234
Rose Elder 342

Rose Geranium 132
Rose hip syrup 288
Rose Moss 273, *273*
Rose of China 170
Rose of Heaven 305
Rose of Jericho *284*
Rose of Sharon 170, 178
Rosebay
  Broad-leaved 321
  East Indian 321
Roselle 140, 143, 170
Rosemary 168, *169*, 289
  Bog 29, *29*
Roseroot 302, *302*
Rosewood 123, **288**
  Bengal 277
  Black 112
  Bombay 112
  Brazilian 112, 329
  East Indian 112, 329
  Honduras 112, 329
  Malabar 112, 329
  Nicaragua 112, 329
  Rio 112
  Senegal 187, 277
  Thailand 112, 329
Rosin 8, 266, **288**
Rosinweed 305
Rotenone 115, 204
Rouge Plant **58**, 287
Roundwood 309
Rowan 308
  Japanese 309
  Kashmire 309
  Vilmarin's 309
Rubber **289**
  African 190
  Ceara 72, 213, 289
  India 72, 289
  Jequie 213, 289
  Lagos 289
  Landolphia 289
  Madagascar 190, **208**, 289
  Mangabeira 163
  Manicoba 289
  Panama 289
  Para 72, 170, *170*, *289*, 289
Rubber Plant 138, *289*
  American 259
  Baby 259
Rubber Tree 299
Rue 143, 168, **292**
  Goat's 150, 324
  Meadow **325**
    Alpine 325
    Yellow 292, 325
  Syrian 220
Rum 56
Runch 282
Rupturewort 167
Rush **183**, **293**
  Chairmaker's 293, 301
  Comb **300**
  Corkscrew 184
  Flowering 66, *66*, 293
  Hard 184
  Japanese-mat 183
  Mud
    Floating 301
  Scouring **174**, 293
  Sharp 183
  Soft 183, *185*
  Spike 125
Rust
  Brown *277*
  Flax 215
Rutabaga 62, 290, **293**
Rye 56, 81, *81*, 145, **293**, *293*, 301, 332
  Wild
    Blue 145
      Smooth 126
    Canada 145
    Siberian 145
Ryegrass 145, **293**
  Australian 293
  Bearded 293
  Italian 293
  Perennial 293, *293*

Sacaline 271
Saddle Fungi **166**
Safflower 75, 122
Saffron 105, 122, *142*, 143
  Meadow 96
Safrole 221
Sage 168, *169*, 223, **294**
  Bethlehem 278

Sage (cont)
  Black 41, 294
  Blue 294
  Clary **89**
  Common 294
  Garden 294
  Green 294
  Jerusalem 262, 278, 294
  Linden-leaf 294
  Lyre-leafed 294
  Meadow 294
  Pineapple-scented 294
  Scarlet 294, 295
  Silver 294
  Yellow 191, 294
Sage Tree 343
Sagebrush
  Common 41
Sago 40, 75, 108, 222, 294
  Portland 42
  Queen 318
Sago Cycas 349
Sago Palm 108, **294**
  Prickly 294
  Spineless 294
Saguaro 67
Sainfoin 246
St. Anthony's Fire 26, 128
St. Johns Bread 74, 195
St. John's Wort **177**
  Creeping 178
  Perforated 178
St. Julien 112
St. Martha Wood 67
St. Peter's Wort 178
St. Thomas Tree 53
Saké 18
Saki 18
Sal Tree 305, 329
Salicin 346
Saligna 329
Sallow 295, 346
  White 9
Salmonberry 292
Salsify 83, *290*, **295**, 330, 338
  Black 295, 301, 338
  Spanish 295, 301
Saltbush 188, **295**
  Four-wing 295
  Mediterranean 295
  Spiny 295
Saltwort
  Black 158
  Prickly 295
Saman 296
Sambong 58
Samphire **105**, **294**
  Golden 179, *179*
  Marsh 294, *294*
Sand Myrtle **193**
Sandalwood
  White 297
Sandalwood Tree **296**
Sandalwood oil 132
Sandarac **296**
Sanderswood, Red 277
Sandwort **40**
  Sea 172, *172*
Sanicle
  Purple 296
  Wood 296
Sanicio, Tree 303
Senecionine 26
Sapele **297**, 329
  Heavy 297, 329
Sap-green 65
Sapida 16
Sapodilla 85, 147, 213
Saponin 221
Sapote 124, **297**
  Black 117, 297
  Colorado 297
  White 297
Sapote Borracho 124
Sappanwood Tree 67
Sapu 289
Sapucaia 239, **297**
Sapucaya 239
Saranta, Black 146
Sarsaparilla 56, 143, **298**, 307
  False 298
  German 73
  Jamaican 307
  Mexican 307
  Tampico 307
  Wild 298, 300, 307
Sarson 62
Sassafras 143, 221, 298
  Chinese 298
Sassafras oil 132
Satal 151
Satan's Bolete 59

Satinwood 86, **298**, 329
  Brazilian 298
  Ceylon 329
  East Indian 298, 329
  Scented 298
  West Indian 298, 329, 351
Satsuma 147, 211, **298**
Sauerkraut 51, **298**
Savanna 48
Savin 184
Savory
  Summer 168, *169*, 298
  Winter 168, 288
Sawbrier 307
Saxifrage **298**
  Magic-carpet 299
  Meadow 299, *298*
  Mossy 299
  Purple Mountain 299
Scab 341
Scabious **299**
  Devil's Bit 299
  Red Indian 299
  Sweet 129, 299
Scabwort 179
Scallops 216
Scammony 97
Scarborough Lily **299**
Scarlet Bush 162
Scarlet fever 51
Scarlet-bugler 259
Schnappes 323
Scholar Tree, Chinese 308
Scopolamine 26, 113, 221
Scouring Rush **174**
  Thatch 252
Screw Pine 252
Screw-augur 312
Scuppernong 344
Scurvy Grass **94**
  Common 94
  Early 94
Sea Coral 248
Sea Eryngium 130
Sea Holly 130
Sea Holm 130
Sea Kale **104**, *104*, 338
Sea Lettuce 21, **301**, *301*
Sea Rocket 67, *67*
  American 67
Sea Urchin 162
Sea Whistles 43
Sea-grass 351
Seakale Beet 83
Sea-pink 40
Seaware 146
Seaweed **301**
Sebil 220
Sedge 73, 145, **302**
  Bottle 73
  Fen 302
  Lesser Pond 73
  Pendulous 73
  Sand 73
  Spring 73
  Tussock 73
Seed Flax 139
Selangan batu 305
Self-heal 19, 276
Seminole-bread 349
Senecio, Tree 303
Senecionine 26
Senna 76, 221
  Aden 76
  Alexandrian 76, 221
  American 76
  Bladder **96**
    Common 96, *97*
  Coffee 76
  Dog 76
  Indian 76
  Italian 76
  Ringworm 76
  Scorpion 101
  Sickle 76
  Smooth 76
  Tinnevelly 76
Sensitive Plant 224, *224*, **303**
  Giant 224
  Wild 303
Sentry 176
Sequoia, Giant 283, 303, 329
Seraya 329
Serradella 145
Service Berry **27**
Service Tree 308
  Bastard 309
  True 309
  Wild 309

Sesame 143, 244, *244*, *304*, 304
Sesame oil 304
Sesban 145, 304
Sevenbark 177
Seville Orange **304**
Shadblow **27**
Shadbush **27**
Shaddock 89, 147, 160, **304**
Shaggy Cap 100, *100*
Shaggy Mane 100, *100*
Shallon 151
Shallot 19, **304**, 338, *339*
Shame Plant 224
Shamrock 251, **304**
Shank, Red 271
She Cabbage 303
She Oak 76
Sheep Berry 342
Shell Flower 226
Shellac 138, **189**
Shellbark, Big 75
Shepherd's Club 227
Shepherd's Purse 72
She-to 220
Shinleaf 279
Shoe Flower 170
Shola 13
Shooting Star 118
Shore Podgrass 331
Shoreweed **203**
Shot, Indian 71
Shrimp Plant *184*, *185*
  Mexican 185
Sickle Medick 205
Sideoats 145
Side-saddle Flower 298
Sierra Leone Butter 321
Silk Cotton Tree 78, 140, 186
  Indian 59
  Red 140
Silk Plant, Chinese 58
Silk Tree 17
Silk tassel Bush 151
Silkweed 43
Silver Bell 162
  Mountain 162
Silver Buffalo **304**
Silver Dollar 206
Silver Rod 45
Silverberry 125
Silverweed 274
Siris Tree 16
  Pink 17
Sisal 140, 141, **306**, *306*
Sisal Agave 306
Sisal hemp **306**
Sissoo Wood, Indian 112
Skullcap, Common 301
Skullcap Flower 301
Slime Fungi *306*
Slipper Flower **68**, 257
Slipperwort **68**
Slivovitz 18, 56, 269
Sloe **58**, 56, 269, **306**
Sloe gin 58
Smartweed **271**
  Water 271
Smilax 43, 44, **307**
Smoke Tree 102, **307**
  American 307
Smut fungi 336, *336*
Snail Flower 261
Snake Bark 10
Snake Plant 140, 297
Snake Wood 64
Snakeroot
  Black 12, 87, 296
  Button 130, **197**
  Indian 221, 283
  Sampson's 156
  Seneca 271
  Virginia 40
Snakeweed 271
Snapdragon **36**
  Chaparral 36
  Common 36, 37
  Withered 36
Snow, red **283**
Snow on the Mountain 134
Snowball Cushion 211
Snowbell
  Fragrant 317
Snowberry 320
Snowdrop **307**, *307*
  Common 307
  Giant 307
Snowdrop Tree 162
Snowflake **197**
  Spring 197

Snowflake (cont)
Summer 197
Water 237
Snowflower 310
Snow-in-summer 80,
80
Soap Root 297
Soap Tree 161
Soapbark Tree 280
Soapberry 304
Soapweed 349
Soapwort **297**, *297*
Rock 297
Soda-apple 307
Solanine 26, 308
Solasodine 308
Solomon's Seal **271**
Common 271
Solomon's Zigzag 307
Sophora, Fourwing 308
Sorghum 81, *81*, **309**
Bicolor 309
Caudatum 309
Common 309
Durha 309
Durra 309
Ghally 309
Guinea 309
Indian 309
Kafir 309
Sweet 318
Sorgo 18, 309, 318
Sorrel 142, *143*, 293
Common 108
French 293
Garden 168, 293
Indian 170
Jamaica 170
Lady's
Creeping 251
Sheep 168, 293
Wood 251, 304
Sorrel Tree 251
Sotol **113**
Souari Nut 75
Soursop 31, 147, *149*
Mountain 31
Wild 31
Sourtop 58
Sourwood 251
Sou-sou 217
Southernwood 42, **309**
Sowthistle 308, *308*
Soya 244
Soybean 158, 195, 244,
309, *310*
Wild 158
Spaghetti **310**
Vegetable 216
Spaniard 11
Spanish Bayonet 349
Spanish-dagger 349
Sparteine 221
Speargrass 11
Spearmint 132, 168,
169 221, 224
Spearmint oil 132
Speedwell **341**
Common 341
Field 342
Germander *341*
Gray 342
Procumbent 342
Rock 342
Wall 342
Spiceberry 40
Spicebush 19, 201
Spider Flower 90
Spider Plant 86
Spiderwort
Common 330
Spikenard 232
American 39
False 307
Spinach 311, *311*, 338,
339
Amaranth 338
New Zealand 311,
338
Prickly-seeded 338
Round-seeded 338
Summer 338
Water 338
Winter 338
Spindle Tree
European 133
Winged 133, *134*
Spiraea 46
Goatsbeard 43
Japanese 312
Reeves 312
Tosa 312
Spirit of turpentine 334
Spleenwort **45**
Mother 45
Sponge

Sponge (cont)
Dishcloth 159
Vegetable 159
Spoon Flower 347
Spotted Dog 278
Spring Beauty 90
Spring Bell 306
Sprouting Leaf 186
Sprouts 62
Spruce 141, **312**, 314
Big-cone 120
Black 313, *313*
Colorado 313
Engelmann 313, 329
Hemlock 122, **167**,
329
Norway 86, 312, 313,
329
Red 312, 329
Rocky Mountain 313
Sitka 313, 329
White 313, *313*, 329
Yeddo 312
Spurge **313**
Caper 134
Ipecacuanha 134
Leafy 134
Slipper 257
Spurge Laurel 112
Squash 107, 216, **313**
Acorn 216
Autumn 216
Banana 216
Bush 216
Butternut 216
Chioggia 216
Cocozelle 216
Hubbard 216
Naples 216
Ohio 216
Pattypan 216
Sea 216
Summer 107, 216,
217
Turban 216
Winter 107, 216, *217*
Squaw Carpet 77
Squaw-bush 342
Squill 221, **313**, 336
Autumn 301, 313
Bell-flowered 177
Blue 313
Dwarf 301
Italian 177
Red 336, *336*
Spring 301
Striped 279
Squirrel Corn 116
Staff Vine 78
Staggerbush 207
Star anise 132
Star Apple **314**
Star of Bethlehem 221,
250
Yellow 150
Star of the Veldt 117
Starfish Plant 106, 314
Star-flower 307, 331
Starleaf 299
Starwort
Yellow 179
Statice **201**, *201*
Steeplebush 312
Stepladder Plant 102
Stepmother's Flower
343
Stevia 267
Stick-tights **55**
Sticky Willie 150
Stinging Nettle 118
Common 336
Stink Tree 314
Stinkhorn 153, 278,
314
Australian 314
Common 314
Stinkwood 79, **314**
Bastard 314
Black 241
Camdeboo 314
Red 314
White 314
Stitchwort 314
Greater 314
Stock **215**
Brompton 215
Garden 215, *215*
Imperial 215
Night-scented 215
Ten Week 215
Virginia 315
Winter 215
Stomach Fungi 278
Stone, Italian 266
Stone-cress
Persian 14

Stonecrop 302
Biting 302
White 302
Stonewort **315**
Storax 221, **315**
American 202
Levant 315
Storesin 221
Storksbill **129**
Stout 18
Strawbell 336
Strawberry 147, *148*,
315
Barren 274, 315
Chiloe 315
Common 274
Hautbois 315
Indian 315
Mexican 124
Mock 315
Musk 315
Wild 315, *315*
Yellow 276
Strawberry Bush 133
Strawberry Tomato 264
Strawberry Tree 39, 147
Hybrid 39
Strawflower 165, *165*
String of Beads 303
String-of-beads Plant
187
Strophanthin 316
Strophanthus 221
Strychnine 26, 221, 316
Strychnine Tree 26
Styfsiekte 105
Stylo 317
Styrax, American 202
Succory 87
Sugar Beet **317**, *317*,
318, *318*
Sugar Berry 79
Mississippi 79
Sugar Cane 56, 225,
294, **317**, *317*, 318,
318
Sugar Maple **319**
Sugarapple 147
Sugarbush 275
Sulfur Flower 129
Sulla Clover 165
Sultana **319**
Sumac 122, 286
Dwarf 286
Elm-leaved 286
Mountain 286
Poison 286
Scarlet 286
Shining 286
Sicilian 286
Smooth 286
Staghorn 286
Tanner's 286
Venetian 307
Wing-rib 286
Summerberry 342
Summer-sweet 90
Sun Plant 273, *273*
Suncup 241, 242
Sundew **121**, *121*
Sundrop 241, 242
Sunfern **158**
Sunflower 244, *244*,
245, **319**, *319*
Common 319
Cucumber-leaved 320
Darkeye 320
Showy 320
Swamp 320
Sun-rose 165, *165*
Suze 156
Swain Flower 40
Swan Flower 320
Swede 62, *63*, 290, 291
320
Swede Rape 320
Swedish Turnip **320**
Sweet Bay 192, 209
Sweet Corn Root 68
Sweet Gale 231
Sweet Gum **202**, 221
American 315
Oriental 221, 315
Sweet Orange 160
Sweet Potato *180*, **320**,
320
Tennessee 216
Sweet Potato Vine 320
Sweet Spire 181
Sweet Sultan 79
Sweet William 115,
115, 305
Sweetbells 197
Sweetberry 342
Sweetcup 255
Sweetleaf 320

Sweetleaf (cont)
Asiatic 320
Common 320
Sweetsop 31, 147
Swiss Chard 54, 311
Swiss-cheese Plant
226, *226*
Swizzlestick 283
Sycamore 10, *286*,
320, 329
American 268
Eastern 320
Egyptian 320
Silver 106
Syphilis 50
Syringa 262
Rose 262

Tacamahac 273
Tagua 264
Tail Flowers 37
T'a-ku-ts'ai 62
Tallow
Vegetable **321**
Tallow Tree
Chinese 122, 321
Sierra Leone 321
Tamarack 191, *191*
Tamarind 143, 147,
195, **321**, 322
Cow 296
Tamarisk *322*
Chinese 322
French 322
Tang, Black 146
Tangelo 211, **322**, 335
Tangerine 147, 211,
250, **322**, 335
Tanier 290, *291*, 318,
347
Tanning 122
Tannins 122
Tansy 122, 168, *169*,
302, 322
Goose 274
Tapa 141
Tapa Cloth 64
Tapioca 75, *76*, *213*
Tara-vine 12
Tare **342**
Taro 290, *291*, 318,
323
Blue 347
Giant 24, 318, 323
Swamp
Giant 318
Tarragon 42, 142, 143,
168, *169*, **323**
Tarragon vinegar 323
Tarweed 208
Tassel Flower 27, *27*,
126
Tatami 184
Tau-saghyz **323**
Tawhiwhi 268, *268*
Tea 56, 57, 221, **323**,
324
Algerian 254
Appalachian 342
Arabian **186**
Assam 323
Beggar's 294
Brown 196
Bush 56
Cape 56
Caucasian 39
China 323
Mexican 84
New Jersey 78
Oswego 168, 226
Paraguay 26, 56, 172
Tansy 322
Tea Bush 50
Teak **324**, 329
African 181, 329
Bastard 66, 277
Bengal 66
Cape 317
Fustic 181
Teak Tree 324
Tea Tree 196, 207
Heath 196
Teaberry 151, 342
Tea-seed oil 324
Teasel 118
Fuller's 118
Teff 128, 145, 224
Telegraph Plant 115
Temple Tree 269
Tendergreen 85
Teosinte 210, **324**
Tequila 15, 18
Terebinth 267, **324**
Tetterbush 207
T-ha-na-sa 220
Thebaine 26

Theine 26, 56, **67**
Theobromine 26, 56,
221
Theol 56
Thimbleberry 282, 292
Thistle **325**
Blessed 91, 325
Bull 88
Cabbage 88
Canadian 325
Carline 74
Stemless 73
Cotton
Common 246
Creeping 88, *88*
Globe **124**
Holy 325
Marsh 88
Melancholy 88
Milk 325
Musk 73, 325
Nodding 73
Picknicker's 88
Russian 295
Scotch 246, 325
Slender 73
Spear 88, *88*
Stemless 88
Torch **80**
Welted 73, 325
Thorn
Babul 9
Branch 129
Camel 9
Cockspur 163
Egyptian 9
Jerusalem 254
Kaffir 207
Kangaroo 9
Rabbit 207
Thorn-apple *113*, *220*,
221
Common 113
Downy 113, 220
Thousand Mothers 327
Three Faces Under a
Hood 253
Thrift 40, *41*
Prickly **10**
Thyme 143, 223, **326**
Cat 325
Caraway 168
Cone Head 326
Common 326
Creeping, Wild 326
Garden 168, *169*,
326
Lemon 108, 326
Wild 326
Thyme oil 132, 326
Thymol 75, 326
Tickseed **55**, **100**, *101*
Tickweed **100**
Tide, red **283**
Tidy-tips 193
Tiger Flower 326
Mexican 326
Peacock 326
Tiger Jaw **135**
Timothy **326**
Alpine 262
Tisswood 260
Toad Plant 313
Giant 313
Hairy 314
Shaggy 314
Toadflax 108, **201**, 311,
327
Alpine 201
Bastard 327
Common 327
Ivy-leaved 108, 327
Pale 201
Purple 201
Striped 201
Toadshade 331
Painted 331
Tobacco 26, 221, 235,
250, **327**, *327*
Common 235
Indian 203, 221
Mountain 41
Tree 235
Toddy 18, 41, 125
Tomatillo 147, 264
Tomato **327**, *327*, 338,
339
Currant 327
Mexican Husk 264
Strawberry 264
Tree **109**, 327
Toon 78, **330**
Toothwort 192
Purple 192, *193*
Topee-tambu 68, *290*,
330

Torch Thistle **80**
Toria 62
Tornillo 275
Torreya
Chinese 330
Japanese 330
Torus Herb 119
Tournesol 86
Tragacanth 46
Traveller's Tree 283
Treacleberry 307
Treasure Flower 156,
156
Tree Celandine 58
Tree fern 330, *330*
Tree of Heaven 16
Tree Poppy 58
Tree Senecio 303
Tree Toad 321
Trefoil **331**
Bird's-foot 205, *205*,
331
Hop 331
Shrubby 331
Sweet 331
Yellow 145, 304, 331
Trigonelline 26
Trillium
Purple 332
Snow 331
Trinidad Devil 283
Triticale 332
Truffle **333**
False **125**
Piedmont
White 333
Summer 333
Trumpet Creeper 71
Trumpet Leaf 298
Trumpet Tree 78
Trumpet Vine 71
Trumpet-flower 55
Chinese 71
Tsamma 345
Tshwala 309
Tuba Root 115
Tuberose 132, 271
Tubocurarine 221, 317
Tulip **333**, *334*, 335
Globe 70
Mariposa 70
Star
Large 70
Small 70
Tulip Breaking Virus
334
Tulip Tree *202*, 311,
334, *334*
Chinese 334
Tulip wood **334**, *334*
Tung 334
Tung oil 244, **334**
Tung-oil Tree 17
Tunka 217
Tupelo 237
Turk's Cap 11
Turmeric 107, 122,
142, 143, **334**
Turmeric Plant, African
71
Turnip 62, *63*, 290,
291, **334**
Dutch 62
Green-top 62
Indian 40, 129, 277
Stubble 62
Swedish 62, **320**
Turnipwood 123
Turpentine 266, **334**
Cyprus 324
Swiss 313
Venice 191
Turpentine Tree 66, 267
Turtle Head **84**
Tutsan 178
Twayblades 202
Twin Flower **201**
Twitch-grass **15**
Two Men-in-a-boat 285
Typhoid fever 51
Typhus 51

Ucuhuba **335**
Ucuhuba butter 335
Uganda Tree 311
Ugli 211, **335**
Ulloco 291, **335**, 338
Ullucu **335**, 338
Umbrella Tree 218, 299
Texas 218
Umzímbeet 335
Giant 335
Unicorn Plant 214
Upas Tree **335**
Urushiol 221, 270
Urn Plant *13*

Usnic acid 199
Utile 329

Valerian 19, **79**
 Cat's 337
 Common 337
 Greek 270
 Red 79
 Spur 79
Vanilla 142, 143, **337**
 Pompona 337
 West Indian 337
Varnish Tree 17, 188, 189
Vegetable Humming Bird 304
Vegetable Oyster 295, 330
Vegetable Sheep 282
Veitchberry 147
Velvet Elephant Ear 186
Velvet Leaf 9, 186
Velvet Plant 161
 Trailing 293
Venus Hair 12
Venus' Necklace *173*
Venus's Fly Trap *32,* 117
Verbena
 Clump 341
 Garden 341
 Lemon 168, 202, 341
 Rose 341
 Sweet-scented 202, 341
Verbena oil 341
Vermouth 18, 56
Vernal Grass
 Sweet 36
Vervain 168, 341
 Creeping 341
 Rose 341
Vetch **342**
 Bitter 342
 Common 342
 Crown 101, *101,* 145
 Hairy 145
 Horse-shoe 342
 Kidney 145, 342
 Milk 342
  Purple 46
 Purple 342
 Russian 342
 Spring 145, 342
 Winter, Hairy 342
Vetchling 160, 342
 Meadow 192
Vetiver 342
Vetiver oil 132, 187
Viburnum *342*
 Sweet 342
Vilea 220
Vinblastine 221
Vincristine 221
Vine 343, **344**
 Balloon 73
 Bamboo 307
 Bay Star 300
 Clock 326
 Cross 55
 Firecracker 212
 Flame *279, 279*
 Grape 344
  Black *344*
 Japanese Hydrangea 300
 Kangaroo 88, *88*
 Kudzu 145
 Lemon
 Link 337
 Matrimony 207
 Moonseed 107
 Pollyanna 308
 Potato 307
 Quarter 55
 Russian 271
 Sky 326
 Staff 78
 Tara 12
 Trumpet
  Blue 326
 Wire 227
 Wonga-wonga 253
 Worm 337
Vinegar Tree 122, 286
Viola 343, *343*
 Horned 343
Violet 132, 343
 African 294, *295*
 Damask 167
 Dame's 167
 Dog 343
 Dog-tooth 131, *131*
  White 131
 English 343
 Florist's 343

Violet (cont)
 Garden 343
 German 134
 Labrador 343
 Northern Blue 343
 Philippine 53
 Sweet 343
 Usambar 294
 Water 173
 Wood 343
Violet Slipper 306
Violetta 343
Virginia Creeper *254,* 255
Virginia-willow 181
Vodka 18, 56

Wahoo Tree 66
Wake-Robin 42, 331
 White 331, *331*
Wallaba 329
Wallfern 272
Wallflower 84, *84,* **344**
 Beach 130
 Canary Islands 84, 344
 Common 344
 Fairy **130**
 Siberian 344
 Western 130
Wallpepper 302
Wallwort 296
Walnut 122, 239, *239,* **344**
 African 344
 Black 239, 344, 329
 Chinese 239, 344
 Common 238, 344
 East India 16
 Eastern 239
 English 238, 344
 Japanese 239, 344, 329
 Persian 238, 344
Wandering Jew 330, 351
Wandflower 116, 310
Wapatu 294
Waratah 324
 Tasmanian 324
Water Aloe *315*
Water Archer 294
Water Chestnut **344,** *344, 345*
 Chinese 125, 345
Water Lily **237,** *237,* **345**
 Amazon 343
 Cape Blue 237
 Fringed 237
 Pygmy 237
 Royal 343
 Santa Cruz 343
 White
  European 237, *237*
 Yellow *236,* 237, 345
Water Melon 216, *217,* **345**
 Chinese 217
Water net 22
Water Soldier 315, *315*
Watercress 62, 232, **345**
 Brown 232
 Fool's 37
 Green 232
 Summer 232
 Wild 63
 Winter 232
Water-platter
 Amazon 343
 Santa Cruz 343
Water-weed
 Canadian 126
Wattle 9, 122
 Australia
  Black 9
 Cootamundra 9
 Golden 9
 Silver 9
 Sydney Golden 9
Wattle and daub 345
Wawa 329
Wax **345**
Wax Flower 176
 Clustered 314
Wax Palm 82
Wax Plant 176, *176*
 Andean 345
 South American 345
Wax Tree 304
Wax Tree, White 275
Waxberry 320
Waxvine 303
Waxwork 78
Weather Clock 73
Weed Blackjack 55

Weld 122, 223
Wheat 81, 139, 318, **332,** *332, 333*
 Bread 81, *81,* 332
  Common 332
 Club 81, 332
 Dinkel 332
 Durum 81, *81,* 332
 Einkorn 81, 332
 Emmer 332
  Cultivated 81
  Wild 81, 332
 Hard 81, *81,* 332
 Inca 159
 Indian 135
 Miracle **230**
 Mummy **230**
 Persian Khorasan 332
 Polish 332
 Soft 332
 Spelt 81, 332
 Strand 293
 Turgidum 81
 Winter 332
Wheatgrass **15**
 Crested 15, 145
 Intermediate 15
 Slender 145
 Western 15
Whin, Common 335
Whisk Fern 92
Whisky 18, 56
Whitebeam 308
 Common 309
 Swedish 309
White leaf 165
White-top 129
Whitewood **326,** 329, 334
 America 329
 Canary 334
Whitlavia 261
Whitten Tree 342
Whooping cough 51
Whorl Flower 226
Whortleberry 57, *337*
 Bushy 176
Widdy 274
Widow's Frill 305
Wild Oat 346, *346*
Wild Spaniard 11
Wild Spice 19
Willow 122, 286, 295
 Basket 295, 346
 Bay-leaved 295
 Black 295
 Crack 295
 Cricket-bat 295
 Dwarf 295
 Florist's 346
 Goat 295, 346
 Pussy 295, 346
 Reticulate 295
 True 295
 White 295
 Woolly 295
 Yellow 295
Willow bark **346**
Willow Leaf 280
Willow Tree, Indian 295
Willowherb
 Rosebay 83, *83,* 127
Wilwilli 130
Windflower **29**
 Grove 30
Window Plant 226
Wine Plant 285
Wineberry 147, 292, **346**
Wing Nut 277
 Caucasian 277
Winter Sweet 11
Winterberry 171
Wintercress 287
Wintergreen 221, 279, 331
 Arctic 279
 Creeping 151
 Flowering 271
 Spicy 151
Wintergreen oil 151
Winter's Bark 121
Wintersweet **85**
Wiregrass 125, 270
Wishbone Flower 330
Wisteria
 Chinese 346, *346*
 Japanese 346
 Scarlet 304
Witch Hazel **162**
 Chinese 162, *163*
 Common 162
 Japanese 162
 Ozark 162
Witches Broom 322

Witloof 85, 87, 126, 338
Woad 122, 181, 346
Wolfbane **11**
Wolfberry 320
 Anderson 207
 Arabian 207
 Chinese 207
Wolffia, Common 347
Wonder Tree 287
Wood Crape 326
Wood oil 132, 282
 China 334
 Japan 244
Woodbine 204, 255
 Italian 204
 Perfoliate 204
Wood-oil Tree
 China 17, 334
 Japan 334
Woodruff **44**
 Dyer's 44
 Sweet 44, 150, 168
Woodrush 206, 293
Woodsia
 Fragrant 347
 Northern 347
 Rocky Mountain 347
 Rusty 347
Wood-vamp 114
Woody Holcus 171
Woolly Sheep 248
Wormseed, American 84
Wormwood 56, *221,* 337
 Common 42
Woundwort
 Hedge *312*
 Marsh 313
 Soldier's 11
Wrack
 Channelled *21,* 258
 Knobbed 43
 Lady 146
 Serrated *150*
 Yellow 43, *43*
Wreath
 Purple 260
 Queen's 260

Yam 81, 221, 291, **348,** *348, 349*
 Aerial 248
 American 290
 Asiatic 248, 290
  Greater 318
 Bitter 290
 Cush-cush 248
 Greater 248, 290
 Guinea
  White 248, 290, 318
  Yellow 248, 290, 318
 Lesser 248
 White 290
 Yellow 290
Yam Bean 195, **348**
Yampee 318
Yangtao 12
Yardgrass 125
Yarrow 11
 Fernleaf 11, *11*
 Musk 11
 Soldier's 315
 Woolly 11
Yaupon 172
Yautia 347
Yeast **294**
Yellow Bark 280
Yellow Wood 45
 African 270
Yerba de maté 26, 56, 172, **214**
Yerba Linda 258
Yew 167, 330, **348**
 American 349
 Canada 349
 Chinese 348
 Common 348, 349
 Irish 349
 Japanese 270, 349
 Plum 79
  Chinese 79
 Plum-fruited 270
 Prince Albert 298
 Stinking 330
Ylang-ylang 132, **349**
 Climbing 349
Yoco 56
Yohimbine 26
Yopo 220
Yorkshire Fog 171
Youngberry 292
Youth-and-old-age 351
Youth-on-age 327

Yucca Datil, Blue 349

Zacatechichi 220
Zamang 296
Zansami 252
Zapote Blanco 297
Zebra Haworthia 163
Zebra Plant 36, 106
Zebra Wood 117
Zedoary 107, 122
 Black 107
 Yellow 107
Zelkova 352
 Japanese 351
 Saw-leaf 351
Zinnia
 Mexican 351
Zucchini 216
Zulu-giant 313

Bibliography

Many hundreds of books and papers have been consulted during the writing and editing of this volume. Much relevant information has been obtained from Floras of many parts of the world and from monographs of hundreds of genera, too numerous to cite here.

BAILEY, L. H. (1949). *Manual of Cultivated Plants* (revised edition). The Macmillan Company, New York.

BAILEY, L. H. & BAILEY, E. Z. *et al.* (1977). *Hortus Third*. Macmillan Publishing Co Inc, New York; Collier Macmillan, London.

BAKER, H. G. (1970). *Plants and Civilization* (edn 2). Wadsworth Publishing Company, Inc., Belmont.

BEAN, W. J. (1970–1976). *Trees and Shrubs Hardy in the British Isles* (8th edn) (vols 1–3). John Murray, London.

BROOKLYN BOTANIC GARDEN (1964). *Dye Plants and Dyeing – a Handbook. Plants and Gardens* (vol. 2, No. 3). Brooklyn Botanic Garden, New York.

BROUK, B. (1975) *Plants Consumed by Man*. Academic Press, London, New York, San Francisco.

CALABRESE, F. (1978). *Frutticoltura Tropicale e Subtropicale*. Cooperativa Libraria Universitaria Editrice, Bologna.

CHITTENDEN, F. J. (ed) (1956–1969). *Dictionary of Gardening* (2nd edn) Clarendon Press, Oxford.

COBLEY, L. S. & STEELE, W. M. (1976). *An Introduction to the Botany of Tropical Crops*. Longman, London, New York.

EMBODEN, W. A. (1972). *Narcotic Plants*. Macmillan, New York.

ENGLER, H. G. A. (1964). *Syllabus der Pflanzenfamilien* (edn 6, by H. Melchior, vol 2). Berlin.

HARRISON, S. G., MASEFIELD, G. B. & WALLIS, M. (1969). *The Oxford Book of Food Plants*. Oxford University Press, Oxford.

HAY, R. & SYNGE, P. M. (1969). *The Dictionary of Garden Plants in Colour in the House and Greenhouse Plants*. Ebury Press and Michael Joseph, London.

HEISER, C. B. (1969). *Nightshades – the Paradoxical Plants*. W. H. Freeman, San Francisco.

HEYWOOD, V. H. (ed) (1978). *Flowering Plants of the World*. Oxford University Press, London, Melbourne.

HEYWOOD, V. H., HARBORNE, J. B. & TURNER, B. L. (1977). *The Biology and Chemistry of the Compositae* (2 vols). Academic Press, London, New York, San Francisco.

HILL, A. F. (1952). *Economic Botany* (edn 2). New York, Toronto & London.

Hillier's Manual of Trees & Shrubs (1977) (4th edn). David & Charles, Newton Abbot.

HOWES, F. N. (1974). *A Dictionary of Useful and Everyday Plants and their Common Names*. Cambridge University Press, London, New York.

HUTCHINSON, J. (1959). *The Families of Flowering Plants* (edn 2, vols 1–2). Oxford University Press, Oxford.

IRVINE, F. R. (1969). *West African Crops*. Oxford University Press, Oxford.

LEWIS, W. & ELVIN-LEWIS, P. F. (1977). *Medical Botany*. John Wiley, London, Sydney, Toronto.

MATHEW, B. (1973). *Dwarf Bulbs*. B. T. Batsford, London.

MATHEW, B. (1978). *The Larger Bulbs*. B. T. Batsford, London.

MENNINGER, E. A. (1962). *Flowering Trees of the World for Tropics and Warm Climates*. Hearthside Press Inc., New York.

MITCHELL, A. F. (1974). *A Field Guide to the Trees of Britain & Northern Europe*. Collins, London.

NATIONAL RESEARCH COUNCIL (1979). *Tropical Legumes; Resources for the Future*. National Academy of Sciences, Washington, D.C.

NICHOLSON, B. E., WALLIS, M., ANDERSON, E., BALFOUR, A. P., FISH, M. & FINNIS, V. (1963). *The Oxford Book of Garden Flowers*. Oxford University Press, Oxford.

PALMER, E. (1977) *A Field Guide to the Trees of Southern Africa*. Collins, London, Johannesburg.

POLUNIN, O. & EVERARD, B. (1976). *Trees & Bushes of Europe*. Oxford University Press, London, New York, Toronto.

PURSEGLOVE, J. W. (1968). *Tropical Crops. Dicotyledons*. Longman, London.

PURSEGLOVE, J. W. (1972). *Tropical Crops. Monocotyledons*. Longman, London.

REHDER, A. (1940). *Manual of Cultivated Trees and Shrubs* (edn 2 – reprint 1956). Macmillan Publishing Company, New York.

RENDLE, A. B. (1904, 1938). *The Classification of Flowering Plants*, (edn 1, vols 1–2). Cambridge University Press, Cambridge.

RENDLE, A. B. (1930). *The Classification of Flowering Plants* (edn 2, vol 1). Cambridge University Press, Cambridge.

RIZZINI, C. T. (1971). *Árvores e Madeiras Úteis do Brasil. Manual de Dendrologia Brasileira*. Editora Edgard Blücher Ltda, São Paulo.

RIZZINI, C. T. & MORS, W. B. (1976). *Botânica Econômica Brasileira*. EPU, Editora da Universidade de São Paulo, São Paulo.

ROWLEY, G. D. (1978). *The Illustrated Encyclopaedia of Succulents and Cacti*. Salamander Books, London.

SANTAPAU, H. (1966). *Common Trees*. National Book Trust, India, New Delhi.

SCHERY, R. W. (1972). *Plants for Man* (edn 2). Prentice-Hall, Englewood Cliffs, N. J.

SCHULTES, R. F. & HOFMANN, A. (1980). *Plants of the Gods*. Hutchinson, London, Melbourne, Sydney, Auckland, Johannesburg.

SIMMONDS, N. W. (ed) (1976). *Evolution of Crop Plants*. Longman, London, New York.

STOBART, T. (1970). *Herbs, Spices and Flavourings*. The International Wine and Food Publishing Company, London.

SYNGE, P. M. (ed) (1969). *Supplement to the Dictionary of Gardening* (edn 2). Clarendon Press, Oxford.

TANAKA, T. (1976). *Tanaka's Cyclopedia of Edible Plants of the World* (edn 5, Nakao). Keigaku Publishing Co., Tokyo.

TUTIN, T. G. & HEYWOOD, V. H. (eds) (1964–1980). *Flora Europaea* (vols 1–5), Cambridge University Press, Cambridge, London, New York, New Rochelle, Melbourne, Sydney.

UPHOF, J. S. (1968). *Dictionary of Economic Plants* (edn 2). Steckert-Hafner, New York.

Urania Pflanzenreich. Höhere Pflanzen. (vol 1 (1971), vol 2 (1973)). Leipzig, Jena & Berlin.

VICKERY, M. L. & B. (1979). *Plant Products of Tropical Africa*. Macmillan, London, Basingstoke.

WETTSTEIN, R. (1933–1935). *Handbuch der Systematischen Botanik* (edn 4). Leipzig-Wien.

WILLIS, J. C. (8th edn by H. K. Airy Shaw 1973). *A Dictionary of Flowering Plants*. Cambridge University Press, Cambridge, New York, Melbourne.

Illustration Credits